SOCIOLOGY IN OUR TIMES

SEVENTH CANADIAN EDITION

SOCIOLOGY IN OUR TIMES

SEVENTH CANADIAN EDITION

JANE LOTHIAN MURRAY
University of Winnipeg

RICK LINDEN
University of Manitoba

DIANA KENDALL
Baylor University

Sociology in Our Times, Seventh Canadian Edition
by Jane Lothian Murray, Rick Linden, and Diana Kendall

VP, Product and Partnership Solutions:
Anne Williams

Publisher, Digital and Print Content:
Leanna MacLean

Marketing Manager:
Terry Fedorkiw

Content Development Manager:
Lisa Berland

Photo and Permissions Researcher:
Julie Pratt

Production Project Manager:
Christine Gilbert

Production Service:
Cenveo Publisher Services

Copy Editor:
Linda Szostak

Proofreader:
Cenveo Publisher Services

Indexer:
Cenveo Publisher Services

Design Director:
Ken Phipps

Managing Designer:
Franca Amore

Interior Design:
Sharon Lucas

Cover Design:
Sharon Lucas

Cover Image:
Preappy/Moment/Getty Images

Compositor:
Cenveo Publisher Services

Printed and bound in the United States of America
1 2 3 4 19 18 17 16

For more information contact Nelson Education Ltd.,
1120 Birchmount Road, Toronto, Ontario, M1K 5G4. Or you can visit our Internet site at
http://www.nelson.com

Library and Archives Canada Cataloguing in Publication

Kendall, Diana Elizabeth, author
Sociology in our times / Jane Lothian Murray (University of Winnipeg), Rick Linden (University of Manitoba), Diana Kendall (Baylor University).

Revision of: Kendall, Diana Elizabeth. Sociology in our times.
Includes bibliographical references and index.
ISBN 978-0-17-655863-5 (paperback)

1. Sociology—Textbooks.
2. Sociology—Canada—Textbooks.
I. Linden, Rick author II. Lothian Murray, Jane, 1960- author III. Title.

HM586.K45 2016
301 C2015-906768-5

ISBN-13: 978-0-17-655863-5
ISBN-10: 0-17-655863-2

To the memory of my brave son Drew Samson Elliot Murray. Your smile lit up the life of everyone around you. Your courage held us strong. Your spirit will guide us always. I will hold you tight forever in my heart.

—Jane Lothian Murray (Mom)

To the memory of my mother-in-law, Mildred Cormack. She was a great friend who taught me the value of storytelling as a way of understanding the world (as well as keeping her grandchildren amused).

—Rick Linden

BRIEF CONTENTS

CONTENTS

BOXES

SOCIOLOGY AND EVERYDAY LIFE

POINT/COUNTERPOINT

SOCIOLOGY IN GLOBAL PERSPECTIVE

SOCIOLOGY AND NEW MEDIA

Welcome to the seventh Canadian edition of *Sociology in Our Times*. Each time we write a new edition, we become acutely aware of how quickly our country and the world change. Even as some things change, however, others remain the same. One thing that has not changed is the significance of education and the profound importance of understanding how and why people act the way they do, how societies grapple with issues and major problems, and why many of us are reassured by social institutions—including family, religion, education, government, and the media—even at times when we might like to see certain changes occur in these institutions.

Like previous editions of this widely read text, this seventh Canadian edition is a cutting-edge book in two ways: (1) by including a diversity of classical and contemporary theory, interesting and relevant research, and lived experiences that accurately mirror the diversity in society itself, and (2) by showing students that sociology involves important questions and issues that they confront both personally and indirectly through the media and other sources. This text captures the interest of a wide variety of Canadian students by taking into account their concerns and perspectives. As the title suggests, we have selected topics most pertinent to "*Our Times*," including the widening income gap between the rich and poor, new and emerging definitions of family life, changing constructions of sexuality, and ongoing concerns related to our environment. In doing so, we hope that students will learn to critically examine their social world, and further to contemplate the social world they hope to live in moving forward.

The research presented in the book includes the best work of classical and established contemporary sociologists, and an inclusive treatment of all people is woven throughout the text. By using the latest theory and research, *Sociology in Our Times* not only provides students with the most relevant information about sociological thinking, but also helps students consider the significance of the interlocking nature of class, race, and gender in all aspects of life.

We have sought to make the research accessible and engaging for both students and instructors. Concepts and theories are presented in a straightforward and understandable way, and the wealth of concrete examples and lived experiences woven throughout the chapters makes the relevance of sociological theory and research abundantly clear to students. We know that people learn best through stories, and we have tried throughout to tell stories about our social world.

NEW FEATURES

Changes in the Seventh Canadian Edition

The seventh Canadian edition of *Sociology in Our Times* provides us with the opportunity to further improve a text that has been well received by students and educators. We have added several hundred new references that incorporate the most recent new developments in sociological research, including the latest census data from Statistics Canada. We have added a new chapter on "Sociology and the Environment" because of the importance of this growing field to the discipline and to the future of society.

- Chapter 1 ("The Sociological Perspective") introduces students to the main theoretical perspectives used in sociology and includes a new opening narrative from a student discussing the challenges faced by millennials as a result of the excesses of the boomer generation. Students are also introduced to the importance of having a global sociological imagination.
- Chapter 2 ("Sociological Research") describes how social scientists conduct their research and the links between research and theory. The chapter now includes a section on big data, perhaps the most exciting new development in sociological methodology in the past several decades. This section discusses some of the important ethical implications of using big data. The Sociology and New Media box on "Methods and the New Media," formerly online, has been integrated into the body of the chapter.
- Chapter 3 ("Culture") has the 2011 census data on language diversity, ethnicity, and Indigenous peoples in Canada. This chapter has an expanded focus on multiculturalism as demonstrated with the new opening narrative, which highlights the challenges of negotiating identity in an increasingly multicultural society. In keeping with this theme, a new Point/Counterpoint box ("Multiculturalism, Reasonable Accommodation, and 'Veiled' Hostility") explores various debates around Muslim women wearing religious head covering and clothing.
- Chapter 4 ("Socialization") has shifted focus somewhat to examine the effects of positive socialization and interesting new issues relating to early childhood and adolescent socialization that are particularly relevant to students attending university for the first time. A new Point/Counterpoint box ("The Issue of Excessive Praise") examines the effects of excessive praise on early childhood socialization.
- Chapter 5 ("Society, Social Structure, and Interaction") includes new research on homelessness in Canada. In the new opening narrative, a young woman describes trying to survive living "on the street" in Winnipeg.
- Chapter 6 ("Families") has the most recent data (2011) on changes in Canadian families, including an increase in the number of same-sex families, stepfamilies, and common-law families, while maintaining its focus on family diversity and change.

- **Chapter 7** ("Groups and Organizations") examines how organizations affect our behaviour. The section on the famous Asch experiments has been updated by describing a study designed to see if the judgments of fingerprint experts would be affected by peer pressure. Also, Milgram's work on obedience was used to help explain the behaviour of U.S. government officials who were involved in torturing prisoners following the 9/11 attacks. The section on gender issues within organizations has been updated to include the recent sexual harassment issues at Dalhousie University's Faculty of Dentistry.
- **Chapter 8** ("Crime and Deviance") has a new chapter introduction, which discusses the lives of British Columbia's notorious Bacon Brothers. These gang leaders grew up in a middle-class household and became the leaders of one of Canada's most violent organized crime groups. The chapter also includes a new section on surveillance, which is becoming an important topic within the field of criminology. Crime statistics have been updated to include 2014 crime rates.
- **Chapter 9** ("Social Class and Stratification in Canada") has the most recent data on the distribution of wealth and income, as well as poverty, in today's society. This chapter highlights the effects of growing income and wealth inequality.
- **Chapter 10** ("Global Stratification") has a new chapter introduction, which tells the inspiring story of Malawi's William Kamkwamba—The Boy Who Harnessed the Wind. There is a new Sociology in Global Perspective box ("The Missing Women") dealing with the world's missing women. These are women who were never born because of sex-selective abortion, or who died early because of infanticide, the denial of healthcare and proper nutrition to young girls, and the impact of death in childbirth and through HIV/AIDS for adult women. The section on the health and safety hazards associated with offshore production now includes a discussion of the deadly collapse of a clothing factory at Bangladesh's Rana Plaza.
- **Chapter 11** ("Ethnic Relations and Race") maintains its emphasis on racism with a revised discussion of different types of racism, new personal narratives from racial minorities who have experienced racism, and a new Point/Counterpoint box ("Explaining White Privilege to the Deniers and the Haters") that challenges students to think about white privilege. This chapter also includes the newest (2011) census data on ethnic origins, language diversity, and visible minorities.
- **Chapter 12** ("Gender") explores our understanding of gender and challenges students to move past binary definitions of gender to an understanding of gender as complex and encompassing more than just two possibilities. The opening narrative tells the story of 11-year-old Wren who is transgender. A Point/Counterpoint box ("How Many Genders: 56 or 2?") defines various new ways that individuals can define their gender.
- **Chapter 13** ("Sex, Sexualities, and Intimate Relationships") examines a range of controversial issues related to sexuality today. It includes updated information on sexual health, diverse sexualities, the sexual double standard, and "hookup cultures."
- **Chapter 14** ("Aging") includes two important additions to the discussion of theories of aging: The aging and society paradigm is now included as a functionalist theory of aging, and cumulative advantage theory is now part of the discussion of conflict theories. The chapter also increases the cross-cultural emphasis by discussing aging in Japan and China. Statistics in the chapter have been extensively updated to reflect the aging of Canadian society. Finally, the chapter presents the innovative ideas of the group called the Committee for Retirement Alternatives for Women.
- **Chapter 15** ("Health, Healthcare, and Disability") contains updated statistics on health and healthcare. There is a discussion of discrimination against individuals as well as against several African nations during the Ebola outbreak in 2014–2015. The chapter also includes a new emphasis on the social determinants of health. The Sociology and New Media box ("Dr. Google: Health on the Web") on the role of new media in health and healthcare is now included in the body of the chapter.
- **Chapter 16** ("Education") has a focus on issues of postsecondary education, including rising tuition costs, increasing student debt, and a shrinking job market. A new opening narrative explores the decision to go to college or university directly out of high school or to take a "gap year." The chapter also includes an examination of home-schooling and dropping out.
- **Chapter 17** ("Religion") has a new chapter introduction discussing the controversy over Trinity Western University's Christian Covenant, which requires that students follow a conduct code prescribing that sexual relations are permissible only within a marital relationship and between a man and a woman. Several provincial law societies have refused to accredit this program. The chapter also includes new material on women and religion, and on Indigenous people and religion. The

theory section now includes a discussion of rational choice theories of religion. New data from the 2011 National Household Survey have been used to update the discussion of the religious affiliation of Canadians.

- **Chapter 18** ("Mass Media") was new in the last edition, so it has not been extensively revised in this edition.
- **Chapter 19** ("The Economy and Work") has been updated to discuss the changes in the economy brought about by new media (including blogs, Wikipedia, crowdsourcing, the sharing economy, etc.), and by new technology such as 3D printing. These changes are occurring rapidly and will have a profound impact on the future economy. There is a more extensive discussion of the problems that unions may face in the future and an expanded section on the difficulties facing women who are seeking careers in the high-tech industry. The chapter also deals with the issue of the hollowing out of the middle class.
- **Chapter 20** ("Sociology and the Environment") is a new chapter that examines a number of important issues related to environmental sociology, including climate change, environmental effects on health, and environmental justice and environmental racism. In this chapter, students are introduced to the primary theoretical frameworks in environmental sociology. This chapter also focuses on the health implications of climate change and other environmental issues. Several controversial cases, both in Canada ("Mercury Poisoning in Grassy Narrows") and internationally ("Bhopal Tragedy"), are used to challenge students to explore the complex interplay between human activity and environmental destruction.
- **Chapter 21** ("Collective Behaviour, Social Movements, and Social Change") includes a new chapter introduction that looks at former Senate Page Brigette Depape, whose "Stop Harper" sign disrupted the throne speech. The chapter also features updated examples of mass behaviour, such as rumours, gossip, fads, and fashion; an updated discussion on revolutionary movements, such as Tunisia in 2013; and interesting new examples of protest movements in Canada.
- **Chapter 22** ("Power, Politics, and Government") has a new chapter introduction looking at the impact of the Idle No More movement. There is additional material on current Canadian political issues, including why Canadians believe there is a democratic deficit and the problem of domestic terrorism. The Sociology and New Media box on social media and politics ("The Political Impact of Social Media") has now been included in the body of the chapter. There is also additional material on social media, including the Facebook political participation study. (This chapter is available online at **www.nelson.com/student**.)
- **Chapter 23** ("Population and Urbanization") has a new chapter introduction showing how a new immigrant to Canada learned about some practices that Canadians take for granted. The chapter also has new material on cities and gender, and a new section on the second demographic transition. (This chapter is available online at **www.nelson.com/student**.)

UNIQUE FEATURES WALKTHROUGH

The following special features are specifically designed to reflect the themes of relevance and diversity in *Sociology in Our Times*, as well as to support student learning. The enhanced pedagogical framework aims to respect diverse learning preferences and engage today's students.

Chapter Learning Objectives

A list of objectives at the beginning of each chapter gives students an overview of major topics and a convenient aid for reviewing the central points of each chapter.

Chapter Focus Questions

Each chapter begins with an open-ended question that provides a starting point for students to think about the material covered in the chapter.

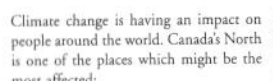

Interesting and Engaging Lived Experiences

Authentic first-person accounts are used as opening vignettes and throughout each chapter to create interest and give concrete meaning to the topics being discussed. Lived experiences, including racism, transgender marriage, environmental activism, transgender issues, disability, and homelessness, provide opportunities for students to consider social life beyond their own experiences and to examine class, ethnicity, gender, and age from diverse perspectives. An unusually wide range of diverse experiences—both positive and negative—is systematically incorporated to expose students to a multiplicity of viewpoints.

Critical Thinking Questions

After the opening lived experience in each chapter, a series of introductory questions invites students to think about the major topics discussed in the vignette and in the chapter.

Sociology and Everyday Life

Each chapter has a brief Sociology and Everyday Life quiz that relates the sociological perspective to the pressing social issues presented in the opening vignette. (Answers are provided online at **www.nelson.com/student**.) Do official statistics accurately reflect crime rates in Canada? Does increasing cultural diversity lead to an increasing incidence of hate crimes and racism? Do individuals over the age of 65 have the highest rate of poverty? Topics such as these will pique the interest of students.

BOX 18.1 SOCIOLOGY AND EVERYDAY LIFE

How Much Do You Know About the Media?

True	False	
T	F	1. You do not need to be concerned about your privacy when using social media sites such as Facebook.
T	F	2. Canadians spend more time watching television than using the Internet.
T	F	3. Canadian radio stations can broadcast any songs they wish.
T	F	4. When people design their online avatars in virtual worlds such as Second Life, they are not bound by our real-life cultural preferences about body size, hairstyles, and dress.
T	F	5. While Internet dating sites are becoming more common, most people still meet their partners through traditional means such as family, school, and church.

For answers to the quiz about the media, go to **www.nelson.com/student.**

Emphasizing the Importance of a Global Perspective

In our interconnected world, the sociological imagination must extend beyond national borders. The global implications of topics are examined throughout each chapter and in Sociology in Global Perspective boxes. Topics include commercial surrogacy in India, missing women around the world, the relationship between long-term environmental pollution and new social pressures in China, aging in Russia, and religious terrorism.

BOX 17.2 SOCIOLOGY IN GLOBAL PERSPECTIVE

Religious Terrorism

Religious terrorism has become a serious threat in postmodern societies. While there is a long history of religious wars among states and many earlier instances of religious terrorism, this type of terrorism has intensified over the past three decades. Following the September 11, 2001, attacks on the United States and subsequent bombings in Madrid, Bali, London, Mumbai, and elsewhere, much of the world's attention is now focused on Islamic terrorists, including members of groups like al-Qaeda and ISIS, and Nigeria's Boko Haram. However, all of the world's major religious traditions—as well as many minor religious movements—have been linked with terrorism. Among the questions that interest sociologists are: What are the causes of religious terrorism? How does it differ from other types of terrorist activities?

Violent extremism is not limited to any one faith. In Northern Ireland, the Catholic Irish Republican Army (IRA) exploded hundreds of bombs and killed hundreds of civilians in an attempt to free Northern Ireland from British rule. In 1994, a Jewish right-wing settler, Dr. Baruch Goldstein, shot and killed more than 30 Palestinians who were praying at the Tomb of the Patriarchs in Hebron. On the other side of the Israeli-Palestinian conflict, hundreds of Israelis have been killed by Palestinian suicide bombers. In Canada and the United States, there have been numerous bombings of abortion clinics, and several doctors who perform abortions have been killed or wounded—some of these attacks were carried out by Christian ministers, and other attacks were supported by militant Christian groups.

There are differences in the motivation behind these different attacks. The IRA bombing campaign had a strong political component, while members of the Japanese sect had few identifiable political goals. In each of the instances, however, the religious ideology of the terrorists defines the enemy and provides a justification for killing innocents. According to Bruce Hoffman, there are important differences between religious and secular terrorism:

> For the religious terrorist, violence first and foremost is a sacramental act or divine duty executed in direct response to some theological demand or imperative. Terrorism assumes a transcendental dimension, and its perpetrators are thereby unconstrained by the political, moral, or practical constraints that seem to affect other terrorists . . . Thus, religion serves as a legitimizing force—conveyed by sacred text or imparted via clerical authorities claiming to speak for the divine. (1995:272)

The acts of secular terrorists may be restrained by their fear of alienating potential supporters. Religious terrorists must please only themselves and their god, and can justify attacks against all "nonbelievers." Finally, purely religious terrorists are not trying to change an existing system, such as a particular government. Rather, they wish to transform the social order. For example, according to former al Qaeda leader Osama bin Laden:

Point/Counterpoint Boxes

Point/Counterpoint boxes encourage students to use their sociological knowledge to grapple with some of today's most hotly contested issues, such as how much accommodation is reasonable for multicultural minorities, the impact of white privilege, and the corporatization of medical charities. The topics covered can be used as springboards for in-class debate or online discussion forums.

BOX 18.3 POINT/COUNTERPOINT

New Media and Privacy

You have just tweeted a friend that you're going for coffee. As you pass a coffee shop, a coupon arrives on your phone offering 50 cents off a large cup of coffee. A friend who is travelling to Paris uses an online site to book a hotel room. Because she is using a Mac computer, the hotels that come up on the booking list are more expensive than if she had used a PC. A new college graduate has submitted a resumé for a job. The potential employer looks at the applicant's Facebook site, finds photos of the applicant using soft drugs at parties, and decides not to hire the person. In each of these cases, information that a person might expect to be private has been used by a third party. In the first two cases, the information was sold to an advertiser.

Online sites such as Facebook, Google, and Twitter provide a useful service for hundreds of millions of users. However, from the perspective of those who own users that allows advertisers to carefully target their ad campaigns. Advertisers are interested in knowing your location, relationship status, travel plans, musical tastes, occupation, and other interests.

Search engines such as Google make billions of dollars from tracking the key words you use. If you search for terms such as *headache* or *upset stomach*, you may receive ads or coupons for remedies for these maladies. Google also tracks your information across its different products, such as Gmail and YouTube, to develop more complete profiles of users in order to personalize the service.

Facebook and other networking sites frequently change their privacy policies with little notice. In 2009, Facebook suddenly made lists of friends publicly available. This change had serious consequences for many

Census Profile

The Census Profiles provide information that highlights changes in Canadian society based on census data. Each unique box uses recent statistics, ensuring students are up-to-date and informed about the topics discussed.

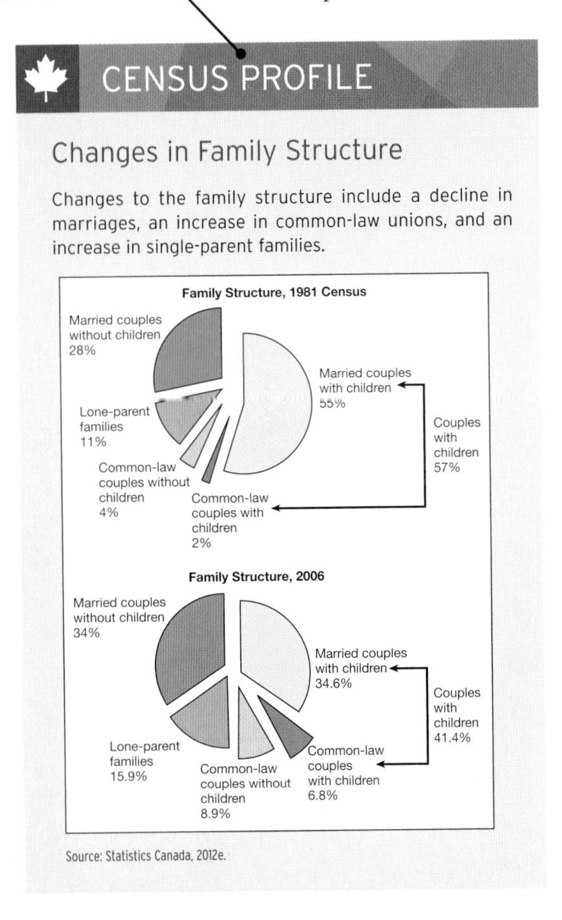

Concept Snapshot

A brief summary of all major perspectives covered in the chapter and the key people connected to those theories are presented in a table format that is efficient for studying.

CONCEPT SNAPSHOT

FUNCTIONALIST PERSPECTIVES Key thinker: Talcott Parsons	In modern societies, families serve the functions of sexual regulation, socialization, economic and psychological support, and provision of social status.
CONFLICT PERSPECTIVES Key thinker: Friedrich Engels	Families both mirror and help perpetuate social inequalities based on class and gender.
FEMINIST PERSPECTIVES Key thinkers: Nancy Mandell, Ann Duffy	Women's subordination is rooted in patriarchy and men's control over women's labour power.
SYMBOLIC INTERACTIONIST PERSPECTIVES Key thinker: Jessie Bernard	Family dynamics, including communication patterns and the subjective meanings that people assign to events, mean that interactions within families create a shared reality.
POSTMODERN PERSPECTIVES Key thinker: David Elkind	In postmodern societies, families are diverse and fragmented. Boundaries between the workplace and home are also blurred.

Time to Review Questions

Time to Review questions help students to review and retain key information from the preceding paragraphs.

TIME TO REVIEW

- Why does the country you are born in play such an important role in determining your life chances?
- How does excessive consumption in high-income countries affect people living in poverty in low-income countries?
- How have experts conceptualized world poverty and global stratification?
- Discuss the role of debt in determining the economic future of low-income nations.
- Why does foreign aid not always work to the benefit of people in low-income countries?

Visual Summary

The Visual Summary provides a concise summary of key points and theoretical perspectives. Summarized learning objectives are illustrated by a relevant image from the chapter. A list of **Key Terms** with page references provides a helpful study aid. The **Key Figures** feature reintroduces students to the major players in each chapter with a few important points and a portrait. Additionally, **Application Questions** encourage students to assess their knowledge of the chapter and apply insights they have gained to other issues.

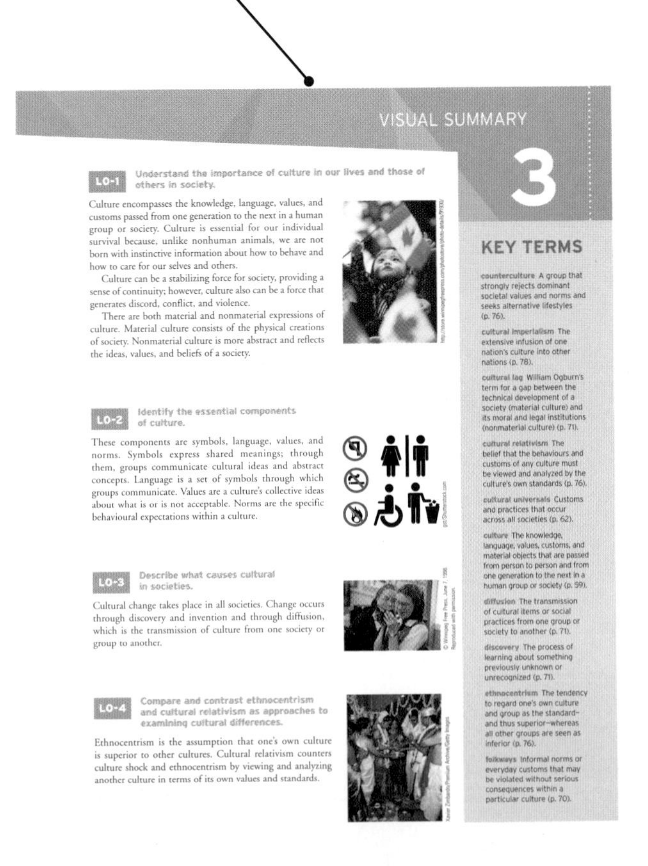

ANCILLARIES MindTap®

For Students

Stay organized and efficient with **MindTap**—a single destination with all the course material and study aids you need to succeed. Built-in apps leverage social media and the latest learning technology. For example:

- ReadSpeaker will read the text to you.
- Flashcards are pre-populated to provide you with a jump-start for review—or you can create your own.
- You can highlight text and make notes in your MindTap Reader. Your notes will flow into Evernote, the electronic notebook app that you can access anywhere when it's time to study for the exam.
- Self-quizzing allows you to assess your understanding of the material in the text.

Visit **www.nelson.com/student** to start using **MindTap**. Enter the Online Access Code from the card included with your text. If a code card is *not* provided, you can purchase instant access at **NELSONbrain.com**.

FOR INSTRUCTORS

The **Nelson Education Teaching Advantage (NETA)** program delivers research-based instructor resources that promote student engagement and higher-order thinking to enable the success of Canadian students and educators. Visit Nelson Education's Inspired Instruction website at www.nelson.com/inspired/ to find out more about NETA.

The following instructor resources have been created for *Sociology in Our Times,* Seventh Canadian Edition. Access these ultimate tools for customizing lectures and presentations at **www.nelson.com/instructor**.

NETA Test Bank

This resource includes over 2500 multiple-choice questions written according to NETA guidelines for effective construction and development of higher-order questions. Also included are approximately 650 true/false questions and more than 250 short-answer and essay questions.

The NETA Test Bank is available in a new, cloud-based platform. **Nelson Testing Powered by Cognero®** is a secure online testing system that allows instructors to author, edit, and manage test bank content from anywhere Internet access is available. No special installations or downloads are needed, and the desktop-inspired interface, with its drop-down menus and familiar, intuitive tools, allows instructors to create and manage tests with ease. Multiple test versions can be created in an instant, and content can be imported or exported into other systems. Tests can be delivered from a learning management system, the classroom, or wherever an instructor chooses. Nelson Testing Powered by Cognero for *Sociology in Our Times* can also be accessed through **www.nelson.com/instructor**.

NETA PowerPoint

Microsoft® PowerPoint® lecture slides for every chapter have been created by Liam Kilmurray of the University of Ottawa. There is an average of 35 slides per chapter, many featuring key figures, tables, and photographs from *Sociology in Our Times.* NETA principles of clear design and engaging content have been incorporated throughout, making it simple for instructors to customize the deck for their courses.

Image Library

This resource consists of digital copies of figures, short tables, and photographs used in the book. Instructors may use these jpegs to customize the NETA PowerPoint or create their own PowerPoint presentations.

ACKNOWLEDGMENTS

This edition of *Sociology in Our Times* would not have been possible without the insightful critiques of these colleagues, who have reviewed some or all of this book or its previous editions. Our profound thanks to each reviewer for engaging in this time-consuming process:

Dawn Anderson, University of Regina

Sean Ashley, Simon Fraser University

Christian Caron, University of Toronto

Choon-Lee Chai, Red Deer College

Bede Eke, University of Alberta

Sandra Enns, Langara College and Capilano University

Anthony Iafrate, Lambton College

Anton Oleinik, Memorial University

Christopher Schneider, University of British Columbia

We express our deep appreciation to Krista Robson of Red Deer College and Anthony Iafrate of Lambton College and the University of Windsor, who coordinated feedback from their students and whose feedback has greatly informed the pedagogical enhancements to the textbook.

We would also like to express our appreciation to the many individuals at Nelson Education involved in the development and production of *Sociology in Our Times*. Among them, Joanna Cotton and Cara Yarzab, who gave us encouragement and sound advice on several earlier editions. The seventh edition included many changes, including a new environment chapter and a new design. We have very much enjoyed the friendship and guidance given throughout this process by Maya Castle, our Publisher, and Content Development Manager Lisa Berland, who have worked tirelessly with us on this project. Linda Szostak has helped keep our prose legible and facts straight as the Copy Editor, and Christine Gilbert has overseen the production process. We also thank Terry Fedorkiw and the sales and marketing staff for their great work in ensuring that there would be a seventh edition of this book, and Julie Pratt and Daniela Glass, who managed the permissions research and clearances for this edition. As always, the leadership, good humour, and hard work of the Nelson team have made this an enjoyable experience.

We would both like to thank our families, who—after an exasperated "are you *still* working on that book?!?!?!?"—continue to provide encouragement and support.

CHAPTER 1

THE SOCIOLOGICAL PERSPECTIVE

LEARNING OBJECTIVES

AFTER READING THIS CHAPTER, YOU SHOULD BE ABLE TO

LO-1 Explain what sociology can contribute to our understanding of social life.

LO-2 Explain why the sociological imagination is important for studying society.

LO-3 Discuss the major contributions of early sociologists.

LO-4 Describe the key assumptions behind each of the contemporary theoretical perspectives.

CHAPTER FOCUS QUESTION

How does sociology add to our knowledge of human societies and of social issues such as consumerism?

The Canadian Press/Nathan Denette

Consider the description by *National Post* writer Danielle Kubes describing some of the challenges of her generation, often referred to as "Millennials," Generation Y, or the "Entitlement Generation" (born between 1981 and 1999). In her article entitled "The spending diaries: What three millennials spend their money on and why," Kubes explains why her generation has it so much harder than the boomers (born between 1946 and 1964):

I'm a millennial. I hate that word: "Millennial."

Before it came into vogue, the nine million Canadians born between 1980 and 2000 were called the echo boom, or gen-Y, which I preferred, since it seemed to denote we came from somewhere—as opposed to a spontaneous wave of children that appeared with genetic mutations that gave them super opposable thumbs, perfect for texting.

Calling us millennials seems an attempt by previous generations to distance themselves from our propensities and our circumstances, as if they had no part. And the widest division seems to be in the economic sphere—how much we earn, the way we earn, and what we choose to buy.

Naturally, these economics have informed every other aspect of our life, like how we mate, the shelter we choose, and the way we transport ourselves.

I spoke recently with David Coletto, who, as the 32-year-old CEO of Abacus Data, an Ottawa-based research firm, is a millennial that studies other millennials.

"We asked 18- to 35-year-olds in Canada, a representative sample, have you achieved a number of big milestones in your life?" Coletto says.

"Have you moved out of your parents' home, do you have children, have you bought a house, are you financially independent, have you got a job in the field that you studied for, or a career, have you started your career?" he says.

"What's unique is that millennials are achieving many of these milestones far later in their life."

It's easy to see why—no generation of Canadians has ever spent so many years being educated, while ending up so poor.

The economy before us had been growing almost steadily for 70 years, with a slight blip in the 1990s (see the 1994 movie Reality Bites for details), and then it contracted at the precise moment the first cohort of millennials graduated and started job-hunting.

What followed was a dramatic rise in precarious employment . . . Wages were stagnant for those who did manage to secure employment—in 20 years they've risen only 35 per cent for the average Canadian.

*In the same 20 years, the cost of undergraduate tuition rose 334 per cent, leading to an average student debt load of around $25,000, according to the Canadian Federation of Students. ... The double whammy of a stunted job market and late adulthood, started deep in the red, has shaped this generation's spending—from choosing to live at home, to outrage at spending $70 a month on cable, to ignoring the boomer markers of success.**

Source: Kubes, D. 2015

Without question, we live in a "consumer society" where many of us rely on credit cards, loans, and lines of credit to pay for items we want to purchase or services we need. However, the younger generation has had to wrestle with financial challenges that older generations did not have to worry about, like paying off their student loans while trying to save and manage spending on lower salaries. A recent survey found that more than one-third of Generation Y'ers find it almost impossible to save (Kubes, 2015). The consequences for Generation Y'ers living in a consumer society is that it is expensive to live, easy to spend, and a struggle to save.

Why are sociologists interested in studying consumerism? Sociologists study the *consumer society*—a society in which discretionary consumption is a mass phenomenon among people across diverse income categories—because it provides interesting and important insights into many aspects of social life and our world. In the consumer society, for example, purchasing goods and services is not limited to the wealthy or even the middle class; people in all but the lowest income brackets spend time, energy, and money on shopping, and some amass large debts in the process. According to sociologists, shopping and consumption—in this instance, the money that people spend on goods and services—are processes that extend beyond our individual choices and are rooted in larger structural conditions in the social, political, and economic order in which we live. In the second decade of the 21st century, many people have had financial problems not only because of their own consumerism but also because of national and global economic instability. In addition, the process of globalization has dramatically affected consumerism and shifted the worldwide production and distribution of goods and services.

Why have shopping, spending, and credit card debt become major problems for some people? How are social relations and social meanings shaped by what people in a given society produce and how they consume? What national and worldwide social processes shape the production and consumption of goods, services, and information? In this chapter, we see how the sociological perspective helps us examine complex questions such as these, and we wrestle with some of the difficulties of attempting to study human behaviour. Before reading on, take the Sociology and Everyday Life quiz in Box 1.1, which lists a number of commonsense notions about consumption and consumer debt.

In this chapter, we will see how the sociological perspective helps us examine social issues, such as debt accumulation and overspending, and wrestle with some of the difficulties of attempting to study human behaviour. Throughout this text, you will be invited to use the sociological perspective and to apply your sociological imagination to reexamine your social world and explore important social issues and problems you may not have considered before.

* Danielle Kubes, "The spending diaries: What three millennials spend their money on and why," *Financial Post*, June 15, 2015, http://business.financialpost.com/personal-finance/young-money/the-spending-diaries-what-three-millennials-spend-their-money-on-and-why?__lsa=3dd8-793b. Material republished with the express permission of: **National Post**, a division of Postmedia Network Inc.

CRITICAL THINKING QUESTIONS

1. Why have shopping, spending, credit card debt, and bankruptcy become major problems for some people?
2. How are social relations and social meanings shaped by what people in a given society produce and how they consume?
3. The millennial generation (those born after 1981) has often been described as the "Entitlement Generation." Is credit card debt an example of this entitlement or of other social factors? How do you respond to this label?

LO-1 PUTTING SOCIAL LIFE INTO PERSPECTIVE

▶ **sociology** The systematic study of human society and social interaction.

Sociology is the systematic study of human society and social interaction. It is a *systematic* study because sociologists apply both theoretical perspectives and research methods (or orderly approaches) to examinations of social behaviour. Sociologists study human societies and their social interactions in order to develop theories of how human behaviour is shaped by group life and how, in turn, group life is affected by individuals.

To better understand the scope of sociology, you might compare it to other social sciences, such as anthropology, psychology, economics, and political science. Like anthropology, sociology studies many aspects of human behaviour; however, sociology is particularly interested in contemporary social organization, relations, and social change. Anthropology primarily concentrates on human existence over geographic space and evolutionary time, meaning that it focuses more on traditional societies and the development of diverse cultures. Cultural anthropology most closely overlaps sociology. Unlike psychology, sociology examines the individual in relation to external factors, such as the effects of groups, organizations, and social institutions on individuals and social life; psychology primarily focuses on internal factors relating to the individual in explanations of human behaviour and mental processes—what occurs in the mind. Social psychology is similar to sociology in that it emphasizes how social conditions affect individual behaviour. Although sociology examines all major social institutions, including the economy and politics, the fields of economics and political science concentrate primarily on a single institution—the economy or the political system. Topics of mutual interest to economics and sociology include issues such as consumerism and debt, which can be analyzed at global, national, and individual levels. Topics of mutual interest to political science and sociology are how political systems are organized and how power is distributed in society. As you can see, sociology shares similarities with other social sciences but offers a comprehensive approach to understanding many aspects of social life.

WHY STUDY SOCIOLOGY?

▶ **society** A large social grouping that shares the same geographical territory and is subject to the same political authority and dominant cultural expectations.

▶ **global interdependence** A relationship in which the lives of all people are closely intertwined and any one nation's problems are part of a larger global problem.

Sociology helps us gain a better understanding of our selves and our social world. It enables us to see how behaviour is largely shaped by the groups to which we belong and by the society in which we live.

Most of us take our social world for granted and view our lives in personal terms. Because of our culture's emphasis on individualism, we often do not consider the complex connections between our own lives and the larger, recurring patterns of the society and world in which we live. Sociology helps us look beyond our personal experiences and gain insights into society and the larger world order. A **society** is a large social grouping that shares the same geographical territory and is subject to the same political authority and dominant cultural expectations, such as Canada, the United States, or Mexico. Examining the world order helps us understand that each of us is affected by **global interdependence**—a relationship in which the lives of all people are closely intertwined and any one nation's problems are part of a larger global problem.

Individuals can make use of sociology on a more personal level. Sociology enables us to move beyond established ways of thinking, thus allowing us to gain new insights into ourselves and to develop a greater awareness of the connection between our own "world" and that of other people. According to sociologist Peter Berger (1963:23), sociological inquiry helps us see that "things are not what they seem." Sociology provides new ways of approaching problems and making decisions in everyday life. It promotes understanding and tolerance by enabling each of us to look beyond our personal experiences (see Figure 1.1).

Many of us rely on intuition or common sense gained from personal experience to help us understand our daily lives and other people's behaviour. **Commonsense knowledge** guides ordinary conduct in everyday life. We often rely on common sense—or "what everybody knows"—to answer key questions about behaviour: Why do people behave the way they do? Who makes the rules? Why do some people break rules and why do others follow them?

commonsense knowledge A form of knowing that guides ordinary conduct in everyday life.

Many commonsense notions are myths. A *myth* is a popular but false notion that may be used, either intentionally or unintentionally, to perpetuate certain beliefs or "theories" even in the light of conclusive evidence to the contrary. For example, one widely held myth is that "money can buy happiness." By contrast, sociologists strive to use scientific standards, not popular myths or hearsay, in studying society and social interaction. They use systematic research techniques and are accountable to the scientific community for their methods and the presentation of their findings. Although some sociologists argue that sociology must be completely value free—without distorting subjective (personal or emotional) bias—others do not think that total objectivity is an attainable or desirable goal when studying human behaviour. However, all sociologists attempt to discover patterns or commonalities in human behaviour. For example, when they study shopping behaviour or credit card abuse, sociologists look for recurring patterns of behaviour and for larger, structural factors that contribute to people's behaviour. Women's studies scholar Juliet B. Schor refers to consumption as the "see–want–borrow–buy" process, which she believes is a comparative process in which desire is structured by what we see around us (1999:68). As sociologists examine patterns such as these, they begin to use their sociological imagination.

TIME TO REVIEW

- What commonsense understandings do you take for granted in everyday life?
- Which of these (if any) are myths?

FIGURE 1.1 : FIELDS THAT USE SOCIAL SCIENCE RESEARCH

In many careers, including jobs in academia, business, communications, health and human services, and law, the ability to analyze social science research is an important asset.

Health and Human Services	Business	Communication	Academia	Law
Medicine Nursing Physical Therapy Occupational Therapy Counselling Education Social Work	Advertising Labour Relations Management Marketing	Broadcasting Public Relations Journalism Social Media	Anthropology Economics Geography History Information Studies Media Studies/ Communication Political Science Psychology Sociology	Law Criminal Justice Mediation Conflict Resolution

Source: Based on Katzer, Cook, and Crouch, 1991.

LO-2 THE SOCIOLOGICAL IMAGINATION

▶ **sociological imagination** C. Wright Mills's term for the ability to see the relationship between individual experiences and the larger society.

How can we make a connection between our personal experiences and what goes on in the larger society? Sociologist C. Wright Mills (1959a) described sociological reasoning as the **sociological imagination**—the ability to see the relationship between individual experiences and the larger society. This awareness enables us to understand the link between our personal experiences and the social contexts in which they occur. The sociological imagination helps us distinguish between personal troubles and social (or public) issues. *Personal troubles* are private problems of individuals and the networks of people with whom they associate regularly. As a result, those problems must be solved by individuals within their immediate social settings. For example, one person being unemployed or running up a high credit card debt could be identified as a personal trouble. *Public issues* are problems that affect large numbers of people and often require solutions at the societal level. Widespread unemployment and massive, nationwide consumer debt are examples of public issues. The sociological imagination helps us place seemingly personal troubles, such as losing one's job or overspending on credit cards, into a larger social context, where we can distinguish whether and how personal troubles may be related to public issues.

OVERSPENDING AS A PERSONAL TROUBLE Although individual behaviour can contribute to social problems, our individual experiences are influenced and in some situations determined by the society as a whole—by its historical development and its organization. In everyday life, we often blame individuals for "creating" their own problems. If a person sinks into debt because of overspending or credit card abuse, many people consider it to be the result of his or her own personal failings. However, this approach overlooks debt among people who are in low-income brackets, having no way other than debt to gain the basic necessities of life. By contrast, at middle- and upper-income levels, overspending takes on a variety of other meanings.

At the individual level, people may accumulate credit cards and spend more than they can afford, thereby affecting all aspects of their lives, including health, family relationships, and employment stability. Sociologist George Ritzer (1999:29) suggests that people may overspend through a gradual process in which credit cards "lure people into consumption by easy credit, and then entice them into still further consumption by offers of 'payment holidays,' new cards, and increased credit limits."

OVERSPENDING AS A PUBLIC ISSUE We can use the sociological imagination to look at the problem of overspending and credit card debt as a public issue—a societal problem.

BOX 1.1 SOCIOLOGY AND EVERYDAY LIFE

How Much Do You Know About Consumption and Debt Accumulation?

True	False	
T	F	1. The average Canadian household has just over $100,000 in debt.
T	F	2. Generation Y'ers are more likely to overspend than previous generations.
T	F	3. Student debt in Canada has declined in recent years.
T	F	4. Overspending is primarily a problem for people in the higher-income brackets in Canada and other affluent nations.
T	F	5. Generation Y'ers are much more inclined to impulse buy and then later regret their purchases than are their baby boomer parents.

For answers to quiz on consumption and credit cards, go to **www.nelson.com/student.**

For example, Ritzer (1998) suggests that the relationship between credit card debt and the relatively low savings rate constitutes a public issue. In 2014, Canadian credit card debt was estimated to be at more than $73 billion, while the savings rate continued to diminish. Because savings is money that governments, businesses, and individuals can borrow for expansion, a lack of savings often creates problems for future economic growth. Some practices of the credit card industry are also a public issue because they harm consumers. Credit card companies may encourage overspending, and then substantially increase interest rates and other fees, making it more difficult for consumers to pay off debts. Mills's *The Sociological Imagination* (1959a) is useful for examining issues because it helps integrate microlevel (individual and small-group) troubles with compelling public issues of our day. Recently, his ideas have been applied at the global level as well.

THE IMPORTANCE OF A GLOBAL SOCIOLOGICAL IMAGINATION Although existing sociological theory and research provide the foundation for sociological thinking, we must reach beyond past studies that have focused primarily on North America to develop a more comprehensive *global* approach for the future. In the 21st century, we face unprecedented challenges, ranging from global political and economic instability to environmental concerns and natural disasters and terrorism. All of the nations of the world are not on equal footing when it comes to economics and politics. The world's **high-income countries** are nations with highly industrialized economies; technologically advanced industrial, administrative, and service occupations; and relatively high levels of national and personal income. Examples include the United States, Canada, Australia, New Zealand, Japan, and the countries of Western Europe (see Map 1.1).

high-income countries Nations with highly industrialized economies; technologically advanced industrial, administrative, and service occupations; and relatively high levels of national and personal income.

MAP 1.1 : THE WORLD'S ECONOMIES IN THE 21ST CENTURY

High-income, middle-income, and low-income countries.

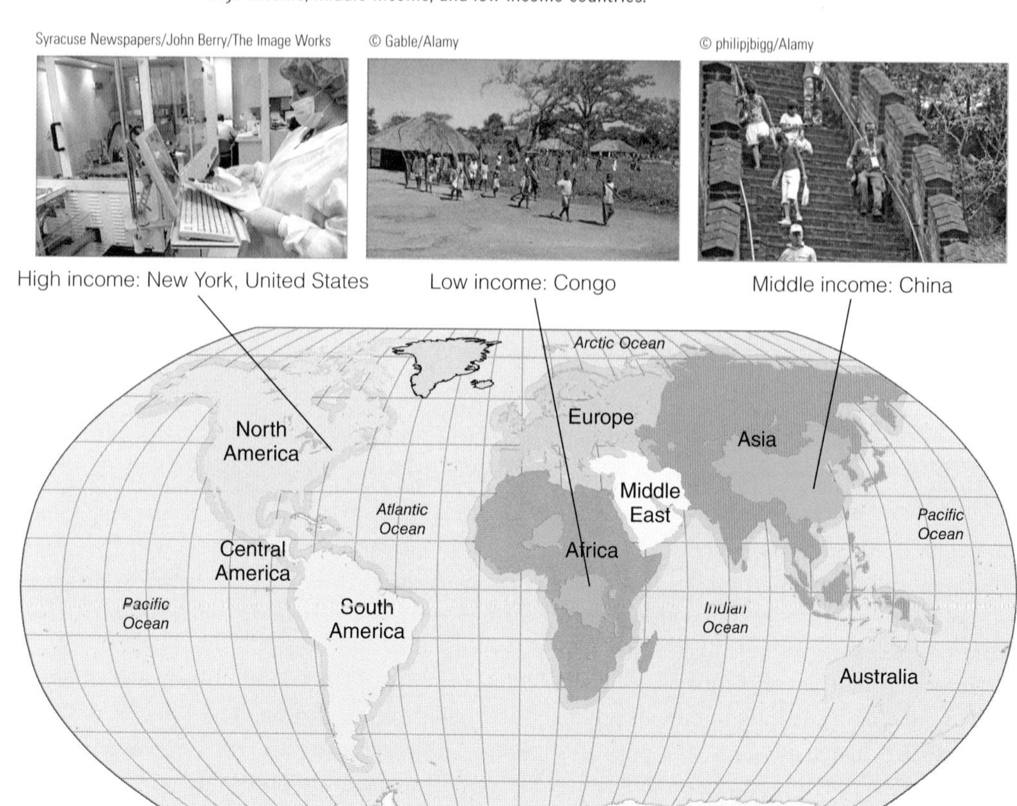

As compared with other nations of the world, many high-income nations have a high standard of living and a lower death rate due to advances in nutrition and medical technology. However, everyone living in a so-called high-income country does not necessarily have a high income or an outstanding quality of life. Even among middle- and upper-income people, problems such as personal debt may threaten economic and social stability. This may increasingly be the case as the effects of the recent global economic crisis take hold.

▶ **middle-income countries** Nations with industrializing economies, particularly in urban areas, and moderate levels of national and personal income.

▶ **low-income countries** Countries that are primarily agrarian, with little industrialization and low levels of national and personal income.

In contrast, **middle-income countries** are nations with industrializing economies, particularly in urban areas, and moderate levels of national and personal income. Examples of middle-income countries include Brazil and Mexico, which are experiencing rapid industrialization. **Low-income countries** are primarily agrarian, with little industrialization and low levels of national and personal income. Examples of low-income countries include many of the nations of Africa and Asia, where people typically work the land and are among the poorest in the world (see Chapter 10).

Throughout this text, we will continue to develop our sociological imaginations by examining social life in Canada and other nations. The future of this country is deeply intertwined with the future of all other nations of the world on economic, political, environmental, and humanitarian levels. We buy many goods and services that were produced in other nations, and we sell much of what we produce to the people of other nations (see Box 1.2).

Peace in other nations is important if we are to ensure peace within our borders. Famine, unrest, and brutality in other regions of the world must be of concern to people in Canada. Moreover, fires, earthquakes, famine, or environmental pollution in one nation typically has an adverse influence on other nations as well. Global problems contribute to the large influx of immigrants who arrive in Canada annually. These immigrants bring with them a rich diversity of language, customs, religions, and previous life experiences; they also contribute to dramatic population changes that will have a long-term effect on this country.

Whatever your race or ethnicity, class, sex, or age, are you able to include in your thinking the perspectives of people who are quite different from you in experiences and points of view? Before you answer this question, a few definitions are in order. *Race* is a term used by many people to specify groups of people distinguished by physical characteristics such as skin colour, but no "pure" racial types exist and most sociologists consider the concept of race to be a social construction used to justify existing social inequalities. *Ethnicity* refers to a group's cultural heritage or identity; it is based on factors such as language or country of origin. *Class* is the relative location of a person or group within the larger society; it is based on wealth, power, prestige, or other valued resources. *Sex* refers to the biological and anatomical differences between females and males. By contrast, *gender* refers to the meanings, beliefs, and practices associated with sex differences, referred to as *femininity* and *masculinity*.

In forming your own global sociological imagination and in seeing the possibilities for sociology in the 21st century, it will be helpful for you to understand the development of the discipline.

LO-3 THE DEVELOPMENT OF SOCIOLOGICAL THINKING

Throughout history, social philosophers and religious authorities have made countless observations about human behaviour. However, early thinkers focused their thoughts on what they believed society *ought* to be like, rather than describing how society *was*.

Several revolutions that took place in the 18th century had a profound influence on the origins of sociology. The Enlightenment produced an intellectual revolution in how people thought about social change, progress, and critical thinking. The optimistic views of the *philosophes* and other social thinkers regarding progress and equal opportunity (at least for some people) became

BOX 1.2 SOCIOLOGY IN GLOBAL PERSPECTIVE

Global Walmartization: From Big-Box Stores to Online Supermarkets in China

Did you know that:

- Walmart has more than 10,000 stores in 27 countries and that more than half of all Walmart stores worldwide are located outside the United States?
- Walmart operates nearly 300 stores in China, including supercentres, neighborhood markets, and Sam's Clubs?
- Walmart is a major player in the credit card business in China, where people in the past were opposed to buying anything on credit?

Although most of us are aware that Walmart stores are visible in virtually every city in North America, we are less aware of the extent to which Walmart and other big-box stores are changing the face of the world economy as megacorporations expand their operations into other nations and into the credit card business.

The strategic placement of Walmart stores both here and abroad accounts for part of the financial success of this retailing giant, but another U.S. export—credit cards—is also part of the company's business plan. Credit cards are changing the way that people shop and how they think about spending money in emerging nations such as China. For example, Walmart China is aggressively seeking both shoppers and credit card holders.

An exciting aspect of studying sociology is comparing our own lives with those of people around the world. Global consumerism, as evidenced by the opening of a Walmart Supercenter in Shanghai, China, provides a window through which we can observe how issues such as shopping and credit affect all of us. Which aspects of this photo reflect local culture? Which aspects reflect a global cultural phenomenon?

By encouraging people to spend money now rather than save it for later, corporations such as Walmart that issue "co-branded" credit cards gain in two ways: (1) people buy more goods than they would otherwise, thus increasing sales; and (2) the corporation whose "brand" is on the credit card increases its earnings as a result of the interest the cardholder pays on credit card debt.

The motto for the Walmart credit card in China is "Maximizing value, enjoying life," and this idea encourages a change in attitude from the past, when—regardless of income level—most residents of that country did not possess credit cards. This has brought a corresponding surge in credit card debt, which can be partly attributed to aggressive marketing by transnational retailers, but also to credit card companies encouraging consumers to buy now, pay later. But Walmart is not stopping there: the company also now owns a controlling (51 percent) interest in Yihaodian, an online Chinese supermarket that sells food, cosmetics, clothing, and consumer electronics to more than one million registered users in five major cities in China.

Throughout this course, as we study the social effects of major changes in societies, such as industrialization, urbanization, and the progression of the digital age, we will see that many of the issues we discuss, such as consumerism and globalization, have both positive and negative effects. Global consumerism, whether in big-box stores or through credit cards or electronic commerce, provides a window through which we can observe how an issue such as shopping affects all of us. Among the poor and those most hard-hit by difficult economic times, the lack of ability to purchase basic necessities is a central litmus test for analyzing quality of life and social inequality. Among persons in the middle class, purchasing power is often used to determine social mobility (the ability to move into) or social stability (the ability to stay on) the middle rungs of a society's ladder of income and wealth. Among persons in the upper class, high rates of luxury consumerism are often seen as an outward sign of "having it all." As we will see, ideas related to consumerism and globalization vary widely across nations.

Are people in North America unique in how we view consumerism? in how we view Walmart and other big-box stores? What do you think?

Sources: Based on Lemaire, 2012; Walmart.com, 2012; WalMart Corporation, 2012.

part of the impetus for political and economic revolutions, first in America and then in France. The Enlightenment thinkers had emphasized a sense of common purpose and hope for human progress; the French Revolution and its aftermath replaced these ideals with discord and overt conflict (see Arendt, 1973; Schama, 1989).

industrialization The process by which societies are transformed from dependence on agriculture and handmade products to an emphasis on manufacturing and related industries.

During the 19th and early 20th centuries, another form of revolution also occurred: the Industrial Revolution. **Industrialization** is the process by which societies are transformed from dependence on agriculture and handmade products to an emphasis on manufacturing and related industries. This process first occurred during the Industrial Revolution in Britain between 1760 and 1850, and was soon repeated throughout Western Europe. By the mid-19th century, industrialization was well under way in Canada and the United States. Massive economic, technological, and social changes occurred as machine technology and the factory system shifted the economic base of these nations from agriculture to manufacturing. A new social class of industrialists emerged in textiles, iron smelting, and related industries. Many people who had laboured on the land were forced to leave their tightly knit rural communities and sacrifice well-defined social relationships to seek employment as factory workers in the emerging cities, which became the centres of industrial work.

urbanization The process by which an increasing proportion of a population lives in cities rather than in rural areas.

Urbanization accompanied modernization and the rapid process of industrialization. **Urbanization** is the process by which an increasing proportion of a population lives in cities rather than in rural areas. Although cities existed long before the Industrial Revolution, the development of the factory system led to a rapid increase in both the number of cities and the size of their populations. People from diverse backgrounds worked together in the same factory. At the same time, many people shifted from being *producers* to being *consumers*. For example, families living in the cities had to buy food with their wages because they could no longer grow their own crops to consume or to barter for other resources. Similarly, people had to pay rent for their lodging because they could no longer exchange their services for shelter.

These living and working conditions led to the development of new social problems: inadequate housing, crowding, unsanitary conditions, poverty, pollution, and crime. Wages were so low that entire families—including young children—were forced to work, often under

As the Industrial Revolution swept through North America beginning in the 19th century, children being employed in factories became increasingly common. Soon social thinkers began to explore such new social problems brought about by industrialization.

hazardous conditions and with no job security. As these conditions became more visible, a new breed of social thinkers turned its attention to trying to understand why and how society was changing.

TIME TO REVIEW

- What were the primary social factors that contributed to the development of sociological thinking?

EARLY THINKERS: A CONCERN WITH SOCIAL ORDER AND STABILITY

At the same time that urban problems were worsening, natural scientists had been using reason, or rational thinking, to discover the laws of physics and the movement of the planets. Social thinkers began to believe that by applying the methods developed by the natural sciences, they might discover the laws of human behaviour and apply these laws to solve social problems. Historically, the time was ripe for such thoughts because the Age of Enlightenment had produced a belief in reason and humanity's ability to perfect itself.

Early social thinkers—such as Auguste Comte, Harriet Martineau, Herbert Spencer, and Émile Durkheim—were interested in analyzing social order and stability, and many of their ideas had a dramatic influence on modern sociology.

AUGUSTE COMTE French philosopher Auguste Comte (1798–1857) coined the term *sociology* from the Latin *socius* ("social, being with others") and the Greek *logos* ("study of") to describe a new science that would engage in the study of society. Even though he never conducted sociological research, Comte is considered by some to be the "founder of sociology." His theory that societies contain *social statics* (forces for social order and stability) and *social dynamics* (forces for conflict and change) continues to be used, although not in these exact terms, in contemporary sociology.

▶ **positivism** A belief that the world can best be understood through scientific inquiry.

Drawing heavily on the ideas of his mentor, Count Henri de Saint-Simon, Comte stressed that the methods of the natural sciences should be applied to the objective study of society. Comte's philosophy became known as **positivism**—a belief that the world can best be understood through scientific inquiry. Comte believed that objective, bias-free knowledge was attainable only through the use of science rather than religion.

The ideas of Saint-Simon and Comte regarding the objective, scientific study of society are deeply embedded in the discipline of sociology. Of particular importance is Comte's idea that the nature of human thinking and knowledge passed through several stages as societies evolved from simple to more complex. Comte described how the idea systems and their corresponding social structural arrangements changed according to what he termed the *law of the three stages:* the theological, metaphysical, and scientific (or positivistic) stages. Comte believed that knowledge began in the *theological stage*—explanations were based on religion and the supernatural. Next, knowledge moved to the *metaphysical stage*—explanations were based on abstract philosophical speculation. Finally, knowledge would reach the *scientific,* or *positive, stage*—explanations are based on systematic observation, experimentation, comparison, and historical analysis. Shifts in the forms of knowledge in societies were linked to changes in the structural systems of society.

Auguste Comte

In the theological stage, kinship was the most prominent unit of society; however, in the metaphysical stage, the state became the prominent unit and control shifted from small groups to the state, military, and law. In the scientific, or positive, stage, industry became the prominent structural unit in society and scientists became the spiritual leaders, replacing in importance the priests and philosophers of the previous stages of knowledge. For Comte, this progression through the three stages constituted the basic law of social dynamics, and, when coupled with the laws of statics (which emphasized social order and stability), constituted the new science of sociology, which could bring about positive social change.

HARRIET MARTINEAU As a woman in a male-dominated discipline and society, Harriet Martineau (1802–1876) received no recognition in the field of sociology until recently; however, the British sociologist made Comte's works more accessible for a wide variety of scholars. Not only did she translate and condense Comte's work, but she was also an active sociologist in her own right. Martineau studied the social customs of Britain and the United States, and analyzed the consequences of industrialization and capitalism. In *Society in America* (1962/1837), she examined religion, politics, child rearing, slavery, and immigration in the United States, paying special attention to social distinctions based on class, race, and gender. Her works explore the status of women, children, and "sufferers" (persons who were considered to be criminal, mentally ill, handicapped, poor, or alcoholic).

Based on her reading of Mary Wollstonecraft's *A Vindication of the Rights of Women* (1974/1797), Martineau advocated racial and gender equality. She was also committed to creating a science of society that would be grounded in empirical observations and widely accessible to people. She argued that sociologists should be impartial in their assessment of society, but that comparing the existing state of society with the principles on which it was founded is entirely appropriate (Lengermann and Niebrugge-Brantley, 1998).

Some scholars have argued that Martineau's place in the history of sociology should be as a founding member of this field of study, not just as the translator of Auguste Comte's work (Hoecker-Drysdale, 1992; Lengermann and Niebrugge-Brantley, 1998). Others have highlighted her influence in spreading the idea that societal progress could be brought about by the spread of democracy and the growth of industrial capitalism (Polanyi, 1944). Martineau believed that a better society would emerge if women and men were treated equally, enlightened reform occurred, and cooperation existed among people in all social classes (but led by the middle class).

Spencer Arnold/Stringer/Getty

Harriet Martineau

In keeping with the sociological imagination, Martineau not only analyzed large-scale social structures in society but also explored how these factors influenced the lives of people, particularly women, children, and those who were marginalized by virtue of being criminal, mentally ill, disabled, poor, or alcoholic (Lengermann and Niebrugge-Brantley, 1998). She remained convinced that sociology, the "true science of human nature," could bring about new knowledge and understanding, enlarging people's capacity to create a just society and live heroic lives (Hoecker-Drysdale, 1992).

HERBERT SPENCER Unlike Comte, who was strongly influenced by the upheavals of the French Revolution, British social theorist Herbert Spencer (1820–1903) was born in a more peaceful and optimistic period in his country's history. Spencer's major contribution to sociology was an evolutionary perspective on social order and social change. Although the term *evolution* has various meanings, evolutionary theory should be taken to mean "a theory to explain the mechanisms of organic/social change" (Haines, 1997:81). According to Spencer's theory of general evolution, society, like a biological organism, has various interdependent parts (such as the family, the economy, and the government) that work to ensure the stability and survival of the entire society.

Spencer believed that societies developed through a process of "struggle" (for existence) and "fitness" (for survival), which he referred to as the "survival of the fittest." Because this phrase is often attributed to Charles Darwin, Spencer's view of society is known as **social Darwinism**—the belief that those species of animals (including human beings) best adapted to their environment survive and prosper, whereas those poorly adapted die out. Spencer equated this process of *natural selection* with progress, because only the "fittest" members of society would survive the competition; the "unfit" would be filtered out of society. Based on this belief, he strongly opposed any social reform that might interfere with the natural selection process and, thus, damage society by favouring its least worthy members.

social Darwinism The belief that those species of animals (including human beings) best adapted to their environment survive and prosper, whereas those poorly adapted die out.

Critics have suggested that many of Spencer's ideas had serious flaws. For one thing, societies are not the same as biological systems; people are able to create and transform the environment in which they live. Moreover, the notion of the survival of the fittest can easily be used to justify class, racial–ethnic, and gender inequalities, and to rationalize the lack of action to eliminate harmful practices that contribute to such inequalities. Not surprisingly, Spencer's hands-off view was applauded by wealthy industrialists of his day. John D. Rockefeller, who gained monopolistic control of much of the U.S. oil industry early in the 20th century, maintained that the growth of giant businesses was merely the "survival of the fittest" (Feagin, Baker, and Feagin, 2006).

Social Darwinism served as a rationalization for some people's assertion of the superiority of the white race. After the American Civil War, it was used to justify the repression and neglect of African Americans as well as the policies that resulted in the annihilation of Native American populations. Although some social reformers spoke out against these justifications, "scientific" racism continued to exist (Turner, Singleton, and Musick, 1984). In both positive and negative ways, many of Spencer's ideas and concepts have been deeply embedded in social thinking and public policy for more than a century.

social facts Émile Durkheim's term for patterned ways of acting, thinking, and feeling that exist outside any one individual.

ÉMILE DURKHEIM French sociologist Émile Durkheim (1858–1917) criticized some of Spencer's views while incorporating others into his own writing. Durkheim stressed that people are the product of their social environment and that behaviour cannot be fully understood in terms of *individual* biological and psychological traits. He believed that the limits of human potential are *socially,* not *biologically,* based. As Durkheim saw religious traditions evaporating in his society, he searched for a scientific, rational way to provide for societal integration and stability (Hadden, 1997).

In *The Rules of Sociological Method* (1964a/1895), Durkheim set forth one of his most important contributions to sociology: the idea that societies are built on social facts. **Social facts** are patterned ways of acting, thinking, and feeling that exist outside any one individual but that exert social control over each person. Durkheim believed that social facts must be explained by other social facts—by reference to the social structure rather than to individual attributes.

Durkheim was concerned with social order and social stability because he lived during the period of rapid social changes in Europe resulting from industrialization and urbanization. His recurring question was this: How do societies manage to hold together? In *The Division of Labor in Society* (1933/1893), Durkheim concluded that preindustrial societies were held together by strong traditions and by members' shared moral beliefs and values. As societies industrialized, more specialized economic activity became the basis of the social bond because people became interdependent.

Émile Durkheim

Durkheim observed that rapid social change and a more specialized division of labour produce *strains* in society. These strains lead to a breakdown in traditional organization, values, and authority and to a dramatic increase in **anomie**—a condition in which social control becomes ineffective as a result of the loss of shared values and of a sense of purpose in society. According to Durkheim, anomie is most likely to occur during a period of rapid social change. In *Suicide* (1964b/1897), he explored the relationship between anomic social conditions and suicide, as discussed in Chapter 2.

anomie Émile Durkheim's term for a condition in which social control becomes ineffective as a result of the loss of shared values and a sense of purpose in society.

Durkheim's contributions to sociology are so significant that he has been referred to as "*the* crucial figure in the development of sociology as an academic discipline [and as] one of the deepest roots of the sociological imagination" (Tiryakian, 1978:187). He has long been viewed as a proponent of the scientific approach to examining social facts that lie outside individuals. He is also described as the founding figure of the functionalist theoretical tradition. Scholars have acknowledged Durkheim's influence on contemporary social theory, including the structuralist and postmodernist schools of thought. Like Comte, Martineau, and Spencer, Durkheim emphasized that sociology should be a science based on observation and the systematic study of social facts rather than on individual characteristics or traits.

Although they acknowledge Durkheim's important contributions, some critics note that his emphasis on societal stability, or the "problem of order"—how society can establish and maintain social stability and cohesiveness—obscures the *subjective meaning* that individuals give to social phenomena, such as religion, work, and suicide. In this view, overemphasis on *structure* and the determining power of "society" resulted in a corresponding neglect of *agency,* the beliefs and actions of the actors involved, in much of Durkheim's theorizing (Zeitlin, 1997).

TIME TO REVIEW

- Why were early thinkers concerned with social order and stability?

class conflict Karl Marx's term for the struggle between the capitalist class and the working class.

bourgeoisie Karl Marx's term for the class comprised of those who own and control the means of production.

means of production Karl Marx's term for tools, land, factories, and money for investment that form the economic basis of a society.

proletariat Karl Marx's term for those who must sell their labour because they have no other means to earn a livelihood.

DIFFERING VIEWS ON THE STATUS QUO: STABILITY VERSUS CHANGE

Henry Guttmann/Getty Images

Karl Marx

Together with Karl Marx, Max Weber, and Georg Simmel, Durkheim established the course for modern sociology. We will look first at Marx's and Weber's divergent thoughts about conflict and social change in societies, and then at Simmel's analysis of society.

KARL MARX In sharp contrast to Durkheim's focus on the stability of society, German economist and philosopher Karl Marx (1818–1883) stressed that history is a continuous clash between conflicting ideas and forces. He believed that conflict—especially class conflict—is necessary to produce social change and a better society. For Marx, the most important changes were economic. He concluded that the capitalist economic system was responsible for the overwhelming poverty that he observed in London at the beginning of the Industrial Revolution (Marx and Engels, 1967/1848).

In the Marxian framework, **class conflict** is the struggle between the capitalist class and the working class. The capitalist class, or **bourgeoisie**, comprises those who own and control the **means of production**—the tools, land, factories, and money for investment that form the economic basis of a society. The working class, or **proletariat**, is composed of those who must sell their labour because they have no other means to earn a livelihood. From Marx's viewpoint, the capitalist class controls and exploits the masses of struggling workers by paying less than the value of their labour. This exploitation results in workers' **alienation**—a feeling of powerlessness and estrangement from other people and from oneself. Marx predicted that the working class would become aware of its exploitation, overthrow the capitalists, and establish a free and classless society.

alienation A feeling of powerlessness and estrangement from other people and from oneself.

Marx's theories provide a springboard for neo-Marxist analysts and other scholars to examine the economic, political, and social relations embedded in production and consumption in historical and contemporary societies (See Box 1.3). But what is Marx's place in the history of sociology? Marx is regarded as one of the most profound sociological thinkers, one who combined ideas derived from philosophy, history, and the social sciences into a new theoretical configuration. However, his social and economic analyses have also inspired heated debates among generations of social scientists. Central to his views was the belief that society should not just be studied but should also be changed because the *status quo* (the existing state of society) involved the oppression of most of the population by a small group of wealthy people. Those who believe that sociology should be value free are uncomfortable with Marx's advocacy of what some perceive to be radical social change. Scholars who examine society through the lens of race, gender, and class believe that his analysis places too much emphasis on class

BOX 1.3 POINT/COUNTERPOINT

Ahead of His Time: Marx, Alienation, and the Occupy Wall Street Movement

Social thinkers have long been fascinated by alienation. This concept is often attributed to the economist and philosopher Karl Marx. As further discussed in Chapter 9, *alienation* is a term used to refer to an individual's feeling of powerlessness and estrangement from other people and from oneself. Marx specifically linked alienation to social relations that are inherent in capitalism; however, more recent social thinkers have expanded his ideas to include social psychological feelings of powerlessness, meaninglessness, and isolation. These may be present because people experience social injustice and vast economic inequalities in contemporary societies. Other analysts believe that rampant consumerism may also be linked to alienation because people spend more than they can afford in hopes of finding happiness, gaining the approval of others, or elevating their social status in society.

How are these concepts of alienation and powerlessness reflected in today's global society? One example can be seen in the recent Occupy Wall Street movement. Jeffrey D. Sachs (2012), an economist and the director of the Earth Institute at Columbia University in New York City, states that the emergence of global capitalism has fostered large economic disparities and other contradictions in society, which have contributed to social activism such as Occupy Wall Street and similar protests. According to Sachs (2012), four factors have contributed to such social unrest:

1. chronic high unemployment rates, especially for young people;
2. high tuition that has put education beyond the reach of many young people;
3. political leaders and governments that do not address the needs and problems of individuals and groups that have been left behind by globalization; and
4. the collapse of the financial bubble that was brought about by a combination of "lax monetary policies, financial deregulation, and flagrant corruption within leading financial companies."

From this perspective, the Occupy Wall Street protesters were an outward indication of the inner frustrations that people were feeling about inadequate political and economic policies and practice. In other words, the protestors were voicing the Marxian concept of alienation that they were individually and collectively feeling.

How do we apply these concepts to better understand our society? For one, we can view pressing social issues as important problems that we must all work together to solve. For many, this might mean that we come together to talk about how we might solve problems rather than continuing to live in our own isolated social worlds, where many individuals feel alienated from others.

Why are we often more concerned about trivial matters, such as who will be the big winner in a sporting event or on a reality TV show, than we are about how to address our most pressing social and economic concerns? What examples can you provide to show that Marx's concept of alienation may still apply to 21st century life?

Mondadori Portfolio/Getty Images

Max Weber

relations, often to the exclusion of issues regarding race/ethnicity and gender. In recent decades, scholars have shown renewed interest in Marx's *social theory,* as opposed to his radical ideology (see Lewis, 1998; Postone, 1997). Throughout this text, we will continue to explore Marx's various contributions to sociological thinking.

MAX WEBER German social scientist Max Weber (pronounced VAY-ber) (1864–1920) was also concerned about the changes brought about by the Industrial Revolution. Although he disagreed with Marx's idea that economics is *the* central force in social change, Weber acknowledged that economic interests are important in shaping human action. Even so, he thought that economic systems are heavily influenced by other factors in a society. As we will see in Chapter 17, one of Weber's most important works, *The Protestant Ethic and the Spirit of Capitalism* (1976/1904–1905), evaluated the role of the Protestant Reformation in producing a social climate in which capitalism could exist and flourish.

Unlike many early analysts who believed that values could not be separated from the research process, Weber emphasized that sociology should be *value free*—that is, research should be conducted in a scientific manner and should exclude the researcher's personal values and economic interests (Turner, Beeghley, and Powers, 1998). Weber realized, however, that social behaviour cannot be analyzed by the objective criteria that we use to measure such things as temperature or weight. Although he recognized that sociologists cannot be totally value free, Weber stressed that they should employ *verstehen* (German for "understanding" or "insight") to gain the ability to see the world as others see it. In contemporary sociology, Weber's idea has been incorporated into the concept of the sociological imagination (discussed earlier in this chapter).

Weber was also concerned that large-scale organizations (bureaucracies) were becoming increasingly oriented toward routine administration and a specialized division of labour, which he believed were destructive to human vitality and freedom. According to Weber, rational bureaucracy, rather than class struggle, was the most significant factor in determining the social relations among people in industrial societies. In this view, bureaucratic domination can be used to maintain powerful (capitalist) interests in society. As we will see in Chapter 7, Weber's work on bureaucracy has had a far-reaching impact.

Weber made significant contributions to modern sociology by emphasizing the goal of value-free inquiry and the necessity of understanding how others see the world. He also provided important insights on the process of rationalization, bureaucracy, religion, and many other topics. In his writings, Weber was more aware of women's issues than were many of the scholars of his day. Perhaps his awareness at least partially resulted from the fact that his wife, Marianne Weber, was an important figure in the women's movement in Germany in the early 20th century (Roth, 1988).

GEORG SIMMEL At about the same time that Durkheim was developing the field of sociology in France, the German sociologist Georg Simmel (pronounced ZIM-mel) (1858–1918) was theorizing about society as a web of patterned interactions among people. The main purpose of sociology, according to Simmel, should be to examine these social interaction processes within groups. In *The Sociology of Georg Simmel* (1950/1902–1917), he analyzed how social interactions vary depending on the size of the social group. He concluded that interaction patterns differed between a *dyad,* a social group with two members, and a *triad,* a social group with three members. He developed *formal sociology,* an approach that focuses attention on the universal recurring social forms that underlie the varying content of social interaction. Simmel referred to these forms as the "geometry of social life." He also distinguished between the *forms*

The Canadian Press/Jonathan Hayward

According to the sociologist Georg Simmel, society is a web of patterned interactions among people. If we focus on the behaviour of individuals in isolation, such as any one of the members of this women's rowing team, we may miss the underlying forms that make up the "geometry of social life."

of social interaction (such as cooperation or conflict) and the *content* of social interaction in different contexts (for example, between leaders and followers).

Like the other social thinkers of his day, Simmel analyzed the impact of industrialization and urbanization on people's lives. He concluded that class conflict was becoming more pronounced in modern industrial societies. He also linked the increase in individualism, as opposed to concern for the group, to the fact that people now had many cross-cutting "social spheres"—membership in a number of organizations and voluntary associations—rather than the singular community ties of the past. Simmel assessed the costs of "progress" on the upper-class city dweller who, he believed, had to develop certain techniques to survive the overwhelming stimulation of the city. Simmel's ultimate concern was to protect the autonomy of the individual in society.

TIME TO REVIEW

- Marx, Weber, and Simmel focused their analysis primarily on social change. Why?

LO-4 CONTEMPORARY THEORETICAL PERSPECTIVES

Given the many and varied ideas and trends that influenced the development of sociology, how do contemporary sociologists view society? Some see it as basically a stable and ongoing entity; others view it in terms of many groups competing for scarce resources; still others describe it as based on the everyday, routine interactions among individuals. Each of these views represents a method of examining the same phenomena. Each is based on general ideas as to how social life is organized and represents an effort to link specific observations in a meaningful way.

theory A set of logically interrelated statements that attempts to describe, explain, and (occasionally) predict social events.

perspective An overall approach to or viewpoint on some subject.

Each utilizes **theory**—a set of logically interrelated statements that attempts to describe, explain, and (occasionally) predict social events. Each theory helps interpret reality in a distinct way by providing a framework in which observations may be logically ordered. Sociologists refer to this theoretical framework as a **perspective**—an overall approach to or viewpoint on some subject. The major theoretical perspectives that have emerged in sociology include the functionalist, conflict, feminist, and interactionist perspectives. Other perspectives, such as postmodernism, have emerged and gained acceptance among some social thinkers more recently. Before turning to the specifics of these perspectives, we should note that some theorists and theories do not fit neatly into any of these perspectives. Nevertheless, these perspectives will be used throughout this book to show you how sociologists try to understand many of the issues affecting Canadian society.

FUNCTIONALIST PERSPECTIVES

functionalist perspectives The sociological approach that views society as a stable, orderly system.

societal consensus A situation whereby the majority of members share a common set of values, beliefs, and behavioural expectations.

Also known as functionalism and structural functionalism, **functionalist perspectives** are based on the assumption that society is a stable, orderly system. This stable system is characterized by **societal consensus**, whereby the majority of members share a common set of values, beliefs, and behavioural expectations. According to this perspective, a society is composed of interrelated parts, each of which serves a function and (ideally) contributes to the overall stability of the society. Societies develop social structures, or institutions, and these persist because they play a part in helping society survive. These institutions include the family, education, government, religion, and the economy. If anything adverse happens to one of these institutions, or parts, all other parts are affected and the system no longer functions properly. As Durkheim noted, rapid social change and a more specialized division of labour produce strains in society that lead to a breakdown in these traditional institutions and may result in social problems, such as increased rates of crime and suicide.

TALCOTT PARSONS AND ROBERT MERTON Talcott Parsons (1902–1979), a founder of the sociology department at Harvard University, was perhaps the most influential advocate of the functionalist perspective. He stressed that, to survive, all societies must make provisions for meeting social needs (Parsons, 1951; Parsons and Shils, 1951). For example, Parsons (1955) suggested that a division of labour (distinct, specialized functions) between husband and wife is essential for family stability and social order. The husband/father performs the *instrumental tasks,* which involve leadership and decision-making responsibilities in the home and employment outside the home to support the family. The wife/mother is responsible for the *expressive tasks,* including housework, caring for the children, and providing emotional support for the entire family. Parsons believed that other institutions, including school, church, and government, must function to assist the family and that all institutions must work together to preserve the system over time (Parsons, 1955). Although his analysis has been criticized for its conservative bias, his work still influences sociological thinking about gender roles and the family.

manifest functions Open, stated, and intended goals or consequences of activities within an organization or institution.

latent functions Unintended functions that are hidden and remain unacknowledged by participants.

dysfunctions A term referring to the undesirable consequences of any element of a society.

Functionalism was refined further by a student of Parsons, Robert K. Merton (1910–2003), who distinguished between manifest and latent functions of social institutions. **Manifest functions** are intended and/or overtly recognized by the participants in a social unit. In contrast, **latent functions** are unintended functions that are hidden and remain unacknowledged by participants. For example, a manifest function of education is the transmission of knowledge and skills from one generation to the next; a latent function is the establishment of social relations and networks. Merton noted that all features of a social system may not be functional at all times; **dysfunctions** are the undesirable consequences of any element of a society. A dysfunction of education can be the perpetuation of gender, racial, and class inequalities. Such dysfunctions may threaten the capacity of a society to adapt and survive (Merton, 1968).

APPLYING A FUNCTIONALIST PERSPECTIVE TO SHOPPING AND CONSUMPTION How might functionalists analyze shopping and consumption? When we examine the part-to-whole relationships of contemporary society in high-income nations, it immediately becomes apparent that each social institution depends on the others for its well-being. For example, a booming economy benefits other social institutions, including the family (members are gainfully employed), religion (churches, mosques, synagogues, and temples receive larger contributions), and education (school taxes are higher when property values are higher). A strong economy also makes it possible for more people to purchase more goods and services. By contrast, a weak economy has a negative influence on people's opportunities and spending patterns. Because of the significance of the economy in other aspects of social life, if people have "extra" money to spend and can afford leisure time away from work, they are more likely to dine out, take trips, and purchase things they might otherwise forgo. However, in difficult economic times, people are more likely to curtail family outings and some purchases.

ValeStock/Shutterstock.com

Shopping malls are a reflection of a consumer society. A manifest function of a shopping mall is to sell goods and services to shoppers; however, a latent function may be to provide a communal area in which people can visit friends and enjoy an event.

Clearly, a manifest function of shopping and consumption is purchasing necessary items such as food, clothing, household items, and sometimes transportation. But what are the latent functions of shopping? Consider shopping malls, for example: Many young people go to the mall to "hang out," visit with friends, and eat lunch at the food court. People of all ages go shopping for pleasure, relaxation, and perhaps to enhance their feelings of self-worth. ("If I buy this product, I'll look younger/beautiful/handsome/sexy!") However, shopping and consuming can also produce problems or dysfunctions. Some people are "shopaholics" or "credit card junkies" who cannot stop spending money; others are kleptomaniacs, who steal products rather than pay for them.

The functionalist perspective is useful in analyzing consumerism because of the way in which it examines the relationship between part-to-whole relationships. How the economy is doing affects individuals' consumption patterns, and when the economy is not doing well, political leaders often encourage us to spend more to help the national economy and keep other people employed.

CONFLICT PERSPECTIVES

According to **conflict perspectives**, groups in society are engaged in a continuous power struggle for control of scarce resources. Conflict may take the form of politics, litigation, negotiations, or family discussions about financial matters. Simmel, Marx, and Weber contributed significantly to this perspective by focusing on the inevitability of clashes between social groups. Today, advocates of conflict perspectives view social life as a continuous power struggle among competing social groups.

conflict perspectives
The sociological approach that views groups in society as engaged in a continuous power struggle for control of scarce resources.

MAX WEBER AND RALF DAHRENDORF As previously discussed, Marx focused on the exploitation and oppression of the proletariat (the workers) by the bourgeoisie (the owners, or capitalist class). Weber recognized the importance of economic conditions in producing inequality and conflict in society but added *power* and *prestige* as other sources of inequality. He defined power as the ability of a person within a social relationship to carry out his or her own will despite resistance from others. Prestige—"status group" to Weber—is a positive or negative social estimation of honour (Weber, 1968/1922).

Other theorists have looked at conflict among many groups and interests (such as employers and employees) as a part of everyday life in any society. Ralf Dahrendorf (1959), for example, observed that conflict is inherent in *all* authority relationships, not just that between the capitalist class and the working class. To Dahrendorf, *power* is the critical variable in explaining human behaviour. People in positions of authority benefit from the conformity of others; those who are forced to conform feel resentment and demonstrate resistance, much as a child may resent parental authority. The advantaged group that possesses authority attempts to preserve the status quo—the existing set of social arrangements—and may use coercion to do so.

APPLYING CONFLICT PERSPECTIVES TO SHOPPING AND CONSUMPTION How might advocates of a conflict approach analyze the process of shopping and consumption? A contemporary conflict analysis of consumption might look at how inequalities based on racism, sexism, and income differentials affect people's ability to acquire the things they need and want. It might also look at inequalities regarding the issuance of credit cards and access to "cathedrals of consumption," such as mega-shopping malls and tourist resorts (see Ritzer, 1999:197–214).

However, one of the earliest social theorists to discuss the relationship between social class and consumption patterns was the U.S. social scientist Thorstein Veblen (1857–1929). In *The Theory of the Leisure Class* (1967/1899), Veblen described early wealthy U.S. industrialists as engaging in conspicuous consumption—the continuous public display of one's wealth and status through purchases such as expensive houses, clothing, motor vehicles, and other consumer goods. According to Veblen, the leisurely lifestyle of the upper classes typically does not provide them with adequate opportunities to show off their wealth and status. In order to attract public admiration, the wealthy often engage in consumption and leisure activities that are both highly visible and highly wasteful. Examples of conspicuous consumption range from Cornelius Vanderbilt's eight lavish mansions and 10 major summer estates in the Gilded Age to the contemporary $85 million Spelling Manor in Los Angeles, which has 56,500 square feet, 14 bedrooms, 27 bathrooms, and a two-lane bowling alley, among many other luxurious amenities. By contrast, however, some of today's wealthiest people engage in inconspicuous consumption, perhaps to maintain a low public profile or out of fear for their own safety or that of other family members.

Conspicuous consumption has become more widely acceptable at all income levels, and many families live on credit in order to purchase the goods and services that they would like to have. According to conflict theorists, the economic gains of the wealthiest people are often at the expense of those in the lower classes, who may have to struggle (sometimes unsuccessfully) to have adequate food, clothing, and shelter for themselves and their children. Chapter 9 and Chapter 10 discuss contemporary conflict perspectives on class-based inequalities.

TIME TO REVIEW

- Evaluate the extent to which you live in a society characterized by consensus, social cohesion, and shared values or by conflict and unequal division of power and wealth.

FEMINIST PERSPECTIVES

In the past several decades, feminists have radically transformed the discipline of sociology. Feminist theory first emerged as a critique of traditional sociological theory and methodology. The primary criticism was that sociology did not acknowledge the experiences of women. Written by men, sociology involved the study of men and not humankind, much less women; sociology examined only half of social reality (Fox, 1989). Feminist scholar Dorothy Smith (1974) argued that sociological methods, concepts, and analyses were products of the "male social universe." If women appeared at all, it was as men saw them and not as they saw themselves. In this way, feminist sociologists argued, sociology furthered the subordination and exploitation of women (Anderson, 1996). The first task of feminist sociology was to provide the missing half of social reality by generating research and theory "by, for, and about women" (Smith, 1987). In doing so, feminist sociology brought the personal problems of women, including violence against women, the poverty of women, and the invisibility of women's reproductive labour, into the public forum.

Feminist perspectives focus on the significance of gender in understanding and explaining inequalities that exist between men and women in the household, in the paid labour force, and in the realms of politics, law, and culture (Armstrong and Armstrong, 1994; Luxton, 1995; Marshall, 1995). Feminism is not one single unified approach. Rather, it is characterized by a variety of perspectives, debates, and approaches among feminist writers—namely, the liberal, radical, and socialist strains (discussed in Chapter 12). Feminist sociology incorporates both microlevel and macrolevel analyses in studying the experiences of women. For example, some feminist theorists, such as Margrit Eichler, have used a structural approach to explain how gender inequality is created and maintained in a society dominated by men (Armstrong and Armstrong, 1994; Eichler, 1988). Other feminist research has focused on the interpersonal relationships between men and women in terms of verbal and nonverbal communication styles, attitudes, and values in explaining the dynamics of power and social control in the private sphere (Mackie, 1995). For example, "Who eats first, sits last, or talks back reflects the micro-politics of gender" (Coltrane, 1992:104).

▶ **feminist perspectives** The sociological approach that focuses on the significance of gender in understanding and explaining inequalities that exist between men and women in the household, in the paid labour force, and in the realms of politics, law, and culture.

All of these approaches share the belief that "women and men are equal and should be equally valued as well as have equal rights" (Basow, 1992). According to feminists (including many men as well as women), we live in a *patriarchy*, a hierarchical system of power in which males possess greater economic and social privilege than females (Saunders, 1999). Feminist perspectives emphasize that gender roles are socially created, rather than determined by one's biological inheritance, and that change is essential for people to achieve their human potential without limits based on gender. Feminism views society as reinforcing social expectations through social learning: What we learn is a social product of the political and economic structure of the society in which we live (Renzetti and Curran, 1995). Feminists argue that women's subordination can end only after the patriarchal system of male dominance is replaced with a more egalitarian system.

Dorothy Smith

SYMBOLIC INTERACTIONIST PERSPECTIVES

The functionalist and conflict perspectives have been criticized for focusing primarily on macrolevel analysis. A **macrolevel analysis** examines whole societies, large-scale social structures, and social systems instead of looking at important social dynamics in individuals' lives. Our final perspective, symbolic interactionism, fills this void by examining people's day-to-day interactions and their behaviour in groups. Thus, symbolic interactionist approaches are based on a **microlevel analysis**, which focuses on small groups rather than large-scale social structures.

▶ **macrolevel analysis** Sociological theory and research that focus on whole societies, large-scale social structures, and social systems.

▶ **microlevel analysis** Sociological theory and research that focus on small groups rather than on large-scale social structures.

We can trace the origins of this perspective to the Chicago School, especially George Herbert Mead (1863–1931), and Herbert Blumer (1900–1987), who is credited with coining the term *symbolic interactionism.* According to **symbolic interactionist perspectives**, society is the sum of the interactions of individuals and groups. Theorists using this perspective focus on the process of *interaction*—defined as immediate, reciprocally oriented communication between two or more people—and the part that *symbols* play in giving meaning to human communication. A **symbol** is anything that meaningfully represents something else. Examples of symbols include signs, gestures, written language, and shared values. Symbolic interaction occurs when people communicate through the use of symbols; for example, a gift of food—a cake or a casserole—to a newcomer in a neighbourhood is a symbol of welcome and friendship. Symbols are instrumental in helping people derive meanings from social situations. In social encounters, each person's interpretation or definition of a given situation becomes a *subjective reality* from that person's viewpoint. We often assume that what we consider to be "reality" is shared; however, this assumption is often incorrect. Subjective reality is acquired and shared through agreed-upon symbols, especially language. If a person shouts, "Fire!" in a crowded movie theatre, for

▶ **symbolic interactionist perspectives** The sociological approach that views society as the sum of the interactions of individuals and groups.

▶ **symbol** Anything that meaningfully represents something else.

example, that language produces the same response (attempting to escape) in all of those who hear and understand it. When people in a group do not share the same meaning for a given symbol, however, confusion results; for example, people who did not know the meaning of the word *fire* would not know what the commotion was about. How people *interpret* the messages they receive and the situations they encounter becomes their subjective reality and may strongly influence their behaviour.

APPLYING SYMBOLIC INTERACTIONIST PERSPECTIVES TO SHOPPING AND CONSUMPTION Sociologists applying a symbolic interactionist framework to the study of shopping and consumption would primarily focus on a microlevel analysis of people's face-to-face interactions and the roles that people play in society. In our efforts to interact with others, we define any situation according to our own subjective reality. This theoretical viewpoint applies to shopping and consumption just as it does to other types of conduct. For example, when a customer goes into a store to make a purchase and offers a credit card to the cashier, what meanings are embedded in the interaction process that takes place between the two of them? The roles that the two people play are based on their histories of interaction in previous situations. They bring to the present encounter symbolically charged ideas, based on previous experiences. Each person also has a certain level of emotional energy available for each interaction. When we are feeling positive, we have a high level of emotional energy, and the opposite is also true. Each time we engage in a new interaction, the situation has to be negotiated all over again, and the outcome cannot be known beforehand.

In the case of a shopper–cashier interaction, how successful will the interaction be for each of them? The answer to this question depends on a kind of social marketplace in which such interactions can either raise or lower one's emotional energy (Collins, 1987). If the customer's credit card is rejected, he or she may come away with lower emotional energy. If the customer is angry at the cashier, he or she may attempt to "save face" by reacting in a haughty manner regarding the rejection of the card. ("What's wrong with you? Can't you do anything right? I'll never shop here again!") If this type of encounter occurs, the cashier may also come out of the interaction with a lower level of emotional energy, which may affect the cashier's interactions with subsequent customers. Likewise, the next time the customer uses a credit card, he or she may say something like "I hope this card isn't over its limit; sometimes I lose track," even if the person knows that the card's credit limit has not been exceeded. This is only one of many ways in which the rich tradition of symbolic interactionism might be used to examine shopping and consumption. Other areas of interest might include the social nature of the shopping experience, social interaction patterns in families regarding credit card debts, and why we might spend money to impress others.

POSTMODERN PERSPECTIVES

postmodern perspectives The sociological approach that attempts to explain social life in contemporary societies that are characterized by post-industrialization, consumerism, and global communications.

According to **postmodern perspectives**, existing theories have been unsuccessful in explaining social life in contemporary societies that are characterized by post-industrialization, consumerism, and global communications. Postmodern social theorists reject the theoretical perspectives we have previously discussed, as well as how those thinkers created the theories (Ritzer, 2011).

Postmodern theories are based on the assumption that the rapid social change that occurs as societies move from modern to postmodern (or post-industrial) conditions has a harmful effect on people. One evident change is the significant decline in the influence of social institutions such as the family, religion, and education on people's lives. Those who live in postmodern societies typically pursue individual freedom and do not want the structural constraints that are imposed by social institutions. However, the collective ties that once bound people together become weakened, placing people at higher levels of risk.

Postmodern (or "post-industrial") societies are characterized by an information explosion and an economy in which large numbers of people either provide or apply information, or they are employed in professional occupations (such as lawyers and physicians) or service jobs (such as fast-food servers and healthcare workers). There is a corresponding rise of a consumer society and the emergence of a global village in which people around the world instantly communicate with one another.

Jean Baudrillard, a well-known French social theorist, has extensively explored how the shift from production of goods to consumption of information, services, and products in contemporary societies has created a new form of social control. According to Baudrillard's approach, capitalists strive to control people's shopping habits, much like the output of factory workers in industrial economies, to enhance their profits and to keep everyday people from rebelling against social inequality (1998/1970). How does this work? When consumers are encouraged to purchase more than they need or can afford, they often sink deeper in debt and must keep working to meet their monthly payments. Instead of consumption being related to our needs, it is based on factors such as our "wants" and the need we feel to distinguish ourselves from others. We will look at this idea in more detail in the next section, where we apply a postmodern perspective to shopping and consumption. We will also return to Baudrillard's general ideas on postmodern societies in Chapter 3.

Postmodern theory opens up broad new avenues of inquiry by challenging existing perspectives and questioning current belief systems. However, postmodern theory has also been criticized for raising more questions than it answers.

APPLYING POSTMODERN PERSPECTIVES TO SHOPPING AND CONSUMPTION

According to some social theorists, the postmodern society is a consumer society. The focus of the capitalist economy has shifted from production to consumption: The emphasis is on getting people to consume more and to own a greater variety of things. As previously discussed, credit cards may encourage people to spend more money than they should, and often more than they can afford (Ritzer, 1998). Television shopping networks, online shopping, and mobile advertising and shopping devices make it possible for people to shop around the clock without having to leave home or encounter "real" people. As Ritzer (1998:121) explains, "So many of our interactions in these settings . . . are simulated, and we become so accustomed to them, that in the end all we have are simulated interactions; there are no more 'real' interactions. The entire distinction between the simulated and the real is lost; simulated interaction is the reality" (see also Baudrillard, 1983).

CONCEPT SNAPSHOT

FUNCTIONALIST PERSPECTIVES **Key thinkers:** Émile Durkheim, Talcott Parsons, Robert Merton	Society is composed of interrelated parts that work together to maintain stability within society. This stability is threatened by dysfunctional acts and institutions.
CONFLICT PERSPECTIVES **Key thinkers:** Karl Marx, Max Weber, Ralf Dahrendorf	Society is characterized by social inequality; social life is a struggle for scarce resources. Social arrangements benefit some groups at the expense of others.
FEMINIST PERSPECTIVES **Key thinkers:** Dorothy Smith, Margrit Eichler, Meg Luxton	Society is based on patriarchy—a hierarchical system of power in which males possess greater economic and social privilege than females.
SYMBOLIC INTERACTIONIST PERSPECTIVES **Key thinkers:** George Herbert Mead, Herbert Blumer	Society is the sum of the interactions of people and groups. Behaviour is learned in interaction with other people; how people define a situation becomes the foundation for how they behave.
POSTMODERNIST PERSPECTIVES **Key thinkers:** Jean-François Lyotard, Jean Baudrillard	Societies characterized by post-industrialization, consumerism, and global communications bring into question existing assumptions about social life and the nature of reality.

For postmodernists, social life is not an objective reality waiting for us to discover how it works. Rather, what we experience as social life is actually nothing more or less than how we think about it, and there are many diverse ways of doing that. According to a postmodernist perspective, the Enlightenment goal of intentionally creating a better world out of some knowable truth is an illusion. Although some might choose to dismiss postmodern approaches, they do give us new and important questions to think about regarding the nature of social life.

The Concept Quick Review reviews all five of these perspectives. Throughout this book, we will be using these perspectives as lenses through which to view our social world.

LO-1 Explain what sociology can contribute to our understanding of social life.

Sociology is the systematic study of human society and social interaction. We study sociology to understand how human behaviour is shaped by group life and, in turn, how group life is affected by individuals. Our culture tends to emphasize individualism, and sociology pushes us to consider more complex connections between our personal lives and the larger world.

Syracuse Newspapers/John Berry/The Image Works

LO-2 Explain why the sociological imagination is important for studying society.

According to C. Wright Mills, the sociological imagination helps us understand how seemingly personal troubles, such as student debt, are related to larger social forces. It allows us to see the relationship between individual experiences and the larger society. It is important to have a global sociological imagination because the future of this country is deeply intertwined with the future of all nations of the world on economic, political, environmental, and humanitarian levels.

The Canadian Press/Nathan Denette

LO-3 Discuss the major contributions of early sociologists.

Comte, considered by many to be the founder of sociology, coined the term *sociology* to describe the new science engaging in the study of society. Others have argued that Harriet Martineau should be viewed as a founding member of sociology due to her enlightened perspective that social progress must involve gender and social equality. The ideas of Émile Durkheim, Karl Marx, and Max Weber helped lead the way to contemporary sociology. Durkheim argued that societies are built on social facts, that rapid social change produces strains in society, and that the loss of shared values and purpose can lead to a condition of anomie. Marx stressed that within society there is a continuous clash between the owners of the means of production and the workers, who have no choice but to sell their labour to others. According to Weber, it is necessary to acknowledge the meanings that individuals attach to their own actions.

Auguste Comte (1798-1857) (oil on canvas), Etex, Louis Jules (1810-1889)/Temple de la Religion de l'Humanite, Paris, France/ The Bridgeman Art Library International

KEY TERMS

alienation A feeling of powerlessness and estrangement from other people and from oneself (p. 15).

anomie Émile Durkheim's term for a condition in which social control becomes ineffective as a result of the loss of shared values and a sense of purpose in society (p. 14).

bourgeoisie Karl Marx's term for the class comprised of those who own and control the means of production (p. 14).

class conflict Karl Marx's term for the struggle between the capitalist class and the working class (p. 14).

commonsense knowledge A form of knowing that guides ordinary conduct in everyday life (p. 5).

conflict perspectives The sociological approach that views groups in society as engaged in a continuous power struggle for control of scarce resources (p. 19).

dysfunctions A term referring to the undesirable consequences of any element of a society (p. 18).

feminist perspectives The sociological approach that focuses on the significance of gender in understanding and explaining inequalities that exist between men and women in the household, in the paid labour force, and in the realms of politics, law, and culture (p. 21).

functionalist perspectives The sociological approach that views society as a stable, orderly system (p. 18).

global interdependence A relationship in which the lives of all people are closely intertwined and any one nation's problems are part of a larger global problem (p. 4).

high-income countries Nations with highly industrialized economies; technologically advanced industrial, administrative, and service occupations; and relatively high levels of national and personal income (p. 7).

industrialization The process by which societies are transformed from dependence on agriculture and handmade products to an emphasis on manufacturing and related industries (p. 10).

latent functions Unintended functions that are hidden and remain unacknowledged by participants (p. 18).

low-income countries Countries that are primarily agrarian, with little industrialization and low levels of national and personal income (p. 8).

macrolevel analysis Sociological theory and research that focus on whole societies, large-scale social structures, and social systems (p. 21).

manifest functions Open, stated, and intended goals or consequences of activities within an organization or institution (p. 18).

means of production Karl Marx's term for tools, land, factories, and money for investment that form the economic basis of a society (p. 14).

microlevel analysis Sociological theory and research that focus on small groups rather than on large-scale social structures (p. 21).

middle-income countries Nations with industrializing economies, particularly in urban areas, and moderate levels of national and personal income (p. 8).

perspective An overall approach to or viewpoint on some subject (p. 18).

positivism A belief that the world can best be understood through scientific inquiry (p. 11).

LO-4 Describe the key assumptions behind each of the contemporary theoretical perspectives.

Courtesy of Dorothy E. Smith

Functionalist perspectives assume that society is a stable, orderly system characterized by societal consensus; however, this perspective has been criticized for overlooking the importance of change in societies. Conflict perspectives argue that society is a continuous power struggle among competing groups, often based on class, race, ethnicity, or gender. Critics of conflict theory note that it minimizes the importance of social stability and shared values in society. Feminist perspectives focus on the significance of gender in understanding and explaining inequalities that exist between men and women in the household, in the paid labour force, and in politics, law, and culture. Symbolic interactionist perspectives focus on how people make sense of their everyday social interactions, which are made possible by the use of mutually understood symbols. However, this approach focuses on the microlevel of society and tends to ignore the macrolevel social context. From an alternative perspective, postmodern theorists believe that entirely new ways of examining social life are needed and that it is time to move beyond functionalist, conflict, and interactionist perspectives.

APPLICATION QUESTIONS

1. What does C. Wright Mills mean when he says the sociological imagination helps us "grasp history and biography and the relations between the two within society" (Mills, 1959a:6)? How might this idea be applied to various trends in consumer spending in today's society?
2. As a sociologist, how would you remain objective and yet see the world as others see it? Would you make subjective decisions when trying to understand the perspectives of others?
3. Early social thinkers were concerned about stability in times of rapid change. In our more global world, is stability still a primary goal? Or is constant conflict important for the well-being of all humans? Use the conflict and feminist perspectives to support your analysis.
4. According to the functionalist perspective, what would happen to society if one of its institutions—say, the education system—were to break down?

postmodern perspectives The sociological approach that attempts to explain social life in contemporary societies that are characterized by post-industrialization, consumerism, and global communications (p. 22).

proletariat Karl Marx's term for those who must sell their labour because they have no other means to earn a livelihood (p. 14).

social Darwinism The belief that those species of animals (including human beings) best adapted to their environment survive and prosper, whereas those poorly adapted die out (p. 13).

social facts Émile Durkheim's term for patterned ways of acting, thinking, and feeling that exist outside any one individual (p. 13).

societal consensus A situation whereby the majority of members share a common set of values, beliefs, and behavioural expectations (p. 18).

society A large social grouping that shares the same geographical territory and is subject to the same political authority and dominant cultural expectations (p. 4).

sociological imagination C. Wright Mills's term for the ability to see the relationship between individual experiences and the larger society (p. 6).

sociology The systematic study of human society and social interaction (p. 4).

symbol Anything that meaningfully represents something else (p. 21).

symbolic interactionist perspectives The sociological approach that views society as the sum of the interactions of individuals and groups (p. 21).

theory A set of logically interrelated statements that attempts to describe, explain, and (occasionally) predict social events (p. 18).

urbanization The process by which an increasing proportion of a population lives in cities rather than in rural areas (p. 10).

CHAPTER 2

SOCIOLOGICAL RESEARCH

Photo by: Sergeant Matthew McGregor, Canadian Forces Combat Camera. © 2012 DND-MDN Canada. Reproduced with the permission of the Department of National Defence, 2015.

CHAPTER FOCUS QUESTION

How does social research add to our knowledge of human societies?

LEARNING OBJECTIVES

AFTER READING THIS CHAPTER, YOU SHOULD BE ABLE TO

LO-1 Understand the relationship between theory and research.

LO-2 Identify the main steps in the sociological research process.

LO-3 Explain why it is important to have different methods of conducting social research and to know something about each of these methods.

LO-4 Discuss how research has contributed to our understanding of altruism.

LO-5 Explain why a code of ethics for sociological research is necessary.

On October 31, 1991, a Canadian Armed Forces Hercules transport plane with 18 passengers and crew was preparing to land at Canadian Forces Station Alert on Ellesmere Island in the Northwest Territories. Just 800 km from the North Pole, Alert is the world's most northerly permanent settlement. In the dark Arctic night, 16 km short of the runway, the left wing of the Hercules struck the peak of a small mountain and the aircraft crashed onto a barren Arctic plateau. Fourteen survivors, many seriously injured, waited for help in the twisted wreckage.

Thirty-two hours later, another Hercules, with a team of search and rescue technicians (SARtechs) on board, was circling the crash site. The SARtechs hoped conditions would allow them to parachute into the site to help the survivors. The temperature was −66°C, ground visibility was limited, and the winds were three times the permissible limit for parachuting. Because of the terrible weather conditions, a U.S. Air Force crew had just cancelled their attempt at a jump onto the crash site. In the back of the Canadian Hercules, Warrant Officer Arnie Macauley, the SARtech team leader, addressed his men:

Okay, guys, you know the situation. The winds are pretty stiff. They've blown away all our marker lights . . . It looks like a snowfield down there, but ground conditions are unknown. We'll be landing at a good clip. We'll try for flare illumination, but I can't promise you anything.

One more thing. Once we're down there, we're down for good. Marv will try for a supply drop, but we can expect the survival gear to be blown away. We'll have no way of extracting ourselves or the survivors . . .

That's the situation, men. We can expect casualties. I have to inform you that the jump involves a knowing risk of life. I can't ask any of you to do this.

Some of the SARtechs studied their boots; others looked out the open door into the howling void. One by one, they looked back at him. "Arnie," one said, "you know how we feel."

Good guys, Arnie thought. I hope like hell I'm doing the right thing.

Arnie checked the closures on his padded orange jumpsuit and the fasteners of his parachute harness. He pulled on his gloves. The jumpmaster clipped their static lines to the overhead cable, and the men crowded around the open door. There were six of them. They squeezed into the opening and grabbed one another by the legs, arms, waist. They would go together. (Mason Lee, 1991:229–231)

Source: From *Death* and *Deliverance*, by Robert Mason Lee (Macfarlane Walter & Ross, Toronto, 1992), pp. 229–231.

The rescuers completed their harrowing jump with relatively minor injuries and immediately began caring for the survivors.

Thirteen people survived because of the heroism of the SARtechs and the other rescue crew members. Why do people like Warrant Officer Macauley and his men risk their lives in order to save others? This question has been asked by sociologists who have studied **altruism**—behaviour intended to help others and done without any expectation of personal benefit.

altruism Behaviour intended to help others and done without any expectation of personal benefit.

In this chapter, you will learn how sociological research methods help us understand social phenomena such as altruism. How do sociologists determine what to study? How do they conduct their research? What factors determine the best method to use in social research? These questions address the process of doing sociological research—an exciting process that challenges us to go as "strangers" into a familiar world. The sociological imagination guides us to examine the social context of acts such as altruism.

Several researchers have conducted studies that try to help us understand altruistic behaviour, and their work will be used to illustrate the different research methods used by sociologists. Before reading on, test your knowledge of altruism by answering the questions in Box 2.1.

CRITICAL THINKING QUESTIONS

1. This chapter presents several studies of altruism. Which aspects of this topic would you be most interested in studying?
2. Can you think of ways in which your fellow students are behaving altruistically by helping others? Do students typically help others by donating time or by contributing money?
3. Why do you think sociologists have spent much more time studying negative behaviours, such as criminality, than studying positive behaviours like altruism?

LO-1 WHY IS SOCIOLOGICAL RESEARCH NECESSARY?

Sociological research results in a body of knowledge that helps us move beyond guesswork and common sense in understanding society. During this course, you will learn that commonsense beliefs about society are often wrong. The sociological perspective incorporates theory and research to arrive at an informed understanding of the "hows" and "whys" of human social interaction.

FIVE WAYS OF KNOWING THE WORLD

Sociologists seek to understand social behaviour. People have always tried to bring order to the chaotic world of experience by trying to understand their social and physical worlds. Understanding is the major goal of science, but this goal is shared by other fields, including philosophy, religion, the media, and the arts. We have several ways of knowing the world:

1. *Personal experience.* We discover for ourselves many of the things we know. If we put our tongue on a frozen doorknob, we learn that removing it can be painful.
2. *Tradition.* People may hold a belief because "everyone knows" it to be true. Tradition tells us that something is right because it has always been done that way. We accept what has always been believed rather than finding the answers by ourselves.
3. *Authority.* Experts tell us that something is true. We do not all need to go to the Moon to discover its mineral composition, but instead accept the judgment of space scientists and geologists. Much of what we know about medicine, crime, and many other phenomena is based on what authorities have told us.
4. *Religion.* Religious authority gives us truths based on our particular scriptures. Beliefs about factors as diverse as morality, diet, dress, and hairstyles are based on religious authority.
5. *Science.* The scientific way of knowing involves controlled, systematic observation. Scientists insist that all statements be tested and that testing procedures be open to public inspection.

The Canadian Press/Adrian Wyld

People are often willing to help others. Barbara Winters is one of the people who rushed to the assistance of Corporal Nathan Cirillo who was fatally wounded at the War Memorial in Ottawa on October 22, 2014.

Personal experience, tradition, authority, and religion are all valid sources of understanding. However, there is no way to resolve disagreements between those who have had different experiences, or who believe in different religions, traditions, or authorities. For example, if religious groups have different views concerning the activities that are permissible on the Sabbath or the role of women in society, there is no institutionalized way of reconciling these contrary positions.

Scientific explanations differ from other ways of knowing in several fundamental ways that allow scientists to resolve differences in their understanding of the world.

First, science is *empirical.* Science is based on the assumption that knowledge is best gained by direct, systematic observation.

Second, scientific knowledge is *systematic* and *public*. The procedures used by scientists are organized, public, and recognized by other scientists. The scientific community will not accept claims that cannot be publicly verified.

▶ **hypotheses** Tentative statements of the relationship between two or more concepts or variables.

Third, science has a built-in mechanism for *self-correction.* Scientists do not claim that their findings represent eternal truths, but rather they present **hypotheses**—tentative statements of the relationship between two or more concepts or variables—that are subject to verification

BOX 2.1 SOCIOLOGY AND EVERYDAY LIFE

How Much Do You Know About Altruism?

True	False	
T	F	1. It is nice to help other people, but altruistic behaviour has little impact on the nature of society.
T	F	2. People who behave altruistically often had parents who also actively helped others.
T	F	3. Helping others after a disaster can be a way of helping oneself recover from the shock of having lived through a traumatic event.
T	F	4. Good Samaritan laws require that Canadians assist those they see in danger or in need of assistance.
T	F	5. Experiments in laboratories and in natural settings have shown that the likelihood of bystanders intervening in situations where someone needs help is reduced as the number of people who are aware of the incident increases.

For answers to the quiz about altruism, go to **www.nelson.com/student.**

by themselves and by others. What is accepted as scientific truth changes over time as more evidence accumulates. By contrast, making changes in understandings based on tradition, authority, or religious belief can be difficult.

Fourth, science is ***objective***. Scientists try to ensure that their biases and values do not affect their research. In some situations, this criterion is easily met. Two scientists measuring the time it takes for a ball to fall 300 metres should arrive at the same answer despite having different backgrounds and values. However, complete objectivity is not possible in the social sciences. Weber pointed out long ago that the researchers' values even influence their selection of research problems. Weber also believed that sociology was fundamentally concerned with the subjective meaning of social action. This means that "the primary task of the sociologist is to understand the meaning an act has for the actor himself, not for the observer" (Natanson, 1963:278). Kirby and McKenna tell us that "our interaction with the social world is affected by such variables as gender, race, class, sexuality, age, physical ability" and conclude that "this does not mean that facts about the social world do not exist, but that what we see and how we go about constructing meaning is a matter of interpretation" (1989:25). For example, marriage may mean different things for men and for women, and people with disabilities may experience the world differently than those without disabilities. The point is not that the observers lack objectivity, but rather that they experience social life in different ways. Researchers must always carefully describe their methods so others can decide for themselves how the researchers' subjectivity has affected their conclusions.

objective Free from distorted subjective (personal or emotional) bias.

DESCRIPTIVE AND EXPLANATORY STUDIES

Sociological studies can be descriptive or explanatory. **Descriptive studies** describe social reality or provide facts about some group, practice, or event. Descriptive studies are designed to find out what is happening to whom, where, and when. For example, the census provides a wealth of descriptive information about the people of Canada, including age, marital status, and place of residence. A descriptive study of altruism might try to determine what percentage of people would return a lost wallet or help a stranger in distress. **Explanatory studies** try to explain relationships and to provide information on why certain events do or do not occur. An explanatory study of altruism might ask, "Why are some people more likely than others to offer help?" or "Why do some countries rely on volunteer blood donations while others pay donors?"

descriptive study Research that attempts to describe social reality or provide facts about some group, practice, or event.

explanatory study Research that attempts to explain relationships and to provide information on why certain events do or do not occur.

THE THEORY AND RESEARCH CYCLE

deductive approach Research in which the investigator begins with a theory and then collects information and data to test the theory.

inductive approach Research in which the investigator collects information or data (facts or evidence) and then generates theories from the analysis of that data.

The relationship between theory and research has been described as a continuous cycle encompassing both the deductive and inductive approaches (see Figure 2.1). In the **deductive approach**, the researcher begins with a theory and uses research to test the theory: (1) Theories generate hypotheses; (2) hypotheses lead to observations; (3) observations lead to generalizations; and (4) generalizations are used to support the theory, to suggest modifications to it, or to refute it. Consider the question, "Why do people help others?" Using the deductive method, we would start by formulating a theory about the "causes" of altruism and then test our theory by collecting and analyzing data. We might conduct experiments on helping behaviour or do surveys that ask why some people help and others don't.

In the **inductive approach**, the researcher collects data and then generates theories from the analysis of those data: (1) Specific observations suggest generalizations; (2) generalizations produce a tentative theory; (3) the theory is tested through the formation of hypotheses; and (4) hypotheses may provide suggestions for additional observations. Using the inductive approach to study altruism, we might start by collecting and analyzing data related to helping behaviour and then generating a theory (see Glaser and Strauss, 1967; Reinharz, 1992).

The process of inquiry is rarely as tidy as Figure 2.1 suggests. Instead, sociologists typically move back and forth from theory to research throughout the course of their inquiry. Investigators rarely, if ever, begin with just a theory or with just research data. Inductive theorists need at least rudimentary theories to guide their data collection, and deductive theorists should refer constantly to the real world as they develop their theories. Researchers may break into the cycle at different points depending on what they want to know and what information is available. Theory gives meaning to research; research helps support theory.

FIGURE 2.1 : THE THEORY AND RESEARCH CYCLE

Different researchers do not necessarily start or stop at the same point, but they share a common goal: to understand social life.

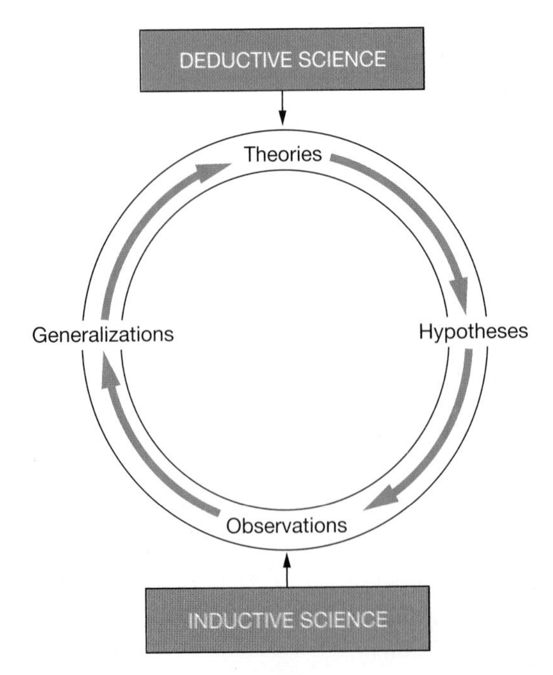

Source: Adapted from Wallace, 1971.

LO-2 THE SOCIOLOGICAL RESEARCH PROCESS

Not all sociologists conduct research in the same way. Some researchers engage primarily in *quantitative* research, whereas others engage in *qualitative* research. With quantitative research, the focus is on data that can be measured numerically. With qualitative research, interpretive description (words) rather than statistics (numbers) is used to analyze underlying meanings and patterns of social relationships. We will trace the steps in both of these research models.

THE QUANTITATIVE RESEARCH MODEL

The steps in the quantitative model are outlined below:

SELECT AND DEFINE THE RESEARCH PROBLEM The first step is to select a research topic. A researcher's personal experience or social policy concerns can trigger interest in a topic. Other researchers select topics to fill gaps or challenge misconceptions in existing research or to test a specific theory.

Once you have selected a topic, ask, "What do I want to know about this topic?" Consider the issue of Good Samaritan laws discussed in Box 2.2. As a researcher, how would you approach the questions raised in the box?

BOX 2.2 POINT/COUNTERPOINT

Does the Law Require Us to Help?

Following Princess Diana's death in a Paris automobile crash in 1997, many Canadians were surprised to learn that France has a law requiring people to help others in distress. Such laws are referred to as Good Samaritan laws after the biblical story of altruistic behaviour. Because of their failure to help the princess, nine photographers and a press motorcyclist were placed under formal investigation—one step short of being charged—for failing to come to the aid of a person in danger.

What about Canada? Does the law require that we intervene in situations where others are in danger? Will we be compensated if we are injured when trying to help others?

In most cases, the answer to these questions is no—Canadian law does little to encourage or to protect Good Samaritans. Most provinces have Good Samaritan laws that protect healthcare professionals from liability if they provide assistance outside a hospital or office setting. Some provinces extend this protection to all citizens who provide emergency medical services or aid. However, only Quebec requires people to assist others. The Quebec Charter of Human Rights and Freedoms states:

> Every human being whose life is in peril has a right to assistance. Every person must come to the aid of anyone whose life is in peril, either personally or calling for aid, by giving him the necessary and immediate physical assistance, unless it involves danger to himself or a third person, or he has another valid reason.

Do you think we should be compelled to help each other, or should we rely on people's altruism? Can the law help to encourage altruism, or is it more a function of our backgrounds and social relationships? Could sociological research help society's understanding of this issue? What does the public think about the need for such laws?

Sources: Nairne, 1998; Quinton, 1989.

REVIEW PREVIOUS RESEARCH Once you have defined your research problem, you should review relevant books and scholarly articles to see what others have learned about the topic. Knowledge of the literature helps refine the research problem, suggests possible theoretical approaches, and identifies the gaps between what is known and what we would like to know.

FORMULATE THE HYPOTHESIS (IF APPLICABLE) After reviewing previous research, you may formulate a hypothesis—a statement of the relationship between two or more concepts. Concepts are the abstract elements representing some aspect of the world in simplified form (social integration and altruism are examples). As you formulate your hypothesis, you will need to convert concepts to variables. A **variable** is any concept with measurable traits or characteristics that can change or vary from one person, time, situation, or society to another. Variables are the observable and/or measurable counterparts of concepts. For example, when altruism is the concept, the percentage of the population who donate blood may be a variable.

Now you are ready to look at the relations between your variables. The most fundamental relationship is between a dependent variable and one or more independent variables. The **independent variable** is presumed to cause or determine a dependent variable. Sociologists often use characteristics such as age, sex, race, and ethnicity as independent variables. The **dependent variable** is assumed to depend on or be caused by the independent variable(s). The dependent variable is also known as the outcome or effect. Several researchers have tested the hypothesis that women are more altruistic than men. In this research, gender is the independent variable and the degree of altruism is the dependent variable. Whether a variable is dependent or independent depends on the context in which it is used, and a variable that is independent in one study may be dependent in another. For example, in a study of the relationship between a family's income and the likelihood of their child graduating from university, the dependent variable is university education. In another study looking at the relationship between university education and voting behaviour, university education is the independent variable.

▶ **variable** In sociological research, any concept with measurable traits or characteristics that can change or vary from one person, time, situation, or society to another.

▶ **independent variable** A variable that is presumed to cause or determine a dependent variable.

▶ **dependent variable** A variable that is assumed to depend on or be caused by one or more other (independent) variables.

When conducting research, sociologists must create operational definitions of each variable. An **operational definition** is an explanation of an abstract concept in terms of observable features that are specific enough to measure the variable. The definition of altruism has been controversial. Earlier, we defined altruism as action intended to help others and done without any expectation of personal benefit. Warrant Officer Macauley and the other SARtechs, however, were paid to rescue people and their heroism was publicly recognized. Blood donors may someday receive the benefits of a transfusion. Mother Teresa received personal satisfaction from working with the poor and may become a saint. How do you think we should define altruism when we conduct research to learn more about it?

▶ **operational definition** An explanation of an abstract concept in terms of observable features that are specific enough to measure the variable.

Some variables are difficult to measure. How do we distinguish between criminals and noncriminals when virtually all of us have broken the law? We might define criminals as those who have been convicted of a crime. However, what do we do about people who have committed crimes but who have been found not guilty at a trial because the evidence against them was obtained illegally? What about corporate offenders whose behaviour may have injured people but who have violated government regulations rather than the criminal law?

▶ **reliability** In sociological research, the extent to which a study or research instrument yields consistent results.

▶ **validity** In sociological research, the extent to which a study or research instrument accurately measures what it is supposed to measure.

▶ **analysis** The process through which data are organized so that comparisons can be made and conclusions drawn.

▶ **replication** In sociological research, the repetition of the investigation in substantially the same way that it originally was conducted.

DEVELOP THE RESEARCH DESIGN Sociologists use several different research methods, including experiments, survey research, field research, and secondary analysis of data, which are described in this chapter. In developing the research design, it is important to consider the advantages and disadvantages of each of these methods.

COLLECT AND ANALYZE THE DATA Next, sociologists collect and analyze their data. Whichever method they choose, researchers must consider the reliability and validity of their data when designing a research project.

Reliability is the extent to which a research instrument yields consistent results when applied to different individuals at one time or to the same individual over time. A ruler is a reliable measure of length because it consistently gives the same results. In the social realm, an IQ test would be considered reliable if a person receives the same score when they take the test more than once. In criminology, a count of the number of people in prison is very reliable, but it is not a valid indicator of criminality because most criminals are not in prison. **Validity** is the extent to which a study or research instrument accurately measures what it is supposed to measure. While IQ tests are quite reliable, their validity as a measure of intelligence is controversial. Proponents believe they are good measures of people's natural abilities. However, others feel the tests measure only some components of intelligence, and criticize their use among people whose language and cultural backgrounds are different from those of the researchers who designed the tests. The number of people arrested is a more valid measure of criminality than the number of people in jail. Do you know why it is still an unsatisfactory measure? (See Chapter 8.)

Once you have collected your data, they must be analyzed. **Analysis** is the process through which data are organized so that comparisons can be made and conclusions drawn.

© Kayte Deioma/PhotoEdit

Surveys and polls are common in many countries. This investigator is conducting her research in Taiwan.

DRAW CONCLUSIONS AND REPORT THE FINDINGS After analyzing the data, your first step in drawing conclusions is to relate the data to your hypotheses. Reporting the findings is the final stage. The report generally reports each step taken in the research process to make the study available for **replication**—the repetition of the investigation in substantially the same way that it was originally conducted. This means that other researchers will see if they get the same results when they repeat your study. If your findings are replicated, the research community will have more faith in your conclusions.

THE QUALITATIVE RESEARCH MODEL

Reuters/Mohammad Ismail/Landov

Natural disasters, such as the May 2014 mudslides in Badakhshan Province, Afghanistan, may be "living laboratories" for sociologists.

Qualitative research is likely to be used where the research question does not easily lend itself to numbers and statistical methods. Compared to a quantitative model, a qualitative approach often involves a different type of research question and a smaller number of cases. Qualitative studies typically provide a detailed picture of some particular social phenomenon or social problem (King, Keohane, and Verba, 1994).

Qualitative researchers typically do not initially define their research problem in as much detail as quantitative researchers. The first step in qualitative research often consists of *problem formulation* to clarify the research question and formulate questions of interest (Reinharz, 1992). Qualitative researchers typically gather data in natural settings, such as places where people live or work, rather than in a laboratory or over the phone. As a result, the qualitative approach can generate new theories and innovative findings that incorporate the perspectives of the research subjects.

The qualitative approach follows the conventional research approach in presenting a problem, asking a question, collecting and analyzing data, and seeking to answer the research question, but it also has several unique features (Creswell, 1998; Kvale, 1996):

1. The researcher begins with a general approach rather than a highly detailed plan. Flexibility is necessary because of the nature of the research question.
2. The researcher decides when the literature review and theory application should take place. Initial work may involve redefining existing concepts or reconceptualizing how existing studies have been conducted. The literature review may take place at an early stage, before the research design is fully developed, or it may occur after development of the research design and after collection of the data.
3. The study presents a detailed view of the topic. Qualitative research usually involves a smaller number of cases and many variables, whereas quantitative researchers typically work with a few variables and many cases. (Creswell, 1998)

TIME TO REVIEW

- What are the different ways people use to understand their social and physical worlds? How does the scientific method differ from the other ways of knowing?
- Explain the cycle of theory and research. How do qualitative and quantitative researchers approach this cycle?

LO-3 RESEARCH METHODS

How do sociologists decide which research method to use? Are some approaches better than others for particular research problems? **Research methods** are specific strategies or techniques for conducting research. *Qualitative* researchers frequently use field observation studies to help them understand the social world from the point of view of the people they are studying. *Quantitative* researchers generally use experimental designs, surveys, and secondary analysis of existing data. We will now look at these research methods.

research methods Specific strategies or techniques for conducting research.

EXPERIMENTS

An **experiment** is a test conducted under controlled conditions in which an investigator tests a hypothesis by manipulating an independent variable and examining its impact on a dependent variable.

experiment A test conducted under controlled conditions in which an investigator tests a hypothesis by manipulating an independent variable and examining its impact on a dependent variable.

experimental group Subjects in an experiment who are exposed to the independent variable.

control group Subjects in an experiment who are not exposed to the independent variable, but later are compared to subjects in the experimental group.

TYPES OF EXPERIMENTS Experiments require that subjects be divided into an experimental group and a control group. **Experimental group** subjects are exposed to an independent variable to study its effect on them. The **control group** contains subjects who are not exposed to the independent variable. The independent variable is manipulated by the researcher and the dependent variable is hypothesized to change, based on the manipulation of the independent variable. Subjects are randomly assigned to each group or matched so that comparisons may be made between the groups. The researcher thereby ensures that the two groups are equivalent at the beginning of the study. In the simplest experimental design, (1) all subjects are pretested (measured in terms of the dependent variable); (2) subjects in the experimental group are then exposed to a stimulus (the independent variable); and (3) all subjects are post-tested (re-measured) in terms of the dependent variable. The experimental and control groups are then compared to see if they differ in relation to the dependent variable, and the hypothesis about the relationship of the two variables is confirmed or rejected.

In a *laboratory experiment,* subjects are studied in a closed setting so researchers can maintain as much control as possible over the research. But not all experiments occur in laboratory settings. Researchers can stage events in natural settings by conducting *field experiments. Natural experiments* are real-life occurrences such as floods or earthquakes that provide researchers with "living laboratories."

EXPERIMENTAL RESEARCH: WOULD YOU HELP ANOTHER PERSON?

At 3 a.m. on March 13, 1964, Kitty Genovese was stabbed to death in the street near her home in New York City. Winston Moseley, her attacker, assaulted her three times over a period of half an hour. At one point, he left her and returned a few minutes later. During the assault, Ms. Genovese screamed, "Oh, my God, he stabbed me! Please help me!" The press reported that she received no help from at least 38 neighbours who saw the attack and heard her cries for

Adisa/Shutterstock.com

What if one of these people suddenly had a heart attack or got stabbed by another person? Under what conditions would others intervene to help? Social research has helped us answer this question.

help and that these neighbours did not turn away or ignore the attack; they continued to watch the murder from their apartment windows without coming to her assistance or calling the police. At Moseley's trial, several of these witnesses said they simply didn't want to get involved. Moseley himself said, "I knew they wouldn't do anything—they never do."

This case received worldwide attention. The killing also raised several questions for researchers: Why did Ms. Genovese's neighbours fail to act altruistically? Under what conditions will people be more or less likely to help others?

Social psychologists Bibb Latané and John Darley (1970) addressed these issues. Their initial field studies found that in routine situations people were very willing to help. The vast majority willingly gave directions, told inquirers the time of day, and provided change for a quarter. However, their willingness to help could be changed by manipulating simple conditions, such as the wording of the request for assistance and the number of people asking for help.

People are so willing to help in non-emergency situations that the question of why they may fail to respond to emergencies is difficult to understand. Latané and Darley rejected the view that this failure is due to apathy or indifference. Instead, they developed a *theoretical model* of the intervention process. Before a bystander will intervene in an emergency, they must notice that something is happening, interpret this as an emergency, and decide that they are going to help. Latané and Darley proposed the hypothesis that the presence of other people will make people less likely to take each of these steps. They predicted the presence of others would inhibit the impulse to help because each of the potential helpers may look to others for guidance rather than acting quickly, potential helpers might be afraid of failing in front of other bystanders, and potential helpers may feel someone else would take care of the problem.

Latané and Darley tested this hypothesis with a series of experiments, each scripted and designed like a short play. One of their studies simulated an emergency. Fifty-two university students were randomly assigned to groups of three different sizes: a two-person group (the subject and the victim), a three-person group, and a six-person group. Each of the students was seated at a table in a room, given a pair of headphones with a microphone, and told to listen for instructions.

Over the intercom, the subjects were told that the study concerned the personal problems facing students in a high-pressure urban environment. They were also told that to maintain their anonymity when discussing personal matters, they had been placed in individual rooms and would talk and listen to others only through an intercom. The discussion would be controlled by a mechanical device, which would turn each student's microphone on for two minutes at a time and then turn it off while the other students were talking. Thus only one student could be heard at a time and students could not talk with each other. The experimenter said he would not listen to the students' discussion but would get their reactions later.

This elaborate script was designed to see how the participants would respond to an emergency—in this case, a seizure in one of the subjects. After receiving the instructions, subjects heard a taped simulation that began with the future seizure victim discussing his difficulties adjusting to university and to big-city life. He also mentioned that he was prone to seizures during studying and exams. Each of the other people in the group, including the subject, then took their turns talking about their own adjustment problems.

The emergency occurred when it was again the victim's turn to talk. After beginning normally, he began to show obvious distress, then asked repeatedly for help, then made choking sounds and said he was going to die. After that, the intercom went quiet.

You will recall that Latané and Darley hypothesized that the presence of others would inhibit a helping response when people were faced with an emergency. The dependent variable was the time that elapsed from the start of the victim's seizure until the participant left the experiment room to get help. The independent variable was the number of other people each participant believed had also heard the victim's distress.

Did the experiment support the hypothesis? Figure 2.2 shows the results. Clearly, the number of bystanders had a significant effect on the likelihood of the student reporting the emergency. All the participants in the two-person groups reported the emergency, compared with 85 percent of the subjects in the three-person groups and only 62 percent of the subjects in the

FIGURE 2.2 : A STUDY OF ALTRUISTIC BEHAVIOUR

Cumulative proportion of subjects reporting seizure who think that they alone hear the victim or that one to four others also are present.

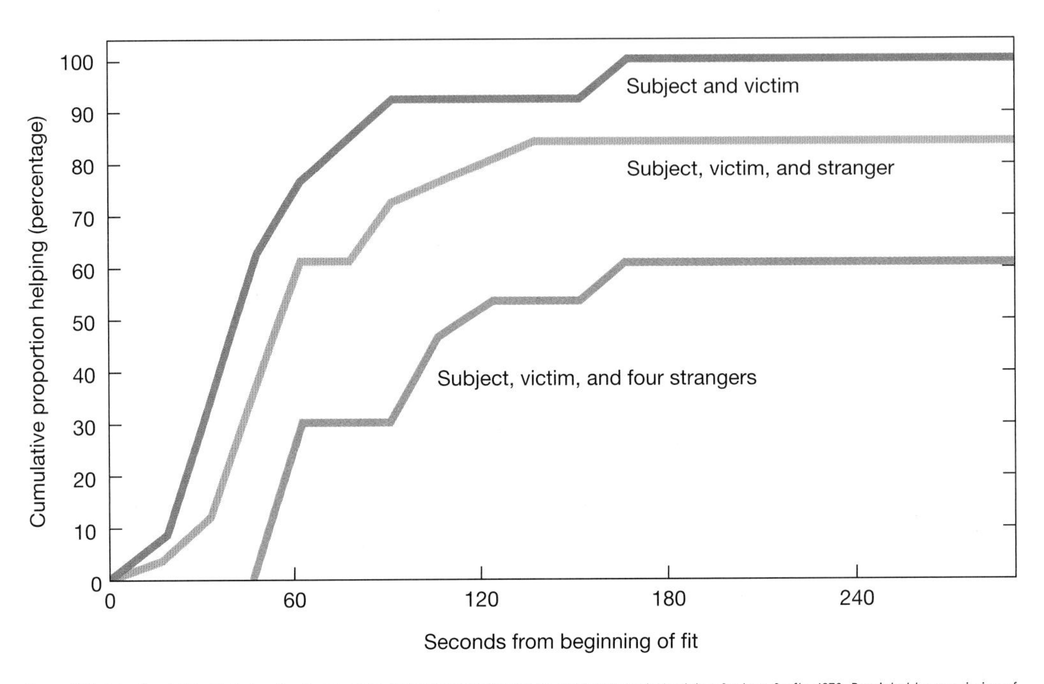

Source: Bibb Latané and John M. Darley, *The Unresponsive Bystander: Why Doesn't He Help?* New York: Appleton Century Crofts, 1970. Reprinted by permission of Prentice Hall.

six-person groups. The participants in the two-person groups also responded more quickly than those in the larger groups.

What about those who failed to respond? They were not apathetic or unconcerned about the victim but were clearly upset by the episode. When the researcher entered the experiment room to end the study, the students who had not reported the seizure were often nervous and emotionally aroused. Many asked if the victim was "all right." While the students in the two-person groups clearly felt they had to intervene because nobody else knew of the victim's distress, those in the larger groups did worry about the victim but were also concerned about making fools of themselves by overreacting and ruining the experiment. Knowledge that others also knew of the emergency made it less likely that they would resolve this conflict by helping the victim.

There is an interesting postscript to the Kitty Genovese story, which had led to this research. A recent book claims that many of the details reported by newspapers, including the *New York Times,* were wrong. According to Cook (2014), while several witnesses did indeed fail to report the crime, there were far fewer than 38 witnesses and many of these did not recognize the seriousness of the event, at least one person did call the police but they did not come, and one of Genovese's neighbours did come out of her apartment to help her as she lay dying.

STRENGTHS AND WEAKNESSES OF EXPERIMENTS The major advantage of the experiment is the researcher's control over the environment and the ability to isolate the experimental variable. This makes experiments the best way of testing cause-and-effect relationships. In order to show that a change in one variable causes a change in another, three conditions must be satisfied.

▶ **correlation** Exists when two variables are associated more frequently than could be expected by chance.

1. *There must be a correlation between the two variables.* **Correlation** exists when two variables are associated more frequently than could be expected by chance. In the Latané and Darley

studies, there was a correlation between the number of people in the group and the likelihood of helping a person in distress.

2. *The independent variable must precede the dependent variable in time.* In an experiment, the variables are manipulated by the researcher, so time order is controlled.
3. *You must ensure that any change in the dependent variable was not because of an extraneous variable*—one outside the research hypothesis. This is referred to as a spurious correlation, or one in which the association between the two variables is caused by a third variable. The classic example of a spurious correlation is the relationship between the number of fire engines at a fire and the cost of the damage caused by the fire. The relationship between these two variables occurs because both are caused by a third variable—the size of the fire.

In addition to these conditions, you also need a theory that links the two variables. Many correlations are simply accidental and the two variables have nothing to do with each other. Figure 2.3 shows one of these taken from the website Spurious Correlations. It is clear that there is no reason for space launches to be related to sociology doctorates (or other correlations on the site, such as the relationship between the divorce rate in Maine and the per capita consumption of margarine in the U.S.), but other accidental correlations can make front page news. Many of these are taken from large medical studies where attempts to find correlations between hundreds of variables lead to accidental correlations, which are debunked by subsequent research.

Artificiality is a major weakness of laboratory experiments. Participants in a laboratory obviously know they are participating in an experiment and may react to what they think the experiment is about or may not react realistically because they do not believe the scenario is real. **Reactivity** is the tendency of participants to change their behaviour in response to the presence of the researcher or to the fact that they know they are being studied. Latané and Darley tried to determine if some respondents chose not to intervene because they did not think the seizure was real. They concluded this was not a problem because the respondents were nervous when reporting the seizure, were surprised when they learned the true nature of the study, and made comments such as "My God, he's having a fit" during the simulated seizure.

reactivity The tendency of experiment participants to change their behaviour in response to the presence of the researcher or to the fact that they know they are being studied.

A second limitation of experiments is that social scientists frequently rely on volunteers or captive audiences, such as students. As a result, the subjects of most experiments may not be representative of a larger population.

Third, experiments are limited in scope, as only a small number of variables can be manipulated.

FIGURE 2.3 : AN EXAMPLE OF AN ACCIDENTAL CORRELATION

Not all correlations are meaningful. This is an example of an accidental relationship.

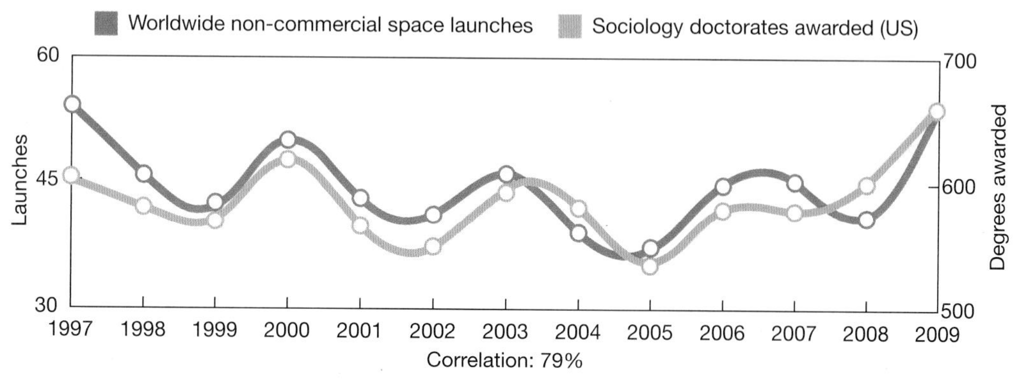

Source: Tyler Vigen, Spurious Correlations, n.d. http://www.tylervigen.com/view_correlation?id=805

SURVEYS

▶ **survey** A research method in which a number of respondents are asked identical questions through a systematic questionnaire or interview.

Survey research is the method perhaps most often associated with sociology. In a **survey**, a number of respondents are asked identical questions through a systematic questionnaire or interview. Researchers may select a sample from a larger population to answer questions about their attitudes, opinions, or behaviour. Surveys are an important research method because they make it possible to study things that are not directly observable—such as people's attitudes and beliefs—and to describe a population too large to observe directly (Babbie, 2004).

▶ **questionnaire** A research instrument containing a series of items to which subjects respond.

TYPES OF SURVEYS Survey data are collected by using self-administered questionnaires, personal interviews, telephone surveys, or Internet surveys. A **questionnaire** is a research instrument containing a series of questions to which subjects respond.

Self-administered questionnaires have certain strengths. They are inexpensive to administer, they allow for rapid data collection and analysis, and they permit respondents to remain anonymous (an important consideration when the questions are personal). A major disadvantage is the low response rate. Mailed surveys sometimes have a response rate as low as 10 percent, and those who respond may not be representative of the larger group. The response rate is usually higher if the survey is handed out to a group, such as a school class, that is asked to fill it out on the spot. In recent years, online surveys have become more common, though response rates for these surveys can also be very low unless they are carefully targeted at specific groups of respondents. It may be difficult to draw a sample using this technique, particularly when the researcher is targeting the general population rather than specific groups where the researcher knows everyone's email address.

▶ **interview** A research method using a data collection encounter in which an interviewer asks the respondent questions and records the answers.

Survey data may also be collected by an **interview** in which an interviewer asks the respondent questions and records the answers. Survey research typically uses structured interviews, in which the interviewer asks questions from a standardized questionnaire.

Interviews are usually more effective than mail or online surveys in dealing with complicated issues, and provide an opportunity for face-to-face communication between the interviewer and respondent. When open-ended questions are used, the researcher may gain more in-depth answers than other techniques. The major disadvantage of interviews is the cost and time involved in conducting them.

Questionnaires may also be administered by *telephone surveys,* which save time and money compared to face-to-face interviews. Some respondents may be more honest over the phone than when they are facing an interviewer. Telephone surveys usually have much higher response rates than mailed questionnaires. However, caller identification systems, cellphones, and do-not-call lists have made telephone surveys more difficult, since many people are now less accessible to researchers.

▶ **population** In a research study, those persons about whom we want to be able to draw conclusions.

▶ **sample** The people who are selected from the population to be studied.

▶ **representative sample** A selection where the sample has the essential characteristics of the total population.

▶ **simple random sample** A selection in which everyone in the target population has an equal chance of being chosen.

SAMPLING CONSIDERATIONS Survey research usually involves some form of sampling. Researchers begin by identifying the population they want to study. They then draw a sample of that population.

The **population** is the group about which we wish to draw conclusions. A **sample** is the people who are selected from that population. A **representative sample** is a selection where the sample has the essential characteristics of the total population. For example, if you have to interview five students selected haphazardly from your sociology class, they would not be representative of your school's total student body. By contrast, if 500 students were selected from the total student body using a random sampling method, they would likely be representative of your school's students. A **simple random sample** is chosen by chance: Every member of an entire population being studied has the same chance of being selected. For example, you might draw a sample of the total student body by placing all the students' names in a rotating drum and drawing names from it.

SURVEY RESEARCH: WHO GIVES IN CANADA?

Many organizations depend upon charitable giving. Many buildings on your campus are likely named after people who have donated large sums of money to your institution; research into the causes of many diseases is funded by donations; food banks depend on donations; and many Canadians contribute to fundraising drives during famines or natural disasters. Because of the importance of charitable giving, we need to know more about the characteristics of those who give their money to help others.

Every five years, Statistics Canada conducts the Canada Survey of Giving, Volunteering and Participating to learn more about charitable giving and volunteer work. In 2010, Statistics Canada drew a sample of 15,482 people 15 years of age and over for this study (Vezina and Crompton, 2012; see Box 2.3). The survey used randomly generated phone numbers to contact respondents. The response rate for the survey was 56 percent, which meant that an interview was conducted with people at 56 percent of the numbers called. The other 44 percent could not be contacted or refused to participate in the survey.

wavebreakmedia/Shutterstock.com

Computer-assisted telephone interviewing is an easy and cost-efficient method of conducting research. The widespread use of voice mail, cellphones, and caller ID has made this form of research more difficult in the 21st century.

BOX 2.3 PRESENTING SOCIOLOGICAL DATA

Sociological data are often presented numerically, and in this text you will find many tables showing the results of research studies. You should learn how to understand these tables and to develop the skill of presenting your own data.

In the study of charitable giving, Statistics Canada researchers wanted to determine if higher-income people were more or less likely to donate than those with lower incomes. Their data are presented below.

TABLE 2.1 DONOR RATE BY HOUSEHOLD INCOME (IN PERCENT)

Household Income	Less Than $20,000	$20,000 To $39,999	$40,000 To $59,999	$60,000 To $79,999	$80,000 To $99,999	$100,000 Or More
DONATED TO CHARITY						
YES	67	81	83	86	89	88
NO	33	19	17	14	11	12
	100%	100%	100%	100%	100%	100%

Source: Statistics Canada, 2010a.

The basic elements of a table are:

1. *Title or heading:* A brief description of the *content* of the table. The title "Donor Rate by Household Income" tells us that the table shows relationships between two variables: donor rate and household income.
2. *Categories of the variables:* The variable "Household Income" is divided into six categories based on reported income, and the variable "Donor Rate" is divided into the categories of Yes or No.
3. *Percentages:* The information in the table is stated in percentages. Using the raw numbers would make comparisons difficult for the reader. By comparing the percentages, we can see that as income increases, so does the likelihood of donating to charities.

This research examined some interesting questions. Which Canadians are most likely to donate their time or money? Are women more likely to donate money than men? Are young people more likely to volunteer their time than seniors?

You may be surprised to learn that the vast majority of Canadians—84 percent—had donated money to a charitable or nonprofit organization in the 12 months preceding the survey. The average person donated $446 and the total amount was $10.6 billion (Turcotte, 2012). The main recipients of these donations were religious organizations (40 percent of the total), health-related organizations (15 percent), and organizations delivering social services (11 percent).

Who was most likely to donate? Women were slightly more likely to donate than men (Statistics Canada, 2010a). Higher-income earners were more likely to donate than those with lower incomes and donated larger amounts of money. However, 67 percent of those in the lowest income bracket (less than $20,000 a year) donated. This compares with 88 percent of those who earned $100,000 or more.

Fewer people volunteered their time than gave their money—47 percent (Vezina and Crompton, 2012) compared with 84 percent who donated funds. The patterns were similar to those for charitable giving, with the exception that young people (15 to 24) were the most likely to volunteer their time: 58 percent of this group had volunteered in the previous year, compared with 54 percent of the next highest group (35 to 44) and 36 percent of those 65 and over. While young people were less able to donate money (a scarce commodity for students and those new to the labour market), they were generous with their time.

One of the survey's most interesting findings was that while there was a broad range of participation in the altruistic activities of charitable giving and volunteering, a core group did much more than others. About 10 percent of respondents were responsible for 63 percent of financial donations and 53 percent of volunteer hours. In the words of researchers reporting on an earlier survey, "a lot do a little, but a little do a lot" (Hall et al., 2006:61).

STRENGTHS AND WEAKNESSES OF SURVEYS Survey research has several strengths. First, it is useful in describing the characteristics of a large population. Second, it enables the researcher to assess the relative importance of a number of variables. In the table in Box 2.3, we can compare the donation rates of people from different income levels.

Survey research also has weaknesses. One is that the use of standardized questions tends to force responses into categories that may not fit well. Moreover, people may be less than truthful, especially on emotionally charged issues or on issues, such as altruism, that have a strong element of social desirability. They may also be unwilling to provide information on sensitive issues, such as sexual activity, income, and criminal behaviour, and may also simply forget relevant information. This can make reliance on self-reported attitudes and behaviour problematic in some surveys. Finally, response rates have become so low for many surveys that their generalizability may be questionable.

FIELD RESEARCH

▶ **field research** The study of social life in its natural setting: observing and interviewing people where they live, work, and play.

▶ **complete observation** Research in which the investigator systematically observes a social process, but does not take part in it.

Field research is the study of social life in its natural setting: observing and interviewing people where they live, work, and play. Some kinds of behaviour are best studied by "being there"; a fuller understanding can be developed through observations, face-to-face discussions, and participation in events. Researchers use these methods to generate *qualitative* data: observations that are best described verbally rather than numerically. Although field research is less structured and more flexible than the other methods we have discussed, it also places many demands on the researcher. For example, field researchers must decide how to approach the target group, how to identify themselves as researchers, and whether to participate in the events they are observing.

OBSERVATION Sociologists interested in observing social interaction may use either complete or participant observation. In **complete observation**, the researcher systematically observes a social process but does not take part in it. Observational research can take place just about anywhere. For example, sociologists David Karp and William Yoels (1976) became interested in why many students do not participate in classroom discussions. Observers sat in on classes and took notes that included the average number of students who participated, the number of times they talked in class, and the sex of the instructor and of the students who talked in class. From these observations, Karp and Yoels found that, on average, a small number of students are responsible for most of the discussion that occurs in any class.

Subjects in observation studies may not realize that they are being studied, especially if the researcher remains unobtrusive. This type of observation helps us view behaviour as it is taking place. However, it provides limited opportunities to learn why people do certain things. One way for researchers to remain unobtrusive is through **participant observation**—collecting systematic observations while being part of the activities of the group they are studying. Participant observation generates more "inside" information than simply asking questions or observing from the outside. As Whyte noted in his classic participant observation study of a Boston low-income neighbourhood, "As I sat and listened, I learned the answers to questions I would not have had the sense to ask" (1957:303).

participant observation A research method in which researchers collect systematic observations while being part of the activities of the group they are studying.

FIELD RESEARCH: RESPONDING TO DISASTER

In 2013, Calgary and many other communities in Southern Alberta suffered heavy flooding. In this and many other natural disasters, thousands of people pitched in to help by giving their time or their money to help out the victims. Natural disasters have brought people together in many parts of the world and have often shown human behaviour at its altruistic best. What motivates people to help after natural disasters? How effective is the aid that is provided to help others? What is the best way to help restore communities after a disaster?

Kamal Kapadia tried to answer these questions when she studied the impact of one of the world's greatest natural disasters. In December 2004, a devastating tsunami struck several countries surrounding the Indian Ocean. Nearly 250,000 people were killed and millions were left homeless. There was a massive global response to the disaster as people around the world donated billions of dollars, and their governments sent ships, aircraft, and disaster response teams to help with the recovery.

Because of the unprecedented global response, those involved in the reconstruction believed they could "build back better"—meaning that infrastructure such as housing and environmental protection to protect coasts could be improved and the poor could be helped to rebuild better lives.

Kapadia, who had lived and conducted research in Sri Lanka, returned there to study the relief effort. As part of her field research, she worked as a volunteer for a Sri Lankan aid agency and so was directly involved with the aid effort. Her research was very thorough: She conducted a document review, analyzing the documents of 49 of the largest organizations involved with the aid effort; conducted 76 interviews with staff from these organizations; participated in dozens of meetings and workshops dealing with reconstruction; kept a database of newspaper clippings; used her volunteer work to learn directly about the aid effort; and moved to the small Sri Lankan village of Muhudupitiya for a month to observe the reconstruction activities, to spend time with people, and to conduct interviews with local residents. She returned to this village several times after moving back to the city of Colombo (Kapadia, 2008).

Kapadia's ethnographic field research with the villagers gave her an invaluable perspective on what was really happening at the community level during the relief effort. For example, when she visited the village as a volunteer representative of an aid group, she was accompanied by village leaders and believed their story that new housing was being allocated with the full participation of community members. However, when she was living in the village as a field researcher, she found that many of the

Keith Morison/Bloomberg/Getty Images

Canadians have always shown a strong willingness to help in natural disasters, such as the 2013 Calgary floods.

villagers were not happy with the way housing was being allocated and felt the leaders were acting unfairly. Housing was not being distributed through participatory decision making, but was being allocated through the village's normal political power structure. One of the villagers complained:

> None of the projects are well managed or based according to actual needs. Those who run these projects show favoritism. If you know somebody powerful then you can get something. Some people are getting four houses even though only one house was damaged. (Kapadia, 2008:268)

Kapadia could not have discovered this information without having developed good relationships with the villagers.

Her fieldwork in the village also helped her to understand why the relief effort did not lead to an increase in Sri Lankans' standard of living. The relief strategy set out a plan for rehabilitating peoples' livelihoods by helping them build upon entrepreneurial activities such as farming and small cottage industries. Her extensive contact with workers in the community showed her that activities that were categorized by aid agencies as "entrepreneurial" were actually just a form of casual labour. Muhudupitiya was considered a good site for entrepreneurial activities, and relief money was used to support businesses manufacturing bricks, to purchase machines to allow local women to reestablish their businesses processing coconut husks into a fibre called coir, and processing coral to make lime that is used in construction. However, Kapadia found that many of the villagers participated in these businesses on a very short-term basis on contracts as short as one day. While the coir business involved workers on a more long-term basis, the women's income was very low, in part because the market for their product was limited. The women also typically only completed one stage in the process, so they were doing piecework for middlemen rather than working as self-employed entrepreneurs. The global market for coir is limited, so the business had little potential to help move villagers out of their lives of poverty.

The aid agencies assumed that "if microfinance, training and assets were made available, people would quickly embrace these opportunities and escape poverty" (Kapadia, 2008:197). However, Kapadia found that relations of production and trade meant that these relief efforts made very little difference in the lives of most individuals. One specific example is the ongoing relationship that many poor villagers have with moneylenders. In many cases, these moneylenders are also employers, so villagers are bound to the people who hire them for casual labour by ties of history, tradition, and debt. The assistance that aid does provide helps those who have already escaped poverty through their business and moneylending activities. But even young people in the village were not attracted to entrepreneurship—most would much rather be employed in a government job.

Kapadia's field research is important because it shows that altruism—in this case, the donation of money, time, and materials to a country ravaged by a natural disaster—is not enough to guarantee that people can move out of poverty or even that their lives will soon return to normal. Without an understanding of the realities of life in the affected country, even the best-funded efforts will have little impact.

▶ **unstructured interview** A research method involving an extended, open-ended interaction between an interviewer and an interviewee.

© Imagestate Media Partners Limited-Impact Photos/Alamy

Case studies of homeless persons have added to our insights on the causes and consequences of this major social concern. Women are often the "invisible homeless."

UNSTRUCTURED INTERVIEWS An **unstructured interview** is an extended, open-ended interaction between an interviewer and an interviewee. The interviewer has a general plan of inquiry rather than a specific set of questions, as is often the case with surveys. Unstructured interviews are essentially conversations in which interviewers establish the general direction by asking open-ended questions. Interviewers can "shift gears" and pursue specific topics raised by interviewees because answers to one question are used to suggest the next question or new areas of inquiry. This technique enables the interviewer to learn more about the interviewees' lives and thoughts.

GROUNDED THEORY Qualitative methods are frequently used to develop theories. The term *grounded theory* was developed by Glaser and Strauss (1967) to describe an inductive method of theory construction. Researchers who use grounded theory collect and analyze data simultaneously. For example, in her study in Sri Lanka, Kapadia did not begin with a theory but developed one as a result of her research:

> It was only when I became more deeply engaged with village life in Muhudupitiya, and started to acknowledge, as opposed to shut out, events and experiences that simply did not fit my conceptual categories, that I was able to see how the managerial and impact models were falling short. I am therefore, in effect, making the argument that one's theories are fundamentally shaped by one's methods, approaches, and experiences, and a self-reflective and engaged ethnographic research approach provides an excellent way to understand the workings of power, [and] the failings of reconstruction (and development) programs. (2008:338–39)

STRENGTHS AND WEAKNESSES OF FIELD RESEARCH Field research provides opportunities for researchers to obtain an inside view of social phenomena. Field methods are useful for understanding attitudes and behaviours within their natural setting and when the researcher wants to study social processes and change over time. They provide detailed information about the reactions of people and let us generate theories from the data collected (Whyte, 1989). Field research is the best way to uncover the meanings that people give to their social lives, the focus of **interpretive sociology**. Weber and others have argued that sociologists should examine the way people construct their social realities and need to understand the meanings people give to various aspects of their social lives. These meanings can best be understood by close contact in natural settings with the people being studied.

interpretive sociology An approach to sociology that examines the meanings that people give to aspects of their social lives.

Participant observation can be difficult. Jack Haas describes his first day in the field in his study of high steel ironworkers:

> My traumatic introduction to the workday realities of high steel ironworking came the day the construction superintendent passed me through the construction gate, gave me a hard hat, and wished me good luck. Directly ahead were five incomplete levels of an emerging 21-storey office building. From my vantage point I observed a variety of workers engaged in the construction process. The most visible and immediately impressive group of workers were those on the upper level who were putting steel beams into place. These were the ironworkers I had come to participate with and observe. This chilling reality filled me with an almost overwhelming anxiety. I began to experience a trepidation that far exceeded any usual observer anxiety encountered in the first days of field research . . . the risks of firsthand observation were profoundly obvious. It was with fearful anticipation that I moved toward the job site. (1977:148)

National Geographic/Getty Images

Participant observers share the same experiences as those who are being observed. Can you think of other observational research projects that would put the researcher in danger?

Haas's understandable fear over the prospect of walking on 10-cm wide steel beams hundreds of metres above city streets helped to focus his research. He eventually learned that while the ironworkers did not talk publicly about their fears, they did share his concerns. However, they were careful not to let their colleagues know they were afraid and often took risks in order to prove themselves to the other workers.

Like others in dangerous occupations, the ironworkers had developed a culture to help them deal with the risks of their work. They tried to control their work environment by making their own decisions about whether it was too dangerous to work rather than going along with the demands of their supervisors and by following a strict informal code of what types of behaviour were appropriate for their co-workers.

Field research rarely involves this much danger, but it can involve spending lengthy periods of time with very different types of people and complicated work schedules. Despite the difficulties, there is really no other way of experiencing the world of your research subjects. It is far different having someone describe what it is like to stand on a steel girder two hundred feet above the ground than actually standing on this piece of steel yourself. Field research provides a richness of data that cannot be obtained in any other way.

LO-4 SECONDARY ANALYSIS OF EXISTING DATA

▶ **secondary analysis** A research method in which researchers use existing material and analyze data that originally was collected by others.

In **secondary analysis**, researchers use existing material and analyze data originally collected by others. Existing data sources include public records, official reports of organizations or government agencies, and surveys conducted by researchers in universities and private corporations. Research data gathered from studies are available in data banks. Other sources of secondary data are books, magazines, newspapers, radio and television programs, personal documents and Internet sites. Box 2.4 shows some of the ways in which new media can be used as a source of sociological data. Secondary analysis is *unobtrusive research* because it includes a variety of nonreactive research techniques—that is, techniques that have no impact on the people being studied. In most cases, they are not even aware they are being studied.

The case study in the next section shows how data collected for other purposes can be used to shed light on the phenomenon of altruism.

SECONDARY ANALYSIS: HELPING AFTER HURRICANE KATRINA

In September 2005, Hurricane Katrina devastated the city of New Orleans and many other communities in Louisiana and Mississippi. More than a thousand people were killed and many more lost their homes and businesses as a result of the hurricane and the resulting flooding. African Americans and the elderly were disproportionately harmed by the storm, and governments at all levels were justifiably blamed for their inadequate planning and response.

While governments failed during Katrina, what about average citizens? Media images of the catastrophe showed New Orleans as a city in chaos, with disorganized rescue attempts, inadequate evacuation plans, and extensive looting and murder. Did members of the community really turn against one another rather than working together and behaving altruistically? Havidan Rodriguez and his colleagues at the University of Delaware's Disaster Research Center (Rodriguez and Dynes, 2006; Rodriguez, Trainor, and Quarantelli, 2006) studied how the media portrayed Hurricane Katrina and its aftermath.

Rodriguez and his colleagues conducted a content analysis of material from several sources:

1. A database of news sources in paper format and/or their website equivalent that were collected over the first month of the response . . . more than 2000 articles have been collected and catalogued by [Disaster Research Center] staff.
2. Reports disseminated by other formal organizations either in printed form or on their websites.
3. Stories from informal sources such as bloggers on the Internet. (Rodriguez, Trainor, and Quarantelli, 2006:86)

They concluded that the media exaggerated many of the negative things that occurred during the catastrophe. This was particularly true of reports of mobs looting and killing people during the flooding. For example:

> When the Louisiana National Guard at the Superdome [an evacuation site where there had been reports of major violence and disorder] turned over the dead to federal authorities, that representative arrived with an 18-wheel refrigerated truck

> since there were reports of 200 bodies there. The actual total was six; of these, four died of natural causes, one from a drug overdose and another had apparently committed suicide. While four other bodies were found in the streets near the Dome, presumably no one had been killed inside as had been previously reported. There were more reports that 30 to 40 bodies were stored in the Convention Center freezers in its basement. Four bodies were recovered: one appeared to have been slain. Prior to this discovery, there had been reports of corpses piled inside the building. (2006a:6)
>
> Source: Rodriguez, Havidán, and Russell Dynes. 2006. *Finding and Framing Katrina: The Social Construction of Disaster*. Understanding Katrina. Social Science Research Council. Retrieved February 1, 2009. Available: http://understandingkatrina.ssrc.org/Dynes_Rodriguez/. Used by permission of the authors.

In contrast to these exaggerated reports of civil disorder, Rodriguez, Trainor, and Quarentelli documented the way in which community members and outsiders helped the hurricane's survivors. In several neighbourhoods, people organized themselves to help others. One group of friends that called itself "the Robin Hood Looters" evacuated their own families, then went back into the community and spent two weeks rescuing people from their flooded homes. Their name came from the practice of taking food and water from abandoned homes to provide it to people who had none. Working with the police and National Guard, they became what sociologists call an "emergent group." They developed a structure and norms during the crisis that helped them operate effectively.

Many existing groups also rose to the occasion. For example, religious organizations already involved in helping the poor and homeless massively expanded their activities to take on many more volunteers to help cope with the demands of Katrina. Even the researchers from the Disaster Research Center benefited from altruistic behaviour, as they were housed and fed by local people. Thus the behaviour of people following the hurricane was not nearly as violent and chaotic as portrayed by the media, and many people worked hard to help restore their communities.

BIG DATA The term **big data** was originally used to describe datasets so large that they could not be analyzed using conventional computers. However, the term has evolved and now also encompasses the fact that governments and corporations have enormous amounts of data that can be accessed in digital form, and that these databases can be linked with one another.

► **big data** Very large datasets that can be accessed in digital form and that can be linked with other large datasets.

These networked data provide unique research opportunities—"[the value of big data] comes from the patterns that can be derived about an individual, about individuals in relation to others, about groups of people, or simply about the structure of information itself" (Boyd and Crawford, 2011:2).

Many disciplines are using big data. Medical scientists conduct large-scale analysis of DNA. Historians use databases such as the Old Bailey Online, which contains digitized transcripts of nearly 200,000 trials held in London over a 250-year period in order to learn about culture, crime, and justice from 1674 to 1913. Researchers from many disciplines are interested in studying social media behaviour. Every day, people post tens of millions of new photos to Facebook and hundreds of millions of new tweets on Twitter, and analysis of this material can reveal a great deal about our social lives. Criminologists use police data to track patterns of crime in our communities in order to plan more effective crime reduction strategies. Urban analysts use GPS data from peoples' phones to track pedestrian and vehicle flows in real time through a city in order to smooth traffic flows and to study the way urban residents experience their neighbourhoods. Perhaps most significantly, corporate marketers use big data to increase their profits by knowing more about their users.

Some have made lofty claims for the power of big data. Pentland says that it provides new tools that "give a view of life in all its complexity—and are the future of social science" (2014:10). Chris Anderson, editor of *Wired* magazine, exemplifies those who feel big data provides unique analytic power: "Who knows why people do what they do? The point is they do it, and we can track and measure it with unprecedented fidelity. *With enough data, the numbers speak for*

BOX 2.4 SOCIOLOGY AND NEW MEDIA

Methods and the New Media

New media are important to social research. Many researchers now conduct their research using the Internet. You have probably been asked to complete an online survey. Corporations often conduct these surveys, and university researchers are increasingly turning to online research. Websites like Survey Monkey make Internet surveys relatively easy, and these surveys will become a more common way of conducting survey research in the future.

The Internet is also a rich source of data. Researchers have studied the comments sent in response to stories on media websites. Elizabeth Comack and Evan Bowness analyzed the anonymous online comments made in response to a CBC report on the fatal shooting of a young Indigenous man by a police officer. Many of the comments were overtly racist despite the CBC's expressed policy of removing such posts:

> ***thepear* wrote:** Posted 2008/08/05 at 10:57 PM ET. . . . And by the way, if there was no white man here to stop this punk, would you be happier to have these types of young thugs running around your teepee villages stabbing you all to death? Probably not.
>
> **21** People recommended this comment Report abuse (2010:42)

Comack and Bowness found that the majority of the comments reinforced a discourse of racial privilege. Only a minority of the comments challenged these racialized comments, and readers were far less likely to recommend these posts than those that expressed hostility toward the man who was killed and toward Indigenous people in general.

Many people live important parts of their lives in online worlds such as Second Life (see Chapter 18). These virtual communities provide a rich setting for sociological analysis, and researchers have begun to study these online communities using the same techniques as they would use to study real-life cultures. Ethnographies usually involve a researcher immersing himself or herself into a setting in the field to learn as much as they can about the culture being studied. "Cyberethnographies" apply this methodology to the online world. Marissa Ashkenaz (2008) wanted to examine the online norms concerning body size, so she asked colleagues to create Second Life avatars that were moderately overweight and found there was a great deal of pressure from others online for the participants to use thinner avatars (you can see the overweight avatars and read some of the comments in Chapter 18, "Creating Avatars in Second Life"). Cyberethnographers can also bring an additional dimension to their research by stepping outside the virtual world and conducting research involving the real-life people behind the avatars to see what their online lives mean to them.

Finally, researchers have begun to use the vast amount of data generated through social media to study social relationships. Kevin Lewis, Marco Gonzalez, and Jason Kaufman (2012) studied the Facebook activities of a group of U.S. college students by recording information from Facebook profiles over a four-year period. One of their studies looked at students who had mutually confirmed "friendships" and examined their stated preferences in music, movies, and books. The researchers found that the students were more likely to form and maintain friendships with peers who liked the same types of music and movies—though this was limited to clusters of "lite/classic rock" and "classical/jazz" music and "dark satire" and "raunchy comedy/gore" movies. Tastes in books had no relationship with friendship. These tastes did not appear to be contagious, in that people's preferences did not become closer to that of their friends over time.

Current studies of Facebook and other social media are only scratching the surface, and it is likely that this kind of research will grow very quickly as researchers learn more about the insights into our society and culture that are available online.

themselves [italics added] (2008:n.p.). While big data do provide tremendous research opportunities, think of researchers who spend years in a particular community talking with residents and observing their behaviour. Their knowledge of that community is not *less* than the digital record captured from smartphones, credit cards, and surveillance cameras. Instead it is *different*—the high level view of big data can give us some idea of what people are doing, but you must talk to the people to find out why they are doing it and to learn what it means to them. At least in the social realm, numbers neither speak for themselves nor do they tell the whole story.

STRENGTHS AND WEAKNESSES OF SECONDARY ANALYSIS One strength of secondary analysis is that data are readily available and may be inexpensive to obtain. Another is that the chances of bias may be reduced because the researcher usually does not collect the data personally. Finally, if records have been kept over time, a researcher can analyze longitudinal data to identify trends or provide a historical context.

However, secondary analysis has inherent problems: the data may be incomplete or inaccurate; if data on certain variables have not been collected, the information will not be available for later research; and, since secondary data are often collected for administrative purposes, the categories may not reflect variables of interest to the researcher. Religion was important to Durkheim's research on suicide, but death records did not contain information on religion—he had to infer that each person who committed suicide belonged to the religious group most common in that community.

MULTIPLE METHODS: TRIANGULATION

What is the best method for studying a particular topic? The Concept Snapshot compares the various social research methods. There is no single best research method; each method has its own strengths and weaknesses, so many sociologists believe that it is best to combine multiple methods in a given study. *Triangulation* is the term used to describe this approach (Denzin, 1989). **Triangulation** refers not only to research methods but also to multiple data sources, investigators, and theoretical perspectives in a study. Multiple data sources include persons, situations, contexts, and times.

▶ **triangulation** Using several different research methods, data sources, investigators, and/or theoretical perspectives in the same study.

For example, in a study of homeless people, Snow and Anderson used as their primary data sources "the homeless themselves and the array of settings, agency personnel, business proprietors, city officials, and neighbourhood activities relevant to the routines of the homeless" (1991:158). They gained a detailed portrait of the homeless and their experiences and institutional contacts by tracking more than 700 homeless individuals through a network of seven institutions with which they had varying degrees of contact. The study also tracked a number of individuals over time and used a variety of methods, including "participant observation and informal, conversational interviewing with the homeless; participant and nonparticipation observation, coupled with formal and informal interviewing in street agencies and settings; and a systematic survey of agency records" (1991:158–169). This study is discussed in depth in Chapter 5.

FEMINIST RESEARCH METHODS

Feminist social scientists have been critical of traditional sociological research methodologies. Margrit Eichler (1988b) has identified several limitations in research that relate to gender, including *androcentricity* (which means approaching an issue from a male perspective or viewing women only in terms of how they relate to men); sexist language or concepts; research methods that are biased in favour of men (for example, in sampling techniques or questionnaire design); and research in which results that focus on members of one sex are used to support conclusions about both sexes.

No one method can be termed *the* feminist methodology. However, qualitative methods, and in particular in-depth interviews, tend to be associated with feminist research. Although feminist research may involve the same basic methods for collecting data as other research, the way in which feminists use these methods is different. First, women's experiences are important and women's lives need to be addressed in their own terms (Edwards, 1993). Feminist research is woman-centred; that is, "It puts women at the center of research that is nonalienating, nonexploitive, and potentially emancipating" (Scully, 1990:2–3). Second, the goal of feminist research is to provide explanations of women's lives that will help them improve their situations.

It is important, therefore, to ensure that women's experiences are not objectified or treated as merely "research data." Indeed, feminist sociologist Dorothy Smith (1987) suggests that "giving voice" to disadvantaged and marginalized groups in society should be a primary goal of sociology. Finally, feminist research methods challenge the traditional role of the researcher as a detached, "value-free," objective observer. Rather, the researcher is seen as central to the research process, and her feelings and experiences should be analyzed as an integral part of the process (Edwards, 1993; Kirby and McKenna, 1989).

CONCEPT SNAPSHOT

STRENGTHS AND WEAKNESSES OF SOCIAL RESEARCH

RESEARCH METHOD	STRENGTHS	WEAKNESSES
Experiments		
Laboratory	Control over research	Artificial by nature
Field	Ability to isolate experimental factors	Frequent reliance on volunteers or captive audiences
Natural	Relatively little time and money required; replication possible, except for natural experiments	Ethical questions of deception; problem of reactivity
Survey Research		
Self-administered questionnaire	Useful in describing features of a large population without interviewing everyone	Potentially forced answers
Interview	Relatively large samples possible	Respondent untruthfulness on emotional issues
Telephone survey	Multivariate analysis possible	Data that are not always "hard facts" presented as such in statistical analyses
Secondary Analysis of Existing Data		
Existing statistics	Data often readily available; inexpensive to collect	Difficulty in determining accuracy of some of the data
Content analysis	Longitudinal and comparative studies possible; replication possible	Failure of data gathered by others to meet goals of current research; questions of privacy when using diaries or other personal documents
Big data	Massive datasets, so lots of information; available in digital form; enables linking diverse types of information	Collected by governments and corporations, so may not be available to public researchers; size of databases may lead to misplaced confidence in validity of findings
Field Research		
Observation	Opportunity to gain insider's view	Problems in generalizing results to a larger population
Participant observation	Useful for studying attitudes and behaviour in natural settings	Data measurements not precise
Case study	Longitudinal/comparative studies possible; documentation of important social problems of excluded groups possible	Inability to demonstrate cause-effect relationships
Unstructured interviews	Access to people's ideas in their words; forum for previously excluded groups	Difficult to make comparisons because of lack of structure; not representative sample

Raquel Bergen's research on marital rape has shown the need for the researcher's personal involvement in the research process. Bergen ensured that the women she interviewed knew that she was supportive and interested in helping them and that she was not simply exploiting their experiences for her own purposes. She shared her own views and experiences with her research subjects and carefully dealt with any emotional distress that was caused by her interviews:

> During the most emotionally difficult interview, I spent a long time offering support to a woman who became extremely upset when she described her husband . . . raping her in front of her child. This experience emphasized the need for researchers . . . to interview with conscious partiality. If I had been a detached and objective researcher merely collecting data, I might have either terminated the interview and discarded the data or possibly suggested that the woman receive outside counseling. As a feminist researcher, however, I was interacting with this woman on a personal level and her distress was deeply affecting. (1993:208)

CRITICAL RESEARCH STRATEGIES

Feminist research practice shows that some sociologists choose not to follow the traditional scientific approach presented earlier in this chapter. The scientific approach involves neutrality—researchers try not to let their values bias their research. However, some researchers feel that this neutral stance implicates the researchers in the dominant power structure that creates and maintains injustice: "To put the matter starkly, in a socially unjust world, knowledge of the social that does not challenge injustice is likely to play a role in reproducing it" (Carroll, 2004:3). By contrast, critical research strategies are rooted in a concern for social justice. William Carroll (2004) sets out three ways in which social inquiry can be considered "critical." First, inquiry can be *oppositional.* Researchers place themselves on the side of those who are victims of injustice and criticize structures that oppress people on the basis of distinctions, such as class, race, and gender. Second, inquiry can be *radical.* In other words, the researcher tries to get at the roots of dominance issues and to explore the interconnections between problems such as capitalism, the environment, and individual suffering. Third, inquiry can be *subversive.* Researchers try to question conventional assumptions in order to open the door for alternative understandings of the social world.

Critical scholars use a variety of research *techniques.* These include the traditional methods you have already learned about, field research, secondary analysis, and surveys among them. As examples, Karl Marx developed a workers' questionnaire in 1880 to investigate working conditions, and Friedrich Engels used ethnographic methods to document the plight of the working class in England in 1844 (Carroll, 2004).

The focus of critical scholars, however, is not so much on techniques but rather on strategies for social inquiry that can lead to a more just society. Among these are dialectical social analysis, institutional ethnography, critical discourse analysis, and participatory action research (Carroll, 2004). The critical approach can be illustrated by looking at institutional ethnography.

At the core of *institutional ethnography* is the view presented by feminist researchers that the lived experience of marginalized people should be at the centre of the research process. This focus sets the researcher apart from the institutionalized power structure and puts the researcher on the side of the oppressed. However, the strategy does not just consider the individual but rather looks at the social organization of the institutions that surround individual actors. These institutions are both local—the settings that people experience in their daily lives—and extra-local—"outside the boundaries of one's everyday experience" (Campbell and Gregor, 2006:170). Feminist scholar Dorothy Smith has done much of the work developing this research strategy. She set out three tasks that define how to conduct an institutional ethnography:

> The first task centers on ideology and involves addressing the ideological practices which are used to make an institution's processes accountable. The second task centers on work in a broad sense (not just paid employment), and involves studying

> the work activities through which people are themselves involved in producing the world they experience in daily life. The third task centers on social relations, and involves discovering the ways in which a localized work organization operates as part of a broader set of social relations which link multiple sites of human activity. (Grahame, 2004:185)

The results of this research will be very different from those of more conventional sociology in the manner in which it shows how the everyday experiences of marginalized people are shaped by larger social and economic factors.

TIME TO REVIEW

- How do researchers design experiments? What are the strengths and weaknesses of experimental design?
- What are the different ways in which researchers can conduct survey research? What are the strengths and weaknesses of survey research?
- Describe how a researcher would carry out an observational study. What are the strengths and weaknesses of observational research?
- Why would a researcher choose to use secondary analysis to help understand society? What are the strengths and weaknesses of secondary analysis?
- Discuss how feminist and critical methods differ from more mainstream research techniques.

LO-5 ETHICAL ISSUES IN SOCIOLOGICAL RESEARCH

Researchers are required by a professional code of ethics to weigh the societal benefits of research against the potential physical and emotional costs to participants.

Sociology has several basic ethical principles. Participation in research must be voluntary. No one should be enticed or forced to participate, and everyone must be told about the study and any possible risks so they can give their informed consent. Deception must not be used to obtain consent. Researchers must ensure that subjects are not harmed. For example, the researcher must be careful not to reveal information that would embarrass the participants or damage their personal relationships. Researchers must respect the rights of research subjects to anonymity and confidentiality. A respondent is *anonymous* when the researcher cannot link a given response to a given respondent. Anonymity is often extremely important in terms of obtaining information on "deviant" or illegal activities. Maintaining *confidentiality* means that the researcher can identify a given person's responses but promises not to do so.

Sociologists are committed to adhering to ethical standards and to protecting research participants. However, research ethics can be ambiguous and researchers may disagree about ethical issues. For example, is it ethical to give students extra marks in a psychology course if they take part in an experiment? Is it ethical to persuade institutionalized young offenders to be interviewed about their crimes by offering payment? Different researchers might have different answers for each of these questions.

How honest do researchers have to be with potential participants? A well-known study raises issues about where the right to know ends and the right to privacy begins.

THE HUMPHREYS RESEARCH

Laud Humphreys' (1970) research focused on homosexual acts between strangers meeting in "tearooms," or public restrooms in parks. He did not ask permission of his subjects, nor did he inform them that they were being studied. Instead, he took advantage of the typical tearoom encounter, which involved three men: two who engaged in homosexual acts and a third who kept a lookout for police and other unwelcome strangers. To conduct his study, Humphreys showed up at public restrooms that were known to be tearooms and offered to be the lookout. Then, he systematically recorded details of the sexual encounters.

Humphreys decided to learn about the everyday lives of these tearoom participants. He recorded their car licence numbers and tracked down their names and addresses. Later, he arranged for these men to be included in a medical survey so that he could go out and interview them personally. He wore disguises and drove a different car so that they would not recognize him. From these interviews, he collected personal information and found that most of the men were married and lived conventional lives.

Humphreys would not likely have gained access to these subjects if he had identified himself as a researcher. His failure to do so was widely criticized. The police became very interested in his notes, but he refused to turn any information over to the authorities. His award-winning study, *Tearoom Trade* (1970), dispelled many myths about homosexual behaviour; however, his study remains controversial.

Do you think Humphreys did his research ethically? Would these men willingly have agreed to participate in his research if he had identified himself as a researcher? What psychological harm might have come to these married men if people, outside of those involved in the encounters, knew about their homosexual behaviour? Today's university ethics committees would never permit this type of research to be conducted.

Ethical issues continue to arise in sociological research. Two Canadian cases deal with a very different sort of question than the Humphreys research.

THE OGDEN AND MAGNOTTA CASES

What should social scientists do when the ethical principles of confidentiality and not harming subjects conflict with the law? In 1992, Simon Fraser University student Russel Ogden began work on his master's thesis, a study of euthanasia (mercy killing) and assisted suicide involving AIDS patients (Ogden, 1994). Both euthanasia and assisted suicide are crimes in Canada. The university's ethics committee approved his research proposal, which included a promise to maintain the "absolute confidentiality" of any information provided by his interview subjects (Palys, 1997).

In 1994, Ogden was subpoenaed to give evidence at a coroner's inquest investigating the possible assisted suicide of an AIDS victim. Ogden refused to testify, citing the guarantee of confidentiality he had given to his respondents. The coroner charged Ogden with contempt of court. After a lengthy legal battle, the coroner agreed that Ogden's guarantee of confidentiality was in the public good and dropped the charges.

Despite this precedent in Ogden's favour, researchers do not know if other courts will support their right to maintain confidentiality. Academics do not have any legal exemption similar to that which exists between a lawyer and client, so without this exemption, decisions are made on a case-by-case basis.

This issue recently arose again in the case of Luka Magnotta, who was accused of murdering and dismembering Lin Jun. Two University of Ottawa criminology professors, Christine Bruckert and Colette Parent, went to court to try to ensure that an interview they conducted with a research subject named "Jimmy"—a name sometimes used by Magnotta when he had worked as a male escort—was kept confidential and was not allowed as evidence in Magnotta's trial (Solyom, 2012). In 2014, the Quebec Superior Court ruled that the police could not review the taped interview. Justice Sophie Bourque said, "much of the research involving vulnerable people can only be conducted if human participants are given a guarantee that their identities and the information they share will remain confidential" (Wilton, 2014:n.p.). The judge said that this privilege is not absolute but that based on her review, the interview was not relevant to the case against Magnotta. This case has established an important precedent concerning the rights of researchers to keep their research material private, but academics must recognize that under some circumstances they still cannot guarantee their subjects' confidentiality.

ETHICS AND BIG DATA

Ethical issues change over time, and the rise of big data has created some new issues. For example, Facebook researchers wanted to learn more about the ways in which positive and negative emotions could be transferred to others (Kramer, Guillory, and Hancock, 2014). They conducted an experiment in which the emotional content of the News Feed posts of nearly 700,000 users was manipulated by Facebook. Some subscribers received reduced positive emotional content in News Feed while others received reduced negative emotional content. The researchers analyzed the emotional content of users' subsequent status updates and found that changing the tone of news posts did lead to more positive and more negative updates, though the differences were extremely small.

While all users agree to Facebook's Data Use Policy, which includes consenting to research, most academics would not consider this to be informed consent for any particular study, especially one that was attempting to manipulate their emotions.

The dating site OkCupid was responsible for an experiment that was an even more egregious violation of research ethics than the Facebook study. Researchers at OkCupid intentionally mismatched people. They told users who had a 30 percent match that they were actually a 90 percent match (Rudder, 2014). As expected, those users sent more first messages than did users who were correctly told they were a 30 percent match. They also continued to exchange more messages. In fact, those who were told they were a 90 percent match were nearly twice as likely to exchange more messages as those who had been told they were 30 percent matches.

While users of social media sites might reasonably expect that the sites will conduct research on topics such as the effectiveness of different kinds of advertising, most of us do not expect that we will be unwitting subjects of experiments such as those conducted by Facebook and OkCupid. However, corporations are not subject to ethics review, so they will do whatever their subscribers will allow. As more corporations collect more data on all of us, we can anticipate that unless users stop using sites that behave unethically, we will all be more likely to become unwitting experimental subjects.

These cases reflect broader ethical concerns with conducting research online. Informed consent is critical to codes of research ethics. However, how do you get consent from people who may be hiding their identities online? Is it feasible to obtain consent for specific projects if research like the Facebook study involves hundreds of thousands of subjects? Are data that are available online 'public' data or should the person providing the data have a say in how the data are used? Will companies such as Google and Facebook monopolize research on many types of social phenomena because they own the data?

In this chapter, we have looked at the research process and the methods used to pursue sociological knowledge. The important thing to realize is that research is the lifeblood of sociology. Without research, sociologists would be unable to test existing theories and develop new ones. Research takes us beyond common sense and provides opportunities for us to use our sociological imagination to generate new knowledge.

TIME TO REVIEW

- Why is a code of research ethics needed?
- What are the key ethical principles that guide social research?

LO-1 Understand the relationship between theory and research.

Photo by: Sergeant Matthew McGregor, Canadian Forces Combat Camera. © 2012 DND-MDN Canada. Reproduced with the permission of the Department of National Defence, 2015.

Sociologists typically move back and forth from theory to research throughout the course of their inquiry. Investigators rarely, if ever, begin with either just a theory or with research data. Inductive theorists need at least rudimentary theories to guide their data collection, and deductive theorists must refer constantly to the real world as they develop their theories. Researchers may break into the cycle at different points, depending on what they want to know and what information is available. Theory gives meaning to research; research helps support theory.

LO-2 Identify the main steps in the sociological research process.

Adisa/Shutterstock.com

The four stages are: (1) Theories generate hypotheses; (2) these hypotheses lead to observations; (3) observations lead to generalizations; and (4) generalizations are used to support, refute, or modify the theory. Researchers following a deductive model will begin with theory, while inductive researchers will begin with their observations of the social world.

LO-3 Explain why it is important to have different methods of conducting social research and know something about each of these methods

wavebreakmedia/Shutterstock.com

Through experiments, researchers study the impact of certain variables on their subjects. Surveys are polls used to gather facts about people's attitudes, opinions, or behaviours; a sample of respondents provides data through questionnaires or interviews. In secondary analysis, researchers analyze existing data, such as a government census, or cultural artifacts, such as a diary. In field research, sociologists study social life in its natural setting through participant and complete observation, case studies, unstructured interviews, and ethnography. Feminist and critical research methods bring a different perspective to sociological research by focusing on social justice for marginalized people.

KEY TERMS

altruism Behaviour intended to help others and done without any expectation of personal benefit (p. 29).

analysis The process through which data are organized so that comparisons can be made and conclusions drawn (p. 34).

big data Very large datasets that can be accessed in digital form and that can be linked with other large datasets (p. 47).

complete observation Research in which the investigator systematically observes a social process, but does not take part in it (p. 42).

control group Subjects in an experiment who are not exposed to the independent variable, but later are compared to subjects in the experimental group (p. 36).

correlation Exists when two variables are associated more frequently than could be expected by chance (p. 38).

deductive approach Research in which the investigator begins with a theory and then collects information and data to test the theory (p. 32).

dependent variable A variable that is assumed to depend on or be caused by one or more other (independent) variables (p. 33).

descriptive study Research that attempts to describe social reality or provide facts about some group, practice, or event (p. 31).

experiment A test conducted under controlled conditions in which an investigator tests a hypothesis by manipulating an independent variable and examining its impact on a dependent variable (p. 35).

experimental group Subjects in an experiment who are exposed to the independent variable (p. 36).

explanatory study Research that attempts to explain relationships and to provide information on why certain events do or do not occur (p. 31).

field research The study of social life in its natural setting: observing and interviewing people where they live, work, and play (p. 42).

hypotheses Tentative statements of the relationship between two or more concepts or variables (p. 30).

independent variable A variable that is presumed to cause or determine a dependent variable (p. 33).

inductive approach Research in which the investigator collects information or data (facts or evidence) and then generates theories from the analysis of that data (p. 32).

interpretive sociology An approach to sociology that examines the meaning that people give to aspects of their social lives (p. 45).

interview A research method using a data collection encounter in which an interviewer asks the respondent questions and records the answers (p. 40).

objective Free from distorted subjective (personal or emotional) bias (p. 31).

operational definition An explanation of an abstract concept in terms of observable features that are specific enough to measure the variable (p. 34)

participant observation A research method in which researchers collect systematic observations while being part of the activities of the group they are studying (p. 43).

population In a research study, those persons about whom we want to be able to draw conclusions (p. 40).

questionnaire A research instrument containing a series of items to which subjects respond (p. 40).

LO-4 Discuss how research has contributed to our understanding of altruism.

Despite the media emphasis on bad-news stories, research on altruism has shown that a high proportion of people are helpful to others. The vast majority of Canadians donate time and money to help others, and in disasters or emergencies many people will pitch in to help others. Experimental research has also found high rates of helping, though this could be suppressed if other people were also present.

The Canadian Press/Adrian Wyld

LO-5 Explain why a code of ethics for sociological research is necessary.

Researchers are required by a professional code of ethics to weigh the societal benefits of research against the potential physical and emotional costs to participants. Ethical principles include ensuring that research subjects provide informed consent; ensuring that subjects are not harmed; maintaining confidentiality unless the respondent waives this right; and ensuring that participation in research is voluntary.

© Imagestate Media Partners Limited-Impact Photos/Alamy

APPLICATION QUESTIONS

1. A university has implemented a program limiting first-year classes to 30 students and wishes to evaluate the impact of this policy on students' subsequent performance. You have been asked to plan this evaluation. How would you proceed? What different research methods would you use?
2. Working with a group of your fellow students, conduct a content analysis of the way in which photographs in several of your textbooks portray people of different races and genders. Try to follow the steps in the sociological research process.
3. Feminist and critical researchers believe that researchers should not be value free in their research, but should be advocates for social justice. Do you agree with this position, or do you feel that sociologists should maintain their objectivity and remain neutral about the way in which their findings are used?
4. For a class project, you want to study the relationship between students' grades and their willingness to cheat on examinations. What are some of the ethical issues to consider before you administer a survey to the other students in your class?
5. Have you ever participated in a behavioural experiment? If you have, do you think your responses were affected by the fact that you knew you were participating in a study?

reactivity The tendency of experiment participants to change their behaviour in response to the presence of the researcher or to the fact that they know they are being studied (p. 39).

reliability In sociological research, the extent to which a study or research instrument yields consistent results (p. 34).

replication In sociological research, the repetition of the investigation in substantially the same way that it originally was conducted (p. 34).

representative sample A selection where the sample has the essential characteristics of the total population (p. 40).

research methods Specific strategies or techniques for conducting research (p. 35).

sample The people who are selected from the population to be studied (p. 40).

secondary analysis A research method in which researchers use existing material and analyze data that originally was collected by others (p. 46).

simple random sample A selection in which everyone in the target population has an equal chance of being chosen (p. 40).

survey A research method in which a number of respondents are asked identical questions through a systematic questionnaire or interview (p. 40).

triangulation Using several different research methods, data sources, investigators, and/or theoretical perspectives in the same study (p. 49).

unstructured interview A research method involving an extended, open-ended interaction between an interviewer and an interviewee (p. 44).

validity In sociological research, the extent to which a study or research instrument accurately measures what it is supposed to measure (p. 34).

variable In sociological research, any concept with measurable traits or characteristics that can change or vary from one person, time, situation, or society to another (p. 33).

CHAPTER 3

CULTURE

LEARNING OBJECTIVES

AFTER READING THIS CHAPTER, YOU SHOULD BE ABLE TO

LO-1 Understand the importance of culture in our lives and those of others in society.

LO-2 Identify the essential components of culture.

LO-3 Describe what causes cultural change in societies.

LO-4 Compare and contrast ethnocentrism and cultural relativism as approaches to examining cultural differences.

LO-5 Explain how the various sociological perspectives view culture.

CHAPTER FOCUS QUESTION

What part does culture play in shaping people and the social relations in which they participate?

Ruth Bonneville/Winnipeg Free Press

Since Canada's multiculturalism policy was first introduced more than 40 years ago, supporters and critics have debated the effects on the social, economic, and political integration of immigrants. Although multiculturalism remains a significant part of our Canadian identity and a source of national pride, it also presents challenges for both new Canadians (who must fit in and succeed) and multi-generational Canadians (to accept and become comfortable with increasing diversity). For example, a 2012 opinion poll found that although Canadians remain supportive of multiculturalism, they express concerns about the extent to which new immigrants are integrating culturally into Canada (Environics, 2012). However, the following comments by Moran highlight just how difficult and confusing this can be:

I am a twenty-five-year-old, second generation Chinese Canadian female. . . . I was born in Belleville, Ontario. My parents are naturalized Canadian citizens originally from People's Republic of China, who met in Hong Kong. Practically my entire life has been spent in the big city [and] I have never left North America. And luckily for me, there is a stable community of my own ethnic background [in my city]. I take great comfort in this.

. . . It has been hard for me to fit into the Canadian culture and be accepted by other Chinese. I sometimes become paranoid and think that my cultural peers perceive me as a "wacko." This is probably because they may think of me as over-assimilated and assertive beyond their comfort zone. I don't fit the stereotype of the submissive East Asian woman. My dress and accent do not give me away. In fact, I am probably not living my culture in many ways, because I have assimilated much more than first-generation immigrants. On the other hand, even though I do not readily mesh with my own culture, I do not possess the privileges that many Canadians of European descent possess. They can mix right into the predominantly white culture at their whim: I will always have my skin colour and physical characteristics to set me apart. Metaphorically I sit on a fence and cannot be categorized or ordered into any group. First generation, lower income immigrants naturally assume that I am totally Westernized, and whites think I am totally Chinese in my ways of thinking. In some ways, both points of view have grains of truth. But where do I fit? And where do I belong? (James, 2010:115)

Source: Carl E. James, *Seeing Ourselves: Exploring Race, Ethnicity and Culture* (4th ed.). Toronto: Thompson Educational Publishing, Inc., 2010.

What is culture? **Culture** is the knowledge, language, values, customs, and material objects that are passed from person to person and from one generation to the next in a human group or society. As previously defined, a *society* is a large social grouping that occupies the same geographic territory and is subject to the same political authority and dominant cultural expectations. Whereas a society is made up of people, a culture is made up of ideas, behaviour, and material possessions. Society and culture are interdependent; neither could exist without the other.

culture The knowledge, language, values, customs, and material objects that are passed from person to person and from one generation to the next in a human group or society.

In this chapter, we will examine society and culture, with special attention to our unique Canadian multicultural society. We will also analyze culture from functionalist, conflict, feminist, interactionist, and postmodern perspectives.

Before reading on, test your knowledge of multiculturalism in Canada by answering the questions in Box 3.1 and referring to Table 3.1.

CRITICAL THINKING QUESTIONS

1. To what extent does our own culture keep us from understanding, accepting, or learning from other cultures?
2. Is intolerance toward "outsiders"—people who are viewed as being different from one's own group or way of life—accepted by some people in Canada? Why?
3. It has been suggested that the cultural freedom legislated by the *Multicultural Act* is more "symbolic" than real. Do you agree?

LO-1 CULTURE AND SOCIETY

How important is culture in determining how people think and act daily? Simply stated, culture is essential for our individual survival and for our communication with other people. We rely on culture because we are not born with the information we need to survive. We do not know how to take care of ourselves, how to behave, how to dress, what to eat, which gods to worship, or how to make or spend money. We must learn about culture through interaction, observation, and imitation in order to participate as members of the group. Sharing a common culture with others simplifies day-to-day interactions. We must, however, also understand other cultures and the worldviews therein.

Just as culture is essential for individuals, it is also fundamental for the survival of societies. Culture has been described as "the common denominator that makes the actions of individuals intelligible to the group" (Haviland, 1993:30). Some system of making and enforcing rules necessarily exists in all societies. What would happen, for example, if *all* rules and laws in Canada suddenly disappeared? At a basic level, we need rules in order to navigate our bicycles and cars through traffic. At a more abstract level, we need laws to establish and protect our rights.

To survive, societies need rules about civility and tolerance toward others. We are not born knowing how to express kindness or hatred toward others, although some people may say, "Well, that's just human nature," when explaining someone's behaviour. Such a statement is built on the assumption that what we do as human beings is determined by *nature* (our biological and genetic makeup) rather than *nurture* (our social environment)—in other words, that our behaviour is instinctive. An *instinct* is an unlearned, biologically determined behaviour pattern common to all members of a species that predictably occurs whenever certain environmental conditions exist. For example, spiders do not learn to build webs; they build webs because of instincts that are triggered by basic biological needs, such as protection and reproduction.

Culture is similar to instincts in animals because it helps us deal with everyday life. Although people may have some instincts, what we most often think of as instinctive behaviour can actually be attributed to reflexes and drives. A *reflex* is an unlearned, biologically determined involuntary response to a physical stimulus (such as a sneeze after breathing some pepper in through the nose or the blinking of an eye when a speck of dust gets in it). *Drives* are unlearned, biologically determined impulses common to all members of a species that satisfy needs, such as for sleep, food, water, and sexual gratification. Reflexes and drives do not determine how people will behave in human societies; even the expression of these biological characteristics is channelled by culture. For example, we may be taught that the "appropriate" way to sneeze (an involuntary response) is to use a tissue or turn our head away from others (a learned response). Most contemporary sociologists agree that culture and social learning—not nature—account for virtually all of our behaviour patterns.

BOX 3.1 SOCIOLOGY AND EVERYDAY LIFE

How Much Do You Know About Multiculturalism in Canada?

True	False	
T	F	1. Canada is one of the most multicultural countries in the world.
T	F	2. A 2012 public opinion poll asked Canadians to describe what made them most proud of their country. Multiculturalism ranked fourth on the list.
T	F	3. Recent high levels of illegal immigration have led an increasing number of Canadians to reject multiculturalism.
T	F	4. The majority of Canadians regard multiculturalism as good for Canada.
T	F	5. Multiculturalism and social integration are mutually exclusive goals.

For answers to the quiz about multiculturalism in Canada, go to **www.nelson.com/student.**

TABLE 3.1 BASIS OF PRIDE IN BEING CANADIAN: TOP MENTIONS, 1994-2012

	1994	2003	2006	2012
Free country/freedom/democracy	31	28	27	26
Quality of life	5	6	3	5
Humanitarian/caring people	9	13	9	9
Multiculturalism	3	6	11	7
Healthcare system	3	2	6	4
Peaceful country	7	5	6	3
Beauty of the land	7	4	4	6
Respected by other countries	4	3	4	4
Social programs	2	1	—	2

Source: Adapted from The Environics Institute for Survey Research, *Focus Canada 2012*. Found at http://environicsinstitute.org/institute-projects/completed-projects/focus-canada-2012

Since humans cannot rely on instincts to survive, culture is a "tool kit" for survival. According to the sociologist Ann Swidler, culture is a "tool kit of symbols, stories, rituals, and world views, which people may use in varying configurations to solve different kinds of problems" (1986:273). The tools we choose will vary according to our own personality and the situations we face. We are not puppets on a string; we make choices from among the items in our own "toolkit."

MATERIAL CULTURE AND NONMATERIAL CULTURE

Our cultural tool kit is divided into two major parts: *material* and *nonmaterial* culture (Ogburn, 1966/1922). **Material culture** consists of the physical or tangible creations that members of a society make, use, and share. Initially, items of material culture begin as raw materials or resources, such as water, trees, and oil. Through technology, these raw materials are transformed into usable items such as books and computers Sociologists define **technology** as the knowledge, techniques, and tools that make it possible for people to transform resources into usable forms, as well as the knowledge and skills required to use them after they are developed. From this standpoint, technology is both concrete and abstract. For example, technology includes computers, iPads and other tablets, and the knowledge and skills necessary to use them. At the most basic level, material culture is important because it is our buffer against the environment. For example, we create shelter to protect ourselves from the weather and provide ourselves with privacy. Beyond the survival level, we make, use, and share objects that are interesting and important to us. Why are you wearing the particular clothes you have on today? Perhaps you're communicating something about yourself, such as where you attend school, what kind of music you like, or where you went on vacation.

Nonmaterial culture consists of the abstract or intangible human creations of society that influence people's behaviour. Language, beliefs, values, rules of behaviour, family patterns, and political systems are examples of nonmaterial culture. A central component of nonmaterial culture is *beliefs*—the mental acceptance or conviction that certain things are true or real. Beliefs may be based on tradition, faith, experience, scientific research, or some combination of these. Faith in a supreme being, conviction that education is the key to success, and the opinion that smoking causes cancer are examples of beliefs. We also have beliefs in items of material culture. For example, most students believe that computers are the key to technological advancement and progress.

material culture A component of culture that consists of the physical or tangible creations—such as clothing, shelter, and art—that members of a society make, use, and share.

technology The knowledge, techniques, and tools that make it possible for people to transform resources into usable forms, as well as the knowledge and skills required to use them after they are developed.

nonmaterial culture A component of culture that consists of the abstract or intangible human creations of society—such as attitudes, beliefs, and values—that influence people's behaviour.

Celia Peterson/Getty Images

© Eddie Gerald/Alamy

Food is a universal type of material culture, but what people eat and how they eat it vary widely, as shown in these cross-cultural examples from the United Arab Emirates (left) and China (right). What might be some of the reasons for the similarities and differences you see in these photos?

CULTURAL UNIVERSALS

▶ **cultural universals** Customs and practices that occur across all societies.

Because all humans face the same basic needs (such as food, clothing, and shelter), we engage in similar activities that contribute to our survival. Anthropologist George Murdock (1945:124) compiled a list of more than 70 **cultural universals**—customs and practices that occur across all societies. His categories included appearance (such as bodily adornment and hairstyles), activities (such as sports, dancing, games, joking, and visiting), social institutions (such as family, law, and religion), and customary practices (such as cooking, folklore, gift giving, and hospitality). These general customs and practices may be present in all cultures, but their specific forms vary from one group to another and from one time to another within the same group. For example, while telling jokes may be a universal practice, what is considered a joke in one society may be an insult in another.

Greg Elms/Lonely Planet Images/Getty Images

The customs and rituals associated with weddings are one example of nonmaterial culture. What can you infer about beliefs and attitudes about marriage represented by this photograph?

How do sociologists view cultural universals? In terms of their functions, cultural universals are useful because they ensure the smooth and continual operation of society (Radcliffe-Brown, 1952). A society must meet basic human needs by providing food, shelter, and some degree of safety for its members so that they will survive. Children and other new members (such as immigrants) must be taught the ways of the group. A society also must settle disputes and deal with people's emotions. All the while, the self-interest of individuals must be balanced with the needs of society as a whole. Cultural universals help to fulfill these important functions of society.

From another perspective, however, cultural universals are not the result of functional necessity; these practices may have been *imposed* by members of one society on members of another. Similar customs and practices do not necessarily constitute cultural universals. They may be an indication that a conquering nation used its power to enforce certain types of behaviour on those who were defeated (Sargent, 1987). Sociologists might ask, who determines the dominant cultural patterns? For example, although religion is a cultural universal, traditional religious practices of indigenous peoples (those who first live in an area) have often been repressed

Steve Russell/GetStock.com

Axel Bueckert/Shutterstock.com

Although body adornment is a cultural universal, the specific form it takes varies from one group to another and from one time to another within the same group.

and even stamped out by subsequent settlers or conquerors who hold political and economic power over them.

TIME TO REVIEW

- What are cultural universals?
- Explain how functionalists and conflict theorists view cultural universals.

LO-2 COMPONENTS OF CULTURE

Even though the specifics of individual cultures vary widely, all cultures have four common nonmaterial cultural components: symbols, language, values, and norms. These components contribute to both harmony and conflict in a society.

SYMBOLS

A symbol is anything that meaningfully represents something else. Culture could not exist without symbols because there would be no shared meanings among people. Symbols can simultaneously produce loyalty and animosity, love and hate. They help us communicate ideas, such as love or patriotism, because they express abstract concepts with visible objects. To complicate matters, however, the interpretation of symbols varies in different cultural contexts. For some Indo-Canadians, for example, the colour green rather than white symbolizes purity or virginity. Similarly, although a swastika represents hate to most Canadians, to a member of the Church of Jesus Christ Christian/Aryan Nations, a swastika represents love.

Flags can stand for patriotism, nationalism, school spirit, or religious beliefs held by members of a group or society. In our technology-oriented society, *emoticons* are systems of symbols used

FIGURE 3.1 : EMOTICONS

The symbols shown here are examples of emoticons, or "smileys," a symbolic way to express moods in email or text messages.

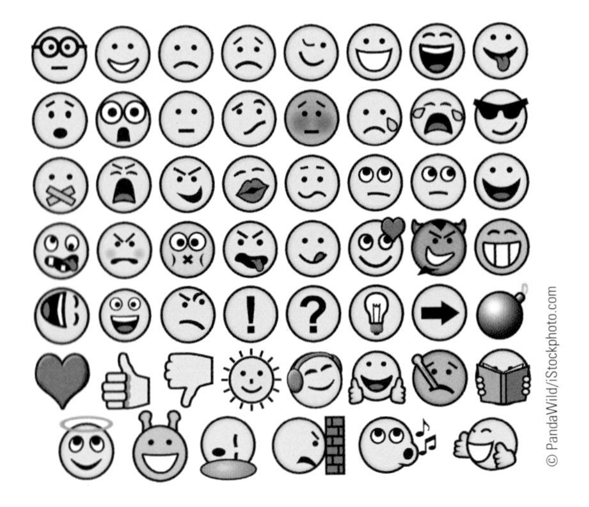

to express emotions when people are communicating on their smartphones or computers, and texting abbreviations have created a new language that may be foreign to older members of society (see Figure 3.1).

Symbols can stand for love (a heart or a valentine), peace (a dove), or hate (a Nazi swastika), just as words can be used to convey meanings. Symbols also can transmit other types of ideas. A siren is a symbol that denotes an emergency situation and sends the message to clear the way immediately. Gestures are also a symbolic form of communication—a movement of the head, body, or hands can express ideas or feelings to others. For example, in Canada, pointing toward your chest with your thumb or finger is a symbol for *me.* We are also all aware of how useful our middle finger can be in communicating messages to inconsiderate drivers!

Symbols affect our thoughts about class. For example, how a person is dressed or the kind of car he or she drives is often at least subconsciously used as a measure of that individual's economic standing or position. With regard to clothing, although many people wear casual clothes on a daily basis, where the clothing was purchased is sometimes used as a symbol of social status. Were the items purchased at Walmart, Old Navy, Club Monaco, or Banana Republic? What indicators on the clothing—such as the Nike swoosh, some other logo, or a brand name—say something about the product's status? Automobiles and their logos are also symbols that have cultural meaning beyond the shopping environment in which they originate.

LANGUAGE

▶ **language** A system of symbols that expresses ideas and enables people to think and communicate with one another.

Language is a system of symbols that expresses ideas and enables people to think and communicate with one another. Verbal (spoken) and nonverbal (written or gestured) language help us describe reality. One of our most important human attributes is the ability to use language to share our experiences, feelings, and knowledge with others. Language can create visual images in our head, such as "the kittens look like little cotton balls" (Samovar and Porter, 1991). Language can also allow people to distinguish themselves from outsiders and maintain group boundaries and solidarity.

"Copy and study this list of text messaging spelling words.

Language is not solely a human characteristic. Other animals use sounds, gestures, touch, and smell to communicate with one another, but they use signals with fixed meanings that are limited to the immediate situation (the present) and cannot encompass past or future situations. For example, chimpanzees can use elements of Standard American Sign Language and manipulate physical objects to make "sentences," but they are not physically endowed with the vocal apparatus needed to form the consonants required for verbal language. As a result, nonhuman animals cannot transmit the more complex aspects of culture to their offspring. Humans have a unique ability to manipulate symbols to express abstract concepts and rules, and thus to create and transmit culture from one generation to the next.

LANGUAGE AND SOCIAL REALITY One key issue in sociology is whether language *creates* or simply *communicates* reality. Consider, for example, the terms used by organizations involved in the abortion debate: pro-life and pro-choice. Do such terms create or simply express a reality?

Anthropological linguists Edward Sapir and Benjamin Whorf have suggested that language not only expresses our thoughts and perceptions but also influences our perception of reality. According to the **Sapir-Whorf hypothesis**, language shapes its speakers' view of reality (Sapir, 1961; Whorf, 1956). If people are able to think only through language, language must precede thought. If language shapes the reality we perceive and experience, some aspects of the world are viewed as important and others are virtually neglected because people know the world only in terms of the vocabulary and grammar of their own language. For example, most Indigenous languages focus on describing relationships between things rather than using language to judge or evaluate. One Indigenous author explains, "No, we don't have any gender. It's a relationship . . . The woman who cares for your heart—that's your wife. Your daughters are the ones who enrich your heart. Your sons are the ones that test your heart!" (Ross, 1996:116). Consequently, many Indigenous languages do not have any personal pronouns based on gender (such as words for *she* or *he*). As writer Rupert Ross explains:

Sapir-Whorf hypothesis The proposition that language shapes its speakers' view of reality.

> Because they don't exist there, searching for the correct ones often seems an artificial and unreasonable exercise. As a result, Indigenous people are often as careless about getting them right as I am when speaking French and trying to remember whether a noun has "le" or "la" in front of it . . . On the more humorous side, my Indigenous friends appear heartily amused by the frenzied Western debate over whether God is a "He" or a "She." (1996:117)

If language does create reality, are we trapped by our language? Many social scientists agree that the Sapir–Whorf hypothesis overstates the relationship between language and our thoughts and behaviour patterns. While acknowledging that language has many subtle meanings and that the words used by people reflect their central concerns, most sociologists contend that language may *influence* our behaviour and interpretation of social reality, but it does not *determine* it.

LANGUAGE AND GENDER What is the relationship between language and gender? What cultural assumptions about women and men does language reflect? Scholars have suggested several ways in which language and gender are intertwined:

- The English language ignores women by using the masculine form to refer to human beings in general. For example, the word *man* is used generically in words like *chairman* and *mankind*, which allegedly include both men and women.
- Use of the pronouns *he* and *she* affects our thinking about gender. Pronouns show the gender of the person we *expect* to be in a particular occupation. For instance, nurses, secretaries, and schoolteachers are usually referred to as *she*, while doctors, engineers, electricians, and presidents are referred to as *he*.
- Words have positive connotations when relating to male power, prestige, and leadership; when related to women, they carry negative overtones of weakness, inferiority, and immaturity (Epstein, 1988).
- A language-based predisposition to think about women in sexual terms reinforces the notion that women are sexual objects. Women are often described by terms such as *fox, broad, bitch, babe,* or *doll,* which ascribe childlike or even petlike characteristics to them. By contrast, performance pressures are placed on men. Words such as *dude, stud,* and *hunk* define them in terms of their sexual prowess (Baker, 1993).

These are some of the most universally recognized signs.

Gender in language has been debated and studied extensively in recent years, and greater awareness and some changes have been the result. Many organizations and publications have established guidelines for the use of nonsexist language and have changed titles such as *chairman* to *chair* or *chairperson*. To develop a more inclusive and equitable society, many scholars suggest that a more inclusive language is needed.

LANGUAGE, RACE, AND ETHNICITY Language may create and reinforce our perceptions about race and ethnicity by transmitting preconceived ideas about the superiority of one category of people over another. Let's look at a few images conveyed by words in the English language in regard to race and ethnicity.

- Words may have more than one meaning and create and reinforce negative images. Terms such as *blackhearted* (malevolent) and expressions such as *a black mark* (a detrimental fact) and *a Chinaman's chance of success* (unlikely to succeed) give the words *black* and *Chinaman* negative associations and derogatory imagery. Although these terms are seldom used today, they are occasionally referenced in popular culture and film.
- Words are frequently used to create or reinforce perceptions about a group. For example, Indigenous peoples have been referred to as *savages* and described as *primitive*, while blacks have been described as *uncivilized, cannibalistic,* and *pagan*.
- The "voice" of verbs may minimize or incorrectly identify the activities or achievements of members of various minority groups. For example, use of the passive voice in the statement "Chinese Canadians *were given* the right to vote" ignores how Chinese Canadians *fought* for that right. Active-voice verbs also may inaccurately attribute achievements to people or groups. Some historians argue that cultural bias is shown by the very notion that "Cabot discovered Canada," given that Canada was already inhabited by the Indigenous population.

In addition to these concerns about the English language, problems also arise when more than one language is involved.

Tomasz Szymanski/Shutterstock.com

Does language influence our perception of reality?

LANGUAGE DIVERSITY IN CANADA The existence of Indigenous languages, the presence of French- and English-speaking populations, and the increasing number of other languages commonly spoken are all evidence of the linguistic diversity in Canadian culture. Over 200 languages were reported as a mother tongue in the 2011 census (Statistics Canada, 2012a).

In 1969, the federal government passed the *Official Languages Act,* making both French and English the country's official languages. In doing so, Canada officially became a bilingual society. However, this action by no means resolved the complex issues regarding language in our society. According to the most recent census, 68 percent of Canadians speak English only, another 13 percent speak French only, and 17 percent are bilingual. Less than 2 percent, just under 600,000 Canadians, indicated that they lacked the skills to converse in either French or English (Statistics Canada, 2012b). Immigrant languages, which are defined as languages other than English, French and Indigenous languages, were the mother tongues of one-fifth of the population (more than 6.8 million people). The immigrant languages spoken most often at home were Punjabi, Chinese (not otherwise specified), Cantonese, Spanish, Tagalog, Arabic, Mandarin, Italian, Urdu and German (Dewing, 2013) (see also Figure 3.2). Although French versus English language issues have been a significant source of conflict over the years, bilingualism remains a distinct component of Canadian culture.

Canada's Indigenous languages are many and diverse. The languages reflect distinctive histories, cultures, and identities linked to family, community, the land, and traditional knowledge. Indigenous peoples' cultures are *oral cultures,* or cultures that are transmitted through speech rather than the written word. Many Indigenous stories can be passed on only in the Indigenous language in which they originated. Language is not only a means of communication, but also a link that connects people with their past and grounds their social, emotional, and spiritual vitality. For Indigenous peoples, huge losses have already occurred as a result of the assimilationist strategies. Indigenous children who attended residential schools were forbidden to speak their language. An Ojibwa woman from northwestern Ontario describes her experience:

> Boarding school was supposed to be a place where you forgot everything about being Anishinabe. And our language too. But I said, "I'm going to talk to myself"—and that's what I did, under my covers—talked to myself in Anishinabe. If we were caught, the nuns would make us stand in a corner and repeat over and over, "I won't speak my language." (Ross, 1996:122)

Despite the efforts of Canadian Indigenous peoples to maintain their languages, these languages are among the most endangered in the world. Only three of the more than 50 Indigenous languages in Canada are in a healthy state; many have already disappeared or are near extinction. In the 2011 Canadian Census, only 17 percent of Indigenous persons reported an Indigenous language as their first language, and even fewer spoke it at home

CENSUS PROFILE

Language Diversity in Canada

Among the categories of information gathered in the 2011 Census are data on the languages spoken in Canadian households. As shown below, two-thirds of Canadians speak English most often at home and just over one-fifth of the population speak French most often at home.

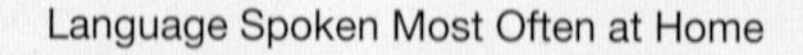

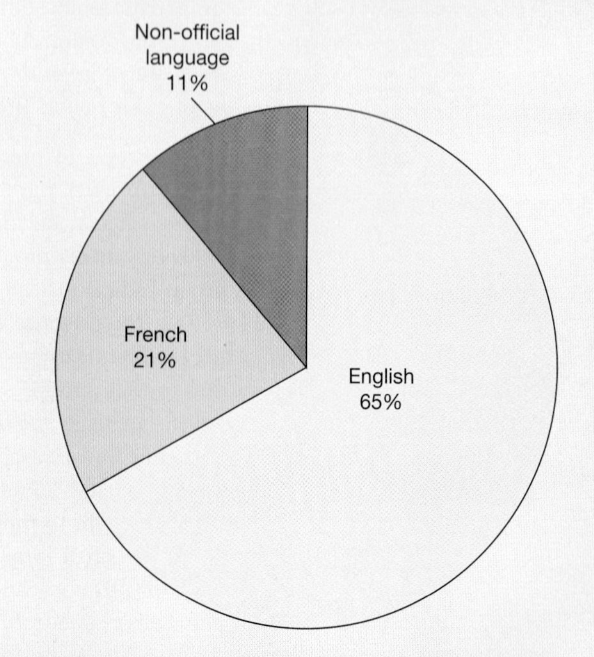

Source: Adapted from Statistics Canada, 2012, Population by language spoken most often and regularly at home, age groups (total), for Canada, provinces and territories, Highlight Tables. Found at: http://www12.statcan.gc.ca/census-recensement/2011/dp-pd/hlt-fst/lang/Pages/Highlight.cfm?TabID=1&Lang=E&PRCode=01&Age=1&tableID=403&queryID=1

FIGURE 3.2 : TOP TEN IMMIGRANT LANGUAGES IN CANADA

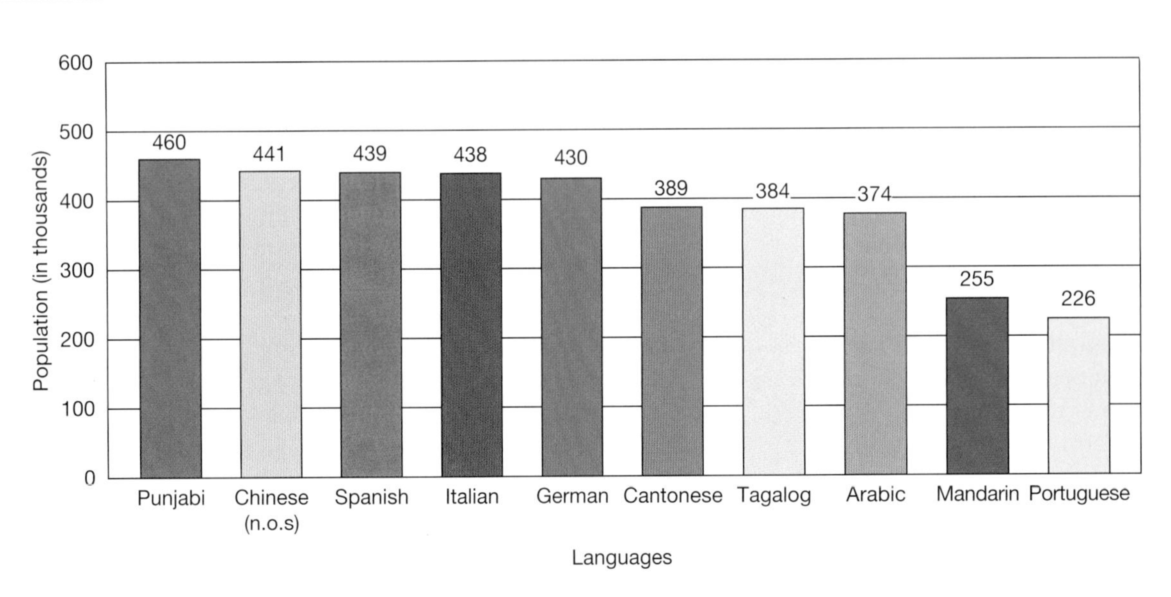

Source: Adapted from Statistics Canada. 2011a. Immigrant Languages in Canada 98-314-X2011003. http://www12.statcan.gc.ca/census-recensement/2011/as-sa/98-314-x/98-314-x2011003_2-eng.pdf

(Statistics Canada, 2011b). According to Indigenous elders, language is the lifeblood of their culture:

> Our native language embodies a value system about how we ought to live and relate to each other . . . Now if you destroy our language, you not only break down these relationships, but you also destroy other aspects of our Indian way of life and culture, especially those that describe man's connection with nature, the Great Spirit, the order of things. Without our language, we will cease to exist as a separate people. (Eli Taylor, Sioux Nation, quoted in Speilmann, 2002:51)

Recent measures to preserve indigenous languages have created optimism. First Nations across North America have worked hard to stop their languages by implementing culturally appropriate curricula, the introduction of Indigenous language courses in schools and universities, Indigenous media programming, and the recording of elders' stories, songs, and accounts of history in Indigenous language (Bougie, 2010).

How does the presence of all these different languages affect Canadian culture? Through language, children learn about their cultural heritage and develop a sense of personal identity in relation to their group.

However, language is also a source of power and social control; it perpetuates inequalities between people and between groups because words are used (intentionally or not) to "keep people in their place." As linguist Deborah Tannen has suggested, "The devastating group hatreds that result in so much suffering in our own country and around the world are related in origin to the small intolerances in our everyday conversations—our readiness to attribute good intentions to ourselves and bad intentions to others" (1993:B5). Furthermore, different languages are associated with inequalities. Consider this Indigenous language instructor's comments on the lure of the English language: "It's to do with the perception of power. People associate English with prestige and power. We don't have movies in [Indigenous language], we don't have hardcover books . . . or neon signs in our language" (Martin, 1996:A8). Language, then, is a reflection of our feelings and our values.

VALUES

Values are collective ideas about what is right or wrong, good or bad, and desirable or undesirable in a particular culture (Williams, 1970). Values do not dictate which behaviours are appropriate and which are not, but they provide us with the criteria by which we evaluate people, objects, and events. Values typically come in positive and negative pairs, such as being brave or cowardly, hardworking or lazy. Since we use values to justify our behaviour, we tend to defend them staunchly.

▶ **values** Collective ideas about what is right or wrong, good or bad, and desirable or undesirable in a particular culture.

VALUE CONTRADICTIONS All societies have value contradictions. **Value contradictions** are values that conflict with one another or are mutually exclusive (achieving one makes it difficult, if not impossible, to achieve another). For example, core values of morality and humanitarianism may conflict with values of individual achievement and success. Similarly, although the majority of Canadians feel that people who are poor have a right to social assistance, they have also shown strong support for governments that have dramatically cut budgets to reduce financial deficits.

▶ **value contradiction** Values that conflict with one another or are mutually exclusive.

IDEAL VERSUS REAL CULTURE What is the relationship between values and human behaviour? Sociologists stress that a gap always exists between ideal culture and real culture in a society. **Ideal culture** refers to the values and standards of behaviour that people in a society profess to hold. **Real culture** refers to the values and standards of behaviour that people actually follow. For example, we may claim to be law-abiding (ideal cultural value) but smoke marijuana (real cultural behaviour), or we may regularly drive over the speed limit but think of ourselves as "good citizens." We may believe in the value of honesty and, at the same time, engage in deception.

▶ **ideal culture** The values and standards of behaviour that people in a society profess to hold.

▶ **real culture** The values and standards of behaviour that people actually follow (as contrasted with *ideal culture*).

NORMS

Values provide ideals or beliefs about behaviour but do not state explicitly how we should behave. Norms, on the other hand, have specific behavioural expectations. **Norms** are established rules of behaviour or standards of conduct. *Prescriptive norms* state what behaviour is appropriate or acceptable. For example, persons making a certain amount of money are expected to file a tax return and pay any taxes they owe. Norms based on custom direct us to open a door for a person carrying a heavy load. By contrast, *proscriptive norms* state what behaviour is inappropriate or unacceptable. Laws that prohibit us from driving over the speed limit and "good manners" that preclude you from texting are examples. Prescriptive and proscriptive norms operate at all levels of society, from our everyday actions to the formulation of laws.

▶ **norms** Established rules of behaviour or standards of conduct.

FORMAL AND INFORMAL NORMS Not all norms are of equal importance; those that are most crucial are formalized. *Formal norms* are written down and involve specific punishments for violators. Laws are the most common type of formal norms; they have been codified and may be enforced by sanctions. **Sanctions** are rewards for appropriate behaviour or penalties for inappropriate behaviour. Examples of *positive sanctions* include praise, honours, or medals for conformity to specific norms. *Negative sanctions* range from mild disapproval to life imprisonment.

▶ **sanctions** Rewards for appropriate behaviour or ' for inappropriate behaviour.

Less important norms are referred to as *informal norms*—unwritten standards of behaviour understood by people who share a common identity. When individuals violate informal norms, other people may apply informal sanctions. *Informal sanctions* are not clearly defined and can be applied by any member of a group. Examples are frowning at someone or making a negative comment or gesture.

▶ **folkways** Informal norms or everyday customs that may be violated without serious consequences within a particular culture.

FOLKWAYS Norms are also classified according to their relative social importance. **Folkways** are informal norms or everyday customs that may be violated without serious consequences within a particular culture (Sumner, 1959/1906). They provide rules for conduct but are not considered essential to society's survival. In Canada, folkways include using underarm deodorant, brushing one's teeth, and wearing appropriate clothing for a specific occasion. Folkways are not often enforced, and when they are, the resulting sanctions tend to be informal and relatively mild.

▶ **mores** Strongly held norms with moral and ethical connotations that may not be violated without serious consequences in a particular culture.

▶ **taboos** Mores so strong that their violation is considered extremely offensive and even unmentionable.

MORES Other norms are considered highly essential to the stability of society. **Mores** (pronounced MOR-ays) are strongly held norms with moral and ethical connotations that may not be violated without serious consequences in a particular culture. Since mores are based on cultural values and are considered crucial for the well-being of the group, violators are subject to more severe negative sanctions (such as ridicule or loss of employment) than are those who fail to adhere to folkways. The strongest mores are referred to as taboos. **Taboos** are mores so strong that their violation is considered extremely offensive and even unmentionable. Violation of taboos is punishable by the group or even, according to certain belief systems, by a supernatural force. The incest taboo, which prohibits sexual or marital relations between certain categories of kin, is an example of a nearly universal taboo.

▶ **laws** Formal, standardized norms that have been enacted by legislatures and are enforced by formal sanctions.

LAWS **Laws** are formal, standardized norms that have been enacted by legislatures and are enforced by formal sanctions. Laws may be either civil or criminal. *Civil law* deals with disputes among persons or groups. Persons who lose civil suits may encounter negative sanctions, such as having to pay compensation to the other party or being ordered to stop certain conduct. *Criminal law,* on the other hand, deals with public safety and well-being. When criminal laws are violated, fines and prison sentences are the most likely negative sanctions.

TIME TO REVIEW

- What are the main types of norms?

LO-3 TECHNOLOGY, CULTURAL CHANGE, AND DIVERSITY

Cultures do not generally remain static. There are many forces working toward change and diversity. Some societies and individuals adapt to this change, whereas others suffer culture shock and succumb to ethnocentrism.

CULTURAL CHANGE

Societies continually experience cultural change at both material and nonmaterial levels. Changes in technology continue to shape the material culture of society. Although most technological changes are primarily modifications of existing technology, *new technologies* are changes that make a significant difference in many people's lives. Examples of new technologies include the introduction of the printing press more than 500 years ago and the advent of computers and electronic communications in the 20th century. The pace of technological change has increased rapidly in the past 150 years, as contrasted with the 4000 years before that, during which humans advanced from digging sticks and hoes to the plow.

All parts of a culture do not change at the same pace. When a change occurs in the material culture of a society, nonmaterial culture must adapt to that change. Frequently, this rate of change is uneven, resulting in a gap between the two. Sociologist William F. Ogburn (1966/1922) referred to this disparity as **cultural lag**—a gap between the technical development of a society and its moral and legal institutions. In other words, cultural lag occurs when material culture changes faster than nonmaterial culture, thus creating a lag between the two cultural components. For example, at the material cultural level, the personal computer and electronic coding have made it possible to create a unique health identifier for each person in Canada. Based on available technology (material culture), it would be possible to create a national data bank that includes everyone's individual medical records from birth to death. Using this identifier, health providers and insurance companies could rapidly transfer medical records around the globe and researchers could access unlimited data on people's diseases, test results, and treatments. The availability of this technology, however, does not mean that it will be used because, from a nonmaterial culture perspective, people may believe that such a national data bank would constitute an invasion of privacy and could easily be abused by others. Social conflict may arise between nonmaterial culture and the capabilities of material culture, often set in motion by discovery, invention, and diffusion.

▶ **cultural lag** William Ogburn's term for a gap between the technical development of a society (material culture) and its moral and legal institutions (nonmaterial culture) (p. 74).

Discovery is the process of learning about something previously unknown or unrecognized. Historically, discovery involved unearthing natural elements or existing realities, such as "discovering" fire or the true shape of Earth. Today, discovery most often results from scientific research. For example, the discovery of a polio vaccine virtually eliminated one of the major childhood diseases. A future discovery of a cure for cancer or the common cold could result in longer and more productive lives for many people.

▶ **discovery** The process of learning about something previously unknown or unrecognized.

As more discoveries have occurred, people have been able to reconfigure existing material and nonmaterial cultural items through invention. **Invention** is the process of reshaping existing cultural items into a new form. Video games, digital devices, reproductive technologies and stem cell research are examples of inventions that positively or negatively affect our lives today.

▶ **invention** The process of reshaping existing cultural items into a new form.

When diverse groups of people come into contact, they begin to adapt one another's discoveries, inventions, and ideas for their own use. **Diffusion** is the transmission of cultural items or social practices from one group or society to another through such means as exploration, military endeavours, the media, tourism, and immigration. Today, cultural diffusion moves at a very rapid pace in our global economy.

▶ **diffusion** The transmission of cultural items or social practices from one group or society to another.

CULTURAL DIVERSITY

Cultural diversity refers to the wide range of cultural differences found between and within nations. Cultural diversity between countries may be the result of natural circumstances (such as climate and geography) or social circumstances (such as level of technology and composition of the population). Some countries—such as Sweden—are referred to as homogeneous societies, meaning they include people who share a common culture and are typically from similar social, religious, political, and economic backgrounds. By contrast, other countries—including Canada—are referred to as heterogeneous societies, meaning they include people who are dissimilar in regard to social characteristics, such as nationality, race, ethnicity, class, religion, or education (see Figure 3.3).

Canada has always been characterized by at least three main cultures. Although cultural diversity in our country is not only the result of immigration, immigration has certainly had a significant impact on the development of our culturally diverse society. Over the past 150 years, more than 13 million "documented," or legal, immigrants have arrived here; innumerable people have also entered the country as undocumented immigrants. Immigration can cause feelings of frustration and hostility, especially in people who feel threatened by the changes that large numbers of immigrants may produce. Often, people are intolerant of those who are different from themselves (See Box 3.2 for a detailed discussion of both sides of the debate around the wearing of religious head cover.)

Have you ever been made to feel like an "outsider"? Each of us receives cultural messages that may make us feel good or bad about ourselves, or may give us the perception that we belong

FIGURE 3.3 HETEROGENEITY OF CANADIAN SOCIETY

Throughout history, Canada has been heterogeneous. Today, Canada is represented by a wide variety of social categories, including our religious affiliations and ethnic origins.

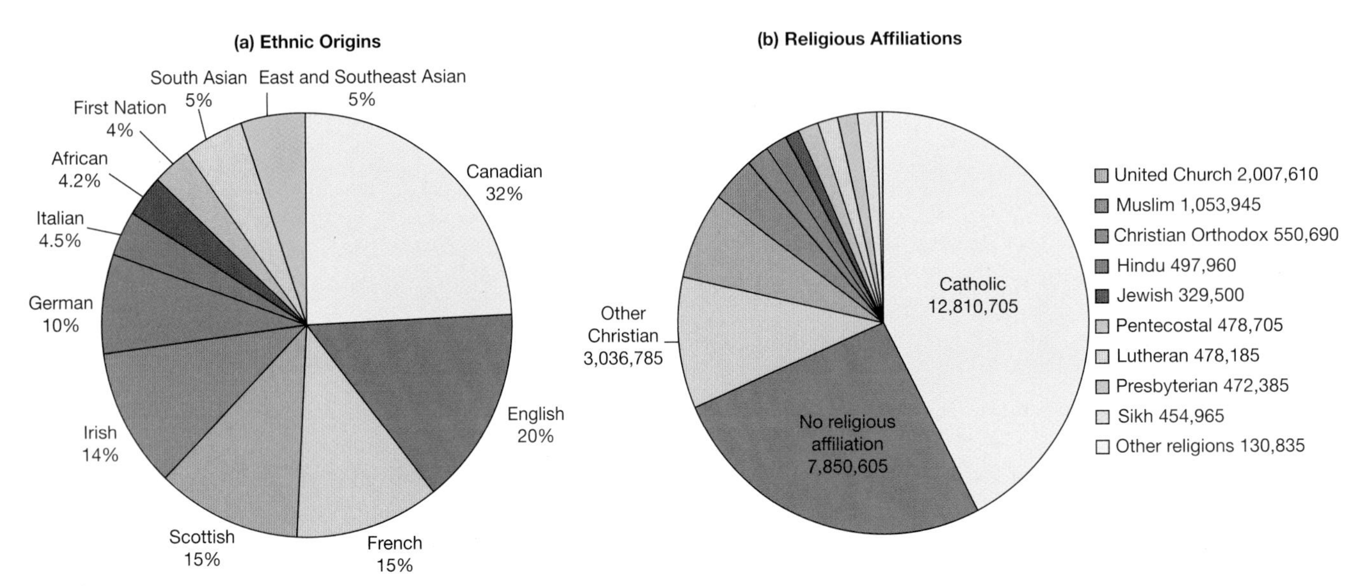

Source: Statistics Canada, 2013, Immigration and Ethnocultural Diversity in Canada National Household Survey, 2011 Catalogue no. 99-010-X2011001 ISBN: 978-1-100-22197-7 http://www12.statcan.gc.ca/nhs-enm/2011/as-sa/99-010-x/99-010-x2011001-eng.pdf

Source: Statistics Canada, 2013, National Household Survey: Immigration and Ethnocultural Diversity - Religion (108), Immigrant Status and Period of Immigration (11), Age Groups (10) and Sex (3) for the Population in Private Households of Canada, Provinces, Territories, Census Metropolitan Areas and Census Agglomerations, 2011 National Household Survey, National Household Survey year 2011 http://www12.statcan.gc.ca/nhs-enm/2011/dp-pd/dt-td/Rp-eng.cfm?LANG=E&APATH=3&DETAIL=0&DIM=0&FL=A&FREE=0&GC=0&GID=0&GK=0&GRP=1&PID=105399&PRID=0&PTYPE=105277&S=0&SHOWALL=0&SUB=0&Temporal=2013&THEME=95&VID=0&VNAMEE=&VNAMEF=

or do not belong. However, in heterogeneous societies such as Canada, cultural diversity is inevitable. In Canada, this diversity has created some unique problems in terms of defining and maintaining our distinct Canadian culture. In fact, what is unique to Canada is the number of distinct subcultures that together make up our Canadian culture.

It has been suggested that complex societies are more likely to produce subcultures. This is certainly the case in Canada, where regional, ethnic, class, language, and religious subcultures combine to produce a highly diverse society.

▶ **subculture** A group of people who share a distinctive set of cultural beliefs and behaviours that differ in some significant way from those of the larger society.

SUBCULTURES A **subculture** is a group of people who share a distinctive set of cultural beliefs and behaviours that differ in some significant way from those of the larger society. Emerging from the functionalist tradition, this concept has been applied to categories ranging from ethnic, religious, regional, and age-based categories to those categories presumed to be "deviant" or marginalized from the larger society. In the broadest use of the concept, thousands of categories of people residing in Canada might be classified as belonging to one or more subcultures, including Muslims, Italian Canadians, Orthodox Jews, Generation Xers, and bikers. However, many sociological studies of subcultures have limited the scope of inquiry to more visible distinct subcultures, such as the Hutterites, to see how subcultural participants interact with the dominant culture.

The Hutterites As a subculture, the Hutterities have fought for many years to maintain their distinct identity. The Hutterites are the largest family-type communal grouping in the Western world, with close to 30,000 members living in approximately 300 settlements. They live in colonies of about 15 families, but each family usually has its own home or apartment. Colonies range in size from about 60 to 150 people (CBC News, 2006).

BOX 3.2 POINT/COUNTERPOINT

Multiculturalism, Reasonable Accommodation, and "Veiled" Hostility

A 2014 survey by the Canadian Race Relations Foundation found overwhelming support for multiculturalism. The majority of respondents agreed that multiculturalism allows people to preserve their ethnic origins, contributes to social cohesion, assists immigrants to adopt shared values, and promotes reasonable accommodation of religious differences, even if those accommodations make some people uncomfortable.

However, a majority of respondents also indicated that multiculturalism appears to open the door to people pursuing certain cultural practices that are not compatible with Canadian laws and norms. When asked to define these 'incompatible' practices, one in three respondents indicated religious head covering and clothing including burqas [a loose garment that covers the hands and face or the entire body and includes a mesh screen over the eyes], hijabs [headscarf or veil covering head, ears, and neck], nijab [face veil], and turbans in public or security settings. (Canadian Race Relations Foundation, 2014)

In the past decade, the term "reasonable accommodation" has been used to describe the guiding principle by which religious diversity could be *reasonably* managed within a society governed by a principle of multiculturalism. How much accommodation is reasonable? For some, there is too much "accommodation" occurring, for others not enough, and for yet others, accommodation is not the appropriate language for assessing claims based on religious identity. Until recently, the discussions, which focused predominantly on Muslim headdress, were most intense in Quebec (Beaman, 2012:1). However, two recent incidents have intensified the debate across the country. In February 2015, a Quebec judge refused to hear a Muslim woman's case unless she removed her hijab. The judge said she wanted her courtroom to be secular, and the woman's headscarf did not conform to courtroom dress codes, stating:

> I will therefore not hear you if you are wearing a scarf on your head, just as I would not allow a person to appear before me wearing a hat or sunglasses on his or her head, or any other garment not suitable for a court proceeding. (Rukavina, 2015)

Shortly after, former Prime Minister Stephen Harper announced he would be appealing a Federal Court decision that allowed women to wear Niqabs during Citizenship swearing-in ceremonies, arguing that the practice of veiling was 'offensive' to Canadian values:

© Darren Ell/Demotix/Corbis

Recent cases involving religious head covering and clothing in public or security settings have generated considerable debate around 'reasonable accommodation.'

> "Almost all Canadians oppose the wearing of face covering during citizenship ceremonies. Why would Canadians, contrary to our own values, embrace a practice at that time that is not transparent, that is not open and, frankly, is rooted in a culture that is anti-women?" (Strapagiel, 2015)

This is certainly not the first time the topic of wearing religious items has sparked controversy and debate. Here are some of the most noteworthy examples:

In 2001, a 12-year-old Muslim boy was banned from carrying his kirpan (a ceremonial dagger worn under his clothing), to school. It was not until five years later that the Supreme Court upheld his right to do so. In their decision, they wrote:

> Allowing him to wear his kirpan under certain conditions demonstrates the importance that our society attaches to protecting freedom of religion and to showing respect for its minorities. (Multani v. Commission scolaire Marguerite-Bourgeoys, [2006] 1 S.C.R. 256, 2006 SCC 6)

In 2007, in a small rural town in Quebec, Herouxville, members of the town council passed 'a code of conduct' for new immigrants outlining behaviours that were 'incompatible' with Quebec culture. They felt it necessary to do so despite the fact that there was only one immigrant family in the town of 1300 residents. The code included bans on beating or burning women alive, veiling one's face, and children carrying symbolic weapons to school. The code was later revised.

(Continued)

In 2009, an Egyptian-born Muslim woman was expelled when she refused to remove her niqab in her French language class, after her teacher requested that she do so in order to correct her pronunciation by seeing her lips. The Quebec Immigration Minister at the time supported the decision stating, "if you want to integrate into Quebec society, here are our values. We want to see your face" (Kay, 2011). The woman subsequently filed a complaint with the Quebec Human Rights Commission.

During the 2013 election, the Parti Québécois proposed modifying the Quebec Charter of Rights and Freedoms to prohibit public employees from wearing conspicuous religious symbols. Moreover, Bill 60 would make it mandatory to provide and receive state services with uncovered faces. The bill died in 2014 when the Parti Québécois lost the election.

As these incidents demonstrate, the debate was played out almost exclusively in Quebec where the discussion was most intense, but this was not and is not solely a Quebec issue.

What do you think? Would you support a ban prohibiting the wearing of Muslim head wear? Under what circumstances? If not, what are your primary objections?

Arguments for a Ban on the Burqa, Niqab, or Hijab in Public Places

Those in support of a ban have raised the following objections to public displays of religion by Muslim women:

- It demonstrates a refusal by Muslims to integrate into Canadian society.
- Traditional Muslim headdress is a testament to the oppression of women.
- The veil is intimating or off-putting, particularly in certain circumstances.
- Covering one's body is a safety or security concern.
- Not showing one's face is impolite.
- Covering one's face is a barrier to communication, which interferes with social relations.
- Covering of faces interferes with identification.
- It is incompatible with the values of tolerance and respect for others. (Bakht 2012:76)

Arguments in Support of Veiling

Until recently, what has been lacking in debates about the hijab and niqab and burqa are the voices of women who wear them. In her article "Veiled Objections: Public Opposition to the Niqab," legal scholar, Natasha Bakht, summarized many of the reasons why Muslim women choose to cover their bodies. Although there are obviously multiple and overlapping reasons why Muslim women choose to veil, a common theme is the value they attached to dressing the way they do. These are some of the explanations:

- The veil is a symbol of their religious conviction. It plays a critical role in advancing in the integration of Muslim women into Canadian society by allowing them to fully participate without giving up their religious beliefs.
- The veil is a sign of a positive religious and cultural identity in contrast to being defined as the Other. "My dress tells you that I am Muslim and that I expect to be treated respectfully."
- The veil symbolizes a commitment to Islamic morality, such as female modesty. Some Muslim women literally 'wear their religious convictions for all to see.'
- The veil allows Muslim women to engage in practices that were unconventional for Muslim women, such as going to university or living alone.
- The veil allows Muslim women to resist being sexualized by men. Muslim women commented: "I feel more free. Especially men, they don't look at your appearance—they appreciate your intellectual abilities. They respect you".
- The veil minimizes social and economic differences among its wearers, which is consistent with Islamic beliefs.
- The veil is a form of resistance to materialism demonstrated by Western adherence to fashion. (Bakht, 2012:74)

As you have learned in this chapter, symbols mean different things to different people and their meaning can transform over time, place, and certainly culture. Although the veil may well have originated in patriarchal societies as a symbol of domination or control of women, if we listen to the voices of the women who wear them, it appears that today, wearing a niqab, hijab, or burqas may be a symbol of choice, identity, and empowerment.

The Hutterites are considered a subculture because their values, norms, and appearance differ significantly from those of members of the dominant culture. They have a strong faith in God and reject worldly concerns. Their core values include the joy of work, the primacy of the home, faithfulness, thriftiness, tradition, and humility. Hutterites hold conservative views of the family, believing that women are subordinate to men, birth control is unacceptable, and wives should remain at home. Children are cherished and seen as an economic asset: They help with the farming and other work.

Hutterite life is centred on the community rather than on the individual. All aspects of day-to-day life are based on sharing, right down to eating every meal in a community hall. Members of this group also have communal rather than private property; nobody is permitted to own as much as a pair of shoes (Curtis and Lambert, 1994). They have a "community of goods."

A predominant tenet of Hutterite faith is *nonassimilation*; that is, the Hutterites wish to maintain their separate status and not be absorbed into the dominant culture. The fact that their colonies are usually located far from towns, cities, and highways emphasizes this. However, the Hutterites do not seek complete social isolation from the wider society. Although they strictly adhere to centuries-old traditions, the Hutterites do not hesitate to take advantage of 21st-century advancements (Lyons, 1998). They are successful farmers who trade with people in the surrounding communities, and they buy modern farm machinery. They also read newspapers, use home computers and telephones, and utilize the services of non-Hutterite professionals.

Subcultures can provide opportunities for the expression of distinctive lifestyles, as well as sometimes helping people adapt to abrupt cultural change. Subcultures can also serve as a buffer against discrimination experienced by many ethnic or religious groups in Canada. However, some people may be forced by economic or social disadvantage to remain in such ethnic enclaves.

Technology and tradition meet at the Fairholme Hutterite Colony as these young women try out a new digital camera at school.

counterculture A group that strongly rejects dominant societal values and norms and seeks alternative lifestyles.

COUNTERCULTURES Some subcultures actively oppose the larger society. A **counterculture** is a group that strongly rejects dominant societal values and norms and seeks alternative lifestyles (Yinger, 1960, 1982). Young people are most likely to join countercultural groups, perhaps because younger persons generally have less invested in the existing culture. Examples of countercultures include the beatniks of the 1950s, the flower children of the 1960s, the drug enthusiasts of the 1970s, and members of non-mainstream religious sects, or cults. Occupy Wall Street and its counterparts throughout North America and other nations began as a counterculture because participants took a stand against dominant cultural values of wealth, power, and political privilege.

TIME TO REVIEW

- How is cultural diversity reflected in society?

LO-4 ETHNOCENTRISM AND CULTURAL RELATIVISM

ethnocentrism The tendency to regard one's own culture and group as the standard—and thus superior—whereas all other groups are seen as inferior.

cultural relativism The belief that the behaviours and customs of any culture must be viewed and analyzed by the culture's own standards.

When observing people from other cultures, many of us use our own culture as the yardstick by which we judge the behaviour of others. Sociologists refer to this approach as **ethnocentrism**—the tendency to regard one's own culture and group as the standard, and thus superior, whereas all other groups are seen as inferior. Ethnocentrism is based on the assumption that one's own way of life is superior to all others. For example, most schoolchildren are taught that their own school and country are the best. The school song and the national anthem are forms of *positive ethnocentrism*. However, *negative ethnocentrism* can also result from constant emphasis on the superiority of one's own group or nation. Negative ethnocentrism is manifested in derogatory stereotypes that ridicule recent immigrants whose customs, dress, eating habits, or religious beliefs are markedly different from those of dominant group members. Long-term Canadian residents who are members of racial and ethnic minority groups, such as First Nations and Indo-Canadians, have also been the target of ethnocentric practices by other groups.

An alternative to ethnocentrism is **cultural relativism**—the belief that the behaviours and customs of any culture must be viewed and analyzed by the culture's own standards. For example, the anthropologist Marvin Harris (1974, 1985) uses cultural relativism to explain why cattle, which are viewed as sacred, are not killed and eaten in India, where widespread hunger and malnutrition exist. From an ethnocentric viewpoint, we might conclude that cow worship is the cause of the hunger and poverty in India. However, according to Harris, the Hindu taboo against killing cattle is very important to their economic system. Live cows are more valuable than dead ones because they have more important uses than as a direct source of food. As part of the ecological system, cows consume grasses of little value to humans. Then, they produce two valuable resources—oxen (the neutered offspring of cows), to power the plows, and manure, for fuel and fertilizer—as well as milk, floor covering, and leather. As Harris's study reveals, culture must be viewed from the standpoint of those who live in a particular society.

Cultural relativism also has a downside. It may be used to excuse customs and behaviour (such as cannibalism) that may violate basic human rights. Cultural relativism is a part of the sociological imagination; researchers must be aware of the customs and norms of the society they are studying, and then spell out their background assumptions so that others can spot possible biases in their studies. According to some social scientists, however, issues surrounding ethnocentrism and cultural relativism may become less distinct in the future as people around the globe increasingly share a common popular culture.

HIGH CULTURE AND POPULAR CULTURE

Before taking this course, what was the first thing you thought about when you heard the term *culture?* In everyday life, culture is often used to describe the fine arts, literature, and classical music. When people say that a person is "cultured," they may mean that the individual has a highly developed sense of style or aesthetic appreciation of the "finer" things.

What is the difference between high culture and popular culture? *High culture* consists of classical music, opera, ballet, live theatre, and other activities usually patronized by elite audiences, composed primarily of members of the upper-middle and upper classes, who have the time, money, and knowledge assumed to be necessary for its appreciation. *Popular culture* consists of activities, products, and services that are assumed to appeal primarily to members of the middle and working classes. These include rock concerts, spectator sports, movies, television soap operas, situation comedies, and, more recently, the Internet. Although we will distinguish between "high" and "popular" culture in our discussion, it is important to note that some social analysts believe that the rise of a consumer society in which luxury items have become more widely accessible to the masses has greatly reduced the great divide between activities and possessions associated with wealthy people or a social elite.

Overall, most sociologists believe that culture and social class are intricately related. French sociologist Pierre Bourdieu's (1984) *cultural capital theory* views high culture as a device used by the dominant class to exclude the subordinate classes. According to Bourdieu, people must be trained to appreciate and understand high culture. Individuals learn about high culture in upper-middle and upper-class families and in elite education systems, especially higher education (university). Once they acquire this trained capacity, they possess a form of symbolic currency, or "cultural capital," that can be exchanged for employment and promotional opportunities in the workplace. The knowledge and skills acquired while earning a university degree (e.g., reading, writing, communication skills, logical reasoning) are valued resources on the job market, and people who possess this form of cultural capital are more likely to secure employment than people who do not. Persons from poor and working-class backgrounds typically do not acquire this cultural capital. Since knowledge and appreciation of high culture is considered a prerequisite for access to the dominant class, its members can use their cultural capital to deny access to subordinate group members and thus preserve and reproduce the existing class structure. Unlike high culture, popular culture is presumed to be available to everyone.

FORMS OF POPULAR CULTURE

Three prevalent forms of popular culture are fads, fashions, and leisure activities.

A *fad* is a temporary but widely copied activity followed enthusiastically by large numbers of people. Most fads are short-lived novelties. According to sociologist John Lofland (1993), fads can be divided into four major categories. First, *object fads* are items that people purchase despite the fact that they have little use value, such as wristbands that make a statement or support a cause. Second, *activity fads* include pursuits such as body piercing or flash mobs. Third are *idea fads,* such as New Age ideologies. Fourth are *personality fads*—for example, those surrounding celebrities such as Lady Gaga, Justin Bieber, and Miley Cyrus.

A *fashion* is a currently valued style of behaviour, thinking, or appearance that is longer lasting and more widespread than a fad. Examples of fashion are found in many areas, including child rearing, education, arts, clothing, music, and sports. Soccer is an example of a fashion in sports. Until recently, only schoolchildren played soccer in Canada, but now soccer has become a really popular sport, perhaps in part because of immigration from European countries and other areas of the world where soccer is widely played.

Michael Stuparyk/GetStock.com

Activity fads, such as flash mobs, are particularly popular with young people.

Is the proliferation of massive shopping malls in China—containing stores from the United States and Western Europe as well as local entities—an example of cultural diffusion? Or is the malling of China an example of cultural imperialism? Can "culture" be sold?

Like soccer, other forms of popular culture move across nations. In Canada, we often assess the quality of popular culture on the basis of whether it is a Canadian or American product. Canadian artists, musicians, and entertainers often believe they have "made it" only when they become part of American popular culture. Music, television shows, novels, and street fashions from the United States have become a part of our Canadian culture. People in this country continue to be strongly influenced by popular culture from nations other than the United States, too. For example, Canada's contemporary music and clothing reflect African, Caribbean, and Asian cultural influences, among others.

Will the spread of popular culture produce a homogeneous global culture? Critics argue that the world is not developing a global culture; rather, other cultures are becoming Westernized. Political and religious leaders in some nations oppose this process, which they view as **cultural imperialism**—the extensive infusion of one nation's culture into other nations. As discussed in Chapter 18, powerful countries often use the media to spread values and ideas that dominate and even destroy other cultures. For example, some view the widespread infusion of the English language into countries that speak other languages as a form of cultural imperialism. On the other hand, the concept of cultural imperialism may fail to take into account various cross-cultural influences. For example, cultural diffusion of literature, music, clothing, and food has occurred on a global scale. A global culture, if it comes into existence, will most likely include components from many societies and cultures.

▶ **cultural imperialism**
The extensive infusion of one nation's culture into other nations.

TIME TO REVIEW

- To what degree are we shaped by popular culture?

LO-5 SOCIOLOGICAL ANALYSIS OF CULTURE

Sociologists regard culture as a central ingredient in human behaviour. Although all sociologists share a similar purpose, they typically see culture through somewhat different lenses because they are guided by different theoretical perspectives in their research. What do these perspectives tell us about culture?

FUNCTIONALIST PERSPECTIVES

As previously discussed, functionalist perspectives are based on the assumption that society is a stable, orderly system with interrelated parts that serve specific functions. Anthropologist Bronislaw Malinowski (1922) suggested that culture helps people meet their *biological needs* (including food and procreation), *instrumental needs* (including law and education), and *integrative needs* (including religion and art). Societies in which people share a common language and core values are more likely to have consensus and harmony.

How might functionalist analysts view popular culture? According to many functionalist theorists, popular culture serves a significant function in society in that it may be the "glue" that holds society together. Regardless of race, class, sex, age, or other characteristics, many people are brought together (at least in spirit) to cheer teams competing in major sporting events, such as the Stanley Cup or the Olympic Games. Television, the Internet, and social media help integrate recent immigrants into the mainstream culture, whereas longer-term residents may become more homogenized as a result of seeing the same images and being exposed to the same beliefs and values. Functionalists acknowledge, however, that all societies have dysfunctions that produce a variety of societal problems. When many subcultures are present within a society, discord results from a lack of consensus about core values. In fact, popular culture may undermine core cultural values rather than reinforce them. For example, movies may glorify crime rather than hard work as the quickest way to get ahead. According to some analysts, excessive violence in music videos, movies, and television programs may be harmful to children and young people. From this perspective, popular culture may be a factor in antisocial behaviour as seemingly diverse as hate crimes and fatal shootings in public schools.

julibbb/Thinkstock

Rapid changes in language and culture in Canada are reflected in this sign. How do functionalist and conflict theorists' views regarding language differ?

A strength of the functionalist perspective on culture is its focus on the needs of society and the fact that stability is essential for society's continued survival. A shortcoming is its overemphasis on harmony and cooperation. This approach also fails to fully account for factors embedded in the structure of society—such as class-based inequalities, racism, and sexism—that may contribute to conflict among people in Canada or to global strife.

CONFLICT PERSPECTIVES

Conflict perspectives are based on the assumption that social life is a continuous struggle in which members of powerful groups seek to control scarce resources. According to this approach, values and norms help to create and sustain the privileged position of the powerful in society while excluding others. As early conflict theorist Karl Marx stressed, ideas are *cultural creations* of a society's most powerful members. Thus, it is possible for political, economic, and social leaders to use *ideology*—an integrated system of ideas that is external to, and coercive of, people—to maintain their positions of dominance in a society. As Marx stated:

> The ideas of the ruling class are in every epoch the ruling ideas, i.e., the class which is the ruling material force in society, is at the same time, its ruling intellectual force. The class, which has the means of material production at its disposal, has control at the same time over the means of mental production . . . The ruling ideas are nothing more than the ideal expression of the dominant material relationships, the dominant material relationships grasped as ideas. (Marx and Engels, 1970/1845–1846:64)

Many contemporary conflict theorists agree with Marx's assertion that ideas, a nonmaterial component of culture, are used by agents of the ruling class to affect the thoughts and actions of members of other classes.

How might conflict theorists view popular culture? Some conflict theorists believe that popular culture, which originated with everyday people, has been largely removed from their domain and has become nothing more than a part of the capitalist economy in North America (Gans, 1974; Cantor, 1980, 1987). From this approach, media conglomerates such as Time Warner and ABC/Disney create popular culture, such as films, television shows, and amusement parks, in the same way that they would produce any other product or service. Creating new popular culture also promotes consumption of *commodities*—objects outside ourselves that we purchase to satisfy our human needs or wants (Fjellman, 1992). According to contemporary social analysts, consumption—even of things that we do not necessarily need—has become prevalent at all social levels, and some middle- and lower-income individuals and families now use as their frame of reference the lifestyles of the more affluent in their communities. As a result, many families live on credit in order to purchase the goods and services that they would like to have or that keep them on the competitive edge with their friends, neighbours, and coworkers (Schor, 1999).

A strength of the conflict perspective is that it stresses how cultural values and norms may perpetuate social inequalities. It also highlights the inevitability of change and the constant tension between those who want to maintain the status quo and those who desire change. A limitation is its focus on societal discord and the divisiveness of culture.

© Scott Larson/Splash News/Newscom

Is this Japanese amusement park a sign of a homogeneous global culture or of cultural imperialism? Discuss.

SYMBOLIC INTERACTIONIST PERSPECTIVES

Unlike functionalists and conflict theorists, who focus primarily on macrolevel concerns, symbolic interactionists engage in a microlevel analysis that views society as the sum of all people's interactions. From this perspective, symbols make communication with others possible because they provide people with shared meanings, and people create, maintain, and modify culture as they go about their everyday activities.

According to some symbolic interactionists, people continually negotiate their social realities. Values and norms are not independent realities that automatically determine our behaviour; instead, we reinterpret them in each social situation we encounter. However, the classical sociologist Georg Simmel warned that the larger cultural world—including both material and nonmaterial culture—eventually takes on a life of its own apart from the actors who daily recreate social life. As a result, individuals may be more controlled by culture than they realize.

Simmel (1990/1907) suggested that money is an example of how people may be controlled by their culture. According to Simmel, people initially create money as a means of exchange, but then money acquires a social meaning that extends beyond its purely economic function. Money becomes an end in itself, rather than a means to an end. Today, we are aware of the relative "worth" not only of objects but also of individuals. Many people revere wealthy entrepreneurs and highly paid celebrities, entertainers, and sports figures for how much money they make, not for their intrinsic qualities. According to Simmel, money makes it possible for us to *relativize* everything, including our relationships with other people. When social life can be reduced to money, people become cynical, believing that anything—including people, objects, beauty, and truth—can be bought if we can pay the price. Although Simmel acknowledged the positive functions of money, he believed that the social interpretations people give to money often produce individual feelings of cynicism and isolation.

A symbolic interactionist approach highlights how people maintain and change culture through their interactions with others. However, interactionism does not provide a systematic framework for analyzing how we shape culture and how it, in turn, shapes us. It also does not provide insight into how shared meanings are developed among people, and it does not take into account the many situations in which there is disagreement on meanings. Whereas the functional and conflict approaches tend to overemphasize the macrolevel workings of society, the interactionist viewpoint often fails to take these larger social structures into account.

POSTMODERN PERSPECTIVES

Postmodern theorists believe that much of what has been written about culture in the Western world is Eurocentric—that it is based on the uncritical assumption that European culture (including its dispersed versions in countries such as Canada, the United States, Australia, and South Africa) is the true, universal culture in which all the world's people ought to believe (Lemert, 1997). By contrast, postmodernists believe that we should speak of *cultures* rather than *culture.*

However, Jean Baudrillard, one of the best-known French social theorists, believes that the world of culture today is based on *simulation,* not reality. According to Baudrillard, social life is much more a spectacle that simulates reality than reality itself. Many people gain "reality" from the media or cyberspace. For example, consider the many North American children who, upon entering school for the first time, have already watched more hours of television than the total number of classroom instruction hours they will encounter in their entire school careers (Lemert, 1997). Add to this the number of hours that some will have spent playing computer games or surfing the Internet. Baudrillard refers to this social creation as *hyperreality*—a situation in which the *simulation* of reality is more real than the thing itself. For Baudrillard, everyday life has been captured by the signs and symbols generated to represent it, and we ultimately relate to simulations and models as if they were reality.

Baudrillard (1983) uses Disneyland as an example of a simulation that conceals the reality that exists outside rather than inside the boundaries of the artificial perimeter. According to Baudrillard, Disney-like theme parks constitute a form of seduction that substitutes symbolic (seductive) power for real power, particularly the ability to bring about social change. From this perspective, amusement park "guests" may feel like "survivors" after enduring the rapid speed and gravity-defying movements of the roller coaster rides, or see themselves as "winners" after surviving fights with hideous cartoon villains on the "dark rides," when they have actually experienced the substitution of an *appearance* of power over their lives for the *absence* of real power.

In their examination of culture, postmodern social theorists make us aware of the fact that no single perspective can grasp the complexity and diversity of the social world. They also make us aware that reality may not be what it seems. According to the postmodern view, no one authority can claim to know social reality and we should *deconstruct*—take apart and subject to intense critical scrutiny—existing beliefs and theories about culture in hopes of gaining new insights (Ritzer, 1997).

Purestock/Thinkstock

People of all ages are spending many hours each week using computers, playing video games, and watching television. How is this behaviour different from the ways in which people enjoyed popular culture in previous generations?

Although postmodern theories of culture have been criticized on a number of grounds, we will mention only three. One criticism is postmodernism's lack of a clear conceptualization of ideas. Another is the tendency to critique other perspectives as being "grand narratives," whereas postmodernists offer their own varieties of such narratives. Finally, some analysts believe that postmodern analyses of culture lead to profound pessimism about the future.

CONCEPT SNAPSHOT

COMPONENTS OF CULTURE	**Symbol:** Anything that meaningfully represents anything else. **Language:** A set of symbols that express ideas and enable people to think and communicate with one another. **Values:** Collective ideas about what is right or wrong, good or bad, and desirable or undesirable in a particular culture. **Norms:** Established rules of behaviour or standards of conduct.
FUNCTIONALIST PERSPECTIVES	A functionalist analysis of culture assumes that a common language and shared values help produce consensus and harmony. Conversely, in a society that contains numerous subcultures, discord results from a lack of consensus and shared core values.
CONFLICT PERSPECTIVES	Conflict theorists suggest that values and norms help create and sustain a position of privilege for those in power in a society. Ideas are a cultural creation of society's most powerful members and can be used by the ruling class to affect the thoughts and actions of members of other classes.
SYMBOLIC INTERACTIONIST PERSPECTIVES	According to symbolic interactionists, people create, maintain, and modify culture during their everyday activities. Symbols assist in our communication with others by providing shared meanings.
POSTMODERN PERSPECTIVES	Postmodern theorists believe that culture today is based on a simulation of reality (e.g., what we see on television) rather than reality itself. According to the postmodern perspective, we should deconstruct existing beliefs and theories about culture in order to gain new insights.

LO-1 Understand the importance of culture in our lives and those of others in society.

Culture encompasses the knowledge, language, values, and customs passed from one generation to the next in a human group or society. Culture is essential for our individual survival because, unlike nonhuman animals, we are not born with instinctive information about how to behave and how to care for our selves and others.

Culture can be a stabilizing force for society, providing a sense of continuity; however, culture also can be a force that generates discord, conflict, and violence.

There are both material and nonmaterial expressions of culture. Material culture consists of the physical creations of society. Nonmaterial culture is more abstract and reflects the ideas, values, and beliefs of a society.

Ruth Bonneville/Winnipeg Free Press

LO-2 Identify the essential components of culture.

These components are symbols, language, values, and norms. Symbols express shared meanings; through them, groups communicate cultural ideas and abstract concepts. Language is a set of symbols through which groups communicate. Values are a culture's collective ideas about what is or is not acceptable. Norms are the specific behavioural expectations within a culture.

gst/Shutterstock.com

LO-3 Describe what causes cultural change in societies.

Cultural change takes place in all societies. Change occurs through discovery and invention and through diffusion, which is the transmission of culture from one society or group to another.

© Winnipeg Free Press, June 7, 1998. Reproduced with permission.

LO-4 Compare and contrast ethnocentrism and cultural relativism as approaches to examining cultural differences.

Ethnocentrism is the assumption that one's own culture is superior to other cultures. Cultural relativism counters culture shock and ethnocentrism by viewing and analyzing another culture in terms of its own values and standards.

Greg Elms/Lonely Planet Images/Getty Images

KEY TERMS

counterculture A group that strongly rejects dominant societal values and norms and seeks alternative lifestyles (p. 76).

cultural imperialism The extensive infusion of one nation's culture into other nations (p. 78).

cultural lag William Ogburn's term for a gap between the technical development of a society (material culture) and its moral and legal institutions (nonmaterial culture) (p. 71).

cultural relativism The belief that the behaviours and customs of any culture must be viewed and analyzed by the culture's own standards (p. 76).

cultural universals Customs and practices that occur across all societies (p. 62).

culture The knowledge, language, values, customs, and material objects that are passed from person to person and from one generation to the next in a human group or society (p. 59).

diffusion The transmission of cultural items or social practices from one group or society to another (p. 71).

discovery The process of learning about something previously unknown or unrecognized (p. 71).

ethnocentrism The tendency to regard one's own culture and group as the standard—and thus superior—whereas all other groups are seen as inferior (p. 76).

folkways Informal norms or everyday customs that may be violated without serious consequences within a particular culture (p. 70).

ideal culture The values and standards of behaviour that people in a society profess to hold (p. 69).

invention The process of reshaping existing cultural items into a new form (p. 71).

language A system of symbols that expresses ideas and enables people to think and communicate with one another (p. 64).

laws Formal, standardized norms that have been enacted by legislatures and are enforced by formal sanctions (p. 70).

material culture A component of culture that consists of the physical or tangible creations—such as clothing, shelter, and art—that members of a society make, use, and share (p. 61).

mores Strongly held norms with moral and ethical connotations that may not be violated without serious consequences in a particular culture (p. 70).

nonmaterial culture A component of culture that consists of the abstract or intangible human creations of society—such as attitudes, beliefs, and values—that influence people's behaviour (p. 61).

norms Established rules of behaviour or standards of conduct (p. 69).

real culture The values and standards of behaviour that people actually follow (as contrasted with *ideal culture*) (p. 69).

sanctions Rewards for appropriate behaviour or penalties for inappropriate behaviour (p. 69).

Sapir-Whorf hypothesis The proposition that language shapes its speakers' view of reality (p. 65).

subculture A group of people who share a distinctive set of cultural beliefs and behaviours that differ in some significant way from those of the larger society (p. 72).

LO-5 Explain how the various sociological perspectives view culture.

A functional analysis of culture assumes that a common language and shared values help produce consensus and harmony. According to some conflict theorists, culture may be used by certain groups to maintain their privilege and exclude others from society's benefits. Symbolic interactionists suggest that people create, maintain, and modify culture as they go about their everyday activities. Postmodern thinkers believe that there are many cultures in Canada alone. To gain a better understanding of how popular culture may simulate reality rather than being reality, postmodernists believe that we need a new way of conceptualizing culture and society.

© Scott Larson/Splash News/Newscom

APPLICATION QUESTIONS

1. Would it be possible today to live in a totally separate culture in Canada? In what ways could you avoid all influences from the mainstream popular culture or from the values and norms of other cultures? How would you avoid any change in your culture?
2. Do fads and fashions in popular culture reflect and reinforce or challenge and change the values and norms of a society? Consider a wide variety of fads and fashions: musical styles; computer and video games and other technologies; literature; and political, social, and religious ideas.
3. Make a list of three or four uniquely Canadian symbols. Then identify examples of symbols that represent other countries.
4. In what ways do we see cultural differences in our everyday life situations and experiences? Which different cultural groups are you a part of, and how do they intersect or interact?

taboos Mores so strong that their violation is considered extremely offensive and even unmentionable (p. 70).

technology The knowledge, techniques, and tools that make it possible for people to transform resources into usable forms, as well as the knowledge and skills required to use them after they are developed (p. 61).

value contradiction Values that conflict with one another or are mutually exclusive (p. 69).

values Collective ideas about what is right or wrong, good or bad, and desirable or undesirable in a particular culture (p. 69).

CHAPTER 4

SOCIALIZATION

© ACE STOCK LIMITED/Alamy Stock Photo

CHAPTER FOCUS QUESTION

How does socialization occur throughout our lives, including our university years?

LEARNING OBJECTIVES

AFTER READING THIS CHAPTER, YOU SHOULD BE ABLE TO

LO-1 Discuss the degree to which our unique physical and human characteristics are based on heredity and to what degree they are based on our social environment.

LO-2 Debate the extent to which people would become human beings without adequate socialization.

LO-3 Identify the key agents of socialization.

LO-4 Describe how sociologists explain our development of a self-concept.

LO-5 Review the main social psychological theories on human development.

LO-6 Describe socialization throughout various stages in the life course.

Scott Dobson-Mitchell studied at the University of Waterloo. He offers several suggestions to new university students as they struggle through the first-year experience.

Assuming I couldn't accidentally cause some sort of butterfly effect that would prevent me being born, I wish I could travel back in time and tell my Freshman Self a few things about university. Considering I've already forgotten the answers to every exam, this is what I'd tell the younger me . . .

Plan ahead. WAY ahead.

It happens to me every semester. Searching through the course calendar, I find the perfect class. It sounds interesting, it fits perfectly into my schedule and it fulfills my upper-year science requirement. The prof checks out on RateMyProfessors and the course has a high score on Bird Courses. But I don't have one of the prerequisites! If I'd been smart enough to plan, I would have that first year zoology credit that's mandatory for nearly everything. Instead, I'm stuck with phytochemical biosystems.

. . . It's going to get easier.

The first year is the worst year. It's sort of like the first 20 minutes of the movie Inception, *when you have no idea what the hell is going on. But if you hang in there, things will start making sense. You'll realize that university isn't impossibly more difficult than high school. In fact, once you've acclimatized, it's easier in some ways. And it only gets better. Once some of those nasty prerequisites are out of the way, you can take courses that truly interest you. My interests happen to coincide with those listed on Bird Courses. (Maclean's Campus Online, 2012)*

Source: Scott Dobson-Mitchell, "The Many Regrets of a Fourth-Year Student," from http://oncampus.macleans.ca/education/2012/01/27/the-many-regrets-of-a-fourth-year-student. Reproduced by permission of the author.

Socialization is a lifelong process that includes socialization in early childhood, in adolescence, and in early and late adulthood. Each of these phases, or stages, of life presents its own special challenges as we are socialized or "resocialized" to new experiences, environments, and expectations. In fact, many of you are being socialized to university or college life with new practices of learning and new social dynamics both inside and outside the classroom.

Many of us experience stress when we take on new and seemingly unfamiliar roles and find that we must learn the appropriate norms regarding how people in a specific role should think, act, and communicate with others. This is especially true when we start university or college and find ourselves in a new setting surrounded by many people we do not know.

Look around in your classes at the beginning of each semester, and you will probably see other students who are trying to find out what is going to be expected of them as a student in a particular course. What is the course going to cover? What are the instructor's requirements? How should students communicate with the instructor and other students in the class? Some information of this type is learned through formal instruction, such as in a classroom, but much of what we know about school is learned informally through our observations of other people, by listening to what they say when we are in their physical presence, or through interacting with them by cellphone, email, or text messaging when we are apart. Sociologists use the term *socialization* to refer to both the formal and informal processes by which people learn a new role and find out how to be a part of a group or organization. Moreover, the process of socialization continues throughout an individual's life.

In this chapter, we will examine the process of socialization and identify reasons why socialization is crucial to the well-being of individuals, groups, and societies. Throughout the chapter, we will focus on positive and negative aspects of the socialization process, including the daily stresses that may be involved in this process. Before reading on, test your knowledge about socialization and the university experience by taking the quiz in Box 4.1.

CRITICAL THINKING QUESTIONS

1. What socialization issues did you face during your first term in higher education?
2. What strategies did you use in acquiring information about your new role as a university student?
3. What suggestions would you offer to students struggling to learn new behaviours, attitudes, and norms for a university student, both inside and outside the classroom?

WHY IS SOCIALIZATION IMPORTANT?

▶ **socialization** The lifelong process of social interaction through which individuals acquire a self-identity and the physical, mental, and social skills needed for survival in society.

Socialization—the lifelong process of social interaction through which individuals acquire a self-identity and the physical, mental, and social skills needed for survival in society—is the essential link between the individual and society. It enables each of us to develop our human potential and learn the ways of thinking, talking, and acting that are essential for social living.

Socialization is most crucial during childhood because it is essential for the individual's survival and for human development. The many people who met the early material and social needs of each of us were central to our establishing our own identity. During the first three years of our life, we begin to develop both a unique identity and the ability to manipulate things and to walk. We acquire sophisticated cognitive tools for thinking and for analyzing a wide variety of situations, and we learn effective communication skills. In the process, we begin a socialization process that takes place throughout our lives and through which we also have an effect on other people who watch us.

Socialization is also essential for the survival and stability of society. Members of a society must be socialized to support and maintain the existing social structure. From a functionalist perspective, individual conformity to existing norms is not taken for granted; rather, basic individual needs and desires must be balanced against the needs of the social structure. The socialization process is most effective when people conform to the norms of society because they believe that doing so is the best course of action. Socialization enables a society to "reproduce" itself by passing on this cultural content from one generation to the next.

The kind of person we become depends greatly on the people who surround us. How will this boy's life be shaped by his close and warm relationship with his mother?

While ways in which people learn beliefs, values, and rules of behaviour are somewhat similar in many countries, the content of socialization differs greatly from society to society. How people walk, talk, eat, make love, and wage war are all functions of the culture in which they are raised. At the same time, we are also influenced by our exposure to subcultures of class, ethnicity, religion, and gender. In addition, each of us has unique experiences in our families and friendship groupings. The kind of human being that we become depends greatly on the particular society and social groups that surround us at birth and during early childhood. What we believe about ourselves, our society, and the world is largely a product of our interactions with others.

BOX 4.1 SOCIOLOGY AND EVERYDAY LIFE

How Much Do You Know About Socialization and the University Experience?

True	False	
T	F	1. Professors are the primary agents of socialization for university students.
T	F	2. In recent studies, few students report that they spend time studying with other students.
T	F	3. Many students find that taking university courses is stressful because it is an abrupt change from high school.
T	F	4. University students typically find the socialization process in higher education less stressful than the one they experience when they enter an occupation or profession.
T	F	5. Getting good grades and completing schoolwork are the top sources of stress reported by university students.

For answers to the quiz about socialization and the university experience, go to **www.nelson.com/student.**

LO-1 HUMAN DEVELOPMENT: BIOLOGY AND SOCIETY

What does it mean to be "human"? To be human includes being conscious of ourselves as individuals with unique identities, personalities, and relationships with others. As humans, we have ideas, emotions, and values. We have the capacity to think and to make rational decisions. But what is the source of "humanness"? Are we born with these human characteristics, or do we develop them through our interactions with others?

When we are born, we are totally dependent on others for our survival. We cannot turn ourselves over, speak, reason, plan, or do many of the things associated with being human. Although we can nurse, wet, and cry, most small mammals can also do those things. As discussed in Chapter 3, we humans differ from nonhuman animals because we lack instincts and must rely on learning for our survival. Human infants have the potential for developing human characteristics if they are exposed to an adequate socialization process.

Every human being is a product of biology, society, and personal experiences—that is, of heredity and environment, or, in even more basic terms, "nature" and "nurture." How much of our development can be explained by socialization? How much by our genetic heritage? Sociologists focus on how humans design their own culture and transmit it from generation to generation through socialization. By contrast, sociobiologists assert that nature, in the form of our genetic makeup, is a major factor in shaping human behaviour. **Sociobiology** is the systematic study of how biology affects social behaviour. According to zoologist Edward O. Wilson (1975), who pioneered sociobiology, genetic inheritance underlies many forms of social behaviour, such as war and peace, envy and concern for others, and competition and cooperation. Most sociologists disagree with the notion that biological principles can be used to explain all human behaviour. Obviously, however, some aspects of our physical makeup—such as eye colour, hair colour, height, and weight—are determined largely by our heredity.

▶ **sociobiology** The systematic study of how biology affects social behaviour.

How important is social influence ("nurture") in human development? There is hardly a behaviour that is not influenced socially. Except for simple reflexes, most human actions are social, either in their causes or in their consequences. Even solitary actions, such as crying or brushing our teeth, are ultimately social. We cry because someone has hurt us. We brush our teeth because our parents (or dentist) told us it was important. Social environment probably has a greater effect than heredity on the way we develop and the way we act. However, heredity does provide the basic material from which other people help to mould an individual's human characteristics.

Our biological and emotional needs are related in a complex equation. Children whose needs are met in settings characterized by affection, warmth, and closeness see the world as a safe and comfortable place and other people as trustworthy and helpful. By contrast, infants and children who receive less than adequate care or who are emotionally rejected or abused often view the world as hostile and have feelings of suspicion and fear.

TIME TO REVIEW

- Why is healthy socialization so important?

LO-2 SOCIAL ISOLATION AND MALTREATMENT

Social environment, then, is a crucial part of an individual's socialization. Even nonhuman primates, such as monkeys and chimpanzees, need social contact with others of their species to develop properly. As we will see, appropriate social contact is even more important for humans.

Martin Rogers/Stone/Getty Images

As Harry and Margaret Harlow discovered, humans are not the only primates that need contact with others. Deprived of its mother, this infant monkey found a substitute.

ISOLATION AND NONHUMAN PRIMATES Researchers have attempted to show the effects of social isolation on nonhuman primates raised without contact with others of their own species. In a series of laboratory experiments, psychologists Harry and Margaret Harlow (1962, 1977) took infant rhesus monkeys from their mothers and isolated them in separate cages. Each cage contained two nonliving "mother substitutes" made of wire, one with a feeding bottle attached and the other covered with soft terry cloth but without a bottle. The infant monkeys instinctively clung to the cloth "mother" and would not abandon it until hunger drove them to the bottle attached to the wire "mother." As soon as they were full, they went back to the cloth "mother," seeking warmth, affection, and physical comfort.

The Harlows' experiments show the detrimental effects of isolation on nonhuman primates. When the young monkeys were later introduced to other members of their species, they cringed in the corner. Having been deprived of social contact with other monkeys during their first six months of life, they never learned how to relate to other monkeys or to become well-adjusted adult monkeys—they were fearful of or hostile toward other monkeys (Harlow and Harlow, 1962, 1977). Because humans rely more heavily on social learning than do monkeys, the process of socialization is even more important for us.

ISOLATED CHILDREN Of course, sociologists would never place children in isolated circumstances so that they could observe what the effects were. However, there are cases in which parents or other caregivers failed to fulfill their responsibilities, leaving children alone or placing them in isolated circumstances. From analysis of these situations, social scientists have documented cases in which children were deliberately raised in isolation. A look at the lives of two children who suffered such emotional abuse provides insights into the importance of a positive socialization process and the negative effects of social isolation.

Anna Born in 1932 to an unmarried, mentally impaired woman, Anna was an unwanted child. She was kept in an attic-like room in her grandfather's house. Her mother, who worked on the farm all day and often went out at night, gave Anna just enough care to keep her alive; she received no other care. Sociologist Kingsley Davis described her condition when she was found in 1938:

> [Anna] had no glimmering of speech, absolutely no ability to walk, no sense of gesture, not the least capacity to feed herself even when the food was put in front of her, and no comprehension of cleanliness. She was so apathetic that it was hard to tell whether or not she could hear. And all of this at the age of nearly six years. (1940)

When she was placed in a special school and given the necessary care, Anna slowly learned to walk, talk, and care for herself. Just before her death at the age of 10, Anna reportedly could follow directions, talk in phrases, wash her hands, brush her teeth, and try to help other children (Davis, 1940).

Genie Almost four decades after Anna was discovered, Genie was found in 1970 at the age of 13. She had been locked in a bedroom alone, alternately strapped down to a child's potty chair or straitjacketed into a sleeping bag, since she was 20 months old. She had been fed baby food and beaten with a wooden paddle when she whimpered. She had not heard the sounds of human speech because no one talked to her, and there was no television or radio in her home

(Curtiss, 1977; Pines, 1981). Genie was placed in a pediatric hospital, where one of the psychologists described her condition:

> At the time of her admission she was virtually unsocialized. She could not stand erect, salivated continuously, had never been toilet-trained and had no control over her urinary or bowel functions. She was unable to chew solid food and had the weight, height and appearance of a child half her age. (Rigler, 1993:35)

In addition to her physical condition, Genie showed psychological traits associated with neglect, as described by one of her psychiatrists:

> If you gave [Genie] a toy, she would reach out and touch it, hold it, caress it with her fingertips, as though she didn't trust her eyes. She would rub it against her cheek to feel it. So when I met her and she began to notice me standing beside her bed, I held my hand out and she reached out and took my hand and carefully felt my thumb and fingers individually, and then put my hand against her cheek. She was exactly like a blind child. (Rymer, 1993:45)

Extensive therapy was used in an attempt to socialize Genie and develop her language abilities (Curtiss, 1977; Pines, 1981). These efforts met with limited success: In the early 1990s, Genie was living in a board-and-care home for mentally challenged adults (see Angier, 1993; Rigler, 1993; Rymer, 1993).

CHILD MALTREATMENT What do the terms *child maltreatment* and *child abuse* mean to you? When asked what constitutes child maltreatment, many people first think of cases that involve severe injuries or sexual abuse. In fact, these terms refer to the violence, mistreatment, or neglect that a child may experience while in the care of someone he or she trusts or depends on, such as a parent, relative, caregiver, or guardian. There are many different forms of abuse, including physical abuse, sexual abuse or exploitation, neglect, emotional abuse, and exposure to intimate partner violence (see Figure 4.1). A child who is abused often experiences more than one form of abuse.

FIGURE 4.1 TYPES OF SUBSTANTIATED CHILD MALTREATMENT IN CANADA IN 2008*

*Total estimated number of substantiated investigations is 85,440, based on a sample of 6163 substantiated investigations.

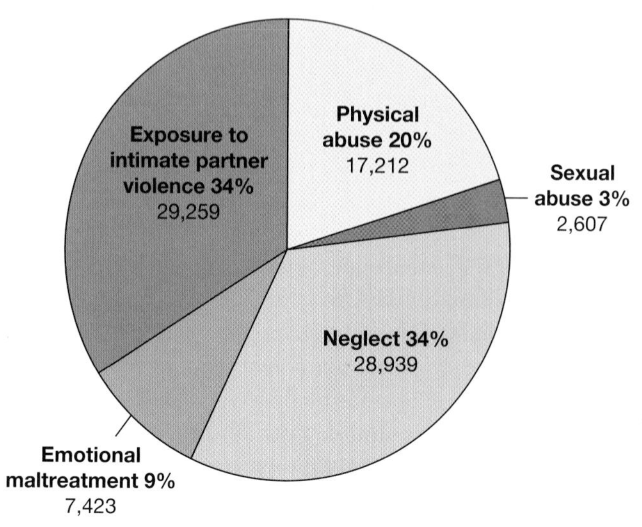

Source: © All rights reserved. Canadian Incidence Study of Reported Child Abuse and Neglect. Public Health Agency of Canada, 2008. Adapted and reproduced with permission from the Minister of Health, 2015.

Recent studies indicate that exposure to intimate partner violence and neglect are the most frequent forms of child abuse. Child neglect occurs when a child's basic needs—including emotional warmth and security, adequate shelter, food, healthcare, education, clothing, and protection—are not met, regardless of the cause (Trocme, 2010). The neglect usually involves repeated incidents over a lengthy time.

Any child—regardless of age, gender, race, ethnicity, socioeconomic status, sexual orientation, physical or mental abilities, and personality—may be at risk of being abused. Sociologists argue that child abuse is linked to inequalities in our society and the power imbalance that exists between adults and children. A child is usually dependent on his or her abuser and has little power to control the abusive circumstances.

TIME TO REVIEW

- Why is healthy socialization so important?
- Explain what problems can develop when children receive inadequate socialization.

LO-3 AGENTS OF SOCIALIZATION

▶ **agents of socialization** The persons, groups, or institutions that teach us what we need to know to participate in society.

Agents of socialization are the persons, groups, or institutions that teach us what we need to know to participate in society. We are exposed to many agents of socialization throughout our lifetime; in turn, we have an influence on those socializing agents and organizations. Here, we will look at those that are most pervasive in childhood—the family, the school, peer groups, and the mass media.

THE FAMILY

The family is the most important agent of socialization in all societies. As the discussion of child maltreatment has demonstrated, the initial love and nurturance we receive from our families are central to our cognitive, emotional, and physical development. As soon as we are born, our families begin to transmit cultural and social values to us. As we will discuss in Chapter 6, families in Canada vary in size and structure. Some families consist of two parents and their biological children, while others consist of a single parent and one or more children. Still other families reflect changing patterns of divorce and remarriage, and an increasing number are made up of same-sex partners and their children. Over time, patterns have changed in some two-parent families so that fathers, rather than mothers, are the primary daytime agents of socialization for their young children.

Theorists using a functionalist perspective emphasize that families serve important functions in society because they are the primary focus for the procreation and socialization of children. Most of us form an emerging sense of self and acquire most of our beliefs and values within the family context (see Box 4.2 for a discussion on the effects of excessive parental praise on children). We also learn about the larger dominant culture (including language, attitudes, beliefs, values, and norms) and the primary subcultures to which our parents and other relatives belong.

Families are also the primary source of emotional support. Ideally, people receive love, understanding, security, acceptance, intimacy, and companionship within families. The role of the family is especially significant because young children have little social experience beyond its boundaries; they have no basis for comparison or for evaluating how their family treats them.

To a large extent, the family is where we acquire our specific social position in society. From birth, we are a part of the specific racial, ethnic, class, religious, and regional subcultural grouping of

BOX 4.2 POINT/COUNTERPOINT

The Issue of Excessive Praise

> I sometimes say that praise is fine "when praise is due." We get into the habit of praising when it isn't praise that is appropriate but encouragement. For example, we're always saying to young children: "Oh, what a beautiful picture," even when their pictures aren't necessarily beautiful. So why not really look at each picture? Maybe a child has painted a picture with many wonderful colors. Why don't we comment on that–on the reality of the picture? (Docia Zavitkovsky, quoted in *Scholastic Parent & Child*, 2007)

Child development specialist Docia Zavitkovsky points out that excessive praise or unearned compliments may be problematic for children because, rather than bolstering their self-esteem, such praise may increase a child's dependence on adults. As children increasingly rely on constant praise and on significant others to identify what is good or bad about their performance, they may not develop the ability to make meaningful judgments about what they have done. Excessively praising children may make it more difficult for them to develop a positive self-concept and visualize an accurate picture of what is expected of them as they grow into young adulthood.

Zavitkovsky's ideas remind us of the earlier sociological insights of sociologist George Herbert Mead, who described how children learn to take into account the expectations of the larger society and to balance the "I" (the subjective element of the self: the spontaneous and unique traits of each person) with the "me" (the objective element of the self: the internalized attitudes and demands of other members of society and the individual's awareness of those demands).

Does this mean that children should not be praised? Definitely not! As noted above, what children may need sometimes is not praise but *encouragement*. From this perspective, positive feedback can have a very important influence on a child's self-esteem because he or she can learn how to do a "good job" when engaging in a specific activity or accomplishing a task rather than simply being praised for any effort expended. Mead's concept of the generalized other makes us aware of the importance of other people's actions in how self-concept develops.

our family. Many parents socialize their children somewhat differently based on race, ethnicity, and class. Some families instruct their children about the unique racial–ethnic and/or cultural backgrounds of their parents and grandparents so that they will have a better appreciation of their heritage. Other families teach their children primarily about the dominant, mainstream culture in hopes that this will help their children get ahead in life.

In terms of class, some upper-class parents focus on teaching their children about the importance of wealth, power, and privilege; however, many downplay this aspect of life and want their children to make their own way in life, fearing that "spoiling them" will not be in their best interest. Middle-class parents have typically focused on academic achievement and the importance of hard work to achieve. However, with the recent economic recession, some optimism that the middle class once passed on to their children may have diminished. Parents in the working-class and lower-income levels often have to struggle to keep shelter overhead and food on the table, and they have little time to help their children learn about things that will help them succeed in school and life (Kendall, 2002, 2011). This problem contributes to and reinforces social inequality, and this is one reason why conflict theorists are concerned about the long-term effects of the socialization process.

However, we should note that socialization is a bidirectional process in which children and young people socialize their agents of socialization, including parents, teachers, and others, as well as receiving socialization from these important agents. **Reciprocal socialization** is the process by

Daycare centres have become important agents of socialization for increasing numbers of children. Today, approximately 54 percent of all Canadian preschool children are in daycare of one kind or another (Sinha, 2014).

▶ **reciprocal socialization** The process by which the feelings, thoughts, appearance, and behaviour of individuals who are undergoing socialization also have a direct influence on those agents of socialization who are attempting to influence them.

which the feelings, thoughts, appearance, and behaviour of individuals who are undergoing socialization also have a direct influence on those agents of socialization who are attempting to influence them. Examples of this process include parents whose preferences in music, hairstyles, and clothing are influenced by their children, and teachers whose choice of words ("like, you know," "LOL," and other slang terms) is similar to that of their students.

THE SCHOOL

As the amount of specialized technical and scientific knowledge has expanded rapidly and the amount of time children spend in educational settings has increased, schools continue to play an enormous role in the socialization of young people. For many people, the formal education process is an undertaking that lasts up to 20 years.

As the numbers of one-parent families and families in which both parents work outside the home has increased dramatically, the number of children in daycare and preschool programs has grown rapidly as well (Sinha, 2014). Generally, studies have found that daycare and preschool programs may have a positive effect on the overall socialization of children. These programs provide children with the opportunity to have frequent interactions with teachers and to learn how to build their language and literacy skills. High-quality programs also have a positive effect on the academic performance of children, particularly those from low-income families. Today, however, the cost of child-care programs has become a major concern for many families.

Although schools teach specific knowledge and skills, they also have a profound effect on children's self-image, beliefs, and values. As children enter school for the first time, they are evaluated and systematically compared with one another by the teacher. Staff keep a permanent, official record of each child's personal behaviour and academic activities. From a functionalist perspective, schools are responsible for (1) socialization, or teaching students to be productive members of society; (2) transmission of culture; (3) social control and personal development; and (4) the selection, training, and placement of individuals on different rungs in the society (Ballantine and Hammack, 2012).

In contrast, conflict theorists assert that students have different experiences in the school system depending on their social class, their ethnic background, the neighbourhood in which they live, their gender, and other factors. As a result, much of what happens in school amounts to teaching a *hidden curriculum* in which children learn to value competition, materialism, work over play, obedience to authority, and attentiveness. Thus, schools do not socialize children for their own well-being but rather for their later roles in the workforce, where it is important to be well-behaved and "know your place." Students who are destined for leadership or elite positions acquire different skills and knowledge than those who will enter working-class and middle-class occupations (Davies & Guppy, 2013).

PEER GROUPS

▶ **peer group** A group of people who are linked by common interests, equal social position, and (usually) similar age.

As soon as we are old enough to have acquaintances outside the home, most of us begin to rely heavily on peer groups as a source of information and approval about social behaviour. A **peer group** is a group of people who are linked by common interests, equal social position, and (usually) similar age. In early childhood, peer groups are composed of classmates in daycare, preschool, and elementary school. Studies have found that pre-adolescence—the latter part of the elementary school years—is a time in which the children's peer culture has an important effect on how they perceive themselves and on how they internalize society's expectations (Robnett and Susskind, 2010). In adolescence, peer groups are typically composed of people with similar interests and social activities. As adults, we continue to participate in peer groups of people with whom we share common interests and comparable occupations, income, and/or social position.

Peer groups function as agents of socialization by contributing to our sense of belonging and our feelings of self-worth. Unlike families and schools, peer groups provide children and adolescents with some degree of freedom from parents and other authority figures (Corsaro, 2011). Peer groups also teach and reinforce cultural norms while providing important information about "acceptable" behaviour. As a result, the peer group is both a product of culture and one of its major transmitters.

Is there such a thing as "peer pressure"? Individuals must earn their acceptance with their peers by conforming to a given group's norms, attitudes, speech patterns, and dress codes. When we conform to our peer group's expectations, we are rewarded; if we do not conform, we may be ridiculed or even expelled from the group. Conforming to the demands of peers frequently places children and adolescents at cross-purposes with their parents. For example, children are frequently under pressure to obtain certain valued material possessions (such as clothing, sporting goods or athletic shoes); they then pass this pressure on to their parents through emotional pleas to purchase the desired items.

Image Source/Getty Images

Texting, social networking, and using smartphones now provide us with instant access to friends, information, and entertainment around the clock. How does this compare to the socialization process when your parents or grandparents were children?

MASS MEDIA

An agent of socialization that has a profound impact on both children and adults is the mass media, composed of large scale organizations that use print and electronic means (such as radio, television, film, and the Internet to communicate with large numbers of people (see Chapter 18). Today, the term *media* also includes the many forms of Web-based and mobile technologies that we refer to as "social media," such as Facebook, Snapchat, Twitter, and YouTube. For many years, the media have functioned as socializing agents in several ways: (1) They inform us about events; (2) they introduce us to a wide variety of people; (3) they provide an array of viewpoints on current issues; (4) they make us aware of products and services that, if we purchase them, supposedly will help us to be accepted by others; and (5) they entertain us by providing the opportunity to live vicariously through other people's experiences. Although most of us take for granted that the media play an important part in contemporary socialization, we frequently underestimate the enormous influence this agent of socialization may have on children's attitudes and behaviour.

The use of social media, such as Facebook and Instagram and Twitter, has grown exponentially in recent years. In a recent survey of students from Grade 4 to Grade 11, 99 percent reported that they used the Internet on a regular basis (Johnson, 2013). Although boys are more likely to play video games, girls lead the charge in the teen blogosphere. Social networking has been added as an additional layer on top of existing layers of media use, particularly among young people.

Consider children, for example. Research has shown that daily media use among Canadian children between the ages of 8 and 18 has increased dramatically. Studies indicate that children in this age range spend an average of 7 hours each day (which adds up to approximately 50 hours a week) using entertainment media. Cellphones, iPods, and other mobile media make it possible for young people to have access to media 24 hours per day, 7 days per week, with little time for other influences or activities in their life (Purcell, 2012; Johnson, 2013). All of this adds up to thousands of hours per year where children are interacting with these media influences; by contrast, Canadian children spend about 1200 hours per year in school. This means that the average 16-year-old will have spent more time in front of a television or computer than attending school.

Parents, educators, social scientists, and public officials have widely debated the consequences of young people watching violence on television. In addition to concerns about violence in television programming, motion pictures, and electronic games, television shows have been criticized for projecting negative images of women and visible minorities. Although the mass media have changed some of the roles that they depict women as playing, some newer characters tend to reinforce existing stereotypes of women as sex objects, even when they are in professional roles such as doctors or lawyers. Often, women are shown in traditional roles as homemakers or caretakers who do not receive respect from other family members unless they demand it (Collins, 2011). What effect does this have on children and young people as they develop their own ideas about the "adult world"? Throughout this text, we will look at additional examples of how the media socialize us in ways that we may or may not realize.

TIME TO REVIEW

- Which of the primary socialization agents have the most significant influence on youth between the ages of 13 and 18? Why?

GENDER SOCIALIZATION

▶ **gender socialization** The aspect of socialization that contains specific messages and practices concerning the nature of being female or male in a specific group or society.

Gender socialization is the aspect of socialization that contains specific messages and practices concerning the nature of being female or male in a specific group or society. Through the process of gender socialization, we learn about what attitudes and behaviours are considered to be appropriate for girls and boys, men and women, in a particular society. Different sets of gender norms are appropriate for females and males in Canada and most other nations.

One of the primary agents of gender socialization is the family. In some families, this process begins even before the child's birth. Parents who learn the sex of the fetus through ultrasound or amniocentesis often purchase colour-coded and gender-typed clothes, toys, and nursery decorations in anticipation of their daughter's or son's arrival. After birth, parents may respond differently toward male and female infants; they often play more roughly with boys and talk more lovingly to girls. Throughout childhood and adolescence, boys and girls are typically assigned different household chores and given different privileges, such as boys being given more latitude to play farther away from home than girls and being allowed to stay out later at night (Lareau, 2011).

Like the family, schools, peer groups, and the media also contribute to our gender socialization. From kindergarten through university, teachers and peers reward gender-appropriate attitudes and behaviour. Sports reinforce traditional gender roles through a rigid division of events into male and female categories. The media are also a powerful source of gender socialization; starting very early in childhood, children's books, television programs, movies, and music provide subtle and not-so-subtle messages about how boys and girls should act. (For a more detailed discussion on gender socialization, see Chapter 12.)

LO-4 SOCIOLOGICAL THEORIES OF HUMAN DEVELOPMENT

Although social scientists acknowledge the contributions of social–psychological explanations of human development, sociologists believe that it is important to bring a sociological perspective to bear on how people develop an awareness of self and learn about the culture in which they live. Let's look at symbolic interactionist, functional, and conflict approaches to describing the socialization process and its outcomes.

SYMBOLIC INTERACTIONIST PERSPECTIVES

According to a symbolic interactionist approach to socialization, we cannot form a sense of self or personal identity without intense social contact with others. How do we develop ideas about who we are? How do we gain a sense of self? The self represents the sum total of perceptions and feelings that an individual has of being a distinct, unique person—a sense of who and what one is. When we speak of the "self," we typically use words such as *I, me, my, mine,* and *myself* (Cooley, 1998/1902). This sense of self (also referred to as *self-concept*) is not present at birth; it arises in the process of social experience. **Self-concept** is the totality of our beliefs and feelings about ourselves. Four components make up our self-concept: (1) the physical self ("I am tall"), (2) the active self ("I am good at soccer"), (3) the social self ("I am nice to others"), and (4) the psychological self ("I believe in world peace"). Between early and late childhood, a child's focus tends to shift from the physical and active dimensions of self toward the social and psychological aspects. Self-concept is the foundation for communication with others; it continues to develop and change throughout our lives.

▶ **self-concept** The totality of our beliefs and feelings about ourselves.

Our *self-identity* is our perception about what kind of person we are and our awareness of our unique identity. Self-identity emerges when we ask the question, "Who am I?" Factors such as individuality, uniqueness, and personal characteristics and personality are components of self-identity. As we have seen, socially isolated children do not have typical self-identities because they have had no experience of "humanness." According to symbolic interactionists, we do not know who we are until we see ourselves as we believe that others see us. The perspectives of symbolic interactionists Charles Horton Cooley and George Herbert Mead help us understand how our self-identity is developed through our interactions with others.

COOLEY AND THE LOOKING-GLASS SELF Charles Horton Cooley (1864–1929) was one of the first sociologists to describe how we learn about ourselves through social interaction with other people. Cooley used the concept of the *looking-glass self* to describe how the self emerges. The **looking-glass self** refers to the way in which a person's sense of self is derived from the perceptions of others. Our looking-glass self is based on our perception of *how* other people think of us (Cooley, 1998/1902). As Figure 4.2 shows, the looking-glass self is a self-concept derived from a three-step process.

▶ **looking-glass self** Charles Horton Cooley's term for the way in which a person's sense of self is derived from the perceptions of others.

Since the looking-glass self is based on how we *imagine* other people view us, we may develop self-concepts based on an inaccurate perception of what other individuals think about us. Consider, for example, the individual who believes that other people see him or her as "fat" when, in actuality, he or she is a person of an average height, weight, and build. The consequences of such a false perception may lead to excessive dieting or health problems such as anorexia, bulimia, and other eating disorders.

Mead and role-taking George Herbert Mead (1863–1931) extended Cooley's insights by linking the idea of self-concept to **role-taking**—the process by which a person mentally assumes the role of another person in order to understand the world from that person's point of view. Role-taking often occurs through play and games, as children try out different roles (such as being mommy, daddy, doctor, or teacher) and gain an appreciation of them. By taking the roles of others, the individual hopes to ascertain the intention or direction of the acts of others. Then, the person begins to construct his or her own roles (role-making) and to anticipate other individuals' responses. Finally, the person plays at her or his particular role (role-playing) (Marshall, 1998).

▶ **role-taking** The process by which a person mentally assumes the role of another person in order to understand the world from that person's point of view.

According to Mead (1962/1934), in the early months of life, children do not realize that they are separate from others. However, they do begin early on to see a mirrored image of themselves in others. Shortly after birth, infants start to notice the faces of those around them, especially the significant others, whose faces begin to have meaning because they are associated with

FIGURE 4.2 HOW THE LOOKING-GLASS SELF WORKS

1. We imagine how our personality and appearance will look to other people.
2. We imagine how other people judge the appearance and personality that we think we present.
3. We develop a self-concept. If we think the evaluation of others is favourable, our self-concept is enhanced. If we think the evaluation is unfavourable, our self-concept is diminished. (Cooley, 1998/1902)

RubberBall Productions/Brand X Pictures/Getty Images

▶ **significant others** Those persons whose care, affection, and approval are especially desired and who are most important in the development of the self.

experiences, such as feeding and cuddling. **Significant others** are those persons whose care, affection, and approval are especially desired and who are most important in the development of the self. Gradually, we distinguish ourselves from our caregivers and begin to perceive ourselves in contrast to them. As we develop language skills and learn to understand symbols, we begin to develop a self-concept. When we can represent ourselves in our own minds as objects distinct from everything else, our self has been formed.

Mead divided the self into the "I" and the "me." The "I" is the subjective element of the self that represents the spontaneous and unique traits of each person. The "me" is the objective element of the self, which is composed of the internalized attitudes and demands of other members of society and the individual's awareness of those demands. Both the "I" and the "me" are needed to form the social self. The unity of the two constitutes the full development of the individual. According to Mead, the "I" develops first and the "me" takes form during the three stages of self development:

1. During the *preparatory stage,* up to about age three, interactions lack meaning and children largely imitate the people around them. At this stage, children are preparing for role-taking.
2. In the *play stage,* from about age three to five, children learn to use language and other symbols, which enable them to pretend to take the roles of specific people. At this stage, children begin to see themselves in relation to others but do not see role-taking as something that they have to do.

© Walter Hodges/CORBIS

© Rubberball Productions/Index Stock Imagery

muzsy/Shutterstock.com

According to Mead the self develops through three stages. In the preparatory stage, children imitate others; in the play stage, children pretend to take the roles of specific people; and in the game stage, children become aware of the "rules of the game" and the expectations of others.

3. During the *game stage*, which begins in the early school years, children understand not only their own social position but also the positions of others around them. In contrast to play, games are structured by rules, are often competitive, and involve a number of other "players." At this time, children become concerned about the demands and expectations of others and of the larger society.

Mead's concept of the **generalized other** refers to the child's awareness of the demands and expectations of the society as a whole or of the child's subculture. According to Mead, the generalized other is evident when a person takes into account other people and groups when he or she speaks or acts. In sum, both the "I" and the "me" are needed to form the social self. The unity of the two (the "generalized other") constitutes the full development of the individual and a more thorough understanding of the social world.

generalized other George Herbert Mead's term for the child's awareness of the demands and expectations of the society as a whole or of the child's subculture.

MORE-RECENT SYMBOLIC INTERACTIONIST PERSPECTIVES The symbolic interactionist approach emphasizes that socialization is a collective process in which children are active and creative agents, not just passive recipients of the socialization process. From this view, childhood is a *socially constructed* category. As children acquire language skills and interact with other people, they begin to construct their own shared meanings. Sociologist William A. Corsaro (2011) refers to this as the "orb web model," whereby the cultural knowledge that children possess consists not only of beliefs found in the adult world but also of unique interpretations from the children's own peer culture. According to Corsaro, children create and share their own *peer culture,* which is an established set of activities, routines, and beliefs that are in some ways different from adult culture. This peer culture emerges through interactions as children "borrow" from the adult culture but transform it so that it fits their own situation. In fact, according to Corsaro, peer culture is the most significant arena in which children and young people acquire cultural knowledge.

According to sociologist George Herbert Mead, the self develops through three stages. In the preparatory stage, children imitate others; in the play stage, children pretend to take the roles of specific people; and in the game stage, children become aware of the "rules of the game" and the expectations of others.

LO-5 SOCIAL PSYCHOLOGICAL THEORIES OF HUMAN DEVELOPMENT

Up to this point, we have discussed sociologically oriented theories; we now turn to psychological theories that have influenced our understanding of how the individual personality develops. Although these are not sociological theories, it is important to be aware of the contributions of Freud, Piaget, Kohlberg, and Gilligan, because knowing about them provides us with a framework for comparing various perspectives on human development.

FREUD AND THE PSYCHOANALYTIC PERSPECTIVE

The basic assumption in Sigmund Freud's (1924) psychoanalytic approach is that human behaviour and personality originate from unconscious forces within individuals. Freud (1856–1939), who is known as the founder of psychoanalytic theory, developed his major theories in the Victorian era, when biological explanations of human behaviour were prevalent. For example, Freud based his ideas on the belief that people have two basic tendencies: the urge to survive and the urge to procreate.

According to Freud (1924), human development occurs in three states that reflect different levels of the personality, which he referred to as the *id*, *ego*, and *superego*. The **id** is the component of personality that includes all of the individual's basic biological drives and needs that demand immediate gratification. For Freud, the newborn child's personality is all id, and from birth the child finds that urges for self-gratification—such as wanting to be held, fed, or changed—are not going to be satisfied immediately. However, id remains with people throughout their life in the form of *psychic energy*, the urges and desires that account for behaviour.

▶ **id** Sigmund Freud's term for the component of personality that includes all of the individual's basic biological drives and needs that demand immediate gratification.

▶ **ego** According to Sigmund Freud, the rational, reality-oriented component of personality that imposes restrictions on the innate pleasure-seeking drives of the id.

▶ **superego** Sigmund Freud's term for the human conscience, consisting of the moral and ethical aspects of personality.

By contrast, the second level of personality—the *ego*—develops as infants discover that their most basic desires are not always going to be met by others. The **ego** is the rational, reality-oriented component of personality that imposes restrictions on the innate pleasure-seeking drives of the id. The ego channels the desire of the id for immediate gratification into the most advantageous direction for the individual. The third level of personality—the *superego*—is in opposition to both the id and the ego. The **superego**, or conscience, consists of the moral and ethical aspects of personality. It is first expressed as the recognition of parental control and eventually matures as the child learns that parental control is a reflection of the values and moral demands of the larger society. When a person is well adjusted, the ego successfully manages the opposing forces of the id and the superego. Figure 4.3 illustrates Freud's theory of personality.

Although subject to harsh criticism, Freud's theories made people aware of the significance of early childhood experiences, including abuse and neglect. His theories have also had a profound influence on contemporary mental health practitioners and on other human development theories.

PIAGET AND COGNITIVE DEVELOPMENT

Jean Piaget (1896–1980), a Swiss psychologist, was a pioneer in the field of cognitive (intellectual) development. Cognitive theorists are interested in how people obtain, process, and use information—that is, in how we think. Cognitive development relates to changes over time in how we think.

Piaget (1954) believed that in each stage of development (from birth through adolescence), children's activities are governed by their perception of the world around them. His four stages of cognitive development are organized around specific tasks that, when mastered, lead to the acquisition of new mental capacities, which then serve as the basis for the next level of development. Piaget emphasized that all children must go through each stage in sequence before moving on to the next one, although some children move through them faster than others.

1. *Sensorimotor stage* (birth to age two). During this period, children understand the world only through sensory contact and immediate action because they cannot engage in symbolic thought or use language. Toward the end of the second year, children comprehend *object permanence;* in other words, they start to realize that objects continue to exist even when the items are out of sight.
2. *Preoperational stage* (age two to seven). In this stage, children begin to use words as mental symbols and to form mental images. However, they are still limited in their ability to use

FIGURE 4.3 FREUD'S THEORY OF PERSONALITY

This illustration shows how Freud might picture a person's internal conflict over whether to commit an antisocial act, such as stealing a candy bar. In addition to dividing personality into three components, Freud theorized that our personalities are largely unconscious–hidden away outside our normal awareness. To dramatize his point, Freud compared conscious awareness (portions of the ego and superego) to the visible tip of an iceberg. Most of personality–including all of the id, with its raw desires and impulses–lies in the subconscious.

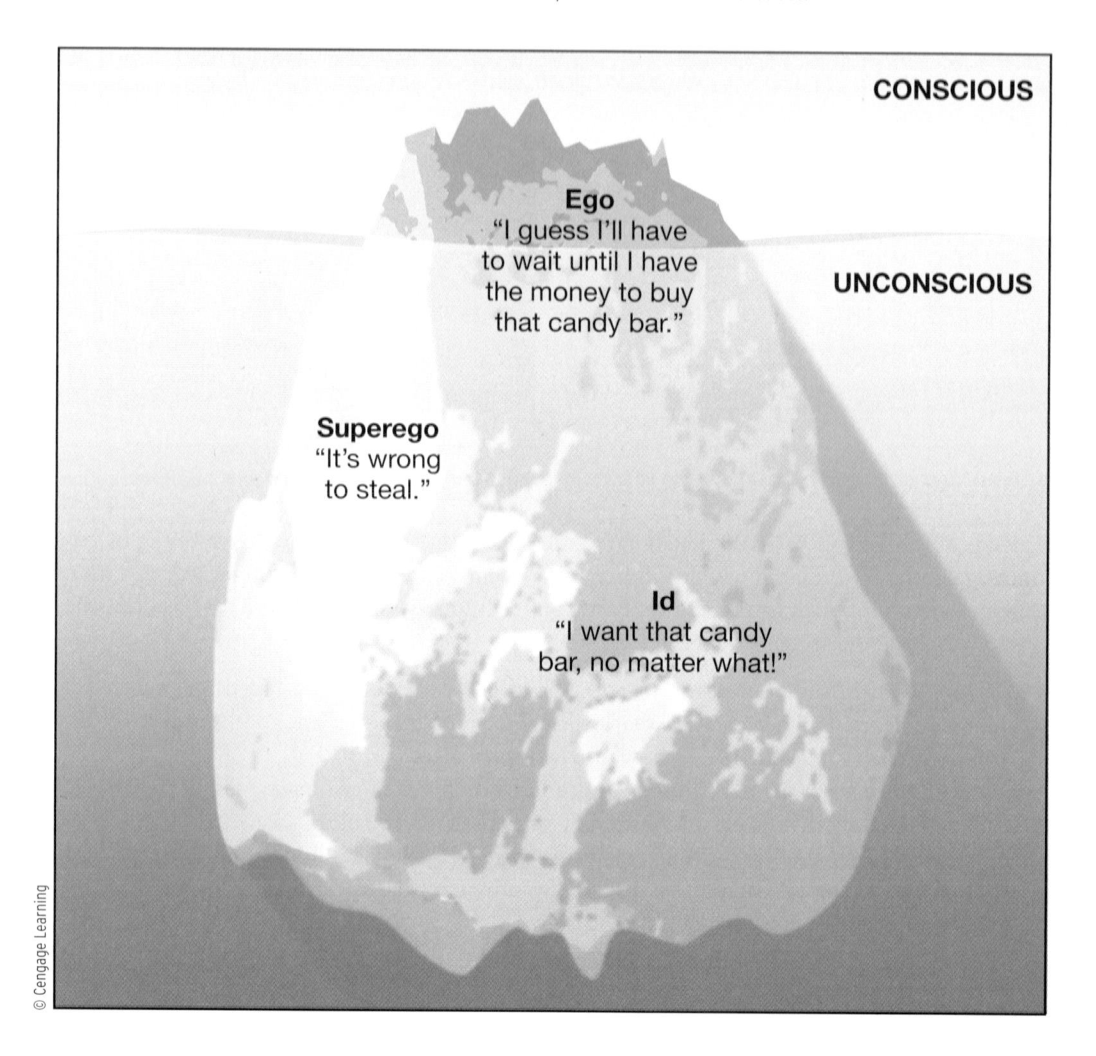

logic to solve problems or to realize that physical objects may change in shape or appearance while still retaining their physical properties (see Figure 4.4).

3. *Concrete operational stage* (age seven to eleven). During this stage, children think in terms of tangible objects and actual events. They can draw conclusions about the likely physical consequences of an action without always having to try out the action. Children begin to take the role of others and start to empathize with the viewpoints of others.
4. *Formal operational stage* (age twelve through adolescence). By this stage, adolescents are able to engage in highly abstract thought and understand places, things, and events they have never seen. They can think about the future and evaluate different options or courses of action.

FIGURE 4.4 THE PREOPERATIONAL STAGE

Psychologist Jean Piaget identified four stages of cognitive development, including the preoperational stage, in which children have limited ability to realize that physical objects may change in shape or appearance. Piaget showed children two identical beakers filled with the same amount of water. After the children agreed that both beakers held the same amount of water, Piaget poured the water from one beaker into a taller, narrower beaker and then asked them about the amounts of water in each beaker. Those still in the preoperational stage believed that the taller beaker held more water because the water line was higher than in the shorter, wider beaker.

Tony Freeman/PhotoEdit

KOHLBERG AND THE STAGES OF MORAL DEVELOPMENT

Lawrence Kohlberg (1927–1987) elaborated on Piaget's theories of cognitive reasoning by conducting a series of studies in which children, adolescents, and adults were presented with moral dilemmas that took the form of stories. Based on his findings, Kohlberg (1969, 1981) classified moral reasoning into three sequential levels:

1. *Preconventional level* (age seven to ten). Children's perceptions are based on punishment and obedience. Evil behaviour is that which is likely to be punished; good conduct is based on obedience and avoidance of unwanted consequences.
2. *Conventional level* (age ten through adulthood). People are most concerned with how they are perceived by their peers and with how one conforms to rules.
3. *Postconventional level* (few adults reach this stage). People view morality in terms of individual rights; "moral conduct" is judged by principles based on human rights that transcend government and laws.

GILLIGAN'S VIEW ON GENDER AND MORAL DEVELOPMENT

Psychologist Carol Gilligan (b. 1936) noted that both Piaget and Kohlberg did not take into account how gender affects the process of social and moral development. According to Gilligan (1982), Kohlberg's model was developed solely on the basis of research with male respondents, who often have different views from women on morality. Gilligan believes that men become more concerned with law and order, but that women tend to analyze social relationships and the social consequences of behaviour. Gilligan argues that men are more likely to use *abstract standards* of right and wrong when making moral decisions, whereas women are more likely to be concerned about the *consequences* of behaviour. Does this constitute a "moral deficiency" on the part of either women or men? Not according to Gilligan, who believes that people make moral decisions according to both abstract principles of justice and principles of compassion and care.

Although the sociological and psychological perspectives we have examined have often been based on different assumptions and reached somewhat different conclusions, an important theme emerges from these models of cognitive and moral development—through the process of socialization, people learn how to take into account other people's perspectives.

LO-6 SOCIALIZATION THROUGH THE LIFE COURSE

Why is socialization a lifelong process? Throughout our lives, we continue to learn. Each time we experience a change in status (such as becoming a university student or getting married), we learn a new set of rules, roles, and relationships. Even before we achieve a new status, we often participate in **anticipatory socialization**—the process by which knowledge and skills are learned for future roles. Many societies organize social experience according to age. Some have distinct *rites of passage,* based on age or other factors that publicly dramatize and validate changes in a person's status. In Canada and other industrialized societies, the most common categories of age are infancy, childhood, adolescence, and adulthood (often subdivided into young adulthood, middle adulthood, and older adulthood). See the Census Profile, which shows three categories as revised by Statistics Canada for the 2006 Census.

anticipatory socialization The process by which knowledge and skills are learned for future roles.

CENSUS PROFILE

Age of the Canadian Population

Just as age is a crucial variable in the socialization process, Statistics Canada gathers data about people's age so that the government and other interested parties will know how many individuals residing in this country are in different age categories. This chapter examines how a person's age is related to socialization and life experiences. The table below shows Canada's senior population in the years 1951, 1981, 2011, and projected for 2061.

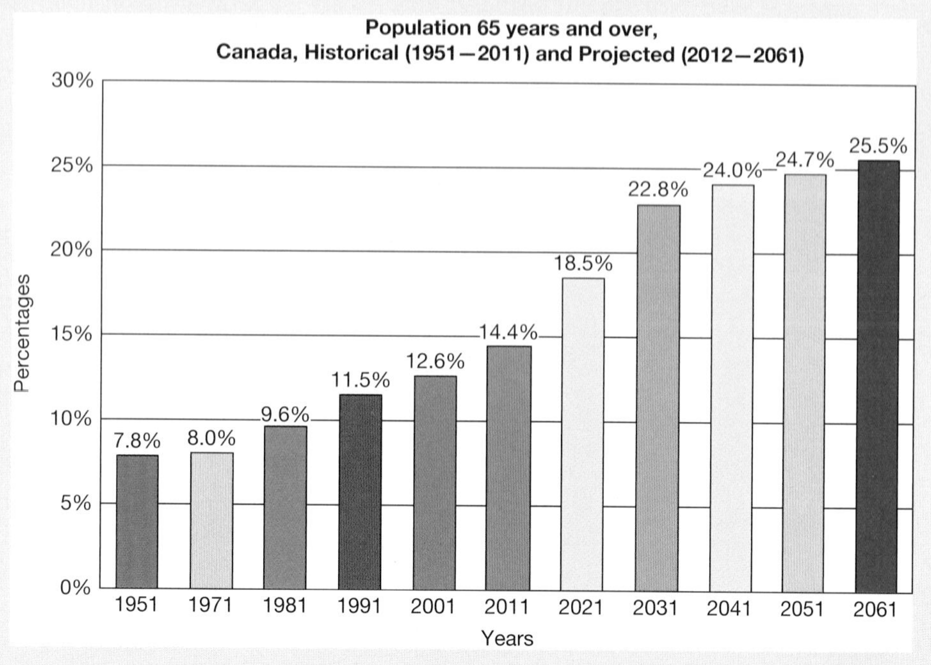

Note: Population projections use a medium-growth scenario (M1) based on interprovincial migration trends from 1981 to 2008. For further information see: Statistics Canada. *Population Projections for Canada, Provinces and Territories (2009–2036).* (Cat. No. 91-520 XIE).

Source: Statistics Canada. 2011c. HRSDC calculations based on Statistics Canada. Estimates of population, by age group and sex for July 1, Canada, provinces and territories, annual (CANSIM Table 051-0001); and Statistics Canada. Projected population, by projection, scenario, sex and age group as of July 1, Canada, provinces and territories, annual (CANSIM Table 052-0005). Ottawa: Statistics Canada, 2011.

Can age be a source of social cohesion among people? Why might age differences produce conflict among individuals in difference age groups? What do you think?

CHILDHOOD

Some social scientists believe that a child's sense of self is formed at a very early age and that it is difficult to change this sense later in life. Interactionists emphasize that during infancy and early childhood, family support and guidance are crucial to a child's developing self-concept. In some families, children are provided with emotional warmth, feelings of mutual trust, and a sense of security. These families come closer to our ideal cultural belief that childhood should be a time of carefree play, safety, and freedom from economic, political, and sexual responsibilities. However, other families reflect the discrepancy between cultural ideals and reality—children grow up in a setting characterized by fear, danger, and risks that are created by parental neglect, emotional abuse, or premature economic and sexual demands Abused children often experience low self-esteem, an inability to trust others, feelings of isolation and powerlessness, and denial of their feelings (UNICEF, 2014.)

ADOLESCENCE

In industrialized societies, the adolescent (or teenage) years represent a buffer between childhood and adulthood. It is a time during which young people pursue their own routes to self-identity and adulthood. Anticipatory socialization is often associated with adolescence, with many young people spending time planning or being educated for future roles they hope to occupy. Although no specific rites of passage exist in Canada to mark *every* child's transition between childhood and adolescence or between adolescence and adulthood, some rites of passage are observed. For example, a celebration known as a Bar Mitzvah is held for some Jewish boys on their thirteenth birthday, and a Bat Mitzvah is held for some Jewish girls on their twelfth birthday; these events mark the occasion upon which young people accept moral responsibility for their own actions and the fact that they are now old enough to own personal property. Although it is not officially designated as a rite of passage, many of us think of the time when we get our first driver's license or graduate from high school as another way in which we mark the transition from one period of our life to the next.

Adolescence is often characterized by emotional and social unrest. In the process of developing their own identities, some young people come into conflict with parents, teachers, and other authority figures who attempt to restrict their freedom. Adolescents may also find themselves caught between the demands of adulthood and their own lack of financial independence and experience in the job market.

The experiences of individuals during adolescence vary according to their ethnicity, class, and gender. Based on their family's economic situation and their own personal choices, some young people leave high school and move directly into the world of work, while others pursue a university education and may continue to receive advice and financial support from their parents. Others are involved in both the world of work and the world of higher education as they seek to support themselves and acquire more years of formal education or vocational/career training.

An important rite of passage for many Jewish Canadians is the bar mitzvah or bat mitzvah—a celebration of the adolescent's passage into manhood or womanhood. Can you see how this might be a form of anticipatory socialization?

ADULTHOOD

One of the major differences between child and adult socialization is the degree of freedom of choice. If young adults are able to support themselves financially, they gain the ability to make more choices about their own lives. In early adulthood (usually until about age 40), people work toward their own goals of creating meaningful relationships with others, finding employment, and seeking personal fulfillment. Of course, young adults continue to be socialized by their parents, teachers, peers, and the media, but they also learn new attitudes and behaviours. For example, when we marry or have children, we learn new roles as partners or parents.

Workplace (occupational) socialization is one of the most important types of early adult socialization. This type of socialization tends to be most intense immediately after a person makes the transition from school to the workplace; however, many people experience continuous workplace socialization as a result of having more than one career in their lifetime.

Between the ages of 40 and 60, people enter middle adulthood, and many begin to compare their accomplishments with their earlier expectations. At this point, people either decide that they have reached their goals or recognize that they have attained as much as they are likely to achieve.

Late adulthood may be divided into three categories: (1) the "young-old" (ages sixty-five to seventy-four), (2) the "old-old" (ages seventy-five to eighty-five), and (3) the "oldest-old" (over age eighty-five). Although these are somewhat arbitrary divisions, the "young-old" are less likely to suffer from disabling illnesses, whereas some of the "old-old" are more likely to suffer such illnesses. Increasingly, studies in gerontology and the sociology of medicine have come to question these arbitrary categories and show that some persons defy the expectations of their age grouping based on individual genetic makeup, lifestyle choices, and a zest for living. Perhaps "old age" is what we make it!

In older adulthood, some people are quite happy and content; others are not. Erik Erikson noted that difficult changes in adult attitudes and behaviour occur in the last years of life, when people experience decreased physical ability, lower prestige, and the prospect of death. Older adults in industrialized societies have experienced **social devaluation**, wherein a person or group is considered to have less social value than other individuals or groups. Social devaluation is especially acute when people are leaving roles that have defined their sense of social identity and provided them with meaningful activity.

Negative images regarding older persons reinforce **ageism**—prejudice and discrimination against people on the basis of age, particularly against older persons. Ageism is reinforced by stereotypes, whereby people have narrow, fixed images of certain groups. Older persons are often stereotyped as thinking and moving slowly; as being bound to themselves and their past, unable to change and grow; as being unable to move forward and often moving backward.

Many older people buffer themselves against ageism by continuing to view themselves as being in middle adulthood long after their actual chronological age would suggest otherwise. Other people begin a process of resocialization to redefine their own identity as mature adults.

▶ **social devaluation** A situation in which a person or group is considered to have less social value than other individuals or groups.

▶ **ageism** Prejudice and discrimination against people on the basis of age, particularly against older persons.

RESOCIALIZATION

Resocialization is the process of learning a new and different set of attitudes, values, and behaviours from those in one's previous background and experience. It may be voluntary or involuntary. In either case, people undergo changes that are much more rapid and pervasive than the gradual adaptations that socialization usually involves. For many new parents, the process of resocialization involved in parenting is the most dramatic they will experience in their lifetimes.

▶ **resocialization** The process of learning a new and different set of attitudes, values, and behaviours from those in one's previous background and experience.

VOLUNTARY RESOCIALIZATION

Resocialization is voluntary when we assume a new status (such as becoming a university student, an employee, or a retiree) of our own free will (see Figure 4.5). Sometimes, voluntary resocialization involves medical or psychological treatment or religious conversion, in which case the person's existing attitudes, beliefs, and behaviours must undergo

FIGURE 4.5 : TIMELINE FOR FIRST-TERM UNIVERSITY RESOCIALIZATION

EARLY FALL

Joy Brown/Shutterstock.com

- Adapting to new people and new situations
- Anticipation and excitement about studying in a new setting
- Insecurity about academic demands
- Homesickness
- If employed, trying to balance school and work life

MID FALL

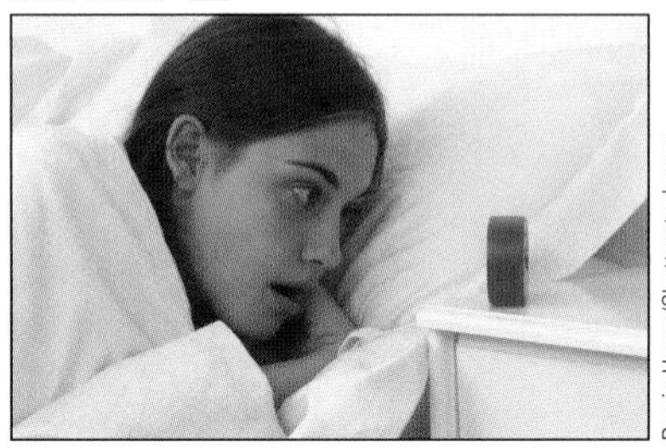

Darrin Henry/Shutterstock.com

- Social pressures from others: What would my parents think?
- Anticipation (and dread) of midterm exams and major papers
- Time-management problems between school and social life
- Intense need for a break
- Concerns about role conflict between school and work

LATE FALL

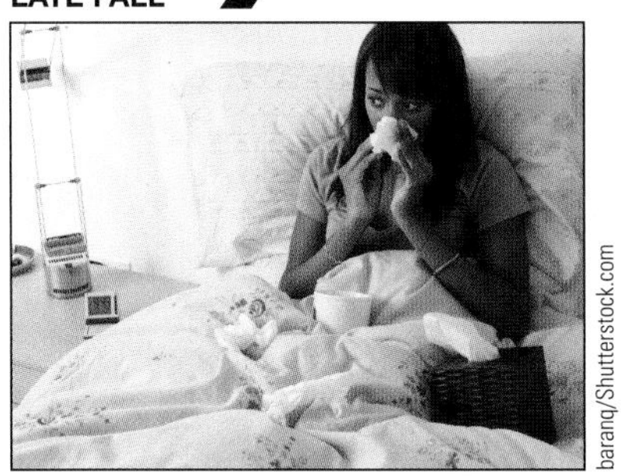

barang/Shutterstock.com

- Positive or negative assessment of grades so far
- Pre-final studying and jitters
- Making up for lost time and procrastination
- First university illnesses likely to occur because of late hours, poor eating habits, and proximity to others who become ill
- Potential problems with roommates or others who make excessive demands on one's time and/or personal space

END OF TERM

fizkes/Shutterstock.com

- Final exams: late nights, extra effort, and stress
- Concerns about leaving new friends and university setting for winter break
- Anticipation (and tension) associated with going home for break for those who have been away
- Reassessment of university choice, major, and career options: Am I on the right track?
- Acknowledgment that growth has occurred and much has been learned, both academically and otherwise, during the first university term

Sources: Based on the author's observations of student life and on Kansas State University, 2010. © Cengage Learning.

strenuous modification to a new regime and a new way of life. For example, resocialization for adult survivors of emotional or physical child abuse includes extensive therapy to form new patterns of thinking and action, somewhat like Alcoholics Anonymous and its 12-step program that has become the basis for many other programs dealing with addictive behaviour (Parrish, 1990).

INVOLUNTARY RESOCIALIZATION

Involuntary resocialization occurs against a person's wishes and generally takes place within a **total institution**—a place where people are isolated from the rest of society for a set period of time and come under the control of the officials who run the institution (Goffman, 1961a). Military boot camps, jails and prisons, concentration camps, and some mental hospitals are total institutions. In these settings, people are totally stripped of their former selves—or depersonalized—through a *degradation ceremony* (Goffman, 1961a). Inmates entering prison, for example, are required to strip, shower, and wear assigned institutional clothing. In the process, they are searched, weighed, fingerprinted, photographed, and given no privacy even in showers and restrooms. Their official identification becomes not a name but a number. In this abrupt break from their former existence, they must leave behind their personal possessions and their family and friends. The depersonalization process continues as they are required to obey rigid rules and to conform to their new environment.

total institution Erving Goffman's term for a place where people are isolated from the rest of society for a set period of time and come under the control of the officials who run the institution.

After stripping people of their former identities, the institution attempts to build a more compliant person. A system of rewards and punishments (such as providing or withholding cigarettes and television or exercise privileges) encourages conformity to institutional norms. Some individuals may be rehabilitated; others become angry and hostile toward the system that has taken away their freedom. Although the assumed purpose of involuntary resocialization is to reform persons so that they will conform to societal standards of conduct after their release, the ability of total institutions to modify offenders' behaviour in a meaningful way has been widely questioned. In many prisons, for example, inmates may conform to the norms of the prison or of other inmates, but little relationship exists between those norms and the laws of society.

CONCEPT SNAPSHOT

SYMBOLIC INTERACTIONIST PERSPECTIVES

Key thinkers: Charles Horton Cooley, George Herbert Mead, William A. Corsaro

According to Cooley, our sense of self is based on how others perceive and treat us. Mead extended Cooley's insights by linking the idea of self-concept to role-playing. According to Mead, our self-concept is developed through role-playing and learning the rules of social interaction through others. More recently, Corsaro developed the "orb web model," arguing that children's socialization reflects not only knowledge from the adult world but also the unique interpretations of children's peer culture.

SOCIAL PSYCHOLOGICAL PERSPECTIVES

Key thinkers: Sigmund Freud

According to Freud, the founder of psychoanalytic theory, the self is comprised of three interrelated components: id, ego, and superego. When a person is well adjusted, the three forces are in balance.

COGNITIVE MORAL PERSPECTIVES

Key thinkers: Jean Piaget, Lawrence Kohlberg, Carol Gilligan

According to Swiss psychologist Jean Piaget, from birth through adolescence, children move through four states of cognitive development, which are organized around acquisition and mastery of specific tasks. Kohlberg classified moral development into six stages. Certain levels of cognitive development must occur before moral reasoning can develop. Gilligan critiqued Kohlberg's research as male-centred and suggested that men and women have different views on morality based on differences in socialization.

VISUAL SUMMARY

4

KEY TERMS

ageism Prejudice and discrimination against people on the basis of age, particularly against older persons (p. 105).

agents of socialization The persons, groups, or institutions that teach us what we need to know to participate in society (p. 92).

anticipatory socialization The process by which knowledge and skills are learned for future roles (p. 103).

ego According to Sigmund Freud, the rational, reality-oriented component of personality that imposes restrictions on the innate pleasure-seeking drives of the id (p. 100).

gender socialization The aspect of socialization that contains specific messages and practices concerning the nature of being female or male in a specific group or society (p. 96).

generalized other George Herbert Mead's term for the child's awareness of the demands and expectations of the society as a whole or of the child's subculture (p. 99).

id Sigmund Freud's term for the component of personality that includes all of the individual's basic biological drives and needs that demand immediate gratification (p. 100).

looking-glass self Charles Horton Cooley's term for the way in which a person's sense of self is derived from the perceptions of others (p. 97).

peer group A group of people who are linked by common interests, equal social position, and (usually) similar age (p. 94).

LO-1 Discuss the degree to which our unique physical and human characteristics are based on heredity and to what degree they are based on our social environment.

As individual human beings, we have unique identities, personalities, and relationships with others. Individuals are born with some of their unique physical and human characteristics; other characteristics and traits are gained during the socialization process. Each of us is a product of two forces: (1) heredity, referred to as "nature," and (2) the social environment, referred to as "nurture." While biology dictates our physical makeup, the social environment largely determines how we develop and behave.

© Rubberball Productions/Index Stock Imagery

LO-2 Debate the extent to which people would become human beings without adequate socialization.

Socialization is crucial to healthy human development. Even nonhuman primates such as monkeys and chimpanzees need social contact with others of their species in order to develop properly. The tragic cases of two children deliberately raised in isolation—Anna and Genie—provide insights into the importance of positive socialization and the negative effects of social isolation.

Martin Rogers/Stone/Getty Images

LO-3 Identify the key agents of socialization.

The people, groups, and institutions that teach us what we need to know to participate in society are called agents of socialization. The agents include the family, schools, peer groups, the media, and the workplace. Families, which transmit cultural and social values to us, are the most important agents of socialization in all societies and have these roles: (1) procreating and socializing children, (2) providing emotional support, and (3) assigning social position. Schools are another key agent of socialization; they not only teach knowledge and skills but also deeply influence the self-image, beliefs, and values of children. Peer groups contribute to our sense of belonging and self-worth; they teach and reinforce cultural norms and are a key source of information about acceptable behaviour. The media function as socializing agents by (1) informing us about world events, (2) introducing us to a wide variety of people, and (3) providing an opportunity to live vicariously through other people's experiences.

© Gaetano/CORBIS

LO-4 Describe how sociologists explain our development of a self-concept.

Charles Horton Cooley developed the image of the looking-glass self to explain how people see themselves through the perceptions of others. Our initial sense of self is typically based on how families perceive and treat us. George Herbert Mead linked the idea of self-concept to role-playing and to learning the rules of social interaction. According to Mead, the self is divided into the "I" and the "me." The "I" represents the spontaneous and unique traits of each person. The "me" represents the internalized attitudes and demands of other members of society.

muzsy/Shutterstock.com

LO-5 Review the main social psychological theories on human development.

According to Sigmund Freud, the self emerges from three interrelated forces: id, ego, and superego. When a person is well adjusted, the three forces act in balance. Jean Piaget identified four cognitive stages of development; at each stage, children's activities are governed by how they understand the world around them. Lawrence Kohlberg classified moral development into six stages; certain levels of cognitive development are essential before corresponding levels of moral reasoning may occur. Carol Gilligan suggested that there are male–female differences regarding morality and identified three stages in female moral development.

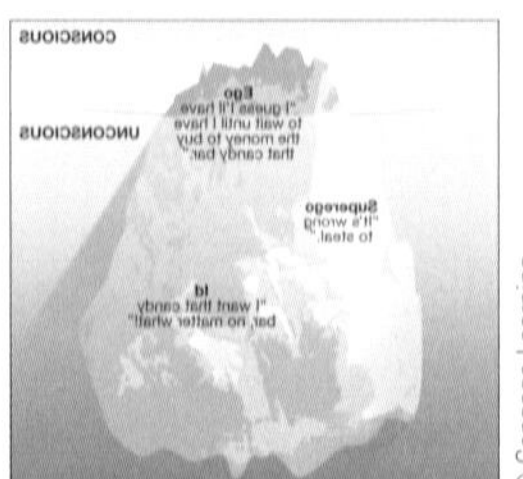

© Cengage Learning

LO-6 Describe socialization throughout various stages in the life course.

Socialization is ongoing throughout the life course. We learn knowledge and skills for future roles through anticipatory socialization. Parents are socialized by their own children, and adults learn through workplace socialization. Resocialization is the process of learning new attitudes, values, and behaviours, either voluntarily or involuntarily.

© PhotoStock-Israel/Alamy

reciprocal socialization The process by which the feelings, thoughts, appearance, and behaviour of individuals who are undergoing socialization also have a direct influence on those agents of socialization who are attempting to influence them (p. 93).

resocialization The process of learning a new and different set of attitudes, values, and behaviours from those in one's previous background and experience (p. 105).

role-taking The process by which a person mentally assumes the role of another person in order to understand the world from that person's point of view (p. 97).

self-concept The totality of our beliefs and feelings about ourselves (p. 97).

significant others Those persons whose care, affection, and approval are especially desired and who are most important in the development of the self (p. 98).

social devaluation A situation in which a person or group is considered to have less social value than other individuals or groups (p. 105).

socialization The lifelong process of social interaction through which individuals acquire a self-identity and the physical, mental, and social skills needed for survival in society (p. 88).

sociobiology The systematic study of how biology affects social behaviour (p. 89).

superego Sigmund Freud's term for the human conscience, consisting of the moral and ethical aspects of personality (p. 100).

total institution Erving Goffman's term for a place where people are isolated from the rest of society for a set period of time and come under the control of the officials who run the institution (p. 107).

KEY FIGURES

© Keystone Pictures USA/Alamy

Sigmund Freud (1856-1939) As the founder of psychoanalytic theory, Freud posited that the personality is comprised of the id, ego, and superego. Although subject to harsh criticism, Freud's analysis drew attention to the importance of early childhood experiences in healthy socialization.

The Granger Collection, New York

George Herbert Mead (1863-1931) Mead divided the self into the "I" (the subjective component of the self) and the "me" (the objective element of the self). According to Mead, the "I" develops first and the "me" develops during three stages.

American Sociological Association, asanet.org

Charles Horton Cooley (1864-1929) According to Cooley, how we see ourselves is based on how we think others see us. He referred to this as our looking-glass self.

APPLICATION QUESTIONS

1. Consider the concept of the looking-glass self. How do you think others perceive you? Do you think most people perceive you correctly? Why, or why not?
2. What are your "I" traits? What are your "me" traits? Which ones are stronger?
3. Is the attempted rehabilitation of a criminal offender—through boot camp programs, for example—a form of socialization or resocialization? Explain.
4. How might functionalist, conflict, symbolic interactionist, and postmodernist analysts view the role of television and computers in childhood socialization? What influence do you think television and computers have on your own socialization?

CHAPTER 5

SOCIETY, SOCIAL STRUCTURE, AND INTERACTION

Photographee.eu/Shutterstock.com

CHAPTER FOCUS QUESTION

How is homelessness related to the social structure of a society?

LEARNING OBJECTIVES

AFTER READING THIS CHAPTER, YOU SHOULD BE ABLE TO

LO-1 Identify the key components of social structure.

LO-2 Compare and contrast functionalist and conflict perspectives on social institutions.

LO-3 Explain how societies maintain stability in times of social change.

LO-4 Define and distinguish between *Gemeinschaft* and *Gesellschaft* societies.

LO-5 Understand Erving Goffman's dramaturgical perspective and the concepts of impression management, and front stage/back stage behaviours.

All activities in life—including scavenging in garbage bins (known as dumpster diving, or binning and living "on the streets" (a rough sleeper)—are social in nature. Here, a young woman recalls her experiences of trying to survive on the street in Winnipeg. Meeting basic needs such as food and shelter was described as a daily struggle:

To find food, we would do a couple of things. We would dumpster dive and sometimes we would know what time restaurants would be throwing out their food you know like even Tim Horton's when they're throwing out their donuts . . . Sometimes shelters, but not all the time because it depends on what city you're in cause you usually don't want to eat in soup kitchens. You would get food poisoning . . . because where they get their food from it's all like expired food like from Safeway or you know any grocery store . . . The soup kitchens would make slop.

I would eat once a day at 7-eleven. I think it was $2.50 or something to get a hot dog, a drink, and chips. People would come and you would ask them for spare change but you know they would think you would go and spend it on something stupid . . . so they would go and buy us like a whole bunch of food, drinks or whatever like at a hot dog stand or whatever. That was really nice you know. We would actually get a meal. (Wingert, Higgitt, & Ristock, 2005:66)

Source: p. 66 in Wingert, S., Higgitt, N., & Ristock, J. (2005). Voices from the Margins: Understanding Street Youth in Winnipeg. *Canadian Journal of Urban Research*, 14(1), 54–80.

These "survival" activities reflect a specific pattern of social behaviour. Homeless persons and domiciled persons (those with homes) live in social worlds that have predictable patterns of social interaction. In this chapter, we will look at the relationship between **social structure** and social interaction. Homelessness is used as an example of how social problems occur and may be perpetuated within social structures and patterns of interaction.

Let's start by defining social interaction and social structure. Although we frequently are not aware of it, our daily interactions with others and the larger patterns found in the social world of which we are a part are important ingredients in the framework of our individual daily lives. **Social interaction** is the process by which people act toward or respond to other people and is the foundation for all relationships and groups in society. As discussed in Chapter 4, we learn virtually all of what we know from our interactions with other people.

Socialization is a small-scale process, whereas social structure is a much more encompassing framework. Social structure is the complex framework of societal institutions (such as the economy, politics, and religion) and the social practices (such as rules and social roles) that make up a society and that organize and establish limits on people's behaviour.

This structure is essential for the survival of society and for the well-being of individuals because it provides a social web of familial support and social relationships that connects each of us to the larger society. Many homeless people have lost this vital linkage. As a result, they often experience a loss of personal dignity and sense of moral worth because of their "homeless" condition (Gaetz, 2014). Although there have always been homeless people, there has been a significant increase in the number of Canadians without homes. The homeless category now includes people who have never before had to depend on social assistance for food, clothing, and a roof over their head. Before reading on, take the quiz on homelessness in Box 5.1.

CRITICAL THINKING QUESTIONS

1. Sociologists suggest that all activities are social in nature. Do you agree?
2. Identify predictable patterns of social interaction with a homeless person. How do you interact with a homeless person you may encounter in your day-to-day life?
3. It was suggested above that homeless persons may not have stable familial and social relationships that are vital to well-being. What are the stable familial and social relationships in your life?

▶ **social structure**
The complex framework of societal institutions (such as the economy, politics, and religion) and the social practices (such as rules and social roles) that make up a society and that organize and establish limits on people's behaviour.

▶ **social interaction**
The process by which people act toward or respond to other people.

SOCIAL STRUCTURE: THE MACROLEVEL PERSPECTIVE

Social structure provides the framework within which we interact with others. This framework is an orderly, fixed arrangement of parts that together comprise the whole group or society (see Figure 5.1). As defined in Chapter 1, a *society* is a large social grouping that shares the same geographical territory and is subject to the same political authority and dominant cultural expectations. At the macrolevel, the social structure of a society has several essential elements: social institutions, groups, statuses, roles, and norms.

Functional theorists emphasize that social structure is essential because it creates order and predictability in a society. Social structure is also important for our human development. As discussed in Chapter 3, we develop a self-concept as we learn the attitudes, values, and behaviours of the people around us. When these attitudes and values are part of a predictable structure, it is easier to develop that self-concept.

Social structure gives us the ability to interpret the social situations we encounter. For example, we expect our families to care for us, our schools to educate us, and our police to protect us. When our circumstances change dramatically, most of us feel an acute sense of anxiety because we do not know what to expect or what is expected of us. For example, newly homeless individuals may feel disoriented because they do not know how to function in their new setting. The person is likely to ask questions: "How will I survive on the streets?" "Where do I go to get help?" "Should I stay at a shelter?" "Where can I get a job?" Social structure helps people make sense out of their environment even when they find themselves on the streets.

However, conflict theorists maintain that there is more to social structure than is readily visible and that we must explore the deeper, underlying structures that determine social relations in a society. For example, Karl Marx suggested that the way economic production is organized is the most important structural aspect of any society. In capitalistic societies, where a few people control the labour of many, the social structure reflects a system of relationships of domination among categories of people (for example, owner–worker and employer–employee).

FIGURE 5.1 : SOCIAL STRUCTURE FRAMEWORK

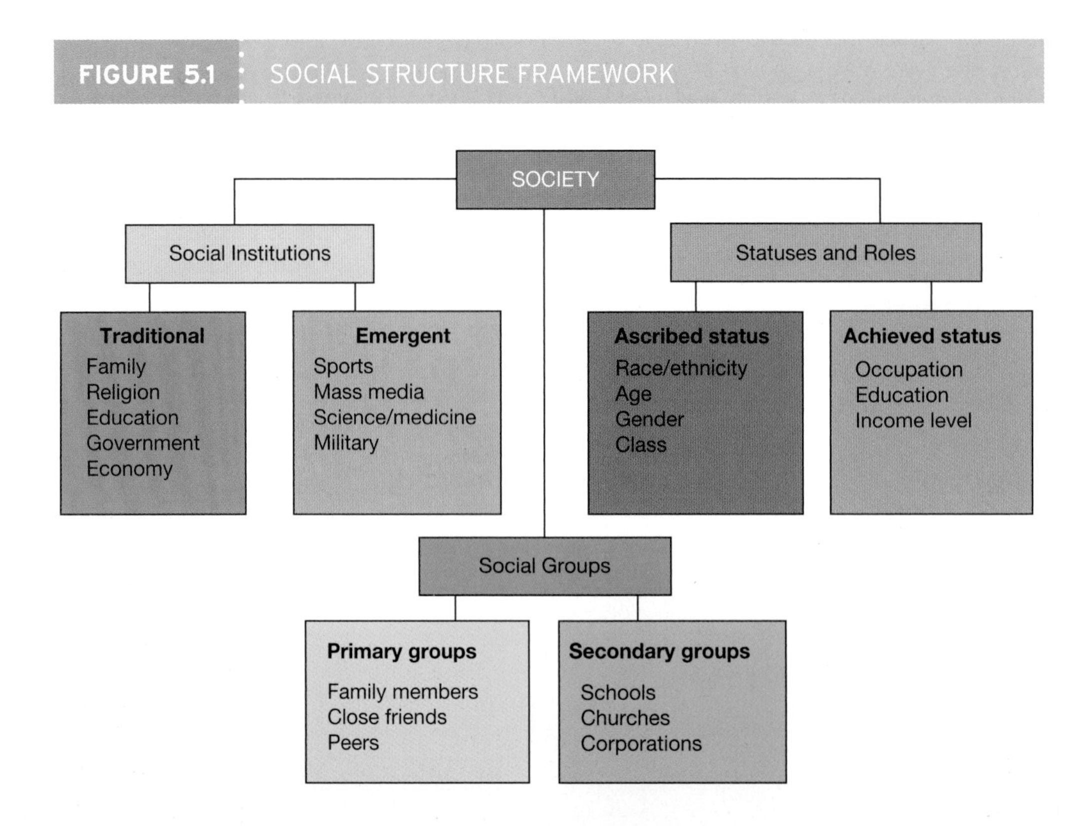

BOX 5.1 SOCIOLOGY AND EVERYDAY LIFE

How Much Do You Know About Homelessness?

True	False	
T	F	1. Most homeless people choose to be homeless.
T	F	2. The number of homeless persons in Canada has gradually declined over the past 30 years.
T	F	3. Most homeless people are mentally ill.
T	F	4. The number of homeless adolescents has increased in the past decade.
T	F	5. One out of every four homeless people is a child.

For answers to the quiz about homelessness, go to **www.nelson.com/student**.

Social structure creates boundaries that define which persons or groups will be the "insiders" and which will be the "outsiders." **Social marginality** is the state of being part insider and part outsider in the social structure. Sociologist Robert Park (1928) coined this term to refer to persons (such as immigrants) who simultaneously share the life and traditions of two distinct groups. Social marginality results in stigmatization. A **stigma** is any physical or social attribute or sign that so devalues a person's social identity that it disqualifies that person from full social acceptance (Goffman, 1963b). The stigmatization of homelessness is clearly revealed in the following interview with a street youth living in Halifax:

> I've had people just go off on rants, like "fuck you, you stupid worthless piece of Shit, blah, blah, blah, get a fucking job, why don't you just fucking die," and like, try to spit on me. I've been kicked and stuff. . . . I've had people throw stuff at me out of cars and shit. (Karabanow et al., 2010:50)

social marginality The state of being part insider and part outsider in the social structure.

stigma According to Erving Goffman, any physical or social attribute or sign that so devalues a person's social identity that it disqualifies that person from full social acceptance.

LO-1 COMPONENTS OF SOCIAL STRUCTURE

What is included in the social structure of a society? The social structure of a society includes its social positions, the relationships among those positions, and the kinds of resources attached to each of the positions. Social structure also includes all of the groups that make up society and the relationships among those groups (Smelser, 1988). We begin by examining the social positions that are closest to the individual.

STATUS

A **status** is a socially defined position in a group or society characterized by certain expectations, rights, and duties. Statuses exist independently of the specific people occupying them; the statuses of professional athlete, rock musician, professor, university student, and homeless person all exist exclusive of the specific individuals who occupy these social positions. For example, although thousands of new students arrive on university campuses each year to occupy the status of first-year student, the status of university student and the expectations attached to that position have remained relatively unchanged for the past century.

status A socially defined position in a group or society characterized by certain expectations, rights, and duties.

Does the term *status* refer only to high-level positions in society? No, not in a sociological sense. Although many people equate the term with high levels of prestige, sociologists use it to refer to all socially defined positions—high rank and low rank.

Take a moment to answer this question: Who am I? To determine who you are, you must think about your social identity, which is derived from the statuses you occupy and is based on

▶ **status set** A term used to describe all the statuses that a person occupies at a given time.

your status set. A **status set** is made up of all the statuses that a person occupies at a given time. For example, Marie may be a psychologist, a professor, a wife, a mother, a Roman Catholic, a school volunteer, an Alberta resident, and a French Canadian. All of these socially defined positions constitute her status set.

▶ **ascribed status** A social position conferred on a person at birth or received involuntarily later in life.

▶ **achieved status** A social position a person assumes voluntarily as a result of personal choice, merit, or direct effort.

ASCRIBED AND ACHIEVED STATUS Statuses are distinguished by the manner in which we acquire them. An **ascribed status** is a social position conferred on a person at birth or received involuntarily later in life, based on attributes over which the individual has little or no control, such as ethnicity, age, and gender. Marie, for example, is a female born to French Canadian parents; she was assigned these statuses at birth. An **achieved status** is a social position a person assumes voluntarily as a result of personal choice, merit, or direct effort. Achieved statuses (such as occupation, education, and income) are thought to be gained as a result of personal ability or successful competition. Most occupational positions in modern societies are achieved statuses. For instance, Marie voluntarily assumed the statuses of psychologist, professor, wife, mother, and school volunteer. However, not all achieved statuses are positions most people would want to attain: Being a criminal, a drug addict, or a homeless person, for example, is a negative achieved status.

Ascribed statuses have a significant influence on the achieved statuses we occupy. Ethnicity, gender, and age affect each person's opportunity to acquire certain achieved statuses. Those who are privileged by their positive ascribed statuses are more likely to achieve the more prestigious positions in a society. Those who are disadvantaged by their ascribed statuses may more easily acquire negative achieved statuses.

▶ **master status** A term used to describe the most important status a person occupies.

Jeff Brass/Getty Images

In the past, a person's status was primarily linked to his or her family background, education, occupation, and other sociological attributes. Today, some sociologists believe that celebrity status has overtaken the more traditional social indicators of status. Bono, shown here performing at a concert, is an example of celebrity status.

MASTER STATUS If we occupy many different statuses, how can we determine which is the most important? Sociologist Everett Hughes has stated that societies resolve this ambiguity by determining master statuses. A **master status** is the most important status a person occupies; it dominates all the individual's other statuses and is the overriding ingredient in determining a person's general social position (Hughes, 1945). Being poor or rich is a master status that influences many other areas of life, including health, education, and life opportunities. For men, occupation has usually been the most important status, although occupation is increasingly a master status for many women as well. "What do you do?" is one of the first questions most people ask when meeting another. Occupation provides important clues to a person's educational level, income, and family background. An individual's race/ethnicity may also constitute a master status in a society in which dominant-group members single out members of other groups as "inferior" on the basis of real or alleged physical, cultural, or nationality characteristics.

Master statuses confer high or low levels of personal worth and dignity on people. These are not characteristics that we inherently possess; they are derived from the statuses we occupy. For someone who has no residence, being a homeless person readily becomes a master status regardless of the person's other attributes. Homelessness is a stigmatized master status; it confers disrepute on its occupant because domiciled people often believe a homeless person has a "character flaw." The circumstances under which someone becomes homeless determine the extent to which that person is stigmatized.

STATUS SYMBOLS When people are proud of a particular social status they occupy, they often choose to use visible means to let others know about their position. A **status symbol** is a material sign that informs others of a person's specific status. For example, just as wearing a wedding ring proclaims that a person is married, owning a Rolls-Royce announces that one has "made it." In North American society, people who have "made it" frequently want symbols to inform others of their accomplishments.

status symbol A material sign that informs others of a person's specific status.

Status symbols for the domiciled and the homeless may have different meanings. Among affluent persons, a full shopping cart in the grocery store and bags of merchandise from expensive department stores indicate a lofty financial position. By contrast, among the homeless, bulging shopping bags and overloaded grocery carts suggest a completely different status.

TIME TO REVIEW

- Define ascribed status, achieved status, and master status.
- Is being unemployed an ascribed or achieved status?
- When does unemployment become a master status?

ROLES

role A set of behavioural expectations associated with a given status.

role expectation A group's or society's definition of the way a specific role ought to be played.

role performance How a person plays a role.

A role is the dynamic aspect of a status. Whereas we *occupy* a status, we *play* a role (Linton, 1936). A **role** is a set of behavioural expectations associated with a given status. For example, a carpenter (employee) hired to remodel a kitchen is not expected to sit down uninvited and join the family (employer) for dinner.

Role expectation is a group's or society's definition of the way a specific role ought to be played. By contrast, **role performance** is how a person plays the role. Role performance does not always match role expectation. Some statuses have role expectations that are highly specific, such as that of surgeon or university professor. Other statuses, such as friend or significant other, have less structured expectations. The role expectations tied to the status of student are more specific than those for being a friend. Role expectations are typically based on a range of acceptable behaviour rather than on strictly defined standards.

Our roles are relational (or complementary); that is, they are defined in the context of roles performed by others. We can play the role of student because someone else fulfills the role of professor. Conversely, to perform the role of professor, the teacher must have one or more students.

Role ambiguity occurs when the expectations associated with a role are unclear. For example, it is not always clear when the provider–dependant aspect of the parent–child relationship ends. Should it end at age 18? When a person is no longer in school? Different people will answer these questions differently depending on their experiences and socialization, as well as on the parents' financial capability and willingness to continue contributing to the welfare of their adult children.

© Jiang Jin/SuperStock

Parents often experience role conflict when they are trying to balance making a living and having a successful career with fulfilling their role as a good parent.

ROLE CONFLICT AND ROLE STRAIN Most people occupy a number of statuses, each of which has numerous role expectations attached. For example, Charles is a student who attends morning classes at the university and he is an employee at a fast-food restaurant where he works from 3 p.m. to 10 p.m. He is also Stephanie's boyfriend, and she would like to see him more often. On December 7, Charles has a final exam at 7 p.m., when he is supposed to be working. Meanwhile, Stephanie is pressuring him to take her to a movie. To top it off, his mother calls, asking him to fly home because his father is going to have emergency surgery. How can Charles be in all of these places at once? Such experiences of role conflict can be overwhelming.

▶ **role conflict** A situation in which incompatible role demands are placed on a person by two or more statuses held at the same time.

Role conflict occurs when incompatible role demands are placed on a person by two or more statuses held at the same time. When role conflict occurs, we may feel pulled in different directions. To deal with this problem, we may prioritize our roles and first complete the one we consider to be most important. Or we may compartmentalize our lives and "insulate" our various roles (Merton, 1968); that is, we may perform the activities linked to one role for part of the day, and then engage in the activities associated with another role in some other time period or elsewhere. For example, under routine circumstances, Charles would fulfill his student role for part of the day and his employee role for another part of the day. In his current situation, however, he is unable to compartmentalize his roles.

Whereas role conflict occurs between two or more statuses (such as being homeless and being a temporary employee of a social services agency), role strain takes place within one status. **Role strain** occurs when incompatible demands are built into a single status that a person occupies. For example, parents may experience role strain because of the demands of managing their time, unclear expectations, unequal division of unpaid work in the home, and lack of emotional support from the other parent. The concepts of role expectation, role performance, role conflict, and role strain are illustrated in Figure 5.2.

▶ **role strain** The strain experienced by a person when incompatible demands are built into a single status that the person occupies.

Individuals frequently distance themselves from a role they find extremely stressful or otherwise problematic. *Role distancing* occurs when people consciously foster the impression of a lack of commitment or attachment to a particular role and merely go through the motions

FIGURE 5.2 ROLE EXPECTATION, PERFORMANCE, CONFLICT, AND STRAIN

When playing the role of "student," do you sometimes personally encounter these concepts?

The Role of "Student"

Role Expectation: a group's or society's definition of the way a specific role *ought* to be played.

Role Performance: how a person does play a role.

Role Conflict: what occurs when incompatible demands are put on a person by two or more statuses held at the same time.

Role Strain: what occurs when incompatible demands are built into a single status that the person holds.

of role performance (Goffman, 1961b). People use distancing techniques when they do not want others to take them as the "self" implied in a particular role, especially if they think the role is "beneath them." While Charles is working in the fast-food restaurant, for example, he does not want people to think of him as a "loser in a dead-end job." He wants them to view him as a university student who is working there just to "pick up a few bucks" until he graduates. When customers from the university come in, Charles talks to them about what courses they are taking, what they are majoring in, and what professors they have. He does not discuss whether the bacon cheeseburger is better than the chili burger. When Charles is really involved in role distancing, he tells his friends that he "works there but wouldn't eat there." Role distancing is most likely to occur when people find themselves in roles in which the social identities implied are inconsistent with how they think of themselves or how they want to be viewed by others.

ROLE EXIT **Role exit** occurs when people disengage from social roles that have been central to their self-identity. Sociologist Helen Rose Fuchs Ebaugh (1988) studied this process by interviewing ex-convicts, ex-nuns, retirees, divorced men and women, and others who had exited voluntarily from significant social roles. According to Ebaugh, role exit occurs in four stages. The first stage is doubt, in which people experience frustration or burnout when they reflect on their existing roles. The second stage involves a search for alternatives; here, people may take a leave of absence from their work or temporarily separate from their marriage partner. The third stage is the turning point at which people realize that they must take some final action, such as quitting their job or getting a divorce. The fourth and final stage involves the creation of a new identity.

role exit A situation in which people disengage from social roles that have been central to their self-identity.

Exiting the "homeless" role is often very difficult. The longer a person remains on the streets, the more difficult it becomes to exit this role. Personal resources diminish over time.

Los Angeles Times columnist Steve Lopez met a homeless man, Nathaniel Ayers (pictured), and learned that he had been a promising musician studying at the Juilliard School who had dropped out because of his struggle with mental illness. In his 2008 book, *The Soloist*, Lopez chronicles the relationship that he developed with Ayers and how he eventually helped get Ayers off the street and treated for his schizophrenia. This story is an example of role exit, and you can see it in the movie *The Soloist*.

Personal possessions are often stolen, lost, sold, or pawned. Work experience and skills become outdated, and physical disabilities that prevent individuals from working are likely to develop on the streets. As 21-year-old Chris describes, breaking the ties with their street families and communities was often the most challenging aspect to their role exit:

> I found my biggest one [obstacle] was leaving the crowd that I was with, like my friends, the situation with my friends, 'cause they are all like, "No, don't go, stay down here and hang with us, go do that," and that was probably my biggest crutch, was getting away from my friends because I'd been friends with them my whole life and for me to just push them away and say, "No, I'm getting away from this, I'm getting out of this." It was a big step for me (Karabanow, 2008:783).

GROUPS

▶ **social group** A group that consists of two or more people who interact frequently and share a common identity and a feeling of interdependence.

▶ **primary group** A small, less specialized group in which members engage in face-to-face, emotion-based interactions over an extended time.

▶ **secondary group** A larger, more specialized group in which the members engage in more impersonal, goal-oriented relationships for a limited time.

▶ **social network** A series of social relationships that link an individual to others.

▶ **formal organization** A highly structured group formed for the purpose of completing certain tasks or achieving specific goals.

Groups are another important component of social structure. To sociologists, a **social group** consists of two or more people who interact frequently and share a common identity and a feeling of interdependence. Throughout our lives, most of us participate in groups, from our families and childhood friends, to our university classes, to our work and community organizations, and even to society.

Primary and secondary groups are the two basic types of social groups. A **primary group** is a small, less specialized group in which members engage in face-to-face, emotion-based interactions over an extended time. Typically, primary groups include our family, close friends, and school- or work-related peer groups. By contrast, a **secondary group** is a larger, more specialized group in which members engage in more impersonal, goal-oriented relationships for a limited time. Schools, churches, and corporations are examples of secondary groups. In secondary groups, people have few, if any, emotional ties to one another. Instead, they come together for some specific, practical purpose, such as getting a degree or a paycheque. Secondary groups are more specialized than primary ones; individuals relate to one another in terms of specific roles (such as professor and student) and more limited activities (such as course-related endeavours). Primary and secondary groups are further discussed in Chapter 7.

Social solidarity, or cohesion, relates to a group's ability to maintain itself in the face of obstacles. Social solidarity exists when social bonds, attractions, or other forces hold members of a group in interaction over a period of time. For example, if a local church is destroyed by fire and congregation members still worship together in a makeshift setting, then they have a high degree of social solidarity.

Many of us build social networks from our personal friends in primary groups and our acquaintances in secondary groups. A **social network** is a series of social relationships that link an individual to others. Social networks work differently for men and women, for different ethnic groups, and for members of different social classes. Research on homeless youth in Toronto and Vancouver revealed that informal social networks that the youth described as "street families" tended to form around issues of survival and support (Gaetz, 2014). Individuals within these groups often assumed specialized roles that were defined in family terms, including references to street brothers and sisters, and even fathers and mothers.

For many years, capitalism has been dominated by powerful "old-boy" social networks.

A **formal organization** is a highly structured group formed for the purpose of completing certain tasks or achieving specific goals. Many of us spend most of our time in formal organizations, such as universities, corporations, or the government. Chapter 7 analyzes the characteristics of bureaucratic organizations; however, at this point, we should note that these organizations are an important component of social structure in all industrialized societies. We expect such organizations to educate us, solve our social problems (such as crime and homelessness), and provide work opportunities.

LO-2 SOCIAL INSTITUTIONS

At the macrolevel of all societies, certain basic activities routinely occur—children are born and socialized, goods and services are produced and distributed, order is preserved, and a sense of purpose is maintained. Social institutions are the means by which these basic needs are met. A **social institution** is a set of organized beliefs and rules that establishes how a society will attempt to meet its basic social needs. In the past, these needs centred around five basic social institutions: the family, religion, education, the economy, and the government or politics. Today, mass media, sports, science and medicine, and the military are also considered to be social institutions.

▶ **social institution** A set of organized beliefs and rules that establishes how a society will attempt to meet its basic social needs.

What is the difference between a group and a social institution? A group is composed of specific, identifiable people; an institution is a standardized way of doing something. The concept of family helps distinguish between the two. When we talk about your family or my family, we are referring to a specific family. When we refer to the family as a social institution, we are talking about ideologies and standardized patterns of behaviour that organize family life. For example, the family as a social institution contains certain statuses organized into well-defined relationships, such as husband–wife, parent–child, brother–sister, and so forth. Specific families do not always conform to these ideologies and behaviour patterns.

Functional theorists emphasize that social institutions exist because they perform five essential tasks:

1. *Replacing members.* Societies and groups must have socially approved ways of replacing members who move away or die.
2. *Teaching new members.* People who are born into a society or move into it must learn the group's values and customs.
3. *Producing, distributing, and consuming goods and services.* All societies must provide and distribute goods and services for their members.
4. *Preserving order.* Every group or society must preserve order within its boundaries and protect itself from attack by outsiders.
5. *Providing and maintaining a sense of purpose.* To motivate people to cooperate with one another, a sense of purpose is needed.

Although this list of functional prerequisites is shared by all societies, the institutions in each society perform these tasks in somewhat different ways depending on their specific cultural values and norms.

Conflict theorists agree with functionalists that social institutions are originally organized to meet basic social needs; however, they do not agree that social institutions work for the common good of everyone in society. For example, the homeless lack the power and resources to promote their own interests when they are opposed by dominant social groups. This problem for homeless people, especially children and youth, exists not only in Canada, but throughout the world (see Box 5.2 later in this chapter). From the conflict perspective, social institutions, such as the government, maintain the privileges of the wealthy and powerful while contributing to the powerlessness of others (see Domhoff, 2002).

LO-3 STABILITY AND CHANGE IN SOCIETIES

Changes in social structure have a dramatic impact on individuals, groups, and societies. Social arrangements in contemporary societies have grown more complex with the introduction of new technology, changes in values and norms, and the rapidly shrinking "global village." How do societies maintain some degree of social solidarity in the face of such changes? Sociologists Émile Durkheim and Ferdinand Tönnies developed typologies to explain the processes of stability and change in the social structure of societies. A *typology* is a classification scheme containing two or more mutually exclusive categories that are used to compare different kinds of behaviour or types of societies.

DURKHEIM: MECHANICAL AND ORGANIC SOLIDARITY

Émile Durkheim (1933/1893) was concerned with this question: How do societies manage to hold together? Durkheim asserted that preindustrial societies were held together by strong traditions and by the members' shared moral beliefs and values. As societies industrialized and developed more specialized economic activities, social solidarity came to be rooted in the members' shared dependence on one another. From Durkheim's perspective, social solidarity derives from a society's social structure, which, in turn, is based on the society's division of labour. *Division of labour* refers to how the various tasks of a society are divided up and performed. People in diverse societies (or in the same society at different points in time) divide their tasks somewhat differently, however, based on their own history, physical environment, and level of technological development.

To explain social change, Durkheim developed a typology that categorized societies as having either mechanical or organic solidarity. **Mechanical solidarity** refers to the social cohesion in preindustrial societies, in which there is minimal division of labour and people feel united by shared values and common social bonds. Durkheim used the term *mechanical solidarity* because he believed that people in such preindustrial societies feel a more or less automatic sense of belonging. Social interaction is characterized by face-to-face, intimate, primary-group relationships. Everyone is engaged in similar work, and little specialization is found in the division of labour.

▶ **mechanical solidarity** Émile Durkheim's term for the social cohesion that exists in preindustrial societies, in which there is a minimal division of labour and people feel united by shared values and common social bonds.

Organic solidarity refers to the social cohesion found in industrial (and perhaps post-industrial) societies, in which people perform specialized tasks and feel united by their mutual dependence. Durkheim chose the term *organic solidarity* because he believed that individuals in industrial societies come to rely on one another in much the same way that the organs of the human body function interdependently. Social interaction is less personal, more status-oriented, and more focused on specific goals and objectives. People no longer rely on morality or shared values for social solidarity; instead, they are bound together by practical considerations.

▶ **organic solidarity** Émile Durkheim's term for the social cohesion that exists in industrial (and perhaps post-industrial) societies, in which people perform specialized tasks and feel united by their mutual dependence.

LO-4 TÖNNIES: GEMEINSCHAFT AND GESELLSCHAFT

Sociologist Ferdinand Tönnies (1855–1936) used the terms *Gemeinschaft* and *Gesellschaft* to characterize the degree of social solidarity and social control found in societies. He was especially concerned about what happens to social solidarity in a society when a "loss of community" occurs.

The ***Gemeinschaft* (guh-MINE-shoft)** is a traditional society in which social relationships are based on personal bonds of friendship and kinship and on intergenerational stability. These relationships are based on ascribed rather than achieved status. In such societies, people have a commitment to the entire group and feel a sense of togetherness. Tönnies used the German term *Gemeinschaft* because it means commune or community; social solidarity and social control are maintained by the community. Members have a strong sense of belonging, but they also have limited privacy.

▶ ***Gemeinschaft* (guh-MINE-shoft)** A traditional society in which social relationships are based on personal bonds of friendship and kinship and on intergenerational stability.

By contrast, the **Gesellschaft (guh-ZELL-shoft)** is a large, urban society in which social bonds are based on impersonal and specialized relationships, with little long-term commitment to the group or consensus on values. In such societies, most people are "strangers" who perceive that they have little in common with most other people. Consequently, self-interest dominates and little consensus exists regarding values. Tönnies selected the German term Gesellschaft because it means association; relationships are based on achieved statuses, and interactions among people are both rational and calculated.

▶ ***Gesellschaft* (guh-ZELL-shoft)** A large, urban society in which social bonds are based on impersonal and specialized relationships, with little long-term commitment to the group or consensus on values.

SOCIAL STRUCTURE AND HOMELESSNESS

In *Gesellschaft* societies, such as Canada, a prevailing core value is that people should be able to take care of themselves. Thus, many people view the homeless as "throwaways"—as beyond help or as having already had enough done for them by society. Some argue that the homeless made their own bad decisions, which led them into alcoholism or drug addiction, and should be held responsible for the consequences of their actions. In this sense, homeless people serve as a visible example to others to "follow the rules" lest they experience a similar fate.

Alternative explanations for homelessness in *Gesellschaft* societies have been suggested. Elliot Liebow (1993) notes that homelessness is rooted in poverty; homeless people overwhelmingly are poor people who come from poor families. Homelessness is a "social class phenomenon, the direct result of a steady, across-the-board lowering of the standard of living of the working class and lower class" (1993:224). The problem is exacerbated by a lack of jobs and adequate housing. Clearly, there is no simple answer to the question about what should be done to help the homeless. Nor, as discussed in Box 5.2, is there any consensus on what legal rights the homeless have in public areas. The answers we derive as a society and as individuals are often based on our social construction of this reality of life.

BOX 5.2 POINT/COUNTERPOINT

Homeless Rights versus Public Space

I had a bit of a disturbing experience yesterday as I was running errands downtown. First, I was glad to see the south Queen sidewalk east of University open. (Months of construction on the new opera house had blocked it off.) As I continued walking eastward past the acclaimed new structure (where I have enjoyed a performance or two), I wondered why the sidewalk was so narrow. It seems this stretch of Queen should feel a bit grander. When I reached the corner of Queen and Bay, I saw some police officers and city workers "taking action on sidewalk clearance." They were clearing a homeless person's worldly belongings off the sidewalk. Using shovels. And a pickup truck . . .

> I think what I saw yesterday is unacceptable. Sure, the situation is complicated. Yes, there are a lot of stakeholders and stories to appreciate. But it's unfairness I want to see shovelled out of public space. Not people. Not blankets. Not kindness. And I hope I'm not alone. (Sandals, 2007)*

"Protection of public space" has become an issue in many cities. Record numbers of homeless individuals and families seek refuge on the streets and in public parks because they have nowhere else to go. However, this seemingly individualistic problem is actually linked to larger social concerns, including long-term unemployment, lack of education and affordable housing, and cutbacks in government and social service budgets. The problem of homelessness also raises significant social policy issues, including the extent to which cities can make it illegal for people to remain for extended periods of time in public spaces.

Should homeless persons be allowed to sleep on sidewalks, in parks, and in other public areas? This issue has been the source of controversy. As cities have sought to improve their downtown areas and public spaces, they have taken measures to enforce city ordinances controlling loitering (standing around or sleeping in public spaces), "aggressive panhandling," and disorderly conduct. Advocates for the homeless and civil liberties groups have filed lawsuits claiming that the rights of the homeless are being violated by the enforcement of these laws. The lawsuits assert that the homeless have a right to sleep in parks because no affordable housing is available for them. Advocates also argue that panhandling is a legitimate livelihood for some of the homeless and is protected speech under the *Charter of Rights and Freedoms*. In addition, they accuse public and law enforcement officials of seeking to punish the homeless on the basis

* Sandals, Leah. 2007. "'Public Space Protection'—But for Which 'Public'?" Retrieved February 24, 2007. Available: http://spacing.ca/wire/?p=1466. Reproduced by permission of the author.

(*Continued*)

of their "status." According to ethics professor Arthur Schafer, punishing panhandlers is the wrong way to go about the issue:

© Mark Ludak/The Image Works

The "homeless problem" is not a new one for city governments. Of the limited public funding that is designated for the homeless, most has been spent on shelters that are frequently overcrowded and otherwise inadequate. Officials in some cities have given homeless people a one-way ticket to another city. Still others have routinely run them out of public spaces.

> Do we, as a society, really want to rely upon still more laws to deal with the serious social problems of poverty, homelessness, and panhandling? Are we convinced that legal coercion, with its use of physical force backed by weapons, lawyers, courts, and jails, will be effective in addressing what is essentially a social problem? Are we prepared to violate fundamental rights to freedom of expression and add further burdens to the least advantaged members of our society? (1998:1)

What responsibility does society have to the homeless? Are laws restricting the hours that public areas or parks are open to the public unfair to homeless persons? Some critics have argued that if the homeless and their advocates win these lawsuits, what they have won (at best) is the right for the homeless to live on the street under extremely adverse conditions. Others have disputed this assertion and note that if society does not make affordable housing and job opportunities available, the least it can do is stop harassing homeless people who are getting by as best they can.

Sources: Based on Kaufman, 1996; Sandals, 2007; Wood, 2002.

Bruce Ayres/Stone/Getty Images

Contrary to a popular myth that most homeless people are single drifters, an increasing number of families are now homeless.

What do you think? What rights are involved? Whose rights should prevail?

SOCIAL INTERACTION: THE MICROLEVEL PERSPECTIVE

So far in this chapter, we have focused on society and social structure from a macrolevel perspective. We have seen how the structure of society affects the statuses we occupy, the roles we play, and the groups and organizations to which we belong. Functionalist and conflict perspectives provide a macrosociological overview because they concentrate on large-scale events and broad social features. By contrast, the symbolic interactionist perspective takes a microsociological approach, asking how social institutions affect our daily lives. We will now look at society from the microlevel perspective, which focuses on social interaction among individuals, especially face-to-face encounters.

SOCIAL INTERACTION AND MEANING

When you are with other people, do you often wonder what they think of you? If so, you are not alone! Because most of us are concerned about the meanings others ascribe to our behaviour, we try to interpret their words and actions so that we can plan how we will react (Blumer, 1969). We know that others have expectations of us. We also have certain expectations about them. For example, if we enter an elevator that has only one other person in it, we do not expect that individual to confront us and stare into our eyes. As a matter of fact, we would be quite upset if the person did so.

Social interaction within a given society has certain shared meanings across situations. For instance, our reaction would be the same regardless of *which* elevator we rode in *which* building. Sociologist Erving Goffman (1963b) described these shared meanings in his observations about two pedestrians approaching each other on a public sidewalk. He noted that each will tend to look at the other just long enough to acknowledge the other's presence. By the time they are about two and a half metres away from each other, both individuals will tend to look downward. Goffman referred to this behaviour as *civil inattention*—the ways in which an individual shows awareness that others are present without making them the object of particular attention. The fact that people engage in civil inattention demonstrates that interaction does have a pattern, or *interaction order*, that regulates the form and processes (but not the content) of social interaction.

George Doyle/Thinkstock

These people are displaying what Goffman referred to as "civil inattention."

Does everyone interpret social interaction rituals in the same way? No. Ethnicity, gender, and social class play a part in the meanings we give to our interactions with others, including chance encounters on elevators or the street. Our perceptions about the meaning of a situation vary widely based on the statuses we occupy and our unique personal experiences.

Social encounters have different meanings for men and women, and for individuals from different social classes and ethnic groups. For example, sociologist Carol Brooks Gardner (1989) found that women frequently do not perceive street encounters to be "routine" rituals. They fear for their personal safety and try to avoid comments and propositions that are sexual in nature when they walk down the street. In another example, members of the dominant classes regard the poor, unemployed, and working class as less worthy of attention, frequently subjecting them to subtle yet systematic "attention deprivation" (Derber, 1983).

THE SOCIAL CONSTRUCTION OF REALITY

If we interpret other people's actions so subjectively, can we have a shared social reality? Some interaction theorists believe that there is little shared reality beyond that which is socially created. Interactionists refer to this as the **social construction of reality**—the process by which our perception of reality is shaped largely by the subjective meaning that we give to an experience (Berger and Luckmann, 1967). This meaning strongly influences what we "see" and how we respond to situations.

social construction of reality The process by which our perception of reality is shaped largely by the subjective meaning that we give to an experience.

Our perceptions and behaviour are influenced by how we initially define situations: We act on reality as we see it. Sociologists describe this process as the *definition of the situation*, meaning that we analyze a social context in which we find ourselves, determine what is in our best interest, and adjust our attitudes and actions accordingly. This can result in a **self-fulfilling prophecy**—a false belief or prediction that produces behaviour that makes the originally false belief come true (Thomas and Thomas, 1928:72). An example would be a person who has been told repeatedly that she or he is not a good student; eventually, this person might come to believe it to be true, stop studying, and receive failing grades.

self-fulfilling prophecy A situation in which a false belief or prediction produces behaviour that makes the originally false belief come true.

People may define a given situation in very different ways. Consider sociologist Lesley Harman's initial reaction to her field research site, a facility for homeless women in an Ontario city: "The initial shock of facing the world of the homeless told me much about what I took for granted . . . The first day I lasted two very long hours. I went home and woke up severely depressed, weeping uncontrollably" (1989:42). In contrast, a resident typical of many of the women who lived there defined living in a hostel in this way: "This is home to me because I feel so comfortable. I can do what I really want, the staff are

Mike Theiler/Getty Images

© AP Images/Chris Gardner

Sharply contrasting perceptions of the same reality are evident in these people's views on the war in Iraq.

very nice to me, everybody is good to me, it's home, you know?" (1989:91). As these two examples show, we define situations from our own frame of reference, based on the statuses we occupy and the roles we play.

Dominant group members with prestigious statuses may have the ability to establish how other people define "reality" (Berger and Luckmann, 1967:109). Some sociologists have suggested that dominant groups, particularly high-income white males in powerful economic and political statuses, perpetuate a dominant worldview that is frequently seen as "social reality."

ETHNOMETHODOLOGY

How do we know how to interact in a given situation? What rules do we follow? Ethnomethodologists are interested in the answers to these questions. **Ethnomethodology** is the study of the commonsense knowledge that people use to understand the situations in which they find themselves (Heritage, 1984:4). Sociologist Harold Garfinkel (1967) initiated this approach and coined the term: *ethno* for "people" or "folk" and *methodology* for "a system of methods." Garfinkel was critical of mainstream sociology for not recognizing the ongoing ways in which people create reality and produce their own world. Consequently, ethnomethodologists examine existing patterns of conventional behaviour in order to uncover people's *background expectancies*; that is, their shared interpretation of objects and events, as well as their resulting actions. According to ethnomethodologists, interaction is based on assumptions of shared expectancies. For example, when you are talking with someone, what are your expectations about taking turns? Based on your background expectancies, would you be surprised if the other person talked for an hour and never gave you a chance to speak?

To uncover people's background expectancies, ethnomethodologists frequently break "rules" or act as though they do not understand some basic rule of social life so that they can observe other people's responses. In a series of *breaching experiments,* Garfinkel (1967) assigned different activities to his students to see how breaking the unspoken rules of behaviour created confusion. In one experiment, when students were asked, "How are you?" they threw their questioners off balance by responding with detailed accounts rather than polite nothings.

The ethnomethodological approach contributes to our knowledge of social interaction by making us aware of subconscious social realities in our daily lives.

▶ **ethnomethodology** The study of the commonsense knowledge that people use to understand the situations in which they find themselves.

▶ **dramaturgical analysis** The study of social interaction that compares everyday life to a theatrical presentation.

▶ **impression management (or presentation of self)** A term for people's efforts to present themselves to others in ways that are most favourable to their own interests or image.

LO-5 DRAMATURGICAL ANALYSIS

Erving Goffman suggested that day-to-day interactions have much in common with being on stage or in a dramatic production. **Dramaturgical analysis** is the study of social interaction that compares everyday life to a theatrical presentation. Members of our "audience" judge our performance and are aware that we may slip and reveal our true character (Goffman, 1959, 1963a). Consequently, most of us attempt to play our role as well as possible and to control the impressions we give to others. **Impression management, or presentation of self**, refers to people's efforts to present themselves to others in ways that are most favourable to their own interests or image.

For example, suppose that a professor has returned graded exams to your class. Will you discuss the exam and your grade with others in the class? If you are like most people, you probably play your student role differently depending on whom you are talking to and what grade you

Betsie Van der Meer/Stone/Getty Images

According to Erving Goffman, our day-to-day interactions have much in common with a dramatic production.

received on the exam. In a study, researchers analyzed how students "presented themselves" or "managed impressions" when exam grades were returned. Students who all received high grades ("Ace–Ace encounters") willingly talked with one another about their grades and sometimes engaged in a little bragging about how they had "aced" the test. However, encounters between students who had received high grades and those who had received low or failing grades ("Ace–Bomber encounters") were uncomfortable. The Aces felt as if they had to minimize their own grades. Consequently, they tended to attribute their success to "luck" and were quick to offer the Bombers words of encouragement. On the other hand, the Bombers believed that they had to praise the Aces and hide their own feelings of frustration and disappointment. Students who received low or failing grades ("Bomber–Bomber encounters") were more comfortable when they talked with one another because they could share their negative emotions. They often indulged in self-pity and relied on face-saving excuses (such as an illness or an unfair exam) for their poor performances (Albas and Albas, 1988).

In Goffman's terminology, *face-saving behaviour* refers to the strategies we use to rescue our performance when we experience a potential or actual loss of face. When the Bombers made excuses for their low scores, they were engaged in face-saving; the Aces attempted to help them save face by asserting that the test was unfair or that it was only a small part of the final grade. Why would the Aces and Bombers both participate in face-saving behaviour? In most social interactions, all role players have an interest in keeping the "play" going so that they can maintain their overall definition of the situation in which they perform their roles.

Goffman noted that people consciously participate in *studied nonobservance,* a face-saving technique in which one role player ignores the flaws in another's performance to avoid embarrassment for everyone involved. Most of us remember times when we have failed in our role and know that it is likely to happen again; thus, we may be more forgiving of the role failures of others.

Social interaction, like a theatre, has a front stage and a back stage. The *front stage* is the area where a player performs a specific role before an audience. The *back stage* is the area where a player is not required to perform a specific role because it is out of view of a given audience. For example, when the Aces and Bombers were talking with each other at school, they were on the "front stage." When they were in the privacy of their own residences, they were in "back stage" settings—they no longer had to perform the Ace and Bomber roles and could be themselves.

The need for impression management is most intense when role players have widely divergent or devalued statuses. As we have seen with the Aces and Bombers, the participants often play different roles under different circumstances and keep their various audiences separated from one another. If one audience becomes aware of other roles that a person plays, the impression being given at that time may be ruined. For example, homeless people may lose jobs or the opportunity to get them when their homelessness becomes known. One woman, Kim, had worked as a receptionist in a doctor's office for several weeks but was fired when the doctor learned that she was living in a shelter. According to Kim, the doctor told her, "If I had known you lived in a shelter, I would never have hired you. Shelters are places of disease" (Liebow, 1993:53–54). The homeless do not passively accept the roles into which they are cast. For the most part, they attempt—as we all do—to engage in impression management in their everyday lives.

The dramaturgical approach helps us think about the roles we play and the audiences who judge our presentation of self; however, this perspective has also been criticized for focusing on appearances and not the underlying substance. This approach may not place enough emphasis on the ways in which our everyday interactions with other people are influenced by occurrences within the larger society. For example, if some political leaders or social elites in a community deride homeless people by saying they are "lazy" or "unwilling to work," it may become easier for everyday people walking down a street to treat homeless individuals poorly. Overall, however, Goffman's dramaturgical analysis has been highly influential in the development of the sociology of emotions, an important area of contemporary theory and research.

TIME TO REVIEW

- Provide three examples of self-fulfilling prophecies you have experienced in your interactions with others.
- Describe how daily interactions are similar to being onstage.

THE SOCIOLOGY OF EMOTIONS

Why do we laugh, cry, or become angry? Are these emotional expressions biological or social? To some extent, emotions are a biologically given sense (like hearing, smell, and touch), but they also are social in origin. We are socialized to feel certain emotions, and we learn how and when to express (or not express) those emotions (Hochschild, 1983).

How do we know which emotions are appropriate for a given role? Sociologist Arlie Hochschild (1983) suggests that we acquire a set of *feeling rules,* which shape the appropriate emotions for a given role or specific situation. These rules include how, where, when, and with whom an emotion should be expressed. For example, for the role of a mourner at a funeral, feeling rules tell us which emotions are required (sadness and grief, for example), which are acceptable (a sense of relief that the deceased no longer has to suffer), and which are unacceptable (enjoyment of the occasion expressed by laughing out loud) (see Hochschild, 1983:63–68).

Feeling rules also apply to our occupational roles. For example, the truck driver who handles explosive cargos must be able to suppress fear. Although all jobs place some burden on our feelings, *emotional labour* occurs only in jobs that require personal contact with the public or the production of a state of mind (such as hope, desire, or fear) in others (Hochschild, 1983). With emotional labour, employees must display only certain carefully selected emotions. For example, flight attendants are required to act friendly toward passengers, to be helpful and open to requests, and to maintain an "omnipresent smile" to enhance the customers' status. By contrast, bill collectors are encouraged to show anger and make threats to customers, thereby supposedly deflating the customers' status and wearing down their presumed resistance to paying past-due bills. In both jobs, the employees are expected to show feelings that are often not their true ones (Hochschild, 1983).

Social class and race are determinants in managed expression and emotion management. Emotional labour is emphasized in middle- and upper-class families. Because middle- and upper-class parents often work with people, they are more likely to teach their children the importance of emotional labour in their own careers than are working-class parents, many of whom work with things, not people (Hochschild, 1983). Race is also an important factor in emotional labour. Members of visible minorities spend much of their life engaged in emotional labour because racist attitudes and discrimination make it continually necessary to manage one's feelings.

Emotional labour may produce feelings of estrangement from one's "true" self. C. Wright Mills (1956) suggested that when we "sell our personality" in the course of selling goods or services, we engage in a seriously self-alienating process. In other words, the "commercialization" of our feelings may dehumanize our work role performance and create alienation and contempt that spill over into other aspects of our life (Hochschild, 1983; Smith and Kleinman, 1989).

Clearly, the sociology of emotions helps us understand the social context of our feelings and the relationship between the roles we play and the emotions we experience. However, it may overemphasize the cost of emotional labour and the emotional controls that exist outside the individual (Wouters, 1989).

NONVERBAL COMMUNICATION

In a typical stage drama, the players not only speak their lines but also convey information by nonverbal communication. In Chapter 3, we discussed the importance of language; now, we will look at the messages we communicate without speaking. **Nonverbal communication** is the transfer of information between persons without the use of speech. It includes not only visual cues (gestures, appearances) but also vocal features (inflection, volume, pitch) and environmental factors (use of space, position) that affect meanings (Wood, 1999). Facial expressions, head movements, body positions, and other gestures carry as much of the total meaning of our communication with others as our spoken words do (Wood, 1999).

▶ **nonverbal communication** The transfer of information between persons without the use of speech.

FUNCTIONS OF NONVERBAL COMMUNICATION Why is nonverbal communication important to you? We obtain first impressions of others from various kinds of nonverbal communication, such as the clothing they wear and their body positions. Head and facial movements may provide us with information about other people's emotional states, and others receive similar information from us (Samovar and Porter, 1991). Through our body posture and eye contact, we signal that we do or do not wish to speak to someone. For example, we may look down at the sidewalk or off into the distance when we pass homeless persons who look as if they are going to ask for money.

Nonverbal communication establishes the relationship among people in terms of their responsiveness to and power over one another (Wood, 1999). For example, we show that we are responsive toward or like another person by maintaining eye contact and attentive body posture, and perhaps by touching and standing close. We can even express power or control over

a
© Steve Vidler/SuperStock

b
© Roderick Chen/SuperStock

c
© Kwame Zikomo/SuperStock

Nonverbal communication can be thought of as an international language. What message do you receive from the facial expression and gestures of each of these people? Is it possible to misinterpret their messages?

others through nonverbal communication. Goffman (1956) suggested that *demeanour* (how we behave or conduct ourselves) is relative to social power. People in positions of dominance are allowed a wider range of permissible actions than are their subordinates, who are expected to show deference. *Deference* is the symbolic means by which subordinates give a required permissive response to those in power; it confirms the existence of inequality and reaffirms each person's relationship to the other (Rollins, 1985).

FACIAL EXPRESSION, EYE CONTACT, AND TOUCHING Nonverbal communication is symbolic of our relationships with others. Who smiles? Who stares? Who makes and sustains eye contact? Who touches whom? All of these questions relate to demeanour and deference; the key issue is the status of the person who is *doing* the smiling, staring, or touching relative to the status of the recipient (Goffman, 1967).

Facial expressions, especially smiles, also reflect gender-based patterns of dominance and subordination in society. Women typically have been socialized to smile and frequently do so even when they are not happy (Halberstadt and Saitta, 1987). Jobs held predominantly by women (including flight attendant, secretary, elementary school teacher, and nurse) are more closely associated with being pleasant and smiling than are "men's jobs." In addition to smiling more frequently, many women tend to tilt their heads in deferential positions when they are talking or listening to others. By contrast, men tend to display less emotion through smiles or other facial expressions and instead seek to show that they are reserved and in control (Wood, 1999).

Women and men use eye contact differently during conversations. Women are more likely to sustain eye contact during conversations (but not otherwise) as a way of showing their interest in and involvement with others. By contrast, men are less likely to maintain prolonged eye contact during conversations but are more likely to stare at other people (especially other men) to challenge them and assert their own status (Pearson, 1985).

Eye contact can be a sign of domination or deference. For example, in a participant observation study of domestic (household) workers and their employers, sociologist Judith Rollins (1985) found that the domestics were supposed to show deference by averting their eyes when they talked to their employers. Deference also required that they present an "exaggeratedly subservient demeanour" by standing less erect and walking tentatively.

Touching is another form of nonverbal behaviour that has many different shades of meaning. Gender and power differences are evident in tactile communication from birth. Studies have shown that touching has variable meanings to parents: Boys are touched more roughly and playfully, while girls are handled more gently and protectively (Condry, Condry, and Pogatshnik, 1983). This pattern continues into adulthood, with women touched more frequently than men. Sociologist Nancy Henley (1977) attributed this pattern to power differentials between men and women and to the nature of women's roles as mothers, nurses, teachers, and secretaries. Clearly, touching has a different meaning to women than to men (Stier and Hall, 1984). Women may hug and touch others to indicate affection and emotional support, while men are more likely to touch others to give directions, assert power, and express sexual interest (Wood, 1999).

PERSONAL SPACE How much space do you like between yourself and other people? Anthropologist Edward Hall (1966) analyzed the physical distance between people speaking to one another and found that the amount of personal space people prefer varies from one culture to another. **Personal space** is the immediate area surrounding a person that the person claims as private. Our personal space is contained within an invisible boundary surrounding our body, much like a snail's shell. When others invade our space, we may retreat, stand our ground, or even lash out, depending on our cultural background (Samovar and Porter, 1991).

personal space
The immediate area surrounding a person that the person claims as private.

Age, gender, kind of relationship, and social class also have an impact on the allocation of personal space. Power differentials are reflected in personal space and privacy issues. With regard to age, adults generally do not hesitate to enter the personal space of a child (Thorne, Kramarae, and Henley, 1983). Similarly, young children who invade the personal space of an adult tend to elicit a more favourable response than do older uninvited visitors (Dean, Willis, and la Rocco, 1976). The need for personal space appears to increase with age (Aiello and Jones, 1971; Baxter, 1970), although it may begin to decrease at about age 40 (Heshka and Nelson, 1972).

For some people, the idea of privacy or personal space is an unheard of luxury afforded only to those in the middle and upper classes. As we have seen in this chapter, the homeless may have no space to call their own. Some may try to "stake a claim" on a heat grate or on the same bed in a shelter for more than one night, but such claims have dubious authenticity in a society in which the homeless are assumed to own nothing and have no right to lay claim to anything in the public domain.

In sum, all forms of nonverbal communication are influenced by gender, ethnicity, social class, and the personal contexts in which they occur. While it is difficult to generalize about people's nonverbal behaviour, we still need to think about our own nonverbal communication patterns. Recognizing that differences in social interaction exist is important. We should be wary of making value judgments—the differences are simply differences. Learning to understand and respect alternative styles of social interaction enhances our personal effectiveness by increasing the range of options we have for communicating with different people in diverse contexts and for varied reasons (Wood, 1999).

TIME TO REVIEW

- Evaluate your nonverbal communication in an encounter with a police officer. What role would demeanour and deference play in this interaction having a positive or negative outcome?

KEY TERMS

achieved status A social position a person assumes voluntarily as a result of personal choice, merit, or direct effort (p. 116).

ascribed status A social position conferred on a person at birth or received involuntarily later in life (p. 116).

dramaturgical analysis The study of social interaction that compares everyday life to a theatrical presentation (p. 126).

ethnomethodology The study of the commonsense knowledge that people use to understand the situations in which they find themselves (p. 126).

formal organization A highly structured group formed for the purpose of completing certain tasks or achieving specific goals (p. 120).

Gemeinschaft (guh-MINE-shoft) A traditional society in which social relationships are based on personal bonds of friendship and kinship and on intergenerational stability (p. 122).

Gesellschaft (guh-ZELL-shoft) A large, urban society in which social bonds are based on impersonal and specialized relationships, with little long-term commitment to the group or consensus on values (p. 122).

impression management (or presentation of self) A term for people's efforts to present themselves to others in ways that are most favourable to their own interests or image (p. 126).

master status A term used to describe the most important status a person occupies (p. 116).

mechanical solidarity Émile Durkheim's term for the social cohesion that exists in

LO-1 Identify the key components of social structure.

Social structure comprises statuses, roles, groups, and social institutions. A status is a specific position in a group or society and is characterized by certain expectations, rights, and duties. Ascribed statuses, such as gender, class, and ethnicity, are acquired at birth or involuntarily later in life. Achieved statuses, such as education and occupation, are assumed voluntarily as a result of personal choice, merit, or direct effort. We occupy a status, but a role is a set of behavioural expectations associated with a given status. A social group consists of two or more people who interact frequently and share a common identity and sense of interdependence. A formal organization is a highly structured group formed to complete certain tasks or achieve specific goals. A social institution is a set of organized beliefs and rules that establish how a society attempts to meet its basic needs.

Jeff Brass/Getty Images

LO-2 Compare and contrast functionalist and conflict perspectives on social institutions.

According to functionalist theorists, social institutions perform several prerequisites of all societies: to replace members; teach new members; produce, distribute, and consume goods and services; preserve order; and provide and maintain a sense of purpose. Conflict theorists, however, note that social institutions do not work for the common good of all individuals. Institutions may enhance and uphold the power of some groups but exclude others, such as the homeless.

© Bob Daemmrich/PhotoEdit

LO-3 Explain how societies maintain stability in times of social change.

According to Durkheim, although changes in social structure may dramatically affect individuals and groups, societies manage to maintain some degree of stability. Mechanical solidarity refers to social cohesion in preindustrial societies, in which people are united by shared values and common social bonds. Organic solidarity refers to the cohesion in industrial societies, in which people perform specialized tasks and are united by mutual dependence.

Mike Theiler/Getty Images

LO-4 Define and distinguish between *Gemeinschaft* and *Gesellschaft* societies.

According to Ferdinand Tönnies, the *Gemeinschaft* is a traditional society in which relationships are based on personal bonds of friendship and kinship, and on intergenerational stability. The *Gesellschaft* is an urban society in which social bonds are based on impersonal and specialized relationships, with little group commitment or consensus on values.

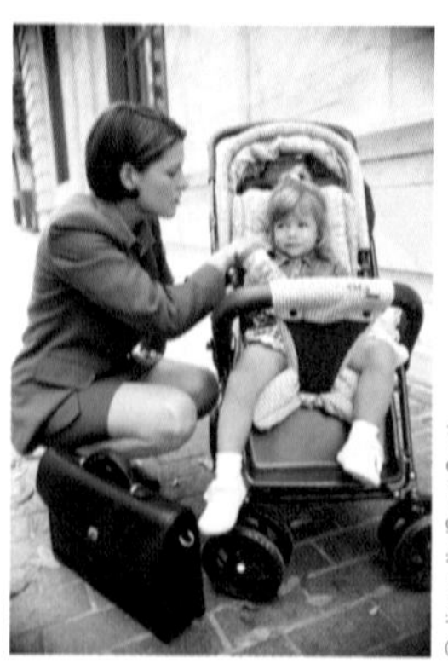

© Jiang Jin/SuperStock

LO-5 Understand Erving Goffman's dramaturgical perspective and the concepts of impression management, and front stage/back stage behaviours.

According to Erving Goffman's dramaturgical analysis, our daily interactions are similar to dramatic productions. *Impression management* refers to efforts to present our self to others in ways that are most favourable to our own interests or self-image. The *front stage* is the area where a player performs a specific role before an audience. The *back stage* is the area where a player is not required to perform a specific role because it is out of view of a given audience.

Betsie Van der Meer/Stone/Getty Images

preindustrial societies, in which there is a minimal division of labour and people feel united by shared values and common social bonds (p. 122).

nonverbal communication The transfer of information between persons without the use of speech (p. 129).

organic solidarity Émile Durkheim's term for the social cohesion that exists in industrial (and perhaps post-industrial) societies, in which people perform specialized tasks and feel united by their mutual dependence (p. 122).

personal space The immediate area surrounding a person that the person claims as private (p. 131).

primary group A small, less specialized group in which members engage in face-to-face, emotion-based interactions over an extended time (p. 120).

role A set of behavioural expectations associated with a given status (p. 117).

role conflict A situation in which incompatible role demands are placed on a person by two or more statuses held at the same time (p. 118).

role exit A situation in which people disengage from social roles that have been central to their self-identity (p. 119).

role expectation A group's or society's definition of the way a specific role ought to be played (p. 117).

role performance How a person plays a role (p. 117).

role strain The strain experienced by a person when incompatible demands are built into a single status that the person occupies (p. 118).

secondary group A larger, more specialized group in which the members engage in more impersonal, goal-oriented relationships for a limited time (p. 120).

self-fulfilling prophecy A situation in which a false belief or prediction produces behaviour that makes the originally false belief come true (p. 125).

social construction of reality The process by which our perception of reality is shaped largely by the subjective meaning that we give to an experience (p. 125).

social group A group that consists of two or more people who interact frequently and share a common identity and a feeling of interdependence (p. 120).

social institution A set of organized beliefs and rules that establishes how a society will attempt to meet its basic social needs (p. 121).

social interaction The process by which people act toward or respond to other people (p. 113).

social marginality The state of being part insider and part outsider in the social structure (p. 115).

social network A series of social relationships that link an individual to others (p. 120).

social structure The complex framework of societal institutions (such as the economy, politics, and religion) and the social practices (such as rules and social roles) that make up a society and that organize and establish limits on people's behaviour (p. 113).

status A socially defined position in a group or society characterized by certain expectations, rights, and duties (p. 115).

status set A term used to describe all the statuses that a person occupies at a given time (p. 116).

status symbol A material sign that informs others of a person's specific status (p. 117).

stigma According to Erving Goffman, any physical or social attribute or sign that so devalues a person's social identity that it disqualifies that person from full social acceptance (p. 115).

KEY FIGURES

Ferdinand Tönnies (1855–1936) German sociologist Ferdinand Tonnies used the terms *Gemeinschaft* (traditional societies) and *Gesellschaft* (large urban societies) to describe the degree of social solidarity and social control in different societies.

American Sociological Association, asanet.org

Erving Goffman (1922–1982) Canadian-born sociologist Erving Goffman, author of *The Presentation of Self in Everyday Life*, used a theatre metaphor in his study of social interaction. According to Goffman, day-to-day interactions are similar to a theatre production—we have a front stage and a back stage and we use "impression management" to ensure that our audience judges our performance favourably.

APPLICATION QUESTIONS

1. Think of a person you know well who often irritates you or whose behaviour grates on your nerves (it could be a parent, friend, relative, or teacher). First, list that person's statuses and roles. Then, analyze his or her possible role expectations, role performance, role conflicts, and role strains. Does anything you find in your analysis help to explain the irritating behaviour? (If not, change your method of analysis!) How helpful are the concepts of social structure in analyzing individual behaviour?
2. How does the structure of Canadian society influence the way in which we understand and respond to homelessness, both individually and collectively?
3. You are conducting field research on gender differences in nonverbal communication styles. How are you going to account for variations in age, ethnicity, and social class?
4. When communicating with other genders, ethnic groups, and ages, is it better to express and acknowledge different styles or to develop a common, uniform style? Why?

CHAPTER 6 FAMILIES

LEARNING OBJECTIVES
AFTER READING THIS CHAPTER, YOU SHOULD BE ABLE TO

LO-1 Explain why it is difficult to define family.

LO-2 Understand the key assumptions of functionalist, conflict, feminist, symbolic interactionist, and postmodernist perspectives on families.

LO-3 Understand the various options available to Canadian families in establishing families.

LO-4 Describe the challenges facing families today.

LO-5 Identify the primary problems facing Canadian families today.

CHAPTER FOCUS QUESTION

How is social change affecting the Canadian family?

Ruaridh Connellan/Barcroft Media/Landov

Madeline Rivers' article, "Shock and Confusion with Love," highlights the complexity of intimate relationships:

For a very private person, this is difficult for me to share. But the lack of stories and information about transgendered partners pushes me to share my own thoughts so that others might learn.

My spouse of 12 years informed me a few months ago of his transgenderedness. I became scared and confused. Yet, celebrating his future goal of happiness, i.e., becoming his true self, eventually became easy for me to accept. It just seemed natural. Many strange things that I look back upon, I can now understand. I just thought that he was gay. Well, my own woman's intuition needs a 30-year check up because I was only half right. I see that most of our lives together has enabled us to get him/her to this point. Brent/Betty is in the safest city, at a job with the most accommodating company policies, and available support groups. Thank goodness for his high intellect that lets him think things through. If I had been approached at the beginning of our relationship, I doubt I'd have written this article.

Information, information, information. I need pamphlets, anything to see what others have felt and done. What happens to the spouse of a M2F?

I've been shoved into the rarest of rare categories. Why? Because most partners leave before the shock wears off. Did I think about it? Of course I did, I'm human; but I'm also in love with a rare and wonderful person. Brent/Betty is still the same person I've loved all these years; he/she just wants to do it in heels!

What a mind-bender! He says he loves me and his attraction is the same. So, boy/girl loves and is attracted to girl. Girl still loves boy/girl but the attraction is not natural to me. Our love for each other is the same. While he/she is dealing with gender issues, it forces me to deal with my sexuality issues. I feel like I'm in a dramatic soap opera. Living like this certainly isn't boring.

Source: Rivers, Madeline. "Shock and confusion with love," *Transforming families: Real stories about Transgendered loved ones.* Ed. Mary Boenke. 2nd ed. Oak Knoll Press, 2003. 59-60. Reproduced by permission.

Today, when we think of families, we think of diversity and change, and exceptions are the rule. The experiences of the family in the above narrative certainly are not unique. Other variations on what has been described as a family are also common. Separation and divorce, remarriage, and blended or reconstituted families are a reality for many Canadians. Regardless of the form it takes, family life continues to be a source of great personal satisfaction and happiness.

In this chapter, we will examine the diversity and complexity of families Pressing social issues, such as same-sex marriage, divorce, child care, and new reproductive technologies, will be used as examples of how families and intimate relationships continue to change. Before reading on, test your knowledge about the changing family by taking the quiz in Box 6.1.

CRITICAL THINKING QUESTIONS

1. It has been suggested that variations of the "traditional" family are common today. Identify some of these alternative family forms.
2. Which, if any, heterosexist institutions continue to place exclusionary pressures on families with same-sex parents today?

LO-1 DEFINING FAMILY

What is a family? Although we all have a family of some form or another and we all understand the concept of family, it is not an easy word to define. More than ever, this term means different things to different people. As the nature of family life and work has changed in high-, middle-, and low-income nations, the issue of what constitutes a family has been widely debated. For example, Hutterite families in Canada live in communal situations in which children from about the age of three spend most of their days in school. In this case, the community is the family, as opposed to a traditional nuclear family.

Some Indigenous families in Canada also tend to have a much broader idea of family membership. Children are often cared for by relatives in the extended family. A social worker may define a family as consisting of parents and children only. Some Indigenous parents may be perceived as neglecting their children when the parents feel they are safe and well cared for by "their family"—that is, by uncles, grandparents, siblings, or other relatives (Vanier Institute of the Family, 2009).

Similarly, gay men and lesbians often form unique families. Many gay men and lesbians have **families we choose**—social arrangements that include intimate relationships between couples and close familial relationships with other couples, as well as with other adults and children (Ambert, 2009).

▶ **families we choose** Social arrangements that include intimate relationships between couples and close familial relationships with other couples, as well as with other adults and children.

In a society as diverse as Canada, talking about "a family" as though a single type of family exists or ever did exist is inaccurate. In reality, different groups will define their family lives in unique ways, depending on a number of factors, such as their socioeconomic background, immigrant status, religious beliefs, or cultural practices and traditions (Baker, 2009).

For many years, a standard sociological definition of family has been a group of people who are related to one another by bonds of blood, marriage, or adoption and who live together, form an economic unit, and bear and raise children. Many people believe that this definition should not be expanded—that social approval should not be extended to other relationships simply because the persons in those relationships wish to be considered a family. Others, however, challenge this definition because it simply does not match the reality of family life in contemporary society. Today's families include many types of living arrangements and relationships, including single-parent households, unmarried couples, lesbian and gay couples, and multiple genera-

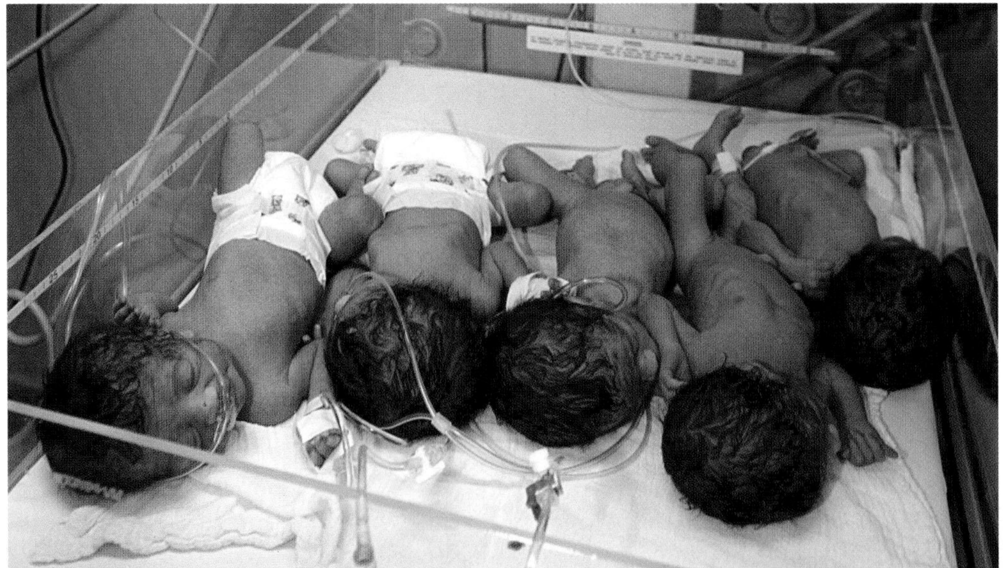

Abid Katib/Getty Images

Despite the idealized image of "the family," North American families have undergone many changes in the past century, as exemplified by the increase in the number of families using assisted reproductive technologies to start a family.

tions (such as grandparents, parents, and children) living in the same household. To accurately reflect these changes in family life, we need an encompassing definition of what constitutes a family. Accordingly, we will define a **family** as a relationship in which people live together with commitment, form an economic unit and care for any young, and consider their identity to be significantly attached to the group. Sexual expression and parent–children relationships are a part of most, but not all, family relationships.

▶ **family** A relationship in which people live together with commitment, form an economic unit and care for any young, and consider their identity to be significantly attached to the group.

In our study of families, we will use our sociological imaginations to see how our personal experiences are related to the larger happenings in our society. At the microlevel, each of us has our own "biography," based on our experience within a family; at the macrolevel, our families are embedded in a specific social context that has a major impact on them. We will examine the institution of the family at both of these levels, beginning with family structure and characteristics.

FAMILY STRUCTURE AND CHARACTERISTICS

In preindustrial societies, the primary form of social organization is through kinship ties. **Kinship** refers to a social network of people based on common ancestry, marriage, or adoption. Through kinship networks, people cooperate so that they can acquire the basic necessities of life, including food and shelter. Kinship systems can also serve as a means by which property is transferred, goods are produced and distributed, and power is allocated.

▶ **kinship** A social network of people based on common ancestry, marriage, or adoption.

In industrialized societies, other social institutions fulfill some of the functions previously taken care of by the kinship network. For example, political systems provide structures of social control and authority, and economic systems are responsible for the production and distribution of goods and services. Consequently, families in industrialized societies serve fewer and more specialized purposes than do families in preindustrial societies. Contemporary families are primarily responsible for regulating sexual activity, socializing children, and providing affection and companionship for family members.

▶ **family of orientation** The family into which a person is born and in which early socialization usually takes place.

▶ **family of procreation** The family that a person forms by having or adopting children.

FAMILIES OF ORIENTATION AND PROCREATION During our lifetime, many of us will be members of two different types of families—a family of orientation and a family of procreation. The **family of orientation** is the family into which a person is born and in which early socialization usually takes place. Although most people are related to members of their family of orientation by blood ties, those who are adopted have a legal tie that is patterned after a blood relationship. The **family of procreation** is the family that a person forms by having or adopting children. Both legal and blood ties are found in most families of procreation. The relationship between a husband and wife is based on legal ties; however, the relationship between a parent and

BOX 6.1 SOCIOLOGY AND EVERYDAY LIFE

How Much Do You Know About the Changing Family in Canada?

True	False	
T	F	1. Today, people in Canada are more inclined to get married than at any time in history.
T	F	2. Men are as likely as women to be single parents.
T	F	3. One out of every two marriages ends in divorce.
T	F	4. Age of first marriage has increased significantly in the past 40 years for both men and women.
T	F	5. In recent years, the number of extended families where members live together in the same home has decreased.

For answers to the quiz about the changing family in Canada, go to **www.nelson.com/student.**

▶ **extended family** A family unit composed of relatives in addition to parents and children who live in the same household.

▶ **nuclear family** A family made up of one or two parents and their dependent children, all of whom live apart from other relatives.

▶ **marriage** A legally recognized and/or socially approved arrangement between two or more individuals that carries certain rights and obligations and usually involves sexual activity.

▶ **monogamy** marriage to one person at a time.

child may be based on either blood ties or legal ties, depending on whether the child has been adopted or is by marriage.

EXTENDED AND NUCLEAR FAMILIES Sociologists distinguish between extended and nuclear families based on the number of generations that live within a household. An **extended family** is a family unit composed of relatives in addition to parents and children who live in the same household. These families often include grandparents, uncles, aunts, or other relatives who live in close proximity to the parents and children, making it possible for family members to share resources. In horticultural and agricultural societies, extended families are extremely important; having a large number of family members participate in food production may be essential for survival. Today, extended families are becoming more common across North America and Britain. This trend is related to an increase in the number of families caring for aging seniors in their homes, an increase in the number of grandparents with children and grandchildren living with them for economic reasons, and an increase in immigration from countries where extended family living is the norm (Milan, Keown, and Robles Urquijo, 2011).

A **nuclear family** is a family composed of one or two parents and their dependent children, all of whom live apart from other relatives. A traditional definition specifies that a nuclear family is made up of a "couple" and their dependent children; however, this definition became outdated as a significant shift occurred in the family structure. A recent trend in Canadian families is that there are now more families without children than families with children. As shown in the Census Profile, in 2011 about 39 percent of all households were composed of couples with children under the age of 18, while almost 45 percent of couples did not have children living at home. This latter group consisted of childless couples and empty nesters, or couples whose children no longer lived at home (Statistics Canada, 2012d).

Nuclear families are smaller than they were 20 years ago; whereas the average family size in 1971 was 3.7 persons, in 2011 it was 2.9 persons. This decrease has been largely attributed to a decline in the overall fertility rate after the baby boom and in a significant increase in the number of lone-parent families.

CENSUS PROFILE

Changes in Family Structure

Changes to the family structure include a decline in marriages, an increase in common-law unions, and an increase in single-parent families.

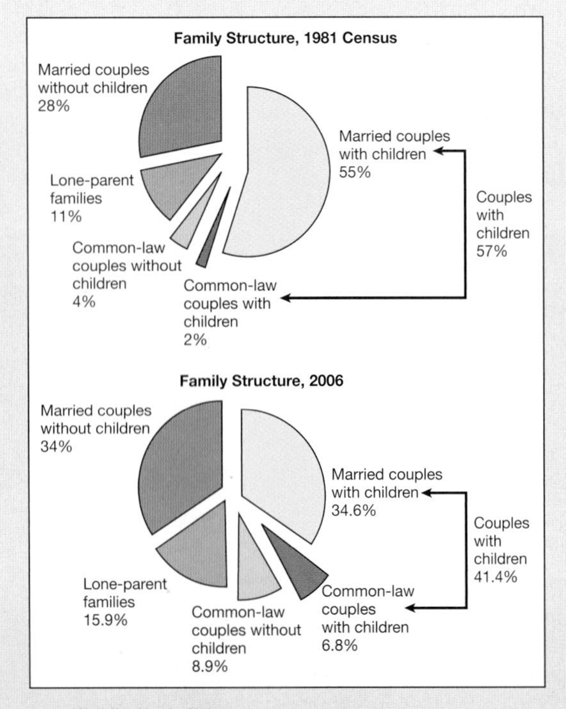

Source: Statistics Canada, 2012a.

MARRIAGE PATTERNS

Across cultures, families are characterized by different forms of marriage. **Marriage** is a legally recognized and/or socially approved arrangement between two or more individuals that carries certain rights and obligations and usually involves sexual activity. In Canada, the only legally sanctioned form of marriage is **monogamy**—marriage to one person at a time. For some people, marriage is a lifelong commitment that ends only with the death of a partner.

Members of some religious groups believe that marriage is "forever"; if one spouse dies, the surviving spouse is precluded from marrying anyone else. For others, marriage is a commitment of indefinite duration. Through a pattern of marriage, divorce, and remarriage, some people practise *serial monogamy*—a succession of marriages in which a person has several spouses over a lifetime but is legally married to only one person at a time.

Polygamy refers to the concurrent marriage of a person of one sex with two or more members of the opposite sex. The most prevalent form of polygamy is **polygyny**—the concurrent marriage of one man with two or more women. Polygyny has been practised in a number of societies, including parts of Europe until the Middle Ages. More recently, some marriages in Islamic societies in Africa and Asia have been polygynous; however, the cost of providing for multiple wives and numerous children makes the practice impossible for all but the wealthiest men. In addition, because roughly equal numbers of women and men live in these areas, this nearly balanced sex ratio tends to limit polygyny.

The second type of polygamy is **polyandry**—the concurrent marriage of one woman with two or more men. Polyandry is rare; when it does occur, it is typically found in societies where men greatly outnumber women because of high rates of female infanticide.

▶ **polygamy** The practice of having more than one spouse at a time.

▶ **polygyny** The concurrent marriage of one man with two or more women.

▶ **polyandry** The concurrent marriage of one woman with two or more men.

PATTERNS OF DESCENT AND INHERITANCE

Even though a variety of marital patterns exist across cultures, virtually all forms of marriage establish a system of descent so that kinship can be determined and inheritance rights established. In preindustrial societies, kinship is usually traced through one parent (unilineally). The most common pattern of unilineal descent is **patrilineal descent**—a system of tracing descent through the father's side of the family. Patrilineal systems are set up in such a manner that a legitimate son inherits his father's property and sometimes his position upon the father's death. In nations such as India, where boys are seen as permanent patrilineal family members and girls are seen only as temporary family members, girls tend to be considered more desirable than boys.

Even with the less common pattern of **matrilineal descent**—a system of tracing descent through the mother's side of the family—women may not control property. However, inheritance of property and position is usually traced from the maternal uncle (mother's brother) to his nephew (mother's son). In some cases, mothers may pass on their property to daughters.

By contrast, in industrial societies, kinship is usually traced through both parents (bilineally). The most common form is **bilateral descent**—a system of tracing descent through both the mother's and father's sides of the family. This pattern is used in Canada for the purpose of determining kinship and inheritance rights; however, children typically take the father's last name.

▶ **patrilineal descent** A system of tracing descent through the father's side of the family.

▶ **matrilineal descent** A system of tracing descent through the mother's side of the family.

▶ **bilateral descent** A system of tracing descent through both the mother's and father's sides of the family.

▶ **patriarchal family** A family structure in which authority is held by the eldest male (usually the father).

▶ **matriarchal family** A family structure in which authority is held by the eldest female (usually the mother).

POWER AND AUTHORITY IN FAMILIES

Descent and inheritance rights are intricately linked with patterns of power and authority in families. A **patriarchal family** is a family structure in which authority is held by the eldest male (usually the father). The male authority figure acts as head of the household and holds power and authority over the women and children as well as over other males. A **matriarchal family** is a family structure in which authority is held by the eldest female (usually the mother). In this case, the female authority figure acts as head of the household. Although there has been a great deal of discussion about matriarchal societies, scholars have found no historical evidence to indicate that true matriarchies ever existed.

The most prevalent pattern of power and authority in families is patriarchy—a hierarchical system of social organization in which cultural, political, and economic structures are controlled by men. Across cultures, men are the primary (and often sole) decision makers regarding domestic, economic, and social concerns facing the family. The existence of patriarchy may give men a sense of power over their

The Kobal Collection at Art Resource, NY

The reality television show *Sister Wives*, which documents the life of a polygamist family that includes a husband with four wives (he divorced one wife in 2014) and 17 children, certainly challenges traditional notions of family.

own lives, but it also can create an atmosphere in which some men feel greater freedom to abuse women and children.

▶ **egalitarian family** A family structure in which both partners share power and authority equally.

An **egalitarian family** is a family structure in which both partners share power and authority equally. In egalitarian families, issues of power and authority may be frequently negotiated as the roles and responsibilities within the relationship change over time. Recently, a trend toward more-egalitarian relationships has been evident in a number of countries as women have sought changes in their legal status and increased educational and employment opportunities. Some degree of economic independence makes it possible for women to delay marriage or to terminate a problematic marriage. Recent cross-national studies have found that larger increases in the proportion of women who have higher levels of education, who hold jobs with higher wages, who have more commitment to careers outside the family, and who have greater interest in gender equality all contribute to the support of egalitarian gender values in the larger society as these ideas eventually spread to others (Pampel, 2011).

TIME TO REVIEW

- Describe how Canadian families have changed in recent years in their structure and characteristics. In particular, comment on changes in nuclear and extended families, patterns of descent and inheritance, and power and authority in families.
- What social factors have driven these changes?

LO-2 THEORETICAL PERSPECTIVES ON FAMILIES

▶ **sociology of family** The subdiscipline of sociology that attempts to describe and explain patterns of family life and variations in family structure.

The **sociology of family** is the subdiscipline of sociology that attempts to describe and explain patterns of family life and variations in family structure. Functionalist perspectives emphasize the functions that families perform at the macrolevel of society, while conflict and feminist perspectives focus on families as a primary source of social inequality. By contrast, symbolic interactionists examine microlevel interactions that are integral to the roles of different family members. Finally, postmodern theorists emphasize the fact that families today are diverse and variable.

FUNCTIONALIST PERSPECTIVES

Functionalists emphasize the importance of the family in maintaining the stability of society and the well-being of individuals. According to Émile Durkheim, marriage is a microcosmic replica of the larger society; both marriage and society involve a mental and moral fusion of physically distinct individuals. Durkheim also believed that a division of labour contributed to greater efficiency in all areas of life—including marriages and families—even though he acknowledged that this division imposed significant limitations on some people.

Talcott Parsons was a key figure in developing a functionalist model of the family. According to Parsons (1955), the husband/father fulfills the *instrumental role* (meeting the family's economic needs, making important decisions, and providing leadership), while the wife/mother fulfills the *expressive role* (running the household, caring for children, and meeting the emotional needs of family members).

Contemporary functionalist perspectives on families derive their foundation from Durkheim and Parsons. Division of labour makes it possible for families to fulfill a number of functions that no other institution can perform as effectively. In advanced industrial societies, families serve four key functions:

1. *Sexual regulation.* Families are expected to regulate the sexual activity of their members and thus control reproduction so that it occurs within specific boundaries. At the macrolevel,

incest taboos prohibit sexual contact or marriage between certain relatives. For example, virtually all societies prohibit sexual relations between parents and their children and between brothers and sisters.
2. *Socialization.* Parents and other relatives are responsible for teaching children the necessary knowledge and skills to survive. The smallness and intimacy of families makes them best suited for providing children with the initial learning experiences they need.
3. *Economic and psychological support.* Families are responsible for providing economic and psychological support for members. In preindustrial societies, families are economic production units; in industrial societies, the economic security of families is tied to the workplace and to macrolevel economic systems. In recent years, psychological support and emotional security have been increasingly important functions of the family.
4. *Provision of social status.* Families confer social status and reputation on their members. These statuses include the ascribed statuses with which individuals are born, such as race and ethnicity, nationality, social class, and sometimes religious affiliation. One of the most significant and compelling forms of social placement is the family's class position and the opportunities (or lack thereof) resulting from that position. Examples of class-related opportunities include access to quality health care, higher education, and a safe place to live.

Functionalist theorists believe that families serve a variety of important functions that no other social institution can adequately fulfill. In contrast, conflict and feminist analysts believe that the functionalist perspective is idealistic and inadequate for explaining problems in contemporary families.

CONFLICT PERSPECTIVES

Both conflict and feminist analysts view functionalist perspectives on the role of the family in society as idealized and inadequate. Rather than operating harmoniously and for the benefit of all members, families are sources of social inequality and conflict over values, goals, and access to resources and power.

According to some classical conflict theorists, families in capitalist economies are similar to the work environment of a factory. Men in the home dominate women in the same manner that capitalists and managers in factories dominate their workers (Engels, 1970/1884). Although childbearing and care for family members in the home contribute to capitalism, these activities also reinforce the subordination of women through unpaid (and often devalued) labour. Other conflict analysts are concerned with the effect that class conflict has on the family. The exploitation of the lower classes by the upper classes contributes to family problems such as high rates of divorce and overall family instability.

FEMINIST PERSPECTIVES

The contributions of feminist theorists have resulted in radical changes in the sociological study of families. Feminist theorists have been primarily responsible for redefining the concept of the family by focusing on the diversity of family arrangements. Some feminist scholars reject the "monolithic model of the family" (Eichler, 1981:368), which idealizes one family form—the family with a male breadwinner and stay-at-home wife and children—as the normal household arrangement. Feminist theorists argue that limiting our concept of family to this traditional form means ignoring or undervaluing diverse family forms, such as single-parent families, childless families, gay or lesbian families, and stepfamilies. Roles within the family are viewed by feminist theorists as primarily socially constructed rather than biologically determined (Smith, 1974). Feminist scholars have challenged a number of common assumptions about family life and the roles women fulfill within families. For example, they question whether all "real" women want to be mothers and whether the inequality between traditional husbands and wives is "natural" (Mandell and Duffy, 2011).

Feminist perspectives on inequality focus on patriarchy. From this viewpoint, men's domination over women existed long before private ownership of property and capitalism (Mann, 1994). Women's subordination is rooted in patriarchy and men's control over women's labour power (Hartmann, 1981). The division of labour by gender, both within the larger society and within households, is a fundamental focus of feminist analysis (Luxton and Corman, 2001, cited in Ambert, 2006:21). Although the division of labour may appear to be an equal pooling of contributions within the family unit, feminist scholars view women as giving much but receiving less in return. According to sociologist Patricia Mann, "Male power in our society is expressed in economic terms even if it does not originate in property relations; women's activities in the home have been undervalued at the same time as their labor has been controlled by men" (1994:42).

Feminist perspectives on families also draw attention to the problems of dominance and subordination inherent in relationships. Specifically, feminist theorists have acknowledged what has been described as the "dark side of the family," focusing research efforts on issues such as child abuse, wife abuse, and violence against the elderly (Ambert, 2001; Johnson and Dawson, 2011; Smith, 1985). The idea that family relations, including wife abuse and child abuse, are private, personal matters has been challenged by feminists and successfully brought into the public domain of social policy and legislative changes. As a result, feminist analysis of families is viewed not only as a theoretical perspective, but also as a broad movement for social change (Johnson and Dawson, 2011).

SYMBOLIC INTERACTIONIST PERSPECTIVES

Early symbolic interactionists, such as Charles Horton Cooley and George Herbert Mead, provided key insights on the roles we play as family members and how we modify or adapt our roles to the expectations of others—especially significant others, such as parents, grandparents, siblings, and other relatives. How does the family influence the individual's self-concept and identity? Contemporary symbolic interactionist perspectives examine the roles of husbands, wives, and children as they act out their own parts and react to the actions of others. From such a perspective, what people think, as well as what they say and do, is very important in understanding family dynamics.

According to sociologists Peter Berger and Hansfried Kellner (1964), interaction between marital partners contributes to a shared reality. Although newlyweds bring separate identities to a marriage, over time they construct a shared reality as a couple. In the process, the partners redefine their past identities to be consistent with new realities. Development of a shared reality is a continuous process, taking place not only in the family but also in any group in which the couple participates together. Divorce is the reverse of this process; couples may start with a shared reality and, in the process of uncoupling, gradually develop separate realities).

Symbolic interactionists explain family relationships in terms of the subjective meanings and everyday interpretations that people give to their lives. As the sociologist Jessie Bernard (1982/1973) pointed out, women and men experience marriage differently. Although the husband may see *his* marriage very positively, the wife may feel less positive about *her* marriage, and vice versa. Researchers have found that husbands and wives may give very different accounts of the same event and that their "two realities" frequently do not coincide.

POSTMODERN PERSPECTIVES

According to postmodern theories, we have experienced a significant decline in the influence of the family and other social institutions. As people have pursued individual freedom, they have been less inclined to accept the structural constraints imposed on them by institutions. Given this assumption, how might a postmodern perspective view contemporary family life? For example, how might this approach answer the question, "How is family life different in the information age"?

The postmodern family has been described as *permeable*—capable of being diffused or invaded in such a manner that the family's original nature is modified or changed (Elkind,

CONCEPT SNAPSHOT

FUNCTIONALIST PERSPECTIVES
Key thinker: Talcott Parsons

In modern societies, families serve the functions of sexual regulation, socialization, economic and psychological support, and provision of social status.

CONFLICT PERSPECTIVES
Key thinker: Friedrich Engels

Families both mirror and help perpetuate social inequalities based on class and gender.

FEMINIST PERSPECTIVES
Key thinkers: Nancy Mandell, Ann Duffy

Women's subordination is rooted in patriarchy and men's control over women's labour power.

SYMBOLIC INTERACTIONIST PERSPECTIVES
Key thinker: Jessie Bernard

Family dynamics, including communication patterns and the subjective meanings that people assign to events, mean that interactions within families create a shared reality.

POSTMODERN PERSPECTIVES
Key thinker: David Elkind

In postmodern societies, families are diverse and fragmented. Boundaries between the workplace and home are also blurred.

1995). From this approach, if the nuclear family is a reflection of the age of modernity, the permeable family reflects the postmodern assumptions of difference and irregularity. This is evident in the fact that the nuclear family is now only one of many family forms. Similarly, under modernity, the idea of romantic love has given way to the idea of consensual love: Individuals agree to have sexual relations with others they have no intention of marrying or, if they marry, do not necessarily see the marriage as having permanence. Maternal love has also been transformed into shared parenting, which includes not only mothers and fathers but also caregivers, who may be either relatives or nonrelatives.

Urbanity is another characteristic of the postmodern family. The boundaries between the public sphere (the workplace) and the private sphere (the home) are becoming more open and flexible. As a result, family life may be negatively affected by the decreasing distinction between what is work time and what is family time. As more people are becoming connected "24/7" (24 hours a day, seven days a week), the boss who before would not have called at 11:30 p.m. or when an employee was on vacation may send an email or text asking for an immediate response to some question that has arisen while the person is away with family members.

Social theorist Jean Baudrillard's idea that the simulation of reality may come to be viewed as "reality" by some people can be applied to family interactions in the Information Age. Does the ability to contact someone anywhere and any time of the day or night provide greater happiness and stability in families?

The Concept Snapshot summarizes these sociological perspectives on the family. Taken together, these perspectives on the social institution of families help us understand both the good and bad sides of familial relationships. Now, we shift our focus to love, marriage, intimate relationships, and family issues in Canada.

LO-3 ESTABLISHING FAMILIES

COHABITATION

Cohabitation refers to the sharing of a household by a couple who live together without being legally married. Attitudes about cohabitation have changed in the last few decades, something that is reflected in Figure 6.1. In Canada, cohabitation (most commonly referred to as a common-law union) has become an increasingly popular alternative to marriage. Since the early 1980s, the number of persons living common-law has doubled, going from 700,000 in

cohabitation The sharing of a household by a couple who live together without being legally married.

FIGURE 6.1 : PROPORTION OF COMMON-LAW FAMILIES GROWS WHILE IT DECLINES FOR MARRIED FAMILIES, CANADA

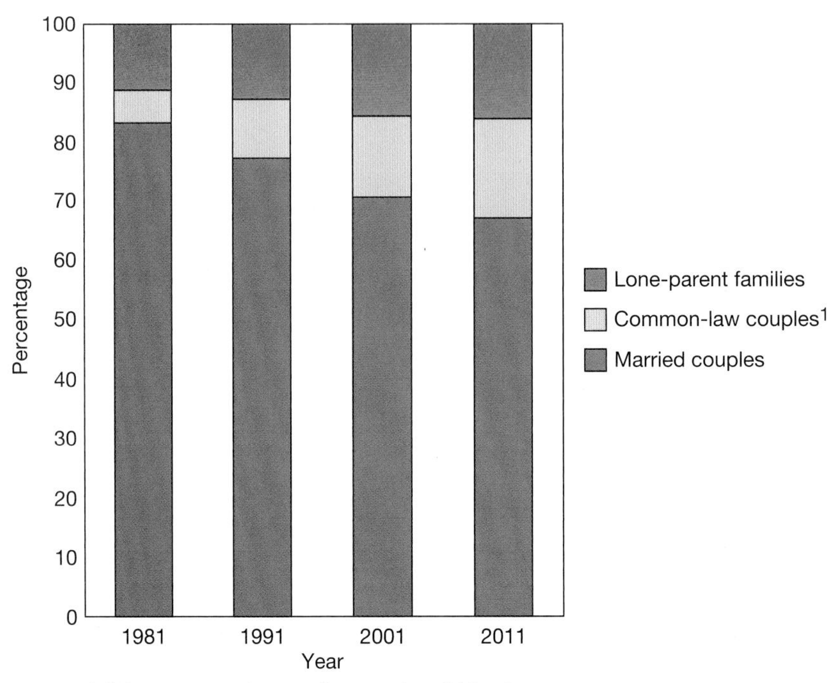

Source: Statistics Canada, 2012a.

1981 to over 1.5 million in 2011 (Statistics Canada, 2012e). In fact, another recent trend is that for the first time, there are more common-law-couple families than lone-parent families in Canada. Almost half of these common-law-couple families included children, whether born to the current union or brought to the family from previous unions. The proportion of people in common-law unions varies considerably by province. In Quebec, one in three couples lives common-law, making it the province with the highest rate of common-law families.

Those most likely to cohabit are young adults between the ages of 25 and 29. Based on the Statistics Canada census data, approximately one in four Canadians in this age group lives in a common-law union. However, the largest growth in common-law unions has occurred among couples in their early 60s, reflecting a growing acceptance among older generations of what was previously a living arrangement of young adults. While "living together" is often a prelude to marriage for young adults, common-law unions are also becoming a popular alternative both to marriage and to remarriage following divorce or separation (Milan, Keown, and Robles Urquijo, 2011).

Today, some people view cohabitation as a form of "trial marriage," but for others, cohabitation is not a first step toward marriage. Some people who have cohabited do eventually marry the person with whom they have been living, whereas others do not. And studies over the past decade have supported the proposition that couples who cohabit before marriage do not necessarily have a stable relationship once they are married (Clark and Crompton, 2006).

MARRIAGE

Despite the prevalence of divorce in our society, marriage continues to be an extremely popular institution. The majority of Canadians will marry at some point in their lives. Furthermore, although marriages today experience many problems, for better or worse the majority of marriages in Canada do last a lifetime (Ward and Belanger, 2011).

Why do people get married? Couples get married for a variety of reasons. Some do so because they are "in love," desire companionship and sex, want to have children, feel social pressure, are attempting to escape from a bad situation in their parents' home, or believe that they will have more money or other resources if married. These factors notwithstanding, the selection of a marital partner is fairly predictable. Most people in Canada tend to choose marriage partners who are similar to themselves. **Homogamy** refers to the pattern of individuals marrying those who have similar characteristics, such as race/ethnicity, religious background, age, education, or social class. However, homogamy provides only the general framework within which people select their partners; people are also influenced by other factors. For example, some researchers claim that people want partners whose personalities match their own in significant ways. As a result, people who are outgoing and friendly may be attracted to people with those same traits. Other researchers, however, claim that people look for partners whose personality traits differ from but complement their own.

homogamy The pattern of individuals marrying those who have similar characteristics, such as race/ethnicity, religious background, age, education, or social class.

Regardless of the individual traits of marriage partners, research indicates that communication and emotional support are crucial to the success of marriages. Common marital problems include lack of emotional intimacy, poor communication, and lack of companionship. One study concluded that for many middle- and upper-income couples, women's paid work was critical to the success of their marriages. People who have a strong commitment to their work have two distinct sources of pleasure—work and family. For members of the working class, however, work may not be a source of pleasure. For all women and men, balancing work and family life is a challenge (Marshall, 2011).

REMARRIAGE

Most people who divorce get remarried. In recent years, over 30 percent of all marriages were between previously married brides and/or grooms. Among individuals who divorce before age 35, about half will remarry within three years of their first divorce. Most divorced people remarry others who have been divorced. However, remarriage rates vary by gender and age. A greater proportion of men than women remarry, often relatively soon after the divorce, regardless of age (Ambert, 2009).

As a result of divorce and remarriage, complex family relationships are often created. Some people become part of stepfamilies or *blended families,* which consist of a husband and wife, children from previous marriages, and children (if any) from the new marriage.

There has been a dramatic increase in the number of blended families in North America over the last 30 years. In Canada, stepfamilies were counted for the first time in the 2011 census. Today, there are just under a half a million stepfamilies in Canada (Statistics Canada, 2012d). At least initially, levels of family stress may be fairly high because of rivalry among the children and hostilities directed toward stepparents or babies born into the family. In spite of these problems, however, many blended families succeed. The family that results from divorce and remarriage is typically a complex, binuclear one in which children may have a biological parent and a stepparent, biological siblings and stepsiblings, and an array of other relatives, including aunts, uncles, and cousins.

The norms governing divorce and remarriage are ambiguous. Because there are no clear-cut guidelines, people must make decisions about family life (such as decisions about Christmas, birthdays, and weddings) based on their own beliefs and feelings about the people involved. Consider the following example of a couple who both brought children to their new marriage:

> The first couple of years after we got married, we had two Christmas trees, one in the living room, and one in the rec room. Once they [Angie's biological children and stepchildren] got older,

Geostock/Getty Images

Dual-earner marriages are a challenge for many children as well as their parents. While parents are at work, latchkey children are often at home alone.

> I could say to them, "It's a really big hassle," but the first couple of years, there was so much decoration and everybody wanted everything: my kids wanted this because we had always had it on our tree, and the other kids wanted that because they had it on their tree. (Church, 2003:66)

Because there are few norms governing these family relationships, it may take years for them to come together, clarify roles and responsibilities, and establish a solid identity as a family unit (Preece, 2004).

TIME TO REVIEW

- Compare and contrast marriage versus cohabitation. Based on what you have learned about these forms of establishing committed relationships, which do you predict will be most common in your generation?

LO-4 CHILD-RELATED FAMILY ISSUES AND PARENTING

Not all couples become parents. Those who decide not to have children often consider themselves "child free," whereas those who do not produce children through no choice of their own may consider themselves "childless."

DECIDING TO HAVE CHILDREN

Cultural attitudes about having children and the ideal family size began to change in North America in the late 1950s. On average, women are now having 1.6 children. However, rates of fertility differ across racial and ethnic categories. For example, Indigenous women have a total fertility rate of 2.6 (O'Donnell and Wallace, 2011; Statistics Canada, 2012e).

Over the past five decades, advances in birth control techniques—including the birth control pill and contraceptive patches and shots—have made it possible for people to decide whether they want to have children and how many they wish to have, and to determine (at least somewhat) the spacing of their births. Sociologists suggest, however, that fertility is linked not only to reproductive technologies but also to women's beliefs that they do or do not have other opportunities in society that are viable alternatives to childbearing (Lamanna and Riedmann, 2012).

The concept of reproductive freedom includes both the desire *to have* or *not to have* one or more children. Women, more often than men, are the first to choose a child-free lifestyle. By age 40, more than 10 percent of Canadian women intend to remain child free (Edmonston, Lee, and Wu, 2008). However, the desire not to have children often comes in conflict with our society's *pronatalist bias,* which assumes that having children is the norm and can be taken for granted, while those who choose not to have children believe they must justify their decision to others (Lamanna and Riedmann, 2012). Many diverse reasons account for why individuals decide not to have children, including never having wanted any, not finding themselves in the right circumstances, and having religious or environmental concerns (Stobert and Kemeny, 2003).

Some couples experience involuntary infertility, whereby they want to have a child but they are physically unable to do so. **Infertility** is defined as an inability to conceive after one year of unprotected sexual relations. Research suggests that fertility problems originate in females in approximately 30 to 40 percent of cases and with males in about 40 percent of cases; in the other approximately 20 percent of cases, the cause is impossible to determine. It is estimated that about half of infertile couples who seek treatments, such as fertility drugs, artificial insemination, and surgery to unblock

▶ **infertility** An inability to conceive after one year of unprotected sexual relations.

fallopian tubes, can be helped; however, some are unable to conceive despite expensive treatments such as in vitro fertilization, which costs as much as $15,000 per attempt (IVF.ca, 2012).

People who are involuntarily childless may choose to become parents by adopting a child.

ADOPTION

Adoption is a legal process through which the rights and duties of parenting are transferred from a child's biological and/or legal parents to new legal parents. This procedure gives the adopted child all the rights of a biological child. In most adoptions, a new birth certificate is issued and the child has no future contact with the biological parents. In Canada, adoption is regulated provincially. Therefore, adopted persons' access to information regarding their "biological parents" varies, as does their desire to access this information.

Matching children who are available for adoption with prospective adoptive parents can be difficult. The available children have specific needs, and the prospective parents often set specifications on the type of child they want to adopt. There are fewer infants available for adoption today than in the past because better means of contraception exist, abortion is more readily available, and more single parents decide to keep their babies. As a result, many prospective parents pursue international adoptions from countries including China, Haiti, South Korea, and India (Mandell & Duffy, 2011)

ASSISTED REPRODUCTIVE TECHNOLOGIES

In recent years, there has been an explosion of research, clinical practice, and experimentation in the area of reproductive technology. These procedures, in particular, have raised some controversial ethical issues in terms of what role medical science should play in the creation of human life (Marquardt, 2006).

Procedures used in the creation of new life, such as artificial insemination and in vitro fertilization, are referred to as "methods of assisted reproduction" (Achilles, 1996). Artificial insemination is the oldest, simplest, and most common type of assisted reproduction. The most common form of artificial insemination is *intrauterine insemination,* which involves a physician inserting sperm directly into the uterus near the time of ovulation. Inseminations may be performed with donor sperm.

Intrauterine insemination with donor sperm raises several complex issues concerning its moral, legal, and social implications. In most cases, the woman is given no information about the donor, and the donor is not told if a pregnancy has occurred. The result of this anonymity is that neither the mother nor the individuals conceived through donor insemination will have access to information regarding the biological father (Achilles, 1996). The term *test-tube baby* is often used incorrectly to describe babies conceived through in vitro fertilization. A real test-tube baby would require conception, gestation, and birth to occur outside of a woman's body. To date, this technology has not been developed (Achilles, 1996). *In vitro* (Latin for "in glass") *fertilization* involves inducing ovulation, removing the egg(s) from a woman, fertilizing the egg(s) with the sperm in a petri dish, and then implanting the fertilized egg(s) (embryos) into the woman.

Another alternative available to couples with fertility problems is the use of a surrogate, or substitute, mother to carry a child for them. There are two types of surrogacy. In *traditional surrogacy,* the surrogate is artificially inseminated with the father's sperm. In this case, the egg is the surrogate's and the child is biologically related to the surrogate and the father. This type of surrogacy is typically used in cases where the woman is infertile or when there is a risk of passing on a serious genetic disorder from mother to child. In the second type of surrogacy, *gestational surrogacy,* the sperm and the eggs from the infertile couple are transferred to the surrogate using an assisted reproductive technology (such as in vitro fertilization). With gestational surrogacy, the surrogate carries the child but is not biologically or genetically related to it. The genetic parents are the man and woman whose eggs and sperm were donated to the surrogate (see Box 6.2, Sociology in Global Perspective).

The availability of a variety of reproductive technologies is having a dramatic impact on traditional concepts of the family and parenthood. In light of all the assisted reproductive technologies available, what does the term *parent* mean? How many "parents" does the child have? How do the children conceived with assisted reproductive technologies define their families? There are now approximately one million donor-conceived children in the world. Now that they are able to speak for themselves, these children have raised some difficult questions about the rights of the child, biology, identity, and families. Some of these issues are highlighted in the following narrative:

> Is it right to deprive people of knowing who their natural parents are? What happens to your sense of identity when one of your biological parents is missing? Is there a difference when you're raised by "social" rather than biological parents? What if those parents are two women, or two men, or perhaps three people? Are children's understandings of parenthood as flexible as we would like to think? How do kids feel about all this? And do their feelings matter? (Wente, 2006:A21)

In 2004, the federal government enacted legislation to monitor and regulate assisted reproductive technologies. This legislation specifies what practices are forbidden: These include human cloning, sex selection, and buying or selling human embryos and sperm. It also outlines allowed practices, including surrogate mothers, donation of human sperm and embryos, and the use of human embryos and stem cells for scientific research (Department of Justice, 2009).

The issues raised by legislation have legal, social, moral, and ethical implications. For example, fertility clinics across the country can send cells from embryos conceived through in vitro fertilization to labs in the United States to test them for disorders the new parents want to avoid. As a result of these new technologies, it is possible to screen for a number of life-threatening conditions. However, it is also possible to screen for other conditions that are less severe and may never present themselves until late adulthood if ever. Most disturbing, however, is the suggestion that these new technologies may lead to the creation of what has been described as "unnatural selection," allowing parents to select for traits such as height, weight, hair and eye colour, and athletic ability (Abraham, 2012).

Despite these concerns and unintended consequences, these new reproductive technologies have enabled some infertile couples to become parents. For them, the benefits far outweigh the costs.

TIME TO REVIEW

- What are the some of the legal, social, moral, and ethical implications of assisted reproductive technologies?

SINGLE-PARENT HOUSEHOLDS

Single parenting is not a new phenomenon in Canada. However, one of the most significant changes in Canadian families is the dramatic increase in single-parent families. Today, there are more than one million single parents in Canada, and just under 80 percent of them are women (Statistics Canada, 2012a). In the past, most single-parent families were created when one parent died. Today, the major causes of single parenthood for women are divorce and separation.

Does living in a single-parent family put children at risk? According to research by sociologist Paul R. Amato (2005), changes in family structure such as an increase in the proportion of single-parent households tends to place children in situations where they experience a lower standard of living, receive less-effective parenting, experience less-cooperative co-parenting, are less emotionally close to both parents, and are subjected to more stressful events and circumstances than children who grow up in stable, two-parent families. Why does this occur? Because of factors such as economic hardships that force single-parent families to do without books, without computers, and without homes in better neighbourhoods and school districts. When this problem is coupled with lack of time for parenting while the single parent struggles to make ends meet, and the fact that many children lose contact with their fathers after separation or divorce, the quality of parenting is often less than that found in supportive, co-parent

BOX 6.2 SOCIOLOGY IN GLOBAL PERSPECTIVE

Wombs-for-Rent: Commercial Surrogacy in India

I wanted to be a surrogate mother because I wanted to deposit money into an account for my children for their future. I also wanted to help parents who cannot have children. I am proud to have given birth to a beautiful baby . . . I feel like part of the family.*

—Thapa, a 31-year-old Indian woman who became a surrogate for an Australian couple, explains why, in addition to her own children, she is willing to help other people become parents (quoted in AFP, 2013). Thapa works with a New Delhi, India, surrogacy centre, where she and other surrogate mothers earn about $6,000 for carrying a child for a foreign couple.

Surrogacy is legal in Canada; however, compensated or commercial (for fee/profit) surrogacy is prohibited. In other words, a surrogate mother in Canada cannot receive any sort of wage or fee for carrying a child. Why do some infertile couples in the Canada, United States, Britain, and elsewhere want to "hire" a woman in India to have their child? Most couples who engage in this practice have made numerous attempts to have a child through in vitro fertilization and other assisted reproductive technologies. If they have been unsuccessful in their efforts, the couple may first attempt to find a surrogate in Canada, but they quickly learn that gestational surrogate costs range from $30,000 to more than $50,000—far more than they would pay for a surrogate in India (Kohl, 2007; Surrogacy in Canada Online, 2015). According to infertility specialist Dr. Nayna Patel, earning money through surrogacy helps uplift Indian women: It provides money for their household and makes them more independent. For example, the typical woman might earn more for one surrogate pregnancy than she would earn in 15 years from other kinds of employment (CBS News, 2007).

Are there any problems with global "rent-a-womb"? If there is an agreement between a surrogate mother and a couple who badly wants a child, some analysts believe that "offer and acceptance" is nothing more than capitalism at work—where there is a demand (for infants by infertile couples), there will be a supply (from low-income surrogate mothers). However, some ethicists raise troubling questions about the practice of commercial surrogacy: A mother should give birth to her child because it is hers and she loves it, not because she is being paid to give birth to someone else's baby. Other social critics are concerned about the potential mistreatment of low-income women who may be exploited or may suffer long-term emotional damage from functioning as a surrogate (Dunbar, 2007). For the time being, in clinics such as the one in India, hopeful parents just provide the egg, the sperm, and the money, and all the rest is done for them by the clinic and the surrogates, who live in a spacious house where they are taken care of by maids, cooks, and doctors.

What are your thoughts on surrogacy? Is there any difference between surrogacy when it occurs in high-income nations such as the United States and Canada as compared to situations in which the parents live in a high-income nation and the surrogate mother lives in a lower-income nation? How might we relate the specific issue of outsourced surrogacy to some larger concerns about families and intimate relationships that we have discussed in this chapter?

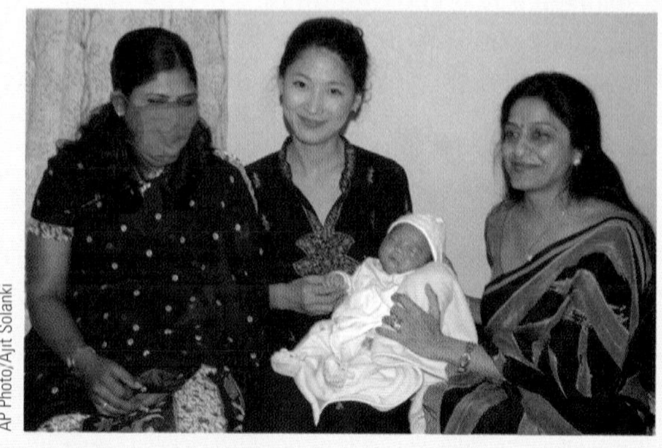

AP Photo/Ajit Solanki

A surrogate mother (left) has delivered a baby for Karen Kim (centre), with the help of infertility specialist Dr. Nayna Patel (right). This practice, sometimes called "rent-a-womb," remains controversial.

relationships. Even for a person with a stable income and a network of friends and family to help with child care, raising a child alone can be an emotional and financial burden.

Currently, men head close to one-fifth of lone-parent families; among many of the men, a pattern of "involved fatherhood" has emerged. While some single fathers remain actively involved in their children's lives, others may become less involved, spending time with their children around recreational activities and on special occasions. Sometimes, this limited role is by choice, but more often it is caused by workplace demands on time and energy, the location of the

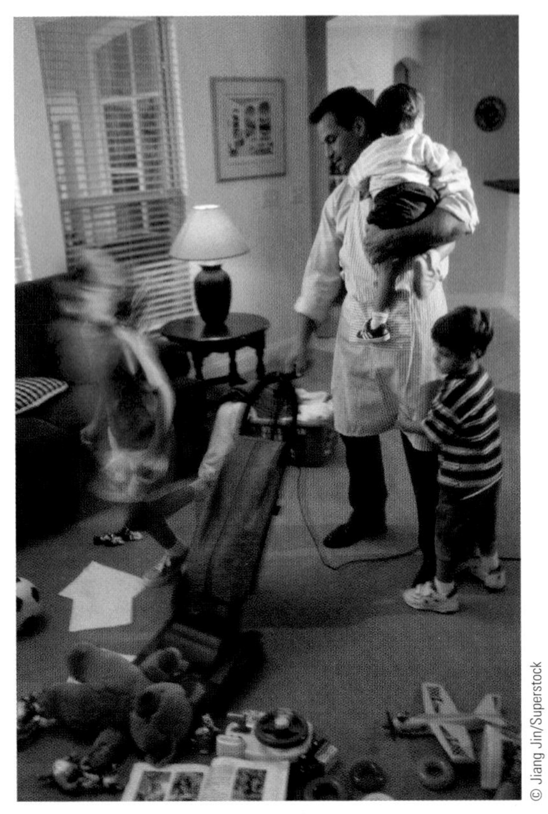
© Jiang Jin/Superstock

In recent years, many more fathers and mothers alike have been confronting the unique challenges of single parenting.

ex-wife's residence, and the limitations placed on the visitation arrangements. Although the courts continue to award mothers custody in divorces, an increasing number of fathers are attempting to gain sole or joint custody of their children. As a result, we can expect to see an increase in the number of single fathers in the future (Ward and Belanger, 2011).

HOUSEWORK AND CHILD-CARE RESPONSIBILITIES

Thirty years ago, most Canadian families relied on one wage earner. Today, approximately 70 percent of all families in Canada are **dual-earner families**—families in which both partners are in the labour force. More than half of all employed women hold full-time, year-round jobs. Even when their children are very young, most working mothers work full time. Moreover, as discussed in Chapter 12, many married women leave their paid employment at the end of the day and go home to perform hours of housework and child care. Difficulty in balancing work and family is the defining feature of family life today. Parents must make difficult decisions—decisions often driven by economic necessity—between the amount of time they spend at work and the amount of time they can be at home with their children (Marshall, 2011). Sociologist Arlie Hochschild (2012) refers to this as the **second shift**—the domestic work that employed women perform at home after they complete their workday on the job. Thus, many women today contribute to the economic well-being of their families and also meet many of the domestic needs of family members by cooking, cleaning, shopping, taking care of children, and managing household routines.

Although a division of labour still exists within families, the hours of paid work, average earnings, and time spent on domestic labour and child care are becoming more similar between

▶ **dual-earner families** Families in which both partners are in the labour force.

▶ **second shift** Arlie Hochschild's term for the domestic work that employed women perform at home after they complete their workday on the job.

Morgan Lane Photography/Shutterstock.com

Juggling housework, child care, and a job in the paid workforce are all part of the average day for many women. Why does sociologist Arlie Hochschild believe that many women work a second shift?

spouses in Canada (Marshall, 2011:13). As women's participation in the paid labour market has increased, men's involvement in housework and child care has also risen.

In Canada, millions of parents rely on child care so they can work and their young children can benefit from early educational experiences that will help in their future school endeavours. Despite this fact, there is still no national child care policy, and there are only regulated child care spaces for approximately 20 percent of Canada's children (Fern & Friendly, 2014).

For millions more parents, after-school care for their school-aged children is an urgent concern because the children need productive and safe activities to engage in while their parents are working. Obtaining child care for children of divorced parents and other young people living in single-parent households is often an especially pressing concern because of the limited number of available adults and a lack of financial resources (Fern & Friendly, 2014).

LO-5 TRANSITIONS AND PROBLEMS IN FAMILIES

Families go through many transitions and experience a wide variety of problems ranging from separation and divorce to family violence. These all-too-common experiences highlight two important facts about families: (1) for good or ill, families are central to our existence, and (2) the reality of family life is far more complicated than the idealized image of families found in the media and in many political discussions. Moreover, as people grow older, transitions inevitably occur in family life.

FAMILY VIOLENCE

Family violence refers to various forms of abuse that take place among family members, including child abuse, spousal abuse, and elder abuse. We will primarily focus on domestic violence—also referred to as spousal abuse or intimate-partner violence—and elder abuse. *Domestic violence* refers to any intentional act or series of acts—whether physical, emotional, or sexual—by one or both partners in an intimate relationship that causes injury to either person. An intimate relationship might include marriage or cohabitation, as well as people who are separated or living apart from a former partner or spouse.

There are numerous causes of domestic violence, and many factors are interrelated. Factors contributing to unequal power relations in families include economic inequality, legal and political sanctions that deny girls and women equal rights, and cultural sanctions that dictate appropriate sex roles and reinforce the belief that males are inherently superior to females. Cultural factors that perpetuate domestic violence include gender-specific socialization that establishes dominant–subordinate sex roles. Economic factors include poverty or limited financial resources within families that contribute to tension and sometimes to violence. Economic factors are intertwined with women's limited access to education, employment, and sufficient income so that they could take care of themselves and their children. Regardless of the factors that contribute to domestic violence, control is central to all forms of abuse: Gaining and maintaining control over the victim is the key factor in abuse. As a result, family violence often involves a cycle of abuse that goes on for extended periods of time.

How much do we know about violence in families? Women, as compared with men, are more likely to be victims of violence perpetrated by intimate partners. Recent statistics indicate that women are five times more likely than men to experience such violence, and that many of these women live in households with children witnessing the violence (Johnson & Dawson, 2011). However, we cannot know the true extent of family violence because much of it is not reported to police. For example, results from the 2009 General Social Survey indicated that just under one-quarter (22 percent) of victims of spousal violence reported the incident to police (Department of Justice, 2011).

Although public awareness of domestic violence has increased in recent years, society is far from finding an effective solution for this pressing social problem.

Although everyone in a household where family violence occurs is harmed psychologically, children are especially affected by household violence. Children who are raised in an environment of violence suffer profoundly, even if they are not the direct targets. Their own physical and emotional needs are often neglected, and they may learn by example to deal with conflict through violence (Johnson and Dawson, 2011). Not surprisingly, the research indicates that domestic violence and child maltreatment often take place in the same household.

It is estimated that approximately one million Canadian children witness some form of domestic violence in their homes each year (Johnson & Dawson, 2011). Long-term effects associated with witnessing violence include aggressive behaviour, emotional problems, and effects on social and academic development (Wathen, 2012). In some situations, family violence can be reduced or eliminated through counselling, the removal of one parent from the household, or other steps that are taken either by the family or by social service agencies or law enforcement officials. However, as noted above, children who witness violence in the home may display certain emotional and behavioural problems that adversely affect their school life and communication with other people. In short, there are no easy solutions to a problem as complex as family violence.

Although differences in power and privilege between women and men do not inevitably result in violence, gender-based inequalities can still produce sustained marital conflicts. In any case, a common consequence of marital strife and unhappiness is divorce.

DIVORCE

Divorce is the legal process of dissolving a marriage that allows former spouses to remarry if they so choose. Prior to 1968, it was difficult to obtain a divorce in Canada. A divorce was granted only on the grounds of adultery. In 1968, the grounds for divorce were expanded to include marital breakdown—that is, desertion, imprisonment, or separation of three or more years—and marital offences (physical or mental cruelty). As shown in Figure 13.3, the divorce rate increased dramatically as a result of the wider grounds for divorce. In 1985, the *Divorce Act* introduced "no fault" provisions that made marital breakdown the sole ground for divorce. Under no-fault divorce laws, proof of "blameworthiness" is no longer necessary. When children are involved, however, the issue of blame may assume greater importance in the determination of parental custody.

Have you heard statements such as "One out of every two marriages ends in divorce"? Statistics might initially appear to bear out this statement. In 2008 (the last year these statistics have been collected in Canada) for example, 147,000 Canadian couples married and 70,000 divorces were granted (Vanier Institute of the Family, 2011). However, comparing the number of marriages with the number of divorces from year to year can be misleading. The couples that are divorced in any given year are unlikely to come from the group that married that year. Some people also may go through several marriages and divorces, thus skewing the divorce rate. The likelihood of divorce goes up with each subsequent marriage in the serial monogamy pattern (Ambert, 2009).

To accurately assess the probability of a marriage ending in divorce, it is necessary to use what is referred to as a *cohort approach*. This approach establishes probabilities based on assumptions about how the various age groups (cohorts) in society might behave, given their marriage rate, their age at first marriage, and their responses to various social, cultural, and economic changes. Canadian estimates based on a cohort approach are that 35 to 40 percent of marriages will end in divorce (Ambert, 2009).

Why do divorces occur? At the macrolevel, societal factors contributing to higher rates of divorce include changes in social institutions such as religion, the family, and the legal system. Some religions have taken a more lenient attitude toward divorce, and the social stigma associated with divorce has lessened. Further, as we have seen in this chapter, the family has undergone a major change that has resulted in less economic and emotional dependency among family members—and thus reduced a barrier to divorce. And, as Figure 6.2 demonstrates, the liberalization of divorce laws in Canada has had a dramatic impact on the divorce rate.

At the microlevel, a number of factors contribute to a couple's "statistical" likelihood of becoming divorced. Social research over the years has identified some factors that closely correlate with divorces. Personal factors that are related to divorce include substance abuse, gambling, infidelity, domestic violence, and mental health issues. The most significant social factor related to divorce is age of marriage. The relationship may be relatively obvious; the younger the couple at marriage the higher the likelihood of divorce. Other social factors correlated to divorce include living common-law before marriage, parental divorce, remarriage, low income and low education, and mixed racial or mixed faith marriages (Ambert, 2009, Sev'er, 2011).

The interrelationship of these and other factors is complicated. For example, the effect of age is intertwined with economic resources: Persons from families at the low end of the income scale tend to marry earlier than those at more affluent income levels. Thus, the question becomes whether age is a factor or whether economic resources are more closely associated with divorce.

CONSEQUENCES OF DIVORCE Divorce may have a dramatic economic and emotional impact on family members. For children, divorce results in the most significant changes they have experienced in their lifetimes—new relationships with each parent, often new residences, changes in schedules to accommodate visitation privileges, and, in some cases, a new parental figure.

FIGURE 6.2 : NUMBER OF MARRIAGES AND DIVORCES, CANADA, 1926 TO 2008

Note: Marriage and Divorce statistics no longer collected in Canada after 2008.

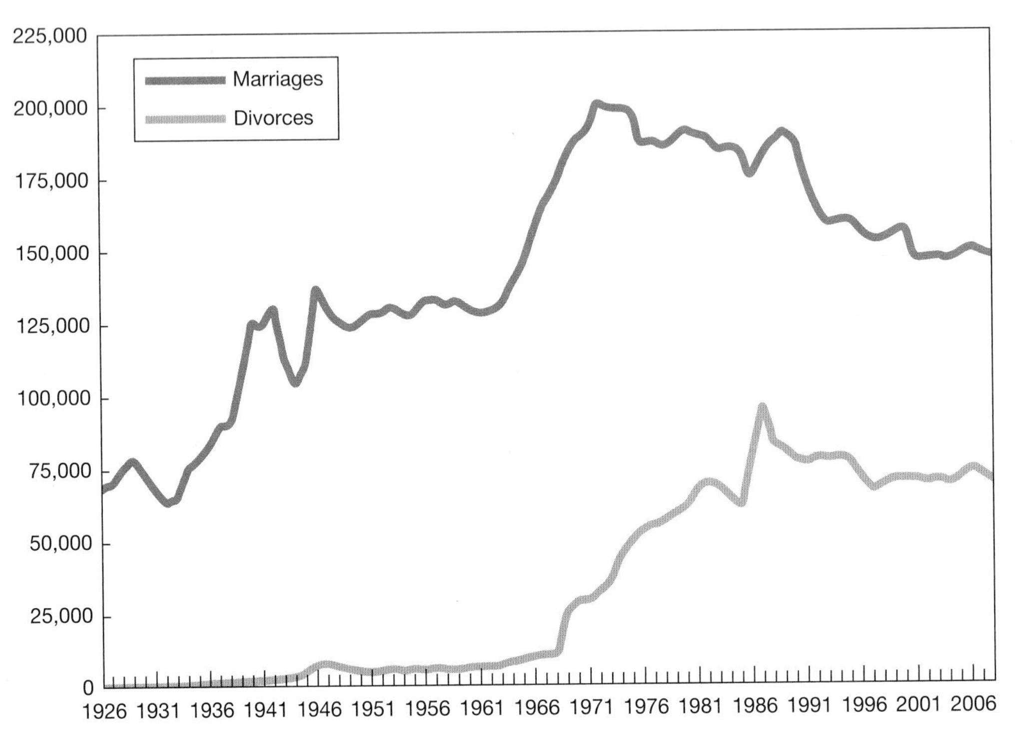

Source: "Marital Status: Overview, 2011," by Ann Milan, Statistics Canada, 2013, http://www.statcan.gc.ca/pub/91-209-x/2013001/article/11788/fig/fig7-eng.htm

The exact number of children affected by divorce in Canada is difficult to determine because no official information is available on out-of-court custody decisions, but approximately 40,000 Canadian children per year are involved in custody disputes. In the majority of these cases, the children reside primarily with their mother, meaning that the mother has *physical* custody of the children. Only about 10 percent of children live with their father despite the fact that joint *legal* custody now represents almost 50 percent of custody orders awarded. Parental joint custody is also an option for some divorcing couples. When joint custody is a voluntary arrangement, and when there is motivation to make it work, it has benefits for both children and parents (Ambert, 2009). However, this arrangement may also create unique problems for children, who must adjust to living in two homes and to the fact that their parents no longer live together.

The consequences of divorce are not entirely negative. There is no doubt that some children are better off after their parents divorce. For some people, divorce may be an opportunity to terminate destructive relationships. For others, it may represent a way to achieve personal growth by enabling them to manage their lives and social relationships and establish their own identity (Sev'er, 2011).

DIVERSITY IN FAMILIES

SAME-SEX FAMILIES

Over the past decade, same-sex families in Canada have changed significantly both in terms of legal recognition and social acceptance. In fact, Canada is recognized as a world leader in same-sex family rights. Today, there is (almost) no legal distinction between heterosexual and same-sex couples when it comes to family life and activities associated with family life. Sex, marriage,

FIGURE 6.3 LGBTQ FAMILY RIGHTS TIMELINE

LGBTQ Family Rights Timeline in Canada	
1967	Pierre Trudeau, then Justice Minister of Canada, declares, "There's no place for the state in the bedrooms of the nation."
1969	Consensual homosexual sex in private between two adults 21 or over is decriminalized.
1977	Quebec becomes the first large jurisdiction in the world to protect citizens from discrimination on the basis of sexual orientation.
1992	Conservative Justice Minister Kim Campbell announces the end of a ban on gay and lesbian people serving in the armed forces.
1995	The Supreme Court of Canada rules that "sexual orientation" should be "read in" to the 1982 *Canadian Charter of Rights and Freedoms.*
1995	Ontario court rules in favour of gay and lesbian couples' adoption rights.
1996	Sexual orientation is added to the *Canadian Human Rights Act.*
1996	The Canadian Psychological Association opposes discrimination against lesbians and gay men, and supports same-sex marriage and same-sex couples parenting.
1999	The Supreme Court of Canada rules that gay and lesbian couples should have the same rights as heterosexual common-law couples.
2005	Canada becomes the fourth country in the world (preceded by the Netherlands, Belgium, and Spain) to legalize same-sex marriage.

divorce, adoption, parenting, custody, hospital visitation rights, inheritance rights—all are now legally protected rights in Canada for same-sex couples.

As shown in Figure 6.3, until recently, same-sex couples were prohibited from sponsoring the immigration of their partners to Canada, from obtaining custody of their children, from jointly adopting children, or from receiving spousal benefits and survivors' pensions, and same-sex couples were prohibited from legally marrying. In 2003, the Ontario Court of Appeal ruled that Canada's legal definition of marriage is unconstitutional and redefined it as "the voluntary union for life of two persons to the exclusion of all others." In her ruling, the judge further explained that "the existing common law rule is inconsistent with the constitutional values in modern Canadian society and offends the equality rights of gays and lesbians" (Kome, 2002:1). In response, the federal government revised the legal definition of marriage to recognize same-sex marriages, and in 2005, Canada became the third country in the world to recognize same-sex marriage.

Even for those who support same-sex marriages, there remain many questions and concerns. Some of the concerns are related to the quality of homosexual unions—for example, are they as long lasting and committed? Others argue that allowing same-sex couples to marry will devalue the institution of marriage. In contrast to stereotypes of same-sex relationships as short-term, promiscuous, and noncommittal, research on homosexual relationships indicates the breakup rates of married or cohabiting heterosexual couples and lesbian and gay couples have been found to be approximately equal.

There is an increasing number of same-sex families with children. In the 2011 Census, just close to 65,000 couples identified themselves as same-sex married or common-law couples. Approximately 10 percent of these same-sex couples had children living with them (Vanier

Institute of the Family, 2013). In many cases, lesbian mothers and gay fathers may have children from a previous marriage or relationship. Lesbians may become pregnant through *alternative insemination* (sexual relations as a means of getting pregnant or artificial insemination) (Meisner, 2014). Lesbian mothers and gay fathers may also form families through fostering or adoption.

Many people believe that being parented by same-sex couples is emotionally unhealthy for children and can cause them confusion about their own sexuality. However, the research has shown that the children of same-sex couples are as well adjusted as children who grow up in heterosexual households. In addition, these children experience no psychological damage, and they are no more likely to be homosexual than are children raised by heterosexual parents (Dufur et al., 2007; Taylor & Ristock, 2011).

TIME TO REVIEW

- What are some of the most significant challenges facing lesbian and gay families today?
- What progress has been made in recent years?

DIVERSITY AMONG SINGLES

While marriage at increasingly younger ages was the trend in Canada during the first half of the 20th century, by the 1960s the trend had reversed and many more adults were remaining single. In 1971, close to half of Canadians aged 20 to 24 were already married. Today, close to 90 percent of Canadians aged 20 to 24 are single. Currently, approximately 25 percent of

Laura Doss/Image Source/Getty Images

Adoption is a complex legal process for most parents; it can be even more complicated for same-sex couples.

households in Canada are one-person households (Statistics Canada, 2012e). This estimate, however, includes people who are divorced, widowed, or have never married. Given the fact that nine out of 10 Canadians marry at some time in their lives, single status is often temporary. Some never-married singles remain single by choice. Reasons include more opportunity for a career (especially for women), the availability of sexual partners without marriage, the belief that the single lifestyle is full of excitement, and the desire for self-sufficiency and freedom to change and experiment. According to some marriage and family analysts, individuals who prefer to remain single hold more individualistic values and are less family-oriented than those who choose to marry. Friends and personal growth tend to be valued more highly than marriage and children (Crompton, 2005).

Other never-married singles remain single out of economic necessity. They cannot afford to marry and set up their own households. Structural changes in the economy have limited the options of many working-class young people. An increasing number of university and college graduates have found that they cannot earn enough money to set up a household separate from that of their parents. Consequently, a growing proportion of young adults are living with one or both parents (Statistics Canada, 2012f).

INDIGENOUS FAMILIES

It is difficult to discuss Indigenous families, given the fact that Indigenous peoples are composed of many distinct nations with different histories, cultures, economic bases, and languages (Ward and Belanger, 2011). In all Indigenous families, however, the extended family was seen as central to both the individual and the community. The concept of family was defined very broadly. For example, to the Ojibwa, *family* referred to individuals who worked together and were bound together by responsibility and friendship as well as by kinship ties. Family size averaged between 20 and 25 persons, and economic cooperation and sharing were essential in the extended family form as this Ojibwas band member explains:

> Trapping kept the family together because everyone in the family had something to do; the man had to lay traps and check them; the woman skinned the animals, cooked, and looked after the kids. The grandparents helped with the kids; they taught them manners, how to behave, and told them stories about our people. The kids, if they were old enough, had work to do. (Shkilnyk, 1985:81)

Under this cooperative family system, Indigenous families were extremely successful in ensuring the survival and well-being of their members.

Tragically, this traditional model of Indigenous family life was profoundly damaged in the 20th century as a result of the assimilation strategies employed by the Canadian church and state. Families were displaced from their traditional lands, moved to reserves, and denied access to the resources that were central to the economic survival of the extended family unit. An estimated 125,000 Indigenous children were forcibly removed from their families and placed in residential schools (where they were often sexually and physically abused). Consider Chief Cinderina Williams's description of the impact of residential schools on the family:

> Later when these children returned home, they were aliens. They did not speak their own language, so they could not communicate with anyone other than their own counterparts. Some looked down on their families because of their lack of English, their lifestyle, and some were just plain hostile. . . . Consequently, when these children became parents, and most did at an early age, they had no parenting skills. (Godin-Beers and Williams, 1994, cited in Castellano, 2002)

Residential schooling of Indigenous children, away from their parents, directly led to the decline of parenting skills because the children were denied parental role models. As a result,

Indigenous parents were seen to be incompetent, their communities socially disorganized, and Indigenous children were deemed to be in 'need of protection.' During the process of phasing out the residential school system during the 1960s and 1970s, Indigenous children were once again forcibly removed from their families (known as the "sixties scoop") and adopted by non-Indigenous families throughout Canada and the United States.

Today, Indigenous families are still recovering from the devastation to Indigenous family life imposed from government assimilation policies as evidenced from the following statistics. Indigenous children represent 40 percent of children in care in Canada; the rate of spousal abuse among Indigenous peoples is five times the national average; the rate of spousal homicide is eight times higher among Indigenous women compared to non-Indigenous women; and the suicide rate of Indigenous peoples is double the rate of the general population (Assembly of First Nations, 2008; Jones and Smith, 2011).

Despite these seemingly insurmountable challenges, significant progress is being made as Indigenous peoples have united behind the goal of self-government, especially in the areas of social services, education, and child welfare (Castellano, 2002). The legacy of colonialism and residential schools is inspiring a generation of First Nation leaders, activists, and Elders to reclaim traditional knowledge and practices, and to reestablish their role as cultural leaders, teachers, and child care providers to the next generation. With a deep commitment to the well-being of their families, their communities and their culture, they have successfully lobbied for recognition of a family form based on extended families (Vanier Institute of the Family, 2012). Indigenous peoples believe in maintaining the ties between children and their natural parents, as well as caring for children within their Indigenous communities. They see this as essential to the rebuilding of Indigenous families in Canada. Many Indigenous communities are striving to return to the practices and values that traditionally nourished Indigenous family life: respect for women and children, mutual responsibility, and, above all, the general creed of sharing and caring (Royal Commission of Aboriginal Peoples, 1995:81).

FAMILY ISSUES IN THE FUTURE

As we have seen, families and intimate relationships have changed dramatically over the last century. What we have also seen in examining families is the resiliency of families and the capacity of family and individual family members to evolve in response to—and maybe, in spite of—constant political, social, and economic change (Vanier Institute of the Family, 2012).

Regardless of the changes in family life, the diversity of families and the problems facing families today, the family remains the central institution in the lives of most Canadians. A national opinion poll found that more than two-thirds of Canadians regard the family as the "greatest joy" in their lives. Moving forward, as sociologist Meg Luxton explains, the challenge for contemporary thinking about families is to focus on functions and practices—on what people do to take care of themselves and each other, to have and raise beloved children, and to ensure as best as possible the well-being of themselves, their households, their communities, and their society (Luxton, 2012:2).

LO-1 Explain why it is difficult to define family.

Families may be defined as relationships in which people live together with commitment, form an economic unit and care for any young, and consider their identity to be significantly attached to the group.

Ruaridh Connellan/Barcroft Media/Landov

LO-2 Understand the key assumptions of functionalist, conflict, feminist, symbolic interactionist, and postmodernist perspectives on families.

Functionalists emphasize the importance of the family in maintaining the stability of society and the well-being of the individuals. Functions of the family include sexual regulation, socialization, economic and psychological support, and provision of social status. Conflict and feminist perspectives view the family as a source of social inequality and an arena for conflict over values, goals, and access to resources and power. Symbolic interactionists explain family relationships in terms of the subjective meanings and everyday interpretations people give to their lives. Postmodern analysts view families as permeable, reflecting the individualism, particularity, and irregularity of social life in the Information Age.

Morgan Lane Photography/Shutterstock.com

LO-3 Understand the various options available to Canadian families in establishing families.

Families are changing dramatically in Canada. Cohabitation has increased significantly in the past two decades. The number of single-parent families has also increased sharply in recent decades. Marriage continues to be an extremely popular institution with the majority of Canadians marrying at some point in their lives. As a result of divorce and remarriage, stepfamilies or *blended families* may be established, which consist of a husband and wife, children from previous marriages, and children (if any) from the new marriage.

© Photos.com

KEY TERMS

bilateral descent A system of tracing descent through both the mother's and father's sides of the family (p. 141).

cohabitation The sharing of a household by a couple who live together without being legally married (p. 145).

dual-earner families Families in which both partners are in the labour force (p. 152).

egalitarian family A family structure in which both partners share power and authority equally (p. 142).

extended family A family unit composed of relatives in addition to parents and children who live in the same household (p. 140).

families we choose Social arrangements that include intimate relationships between couples and close familial relationships with other couples, as well as with other adults and children (p. 138).

family A relationship in which people live together with commitment, form an economic unit and care for any young, and consider their identity to be significantly attached to the group (p. 139).

family of orientation The family into which a person is born and in which early socialization usually takes place (p. 139).

family of procreation The family that a person forms by having or adopting children (p. 139).

homogamy The pattern of individuals marrying those who have similar characteristics, such as race/ethnicity, religious background, age, education, or social class (p. 147).

infertility An inability to conceive after one year of unprotected sexual relations (p. 148).

kinship A social network of people based on common ancestry, marriage, or adoption (p. 139).

marriage A legally recognized and/or socially approved arrangement between two or more individuals that carries certain rights and obligations and usually involves sexual activity (p. 140).

matriarchal family A family structure in which authority is held by the eldest female (p. usually the mother) (p. 141).

matrilineal descent A system of tracing descent through the mother's side of the family (p. 141).

monogamy marriage to one person at a time (p. 140).

nuclear family A family made up of one or two parents and their dependent children, all of whom live apart from other relatives (p. 140).

patriarchal family A family structure in which authority is held by the eldest male (p. usually the father) (p. 141).

patrilineal descent A system of tracing descent through the father's side of the family (p. 141).

polyandry The concurrent marriage of one woman with two or more men (p. 141).

LO-4 Describe the challenges facing families today.

Canadian families are faced with many challenges as a result of the increasing diversity and choice available to individuals establishing intimate relationships. These include choosing among a wide range of reproductive choices and establishing ways to maintain balance between paid work and family responsibilities. With the increase in dual-earner marriages, women increasingly have been burdened by the "second shift"—the domestic work that employed women perform at home after they complete their workday on the job

Geostock/Getty Images

LO-5 Identify the primary problems facing Canadian families today.

Two of the most significant problems facing families are family violence and divorce. Both *spouse abuse* (violence or mistreatment that a woman or man may experience at the hands of a marital, common-law, or same-sex partner) and *child abuse* (physical or sexual abuse and/or neglect by a parent or caregiver) occur at alarming rates in Canadian families. Divorce is the legal process of dissolving a marriage. At the macrolevel, changes in social institutions may contribute to an increase in divorce rates; at the microlevel, factors contributing to divorce include age at marriage, economic resources, religiosity, and parental marital happiness. Divorce has contributed to greater diversity in family relationships, including stepfamilies or blended families and the complex binuclear family.

© 1997–2012 Centre for Children and Families in the Justice System, London Family Court Clinic Inc., http://www.lfcc.on.ca

APPLICATION QUESTIONS

1. In your own thinking, what constitutes an ideal family?
2. Based on your understanding of the term *family*, should the following be considered families? Why, or why not?
   - Man, woman, no children; married but living apart
   - Woman, woman, and child of one woman living together; women a same-sex couple
   - Man, his biological child, and woman (not his wife) with whom he has a sexual relationship living together
   - Four adults sharing household for many years; none a same-sex couple

polygamy The practice of having more than one spouse at a time (p. 141).

polygyny The concurrent marriage of one man with two or more women (p. 141).

second shift Arlie Hochschild's term for the domestic work that employed women perform at home after they complete their workday on the job (p. 152).

sociology of family The subdiscipline of sociology that attempts to describe and explain patterns of family life and variations in family structure (p. 142).

CHAPTER 7

GROUPS AND ORGANIZATIONS

Gemenacom/Shutterstock.com

CHAPTER FOCUS QUESTION

How can we explain the behaviour of people who work in bureaucracies?

LEARNING OBJECTIVES

AFER READING THIS CHAPTER, YOU SHOULD BE ABLE TO

LO-1 Identify the differences among social groups, aggregates, and categories.

LO-2 Understand the effect that size has on the functioning of groups.

LO-3 Explain the impact of groups on people's behaviour.

LO-4 Identify the characteristics that define a bureaucracy and the "other face" of bureaucracies.

LO-5 Discuss the form large organizations may take in the future.

Kenneth Payne describes his journey through a bureaucratic maze:

Since November, I have spent six to eight hours a day trying to persuade the authorities to accommodate me, but it just goes around in a circle . . . It's George Orwell's Big Brother *. . . The bureaucracy is making me prove a negative and it turns "innocent until proven guilty" on its head . . . There's no common sense here. It's an inflexible bureaucracy where nobody takes any responsibility. (Reed, 1998:A11)*

What is Mr. Payne's problem? The former carpenter wants to be a schoolteacher. He has a degree in education and teaching experience. However, he is unable to get a permanent job teaching because he has a disease that causes the skin on his hands to blister and peel.

Why should this disqualify Payne from teaching? California legislators passed a law requiring that all teachers be fingerprinted so their criminal records could be checked. Because of his disease, Payne has never had proper fingerprints, so there is no file to check against. Payne has appealed to the state and offered to prove in other ways that he has no criminal record, but he has been unable to get exempted from the rule.

In a similar case, former Canadian Army Master Corporal Paul Franklin lost both legs in a roadside bombing in Afghanistan. In order to keep receiving his pension, Master Corporal Franklin must have a doctor send a medical report every year certifying that he still has his disability—in other words, that his legs have not grown back in the previous twelve months.

Why do people in organizations behave so inflexibly? Some rules are necessary. Even in small groups, such as families or friendship groups, informal rules help to ensure that people interact smoothly. In a large bureaucracy, an explicit system of rules means that employees and clients know what is expected of them. These rules help to ensure that everyone receives equal treatment from the organization. Unfortunately, adherence to the rules can stifle individual judgment, and some bureaucrats become so inflexible they hurt the organization and its clients. In Payne's case, it made sense for the school system to protect children by establishing background checks for prospective teachers. However, in this case, the bureaucrats focused on fingerprinting, which is just one way of ensuring that people with criminal records are not hired as teachers. An official who was concerned with the *goal* of the policy (protecting children) rather than with one of the *means* of achieving that goal (fingerprinting) would have allowed Payne to prove he had a clean record in other ways.

It also makes sense for Canada's Veterans' Affairs department to ensure the medical condition of those receiving disability pensions, but it could easily eliminate this requirement for those who have lost limbs. Instead, Veterans' Affairs responded to public criticism about this case by requiring Master Corporal Franklin to submit his medical form every three years rather than annually.

While Payne suffered personal hardship, and Master Corporal Franklin endured a lot of aggravation, the consequences of bureaucratic inflexibility can be much more serious. Just over a decade ago, Hurricane Katrina devastated the city of New Orleans. Tens of thousands of evacuees were not properly cared for, and law and order broke down. There were massive failures in disaster planning and in coordinating the response after the city was flooded. The most serious flaw was a lack of coordination among the local, state, and federal agencies responsible for the emergency, but bureaucratic inflexibility was also pervasive.

Even in the face of the largest natural disaster ever to hit North America, some bureaucrats were focused more on rules and regulations than on saving lives. Despite a desperate need for water, many truckloads of water were turned back because the drivers didn't have the proper paperwork (Lipton et al., 2005). A group of doctors were evacuated from their hospital and taken to the New Orleans airport. They offered to help look after the many sick people who had also been evacuated to the airport, but authorities were worried about liability issues and told them they could best help by mopping floors (CNN, 2005). While the doctors cleaned floors, patients died because of the lack of medical care.

Other examples of bureaucratic inflexibility include an incident during the 2013 Alberta floods when Calgary health officials shut down a coffee shop that was serving coffee and sandwiches to emergency responders and evacuees—a decision later reversed by city officials. More recently, a man had to get help from the Ontario Ombudsman to get a birth certificate for his newborn baby because the mother had died shortly after childbirth and could not sign the application form.

We deal constantly with bureaucratic organizations. Most of us are born in hospitals, educated in schools, fed by restaurants and supermarket chains, entertained by communications companies, employed by corporations, and buried by funeral companies. We often think negatively of bureaucracies because of their red tape and impersonality. While they can be inflexible and inhumane, bureaucracies are an essential part of our industrialized society.

In this chapter, you will learn about different types of groups and organizations, including bureaucracies. We live our lives in groups and they constantly affect our behaviour. Before reading on, test your knowledge about bureaucracies by taking the quiz in Box 7.1.

CRITICAL THINKING QUESTIONS

1. What bureaucracies have you recently encountered? What are the benefits and shortcomings of this form of social organization?
2. Have you ever run into the kind of bureaucratic inflexibility described in the chapter introduction? If you have, how did that make you feel about the organization that behaved unreasonably?
3. How could bureaucracies be changed so they treat people more like individuals than many of them do now?

LO-1 SOCIAL GROUPS

We spend most of our lives in groups, including families, friends, and school and work groups, so it is important to understand the characteristics and dynamics of groups ranging from small, informal groups to large bureaucracies.

Consider these situations. Three strangers are standing at a street corner waiting for a traffic light to change. Do they constitute a group? Five hundred people are first-year students at a university. Do they constitute a group? In everyday usage, we use the word *group* to mean any collection of people. According to sociologists, however, the answer to these questions is no; individuals who happen to share a common feature or to be in the same place at the same time do not constitute social groups.

GROUPS, AGGREGATES, AND CATEGORIES

A *social group* is a collection of two or more people who interact frequently with one another, share a sense of belonging, and have a feeling of interdependence. Several people waiting for a traffic light to change constitute an **aggregate**—they happen to be in the same place at the same time but have little else in common. People in aggregates generally do not interact with one another. The first-year students, at least initially, constitute a **category**—a number of people who may never have met one another but who share a similar characteristic (such as education level, age, ethnicity, and gender).

▶ **aggregate** A collection of people who happen to be in the same place at the same time but have little else in common.

▶ **category** A number of people who may never have met one another but who share a similar characteristic.

Social groups can change over time. An informal group may become a formal organization with a specific structure and clear-cut goals. A *formal organization* is a structured group formed to achieve specific goals. Universities, factories, corporations, and the military are examples of formal organizations. Before we examine formal organizations, we need to know more about groups in general and about how they function.

Napoleon's defeat at Waterloo in 1815 showed that massive armies could not be led in the traditional way, by a single commander responsible for everything. Subsequently, armies developed more effective organizational structures.

TYPES OF GROUPS

PRIMARY AND SECONDARY GROUPS Charles H. Cooley (1962/1909) used the term *primary group* to describe a small, less specialized group in which members engage in face-to-face, emotion-based interactions over an extended time. We have primary relationships with other individuals in our primary groups—that is, with our *significant others.*

A *secondary group* is a larger, more specialized group in which members engage in more impersonal, goal-oriented relationships for a limited time. The size of a secondary group may vary. Formal organizations are secondary groups, but they also contain many primary groups within them. There are many thousands of primary groups within the secondary group setting of your university.

INGROUPS AND OUTGROUPS Groups set boundaries by distinguishing between insiders, who are members, and outsiders, who are not. Sumner (1959/1906) coined the terms *ingroup* and *outgroup* to describe people's feelings toward members of their own and other groups. An **ingroup** is a group to which a person belongs and identifies with. An **outgroup** is a group to which a person does not belong and toward which the person may feel a sense of competitiveness or hostility. Distinguishing between our ingroups and our outgroups helps us establish our individual identity.

Group boundaries may be formal, with clearly defined criteria for membership. For example, a country club that requires applicants for membership to be recommended by four current members and pay a $25,000 initiation fee and monthly membership dues has set requirements for its members. The club may post "Members Only" signs and use security personnel to ensure that nonmembers do not encroach on its grounds. Boundary distinctions are often reflected in symbols, such as emblems or clothing. These symbols, including membership cards and shirts bearing the club's logo, denote that the individual is a member of the ingroup.

Many groups do not have these formal boundaries. Friendship groups, for example, usually do not have clear guidelines for membership. Rather, the boundaries tend to be informal and vaguely defined.

Ingroup and outgroup distinctions may encourage social cohesion among members, but they also may promote classism, racism, sexism, and ageism. Ingroup members typically view themselves positively and may view members of outgroups negatively. These feelings of group superiority, or *ethnocentrism,* can be detrimental to groups and individuals not part of the ingroup. Sexual harassment and racial discrimination are two negative consequences of ethnocentrism.

ingroup A group to which a person belongs and identifies with.

outgroup A group to which a person does not belong and toward which the person may feel a sense of competitiveness or hostility.

BOX 7.1 SOCIOLOGY AND EVERYDAY LIFE

How Much Do You Know About Bureaucracy?

True	False	
T	F	1. Large bureaucracies have existed for about a thousand years.
T	F	2. Because of the efficiency and profitability of emerging factory bureaucracies, people were eager to leave farms to work in the factories.
T	F	3. Bureaucracies are deliberately impersonal.
T	F	4. The organizational principles used by McDonald's restaurants have been adopted by other sectors of the global economy.
T	F	5. The rise of Protestantism helped create the social conditions favourable to the rise of modern bureaucracies.

For answers to the quiz about bureaucracy, go to **www.nelson.com/student**.

© Gari Wyn Williams/Alamy

Visiting spectators to the game form an outgroup in relation to the home team fans.

▶ **reference group** A group that strongly influences a person's behaviour and social attitudes, regardless of whether that individual is a member.

▶ **network** A web of social relationships that link a number of people.

REFERENCE GROUPS A **reference group** is a group that influences our attitudes and behaviour, regardless of whether that individual is an actual member. When we evaluate our appearance, ideas, or goals, we automatically refer to the standards of a group. Sometimes, we will refer to our membership groups, such as family or friends. Other times, we will rely on groups to which we do not belong but that we might wish to join in the future, such as a social club, profession, or even an outlaw motorcycle gang.

NETWORKS A **network** is a web of social relationships that link a number of people. Frequently, networks connect people who share common interests but who otherwise might not interact with one another. For example, if A is tied to B and B is tied to C, then a network may be formed among individuals A, B, and C. Think of your experiences looking for summer jobs. If you have a friend who works at a company that needs more people, he or she may recommend you to the potential employer. This recommendation helps you get a job and gives the employer the assurance that you are likely to be a good employee. Research shows that networks play a very important role for graduating students in finding employment (Granovetter, 1994).

It's a Small World: Networks of Acquaintances On September 11, 2001, nearly 3000 people died when terrorists crashed two planes into New York City's World Trade Center and a third into the Pentagon. Many people around the world were surprised to learn that they, or some of their acquaintances, knew someone who had been personally touched by the tragedy. For example, one of the authors of this textbook has a colleague whose partner was killed at the World Trade Center. This is consistent with a fascinating research project done nearly 50 years ago by psychologist Stanley Milgram (1967).

Milgram sent packages of letters to people in the Midwestern United States. The objective was to get the letters to one of two target recipients in Boston using personal contacts. Those originating the chain were given the name of the target recipient and told that the person was either a Boston stockbroker or the wife of a Harvard divinity student. They were asked to mail the letter to an acquaintance who they felt would be able to pass it on to another acquaintance even closer to the intended target. Milgram found that it took an average of five contacts to get the letters to the intended recipient.

The research was popularized through the play and movie *Six Degrees of Separation,* and the popular trivia game *Six Degrees of Kevin Bacon,* in which the objective is to link actors to other actors who have appeared in films with Kevin Bacon. Thus, Angelina Jolie has a Kevin Bacon number of 2, as she appeared in *Taking Lives* with Kiefer Sutherland, who worked with Kevin Bacon in the film *A Few Good Men*. Virtually no North American actor has a Bacon number larger than 4 (See the Oracle of Bacon website at oracleofbacon.org.).

The "small world" research has important implications. Strogatz and Watts (1998) have studied the mathematics behind the phenomenon and have documented the importance of "bridges"—people who bridge very different social worlds. For example, in Milgram's study, the Boston stockbroker received 64 letters, 16 of which were delivered by the owner of a Boston clothing store. Perhaps you have friends or acquaintances who come from other countries or who have unusual hobbies, or jobs that would enable them to bridge vast distances or widely different social groups. The study of networks and of the role of bridges has important implications for researchers in many fields, including *epidemiology*, which is the study of the spread of disease. For example, the spread of HIV/AIDS was hastened by a Canadian flight attendant (Patient X), whose travels meant that he bridged several different networks of gay males (Saulnier, 1998), though he was not as great a factor in the spread of HIV/AIDS as once believed (Tiemeyer, 2013).

While Milgram's research was influential, Kleinfeld (2002) found that most of Milgram's letters never reached their intended destination. We do not know if the connections failed because the participants could not think of anyone who could act as the next link in the chain, or simply because they did not bother moving the letter along toward the intended recipient. However, a study using email contacts had lower failure rates and had similar results to Milgram's. Dodds and his colleagues (2003) found that those who continued the chain needed an average of five to seven contacts to reach their targets, even when in another country. A recent study found that there was an average of just under four degrees of separation among Facebook users around the world, so social media may be bringing people closer together (Backstrom et al., 2011).

Race and gender are an important part of social networks. Korte and Milgram (1970) found a significantly higher number of completed chains when both the sender and the recipient were the same race, and Dodds and his colleagues (2003) found that people most frequently contacted persons of the same gender. They also found that workplace and educational contacts were most likely to be used in completing chains. These findings imply that some categories of people may be systemically disadvantaged in a world that is increasingly dependent upon geographically dispersed social networks.

Why is this important? We live in a world where many things get done through networks. Granovetter (1995) showed the importance of social networks for people looking for work. Most people get their jobs through personal contacts rather than through formal job-search mechanisms. Those with strong networks have an advantage in their search for work, while those without extensive networks or whose networks are not oriented to the labour market will be at a great disadvantage. If most of your friends are unemployed, they cannot help you find a job. This can perpetuate unemployment among groups, including some visible minorities and women whose networks cannot connect them to potential employers.

Network analysis is becoming more important in sociology. For example, email patterns may help us to understand how organizations work. How do ideas spread within an organization? Do email messages frequently pass between different levels of an organization, or are communications restricted to one level? Do women and visible minorities have the same interaction patterns as white males, and are they able to bridge different parts of their organizations? On a broader level, can genuine communities flourish in cyberspace, or does the Internet reduce community by reducing the personal contact between people?

LO-2 GROUP CHARACTERISTICS AND DYNAMICS

What purpose do groups serve? Why do individuals participate in groups? According to functionalists, groups meet peoples' instrumental and expressive needs. *Instrumental,* or task-oriented, needs cannot always be met by one person, so the group works cooperatively to fulfill a specific goal. Groups help members do things that are difficult or impossible to do alone. For example, you could not function as a one-person football team or single-handedly build a skyscraper. Groups also help meet *expressive,* or emotional, needs from family, friends, and peers.

Conflict theorists and symbolic interactionists, of course, have a different understanding of groups. While not disputing that groups ideally perform positive functions, conflict theorists suggest that groups also involve power relationships whereby the needs of individual members may not be equally served. Symbolic interactionists focus on how the characteristics of groups influence the kind of interaction that takes place among members.

▶ **small group** A collectivity small enough for all members to be acquainted with one another and to interact simultaneously.

▶ **dyad** A group consisting of two members.

▶ **triad** A group composed of three members.

To many postmodernists, groups and organizations—like other aspects of postmodern societies—are generally characterized by superficiality and by shallow social relationships. One postmodern thinker who focuses on this issue is the literary theorist Fredric Jameson, whose works have had a significant influence on contemporary sociology. According to Jameson (1984), postmodern organizations (and societies as a whole) are characterized not only by superficial relations, but also by people experiencing a lack of emotion because the world and the people in it have become more fragmented (Ritzer, 1997). For example, Ritzer (1997) examined fast-food restaurants and concluded that restaurant employees and customers interact in extremely superficial ways that are largely scripted by large-scale organizations: The employees learn to follow scripts in taking and filling customers' orders ("Would you like fries with that?"), while customers respond with their own "recipied" action. This can be contrasted with eating at a neighbourhood restaurant where you may know the owners and servers by name and have a friendly relationship with them.

© Roy Morsch/Corbis

Our most intense relationships occur in dyads—groups with two members. How might the interaction of these two people differ if they were with several other people?

GROUP SIZE

The size of a group is important. Interactions are more personal and intense in a **small group**, in which all members are acquainted with one another and interact simultaneously.

In a **dyad**—a group composed of two members—the active participation of both members is crucial for the group's survival. If one member withdraws from interaction, or "quits," the group ceases to exist. Examples of dyads include two people who are best friends, and married couples. Dyads provide an intense bond not found in most larger groups.

The nature of the relationship and interaction patterns change with the addition of a third person to form a **triad**. In a triad, even if one member ignores another or declines to participate, the group can still function. In addition, two members may unite to subject the third member to group pressure to conform.

As group size increases beyond three, members tend to specialize in different tasks and communication patterns change. In groups of more than six or seven people, it becomes increasingly difficult for everyone to participate in the same conversation, so several conversations will likely take place simultaneously. In groups of more than 10 or 12 people, it becomes virtually impossible for all members to participate in a single conversation unless one person serves as moderator and facilitates the discussion. Figure 7.1 shows that when the size of the group increases, the number of possible social interactions increases dramatically.

FIGURE 7.1 GROWTH OF POSSIBLE SOCIAL INTERACTIONS BASED ON GROUP SIZE

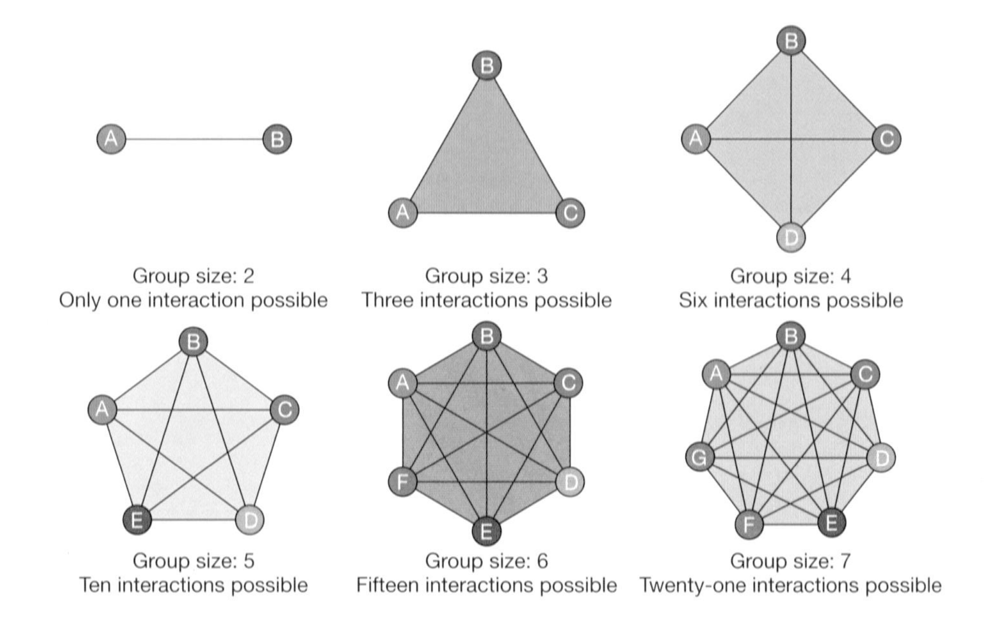

LO-3 GROUP CONFORMITY

Groups exert a powerful influence in our lives. To gain and then retain our membership in groups, most of us are willing to exhibit a high level of conformity to the wishes of other group members. **Conformity** is the process of maintaining or changing behaviour to comply with the norms established by a society, subculture, or other group. We often experience powerful pressure from other group members to conform. In some situations, this pressure may be almost overwhelming.

conformity The process of maintaining or changing behaviour to comply with the norms established by a society, subculture, or other group.

Several researchers have found that the pressure to conform can cause group members to say they see something they don't see or to do something they otherwise would not be willing to do. As you consider two of these studies, ask yourself what you might have done if you had been one of the research subjects.

ASCH'S RESEARCH Solomon Asch (1955, 1956) conducted a series of experiments in which the pressure to conform was so great that participants were willing to contradict their own best judgment rather than disagree with other group members.

One of Asch's experiments involved groups of undergraduate men (seven in each group) who were supposedly recruited for a study of visual perception. All the men were seated in chairs. However, the person in the sixth chair did not know that he was the only actual subject; all of the others were assisting the researcher. The participants were first shown a large card with a vertical line on it and then a second card with three vertical lines (see Figure 7.2). Each of the seven participants was asked to indicate which of the three lines on the second card was identical in length to the "standard line" on the first card.

In the first two trials with each group, all seven men selected the correct matching line. In the third trial, however, the subject became very uncomfortable when all of the others selected the incorrect line. The subject could not understand what was happening and became even more confused as the others continued to give incorrect responses on 11 out of the next 15 trials.

If you had been in the position of the subject, how would you have responded? Would you have given the correct answer, or would you have been swayed by the others? Asch (1955)

FIGURE 7.2 ASCH'S CARDS

Although Line 2 is clearly the same length as the line in the lower card, Asch's research assistants tried to influence "actual" participants by deliberately picking Line 1 or Line 3 as the correct match. Many of the participants went along rather than risk expressing a judgment that did not match the consensus of the "group."

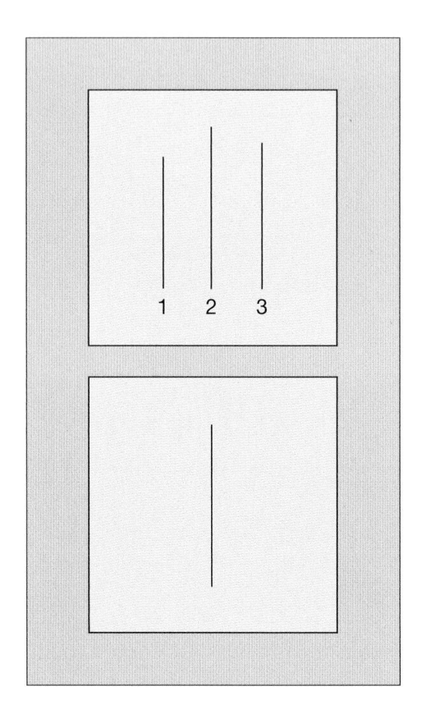

found that about 33 percent of the 'actual' subjects routinely chose to conform to the group by giving the same (incorrect) responses as Asch's assistants. Another 40 percent gave incorrect responses in about half of the trials. Although 25 percent always gave correct responses, even they felt very uneasy and "knew that something was wrong." In discussing the experiment afterward, most of the subjects who gave incorrect responses indicated that they had known the answers were wrong but decided to go along with the group to avoid ridicule or ostracism.

In later studies, Asch found that if even a single assistant did not agree with the others, the subject was reassured by hearing someone else question the accuracy of incorrect responses and was much less likely to give a wrong answer. Figure 7.3 shows how group size was related to conformity. This shows the power that groups have to produce conformity among members.

In a real-life verification of Asch's experiment, Dror and colleagues (2006) asked if the judgment of forensic experts would be affected by the opinions of others. They presented five fingerprint experts with pairs of fingerprints the experts had testified were identical in actual criminal cases several years earlier. Each pair of prints was checked by two other experts who verified that the prints matched.

The subject experts had consented to their participation in the study, but did not know when they would be tested. During a normal work period, the subject experts were asked by one of their work colleagues about a set of prints. The colleagues told them that the prints were those of a well-known case of erroneous matching by the FBI in a case involving a bombing in Madrid in which the wrong person was identified. Given this context, three of the five experts now said the prints definitely did not match, one stuck with his earlier judgment, and one said there was insufficient information. This shows that even expert judgments are subject to contextual influences such as those Asch designed into his experiments.

FIGURE 7.3 : EFFECT OF GROUP SIZE IN THE ASCH CONFORMITY STUDIES

As more people are added to the "incorrect" majority, subjects' tendency to conform by giving wrong answers increases–but only up to a point. Adding more than seven people to the incorrect majority does not further increase subjects' tendency to conform–perhaps because subjects are suspicious about why so many people agree with one another.

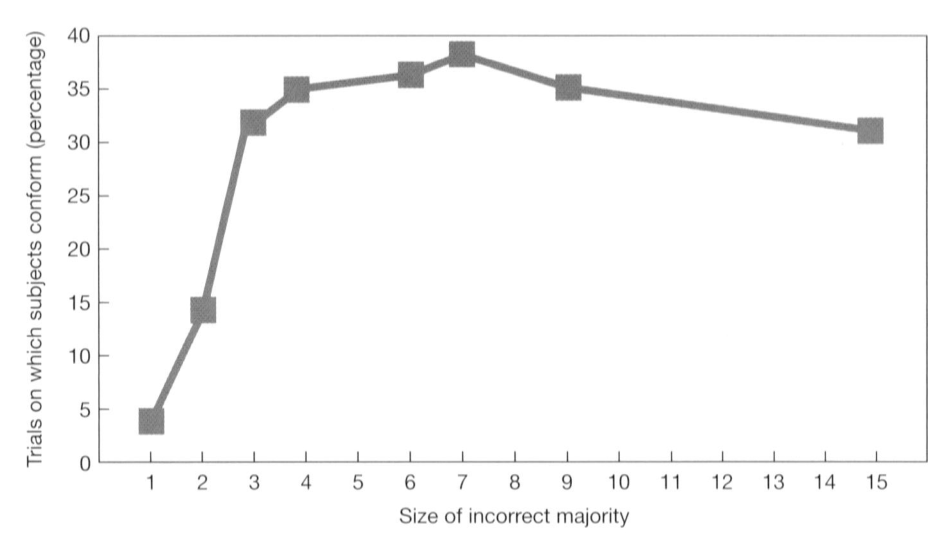

Source: Asch, 1955.

MILGRAM'S RESEARCH ON OBEDIENCE How willing are we to do something because someone in a position of authority has told us to do it? How far are we willing to go in following that individual's demands? Stanley Milgram (1963, 1974) conducted a series of controversial experiments to answer these questions about people's obedience to authority.

Milgram felt that making judgments about the length of a line was a substantively trivial matter, so he decided to conduct an experiment dealing with a more serious matter—how people would respond to pressure to inflict significant pain on another person. Milgram wanted to understand atrocities such as the Holocaust, where ordinary citizens behaved brutally when they were ordered to do so. Of course, the Holocaust is by no means the only example of horrifying behaviour by people working in formal organizations. In 2014, the U.S. Senate released a report on the many different kinds of torture—including waterboarding, rectal feeding, beatings, and sleep deprivation—used by members of the CIA on people suspected of terrorism in the years following the 9/11 attacks on the United States. Dozens of people participated in these activities, the leadership of the CIA misled government overseers about what they were doing, and there was no evidence that the torture resulted in any valuable information being collected.

While Milgram's experiment focused on obedience to authority rather than on conformity to others, it is still very relevant to the behaviour of people in groups, including bureaucracies.

Milgram's subjects were men who had responded to an advertisement for participants in an experiment. In each trial, the first subject was told that the study concerned the effects of punishment on learning. After the second subject (actually an actor hired by Milgram) arrived, the two men were directed to draw slips of paper from a hat to get their assignments as either the "teacher" or the "learner." Because the drawing was rigged, the actual subject always became the teacher and the actor became the learner. The learner was strapped into a chair with protruding electrodes that looked like an electric chair. The teacher was placed in an adjoining room and given a realistic-looking, but nonoperative, shock generator. The "generator's" control panel showed levels that went from "Slight Shock" (15 volts) on the left, to "Intense Shock" (255 volts) in the middle, to "DANGER: SEVERE SHOCK" (375 volts), and finally "XXX" (450 volts) on the right.

Shane Mccoy/The LIFE Images Collection/Getty Images

Following the 9/11 attacks on the World Trade Center and the Pentagon, the U.S. government held hundreds of Muslim prisoners at Guantanamo Bay, Cuba. Members of the Central Intelligence Agency and the U.S. military followed their superiors' orders and used torture methods such as waterboarding and rectal feeding on some of the detainees.

The teacher was instructed to read aloud a pair of words and then repeat the first of the two words. At that time, the learner was supposed to respond with the second of the two words. If the learner could not provide the second word, the teacher was instructed to press the lever on the shock generator so that the learner would be punished for forgetting the word. Each time the learner gave an incorrect response, the teacher increased the shock level by 15 volts. The alleged purpose of the shock was to determine whether punishment improves a person's memory.

What was the maximum level of shock that a "teacher" was willing to inflict on a "learner"? The learner had been instructed (in advance) to beat on the wall between himself and the teacher as the experiment continued, pretending that he was in intense pain. The teacher was told that the shocks might be "extremely painful" but would cause no permanent damage. At about 300 volts, when the learner quit responding to questions, the teacher often turned to the experimenter to see what he should do next. When the experimenter indicated that the teacher should give increasingly painful shocks, 65 percent of the teachers administered shocks all the way up to the "XXX" (450 volt) level (see Figure 7.4). By this point in the process, the teachers were frequently sweating, stuttering, or biting on their lip.

According to Milgram, the "teachers"—who were told they were free to leave whenever they wanted to—continued in the experiment because they were being given directions by a person in a position of authority (a scientist wearing a white coat).

What can we learn from Milgram's study? The study suggests that obedience to authority may be more common than most of us would like to believe. None of the "teachers" challenged the process before they had applied 300 volts. Almost two-thirds went all the way to what could have been a painful jolt of electricity if the shock generator had been real.

Burger (2009) conducted a partial replication of Milgram's work and had results similar to those of the earlier study. Most people went to the end of the experiment. He also found that women were as likely to obey as men.

Milgram's research raises ethical questions. The subjects were deceived about the study. Many found the experiment extremely stressful, and some suffered anxiety so severe that the

FIGURE 7.4 : RESULTS OF MILGRAM'S OBEDIENCE EXPERIMENT

You may be surprised by the subjects' willingness to administer what they thought were severely painful and even dangerous shocks to a helpless "learner."

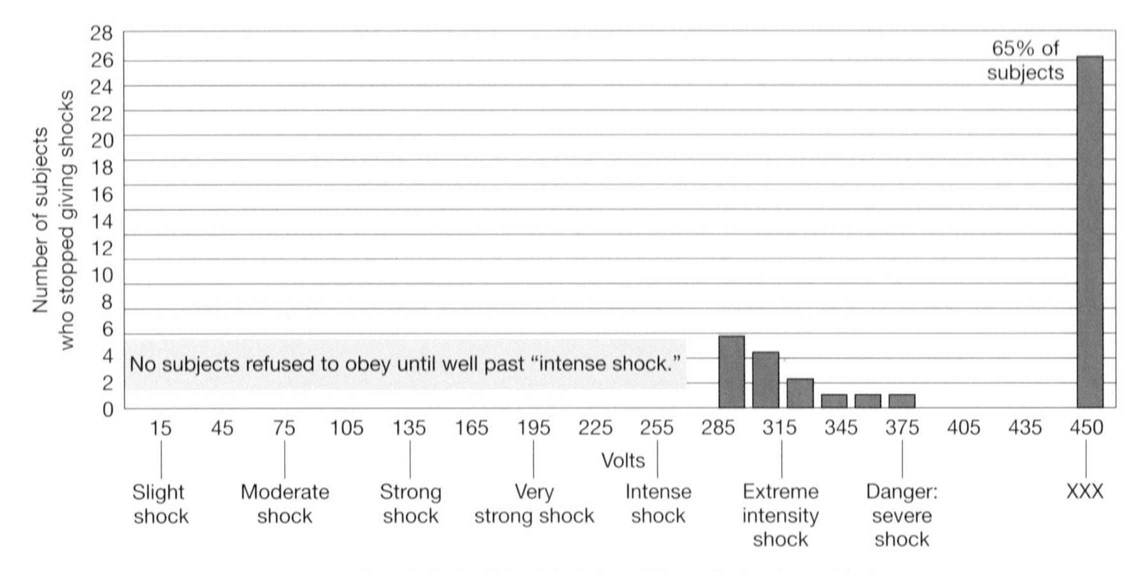

Source: Graph is based on Table 2 in Milgram, 1963: 376.

experimental sessions had to be ended (Milgram, 1963). It would be impossible today to obtain permission to replicate this experiment in a university setting, though such studies were common in the 1960s. Burger's partial replication of the study was approved because he made a number of changes to the original, including stopping the experiment at the 150-volt level. Burger felt this was justified because in Milgram's work, most people who went past this level continued all the way to the end.

In addition to ethical problems, some critics feel that Milgram's study was methodologically flawed. Brannigan (2004) has raised the issue of whether the subjects actually believed that they were hurting people. The more realistic Milgram made the experiment, the more likely the subjects were to refuse to proceed. One critic explains why he does not take the results of Milgram's study seriously:

> Every experiment was basically preposterous . . . the entire experimental procedure from beginning to end could make no sense at all, even to the laymen. A person is strapped to a chair and immobilized and is explicitly told he is going to be exposed to extremely painful electric shocks . . . The task the student is to learn is evidently impossible. He can't learn it in a short time . . . No one could learn it . . . This experiment becomes more incredulous and senseless the further it is carried. (Mantell, 1971:110–111)

Because of the artificiality of the laboratory situation, Brannigan doubts that this experiment tells us anything about why German citizens willingly participated in the atrocities of the Holocaust. The issue of artificiality means that we should always be cautious when we consider the findings of laboratory experiments involving human behaviour.

GROUPTHINK As we have seen, individuals often respond differently in a group context than if they were alone. In many circumstances, it is good for group members to work together and to support each other—think of professional hockey players following their coach's system.

groupthink The process by which members of a cohesive group arrive at a decision that many individual members privately believe is unwise.

However, Janis (1972, 1989) examined group decision making and found that major blunders may be attributed to pressure toward group conformity. To describe this phenomenon, he coined the term **groupthink**—the process by which members of a group arrive at a decision that many individual members privately believe is unwise. Why not speak up at the time? Members usually want to be "team players." They may not want to be the ones who undermine the group's consensus or who challenge the leadership. Some group leaders do not welcome dissent, so those who express disagreement may be punished or ignored. Consequently, members often withhold their opinions and focus on consensus rather than on exploring all the options and making the best decision.

The tragic 2010 explosion of the BP Deepwater Horizon oil rig, owned by British Petroleum (BP) and located in the Gulf of Mexico, is an example of this process. Errors in decision making contributed to one of the worst oil spills and marine and wildlife disasters in U.S. history. Eleven people were killed and 17 were injured in the rig explosion, and massive damage was done to the Gulf Coast and the fishing and tourism industries because of this accident. Figure 7.5 summarizes the dynamics and results of groupthink.

Why is this an example of groupthink? It is one thing to doubt your judgment about the length of a line, as in Asch's experiments, and quite another to send 11 people to their deaths and create an environmental disaster. The event occurred because officials for BP, Transocean, and Halliburton, the major transnational corporations responsible for this error in decision making, closed off their discussions about safety and hid bad news from one another and public officials; because they began to think alike in their assumption about safety, namely that a blowout preventer would keep such a massive disaster from occurring; and because their companies were already behind schedule, had put millions of dollars into production, and did not want to stop to check out reports that a rubber safety seal was broken. Corporate officials risked other people's lives to accomplish their own bureaucratic goals.

TIME TO REVIEW

- Discuss the differences between groups, aggregates, and categories.
- Explain the importance of the difference between primary and secondary groups.
- Discuss how people interact though networks.
- Explain how groups can lead people to make bad decisions. How does "groupthink" affect peoples' behaviour?
- Discuss the impact of group size on interaction patterns within groups.

LO-4 FORMAL ORGANIZATIONS

In earlier times, life was centred in small, informal groups, such as the family and the village. With industrialization and urbanization, people's lives became increasingly dominated by large, formal organizations. A formal organization is a highly structured group formed for the purpose of completing certain tasks or achieving specific goals. Formal organizations (including corporations, schools, and government agencies) usually keep their basic structure for many years.

BUREAUCRACIES

bureaucracy An organizational model characterized by a hierarchy of authority, a clear division of labour, explicit rules and procedures, and impersonality in personnel matters.

The bureaucratic model is the most universal organizational form in government, business, education, and religion. A **bureaucracy** is an organizational model characterized by a hierarchy of authority, a clear division of labour, explicit rules and procedures, and impersonality in personnel matters.

When we think of a bureaucracy, we may think of "buck-passing," such as occurs when we are directed from one office to the next without receiving a solution to our problem.

FIGURE 7.5 JANIS'S DESCRIPTION OF GROUPTHINK

In Janis's model, prior conditions, such as a homogeneous group with committed leadership, can lead to potentially disastrous "groupthink," which short-circuits careful and impartial deliberation. Events leading up to the tragic 2010 explosion of the BP oil rig provides an example of this process.

Process of Groupthink	Example: Deepwater Horizon Explosion
PRIOR CONDITIONS Isolated, cohesive, homogeneous decision-making group Lack of impartial leadership High stress	Millions of dollars had been spent on production of mobile offshore drilling unit; BP was running behind schedule and was under pressure to complete work despite reports of a leak in the rig's blowout preventer.
SYMPTOMS OF GROUPTHINK Closed-mindedness Rationalization Squelching of dissent "Mindguards" Feelings of righteousness and invulnerability	Although rig workers reported pieces of rubber seal coming loose, superiors stated this happened often. Superiors closed off debate by saying blowout-preventer problem would be resolved if something went wrong.
DEFECTIVE DECISION MAKING Incomplete examination of alternatives Failure to examine risks and contingencies Incomplete search for information	In March 2010, when rig workers informed superiors of actual leaks in gasket on blowout preventer on rig, no decision was made to repair the rubber seal or to stop work. Superiors began to hide bad news, and decisions were made without a clear sense of what risks were involved.
CONSEQUENCES Poor decisions	No one stopped production. The BP oil rig exploded, killing 11 and injuring 17. The spill was the largest of its kind in the history of the petroleum industry. BP, Transocean, and others were blamed for making a series of bad decisions based on money, time pressures, and too many people thinking alike.

AP Images/Anonymous/US Coast Guard

We also may view a bureaucracy in terms of red tape because of the situations in which there is so much paperwork and so many incomprehensible rules that no one really understands what to do. However, bureaucracy originally was seen as a way to make organizations *more* productive and efficient. Weber (1968/1922) was interested in the historical trend toward bureaucratization that accelerated during the Industrial Revolution. To Weber, the bureaucracy was the most efficient means of attaining organizational goals because of its coordination and control.

WHY BUREAUCRACY? While much of the rest of this chapter focuses on how bureaucracies work, it is also important to understand why they exist. The simple answer is that they exist because organizations grew too large to be managed in any other way. However, large organizations existed for thousands of years before the birth of bureaucracy, so we must also consider social conditions to explain why the modern bureaucratic form of social organization arose in the 19th century in Europe and North America.

Max Weber suggested that the growth of bureaucratic organizations required cultural and structural changes that did not occur until this time. The cultural change was the rejection of *traditional authority* and the acceptance of *rational-legal authority* as the basis of conduct. This means that people were less willing to accept rules based on tradition and more willing to grant legitimacy to a set of rules intended to achieve certain ends (Weber, 1947). Weber's influential work on the relationship between the rise of Protestantism and the development of capitalism (Weber, 1976) analyzes the factors that led to this change.

The social conditions for factory bureaucracies were established during the Industrial Revolution, when peasants were forced off the farms. These former peasants became the first large labour pool for the factories, as they had no alternative but to work for whatever wages the owners would pay them. The system of wage employment gave the profits from the workers' labour to the factory owner, while the workers were paid only a subsistence wage. This cheap labour provided a tremendous incentive for the factory owners to expand their enterprises. Owners used their profits to mechanize the factories; they also developed systems of specialization and standardization to make their organizations more productive and more profitable. Of course, breaking down production into specialized tasks required managers to coordinate activities, so the factories quickly became hierarchical organizations.

The success of the factory bureaucracy encouraged other organizations to adopt the same principles, and the bureaucratic form quickly spread to governments, schools, and churches.

FORMAL CHARACTERISTICS OF BUREAUCRACY Weber set forth several characteristics of bureaucratic organizations. Although real bureaucracies may not feature all of these ideal characteristics, Weber's model highlights the organizational efficiency and productivity that bureaucracies strive for.

Division of Labour Bureaucracies are characterized by specialization, and each member has a specific role with assigned tasks to fulfill. This division of labour requires the employment of specialized experts. In a university, for example, a distinct division of labour exists between the teaching faculty and the administration.

Hierarchy of Authority Hierarchy of authority, or chain of command, means that each lower office is under the control of a higher one. Hierarchical authority takes the form of a pyramid. Those few individuals at the top have more power and exercise more control than do the many at the lower levels. People lower in the hierarchy report to (and take orders from) those above them. Persons at the upper levels are responsible not only for their own actions but also for those of the individuals they supervise.

Rules and Regulations Weber asserted that rules and regulations establish authority within an organization. These rules are typically standardized and provided in written form. Written rules and regulations can offer clear-cut standards for determining satisfactory performance. They also provide continuity so that each new member does not have to reinvent the rules and regulations.

Qualification-Based Employment Bureaucracies hire staff members and professional employees based on specific qualifications. Favouritism, family connections, and other subjective factors not relevant to organizational efficiency are not acceptable criteria for employment. Individual performance is evaluated against specific standards, and promotions are based on merit as defined by personnel policies.

Impersonality A detached approach should prevail toward clients so that personal feelings do not interfere with organizational decisions. Officials must interact with subordinates based on their status in the organization, not on the officials' personal feelings.

INFORMAL STRUCTURE IN BUREAUCRACIES An organizational chart makes the official, formal structure of a bureaucracy readily apparent. In practice, however, people in bureaucracies behave in ways that do not always follow organizational charts and formal rules. In addition to its formal structure, every bureaucracy has an informal structure, which has been called "bureaucracy's other face" (Page, 1946).

An organization's **informal structure** comprises those aspects of participants' day-to-day activities that do not correspond with the official rules and procedures of the bureaucracy. An example is an informal "grapevine" that spreads information (with varying degrees of accuracy) much faster than do official channels of communication, which are often slow and unresponsive. The informal structure also includes the ideology and practices of workers on the job. Workers create a work culture to help deal with the constraints of their jobs and to guide their interactions with co-workers.

informal structure Those aspects of participants' day-to-day activities and interactions that ignore, bypass, or do not correspond with the official rules and procedures of the bureaucracy.

HAWTHORNE STUDIES AND INFORMAL NETWORKS The Hawthorne studies first made social scientists aware of the effect of informal networks on workers' productivity.

Researchers observed 14 men in the "bank wiring room" who made parts of switches for telephone equipment. Although management offered financial incentives to encourage the men to work harder, the men continued to work according to their own informal rules. They tended to work rapidly in the morning and to ease off in the afternoon. They frequently stopped their own work to help colleagues who had fallen behind. When they got bored, they swapped tasks so their work was more varied. They played games and bet on sports events.

Why did these men insist on limiting production even when they were offered financial incentives to work harder? Perhaps they feared that production quotas would increase if they showed that they could do more. Some may have feared that they would lose their jobs if the work were finished more rapidly. One finding stood out: their productivity level was clearly related to the pressure they received from other members of their informal networks. Those who worked too hard were called "speed kings" and "rate busters"; individuals who worked too slowly were referred to as "chisellers." Those who broke the informal norm against telling a supervisor about someone else's shortcomings were called "squealers." Negative informal sanctions made the workers adhere to the informal norms of their work group. Ultimately, the productivity level was determined by the workers' informal networks, not by the levels set by management (Blau and Meyer, 1987; Roethlisberger and Dickson, 1939).

POSITIVE AND NEGATIVE ASPECTS OF INFORMAL STRUCTURE Is informal structure good or bad? Should it be controlled or encouraged? Two schools of thought have emerged with regard to these questions. One approach emphasizes control of informal groups; the other suggests that they should be nurtured.

Traditional management theories are based on the assumption that people are basically lazy and motivated by greed. Consequently, informal groups must be controlled (or eliminated)

The Canadian Press/Richard Lam

Corporal Catherine Galliford is one of the members and former members who have launched sexual harassment suits against the RCMP. Sociologists have found that women in male-dominated fields are less likely than men to be included in informal networks and more likely to be harassed on the job. Are these two factors related? What steps could be taken to reduce the problems of harassment and lack of networks?

to ensure greater worker productivity. Proponents of this view cite the bank wiring room study to demonstrate the importance of controlling informal networks.

The other school of thought asserts that people are cooperative. Thus, organizations should foster informal groups that permit people to work more efficiently toward organizational goals. Barnard (1938) suggested that informal groups help organizations by providing understanding and motivation for participants. Research on soldiers in combat has shown that bonds with other soldiers in each small squad or platoon have much more impact on performance than abstract notions of patriotism and love for one's country (Marshall, 1947). Even in huge organizations, close interpersonal relationships provide meaning and a sense of belonging to individual workers.

On the other hand, informal groups can have a negative impact on employees who are excluded from them. Some employees may be excluded from networks that are important for survival and advancement in the organization. For example, women and visible minorities who are employed in positions traditionally held by white men (such as firefighters, police officers, and construction workers) are often excluded from the informal structure. Not only do they lack an informal network to "grease the wheels," they also may be harassed and endangered by their co-workers. For example, in 2012, many female RCMP members and ex-members sued the RCMP, claiming they had suffered sexual harassment, gender discrimination, and exposure to pornography throughout their careers. Many of these women had stress-related medical problems and some had left the RCMP because of the working conditions. While many women have had long and successful careers in the RCMP and other Canadian police departments, others have had their careers ruined by the informal culture within their organizations.

SHORTCOMINGS OF BUREAUCRACIES

Weber's intentionally gave us an idealized model of bureaucracy. However, the characteristics that make up this "rational" model have a dark side that has frequently given bureaucracies a bad name (see Figure 7.6). Three of the major problems of bureaucracies are inefficiency and rigidity; resistance to change; and perpetuation of gender, race, and class inequalities.

FIGURE 7.6 CHARACTERISTICS AND EFFECTS OF BUREAUCRACY

The characteristics that define Weber's idealized bureaucracy can create or worsen the problems that many people associate with this type of organization.

© John Aikins/Corbis

Pavel LPhoto and Video/Shutterstock.com

Characteristics	Effects
• Division of labour • Hierarchy of authority • Rules and regulations • Qualification-based employment • Impersonality	• Inefficiency and rigidity • Resistance to change • Perpetuation of race, class, and gender inequalities

INEFFICIENCY AND RIGIDITY Bureaucracies can be rigid and inefficient. The self-protective behaviour of officials at the top may render the organization inefficient. One type of self-protective behaviour is the monopolization of information. Information is crucial for decision making at all levels of an organization. However, those in positions of authority may guard information because it gives them power—others cannot second-guess their decisions without access to relevant (and often confidential) information (Blau and Meyer, 1987).

Those at the top may use their power and authority to monopolize information, and they may also fail to communicate with workers at the lower levels. This may leave them unaware of potential problems within their organizations. Meanwhile, those at the bottom of the structure hide their mistakes from supervisors, a practice that ultimately may result in problems for the organization.

Policies and procedures also contribute to inefficiency and rigidity. Bureaucratic regulations are often written out in great detail to ensure that almost all conceivable situations are covered (Blau and Meyer, 1987). This can lead to **goal displacement,** which occurs when the rules become an end in themselves rather than a means to an end (Merton, 1968). Some administrators may overconform to the rules because their expertise is knowledge of the regulations and they are paid to enforce them. They also fear that if they bend the rules for one person, they may be accused of engaging in favouritism (Blau and Meyer, 1987).

Bureaucrats may also be inflexible because they fear criticism or liability if they do not follow the rules closely. In the case of Kenneth Payne, the aspiring teacher you read about in the chapter introduction, bureaucrats were afraid to waive the need for fingerprints because of public concern about the possibility of sexual offenders working in the schools. These bureaucrats were able to avoid taking responsibility for their unreasonable decision by saying that they were "just following the rules." Mistakes can be blamed on the bureaucracy rather than on the individuals who run it.

Rigidity can also occur at lower levels. Merton (1968) used the term **bureaucratic personality** to describe workers who are more concerned with following correct procedures than with getting the job done correctly. Such workers are usually able to handle routine situations effectively but may be incapable of handling a unique problem or an emergency. Box 7.2 shows how bureaucratic rigidity contributed to serious terrorist attacks in Canada and the United States.

▶ **goal displacement** A process that occurs in organizations when the rules become an end in themselves rather than a means to an end.

▶ **bureaucratic personality** A psychological construct that describes those workers who are more concerned with following correct procedures than they are with doing the job correctly.

The Canadian Press/AP Photo/Barry Sweet

REUTERS/Claro Cortes IV/Landov

The "organization man" of the computer age varies widely in manner and appearance, as shown in the contrast between the top photo of a casually clad Microsoft programmer and the one of Bill Gates, the formally dressed chairman of Microsoft.

RESISTANCE TO CHANGE Resistance to change occurs in all bureaucracies. This resistance can make it difficult for organizations to adapt to new circumstances. Many workers are reluctant to change because they have adapted their professional and personal lives to the old way of doing their jobs. Some have also seen previous change efforts fail and do not want to commit to the latest effort at transforming their organization. Those trying to implement change can have a difficult task breaking through this resistance.

The hierarchical structure of bureaucracies can make this situation worse. Management is separated from labour, clerical workers from professional workers, and people doing one function from those doing another. This creates structural barriers to communication and to joint problem solving. Information is restricted and problems are dealt with in a segmented way. People are rewarded for not taking risks and punished when they try to make changes. Often, people have no structural way of getting innovative ideas from the bottom to the top, so they give up trying.

BOX 7.2 POINT/COUNTERPOINT

How Bureaucratic Rigidity Contributed to Terrorist Attacks

Bureaucratic rigidity can impede information flow within and between large-scale organizations. Failures of governmental organizations to properly utilize information have contributed to the success of major terrorist strikes in Canada and the United States.

In 1985, an explosion destroyed Air India Flight 182, killing 329 people. The flight had originated in Vancouver and most of the victims were Canadians. After nearly two decades of investigation, two Sikh militants were tried for the crime, but were acquitted in 2005.

A public inquiry into the bombing was critical of both the RCMP and the Canadian Security and Intelligence Service (CSIS). The inquiry concluded that the investigation was seriously flawed and that there was little cooperation between the RCMP and CSIS. Both agencies had information that could have prevented the attack but took no action. Bureaucratic inefficiencies facilitated the attack and hindered the subsequent investigation.

CSIS and the RCMP have different roles. CSIS is responsible for collecting intelligence on possible terrorist activity, while the RCMP investigates criminal matters and helps prosecute accused persons. There has been a history of poor relationships between these agencies, and information collected by one agency has not always been made available to the other. CSIS did not process information on the prime suspect, prematurely erased wiretap evidence, and did not share evidence with the RCMP. The RCMP did not provide CSIS with information that would have allowed them to correctly assess the threat: "Unforgivably, the RCMP did not forward to CSIS the June 1st Telex that set out Air India's own intelligence, forecasting a June terrorist attempt to bomb an Air India flight by means of explosives hidden in checked baggage" (Commission of Inquiry into the Investigation of the Bombing of Air India Flight 182, 2010:23). The failure of these agencies to properly assess and share evidence contributed to the murder of hundreds of people.

These problems are not unique. Following the 2001 terrorist attacks on the United States, the Federal Bureau of Investigation (FBI), the Central Intelligence Agency (CIA), and other U.S. governmental organizations faced similar criticisms. The U.S. Senate Judiciary Committee conducted hearings to determine what information about terrorist activities and possible U.S. targets had been available to federal agencies before September 11, and why the government had not acted on this information to prevent the attacks. The judiciary committee interviewed FBI agent Coleen Rowley, who testified that she believed the culture of the FBI had prevented the organization from acting on what it knew before the attacks (*New York Times*, 2002):

> Agent Rowley: We have a culture in the FBI that there's a certain pecking order, and it's pretty strong. And it's very rare that someone picks up the phone and calls a rank or two above themselves. It would have to be only on the strongest reasons. Typically, you would have to . . . pick up the phone and talk to somebody who is at your rank. So when you have an item that requires review by a higher level, it's incumbent for you to go to a higher-level person in your office and then for that person to make a call . . .
>
> Senator Grassley: In your letter [to the FBI director], you mention a culture of fear, especially a fear of taking action, and the problem of careerism. Could you talk about how this hurts investigations in the field, what the causes are, and what you think might fix these problems?
>
> Agent Rowley: [W]hen I looked up the definition [of careerism], I really said [it's] unbelievable how appropriate that is. I think the FBI does have a problem with that. And if I remember right, it means, "promoting one's career over integrity." So, when people make decisions, and it's basically so that [they] can get to the next level and not rock–either it's not rock the boat or do what a boss says without question. And either way that works, if you're making a decision to try to get to the next level, but you're not making that decision for the real right reason, that's a problem . . .

Kanter provides an example of this kind of blockage in a textile company that had been dealing with frequent and costly yarn breakages for decades:

> A new plant manager interested in improving employee communication and involvement discovered a foreign-born worker with an ultimately successful idea for modifying the machine to reduce breakage—and was shocked to learn that the man had wondered about the machine modification for thirty-two years. "Why didn't you say something before?" the manager asked. The reply: "My supervisor wasn't interested and I had no one else to tell it to." (Kanter, 1983:70)

BOX 7.3 SOCIOLOGY AND NEW MEDIA

Organizations and New Media

Do you enjoy going to the grocery store, standing in line at the checkout counter, and carrying your groceries to the car? What if you could order your groceries online and have them delivered to your home for less than what it would cost to buy them at the store? The Internet is having a major impact on the organizational structure of many businesses, by simplifying product distribution. One company, Grocery Gateway (www.grocerygateway.com), sells groceries over the Internet in Ontario, and other Canadian companies have also entered the business. In 2015, Loblaw is testing a system that allows people to order groceries online and to pick them up at one of their stores.

Shopping on the Internet results in *disintermediation*—removing the middleman. An Internet transaction can be made directly between the producer and the consumer. The Internet allows manufacturers to sell their products faster and more cheaply without going through wholesalers and retailers. This capacity is having a major impact on the structure of organizations that sell consumer goods and services. The book publishing industry is being transformed by e-readers, which enable people to download books rather than buying the physical product. Programs such as the Kindle Direct Publishing service allow people to self-publish and avoid the entire publishing industry—there are no writers' agents, no publishing staffs, no bookstores, and no paper is needed for hard copies of books.

The greeting card business shows how the Internet can affect retailers. For more than a century, companies have designed and produced greeting cards and have sold the cards to wholesalers, who distributed them to retail outlets. Customers bought the cards, bought stamps, and mailed them to the recipient through Canada Post. Jobs are created at each step of this process. A new approach taken by companies such as Blue Mountain (www.bluemountain.com) eliminates the need for printing and paper and for wholesalers and retailers. Blue Mountain cards are selected online and sent electronically. Online cards have contributed to flattened sales of greeting cards over the past decade. Christmas mail volumes have also declined in Canada and the United States over this period as people share greetings electronically rather than by mail.

Music downloads, both legal and illegal, have already put many music stores out of business, and electronic commerce has become common in areas such as banking and investing, travel services, and selling goods such as flowers, clothing, and pizza. What do you think will happen to the jobs of stockbrokers and people who work in banks and travel agencies? How will the change affect the bureaucracies in which they work? Internet commerce represents a growing portion of consumer sales. The use of the Internet is growing; those who use the Internet tend to be well-educated people with high incomes who are attractive to marketers; and secure payment mechanisms are making people more comfortable about providing their credit card numbers. Amazon is now aiming to provide same-day delivery in some cities, which will eliminate one of the major advantages of retailers and encourage more people to shop online. In the future, merchandising organizations will look very different than traditional bureaucracies.

Organizations that resist change, rather than adapt to it, may not survive and will not flourish. Thus leaders of many organizations face the task of developing new organizational models that are better suited to today's environment. Consider the challenges faced by leaders of corporations such as those discussed in Box 7.3, "Organizations and New Media."

PERPETUATION OF GENDER, RACE, AND CLASS INEQUALITIES Bureaucracies can perpetuate inequalities of gender, race, and class. Affluent white men still control most North American bureaucracies. These divisions can be perpetuated by the "dual labour market" in which bureaucracies provide different career paths for different categories of workers. Middle- and upper-class employees are more likely to have careers characterized by higher wages, job security, and opportunities for advancement. By contrast, poor and working-class employees (who are more likely to be women and members of racial minorities) work in occupations characterized by low wages, lack of job security, and few opportunities for promotion. This dual labour market not only reflects class, race, and gender inequities but also perpetuates them (see Box 7.4).

While the situation has improved over the past several decades, women and members of racial minorities have found themselves excluded from informal networks. Kanter (1977) conducted an important study of the difficulties faced by workers who did not fit the white male stereotype. There are enormous pressures on "tokens"—group members who were different from the dominant group members. Tokens were singled out and were often viewed as representatives of their group rather than as individual workers. These pressures led to higher turnover rates and to reduced performance by those in the token groups.

To counteract these pressures, organizations must establish policies that ensure supportive environments for members of disadvantaged groups. Pryor and McKinney (1991) showed how people respond to environments that condone sexist behaviour. Their experiments examined the dynamics of sexual harassment on university campuses. In one experiment, a graduate student (a member of the research team) led research subjects to believe that they would be training undergraduate women to use a computer. The actual purpose of the experiment was to observe whether the trainers (subjects) would harass the women if given the opportunity and encouraged to do so. By design, the graduate student purposely harassed the women (who were also part of the research team), setting an example for the subjects to follow.

Pryor and McKinney found that when the "trainers" were led to believe that sexual harassment was condoned and were then left alone with the women, they harassed the women in 90 percent of the experiments. One of the women on the research team felt vulnerable because of the permissive environment created by the men in charge:

> So it kind of made me feel a little bit powerless as far as that goes because there was nothing I could do about it. But I also realized that in a business setting, if this person really was my boss, it would be harder for me to send out the negative signals or whatever to try to fend off that type of thing. (1995)

While this was a laboratory experiment, women often face misogynistic attitudes on university and college campuses. In 2014, 13 males in Dalhousie University's fourth-year dentistry program (calling themselves the Class of DDS 2015 Gentlemen) posted a series of sexist comments about their female colleagues on Facebook. These included comments such as polling each other about which female student they would most like to 'hatef—k' and frequent references to using chloroform and nitrous oxide on bikini-clad women. Think of how stressful it would be for the women to sit in classes with these men for the rest of the academic year, and the possible consequences on their grades and future employment.

MCDONALDIZATION

Weber's views of bureaucracy were based on his belief that rationalization was an inevitable part of the social world. George Ritzer has updated Weber's work by looking at what he calls *McDonaldization*—"the process by which the principles of the fast-food restaurant are coming to dominate more and more sectors of American society, as well as of the rest of the world" (2004:1). McDonald's restaurants embody the principles of rationalization and establish a model that is emulated by many other types of organizations. To Ritzer, fast-food restaurants go beyond Weber's model of bureaucracy. The basic elements of McDonaldization are as follows:

- *Efficiency.* Fast-food restaurants operate like an assembly line. Food is cooked, assembled, and served according to a standardized procedure. Customers line up or move quickly past a drive-through window. Despite the McDonald's slogan, "We do it all for you," it is the customer who picks up the food, takes it to the table, and cleans up the garbage at the end of the meal.
- *Calculability.* The emphasis is on speed and quantity rather than quality. Cooking and serving operations are precisely timed, and the emphasis on speed often results in poor employee morale and high turnover rates. Restaurants are designed to encourage customers to leave quickly.
- *Predictability.* Standard menus and scripted encounters with staff make the experience predictable for customers. The food is supposed to taste the same wherever it is served.
- *Control.* Fast-food restaurants have never allowed individual employees much discretion—employees must follow detailed procedures. The level of control has been enhanced

through technology. For example, automatic french fry cookers and other devices ensure a standardized product. There are no chefs in a fast-food restaurant.
- *Irrationalities of rationality.* Fast-food restaurants are dehumanizing for both customers and employees. The examples of bureaucratic inflexibility used at the beginning of this chapter demonstrate this dehumanization, or **rationality**—the process by which traditional methods of social organization, characterized by informality and spontaneity, are gradually replaced by formal rules and procedures (bureaucracy). Kenneth Payne was denied a teaching career because of an inflexible interpretation of the rules, and the real human concerns of the victims of the New Orleans hurricane were subordinated to organizational rules.

rationality The process by which traditional methods of social organization, characterized by informality and spontaneity, are gradually replaced by formal rules and procedures (bureaucracy).

McDonaldization is expanding to other parts of our lives. Many universities process huge numbers of students by teaching them in large lecture theatres and testing them using machine-graded, multiple-choice exams. The questions on these examinations are often taken from test banks provided by the textbook publishers, who also provide instructors with many of their teaching aids. Students who are more interested in efficiency than in learning can purchase their term papers online so they don't have to spend time writing them.

Recent increases in the number of babies born via surgery, using Cesarean sections rather than waiting for a natural birth, show that even the birth process is being rationalized. Families and doctors may welcome the predictability associated with scheduling birth on a specific day during normal working hours rather than waiting for nature to take its course.

TIME TO REVIEW

- Why have bureaucracies become the most universal organizational form in modern society?
- Describe the five characteristics that Weber believed characterized bureaucracies.
- How can the informal structure affect the way bureaucracies operate?
- Why do bureaucracies sometimes become rigid, inefficient, and resistant to change?
- How do bureaucracies perpetuate inequalities of gender, race, and class?

LO-5 ORGANIZATIONS OF THE FUTURE: THE NETWORK ORGANIZATION

The form of organizations has changed over time. Broad social trends including globalization, technological innovation, and the transition to a service-based economy suggest that *networks* will become the dominant organization of the future. Manuel Castells argues that "the old order, governed by discrete individual units in the pursuit of money, efficiency, happiness, or power, is being replaced by a novel one in which motives, decisions, and actions flow from ever more fluid, yet ever-present networks. It is networks, not the firm, bureaucracy, or the family, that gets things done" (Esping Anderson, 2000:68).

You can learn about global networks by reading about the structure of terrorist organizations in Box 7.5, or by thinking about how illicit drugs get from the coca fields of Colombia and the poppy fields of Afghanistan to users on the streets of Halifax, Toronto, and Victoria. Large bureaucracies are not involved in either of these complex global enterprises, as terrorists and drug dealers operate very effectively through decentralized global networks. One reason that drug suppression strategies have not succeeded is because there is no company called Global Drugs Incorporated that can be located and destroyed by law enforcement agencies. Instead, there are shifting, fluid networks of people who are difficult to identify and who are easily replaced when the legal system (or their competitors) take them out of the network. Similar problems face those who are trying to deal with the threat of terrorism.

BOX 7.4 POINT/COUNTERPOINT

Dilbert and the Bureaucracy

Our experiences with red tape and other bureaucratic inefficiencies have been satirized by cartoonist (and disillusioned bureaucrat) Scott Adams. In the late 1980s, Adams began passing his cartoons around the office at Pacific Bell. Since then, *Dilbert* has become a phenomenal success and is read in more than 2000 newspapers in 70 countries. *Dilbert* ridicules many of the worst features of bureaucracy, including stupid bosses, cubicles, management consultants, pointless meetings, and inflexibility. (For examples of the cartoon, go to www.unitedmedia.com/comics/dilbert).

Workers enjoy *Dilbert*. The cartoons are posted on office doors, walls, and desks, and many of Adams's ideas come from readers' suggestions. *The Economist* magazine attributes *Dilbert's* popularity to the fact that the comic strip taps into three trends that are troubling workers:

1. Employees are forced to work harder because of corporate downsizing.
2. Workers are afraid of being laid off and see their wage increases falling far behind those of their managers.
3. New management fads have led to constant reorganization but have had little impact on efficiency or on job satisfaction.

Ironically, while many workers feel *Dilbert* says what they are thinking about ineffective and uncaring managers, the leaders of many of North America's largest corporations have used the cartoons for training and corporate communications.

Sources: The Economist, 1997; Merton, 1968; Whitaker, 1997.

Another example of a flexible global network is the production of open source software, such as the Linux operating system and the Firefox Internet browser. This software was not produced by large profit-making corporations, but by networks of people working together with no expectation of profit. The product is available freely to anyone who wishes to download it, and programmers all over the world can make improvements in the software. While some coordination is necessary to develop a product that can be used by the public, no large bureaucracy is required and individual users are free to modify programs to suit their own needs.

Networks have always had an advantage over other organizational forms because they are agile and can quickly adapt to new circumstances. However, the ability to coordinate network activities has been weak compared to hierarchical bureaucratic organizations that have defined lines of communication and coordination. This has meant that bureaucracies have had a competitive advantage in handling complex tasks (Castells, 2000b). However, modern information and communication technology has now provided networks with a competitive advantage. Each part of a network can communicate instantly with other parts, and those leading the network can constantly monitor performance even if the network is globally distributed. Technology enables networks to quickly shift and change, as pieces can be eliminated if they are no longer useful or can be temporarily set aside if they are not needed for a particular project (Castells, 2000b).

It is more difficult to control a network than a traditional hierarchical organization because, once the network has been programmed and set in motion, it may be difficult for anyone, even those who started the network, to shut it down. With no central communication and control

BOX 7.5 SOCIOLOGY IN GLOBAL PERSPECTIVE

The Structure of Terrorist Networks

The growth of large armies was important in the development of bureaucratic organizations. While these large armies represented the governments of established countries, many of today's wars are not between two countries and are fought by a very different type of organization.

Terrorist attacks around the globe and the difficulties in fighting insurgencies such as that in Afghanistan have drawn attention to what military planners call *asymmetrical warfare*. This term refers to attacks by small groups of people, who usually do not represent states or governments, upon much larger and stronger opponents. Terrorists do not directly confront their opponents, since they would be quickly defeated in such a confrontation. Rather, they use covert tactics, such as car bombs and suicide bombings, which are difficult to prevent.

To fight successfully against larger and more powerful opponents, terrorist groups must develop organizational structures that are difficult to identify and to fight against. Rather than forming large hierarchical armies, terrorist groups such as al-Qaeda have evolved sophisticated network structures made up of loosely coupled cells, each of which has only a few members. This structure provides a high level of secrecy, flexibility, and innovation. Participants coordinate their activities through relationships with other members of the network, with whom they share a common cause, not through bureaucratic control. Al-Qaeda network members are linked by a common religious background and philosophy, and through the leadership of now-deceased Osama bin Laden, and now of his associates. Figure 7.7 is a simplified diagram of the al-Qaeda network. This structure is very different from the hierarchical organization charts of the military and security organizations that are trying to defeat al-Qaeda. This loose and flexible network structure makes it difficult to defeat terrorist organizations. For example, the network

FIGURE 7.7 SIMPLIFIED REPRESENTATION OF THE AL-QAEDA NETWORK

Subcategories identify the given component's functions; arrows indicate direction of moderately to tightly coupled dependency. Al-Qaeda refers to bin Laden's leadership core. Each component is itself comprised of numerous aggregates linked in loosely to tightly coupled networks of interdependency.

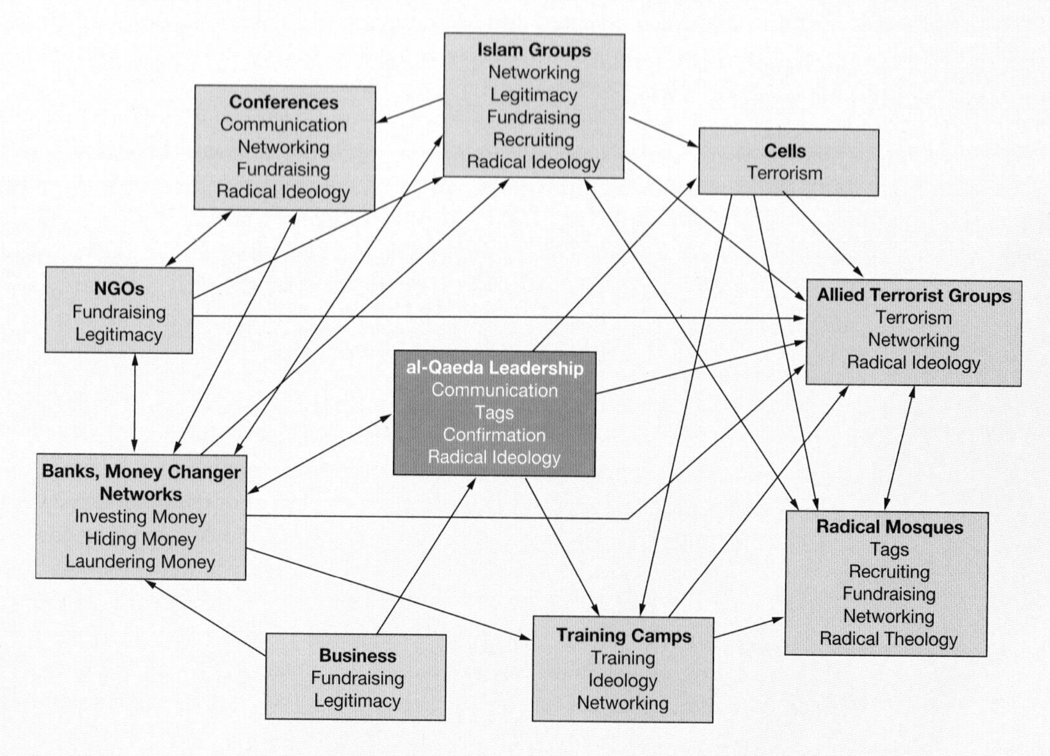

Source: Russ Marion and Mary Uhl-Bien, "A Complexity Theory and Al-Qaeda: Examining Complex Leadership," *Emergence* vol. 5 (1), 2003, pp. 54-76.

(*Continued*)

has roots in many different countries, and information and funds can flow relatively freely from one jurisdiction to another. On the other hand, security and intelligence agencies are based in individual countries and for a variety of reasons find it difficult to work cooperatively (Arquilla and Ronfeldt, 2001). Using modern communications technology, including the Internet, information can flow to all parts of the network much more easily than can information that must be filtered through national governments and their internal bureaucracies (see Box 7.2).

Krebs (2002) analyzed the relationships between the hijackers responsible for the September 11 attacks on the United States. Mohamed Atta, who was the leader of the attacks, had contacts with each of the teams of hijackers. However, most of the others had no contact with teams other than their own. The strategy of minimizing ties among members of the network is deliberate—if security personnel identify or apprehend one or two members of the network, they can provide only limited information about other members so the entire network would not be jeopardized. In a videotape that was found at an al-Qaeda training camp in Afghanistan, Osama bin Laden said, "Those who were trained to fly didn't know the others. One group of people did not know the other group" (Department of Defense, 2001, cited in Krebs, 2002:46). It is very difficult for those opposing such networks to be able to target more than a limited part of the terrorist organization.

Sources: Arquilla and Ronfeldt, 2001; Krebs, 2002.

system, parts of the network can continue to operate even if the central core is eliminated. Thus, opponents of al-Qaeda could not shut down the network by simply closing down some of its pieces. Even Osama bin Laden would have had difficulty closing down the network or changing its goals if other members of al-Qaeda and its affiliated groups around the world wanted to continue with their activities. This was illustrated in 2013 when al-Qaeda's leader, Ayman al-Zawahri, ordered an affiliate group, ISIS, to stop operating in Syria. ISIS leader Abu Bakr al-Baghdad refused to leave (Hubbard, 2014). ISIS continued to fight and became known around the world for its extreme violence, including filmed beheadings of Western prisoners.

▶ **network enterprise** Separate businesses, which may be companies or parts of companies, join together for specific projects that become the focus of the network.

Castells (2000a, 2000b) speaks of a new type of economic organization called the **network enterprise**, in which separate businesses, which may be companies or parts of companies, join together for specific projects that become the focus of the network. This structure gives those responsible for the network considerable flexibility, as they can select and change network partners based upon factors such as cost, efficiency, and technological innovation. The Dell laptop computer that you may be working on is the product of a network enterprise (Friedman, 2005). Dell sells its products over the Internet and by telephone rather than in stores, so your order may be taken by a person in Bangalore, India, rather than by a clerk in your own city. The hardware that makes up the computer was manufactured by companies in Israel, the Philippines, Malaysia, Costa Rica, China, Taiwan, South Korea, Germany, Japan, Mexico, Singapore, Indonesia, India, and Thailand. The computers are assembled in Dell factories located in Ireland, China, Brazil, Malaysia, and the United States. Thomas Friedman describes how the company fills its orders:

> "In an average day, we sell 140,000 to 150,000 computers," explained Dick Hunter, one of Dell's three global production managers. "Those orders come over Dell.com or over the telephone. As soon as these orders come in, our suppliers know about it. They get a signal based on every component in the machine you ordered, so the supplier knows just what he has to deliver. If you are supplying power cords for desktops, you can see minute by minute how many power cords you are going to have to deliver." Every two hours, the Dell factory in Penang [Malaysia] sends an email to the various SLCs [supplier logistics centres] nearby, telling each one what parts and what quantities of those parts it wants delivered within the next 90 minutes—and not one minute later. Within 90 minutes, trucks from the various SLCs around Penang pull up to the Dell manufacturing plant and unload the parts needed for all those notebooks ordered in the last two hours. This goes on all day, every two hours," said Hunter. (2005:415)

This system is a major reason why Dell helped to dramatically reduce computer prices.

Modern communications technology has been a critical factor in the development of dispersed network organizations. Networks are held together by a rapid flow of information rather than by bricks and mortar and a rigid organizational chart like that of the industrial organization. The globalized production processes used by Dell and many other large companies would be impossible without instant global communication. The Internet itself is a decentralized and loosely coupled network. The Internet was originally designed as a way of ensuring that communications systems would survive an attack targeted at central information hubs. Instead of flowing from a central hub, information on the Internet is transmitted in small packets that can follow a wide range of electronic routes and are put together at the destination computer (Castells, 2000a). Nobody owns the Internet, so it is universally accessible to anyone who has a computer and a connection. While the Internet is vulnerable to a variety of threats, including computer hacking, it would be almost impossible to completely shut it down.

The cultural and organizational means of using the technology is also important (Castells, 2004). This means that just having computers does not guarantee access to the networked global economy. Countries with ineffective governments, few trained workers, and no entrepreneurial culture that supports innovation will be excluded from these networks. India has been successful in getting involved in network enterprises because of an entrepreneurial culture, a democratic government, and the presence of a well-educated workforce with English language skills, while many parts of Africa and Latin America have less involvement in the new economy.

What is the impact of network enterprises on people? While network structures can help corporations to become more profitable, the impact on workers has not always been as positive. For example, unions lose their bargaining power when production at one plant can be quickly moved to another part of the network in a different country. Thus a strike may result in the permanent closure of a factory and the movement of jobs offshore. It is likely that the future work lives of today's students will be affected in many ways—some positive but others negative—by the shift to networked organizations. Because network enterprises are fluid and can quickly transform themselves, you should anticipate that your working lives may also change rapidly after you enter the labour market.

Finally, there are inherent dangers in networked organizations. These dangers were illustrated in August 2003 when the power went off for several days in much of Ontario because of a power outage in Ohio that cascaded through the transmission networks covering the northeastern United States and Ontario. A different type of network failure led to the global financial crisis in 2008 in which lax U.S. mortgage practices affected the global economy and billions of people who had not even invested in these mortgages (Watts, 2009). Our reliance on complex physical and social networks places us at risk, and these risks are not always predictable. Global transportation networks facilitate the spread of disease; email viruses threaten our computers; a moment of indiscretion can be spread around the globe through YouTube; and any disruption of traffic between Detroit and Windsor would have a serious impact on North American automobile production and food distribution because of the system of parts manufacture and supply. Governments will be challenged in the future to determine ways of minimizing this risk.

TIME TO REVIEW

- Explain Ritzer's theory that the principles that guide McDonald's operations are expanding to other parts of our lives and to other parts of the world.
- How does the network structure increase the effectiveness of terrorist groups?
- According to Castells, what are the advantages of network enterprises over traditional businesses?
- How is the Internet changing the way businesses operate?

VISUAL SUMMARY

7

KEY TERMS

aggregate A collection of people who happen to be in the same place at the same time but have little else in common (p. 166).

bureaucracy An organizational model characterized by a hierarchy of authority, a clear division of labour, explicit rules and procedures, and impersonality in personnel matters (p. 176).

bureaucratic personality A psychological construct that describes those workers who are more concerned with following correct procedures than they are with doing the job correctly (p. 181).

category A number of people who may never have met one another but who share a similar characteristic (p. 166).

conformity The process of maintaining or changing behaviour to comply with the norms established by a society, subculture, or other group (p. 171).

dyad A group consisting of two members (p. 170).

goal displacement A process that occurs in organizations when the rules become an end in themselves rather than a means to an end (p. 181).

groupthink The process by which members of a cohesive group arrive at a decision that many individual members privately believe is unwise (p. 176).

informal structure Those aspects of participants' day-to-day activities and interactions that ignore, bypass, or do not correspond with the official rules and procedures of the bureaucracy (p. 179).

LO-1 Identify the differences among social groups, aggregates, and categories.

A social group is a collection of two or more people who interact frequently, share a sense of belonging, and depend on one another. People who happen to be in the same place at the same time are considered an aggregate. Those who share a similar characteristic are considered a category. Neither aggregates nor categories are considered social groups.

© Gari Wyn Williams/Alamy

LO-2 Understand the effect that size has on the functioning of groups.

In small groups, all members know one another and interact simultaneously. In groups with more than three members, the dynamics of communication change and members tend to assume specialized tasks. As groups grow larger, keeping them operating effectively becomes increasingly challenging.

© Roy Morsch/Corbis

LO-3 Explain the impact of groups on people's behaviour.

Groups have a significant influence on our values, attitudes and behaviour. Most of us are willing to conform to the wishes of other group members. This sometimes leads to groupthink—the process by which members of a group arrive at a decision that many individual members privately believe is unwise.

AP Images/Anonymous/US Coast Guard

LO-4 Identify the characteristics that define a bureaucracy and the "other face" of bureaucracies.

A bureaucracy is a formal organization characterized by hierarchical authority, division of labour, explicit procedures, and impersonality. Bureaucracy's "other face" is the informal structure of daily activities and interactions that bypass the official rules and procedures. Informal networks may enhance productivity or may be counterproductive to the organization. They also may be detrimental to those who are excluded from them.

© Lebrecht Music and Arts Photo Library/Alamy

LO-5 Discuss the form large organizations may take in the future.

Broad social trends, such as globalization, technological innovation, and the increased prevalence of a service economy, make it likely that networks will be the dominant organization of the future. Networked organizations, which are made possible by modern communications technology, are flexible and can respond quickly to social change.

REUTERS/Claro Cortes IV/Landov

KEY FIGURES

Used courtesy of George Ritzer

George Ritzer (b. 1940) Ritzer is a prolific social theorist who is perhaps best known for his work on the concept of "McDonaldization." This work is a contemporary extension of Weber's study of rationalization.

Used courtesy of Manuel Castells

Manuel Castells (b. 1942) One of the foremost theorists of network theory, Castells is a Spanish sociologist who has taught for many years in the United States. His other work has focused on the influence of the media, globalization, and the processes of urban life.

Stanley Milgram (1933-1984) Milgram was a social psychologist whose work contributed to the study of social organization. His best-known research involves his study of the "small world" and his controversial obedience studies.

Courtesy of Alexandra Milgram

Rosabeth Moss Kanter (b. 1943) A professor at Harvard Business School, much of Kanter's work has focused on change management. Her best-known book, *Men and Women of the Corporation* (1977), examined the difficulties faced by members of token groups, including women, in formal organizations.

APPLICATION QUESTIONS

1. Do you think the insights gained from Milgram's research on obedience outweigh the elements of deception and stress that were forced on his subjects?
2. Many students have worked at a McDonald's or at some other fast-food restaurant. Relate your experience (or that of your friends) to George Ritzer's analysis of "McDonaldization."
3. Technology is changing the way organizations operate. Consider the entertainment industry. Downloading music and movies from the Internet has become very popular, and the film and music industries are trying hard to convince the Canadian government to pass new legislation to combat this downloading. What are the arguments of those who think that this material should be widely available on the Internet? What are the counterarguments of those who wish to see downloading regulated? Which side do you support in this debate?
4. What happens to people when they violate bureaucratic regulations? What range of sanctions do bureaucracies have to enforce these regulations?
5. Networks are important organizational forms. How have social networking sites changed the way in which people stay connected with their personal and professional networks?

ingroup A group to which a person belongs and identifies with (p. 167).

network A web of social relationships that link a number of people (p. 168).

network enterprise Separate businesses, which may be companies or parts of companies, join together for specific projects that become the focus of the network (p. 188).

outgroup A group to which a person does not belong and toward which the person may feel a sense of competitiveness or hostility (p. 167).

rationality The process by which traditional methods of social organization, characterized by informality and spontaneity, are gradually replaced by formal rules and procedures (bureaucracy) (p. 185).

reference group A group that strongly influences a person's behaviour and social attitudes, regardless of whether that individual is a member (p. 168).

small group A collectivity small enough for all members to be acquainted with one another and to interact simultaneously (p. 170).

triad A group composed of three members (p. 170).

CHAPTER 8

CRIME AND DEVIANCE

Sam Leung/Vancouver Sun

CHAPTER FOCUS QUESTION

What are the causes and consequences of crime and deviance in Canada?

LEARNING OBJECTIVES

AFTER READING THIS CHAPTER, YOU SHOULD BE ABLE TO

LO-1 Explain the meanings of the terms *crime* and *deviance.*

LO-2 Understand how crime and deviance are explained by functionalist, conflict, interactionist, feminist, and postmodern theories.

LO-3 Describe how sociologists count and classify crimes.

LO-4 Understand how age, gender, class, and race are related to deviance and crime.

LO-5 Describe how the criminal justice system deals with crime.

Many Canadians buy drugs or engage in illegal gambling without recognizing the way these actions contribute to the harm caused by organized crime. The massive profits from these illicit enterprises—particularly drugs—leads to competition among different groups seeking to control the market. Territorial and financial disputes involving organized criminals are typically resolved through violence.

In 2009, Vancouver had 23 gang murders (Beattie and Cotter, 2010). Many occurred in busy parts of the city in broad daylight, endangering the lives of ordinary citizens. This wave of homicides resulted from a war between a number of gangs—among them the Red Scorpions, the United Nations Gang, the Independent Soldiers, and the Hells Angels. Territorial battles over drug sales quickly turned into a cycle of revenge killings. Three of the key players in these activities were the notorious Bacon brothers—Jamie, Jarrod and Jonathan. The Bacon brothers did not fit the traditional stereotype of gang members. They came from a middle-class suburb and they were affiliated with multi-ethnic gangs. Traditionally, organized crime gangs were ethnically based, but several of the major gangs in British Columbia's Lower Mainland recruited members based on their loyalty and skills rather than on their ethnicity.

In the midst of the gang war, the Port Moody Police Department (the suburb where Jonathan Bacon lived) issued this unusual public warning:

As part of the response to gang violence which has gripped many communities throughout B.C., the Port Moody Police Department is taking steps to warn the public, friends, and associates of the BACON brothers, that there are significant threats to their safety.

The Port Moody Police Department through its investigation and information received from other police agencies have learned there are plans to murder Jarrod, Jamie and Jonathan BACON. The BACON brothers are well known to the police and have been linked to violence and weapons in the past. The BACON brothers have been approached by police and advised of these threats, but appear to be unconcerned for their safety or for that of their friends, associates and, particularly the public. Due to the seriousness of the threats made against the BACON brothers, those associated to them in any way are being advised to discontinue their association or interaction immediately. (Langton, 2013).

Source: Courtesy of the Port Moody Police Department.

Not surprisingly, the warning did not work. Hours after it was issued, a close associate of the Bacon brothers was shot to death in the parking lot of a busy shopping mall. Also not surprisingly, the saga of the Bacon brothers—heroes to many because of their money, lifestyles, and ability to avoid prosecution—also did not end well. Jonathan was leaving a luxury resort in downtown Kelowna in a Porsche Cayenne when three men jumped out of an SUV and opened fire with automatic weapons. He was killed and four other passengers were injured. A woman travelling with Bacon was paralyzed from the neck down by a bullet.

Jamie Bacon was charged with murder for his alleged role in a shooting in which six people (four gang members and two innocent bystanders) were killed in a Surrey, B.C. apartment building. Jarrod was sentenced to 12 years in prison for cocaine trafficking. Since 2009, gang homicides have declined in Vancouver because many of the gang members were dead or in jail and their successors were not as willing to risk their own lives by getting involved in gang feuds.

Organized crime is one of a wide range of behaviours that society has defined as deviant and/or criminal. Crime and deviance have been of special interest to sociologists. Many of the issues they have examined include: What is deviant behaviour, and how does it differ from criminal behaviour? Why are some people considered to be "deviants" or "criminals" while others are not? How should society deal with those who break the rules? Before reading on, take the quiz on crime and organized crime in Box 8.1.

CRITICAL THINKING QUESTIONS

1. The violence perpetrated by organized crime groups is fed by the huge amounts of money made by selling illegal drugs. Many of the people who buy these drugs are otherwise respectable lawyers, electricians, and perhaps even students. How do you think these people justify sustaining organized crime?
2. In the *Safe Streets and Communities Act*, passed in 2012, the federal government imposed mandatory minimum penalties for some relatively minor drug offences, including a mandatory minimum of nine months for growing six marijuana plants and selling or sharing the crop. Do you think these mandatory penalties will reduce drug crime in Canada?
3. Should so-called victimless crimes such as recreational drug use be decriminalized? What do you think would be the positive and negative effects of decriminalization?

LO-1 WHAT IS DEVIANCE?

How do societies determine what behaviour is acceptable and what is unacceptable? All societies have norms that govern acceptable behaviour. If we are to live and to work with others, these rules are necessary. We must also have a reasonable expectation that other people will obey the rules. Think of the chaos that would result if each driver decided which side of the road she would drive on or which stop sign he would decide to obey. Most of us usually follow the rules our group prescribes. Of course, not all members of the group obey all the time. All of us have broken many rules, sometimes even important ones. These violations are dealt with through various mechanisms of **social control**—systematic practices developed by groups to encourage conformity and discourage deviance. One form of social control is socialization, whereby individuals *internalize* societal norms and values. A second is the use of *negative sanctions* to punish rule-breakers and nonconforming acts. Later in this chapter, you will read about the legal system, which is a *formal* means of social control.

▶ **social control** Systematic practices developed by social groups to encourage conformity and discourage deviance.

Although the purpose of social control is to ensure some level of conformity, all societies have some degree of **deviance**—any behaviour, belief, or condition that violates cultural norms of the society or group in which it occurs (Adler and Adler, 1994).

▶ **deviance** Any behaviour, belief, or condition that violates cultural norms of the society or group in which it occurs.

DEFINING DEVIANCE

Deviance is relative—an act only becomes deviant when it is socially defined as such. Definitions of deviance vary widely from place to place, from time to time, and from group to group. For example, you may have played the Pick 3 lottery. To win, you must pick a three-digit number matching the one drawn by the government lottery agency. We are encouraged to risk our money on this game, and the government takes the profits. In earlier days, the same game was called the numbers racket and was the most popular form of gambling in many low-income neighbourhoods. The two main differences between now and then are that the game used to be run by organized criminals, and that they paid the winners a higher share of the take than the government now does. While the profits now go to social services rather than into the pockets of criminals, the example illustrates the point that the way societies define behaviour can be more important than the harm caused by that behaviour, as legalized gambling involves far more people suffering losses than was the case when gambling was illegal.

Definitions of deviance are continually changing. Several hundred thousand "witches" were executed in Europe during the Middle Ages; now the crime of witchcraft doesn't exist. Racist comments used to be socially acceptable; now they are not. Tattoos and piercings are now common among students, but 30 years ago they were almost unknown.

BOX 8.1 SOCIOLOGY AND EVERYDAY LIFE

How Much Do You Know About Crime and Organized Crime?

True	False	
T	F	1. Official statistics accurately reflect the amount of crime in Canada.
T	F	2. Most organized criminals are affiliated with the Italian Mafia.
T	F	3. Organized crime exists largely to provide goods and services demanded by "respectable" members of the community.
T	F	4. Rates of murder and other violent crimes have been steadily rising for the past 20 years.
T	F	5. Gang-related killings have been declining at about the same rate as other types of homicides in Canada.

For answers to the quiz about crime and organized crime, go to **www.nelson.com/student**.

The Canadian Press/Darryl Dyck

Violence is an important part of the world of organized crime. This shows another victim of a major gang war in Vancouver.

Deviance can be difficult to define. Good and evil are not two distinct categories, and the line between deviant and nondeviant can be *ambiguous*. For example, how do we decide if someone is mentally ill? What if your brother begins to behave in a strange fashion? He occasionally yells at people for no apparent reason and keeps changing topics when you talk to him. He frequently phones you in the middle of the night to complain about people who are threatening him. How would you respond to this change in behaviour? Would it make any difference if you knew that your brother was drinking heavily at the time or was under a lot of stress at work? Would it make a difference if he behaved this way once a year or twice a week? When would you decide that he had a problem and should seek help? What is the difference between someone who is eccentric and someone who is mentally ill? These questions reflect the difficulty we have in defining deviance.

Deviant behaviours also vary in seriousness, ranging from mild transgressions to serious violations of the law. Have you kept a library book past its due date? If so, you have broken the rules. This infraction is relatively minor; at most, you might have to pay a fine. Violations of other university regulations—such as cheating on an examination—are viewed as more serious infractions and are punishable by stronger sanctions, including academic probation or expulsion. Some types of deviance are defined as crimes. A **crime** is an act that violates criminal law and is punishable with fines, jail terms, and other sanctions. Crimes range from minor—running an illegal bingo game or disorderly conduct—to major offences such as sexual assault and murder.

crime An act that violates criminal law and is punishable with fines, jail terms, and other sanctions.

Sociologists are interested in asking who defines acts as deviant, how and why people become deviants, and how society deals with deviants (Schur, 1983). This chapter presents several sociological explanations of deviance. These theories are quite different from one another, but each contributes in its own way to our understanding of deviance. No single theory provides a comprehensive explanation of all deviance. In many respects, the theories presented in this chapter can be considered complementary.

LO-2 SOCIOLOGICAL PERSPECTIVES ON CRIME AND DEVIANCE

FUNCTIONALIST PERSPECTIVES ON CRIME AND DEVIANCE

STRAIN THEORY: GOALS AND THE MEANS TO ACHIEVE THEM According to Robert Merton (1938, 1968), in a smoothly functioning society, deviance will be limited because most people share common cultural goals and agree upon the appropriate means for reaching them. However, societies that do not provide sufficient opportunities to reach these goals may also lack agreement about how people may achieve their aspirations. Deviance may be common in such societies because people may be willing to use whatever means they can to achieve their goals.

▶ **strain theory** The proposition that people feel strain when they are exposed to cultural goals that they are unable to obtain because they do not have access to culturally approved means of achieving these goals.

According to **strain theory,** people feel strain when they are exposed to cultural goals that they are unable to obtain because they do not have access to culturally approved means of achieving these goals. The goals may be material possessions and money; the approved means may include an education and jobs. Some of those who do not have the opportunity to get a good education and a good job may try to achieve their goals through deviant means.

Strain theory has typically been used to explain the deviance of the lower classes. Denied legitimate access to the material goods that are so important in North American culture, some individuals may turn to illegal activities to achieve their goals. However, not only the poor turn to illegal ways of achieving their goals. Some sociologists feel that strain theory can help explain upper-class deviance as well. In 2007, Conrad Black, one of Canada's wealthiest and most influential businessmen, was convicted of fraud and obstruction of justice and sentenced to three and a half years in prison. Despite his wealth, Black took money that rightfully belonged to shareholders from the Hollinger company. A committee established to investigate Black's activities concluded that "Black and [his partner] Radler were motivated by a ravenous appetite for cash . . . and Hollinger International, under their reign 'lost any sense of corporate purpose, competitive drive or internal ethical concerns' as the two executives looked for ways to 'suck cash' out of the company" (McNish and Stewart, 2004:288).

OPPORTUNITY THEORY: ACCESS TO ILLEGITIMATE OPPORTUNITIES Expanding on Merton's strain theory, Richard Cloward and Lloyd Ohlin (1960) suggested that for deviance to occur, people must have access to **illegitimate opportunity structures** that provide an opportunity for people to acquire through illegitimate activities what they cannot achieve through legitimate channels. For example, members of some communities may have insufficient legitimate means to achieve conventional goals of status and wealth, but have much greater access to illegitimate opportunity structures—such as theft, drug dealing, or robbery—through which they can achieve these goals.

▶ **illegitimate opportunity structures** Circumstances that provide an opportunity for people to acquire through illegitimate activities what they cannot achieve through legitimate channels.

According to Cloward and Ohlin (1960), three different forms of delinquent subcultures—criminal, conflict, and retreatist—emerge based on the type of illegitimate opportunities available in a specific area. The criminal subculture focuses on economic gain and includes acts such as theft and drug dealing. Anderson (1990) suggested that the "drug economy [is an] employment agency superimposed on the existing gang network" for many young men who lack other opportunities. For young men who grow up in a gang subculture, selling drugs on street corners provides illegitimate opportunities. Using the money from these "jobs," they can support themselves and can purchase material possessions to impress others. When illegal economic opportunities are unavailable, gangs may become conflict subcultures that fight over territory and adopt a value system of toughness and courage to gain status from their peers. Those who lack the opportunity or ability to join one of these gangs may turn to retreatist forms of deviance, such as drinking and drug use.

Opportunity theory expands strain theory by pointing out the relationship between deviance and the availability of illegitimate opportunity structures. Some studies of gangs have supported this premise by pointing out that gang membership provides some people in low-income central-city areas with the illegitimate means to acquire money, entertainment, refuge, and physical protection (Esbensen and Huizinga, 1993; Jankowski, 1991).

CONTROL THEORY: SOCIAL BONDING Early social control theories proposed that social structure could affect rates of deviance. Communities characterized by poverty, physical deterioration, and internal conflict were too disorganized to exert effective control over residents' behaviour. These communities often had high rates of suicide, mental illness, substance abuse, and crime.

Although most of the research documenting the correlation between community disorganization and crime has been done in large, urban areas, Deutschmann (2002) has applied the theory to 'frontier' areas. Many small Canadian communities were created solely to develop an economic resource. Towns have grown up around mines, mills, and hydro dams. These towns may be lasting or short-lived depending on the nature of the project or the life of the resource. Deutschmann notes that in these towns' early stages, the absence of controls, such as families and churches, means that deviant behaviour, such as fighting and alcohol abuse, may be common. In later stages of development, the strains of a booming town may also facilitate deviance.

Most of the recent work on control theory has focused on the individual rather than on the community. Control theorists have posed the fundamental question about causes of deviance in a new way. Most theories of deviance ask the question, why do they do it? Control theorists reverse the question. They ask, why don't we *all* do it? Why do some people *not* engage in deviant behaviour? In answer to this question, Travis Hirschi (1969) suggested that deviant behaviour is minimized when people have strong bonds that tie them to social institutions, including families, school, peers, and churches.

Social bond theory holds that deviant behaviour will be more likely if a person's ties to society are weakened or broken. According to Hirschi, social bonding consists of (1) *attachment* to other people; (2) *commitment* to conventional lines of behaviour, such as schooling and job success; (3) *involvement* in conventional activities; and (4) *belief* in the legitimacy of conventional values and norms. Hirschi's research found that the variables of attachment and commitment are much more strongly related to delinquency than involvement and belief.

▶ **social bond theory** The likelihood of deviant behaviour increases when a person's ties to society are weakened or broken.

While Hirschi's theory did not differentiate between bonds to conventional and to deviant others, others have modified the theory and suggested that the probability of crime or delinquency increases when a person's conventional social bonds are weak and when peers promote antisocial values and deviant behaviour. Research suggests this modified theory fits the data better than Hirschi's original version (Linden and Fillmore, 1981).

SYMBOLIC INTERACTIONIST PERSPECTIVES ON CRIME AND DEVIANCE

According to symbolic interactionists, deviance is learned in the same way as conformity—through interaction with others. Differential association and labelling theory are two interactionist theories of deviance.

DIFFERENTIAL ASSOCIATION THEORY Edwin Sutherland (1939) developed a theory to explain how people learn deviance through social interaction. **Differential association theory** states that people have a greater tendency to deviate from societal norms when they frequently associate with persons who favour deviance over conformity. According to Sutherland, people learn the necessary techniques and the motives, drives, rationalizations, and attitudes of deviant behaviour from these people.

▶ **differential association theory** The proposition that individuals have a greater tendency to deviate from societal norms when they frequently associate with persons who favour deviance over conformity.

Glenny described the transition of a "whitehat" computer hacker named Max Vision into a criminal "blackhat" hacker named Iceman. Vision had been working for the U.S. government searching for website security vulnerabilities. For reasons that were unclear, Vision left a U.S. Air Force website vulnerable to later attack and was sentenced to prison. After his release, Vision decided to change his life:

> Abandoned by his wife for another man, forsaken by his erstwhile friends in the FBI, Max Vision tumbled down the abyss, at the bottom of which lay a deep depression. Here he landed next to a fellow inmate, one Jeffrey Normington, who extended a hand of friendship when nobody else would.

> On his release from prison, Vision was unable to find regular work that paid more than the minimum wage. He . . . was offered senior positions in security companies abroad, but as he was on parole, he was not eligible for a passport. In Silicon Valley, nobody wanted to employ someone whose CV included an indelible conviction for computer crime.
>
> His debts mounted as his despair deepened. Then one day friend Normington reappeared, promising a path out of the abyss and back into California's sunshine . . . Normington promised him a top-of-the-line Alienware laptop, a must-have but expensive accessory for hackers. That was just for starters. He said he'd find Vision an apartment and pay for it. Normington would arrange everything.
>
> In exchange for a few favours.
>
> Crime was not Vision's sole option. There were other avenues to explore. He could have gone to friends and family. But he was tired, he felt abandoned and Normington was convincing . . .
>
> Max Vision, all-round good guy, was discarded back into an abyss. In his place, Iceman emerged—all-round bad guy. . . . (2011:102)

Vision is now in prison, serving a 13-year sentence for credit card hacking that cost consumers over $85 million.

Differential association is most likely to result in deviance when a person has extensive interaction with rule-breakers. Ties to other deviants can be particularly important in organized crime, where the willingness of peers to stand up for one another is a response to violent competitors. Daniel Wolf, an anthropologist who rode with an Edmonton biker gang, describes this solidarity:

> For an outlaw biker, the greatest fear is not of the police; rather, it is of a slight variation of his own mirror image: the patch holder [full-fledged member] of another club. Under slightly different circumstances those men would call each other "brother." But when turf is at stake, inter-club rivalry and warfare completely override any considerations of the common bonds of being a biker—and brother kills brother. None of the outlaws that I rode with enjoyed the prospect of having to break the bones of another biker. Nor did they look forward to having to live with the hate–fear syndrome that dominates a conflict in which there are no rules . . .
>
> When a patch holder defends his colours, he defends his personal identity, his community, his lifestyle. When a war is on, loyalty to the club and one another arises out of the midst of danger, out of apprehension of possible injury, mutilation, or worse. Whether one considers this process as desperate, heroic, or just outlandishly foolish and banal does not really matter. What matters is that, for patch holders, the brotherhood emerges as a necessary feature of their continued existence as individuals and as a group. (1996:11)

Group ties are not only important in groups such as motorcycle gangs. Think of the different subcultural groups that are involved in deviant activities in many Canadian high schools. Whether the focus of the group is graffiti, using drugs, or fighting, the encouragement and support of peers is vital to recruiting and teaching new members and to sustaining the group.

Differential association theory contributes to our knowledge of how deviant behaviour reflects the individual's learned techniques, values, attitudes, motives, and rationalizations. However, critics criticize the fact that the theory does not adequately assess possible linkages between social inequality and criminal behaviour.

LABELLING THEORY Two complementary processes are involved in the definition of deviance. First, some people act (or are believed to act) in a manner contrary to the expectations of others. Second, others disapprove of and try to control this contrary behaviour. Part of this social control process involves labelling people as deviants. An important contribution to our understanding of deviance was made by sociologists who asked the question: Why are some people labelled as deviants while others are not? **Labelling theory** suggests that deviants are those people who have been successfully labelled as such by others. The process of labelling is directly related to the power and status of those persons who do the labelling and those who are being labelled. To the labelling theorist, behaviour is not deviant in and of itself; it is defined as deviant by a social audience (Erikson, 1962). Labels are applied most easily to those who lack the power to resist them.

labelling theory The proposition that deviants are those people who have been successfully labelled as such by others.

The concept of secondary deviance suggests that when people accept a negative label or stigma that has been applied to them, the label may contribute to the type of behaviour it was initially meant to control (see Figure 8.1). **Primary deviance** is the initial act of rule breaking (Lemert, 1951). **Secondary deviance** occurs when a person who has been labelled deviant accepts that new identity and continues the deviant behaviour. For example, a person may shoplift, not be labelled deviant, and subsequently decide to forgo such acts in the future. Secondary deviance occurs if the person steals from a store, is labelled a "shoplifter," accepts that label, and then continues to steal.

primary deviance A term used to describe the initial act of rule breaking.

secondary deviance A term used to describe the process whereby a person who has been labelled deviant accepts that new identity and continues the deviant behaviour.

Labelling theorists have contributed to our understanding of the way society defines behaviours and individuals as deviant, and of the consequences of that definition. Let us first look at the impact of labelling on a person who is defined as deviant.

Scott (1969) conducted a fascinating study of the effects of two different ways of treating blind people. One agency defined the blind as helpless, dependent people and developed programs to accommodate them. Clients were driven to the agency's offices, where they worked in sheltered workshops and ate food that had been cut before being served. Not surprisingly, the clients had trouble adapting to life outside the agency. Another agency, which dealt mainly with Vietnam War veterans, used a different approach. Their goal was to reintegrate clients into the community. Instead of being driven places, they were trained to take public transit. They were given confidence training and encouraged to live on their own and work in normal job settings. Scott concluded that these different approaches, with different labels for their visually impaired clients, had a significant impact on the self-image and social adjustment of these clients.

Think of the impact that labels can have on a person's self-concept and life chances. Being labelled a drug addict can lead to serious difficulties in getting a job even after successful treatment. If the label prevents the former addict from reintegrating into the conventional community, that person may accept this deviant status and return to his or her friends in the drug world. Similar problems come with other deviant labels. The impact of the label "mentally ill" is described by Tom, an ex-patient:

FIGURE 8.1 LABELLING THEORY

When an act is labelled as deviant, the label may cause the behaviour it was intended to control. The individual may find that others now respond differently to him or her because of the stigmatization of being labelled as deviant. Some individuals may successfully resist the label, but others may develop a deviant self-image and subsequently get involved in secondary deviance.

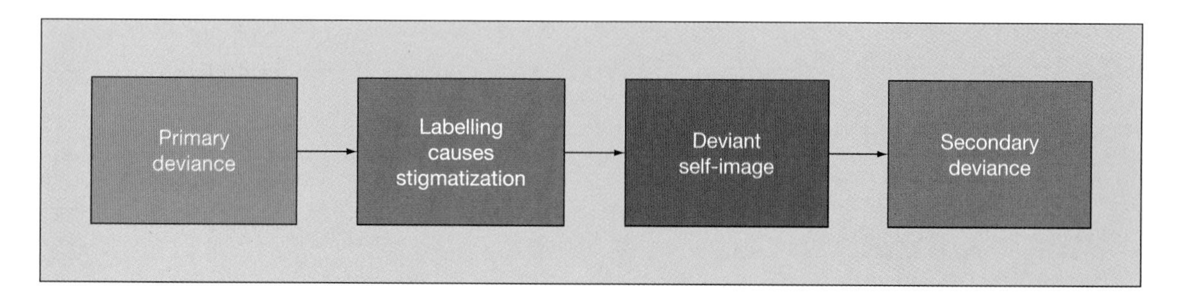

> Having been diagnosed as a psychiatric patient with psychotic tendencies is the worst thing that has ever happened to me. It's shitty to be mentally ill; it's not something to be proud of. It makes you realize just how different you are from everybody else—they're normal and you're not. Things are easy for them; things are hard for you. Life's a ball for them; life's a bitch for you. I'm like a mental cripple! I'm a failure for life! (Herman, 1996:310)

Not everyone passively submits to the labelling process—some people successfully resist the imposition of a label. This can be done individually or by working with others. The leader of an Ontario group of former psychiatric patients described the aims of his group:

> Simply put, we're tired of being pushed around. We reject everything society says about us, because it's just not accurate . . . We don't like the meaning of the words [people] use to describe us—"mentals" and "nuts." We see ourselves differently, just as good and worthy as everybody out there. In our newsletter, we're trying to get across the idea that we're not the stereotypical mental patient you see in movies. We're real people who want to be treated equally under the Charter of Rights. We're not sitting back, we're fighting back! (Herman, 1996:323)

moral entrepreneurs People or groups who take an active role in trying to have particular behaviours defined as deviant.

The view that deviance is socially defined raises the question of why some behaviours are defined as deviant and others are not. Some sociologists have highlighted the role of **moral entrepreneurs**—people or groups who take an active role in trying to have particular behaviours defined as deviant (Becker, 1963). Think of the role that groups such as Mothers Against Drunk Driving (MADD) have played in getting governments to increase the penalties for drunk driving and in educating the public about the dangers of this behaviour. In recent years, health advocates have stigmatized cigarette smoking and many communities have passed legislation banning smoking in most public places. Several decades ago, there was little opposition to smoking; people smoked on buses, in airplanes, in classrooms, and in offices. Because of the health risks of smoking, the antismoking movement was able to overcome tobacco company lobbying and convince governments to impose restrictions on smoking. Not only is smoking banned in public places, but many people now define smokers as deviants who threaten public health (Tuggle and Holmes, 2000). Smoking moved from a normative behaviour to a stigmatized behaviour over a period of about 30 years.

moral crusades Public campaigns that help generate public and political support for moral entrepreneurs' causes.

Moral entrepreneurs often create **moral crusades**—public campaigns that help generate public and political support for their causes. In recent years, we have seen moral crusades against abortion providers, wife abusers, squeegee kids, panhandlers, prostitutes, and a wide variety of other real or perceived threats to society. Some crusades have been more successful than others. The campaign by women's groups for zero-tolerance policies mandating arrest in domestic violence cases has been successful. However, anti-abortion groups have drawn attention to their cause but have been unable to bring about changes in the law.

Some groups succeed in changing perceptions and laws, and others do not. A major reason for the difference is the distribution of power and resources. Those who control the levers of power are much more likely to be able to impose their definitions of what is right and wrong on the rest of society.

Labelling theory has had an impact on the justice system. It has led to an increased use of diversion for minor offences to avoid a formal label. However, critics argue that labelling theory neither explains what causes the original acts that make up primary deviance nor explains why some people accept deviant labels and others do not (Cavender, 1995).

CONFLICT PERSPECTIVES ON CRIME AND DEVIANCE

Who determines what behaviours are deviant or criminal? Conflict theorists feel that people in positions of power use the law to protect their own interests. Conflict theorists suggest that the activities of poor and lower-income individuals are more likely to be defined as criminal than those of persons from middle- and upper-income backgrounds. For example, those who commit welfare fraud are more likely to face criminal charges than are professionals whose misconduct is generally dealt with by disciplinary committees of their peers rather than by the criminal courts. The relative social harm caused by either of these groups seems to have little relevance in the determination of who is defined as criminal; what matters more is the power of some groups to resist sanctions.

THE CONFLICT APPROACH Although Karl Marx wrote very little about deviance and crime, many of his ideas influenced a critical perspective based on the assumption that the criminal justice system protects the power and privilege of the capitalist class.

Marx based his critique of capitalism on the inherent conflict that he believed existed between the capitalists and the working class. According to Marx, social institutions (such as law, politics, and education) make up a superstructure that legitimizes the class structure and maintains the capitalists' dominant position in it. Crime is an expression of the individual's struggle against the unjust social conditions and inequality produced by capitalism.

According to conflict theorists, people with economic and political power define as criminal any behaviour that threatens their own interests or values. Drug laws enacted early in the 20th century were passed and enforced in an effort to control immigrant workers, particularly Chinese workers, who were more inclined than most other residents of Canada to smoke opium. The laws were motivated by racism more than by a real concern with drug use (Cook, 1969). By contrast, while the Canadian government passed anti-combines legislation in 1889 in response to concerns expressed by labour and small-business people about the growing power of monopoly capitalists, the law had no impact. Large companies still engaged in price fixing and other means of limiting competition. Having symbolic anti-combines laws on the

The Canadian Press/Don MacKinnon

This British Columbia accident killed three farm workers and injured 14 when their employer transported 17 people in a 10-passenger van with wooden benches and no seat belts. While the RCMP recommended 33 criminal charges, none were laid and the driver was only fined $2000.

books merely made the government appear responsive to public concerns about big business (Smandych, 1985).

Why do people commit crimes? Some conflict theorists believe that the affluent commit crimes because they are greedy. Corporate and white-collar crimes such as stock market manipulation and price fixing often involve huge sums of money and harm many people. By contrast, street crimes such as robbery and break and enter generally involve small sums of money and cause harm to a limited number of victims. According to conflict theorists, the poor commit street crimes to survive; they cannot afford the necessary essentials, such as food, clothing, and shelter. Thus, some crimes represent a rational response by the poor to the unequal distribution of resources in society (Gordon, 1973). Further, living in poverty may lead to violent crime and victimization *of the poor by the poor.* For example, violent gang activity may be a collective response of young people to seemingly hopeless poverty (Quinney, 1979).

In sum, the conflict approach argues that the law protects the interests of the rich and powerful. The criminal law benefits the capitalist class by ensuring that individuals at the bottom of the class structure do not take the property or threaten the safety of those at the top (Reiman, 1984). However, this theory explains some types of laws but not others. People of all classes share a consensus about the criminality of certain acts. For example, laws that prohibit murder, rape, and armed robbery protect not only middle- and upper-income people but also low-income people, who are frequently the victims of such violent crimes (Klockars, 1979). While some laws do protect the rich and powerful, others reflect the interests of all citizens.

FEMINIST PERSPECTIVES ON CRIME AND DEVIANCE

The few early studies that were conducted on "women's crimes" focused almost exclusively on prostitution and attributed the cause of this crime to women's biological or psychological "inferiority." As late as the 1980s, researchers were still looking for unique predisposing factors that led women to commit crime, which was often believed to reflect individual psychopathology rather than a response to their social environment. These theories, which reinforced existing female stereotypes, have had a negative impact on our understanding and treatment of female offenders.

A new interest in women and deviance developed in 1975 when two books—Freda Adler's *Sisters in Crime* and Rita James Simons's *Women and Crime*—declared that women's crime rates were going to increase significantly as a result of the success of the feminist movement. As women gained access to male-dominated roles, they would also become more involved in criminality and other forms of deviance. Although this 'emancipation' theory of female crime has been strongly criticized by subsequent analysts (Comack, 2012), Adler's and Simons's works encouraged feminist scholars to examine the relationship between gender, deviance, and crime more closely.

Feminist scholars have concluded that the roots of female criminality lie in a social structure that is "characterized by inequalities of class, race, and gender" (Comack, 2012:173). Women's deviance is seen as a rational response to gender discrimination experienced in work, marriage, and interpersonal relationships. Some female crimes are attributed to women's lack of job opportunities and to stereotypical expectations about appropriate roles for women. Other theorists feel that women are exploited by capitalism and patriarchy. Because most females have had relatively low-wage jobs and few economic resources, minor crimes such as prostitution, shoplifting, and passing bad cheques were means to earn money or acquire consumer products. Increases in women's criminality during the 1970s and 1980s reflect the fact that the number of single female parents living in poverty grew significantly during this period.

Some of the most interesting work on female criminality has focused on the simultaneous effects of race, class, and gender on deviant behaviour. Arnold (1990) attributes many of the women's offences to living in families in which sexual abuse, incest, and other violence left them few choices except deviance. Economic marginality and racism also contributed to their problems.

These conclusions are supported by the work of Elizabeth Comack, who examined the relationship between women's victimization and their subsequent involvement in Manitoba's criminal justice system. The incidence of prior victimization was pervasive among the 24 incarcerated women she interviewed. The abuse suffered by the women was connected to their criminal behaviour in several ways. Some women turned to crime as a means of coping. "Meredith" had been sexually abused by her father since the age of four or five. She was in jail for fraud and had been involved in drug use and prostitution:

> Some people are violent, some people take it out in other ways, but that was my only way to release it. It was like, it's almost orgasmic, you know, you'd write the cheques, and you'd get home and you'd go through all these things and it's like, "There's so much there. I have all these new things to keep my mind off. I don't have to deal with the old issues." And so you do it. And it becomes an escape. (1996:86)

Others broke the law while resisting abuse. "Janice" had been raped as a teenager and turned to alcohol as a way to cope. Serving time for manslaughter, she recounts the offence:

> Well I was at a party, and this guy, older, older guy, came, came on to me. He tried telling me, "Why don't you go to bed with me. I'm getting some money, you know." And I said, "No." And then he started hitting me and then he raped me and then [pause] I lost it. Like I just, I went, I got very angry and I snapped. And I started hitting him. I threw a coffee table on top of his head and then I stabbed him, and then I left. (1996:96)

While abuse was strongly related to the women's law violations, Comack also found that race and class were factors contributing to their criminal behaviour—most were Indigenous and poor.

Feminist theorists feel that women who violate the law are not "criminal women" but "criminalized women" (Laberge, 1991). This means they commit crimes and acts of deviance because they have been forced into difficult situations that are not of their own making. The women interviewed by Comack faced many social pressures caused by race, class, and gender, and had few options for escaping their situations and improving their lives.

Feminist scholars have also studied violence against women. Much of this violence was hidden, as sexual assault and domestic violence were rarely reported and were not taken seriously by the justice system. Several studies focusing on how rapes were dealt with by the justice system (Clark and Lewis, 1977) led to new sexual assault legislation in 1983 that changed some of the worst parts of the old law, including a section that gave husbands the right to rape their wives (Comack, 2016).

Feminist research has also helped to change the way domestic violence is dealt with by the justice system. Until the 1970s, little was known about this crime. In 1980, the Canadian Council on the Status of Women released a report showing that wife abuse was a major problem (MacLeod, 1980). Attitudes toward domestic violence at that time were illustrated by the fact that when these findings were released in the male-dominated House of Commons, parliamentarians responded with laughter. Despite this response, some legislators did take the issue seriously and recognized that the police response to domestic violence was inadequate. As a result, many provinces implemented mandatory charging policies for

AP Photo/Brett Coomer

Michel Foucault contends that new means of surveillance would make it possible for prison officials to use their knowledge of prisoners' activities as a form of power over the inmates. These guards are able to monitor the activities of many prisoners without ever leaving their station.

domestic violence complaints. This dramatically increased the number of charges laid for this offence (Ursel, 1996).

POSTMODERN PERSPECTIVES ON CRIME AND DEVIANCE

How do postmodernists view deviance and social control? In his book *Discipline and Punish* (1979), Michel Foucault analyzed the intertwining nature of power, knowledge, and social control. In his study of prisons from the mid-1800s to the early 1900s, Foucault found that many penal institutions stopped torturing prisoners who disobeyed the rules and began using new surveillance techniques to maintain social control. Although prisons appeared to be more humane in the post-torture era, Foucault contends that the new means of surveillance impinged more on prisoners and brought greater power to prison officials. Foucault described the *Panopticon*—a structure that gives prison officials the possibility of complete observation of inmates at all times. The Panopticon might be a tower in the centre of a circular prison from which guards can see all the cells. The prisoners know they can be observed at any time but do not know precisely when their behaviour is being scrutinized. The guards would not even have to be present all the time because prisoners would believe that they were under constant scrutiny by officials in the observation post.

How does Foucault's perspective explain social control outside the prison? New technologies make widespread surveillance and disciplinary power possible in many settings. And current technology has the potential to expand surveillance far more broadly than Foucault could have imagined (see Box 8.2). The computer can act as a modern Panopticon that gives workplace supervisors virtually unlimited capabilities for surveillance. Technological developments have broadened the capacity of governments and corporations to control our behaviour. The Japanese have designed a toilet that companies can use to determine whether employees have recently used illegal drugs. Many people who were responsible for the 2011 Stanley Cup riots in Vancouver were identified and arrested because of facial recognition software that was used to analyze digital photos of the riots.

These technologies can be valuable tools in improving public safety. Closed-circuit television cameras were used to identify the people who carried out the July 2005 bombings that took more than 50 lives in London. Licence number recognition cameras have helped the police to get many auto thieves and suspended drivers off the road. DNA technology has freed many people who had been unjustly convicted of serious crimes and has enabled the justice system to imprison others who have committed serious crimes. A system that would allow us to log on to our computers by scanning the iris of our eyes or our fingerprints would eliminate the confusing number of security passwords that each of us must remember and help protect us against cybercrime.

However, these technologies raise important issues of privacy and individual rights, and society will have to decide whether greater protection is worth our loss of personal privacy. What are your views on this issue? Where should the balance lie between collective security and individual rights?

We have examined functionalist, interactionist, conflict, feminist, and postmodern perspectives on deviance and crime (see the Concept Snapshot). These explanations help us understand the causes and consequences of certain kinds of behaviour and provide us with guidance about how we might reduce crime and deviance.

BOX 8.2 POINT/COUNTERPOINT

Surveillance Studies

Following the work of Foucault, one of the newest areas studied by criminologists is surveillance studies. Governments have always wanted to know more about what citizens were up to, particularly in totalitarian countries where surveillance was often oppressive. The technological capacity for surveillance has grown so quickly that massive amounts of personal information are now available, leading to a greater possibility of violating our privacy rights. Four leading Canadian surveillance studies scholars have outlined the problem:

> Today, our lives are transparent to others in unprecedented ways . . . . many kinds of organizations watch what we do, keep tabs on us, check our details, and track our movements. Almost everything we do generates an electronic record: we cannot go online, walk downtown, attend a university class, pay with a credit card, hop on an airplane, or make a phone call without data being captured. Personal information is picked up, processed, stored, retrieved, bought, sold, exchanged. Our lives—or rather, those traces and trails of data, those fragments of reality to which our lives can be reduced—are visible as never before, to other individuals, to public and private organizations, to machines. (Bennett et al, 2014:3)

What exactly is surveillance? Bennett and colleagues define it as "any systematic focus on personal information in order to influence, manage, entitle, or control those whose information is collected" (2014:6).

Governments are increasingly relying on information to govern us, and corporations use our information for their profit. This can be a good thing. Government tracking of prescription drug use can help to ensure that people are not taking drugs that interact with each other or that addicts are not getting opiate prescriptions written by several different doctors. Police have been able to reduce crime by targeting their efforts on crime hot spots and on monitoring high-risk offenders. Starbucks customers may appreciate having their cellphones receive an electronic discount coupon when they walk near a Starbucks location.

However, surveillance can also be harmful. In 2014, Ontario's privacy commissioner filed a court action against the Toronto Police Service because they refused to stop releasing information about attempted suicides to other agencies. The action was precipitated by a case involving a woman who missed a Caribbean cruise after having been refused admission to the United States because she had attempted suicide several years earlier. The police had given U.S. officials access to this information. A youthful indiscretion captured in a photo and posted online may cost a person a job when a potential employer views it many years later. Today's unprecedented level of surveillance means that each of us has given up power to the governments and corporations that are collecting and analyzing information about us. Once information has been collected, individuals no longer have control over how it is used.

A major issue is not just the presence of so many surveillance technologies, but the increasing capacity of corporations and governments to link these technologies, as commercial companies aggregate data from multiple sources and as governments establish fusion centres that integrate a variety of different databases to enhance security or to monitor and deliver government services. Drones (small airborne vehicles with cameras) are quickly multiplying, and they have unique surveillance capabilities. They are being used globally by military and intelligence agencies; police departments are using them for surveillance and for taking overhead photos of auto accidents and crime scenes; and hydro and pipeline companies are using them to conduct safety inspections on their power lines.

While governments have always tried to exercise surveillance of their citizens, technology such as cellphone cameras has now made it possible for everyone to watch everyone else (Brodeur, 2010). This technology has profound impacts on our privacy—when your neighbor can afford a small drone, what privacy rights do you have if he chooses to hover it above your fenced backyard or beside your second-story window transmitting live video of your activities back to his computer screen? The state is now a target of surveillance. The relationship between police and the public has been profoundly changed by the fact that phone cameras mean that the police are no longer able to conceal actions that may be illegal, and that it is no longer simply a case of a police officer's word against that of a citizen if the officer is accused of misconduct.

Society has not yet worked out an appropriate balance between privacy rights and the need for more surveillance in order to protect the population or to enable companies to provide better service to customers. This means that surveillance issues will continue to be important and that this new field of criminology will become an important source of theory and data about surveillance and privacy issues.

Source: Rick Linden. 2016. *Criminology: A Canadian Perspective*, 8th edition. Toronto: Nelson.

CONCEPT SNAPSHOT

FUNCTIONALIST PERSPECTIVES **Key thinkers:** Robert Merton, Richard Cloward, Lloyd Ohlin	In a smoothly functioning society, deviance will be limited because people will share common culture goals and agree upon the appropriate means for reaching them. However, societies that do not provide sufficient avenues to reach these goals may also lack agreement about how people may achieve their aspirations. Deviance may be common in such societies because people may feel free to use whatever means they can to achieve their goals.
INTERACTIONIST PERPECTIVES **Key thinkers:** Edwin Sutherland, Howard Becker, Edwin Lemert	Deviance is learned in the same way as conformity—through interaction with others. A person becomes deviant when exposure to law-breaking attitudes is more meaningful to them than exposure to law-abiding attitudes. Societal reaction to someone who has been labelled as deviant may also cause people to develop a deviant self-concept.
CONFLICT PERSPECTIVES **Key thinkers:** Karl Marx, Richard Quinney	The powerful use law and the criminal justice system to protect their class interests. The way laws are written and enforced benefits the capitalist class by ensuring that individuals at the bottom of the class structure do not take the property or threaten the safety of those at the top. The poor may also be forced to commit crimes to survive.
FEMINIST PERSPECTIVES **Key thinker:** Elizabeth Comack	The structured inequalities of race, class, and gender lead to the criminalization of women. Women's deviance and crime is seen as a rational response to gender discrimination. Some female crimes are attributed to women's lack of job opportunities and to stereotypical expectations about appropriate roles for women. Other theorists feel that women are exploited by capitalism and patriarchy.
POSTMODERN PERSPECTIVES **Key thinker:** Michel Foucault	Power, knowledge, and social control are intertwined. In prisons, for example, new means of surveillance make prisoners think they are being watched all the time. This gives prison officials power over the inmates. Modern technologies make widespread surveillance and disciplinary power possible in many settings, including the police network, factories, schools, and hospitals.

TIME TO REVIEW

- Discuss how crime differs from deviance.
- Describe how functionalist theorists explain the causes of crime and deviance. How do strain theories differ from control theories of deviance?
- Explain the two interactionist theories described in the text—differential association theory and labelling theory.
- Describe how conflict theorists critically assess the justice system and blame capitalism for causing crime.
- How do feminist theorists link women's criminality with inequalities of class, race, and gender?
- Explain how postmodern theorists such as Foucault view deviance and social control.

LO-3 CRIME CLASSIFICATION AND STATISTICS

There are many different types of crimes. To study them, sociologists have put them into broader categories.

HOW SOCIOLOGISTS CLASSIFY CRIME

We will examine four types of crime: (1) street crime; (2) occupational, or white-collar, and corporate crime; (3) organized crime; and (4) political crime. Box 8.3 also introduces the relatively new category of **cybercrime**—offences where a computer is the object of a crime or the tool used to commit a crime. As you read about these types of crime, ask yourself how you feel about them. Should each be a crime? How severe should the sanctions be against each type?

▶ **cybercrime** Offences where a computer is the object of a crime or the tool used to commit a crime.

STREET CRIME When people think of crime, they most commonly think of **street crime**, which includes most violent crime, certain property crimes, and certain morals crimes. Examples are robbery, assault, and break and enter. These crimes occupy most of the criminal justice system's time. All street crime does not occur on the street; it frequently occurs in the home, workplace, and other locations.

▶ **street crime** most violent crime, certain property crimes, and certain morals crimes.

Violent crime involves force or the threat of force against others, including murder, sexual assault, robbery, and aggravated assault. Violent crime receives the most sustained attention from law enforcement officials and the media. While much attention may be given to the violent stranger, the vast majority of violent crime victims are injured by someone they know: family members, friends, neighbours, or co-workers (Silverman and Kennedy, 1993).

Property crimes include break and enter, theft, motor vehicle theft, and arson. These crimes are far more common than violent crimes.

Morals crimes involve an illegal action voluntarily engaged in by the participants, such as prostitution, illegal gambling, the use of illegal drugs, and illegal pornography. Many believe these activities should not be labelled as a crime. These offences are often referred to as "victimless crimes" because they involve exchanges of illegal goods or services among willing adults (Schur, 1965).

▶ **occupational, or white-collar, crime** A term used to describe illegal activities committed by people in the course of their employment or in dealing with their financial affairs.

▶ **corporate crime** An illegal act committed by corporate employees on behalf of the corporation and with its support.

OCCUPATIONAL AND CORPORATE CRIME **Occupational, or white-collar, crime** consists of illegal activities committed by people in the course of their employment or in dealing with their financial affairs. Many white-collar crimes involve the violation of positions of trust. These include employee theft, soliciting bribes or kickbacks, and embezzling. Some white-collar criminals set up businesses for the sole purpose of victimizing the general public, engaging in activities such as land swindles, securities thefts, and consumer fraud.

In addition to acting for their own profit, some white-collar offenders become involved in criminal conspiracies designed to improve the profitability of their companies. This is known as **corporate crime**—illegal acts committed by corporate employees on behalf of the corporation and with its support. Examples include antitrust violations; false advertising; infringements on patents; price fixing; and financial fraud. These crimes involve deliberate decisions made by corporate personnel to enhance profits at the expense of competitors, consumers, and the general public.

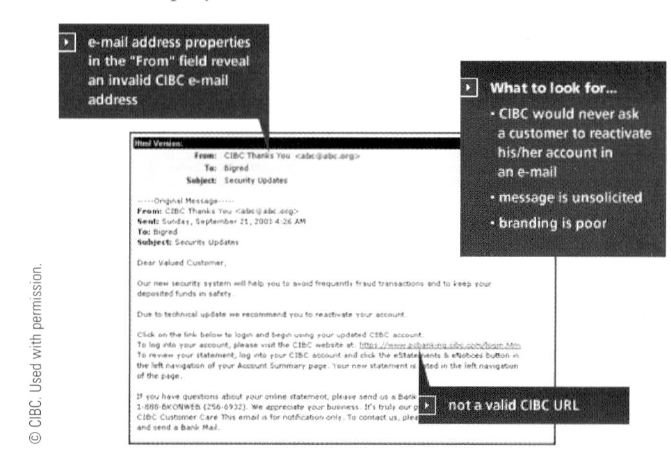

© CIBC. Used with permission.

For some people, the "information superhighway" is a new avenue of illegitimate opportunity. Cybercrime demonstrates how new opportunity structures can elicit new forms of deviance. This is an attempt by CIBC to educate its banking customers to avoid Internet fraud.

Dick Loek/GetStock.com

Conrad Black, one of Canada's most influential businessmen, has completed a three-and-a half-year sentence in a Florida prison for misappropriating millions of dollars from shareholders of his newspaper chain.

The cost of white-collar and corporate crimes far exceeds that of street crime. Tax evasion costs Canadians billions of dollars a year. In one of the world's biggest white-collar crimes, investors in Calgary's Bre-X gold-mining company lost about $5 billion because geologist Michael de Guzman had salted core samples with gold to make a worthless mining property look like the world's biggest gold find. In 2011, investors discovered that Sino-Forest, a Canadian-listed company that claimed to control vast amounts of Chinese forests, may have issued fraudulent reports concerning its ownership of timber reserves in China. Auditing firm Ernst and Young and a number of Canadian investment firms have paid almost $150 million in connection with their role in the Sino-Forest affair, but investors lost $6 billion when the company collapsed.

At the individual level, while bank robbers seldom steal more than a few thousand dollars, Calgary businessmen Gary Sorenson and Milowe Brost defrauded investors of between $100 million and $400 million to support their lavish lifestyles. Many investors trusted the men with their life savings and were financially devastated by this fraud.

Corporate crimes can also be costly in terms of lives lost and injury. Occupational accidents and illnesses—many caused by unsafe and illegal working conditions—have killed thousands of Canadians. Working conditions in the mining industry, for example, have been especially dangerous. Many Canadian miners have died because their employers failed to protect them from mine hazards. Coal miners died of black lung, a condition caused by inhaling coal dust; fluorspar miners died from the effects of inhaling silica dust in unventilated mineshafts; and asbestos miners died of lung disease caused by that mineral. Not only did the mine owners fail to provide safe working conditions, but company doctors were also told not to advise the miners of the seriousness of their illnesses (Leyton, 1997).

One reason why many employers have failed to implement required safety measures is because the penalties for violating workplace health and safety laws are so light. Typically, companies have been fined only a few thousand dollars even when employees died because of their employers' negligence. These events are usually treated as regulatory matters rather than as criminal violations (Bittle, 2014).

Although people who commit occupational and corporate crimes can be arrested, fined, and sent to prison, many people do not regard such behaviour as "criminal." In Canada, punishment for such offences is usually a fine or a relatively brief prison sentence at a minimum-security facility; in the United States, however, penalties have become much more severe.

The concept of white-collar crime also fits some people who wear blue collars. Thus, *occupational crime* may be a more accurate term. Many tradespeople defraud the government by doing work "off the books" in order to avoid sales taxes, and some blue-collar businesses, such as auto repair, have bad records of consumer fraud.

▶ **organized crime** A business operation that supplies illegal goods and/or services for profit.

ORGANIZED CRIME **Organized crime** is a business operation that supplies illegal goods and/or services for profit. Organized crime includes drug trafficking, prostitution, liquor and cigarette smuggling, loan sharking, money laundering, and large-scale theft, such as truck hijacking (Simon and Eitzen, 1993). Many different organized crime groups operate at all levels of society. Organized crime thrives because there is great demand for illegal goods and services. This public demand has produced illicit supply systems with global connections. These activities are highly profitable, since groups that have a monopoly over goods and services the public strongly desires can set their own price. Legitimate competitors are excluded because of the illegality; illegitimate competitors are controlled by force. While gang-related killings peaked in 2008 and have declined since then, they still accounted for 85 homicides in 2013.

Along with their illegal enterprises, organized crime groups have infiltrated the world of legitimate business. Linkages with organized crime exist in many businesses, including immigration consulting, real estate, garbage collection, vending machines, construction, and trucking.

POLITICAL CRIME **Political crime** involves illegal or unethical acts involving the misuse of power by government officials, or illegal or unethical acts perpetrated against a government by outsiders seeking to make a political statement or to overthrow the government. Government officials may use their authority unethically or illegally for material gain or political power. They may engage in graft (taking advantage of political position to gain money or property) through bribery, kickbacks, or "insider" deals that financially benefit them. While Canadian governments have a better record than those of most other countries, there have been a number of scandals. The 2005 Gomery Inquiry exposed serious wrongdoing by some members of Prime Minister Jean Chrétien's Liberal government who illegally funnelled millions of dollars to Quebec advertising agencies in exchange for their political support. Stephen Harper's Conservative government was also involved in several scandals. In 2014, a party official received a prison sentence for making misleading robocalls to voters, telling them their voting locations had changed, and Dean Del Mastro, Prime Minister Harper's former Parliamentary Secretary, was convicted of campaign fraud. Several Conservative senators were accused of expense fraud, and in 2015 Senator Mike Duffy went on trial for a number of charges, including fraud, bribery, and breach of trust.

political crime Illegal or unethical acts involving the usurpation of power by government officials, or illegal or unethical acts perpetrated against a government by outsiders seeking to make a political statement or to overthrow the government.

BOX 8.3 SOCIOLOGY AND NEW MEDIA

The Growth of Cybercrime

The future of crime lies in the Internet and other new media technologies. Cybercrime includes offences where a computer is the object of a crime (hacking, viruses, denial of service) or the tool used to commit a crime (child pornography, human trafficking).

There are many different cybercrimes. Computer hackers spread viruses and carry out attacks intended to shut down specific websites. Many hackers simply intend to create chaos on the Internet, but some have tried to extort money from corporations by threatening to disrupt their communications systems.

Cybercriminals have stolen people's identifying information and credit card numbers in order to obtain money by fraud. The largest identity theft that has been uncovered involved 130 million credit and debit card numbers obtained from the databases of 7-Eleven and two other companies (Shapiro and Block, 2009). More recently, tens of millions of peoples' information has been stolen from large corporations including Target, Home Depot, and JP Morgan Bank. Company websites have also been hacked to find corporate information that can be used by competing businesses.

Individuals can be victimized by cyberstalking and bullying over the Internet. Sites like eBay are vulnerable to auction fraud. Organized crime groups have also seen the opportunities of the Internet. The Russian Mafia has been involved in extorting money by threatening to take down business websites unless they are paid off (Kshetri, 2010). In 2015, the security firm Kaspersky Lab reported that a gang of hackers had stolen between $300 million and $1 billion from banks in 30 countries by taking control of ATM machines and by transferring money to other banks (Sanger and Perlroth, 2015).

Cybercrime can also involve attacks against governments. The U.S. Department of Defense suffers three million attacks annually (Kshetri, 2010), and other agencies face similar threats. In 2007, cyberattacks, likely orchestrated by the Russian government, shut down Estonia's government and its banks. In 2009, a virus called Stuxnet damaged some of the technology being used by Iran to develop its nuclear program. Many experts have attributed this virus to the Israeli and U.S. governments who wished to ensure Iran did not develop a nuclear capability (Broad, Markoff, and Sanger, 2011). Cyberattacks on infrastructure such as power grids could have a catastrophic impact. U.S. President Obama has stated, "We know that cyber-intruders have probed our electrical grid and that in other countries, cyberattackers have plunged cities into darkness" (Kshetri, 2010:6), and we know that many countries are developing the capacity to carry out such attacks.

(Continued)

Sophisticated technology skills are not always required to attack a country's security. In 2013, U.S. Army Private Chelsea Manning received a 35-year sentence for releasing hundreds of thousands of highly classified military documents and diplomatic cables through the Wikileaks website. Manning simply downloaded the documents at her worksite in Iraq and removed them on a CD labelled as a Lady Gaga album.

Cybercrimes are difficult to deal with because technology provides an anonymity that is not present in many other types of offences. The global nature of the Internet also means that cybercrime can be committed from almost anywhere in the world. Some countries, including Russia and China, act as cybercrime havens in that they typically do not pursue cybercriminals who are attacking targets outside their countries. Police agencies are set up to deal with real-world crimes, not cybercrimes, and have not kept pace with the very expensive task of enforcing laws against cybercrime. The laws themselves also lag behind. It can take years to pass laws, and technology and innovation by cybercriminals changes much more rapidly (Brenner, 2010).

The activities of the "hacktivist" group Anonymous provide an example of the difficulties of policing Internet crimes. Anonymous is a collective of people who work together online to protest against those who they feel restrict freedom of speech and access to information. Their tactics include disrupting websites and online services. Among their targets have been the Church of Scientology, Wall Street banks, and companies and governments that have supported more restrictions on Internet sharing. For example, in 2012, Anonymous took over the website of Greece's Department of Justice to protest the Greek government's support of tougher copyright laws (BBC, 2012).

The Internet has also facilitated the actions of 'lone wolf' terrorists. These are people who are not directly affiliated with members of terrorist groups, but who become radicalized through jihadist websites. In October 2014, two Canadian Forces members, Warrant Officer Patrick Vincent and Corporal Nathan Cirillo, were killed in separate incidents by two men who had self-radicalized with the help of jihadist websites.

JEAN-PHILIPPE KSIAZEK/AFP/Getty Images

The "hacktivist" group Anonymous uses disruptive tactics to protest against those who they feel restrict Internet freedom.

Cybercrime is usually not reported to the police, so it does not show up in our crime statistics (Tcherni et al., 2015). When you look at the decline in Canadian crime rates in Figure 8.2, you should be aware that huge numbers of cybercrimes are not shown. As our homes and cars become connected (the Internet of Things), the risk of cybercrime grows.

CRIME STATISTICS

Crime is hard to measure. While citizens, police, and policymakers all wish to know how much crime there is, those committing crimes normally try to conceal their actions. Thus, our information about crime will always be incomplete and we can never be certain of its accuracy. Our main sources of information about crime are police statistics and victimization surveys.

OFFICIAL STATISTICS Our most important source of crime data is the Canadian Uniform Crime Reports (CUCR) system, which summarizes crimes reported to all Canadian police departments. When we read that British Columbia's homicide rate is higher than the national average, or that in 2013 more than 1.8 million offences were reported to the police, the information is usually based on CUCR data. Figure 8.2 shows trends in violent and property crimes, and Figure 8.3 shows Canada's homicide rates. These figures show that crime has declined

FIGURE 8.2 : CANADIAN CRIME RATES, 1962-2014

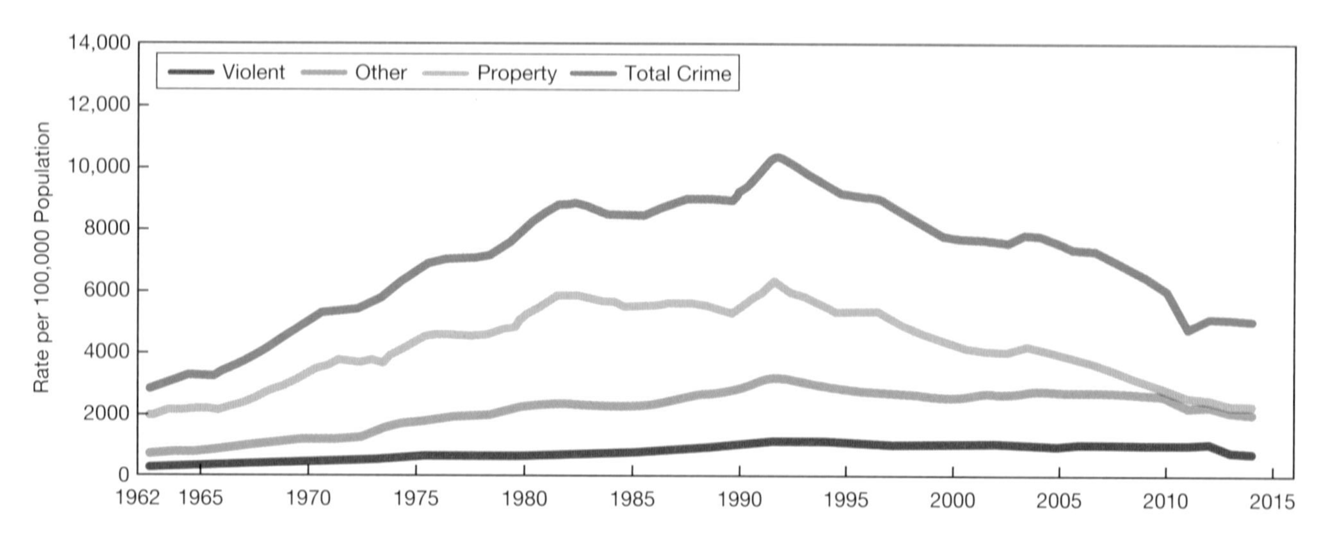

Sources: Adapted from Statistics Canada, Crime Statistics in Canada 2007, Cat. no. 85-002-X, Vol. 28 no. 7, 2008; Boyce, Jillian. 2015 "Police-Reported Crime Statistics in Canada, 2014." *Juristat*. Catalogue no. 85-002-X. Ottawa: Statistics Canada.

FIGURE 8.3 : CANADIAN HOMICIDE RATES, 1961-2014

As of 1971, population estimates were adjusted to reflect new methods of calculation.

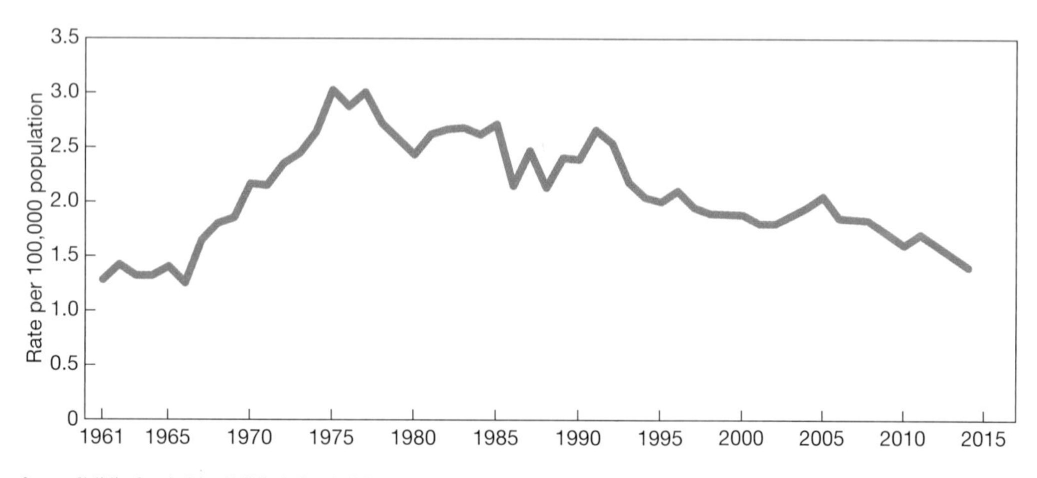

Sources: Statistics Canada, Crime Statistics in Canada 2007, Cat. no. 85-002-X, Vol. 28 no. 7, 2008; Boyce, Jillian. 2015 "Police-Reported Crime Statistics in Canada, 2014." *Juristat*. Catalogue no. 85-002-X. Ottawa: Statistics Canada.

significantly over the past two decades. Homicide rates are now the lowest they have been in almost 50 years.

Crime figures should be interpreted cautiously. While we can have confidence in homicide and auto theft statistics, the accuracy of other crime statistics is less certain and it is important to recognize their limitations.

The major weakness of the CUCR is that police statistics always underreport the actual amount of crime. Victims do not report most crimes to the police. Also, rates of reporting may change over time, and this can lead to an increase or decrease in crime rates. As a result, it can be difficult to analyze crime patterns and trends.

For example, Figure 8.2 shows that rates of reported violent crimes increased significantly in Canada during the late 1980s and early 1990s, almost doubling between 1980 and 1990. A large part of this increase in reported crime was actually due to the fact that violence against women was more likely to be reported than in earlier years (Linden, 1994). In the mid-1980s, many provincial governments directed police to lay charges in all suspected cases of domestic violence. These zero-tolerance policies have been made more effective in subsequent years and meant that crimes previously dealt with informally by the police now showed up in crime statistics. Also, this visible support from the justice system may have encouraged more victims to report spousal assaults. The impact of these changes was seen in Winnipeg, where the police instituted a mandatory charging policy, and where the province established a special family violence court for spouse abuse cases. The number of domestic violence cases dealt with by this court rose from 1444 in 1990, the first year of the court's operation, to 3387 three years later (Ursel, 1996). This increase was likely due to changes in the reporting and recording of domestic assaults rather than to any increase in family violence.

Another weakness of official statistics is that many crimes committed by middle- and upper-class people are routinely handled by administrative bodies or by civil courts. To avoid negative publicity, many companies prefer to deal privately with offences like embezzlement committed by their employees, and these cases may not be reported to the police. As a result, many elite crimes are never classified as "crimes," nor are the businesspeople who commit them labelled as "criminals."

VICTIMIZATION SURVEYS These weaknesses led to the development of the victimization survey. Because many people do not report their victimization to police, governments conduct surveys in which people are asked if they have been victims of crime. In the latest Canadian survey, only 31 percent of the victimizations reported by respondents had been reported to the police (Perreault and Brennan, 2010). Thus, reported crimes are only the tip of the iceberg. People said they did not report crimes because: they considered the incident too minor; they felt it was a personal matter; they preferred to deal with the problem in another way; or they did not feel the police could do anything about the crime.

The additional information provided by victimization surveys has helped to confirm that the rise in violent crime during the 1980s and early 1990s was due to an increase in the reporting and recording of domestic assaults. Assaults did not likely increase during this period—we just did a better job of counting them.

These surveys also have weaknesses: people may not remember minor types of victimization; they may not report honestly to the interviewer; and the surveys do not provide any information about "victimless crimes," such as drug use and illegal gambling. Despite these flaws, victimization surveys have shed new light on the extent of criminal behaviour. They are a valuable complement to other ways of counting crimes.

LO-4 WHO COMMITS CRIMES: CHARACTERISTICS OF OFFENDERS

Age, gender, class, and race are important *correlates of crime.* That is, they are factors associated with criminal activity. One method of testing theories of crime is to see how well they explain these correlates.

AGE AND CRIME

The offender's age is one of the strongest correlates of crime and most other kinds of deviance. Arrests increase through adolescence, peak in young adulthood, and steadily decline with age. There is some variation—for example, violent crimes peak at a later age than property crimes—but the general pattern is consistent. Crime is a young person's game, with rates peaking between the ages of 15 and 18.

The relationship between age and criminality exists in every society for which we have data (Hirschi and Gottfredson, 1983). This is true of most other types of high-risk behaviours, some of which are considered to be deviant. Adolescence and early adulthood are the peak times for both offending and victimization. Possible explanations for the decline in crime and deviance rates after early adulthood are the physical effects of aging, which make some criminal activities more difficult, and the realization by older chronic offenders that further arrests will result in very long jail sentences. Also, older people sometimes do get wiser. Perhaps the best explanation for maturational reform, though, is related to the different social positions of youth and adults. Adolescents are between childhood and adult life. They have few responsibilities and no clear social role. During adolescence, young people are breaking away from parental controls and preparing to live on their own. As we age, we begin to acquire commitments and obligations such as jobs and families that limit our freedom to choose a lifestyle that includes crime and other forms of deviance.

GENDER AND CRIME

Another consistent correlate of deviance is gender. Most crimes are committed by males. For many types of crimes, males are also more likely to be victims. As with age and crime, this relationship has existed in almost all times and cultures. However, while the age distribution is remarkably stable, there is considerably more variation in male/female crime ratios in different places, at different times, and for different types of crime.

Men make up more than 80 percent of those charged with crimes in Canada. As Figure 8.4 shows, the degree of involvement of males and females varies substantially for different crimes. The largest gender differences in charges are reflected in the proportionately greater involvement of men in violent crimes and major property offences.

The difference between male and female involvement in crime narrowed at the end of the last century. Hartnagel (2004) found that the percentage of *Criminal Code* offences committed by females nearly doubled, from 9 percent to 17 percent, between 1968 and 2000. While there was virtually no change in the percentage of homicides committed by women (11 percent versus 10 percent), women's involvement in serious theft (9 percent versus 23 percent), fraud (11 percent versus 30 percent), and minor theft (22 percent versus 28 percent) increased substantially. However, since 2000, the male/female ratio has remained nearly unchanged.

Why did the sex distribution of crime change? One clue comes from cross-cultural data showing very large differences in sex ratios of criminal involvement in different parts of the world. Women's rates of crime are lowest in countries with the greatest differences between the roles of men and women. Where women follow traditional roles in which their lives are centred exclusively on the home, their crime rates are low. On the other hand, where women's lives are more similar to men's, their crime rates will be higher.

While role convergence may explain some of the reduction in the gap between male and female crime rates, the convergence in crime rates had almost ended over two decades ago, and it is unlikely that women will ever adopt male patterns of crime, particularly violent crime. The increase in female crime has been greatest for property crimes, particularly minor offences such as shoplifting, credit card fraud, and passing bad cheques. Comack (2009) concluded that this reflects the feminization of poverty rather than a convergence of gender roles. Thus, much of the increase in female crime may reflect the increased economic marginalization of poor women.

SOCIAL CLASS AND CRIME

Many theories assume that crime is economically motivated and that poverty will lead to criminal behaviour. However, the evidence concerning the impact of economic factors on crime is mixed. Persons from lower socioeconomic backgrounds are more likely to be

FIGURE 8.4 PERCENTAGE OF CHARGES IN ADULT CRIMINAL COURTS, BY GENDER, 2011-12 (SELECTED CRIMINAL OFFENCES)

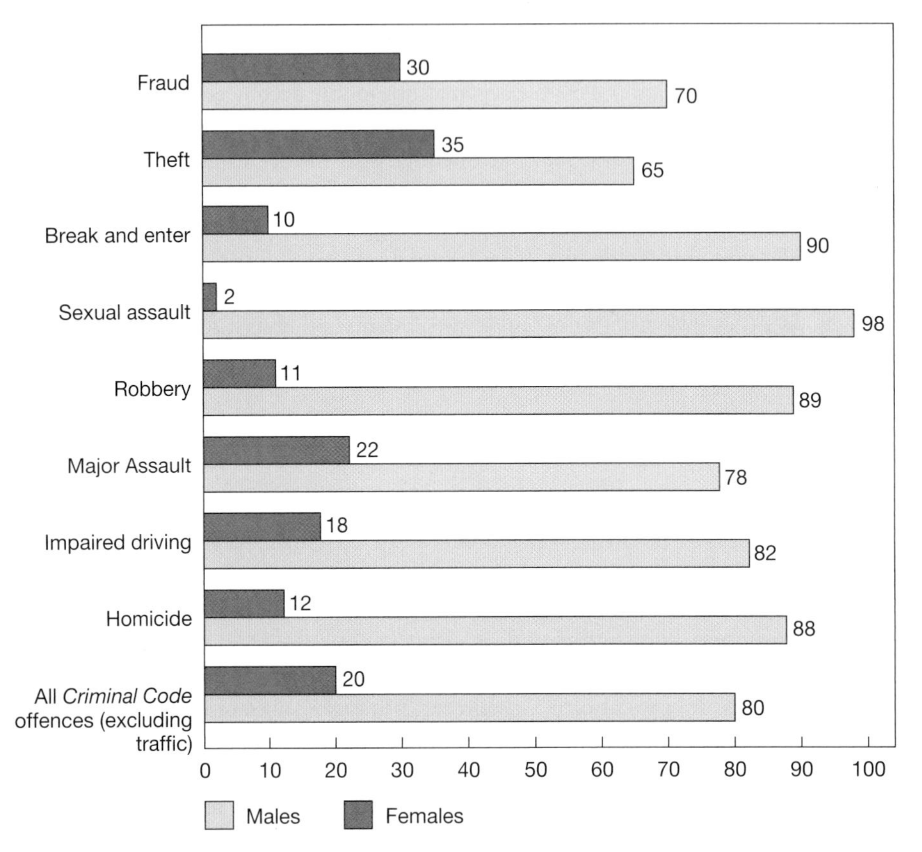

Source: Statistics Canada, CANSIM Table 252-0053. Integrated Criminal Court Survey.

arrested for violent and property crimes. However, we also know that these crimes are more likely to come to the attention of the police than are the white-collar crimes that are more likely to be committed by members of the upper class. Because the vast majority of white-collar crimes are never reported, we lack the data to fully assess the relationship between class and crime.

Before considering some of the data on social class and crime, let us consider several other economic variables. Does unemployment cause crime? Do poor cities, provinces, and countries have higher crime rates than richer communities? The answer to both these questions is no. Historically, crime rates are at least as likely to rise during periods of prosperity as during times of high unemployment. We are also as likely to find high crime rates in rich countries as in poor ones. Within Canada, the poorer provinces of Quebec and New Brunswick have crime rates far lower than the wealthier provinces of British Columbia and Alberta (see Map 8.1). Hartnagel (2012) has concluded that the *degree of inequality*—poverty amid affluence—is a better predictor of crime than is the amount of poverty.

We know that lower-class people are overrepresented in arrest and prison admission statistics; however, we do not know if lower-class people commit more crimes or if the justice system treats them more harshly. To get closer to actual behaviour, researchers developed self-report surveys in which respondents were asked to report the number of deviant acts they had committed during a specified time. There is some disagreement about the conclusions that should be drawn from this research, most of which has used adolescent subjects. However, the most likely conclusion is that for the vast majority of people, class and crime or delinquency are not related. However, the most frequent and serious offenders are most likely to come from the bottom of the class ladder—from

MAP 8.1 2014 CRIME RATES PER 100,000 POPULATION (CRIMINAL CODE EXCLUDING TRAFFIC OFFENCES)

Crime rates are highest in the West and the North, and lowest in Central Canada.

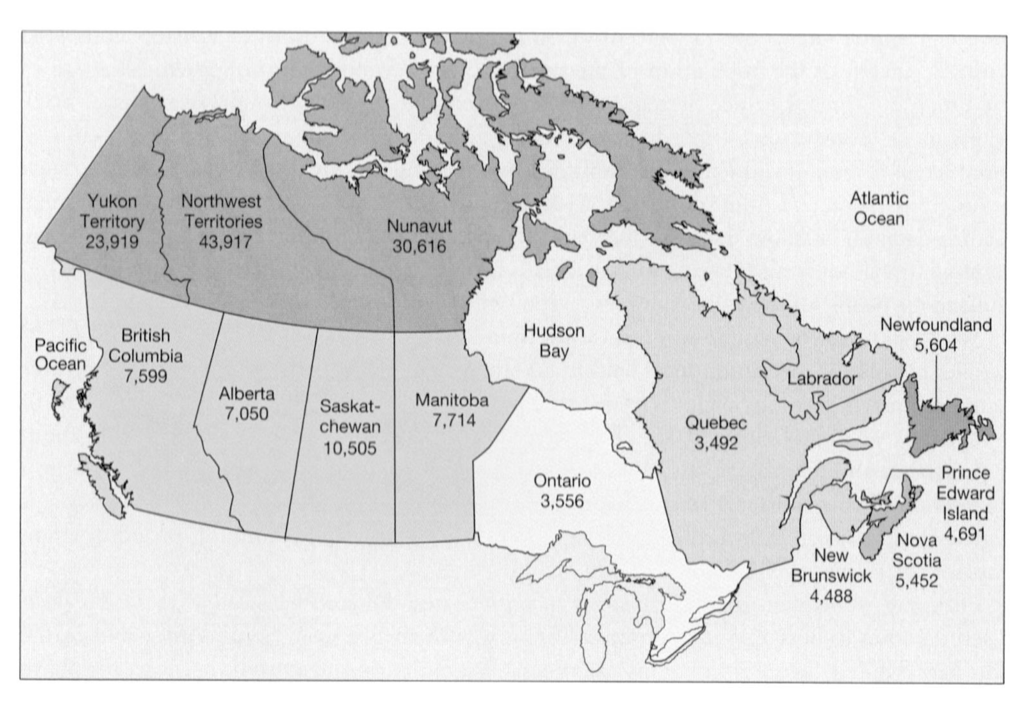

Source: Statistics Canada, "Police-Reported Crime Statistics in Canada, 2014," Catalogue no. 85-002-X.

an underclass that is severely disadvantaged economically, educationally, and socially. There is also some evidence that other forms of deviance, such as suicide, alcoholism, mental illness, and drug addiction, are also more common among the underclass.

A unique victimization survey reinforces this conclusion. More than 12,000 Canadian women were interviewed for the national Violence Against Women Survey (Johnson, 1996). Several findings supported the view that violence is greatest at the very bottom of the class ladder. First, men with high school educations assaulted their wives at twice the rate of men with university degrees. Second, men who were out of work committed assaults at twice the rate of men who were employed. Third, men in the lowest income category (less than $15,000 a year) assaulted their wives at twice the rate of men with higher incomes. Above this $15,000 level, however, there was no relationship between income and crime. This again suggests that the highest crime rates can be found at the bottom of the economic ladder, but that above this level there is no relationship.

RACE AND ETHNICITY AND CRIME

In societies with diverse populations, some ethnic and racial groups will have higher crime rates than others. For example, in the United States, African Americans and Hispanics are overrepresented in arrest data. However, because Statistics Canada does not routinely collect data about racial and ethnic correlates of crime, we know relatively little about the situation in Canada.

Several Canadian studies have examined minorities and crime. The first of these examined race and ethnicity in the federal prison system. Offenders from non-Indigenous visible ethnic minorities were *underrepresented* in the federal correctional system's population (Thomas, 1992). Specifically, the study found that in 1989, 5.2 percent of the federal corrections population were members of ethnic minority groups, while these groups made up more than 6.3 percent of the general population. The second study, which examined provincial youth and adult correctional centres in British Columbia, arrived at similar findings. Only 8.2 percent of the prison population were members of non-Indigenous visible ethnic minorities, yet these groups made up 13.5 percent of the province's population. Contrary to the common view that immigrants have high crime rates, only 11 percent of B.C. inmates were not born in Canada, compared with 22 percent of the population of the province. The Commission on Systemic Racism in the Ontario Criminal Justice System (1995) reported that the rate of imprisonment for black adults in Ontario was five times higher than the rate for white adults. Black adults were also more likely to be imprisoned while awaiting trial, particularly for discretionary charges, such as drug possession and drug trafficking. Most recently, the *Toronto Star* used Canada's criminal records database to show that 16.7 percent of people with a criminal record in Canada were "non-white" (Rankin and Powell, 2008). This is below the percentage of visible minorities and Indigenous people in the Canadian population, which is about 20 percent.

While statistics on visible minorities are limited, there are extensive data on Indigenous peoples because of special inquiries held to determine whether the justice system has discriminated against Indigenous people. Many studies have demonstrated the over-involvement of Indigenous people (Oriola, 2016). For example, while Indigenous people made up about 4 percent of the population in 2012, they made up about 23 percent of the federal prison population (Office of the Correctional Investigator, 2014). It is important to note that there is a great deal of variation in Indigenous crime rates among different communities and different parts of the country (Wood and Griffiths, 1996).

How can we explain these differences in crime rates? One answer is that there has been discrimination against minority groups. The treatment of blacks in South Africa and in the Southern United States are obvious examples. Discrimination against Indigenous people in Canada, Australia, and New Zealand has also been well documented by commissions of inquiry.

Poor people, many of whom may be members of minority groups, may go to prison for minor offences if they are unable to pay fines. While this type of discrimination may be unintentional, it is nonetheless real. The justice system also tends to focus its efforts on the types of crimes that are committed by low-income people rather than on white-collar crimes, so members of poor minority groups may be overrepresented in crime statistics. Discrimination accounts for some, but not all, of the high rates of criminality of some minority groups.

BOX 8.4 POINT/COUNTERPOINT

"If It Bleeds, It Leads": Fear of Crime and the Media

Most Canadians learn about crime through the media, which shape our views about crime and criminals. However, the media do not simply "report" the news. Editors and reporters select the news about crime and construct the way this news is presented to us (McKnight, 2012).

Unfortunately, the media provide us with a distorted picture of crime. For example, while most crime is property crime, most media stories deal with violent crime, particularly murder. Serial killers are the subject of massive media attention, and their names (Clifford Olson, Paul Bernardo, Charles Manson) and their crimes become part of popular culture for generations.

While violent crimes are over-reported, property crimes rarely receive much attention. The portrayal of crime in the fictional media is even more distorted. Video games, television programs, and movies are often extremely violent.

Both the news and fictional media also focus on the drama of criminal events without attempting to explain the context or to provide readers with an understanding of the causes of crime (Doyle, 2006). The media also present distorted views of the racial dimensions of crime. Wortley found that while immigrants actually have lower rates of crime than those who are Canadian-born, the media perpetuate stereotypes: "One popular columnist, for example, wrote that 'our culture is not used to this type of savagery' and that 'this type of crime is the direct result of choosing too many of the wrong immigrants.' . . . Another reporter maintained that 'White Canadians are understandably fed up with people they see as outsiders coming into their country and beating and killing them'" (2009:349).

The media may also contribute to crime. The linkage between media violence and violent behaviour is complex, and researchers disagree about the degree to which television influences behaviour rather than simply reflecting a pre-existing interest in violence. However, there is a body of evidence concluding that children who are exposed to extensive television violence are more likely to be violent themselves (Christakis, et al., 2013).

The news media may also facilitate crime by providing patterns for criminal acts such as rampage shootings. In the spring of 2014, North America saw three mass shootings within a two-week period, including the ambush deaths of three RCMP members in Moncton. While there were differences between these tragic events, there were also similarities, and some researchers believe that these similarities are a result of media reporting of previous events. The shooters carefully plan their actions based on their knowledge of what others have done. This theory helps to explain the similarities between rampage killings, and also why these events often occur in clusters. One of the few researchers who has interviewed rampage shooters is Australian forensic psychiatrist Paul Mullen, who sets out characteristics common to many of these killers (Alberici, 2007, n.p.). Typically, they are young males who have few friends and few intimate relationships. They are resentful and blame others for their unhappiness with their lives. Typically, they issue 'manifestos,' often on social media, outlining their grievances and explaining their killing sprees (Knoll, 2010). Their shootings are a way of getting revenge for the injustices they feel they have suffered. Many plan to commit suicide at the end of their killing sprees. They are also gun-obsessed, and firearms play a central role in their lives. Unlike many offenders, they carefully plan out their killings and use what Mullen has termed "cultural scripts" learned through media reports of previous rampage killings.

Why do the media misrepresent crime? The primary goal of the media is to make profits by selling advertising. Stories about violent crime will boost ratings and circulation, even if these stories give people a false picture of crime. The informal media rule, "If it bleeds, it leads," reflects the public fascination with sensationalized, bloody stories, such as those of mass murders. Commenting on his experience with the media, the executive director of a provincial legal society said, "If there's no blood and gore, or there's no sex, it's not newsworthy. And if it falls into the category of being newsworthy, then they have to show the dead body. They've got to show the corpse" (McCormick, 1995:182).

Some parts of the media also have an ideological agenda, as they favour "tough on crime" policies that are more likely to be supported by members of the public who fear being victimized by violent crime. Boyd and Carter (2014) found that the police (particularly the RCMP) and the media misrepresented the dangers of marijuana grow-ops to convince the public that marijuana cultivation poses a danger to public safety. This campaign has culminated in the very harsh penalties—including mandatory minimum sentences for growing small amounts of marijuana—in the Harper government's recent *Safe Streets and Communities Act* at a time when several U.S. states are legalizing marijuana use.

The media's crime coverage is selective in other ways. Some have blamed the media for failing to cover the story of large numbers of missing women in Vancouver until Robert Pickton was charged with 26 murders. While a missing child from a middle-class home will generate an avalanche of publicity, the stories of dozens of missing lower-class women—many of whom were sex trade workers—were not seen as important. Robert Pickton's trial generated international coverage, but the media focused on the gruesome crimes and did not consider larger social issues, such as legal policies that endanger sex trade workers, the structural reasons why so many of the victims were Indigenous, and the role of the state in producing socially impoverished neighbourhoods such as Vancouver's Downtown Eastside, where Pickton found most of his victims (Hugill, 2010).

Sources: Rick Linden, *Criminology: A Canadian Perspective* (8th ed.), Toronto: Nelson, 2016.

The Canadian Press/Nathan Denette

When sexual assault charges were laid against popular CBC radio host Jian Ghomeshi, the case immediately became an international media sensation.

To provide a further explanation, consider the circumstances of Canada's Indigenous people. Their situation is unique, but the same kinds of factors may apply in other racial contexts. While several theories have been advanced to explain Indigenous over-involvement, consider the following explanation drawn from conflict and social control theories. Canada's Indigenous people have far less power than other Canadians and are economically disadvantaged. They must cope with systems of education and religion that have been imposed on them from outside their cultural communities and that are incompatible with their customs and traditions. In the past, forced attendance at residential schools and forced adoption outside the community weakened family ties. Crippling rates of unemployment in many areas mean no job ties, and school curricula that are irrelevant to the lives of Indigenous students mean that children may not become attached to their schools. Under these conditions, strong social bonds are difficult to develop and high rates of crime can be predicted. Manitoba's Aboriginal Justice Inquiry concluded: "We believe that the relatively high rates of crime among Aboriginal people are a result of the despair, dependency, anger, frustration and sense of injustice prevalent in Aboriginal communities, stemming from the cultural and community breakdown that has occurred over the past century" (Hamilton and Sinclair, 1991:91).

TIME TO REVIEW

- Compare the different categories used by sociologists to classify different types of crime.
- Discuss some of the different types of cybercrime.
- What are the strengths and weaknesses of official crime statistics, self-reported crime statistics, and victimization surveys?
- Explain how and why age, gender, class, and race are correlated with crime.

LO-5 THE CRIMINAL JUSTICE SYSTEM

The criminal justice *system* includes the police, the courts, and prisons. However, the term criminal justice *system* is misleading because these institutions do not work together, and each has considerable autonomy.

THE POLICE

Most people think that the main role of the police is to enforce the law. That is indeed one of their functions, but most of their time is spent doing other things, including order maintenance and the provision of social services. *Order maintenance* refers to keeping the peace—stopping arguments, controlling the areas where skid-row alcoholics drink, and moving boisterous teenagers away from a convenience store parking lot. While the main concern in law enforcement is arresting a suspect, the main concern in order maintenance is restoring peace in the community. The *service* role, also important, consists of many different activities, including finding lost children, counselling crime victims, and notifying next of kin in fatal accidents.

Two important questions about the police are: Why do the police have such a broad range of responsibilities? What ties these diverse activities together? There are several reasons why the police have the broad responsibilities that they do:

1. The police are one of the few public agencies open 24 hours a day.
2. The police often serve clients who other agencies may not be interested in. The poor, the homeless, and the mentally ill may become police clients by default. If no other agency will look after intoxicated people who pass out on downtown streets, the police must do it.
3. The police may not know about, or have access to, other agencies that could handle some of their cases.

The second question concerning what ties these diverse activities together is best answered by considering two dimensions of the police role. First, the police have the *authority* (and often the duty) to intervene in situations where something must be done immediately. This authority is the same whether the incident is an armed robbery in progress, a naked man standing on a busy street screaming at people, or a complaint that someone's pet boa constrictor has just appeared in someone else's apartment. Second, the authority is backed up by *non-negotiable force.* If someone refuses to obey a police officer, the officer can use force (usually arrest) to back up his or her demands. Even professional caregivers may resort to calling the police when clients refuse to cooperate with them. Egon Bittner has summed up the patrol officer's role: "What policemen do appears to consist of rushing to the scene of any crisis whatever, judging its needs in accordance with canons of common sense reasoning, and imposing solutions upon it without regard to resistance or opposition" (1980:137).

The police have the discretion to decide which rules to apply and how to apply them. For example, if a police officer stops a driver for speeding and the driver has alcohol on his or her breath, the police officer may warn the person and tell them to go straight home, write a speeding ticket, or administer a Breathalyzer and lay charges of impaired driving if the limits are exceeded.

Police discretion is unavoidable. If discretion is used fairly and in a manner consistent with community standards and the rule of law, this is not a problem. However, if it is based on extra-legal factors, such as race or class, or if it is used to favour certain individuals over others, it can be *discriminatory.* Issues of racial discrimination have generated a great deal of discussion in Canada in recent years, as inquiries have been held in many provinces following police shootings of minority group members.

We have discussed some of the ways in which the justice system has discriminated against Indigenous people. Similar conclusions may also apply to other groups. Many blacks, particularly young people, feel that they are harassed by the police and that they are stopped and questioned because of their race. Carl James interviewed a number of black youth in several Ontario cities about their experiences. Many feel they have been treated unfairly by the police:

> You can't win. As long as you're Black you are a target.
>
> They drive by. They don't glimpse your clothes, they glimpse your colour. That's the first thing they look at. If they judge the clothes so much why don't they go and stop those white boys that are wearing those same things like us?
>
> No matter what the situation that you're in, what your dress is like . . . they will find a negative way of thinking about you, because you're Black, and secondly because you're Somali, and thirdly because you're an immigrant and you speak a different language. (1998:165–168)

A controversial Toronto Police Service policy of stopping citizens in non-criminal situations and recording detailed information about them on contact cards was widely criticized because black Torontonians were stopped at a rate almost three times higher than their representation in the

The Canadian Press/Halifax Chronicle Herald - Darren Pittman

A Nova Scotia Board of Inquiry found that a member of the Halifax Police Service discriminated against boxer Kirk Johnson when Mr. Johnson was stopped, ticketed, and had his car impounded despite having committed no offence.

population (Logical Outcomes, 2014). The Toronto Police Services Board eventually curtailed this practice, but it was widely criticized by those it affected who felt the policy was discriminatory. These perceptions of racial profiling have been supported by research done in Toronto (Wortley and Tanner, 2003) and Kingston (CTV, 2005), showing that black drivers are more likely to be stopped by the police than drivers of other races. While not all researchers agree that these studies demonstrate racial profiling (see Melchers, 2003), there is little doubt that being more frequently stopped does contribute to the perception of police harassment felt by many black people.

Most recently, Carrington and Fitzgerald reported that minority youth were more likely than white youth to be questioned by the police. This difference could not be explained by the extent of differential involvement in crime by minority youth. They conclude that the evidence suggests that the "disproportionate minority youth contact with the police in Canada is at least partly a result of racially discriminatory policing practices" (2012:195).

THE COURTS

Criminal courts decide whether persons accused of a crime are found guilty or not guilty. In theory, justice is determined in an adversarial process in which the prosecutor (a lawyer representing the state) argues that the accused is guilty, and the defence lawyer presents the accused's defence. Proponents of the adversarial system feel this system best provides a just decision about guilt or innocence.

The essence of the adversarial system can be seen in the defence lawyer's role, which is to defend the accused without concern for the client's actual guilt or innocence. This role was described by Lord Brougham, the defence lawyer in a complex 1821 case that could have had disastrous consequences for the British government had his defence been successful:

> An advocate, in the discharge of his duty, knows but one person in all the world, and that person is his client. To save that client by all means and expedients, and at all hazards and costs to their persons, and amongst them, to himself, is his first and only duty; and in performing this duty he must not regard the alarm, the torments, the destruction which he may bring upon others. Separating the duty of a patriot from that of an advocate, he must go on reckless of the consequences, though it should be his unhappy fate to involve his country in confusion. (cited in Greenspan, 1982:201)

Many view the adversarial system as one of the cornerstones of a democratic society. Many of the procedures that seem to restrict the ability of the court to get at the "truth," such as the rule that accused persons cannot be forced to testify against themselves, were adopted to prevent the arbitrary use of state power against the accused. However, some critics feel that our system does not deal adequately with crime because it places more emphasis on winning than on doing what is best for the accused, the victim, and society.

Not all Western countries use the adversarial court system. In several European countries, the judge takes an active role in the process. At the trial, the judge leads the questioning, and there is much more concern with getting at the 'truth' than in our system.

RESTORATIVE JUSTICE

Another alternative to our adversarial system is the restorative justice approach. For many years, we have relied on the formal justice system to deal with crime. Community members have been discouraged from participating in the system and have had little say in the services they received. After the victim called the police, the police would arrive to take care of the problem, and if an arrest was made, processing the case was left in the hands of the formal justice system. Some of those found guilty were removed from the community and sent away to jail. Professionals controlled each step in the system, with little input from victims and other community members.

While most people have come to accept this as the proper way of dealing with crime, some feel the system has failed them. Victims feel left out, as their injury is forgotten and they are relegated to the role of witnesses. Offenders are also dealt with impersonally and are rarely reminded of the personal harm they have done. The public is often dissatisfied with a justice system that does not respond to their concerns.

Critics have proposed an alternative system that is intended to restore social relationships rather than simply to punish. Advocates of *restorative justice* seek a system that will repair the harm that has been done to the victim and to the community. A key element is the involvement of the victim and other community members in the process in order to reconcile offenders with those they have harmed and to help communities reintegrate victims and offenders.

Restorative justice has its roots in traditional societies where the restoration of order was crucial to the community's survival. In Canada, Indigenous communities are leading the way in the return to traditional restorative justice practices. They have used a variety of different methods, including *sentencing circles,* which bring an offender together with the victims and other community members, to resolve disputes. Two of the most widespread contemporary restorative justice methods are *victim–offender reconciliation* and *family group conferencing.*

Victim–offender reconciliation was devised in Elmira, Ontario, as an initiative to persuade a judge to deal in a positive fashion with two youths who had vandalized property belonging to 22 different victims. Mediators worked with the victims and the offenders to reach an acceptable resolution. As a result, the youths had to deal personally with each of their victims and to make restitution for the damage they had caused. The restorative process gave victims a say in what happened and gave offenders the chance to make amends.

Family group conferencing is similar to sentencing circles. It typically applies to young offenders and normally involves the victim, the offender, and as many of their family and friends as possible. All parties speak and then discuss how to repair the harm done to the victim. Negotiation continues until a plan is agreed on and written down. The coordinator then establishes mechanisms for enforcing the plan. The family and friends of both the victim and the offender are encouraged to offer continuing help to ensure that the resolution arrived at during the conference is carried out in the community.

PRISONS

The incarceration rate in Canada in 2013 was 118 per 100,000 people. This rate is lower than the rates in the United States (716) and England and Wales (148), but higher than the rates in many other countries, including Italy (106), Germany (79), and Sweden (67) (Public Safety Canada, 2013). And prisons are very expensive—it costs nearly $120,000 a year to keep an inmate in a federal penitentiary.

We send people to jail for several reasons:

1. *Retribution.* To punish them for their crimes.
2. *Incapacitation.* To restrain them from committing further crimes.
3. *Rehabilitation.* To change offenders so they return to the community as law-abiding citizens.
4. *Deterrence.* To instill a fear of punishment to reduce future crimes.

These goals often conflict. Those who focus on retribution and deterrence may want to make the prison experience as punitive and harsh as possible. This conflicts with rehabilitation, however, because this goal is best accomplished by providing inmates with the skills to get jobs following release, by helping them deal with the issues that led them into crime, and by carefully reintegrating them into their communities.

BOX 8.5 POINT/COUNTERPOINT

Do Tougher Prison Sentences Reduce Crime?

The law clearly deters. Most people do not deliberately park where they know their car will be towed away and do not speed if they see a police car behind them. However, the more important question concerns the limits of deterrence. How can we make the justice system more effective at reducing crime?

The Conservative government was elected in 2006 promising to "crack down on crime." Since then, Parliament has passed many laws designed to put more people in jail for longer periods of time. Prisons are expensive, so it is important to know whether this policy of sending more people to jail for longer periods reduces crime rates or whether other crime reduction strategies would keep Canadians safer.

The research tells us that longer sentences do *not* reduce crime rates. Durlauf and Nagin reviewed the evidence on the deterrent effect of imprisonment and concluded that long prison sentences "are difficult to justify on a deterrence-based, crime prevention basis" (2011:38). Some of the research they reviewed suggests that imprisonment may actually *increase* an individual's likelihood of future criminal behaviour.

The most conclusive studies reviewed by Durlauf and Nagin are evaluations of laws requiring mandatory minimum prison sentences for particular offences or for offenders with significant prior records. Mandatory minimum sentences have become widely used and there has been much debate about their effectiveness.

The harshest mandatory sentencing law in any Western country is California's three-strikes law. This law provides a mandatory sentence of 25 years in prison for a third felony conviction following two earlier convictions for serious felonies. This has resulted in some bizarre sentences, including cases where two men will each spend 25 years in prison, one for stealing a slice of pizza and the other for shoplifting a small package of meat because these minor offences were their third felony.

The three-strikes law has been costly. The California State Auditor (2010) calculated that the cost was $20 billion more than if the inmates had been sentenced for the crimes they committed rather than for the "strikes" against them. California can no longer afford its prison system, and in 2012, the U.S. Supreme Court ordered the state to release 32,000 inmates because of severe overcrowding. Mandatory minimum sentences also increase court costs because individuals facing long mandatory penalties are more likely to insist on a trial rather than pleading guilty. The law is also very hard on the offenders and their families.

The high social and financial costs of mandatory minimum sentences might be worthwhile if they reduced crime rates. However, they do not. Michael Tonry concluded:

> "Mandatory penalties are a bad idea. They often result in injustice to individual offenders . . . And the clear weight of the evidence is . . . that there is insufficient credible evidence to conclude that mandatory penalties have significant deterrent effects" (2009:100).

Tonry bases this conclusion in part on several evaluations of California's three-strikes laws. While California's crime rate has declined since the passage of three strikes in 1994, this decline was not due to the three-strikes laws. Only one of 15 studies reviewed by Tonry concluded that the legislation reduced crime rates. Several of the studies showed that crime rates in California did not decline faster than in other states, even though the penalties in California were far more severe than in any other state.

Why don't severe penalties like mandatory sentences deter crime? One reason is because offenders may not feel they are at risk of receiving those penalties. And potential offenders are actually correct in believing that their next crime is unlikely to lead to punishment. Most crimes are not reported, most reported offences do not result in arrests, most arrests do not lead to convictions, and most convictions do not result in imprisonment. Nearly 1.8 million crimes were reported to Canadian police in 2014. Victimization surveys have shown that less than one-third of all crimes are reported to the police, so there are likely around 6 million crimes each year in Canada. Despite this huge number of offences, only about 5000 people were sentenced to federal penitentiaries (all sentences of two years or more) and 87,000 to provincial custody each year (Public Safety Canada, 2013). Thus, the likelihood of being arrested, convicted, and punished for any offence is so low that tinkering with the level of punishment makes no difference. The promise to crack down on crime is kept so rarely that it is ignored by potential offenders, who know from their own experience (and from that of their peers) that the odds of avoiding punishment are in their favour. A harsh system like California's is really one of randomized severity in which a small number of offenders receive very harsh sentences, while many others with similar patterns of offending remain on the streets.

Techniques such as improving the lives of high-risk youth and policing targeted toward high-crime locations and high-rate offenders can be very effective at reducing crime (Linden and Koenig, 2016). Why do you think the Conservative government chose to pursue a prison-based strategy they know will not reduce crime?

Canada's latest mandatory minimum sentences require a mandatory sentence of at least six months in prison for growing as few as six marijuana plants if the grower is involved in marijuana trafficking. Do you think this will have any impact on marijuana use in your community?

Source: Rick Linden, *Criminology: A Canadian Perspective* (8th ed.), Toronto: Nelson, 2016.

While prisons are very costly, we do not know as much as we should about their effectiveness. The research discussed in Box 8.5 shows that longer prison sentences are not effective deterrents and that other methods are more effective at ensuring community safety.

COMMUNITY CORRECTIONS

Our relatively high incarceration rate has led some to suggest that Canada should rely more heavily on community corrections. These dispositions include programs such as community probation, community service orders, intensive probation supervision, and bail supervision.

The movement toward community-based sanctions has been driven by three major concerns. First, these programs are much cheaper. It costs about $35,000 for community supervision, which is about one-quarter the cost of imprisonment (Public Safety Canada, 2013). The second concern is humanitarian. Prison life is unpleasant and it can be unfair to send people to jail for relatively minor offences. Finally, an offender may benefit from maintaining ties with family and community, which may make subsequent involvement in crime less likely.

TIME TO REVIEW

- Explain why the police have such a broad range of responsibilities.
- Discuss the use of discretion in the criminal justice system.
- How does the adversarial system affect the way our court system operates?
- Explain how the restorative approach to justice differs from the normal operation of our criminal justice system.
- Why do we send some convicted criminals to jail? What are some of the alternatives to this practice?

VISUAL SUMMARY

8

KEY TERMS

corporate crime An illegal act committed by corporate employees on behalf of the corporation and with its support (p. 207).

crime An act that violates criminal law and is punishable by fines, jail terms, and other sanctions (p. 195).

cybercrime Offences where a computer is the object of a crime or the tool used to commit a crime (p. 207).

deviance Any behaviour, belief, or condition that violates cultural norms in the society or group in which it occurs (p. 194).

differential association theory The proposition that individuals have a greater tendency to deviate from societal norms when they frequently associate with persons who favour deviance over conformity (p. 197).

illegitimate opportunity structures Circumstances that provide an opportunity for people to acquire through illegitimate activities what they cannot achieve through legitimate channels (p. 196).

labelling theory The proposition that deviants are those people who have been successfully labelled as such by others (p. 199).

moral crusades Public campaigns that help generate public and political support for moral entrepreneurs' causes (p. 200).

moral entrepreneurs People or groups who take an active role in trying to have particular behaviours defined as deviant (p. 200).

LO-1 Explain the meanings of the terms *crime* and *deviance*.

Deviant behaviour is any act that violates established norms. Deviance varies from culture to culture and in degree of seriousness. Crime is seriously deviant behaviour that violates written laws and that is punishable by fines, incarceration, or other sanctions.

The Canadian Press/Darryl Dyck

LO-2 Understand how crime and deviance are explained by functionalist, conflict, interactionist, feminist, and postmodern theories.

Sam Leung/Vancouver Sun

Strain theory says that if people are denied legitimate access to cultural goals, some will engage in illegal behaviour to achieve these goals. Social control theory says that our social bonds help to keep us from crime and deviance, while differential association theory focuses on ties to deviant peers. The emphasis of labelling theory is on those who apply a deviant label to people who break the rules, because that label may lead to subsequent deviance. Conflict theories examine the impact of social inequality. The powerful exploit the lower classes and the legal order protects people at the top. Feminist theorists conclude that women's deviance and crime is a response to gender discrimination experienced in work, marriage, and interpersonal relationships. Postmodern theorists have focused on social control and discipline based on the use of knowledge, power, and technology.

LO-3 Describe how sociologists count and classify crimes.

The Canadian Uniform Crime Reports survey is based upon crimes reported to the police. We also use victimization surveys that interview households to determine the incidence of crimes, including those not reported to police.

JEAN-PHILIPPE KSIAZEK/AFP/Getty Images

LO-4 Understand how age, gender, class, and race are related to deviance and crime.

Young people have the highest rates of crime. Women have much lower rates of crime than men. Persons from lower socioeconomic backgrounds are more likely to be arrested for violent and property crimes, while corporate crime is committed by those from the upper classes.

Dick Loek/GetStock.com

LO-5 Describe how the criminal justice system deals with crime.

The criminal justice system includes the police, the courts, and prisons. These agencies have considerable discretion. The police often use discretion in deciding whether to act on a situation. Prosecutors and judges use discretion in deciding which cases to pursue and in determining guilt and penalties.

AP Photo/Brett Coomer

occupational, or white collar, crime A term used to describe illegal activities committed by people in the course of their employment or in dealing with their financial affairs (p. 207).

organized crime A business operation that supplies illegal goods and/or services for profit (p. 208).

political crime Illegal or unethical acts involving the usurpation of power by government officials, or illegal or unethical acts perpetrated against a government by outsiders seeking to make a political statement or to undermine or overthrow the government (p. 209).

primary deviance A term used to describe the initial act of rule breaking (p. 199).

secondary deviance A term used to describe the process whereby a person who has been labelled deviant accepts that new identity and continues the deviant behaviour (p. 199).

social bond theory The proposition that the likelihood of deviant behaviour increases when a person's ties to society are weakened or broken (p. 197).

social control Systematic practices developed by social groups to encourage conformity and discourage deviance. (p. 194).

strain theory The proposition that people feel strain when they are exposed to cultural goals that they are unable to obtain because they do not have access to culturally approved means of achieving these goals (p. 196).

street crime All violent crime, certain property crimes, and certain morals crimes (p 207).

KEY FIGURES

Pictorial Parade/Staff/Getty Images

Robert Merton (1910–2003) Merton was an important U.S. sociologist. His work in criminology was important because it placed the sources of crime in the social structure rather than in individual psychopathology.

American Sociological Association. asanet.org

Edwin Sutherland (1883–1950) Sutherland was also a key figure in criminological theory. He is best known for differential association theory and for the identification of white-collar crime as a serious problem.

Jean Pierre FOUCHET/RAPHO/ Gamma-Rapho/Getty

Michel Foucault (1926–1984) Foucault's work spanned many different areas in the field of social theory. He was one of the first to recognize the impact of new surveillance technologies on all members of society.

Courtesy of Elizabeth Comack

Elizabeth Comack (b. 1952) Comack's work has focused on the intersections of gender, race, and crime, and on the sociology of law. Her most recent research has been on inner-city crime and on racialized policing.

APPLICATION QUESTIONS

1. Using your sociological imagination, how would you deal with the problem of crime in Canada? What programs should be enhanced? What programs would you reduce?
2. Do you ever feel afraid of being a crime victim? How do you think that people like you can best reduce their likelihood of being victimized?
3. Legalized gambling provides Canadian governments with billions of dollars in revenue. What are the positive and negative consequences of this gambling? Do you think that gambling laws should be liberalized, or should gambling be restricted? Why?
4. Look at today's newspaper or an online media site. What types of crimes are in the headlines? How well do these media reports reflect the reality of crime in your community?
5. Legislation passed by the federal government in 2012 specified mandatory penalties for some marijuana offences that were as long or longer than sentences for sexually abusing children. Do you think these mandatory minimum sentences reflect society's attitudes toward these offences? Why do you think the government passed such harsh laws against growing marijuana?

CHAPTER 9

SOCIAL CLASS AND STRATIFICATION IN CANADA

© Shepard Sherbell/CORBIS SABA

CHAPTER FOCUS QUESTION

How are the lives of Canadians affected by social inequality?

LEARNING OBJECTIVES

AFTER READING THIS CHAPTER, YOU SHOULD BE ABLE TO

LO-1 Identify three types of stratification systems.

LO-2 Discuss the extent of social inequality in Canada based on measures of income and wealth.

LO-3 Understand the classical analysis of social class by Karl Marx and Max Weber.

LO-4 Provide an overview of poverty and its effects in Canada.

LO-5 Compare and contrast a functionalist view and a conflict perspective on social inequality.

Financial wealth—or, in more technical terms, net worth—is an important element of Canadian society. Most Canadians aspire to get some of it to help them live long and well. Others want it to help care for the less fortunate. Fair or not, wealth is also one of the ways used to measure people and their achievements. Advertisers push "the dream" of wealth—and what it can bring. For some, the dream and the reality do become one. There are more than one million millionaire families in Canada, and the richest 20% of Canadian households have over two-thirds of all wealth in Canada. For many, however, the dream never does become the reality. Their reality regarding wealth may consist of having enough to live just above the poverty line. The poorest 20 percent of Canadian households have more debts than assets—in fact a negative net wealth of approximately $3500 (Battams, Spinks, & Sauvé, 2014). The following speech, read by Vancouver teacher Anna Chudnovsky at the launch of 2014 Child Poverty Report Card, highlights the tragic effects of poverty on children. The report found that British Columbia has the fifth-highest child poverty rate in the country, with approximately one in five B.C. children living in poverty.

My name is Anna Chudnovsky. I'm a teacher in an inner city school in Vancouver.

I don't have a lot of numbers to share with you; instead, I have a story of a real student and of his family. A story that I hope will tell you a little bit about what it's like to be a student living in poverty. I'm telling you this story not because it's sad, but because I believe that being poor makes it hard to learn. And when it's too hard to learn, lifting yourself out of the poverty you're born into is incredibly hard.

There's a boy in my class. He is Aboriginal, eight years old, has a younger brother and a sister, and his mother is pregnant and due to give birth any day now. The family has just moved here from Smithers because his mum is trying to start fresh, give her children some opportunities that she herself didn't have. The family lives in B.C. housing.

This boy is smart. He is capable. He is kind and friendly. He helps others, he tries his best at school. But his life is too hard. The task he has in front of him is too grand, too monstrous to overcome. You see, he's poor. And when you're poor, as poor as he is, succeeding at school is such a very difficult challenge.

We have a "walking school bus" that picks kids up at home from the housing complexes in the neighbourhood and brings them to school. It helps get kids to school on time, builds a bit of community, and takes a bit of the stress of the before-school-rush off the parents' shoulders. The walking bus picked up this child last week and when the support worker got to the door, the mum, nine months pregnant in a new city trying to build her new life, confided in the worker that she had no groceries. She had only one single jar of olives in the fridge. She started to cry. She was worried she wouldn't have time to get groceries before the baby came.

She's panicked that she won't be ready. Her son, my student, my kind, smart, lovely student, tried to comfort his mum as she stood there crying in the doorway. And then he came to school to try to learn. We're working on adding with regrouping. His family is in utter crisis and he's trying to carry the one. (Chudnovsky, 2014)

Source: Chudnovsky, Anna 2014, "In My Class, Child Poverty Is No Numbers Game," *The Tyee*, November 25th, 2014 http://thetyee.ca/Opinion/2014/11/25/Child-Poverty-No-Numbers-Game/. Reproduced by permission of the author.

In Canadian society, we are socialized to believe that hard work is the key to personal success. In other words, anyone can go from poverty to wealth if she or he works hard enough and plays by the rules. Conversely, we are taught that individuals who fail—who do not achieve success—do so as a result of personal inadequacies. Poverty is attributable to personal defect; it is up to the individual to find a way to break the cycle of poverty. Do you agree? Or do you think that structural factors in Canadian society affect the degree of success that individuals achieve?

In this chapter, we will examine systems of social stratification and how the Canadian class system may make it easier for some individuals to attain (or maintain) top positions in society while others face significant obstacles in moving out of poverty or low-income origins. Before we explore class and stratification, test your knowledge of wealth and poverty in Canada by taking the quiz in Box 9.1.

CRITICAL THINKING QUESTIONS

1. What factors–individual, social, or structural–do you think contribute to a person living in poverty?
2. If a child is raised in poverty, what chance does he or she have of changing his or her economic position later in life? What would be the best way to accomplish this?
3. Is it true that "the rich are getting richer and the poor are getting poorer" in Canadian society?

WHAT IS SOCIAL STRATIFICATION?

▶ **social stratification** The hierarchical arrangement of large social groups based on their control over basic resources.

▶ **life chances** Max Weber's term for the extent to which individuals have access to important societal resources, such as food, clothing, shelter, education, and healthcare.

Social stratification is the hierarchical arrangement of large social groups based on their control over basic resources. Stratification involves patterns of structural inequality that are associated with membership in each of these groups, as well as the ideologies that support inequality. Sociologists examine the social groups that make up the hierarchy in a society and seek to determine how inequalities are structured and persist over time.

Max Weber's term **life chances** refers to the extent to which individuals have access to important societal resources, such as food, clothing, shelter, education, and healthcare. According to sociologists, more affluent people typically have better life chances than the less affluent because they have greater access to quality education, safe neighbourhoods, high-quality nutrition and healthcare, police and private security protection, and an extensive array of other goods and services. In contrast, persons with low- and poverty-level incomes tend to have limited access to these resources. *Resources* are anything valued in a society. They range from money and property to medical care and education; they are considered scarce because of their unequal distribution among social categories. If we think about the valued resources available in Canada, for example, the differences in life chances are readily apparent. Our life chances are intertwined with our class, race, gender, and age.

All societies distinguish among people by age. Young children typically have less authority and responsibility than older persons. Older persons, especially those without wealth or power, may find themselves at the bottom of the social hierarchy. Similarly, all societies differentiate between females and males: Women are often treated as subordinate to men. From society to society, people are treated differently as a result of their religion, race and ethnicity, appearance, physical strength, disabilities, or other distinguishing characteristics. All of these differentiations result in inequality. However, systems of stratification are also linked to the specific economic and social structure of a society and to a nation's position in the system of global stratification, which is so significant for understanding social inequality that we will devote Chapter 10 to this topic.

LO-1 SYSTEMS OF STRATIFICATION

▶ **social mobility** The movement of individuals or groups from one level in a stratification system to another.

▶ **intergenerational mobility** The social movement (upward or downward) experienced by family members from one generation to the next.

▶ **intragenerational mobility** The social movement (upward or downward) experienced by individuals within their own lifetime.

▶ **slavery** An extreme form of stratification in which some people are owned by others.

Around the globe, one of the most important characteristics of systems of stratification is their degree of flexibility. Sociologists distinguish among such systems based on the extent to which they are open or closed. In an *open system,* the boundaries between levels in the hierarchies are more flexible and may be influenced (positively or negatively) by people's achieved statuses. Open systems are assumed to have some degree of social mobility. **Social mobility** is the movement of individuals or groups from one level in a stratification system to another. This movement can be either upward or downward. **Intergenerational mobility** is the social movement experienced by family members from one generation to the next. By contrast, **intragenerational mobility** is the social movement of individuals within their own lifetime. Both intragenerational mobility and intergenerational mobility may be downward as well as upward. In a *closed system,* the boundaries between levels in the hierarchies of social stratification are rigid and people's positions are set by ascribed status.

Open and closed systems are ideal-type constructs; no stratification system is completely open or closed. The systems of stratification that we will examine—slavery, caste, and class—are characterized by different hierarchical structures and varying degrees of mobility.

SLAVERY

Slavery is an extreme form of stratification in which some people are owned by others. It is a closed system in which people designated as "slaves" are treated as property and have little

BOX 9.1 SOCIOLOGY AND EVERYDAY LIFE

How Much Do You Know About Wealth and Poverty in Canada?

True	False	
T	F	1. There is less child poverty in Canada today than there was 20 years ago.
T	F	2. Individuals over the age of 65 have the highest rate of poverty.
T	F	3. Men account for two out of every three impoverished adults in Canada.
T	F	4. Most poor children live in female-headed, single-parent households.
T	F	5. Age plays a key role in wealth accumulation in Canada.
T	F	6. The richest 10 percent of Canadian households account for approximately one-third of all wealth.
T	F	7. Fewer than 1 percent of Canadian households have a net worth of at least a million dollars.
T	F	8. The income gap between poor families and rich families has widened in the past 10 years.

For more questions and the answers to the quiz about wealth and poverty in Canada, go to **www.nelson.com/student**.

or no control over their lives. Many Canadians are not aware of the legacy of slavery in our own country. Beginning in the 1600s, people were forcibly imported to what are now Canada and the United States to serve as slaves and cheap labour. Slavery existed in what are now the provinces of Quebec, New Brunswick, Nova Scotia, and Ontario until the early 19th century (Satzewich, 1998).

As practised in North America, slavery had four primary characteristics: (1) It was for life and was inherited (children of slaves were considered slaves); (2) slaves were considered property, not human beings; (3) slaves were denied rights; and (4) coercion was used to keep slaves "in their place" (Noel, 1972).

THE CASTE SYSTEM

Like slavery, caste is a closed system of social stratification. A **caste system** is a system of social inequality in which people's status is permanently determined at birth based on their parents' ascribed characteristics. Vestiges of caste systems exist in contemporary India and South Africa.

caste system A system of social inequality in which people's status is permanently determined at birth based on their parents' ascribed characteristics.

In India, caste is based in part on occupation, while in South Africa it was based on race and a sense of moral superiority. In India, families have typically performed the same type of work from generation to generation. By contrast, the caste system of South Africa was based on racial classifications and the belief of white South Africans (Afrikaners) that they were morally superior to the black majority. Until the 1990s, the Afrikaners controlled the government, the police, and the military by enforcing *apartheid*—the separation of the races. Blacks were denied full citizenship and restricted to segregated hospitals, schools, residential neighbourhoods, and other facilities. Whites held almost all the desirable jobs; blacks worked as manual labourers and servants.

In a caste system, marriage is *endogamous,* meaning that people are allowed to marry only within their own group. In India, parents have traditionally selected marriage partners for their children. In South Africa, interracial marriage was illegal until 1985.

Cultural beliefs and values sustain caste systems. Hinduism, the primary religion of India, reinforced the caste system by teaching that people should accept their fate in life and work hard as a moral duty. However, caste systems grow weaker as societies industrialize: The values reinforcing the system break down, and people begin to focus on the types of skills needed for industrialization.

As we have seen, in closed systems of stratification, group membership is hereditary and it is almost impossible to move up within the structure. Custom and law frequently perpetuate privilege and ensure that higher-level positions are reserved for the children of the advantaged.

THE CLASS SYSTEM

class system A type of stratification based on the ownership and control of resources and on the type of work people do.

The **class system** is a type of stratification based on the ownership and control of resources and on the type of work people do. At least theoretically, a class system is more open than a caste system because the boundaries between classes are less distinct than the boundaries between castes. In a class system, status comes at least partly through achievement rather than entirely by ascription.

In class systems, people may become members of a class other than that of their parents through both intergenerational and intragenerational mobility, either upward or downward. *Horizontal mobility* occurs when people experience a gain or loss in position and/or income that does not produce a change in their place in the class structure. For example, a person may get a pay increase and a more prestigious title but still not move from one class to another.

By contrast, movement up or down the class structure is *vertical mobility*. For example, Bruce is a physician, but he was the first person in his family to attend college, much less to graduate from medical school. His father was a day labourer picked up each day by contractors outside the local lumberyard to work on various building sites. His mother was a stay-at-home mom who took care of Bruce and his four younger siblings. Bruce's parents did not complete high school and had little support from home because their families had limited economic means. In high school, Bruce became involved in the Upward Bound program, which encourages young people to remain in school and provides mentoring, tutoring, and enrichment activities that ultimately helped him achieve his dream.

Bruce's situation reflects upward mobility; however, people may also experience downward mobility, caused by any number of reasons, including a lack of jobs, low wages and employment instability, marriage to someone with fewer resources and less power than oneself, and changing social conditions. Ascribed statuses, such as race and ethnicity, gender, and religion, also affect

Hulton Archive/Getty Images

© Jake Norton/Alamy

iStockphoto/Thinkstock

Systems of stratification include slavery, caste, and class. As shown in these photos, the life chances of people living in each of these systems differ widely.

people's social mobility. Sometimes, the media portray upward social mobility as something that is easily achieved by a few lucky people regardless of their ascribed or achieved statuses. We will return to the ideals versus the realities of social mobility when we examine the Canadian class structure later in the chapter.

TIME TO REVIEW

- How does social mobility differ in the three systems of stratification?
- Is it possible to have vertical or horizontal mobility in a caste system? Why, or why not?

LO-2 INEQUALITY IN CANADA

Throughout human history, people have argued about the distribution of scarce resources in society. Disagreements often concentrate on whether the share people get is a fair reward for their effort and hard work. Social analysts have recently pointed out that the old maxim "the rich get richer" continues to be valid in Canada. To understand how this happens, let us take a closer look at income and wealth inequality in our country.

income The economic gain derived from wages, salaries, income transfers (governmental aid), and ownership of property.

Money is essential for acquiring goods and services. People without money cannot acquire food, shelter, clothing, legal services, education, and the other things they need or desire. Money—in the form of both income and wealth—is unevenly distributed in Canada.

Among the approximately 20 industrialized nations of North America and Europe, Canada has a poor record pertaining to income inequality, ranking sixth in income inequality among developed countries (Conference Board of Canada, 2011).

INCOME INEQUALITY

Income is the economic gain derived from wages, salaries, income transfers (governmental aid), and ownership of property. One common method of analyzing the distribution of income is the concept of income *quintiles* comprised of five income groups ranging from the lowest-income group to the highest-income group. As shown in Figure 9.1, the top quintile represents the 20 percent of families with the lowest incomes and the bottom quintile represents the 20 percent of families with the highest incomes. Sociologist Dennis Gilbert (2011) compares the distribution of income to a national pie that has been cut into portions ranging from stingy to generous, for distribution among segments of the population. Today, Canada's richest households take home close to 45 percent of the total income "pie," while the poorest receive only 4 percent of all income (Battams, Spinks, & Sauvé, 2014). Analysts further report that incomes have remained remarkably stable over time. In 1990, average family income was just under \$60,000, and by 2012, it had increased to only \$71,000 (Statistics Canada, 2012g). However, a closer examination of the data reveals that focusing on the overall average family income tends to conceal wide variations between different segments of the population and hides increasing inequities in the distribution of income in Canada. Overall, the percentage of the pie of the three lowest income groups has decreased in the past decade. As shown in Figure 9.1, only those in the highest quintile increased their share of the income pie between 1999 and 2011 (Battams, Spinks, & Sauvé, 2014).

Photodisc/Getty Images

People's life chances are enhanced by access to important societal resources, such as education. How will the life chances of students who have the opportunity to pursue a university degree differ from those of young people who do not have the chance to go to university?

FIGURE 9.1 : DISTRIBUTION OF DISPOSABLE INCOME AFTER TAX, CANADA, 1999 AND 2011

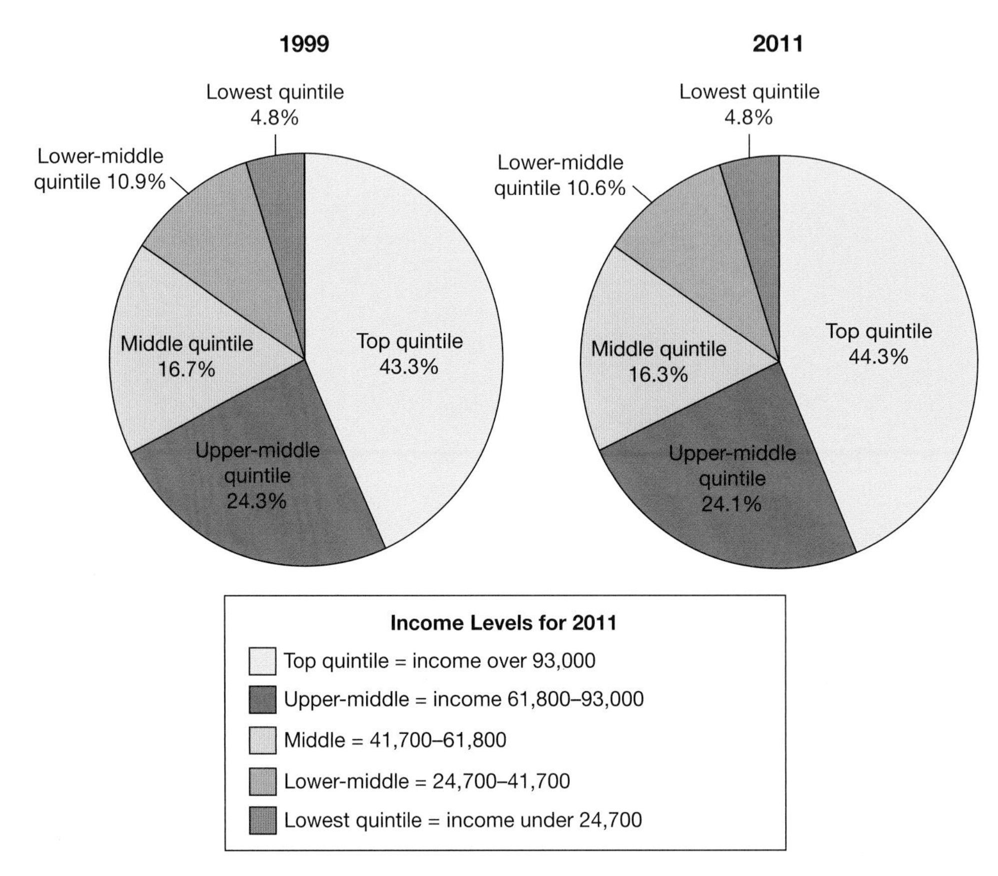

Source: Battams, N., Spinks, N., & Sauvé, R. (2014). *The Current State of Canadian Family Finances: 2013-2014 Report*. Vanier Institute of the Family. http://www.vanierinstitute.ca/include/get.php?nodeid=3773

There is considerable variation on other fronts, too. Consider regional variation in income across the country. As shown in Figure 9.2, family income is highest in Saskatchewan, Ontario, and Alberta, and lowest in the Atlantic provinces and Quebec (Statistics Canada, 2012g). There is also significant income variation among particular racial–ethnic groups. For example, recent statistics indicate that almost 40 percent of visible minorities are in the low-income group, as compared to 20 percent of the general population. Beyond that, the data clearly demonstrate the inequities in income distribution experienced by Indigenous peoples in Canada; their average income is less than two-thirds the average income of the general population (Wilson & MacDonald, 2010; Pendakur & Pendakur, 2013; Statistics Canada, 2013c).

▶ **wealth** The value of all of a person's or family's economic assets, including income and property, such as buildings, land, farms, houses, factories, and cars, as well as other assets, such as money in bank accounts, corporate stocks, bonds, and insurance policies.

WEALTH INEQUALITY

Income is only one aspect of wealth. **Wealth** includes property, such as buildings, land, farms, houses, factories, and cars, as well as other assets, such as money in bank accounts, corporate stocks, bonds, and insurance policies. Wealth is computed by subtracting all debt obligations and converting the remaining assets into cash. The terms *wealth* and *net worth,* therefore, are used interchangeably. For most people in Canada, wealth is invested primarily in property that generates no income, such as a house or a car. In contrast, the wealth of an elite minority is often in the form of income-producing property.

FIGURE 9.2 MEDIAN AFTER-TAX INCOME, FAMILIES OF TWO PERSONS OR MORE, PROVINCES AND TARGETED CENSUS METROPOLITAN AREAS, 2012

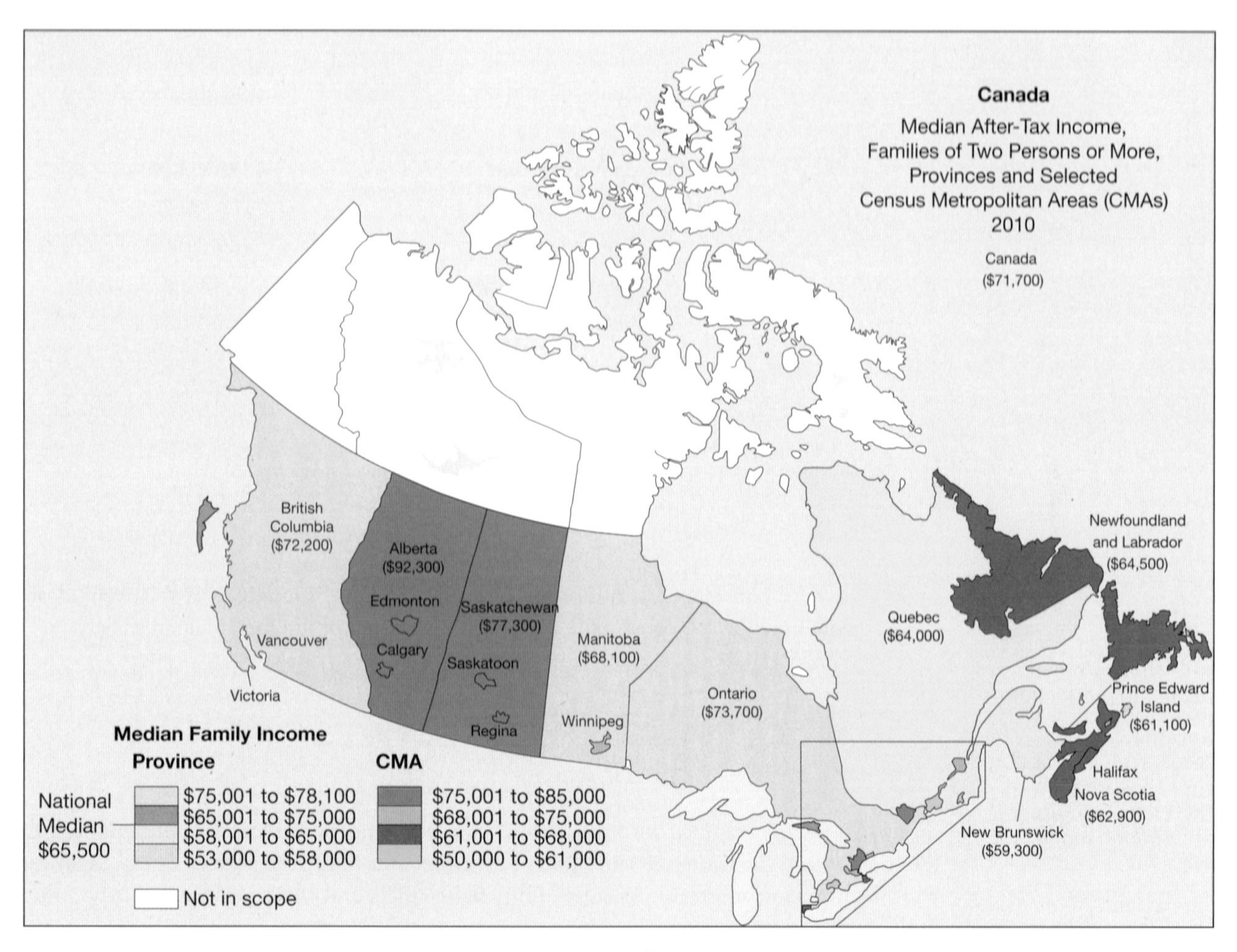

Source: Statistics Canada. 2012. "Canadian Income Survey." *The Daily*. CANSIM table 206-0001. Retrieved from: http://www.statcan.gc.ca/daily-quotidien/141210/t141210a001-eng.htm

Research on the distribution of wealth in Canada reveals that wealth is more unevenly distributed among the Canadian population than is income. Although the term *wealthy*, like the term *poor*, is relative, analysts generally define the wealthy as those whose total assets after debt payments are more than $250,000. Today, there are close to 350,000 Canadians who can claim millionaire status because their net worth is at least $1 million. That's about 1 per cent of Canada's population of 35 million. Their collective wealth is estimated at $979 billion (World Wealth Report, 2015).

Most of the wealthiest people in Canada are inheritors, with some at least three or four generations removed from the original fortune. As shown in Table 9.1, the combined wealth of Canada's richest families totals approximately $97 billion, $31 billion of which is accounted for by Canada's richest person, former newspaper publisher Kenneth Thomson *(Canadian Business,* 2015). It is clear that a limited number of people own or control a huge portion of the wealth in Canada. Indeed, one survey of wealth concluded that there is "gross and persistent inequality in the distribution of wealth in Canada. A surprisingly small number of Canadians have huge slices of the wealth pie.

TABLE 9.1 : WEALTHIEST CANADIANS 2015

Rank	Last Year	Name	2014 net worth
1	1	Thomson Family	$30,738,353,336
2	2	Galen Weston	$11,381,211,658
3	3	Irving Family	$8,228,039,229
4	5	James Pattison	$7,875,807,140
5	4	Rogers Family	$7,404,572,651
6	6	Saputo Family	$6,240,002,796
7	7	Estate of Paul Desmarais Sr.	$5,581,765,394
8	9	Richardson Family	$5,049,518,418
9	8	Jeffrey S. Skoll	$5,013,336,412
10	10	Carlo Fidani	$4,576,963,843

Source: "Canada's Richest People 2015: The Top 100 Richest Canadians." *Canadian Business*, 2015. Retrieved from: http://www.canadianbusiness.com/lists-and-rankings/richest-people/top-100-richest-canadians-2015/

Whether we consider distribution of income or wealth, though, it is relatively clear that social inequality is a real, consistent, and enduring feature of life in Canadian society.

▶ **capitalist class (bourgeoisie)** Karl Marx's term for those who own the means of production.

▶ **working class (proletariat)** Karl Marx's term for those who must sell their labour in order to earn enough money to survive.

▶ **alienation** A feeling of powerlessness from other people and from oneself.

LO-3 CLASSICAL PERSPECTIVES ON SOCIAL CLASS

Early sociologists grappled with the definition of class and the criteria for determining people's location within the class structure. Both Karl Marx and Max Weber viewed class as an important determinant of social inequality and social change, and their works have had a profound influence on contemporary class theory.

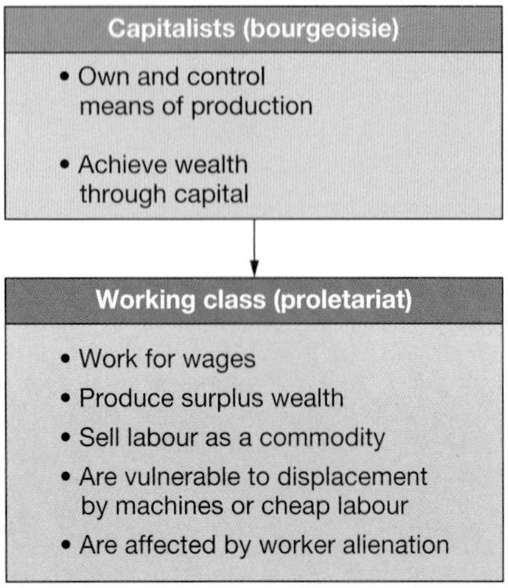

KARL MARX: RELATION TO MEANS OF PRODUCTION

FIGURE 9.3 : MARX'S VIEW OF SOCIAL CLASS

Capitalists (bourgeoisie)
- Own and control means of production
- Achieve wealth through capital

Working class (proletariat)
- Work for wages
- Produce surplus wealth
- Sell labour as a commodity
- Are vulnerable to displacement by machines or cheap labour
- Are affected by worker alienation

According to Karl Marx, class position is determined by people's work situation, or relationship to the means of production. As we have previously seen, Marx stated that capitalistic societies are made up of two classes—the capitalists and the workers. The **capitalist class (bourgeoisie)** consists of those who own the means of production—the land and capital necessary for factories and mines, for example. The **working class (proletariat)** consists of those who must sell their labour to the owners in order to earn enough money to survive (see Figure 9.3).

According to Marx, class relationships involve inequality and exploitation. Capitalists maximize their profits by exploiting workers, paying them less than the resale value of what they produce but do not own. This exploitation results in workers' **alienation**—a feeling of powerlessness and estrangement from other people and from oneself. In Marx's view, alienation develops as workers manufacture goods that embody their creative talents but the goods do not belong to them. Workers are also alienated from the work itself because they are forced to perform it in order to live. Because the workers' activities are not their own, they feel self-estrangement. Moreover, workers are separated from others in the factory because they individually sell their labour power to the capitalists as a commodity.

In Marx's view, the capitalist class maintains its position at the top of the class structure by control of the society's *superstructure,* which is composed of the government, schools, churches, and other social institutions that produce and disseminate ideas perpetuating the existing system of exploitation. Marx predicted that the exploitation of workers by the capitalist class would ultimately lead to **class conflict**—the struggle between the capitalist class and the working class. According to Marx, when the workers realized that capitalists were the source of their oppression, they would overthrow the capitalists and their agents of social control, leading to the end of capitalism. The workers would then take over the government and create a more egalitarian society.

class conflict Karl Marx's term for the struggle between the capitalist class and the working class.

Why has no workers' revolution occurred? Capitalism may have persisted because it has changed significantly since Marx's time. Individual capitalists no longer own and control factories and other means of production; today, ownership and control have largely been separated. For example, contemporary transnational corporations are owned by a multitude of stockholders but run by paid officers and managers. Similarly, many (but by no means all) workers have experienced a rising standard of living, which may have contributed to a feeling of complacency. During the 20th century, workers pressed for salary increases and improvements in the workplace through their activism and labour union membership. They also gained more legal protection in the form of workers' rights and benefits such as workers' compensation insurance for job-related injuries and disabilities. For these reasons, and because of a myriad of other complex factors, the workers' revolution predicted by Marx has not come to pass. However, the failure of his prediction does not mean that his analysis of capitalism and his theoretical contributions to sociology are without validity.

Marx had a number of important insights about capitalist societies. First, he recognized the economic basis of class systems (Gilbert, 2011). Second, he noted the relationship between people's location in the class structure and their values, beliefs, and behaviour. Finally, he acknowledged that classes may have opposing (rather than complementary) interests. For example, capitalists' best interests are served by a decrease in labour costs and other expenses and a corresponding increase in profits; workers' best interests are served by well-paid jobs, safe working conditions, and job security.

MAX WEBER: WEALTH, PRESTIGE, AND POWER

Max Weber's analysis of class builds upon earlier theories of capitalism (particularly those by Marx). Living in the late 19th and early 20th centuries, Weber was in a unique position to see the transformation that occurred as individual, competitive, entrepreneurial capitalism went through the process of shifting to bureaucratic, industrial, corporate capitalism. As a result, Weber had more opportunity than Marx did to see how capitalism changed over time.

Weber agreed with Marx's assertion that economic factors are important in understanding individual and group behaviour. However, he emphasized that no one factor (such as economic divisions between capitalists and workers) was sufficient to define people's location within the class structure. For Weber, the access that people have to important societal resources (such as economic, social, and political power) is crucial in determining their life chances. To highlight the importance of life chances for categories of people, Weber developed a multidimensional approach to *social stratification* that reflects the interplay among wealth, prestige, and power. In his analysis of these dimensions of class structure, Weber viewed the concept of class as an *ideal type* (which can be used to compare and contrast various societies), rather than as a specific category of "real" people (Bourdieu, 1984).

Weber placed categories of people who have a similar level of wealth and income in the same class. For example, he identified a privileged commercial class of *entrepreneurs*—wealthy bankers, ship owners, professionals, and merchants who possess similar financial resources. He also described a class of *rentiers*—wealthy individuals who live off their investments and do not have to work. According to Weber, entrepreneurs and rentiers have much in common.

Both are able to purchase expensive consumer goods, control other people's opportunities to acquire wealth and property, and monopolize costly status privileges (such as education) that provide contacts and skills for their children.

Weber divided those who work for wages into two classes: the middle class and the working class. The middle class consists of white-collar workers, public officials, managers, and professionals. The working class consists of skilled, semiskilled, and unskilled workers.

prestige The respect or regard with which a person or status position is regarded by others.

The second dimension of Weber's system of social stratification is **prestige**—the respect with which others regard a person or a status position. Fame, respect, honour, and esteem are the most common forms of prestige. A person who has a high level of prestige is assumed to receive deferential and respectful treatment from others. Weber suggested that individuals who share a common level of social prestige belong to the same status group regardless of their level of wealth. They tend to socialize with one another, marry within their own group of social equals, spend their leisure time together, and safeguard their status by restricting outsiders' opportunities to join their ranks (Beeghley, 2000).

power According to Max Weber, the ability of people or groups to achieve their goals despite opposition from others.

The other dimension of Weber's system is **power**—the ability of people or groups to achieve their goals despite opposition from others. The powerful shape society in accordance with their own interests and direct the actions of others. According to Weber, social power in modern societies is held by bureaucracies; individual power depends on a person's position within the bureaucracy. Weber suggested that the power of modern bureaucracies was so strong that even a workers' revolution, as predicted by Marx, would not lessen social inequality.

socioeconomic status (SES) A combined measure that attempts to classify individuals, families, or households in terms of indicators, such as income, occupation, and education, to determine class location.

Weber stated that wealth, prestige, and power are separate continuums on which people can be ranked from high to low. As shown in Figure 9.4, individuals may be high in one dimension while being low in another. For example, people may be very wealthy but have little political power (for example, a recluse who has inherited a large sum of money). They also may have prestige but not wealth (for instance, a university professor who receives teaching excellence awards but lives on a relatively low income). In Weber's multidimensional approach, people are ranked in all three dimensions. Sociologists often use the term **socioeconomic status (SES)** to refer to a combined measure that attempts to classify individuals,

FIGURE 9.4 : WEBER'S MULTIDIMENSIONAL APPROACH TO SOCIAL STRATIFICATION

According to Max Weber, wealth, power, and prestige are separate continuums. Individuals may rank high in one dimension and low in another, or they may rank high or low in more than one dimension. They may also use their high rank in one dimension to achieve a comparable rank in another.

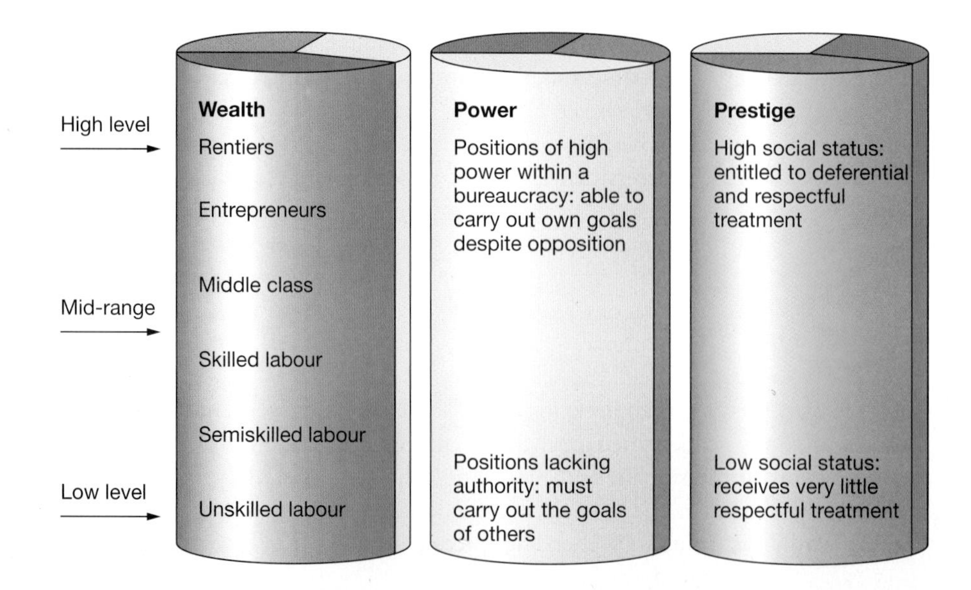

families, or households in terms of indicators, such as income, occupation, and education, to determine class location.

What important insights does Weber provide in regard to social stratification and class? Weber's analysis of social stratification contributes to our understanding by emphasizing that people behave according to both their economic interests and their values. He also added to Marx's insights by developing a multidimensional explanation of the class structure and identifying additional classes. Both Marx and Weber emphasized that capitalists and workers are the primary players in a class society, and both noted the importance of class to people's life chances. However, they saw different futures for capitalism and the social system. Marx saw these structures being overthrown; Weber saw increasing bureaucratization of life even without capitalism.

TIME TO REVIEW

- Explain the statement, "The rich get richer and the poor get poorer." Does this apply to both income and wealth?
- Explain how prestige, power, and wealth determine social class.

CONTEMPORARY SOCIOLOGICAL MODELS OF THE CLASS STRUCTURE IN CANADA

How many social classes exist in Canada today? What criteria are used for determining class membership? No broad consensus exists about how to characterize the class structure of this country. Canadians do not like to talk about social class, and many deny that class distinctions even exist. Most people like to think of themselves as middle class; it puts them in a comfortable middle position—neither rich nor poor.

Sociologists have developed a few models of the class structure. One is based on a Weberian approach, the other on a Marxian approach. We will examine each of these models briefly.

A WEBERIAN MODEL OF CLASS STRUCTURE

Expanding on Weber's analysis of the class structure, sociologist Dennis Gilbert (2011) developed a widely used model of social classes based on three elements: (1) education, (2) occupation of the family head, and (3) family income. This model can be used to describe the social class structure in Canadian society.

THE UPPER CLASS The upper class is the wealthiest and most powerful class in Canada. About 1 percent of the population is included in this class, whose members own substantial income-producing assets and operate on both the national and international levels. According to Gilbert (2011), people in this class have an influence on the economy and society far beyond their numbers.

Some models further divide the upper class into *upper-upper class* ("old money") and *lower-upper class* ("new money"). Because such a small number are members of the upper-upper class, many analysts have referred to it as the *elite* class (Clement, 1975; Clement and Myles, 1994). In the past, family names—such as Thomson, Richardson and Asper—were well known and often held in high esteem. Today, some upper-class family names are well known, but many of the individuals and families involved have often made their fortunes much more recently.

Numerous scholars have examined the distribution of wealth and power in Canada and found one consistent result: A small number of individuals—in the upper-upper class—yield an enormous amount of power. For example, John Porter's classic study identified just over 900 individuals who controlled all of the major corporations in Canada (1965:579). Author Peter Newman identified slightly less than 1000 individuals as members of what he described as "the Canadian establishment." As Newman points out, members of the upper-upper class share more than wealth and power; they tend to have strong feelings of ingroup solidarity. They belong to the same exclusive clubs, share social activities, and support high culture (such as the symphony, opera, ballet, and art museums). Their children are educated at prestigious private schools and universities. In general, children of the upper class are socialized to view themselves as different from others; they may also learn that they are expected to marry within their own class (Domhoff, 2005).

Members of the lower-upper class may be extremely wealthy but have not attained as much prestige as the members of the upper-upper class. The "new rich" have earned most of their money as entrepreneurs, presidents of major corporations, sports or entertainment celebrities, or top-level professionals.

THE UPPER-MIDDLE CLASS Persons in the upper-middle class are often highly educated professionals who have established careers as physicians, lawyers, stockbrokers, or corporate managers. Others have derived their income from family-owned businesses. A combination of three factors qualifies people for the upper-middle class: a university education, authority and independence on the job, and high income. Of all the class categories, the upper-middle class is the one that is most influenced by education.

THE MIDDLE CLASS In past decades, a high school diploma was required to qualify for most middle-class jobs. Today, undergraduate university degrees or college programs have replaced the high school diploma as an entry-level requirement for employment in many middle-class occupations, including medical technicians, nurses, legal and medical secretaries, lower-level managers, semiprofessionals, and nonretail salespersons. Traditionally, most middle-class positions have been relatively secure and provided more opportunities for advancement (especially with increasing levels of education and experience) than working-class positions. Recently, however, several factors have diminished the chances for material success for members of this class: (1) escalating housing prices; (2) a high rate of job loss and instability, with long-term unemployment and blocked mobility on the job; and (3) the cost-of-living squeeze that has penalized younger workers, even when they have more education and better jobs than their parents.

THE WORKING CLASS An estimated 30 percent of the Canadian population is in the working class. The core of this class is made up of semiskilled machine operators who work in factories and elsewhere. Members of the working class also include some workers in the service sector and salespeople whose job responsibilities involve routine, mechanized tasks requiring little skill beyond basic literacy and a brief period of on-the-job training (Gilbert, 2011). Within the working class are also **pink-collar occupations**—relatively low-paying, nonmanual, semiskilled positions primarily held by women, such as daycare workers, checkout clerks, cashiers, and waitresses.

▶ **pink-collar occupation** Relatively low-paying, nonmanual, semiskilled positions primarily held by women.

How does life in the working-class family compare with that of individuals in middle-class families? According to sociologists, working-class families not only earn less than middle-class families, but they also have less financial security, particularly with high rates of layoffs and corporate downsizing in some parts of the country. Few people in the working class have more than a high school diploma, which makes job opportunities increasingly scarce in our "high-tech" society (Gilbert, 2011).

THE WORKING POOR The working poor account for about 20 percent of the Canadian population. Members of the working-poor class live from just above to just below the poverty line. They typically hold unskilled jobs, seasonal jobs, lower-paid factory jobs, and service jobs (such as counter help at restaurants). Employed single mothers often belong to this class; consequently, children are overrepresented in this category. Members of some visible minority groups, Indigenous peoples, and recent immigrants are also overrepresented among the working poor. For the working poor, living from paycheque to paycheque makes it impossible to save money for emergencies, such as periodic or seasonal unemployment, which is a constant threat to any economic stability they may have.

Social critic and journalist Barbara Ehrenreich (2001, 2011) left her upper-middle-class lifestyle for a time to see whether the working poor could live on the wages they were being paid as restaurant servers, salesclerks at discount department stores, aides in nursing homes, house cleaners for franchise maid services, and other similar jobs. She conducted her research by holding those jobs for periods of time and seeing if she could live on the wages she received. Through her research, Ehrenreich persuasively demonstrated that people who work full time, year-round for poverty-level wages must develop survival strategies that include getting help from relatives or constantly moving from one residence to another to have a place to live. Like many other researchers, Ehrenreich found that minimum-wage jobs cannot cover the full cost of living, such as rent, food, and the rest of an adult's monthly needs, even without considering the needs of children or other family members.

THE UNDERCLASS According to Gilbert (2011), people in the underclass are poor, seldom employed, and caught in long-term deprivation that results from low levels of education and income and high rates of unemployment. Some are unable to work because of age or disability; others experience discrimination based on race or ethnicity. Single mothers are overrepresented in this class because of lack of jobs, affordable child care, and many other impediments to their future and that of their children. People without a "living wage" must often rely on public or private assistance programs for their survival. About 5 percent of the population is in this category, and the chances of their children moving out of poverty are about fifty-fifty (Gilbert, 2011).

Gaining work-related skills and having employment opportunities are two critical issues for people on the lowest rungs of the class ladder. Many of the jobs that exist today require specialized knowledge or skills that are inaccessible to people in the underclass. Skills and jobs are essential for people to have the opportunity to earn a decent wage; have medical coverage; live meaningful, productive lives; and raise their children in a safe environment.

A CONFLICT MODEL OF CLASS STRUCTURE

The earliest Marxian model of class structure identified ownership or nonownership of the means of production as the distinguishing feature of classes. From this perspective, classes are social groups organized around property ownership, and social stratification is created and maintained by one group to protect and enhance its own economic interests. Moreover, societies are organized around classes in conflict over scarce resources. Inequality results when the more powerful exploit the less powerful.

Contemporary Marxian (or conflict) models examine class in terms of people's relationships with others in the production process. For example, conflict theorists attempt to determine what degree of control workers have over the decision-making process and the extent to which they are able to plan and implement their own work. They also analyze the type of supervisory authority, if any, that a worker has over other workers. According to this approach, most employees are a part of the working class because they do not control either their own labour or that of others.

Erik Olin Wright (1978, 1979, 1985, 1997, 2010), one of the leading stratification theorists to examine social class from a Marxian perspective, has concluded that Marx's definition of "workers" does not fit the occupations found in advanced capitalist societies. For example,

Bill Truslow/Stone/Getty Images
Rido/Shutterstock.com
Norm Betts/Bloomberg/Getty Images
© Katie Deits/Index Stock Imagery

Erik Olin Wright's conflict model of the class system emphasizes the differing interests of the capitalist class, exemplified by the small-business class (top left); the managerial class (top right); the capitalist class—David Thomson (bottom left); and the working class (bottom right).

many top executives, managers, and supervisors who do not own the means of production (and thus would be "workers" in Marx's model) act like capitalists in their zeal to control workers and maximize profits. Likewise, some experts hold positions in which they have control over money and the use of their own time even though they are not owners. Wright views Marx's category of "capitalist" as being too broad as well. For instance, small-business owners might be viewed as capitalists because they own their own tools and have a few people working for them, but they have little in common with large-scale capitalists and do not share the interests of factory workers. Figure 9.5 compares Marx's model and Wright's model.

Wright (1979) argues that classes in modern capitalism cannot be defined simply in terms of different levels of wealth, power, and prestige, as in the Weberian model. Consequently, he outlines four criteria for placement in the class structure: (1) ownership of the means of production, (2) purchase of the labour of others (employing others), (3) control of the labour of others (supervising others on the job), and (4) sale of one's own labour (being employed by someone else). Wright (1978) assumes that these criteria can be used to determine the class placement of all workers, regardless of race and ethnicity, in a capitalist society.

Let's take a brief look at Wright's (1979, 1985) four classes—(1) the capitalist class, (2) the managerial class, (3) the small-business class, and (4) the working class—so that you can compare them to those found in the Weberian model.

THE CAPITALIST CLASS According to Wright, this class holds most of the wealth and power in society through ownership of capital—for example, banks, corporations, factories, mines, news and entertainment industries, and agribusiness firms. The "ruling elites," or "ruling class," within the capitalist class hold political power and are often elected or appointed to influential political and regulatory positions (Parenti, 1994).

This class is composed of individuals who have inherited fortunes, own major corporations, or have extensive stock holdings or control of company investments because they are top corporate executives. Even though many top executives have only limited *legal ownership*

FIGURE 9.5 COMPARISON OF MARX'S AND WRIGHT'S MODELS OF CLASS STRUCTURE

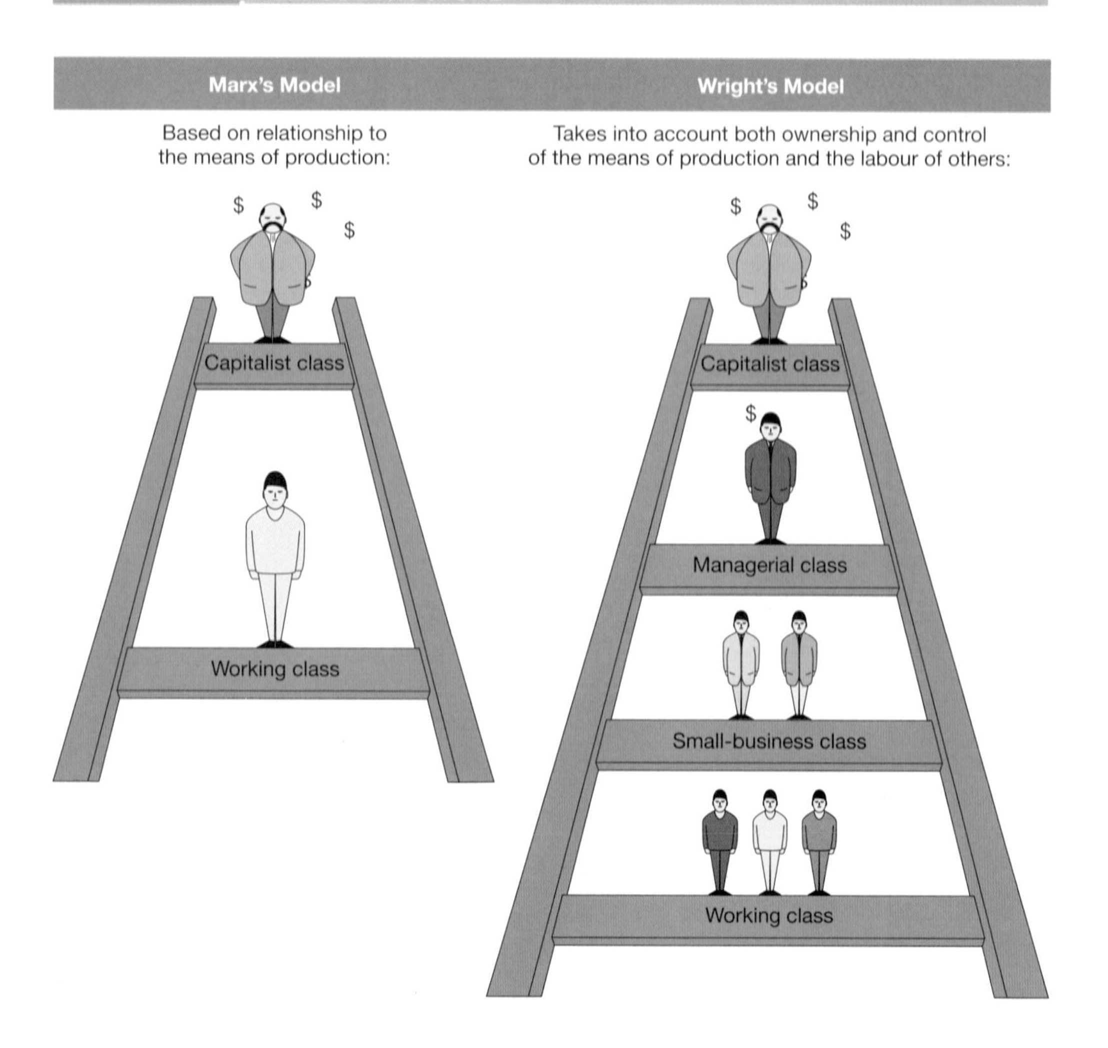

of their corporations, they have substantial economic ownership and exert extensive control over investments, distribution of profits, and management of resources. The major sources of income for the capitalist class are profits, interest, and very high salaries. Members of this class make important decisions about the workplace, including which products and services to make available to consumers and how many workers to hire or fire.

According to *Forbes* magazine's 2015 list of the richest people in the world, Bill Gates (co-founder of Microsoft Corp., the world's largest microcomputer software company) was the richest person in the world with a net worth of $79 billion. Mexican telecom entrepreneur Carlos Slim Helú came in second, followed by investor Warren E. Buffett with $73 billion. Although some of the men who made the *Forbes* list of wealthiest people have gained their fortunes through entrepreneurship or as chief executive officers (CEOs) of large corporations, women who made the list typically acquired their wealth through inheritance, marriage, or both. Of the top 20 richest people in the world, only three are women (Forbes, 2015).

THE MANAGERIAL CLASS People in the managerial class have substantial control over the means of production and over workers; however, these upper-level managers, supervisors, and professionals typically do not participate in key corporate decisions, such as how to invest profits. Lower-level managers may have some control over employment practices, including the hiring and firing of some workers.

Top professionals, such as physicians, lawyers, accountants, and engineers, may control the structure of their own work; however, they typically do not own the means of production and may not have supervisory authority over more than a few people. Even so, they may influence the organization of work and the treatment of other workers. Members of the capitalist class often depend on these professionals for their specialized knowledge.

THE SMALL-BUSINESS CLASS This class consists of small-business owners and craftspeople who may hire a few employees but largely do their own work. Some members own businesses, such as "mom and pop" grocery stores, retail clothing stores, and jewellery stores. Others are doctors and lawyers who receive relatively high incomes from selling their services. Some of these professionals now share attributes with members of the capitalist class because they have formed corporations that hire and control the employees who produce profits for the professionals.

It is in the small-business class that we find many people's hopes of achieving upward mobility. Recent economic trends, including corporate downsizing, telecommuting, and the movement of jobs to other countries, have encouraged more people to think about starting their own business. As a result, more people today are self-employed or own a small business than at any time in the past.

THE WORKING CLASS The working class is made up of a number of subgroups, one of which is blue-collar workers, some of whom are highly skilled and well paid, and others of whom are unskilled and poorly paid. Skilled blue-collar workers include electricians, plumbers, and carpenters; unskilled blue-collar workers include janitors and gardeners.

White-collar workers are another subgroup of the working class. Referred to by some as a "new middle class," these workers are members of the working class because they do not own the means of production, do not control the work of others, and are relatively powerless in the workplace. Administrative assistants, clerical workers, and sales workers are members of the white-collar faction of the working class. They take orders from others and tend to work under constant supervision. Thus, these workers are at the bottom of the class structure in terms of domination and control in the workplace. The working class consists of about half of all employees in Canada.

TIME TO REVIEW

- What are the fundamental differences between Weberian and Marxian models of the class structure?

LO-4 CONSEQUENCES OF INEQUALITY

Income and wealth are not simply statistics; they are intricately related to our individual life chances. Persons with a high income or substantial wealth have more control over their lives. They have greater access to goods and services, and they can afford better housing, more education, and a wider range of medical services. Similarly, those with greater access to economic resources fare better when dealing with the criminal justice system (Linden, 2009; Reiman, 1979). Persons with less income, especially those living in poverty, must spend their limited resources on the basic necessities of life.

PHYSICAL AND MENTAL HEALTH AND NUTRITION

People who are wealthy and well educated and who have high-paying jobs are much more likely to be healthy than are poor people. As people's economic status increases, so does their health status. People who are poor have shorter life expectancies and are at greater risk for chronic illnesses, such as diabetes, heart disease, and cancer, as well as for infectious diseases, such as tuberculosis (see Chapter 15).

A growing body of research suggests that by following a healthy lifestyle, the risk factors for chronic diseases such as cancer, diabetes, heart disease/stroke, and lung disease can be reduced. In low-income populations, however, underlying social and economic factors often make it difficult to initiate lifestyle changes such as quitting smoking, eating healthier food, and increasing physical activity. In short, people with lower incomes have greater health inequities because they have less personal control over where they live, what they eat, and how they work and relax (Health Nexus, 2015a).

Children born into poor families are at much greater risk of dying during their first year of life. Some die from disease, accidents, or violence. Others are unable to survive because they are born with low birth weight, a condition linked to birth defects and increased probability of infant mortality. Low birth weight in infants is attributed, at least in part, to the inadequate nutrition received by many low-income pregnant women. Most of the poor do not receive preventive medical and dental checkups; many do not receive adequate medical care after they experience illness or injury. Furthermore, many high-poverty areas lack an adequate supply of doctors and medical facilities (Campaign 2000, 2014). The higher death rates among Indigenous peoples in Canada are partly attributable to unequal access to medical care and nutrition.

Although the precise relationship between class and health is not known, analysts suggest that people with higher incomes and greater wealth tend to smoke less, exercise more, maintain a healthy body weight, and eat nutritious meals. As a category, affluent people tend to be less depressed and face less psychological stress, conditions that tend to be directly proportional to income, education, and job status (Canadian Mental Health Association, 2015).

Good health is basic to good life chances, and adequate amounts of nutritious food are essential for good health. Hunger is related to class position and income inequality. After spending 60 percent of their income on housing, low-income families are often unable to provide enough food for their children. The following comments of this mother living in poverty reveals it is particularly challenging to provide healthy food, which is generally more expensive than junk food:

> Food—buying food period, and making healthier choices, when you know that pasta goes this far [gestures with hands] and your salad goes this far [gestures a smaller amount with her hands]. Especially in the last five days of the month before my next OW cheque arrives, we have a more vegetarian diet. Meat is more expensive, and so we have filler foods. We have limited fruits, limited vegetables, a lot of bread, carbohydrates. Filler food. (Health Nexus, 2015b:35)

Most recent statistics confirm that more than 800,000 people use food banks each month. Children represent more than one in three (37 percent) of those users in Canada (Campaign 2000, 2014). These numbers clearly indicate that many Canadians are unable to meet their nutritional needs.

EDUCATION

Educational opportunities and life chances are directly linked. Some functionalist theorists view education as the "elevator" to social mobility. Improvements in the educational achievement levels (measured in number of years of schooling completed) of the poor, visible minorities, and women have been cited as evidence that students' abilities are now more important than their class, race, or gender. From this perspective, inequality in education is declining and students have an opportunity to achieve upward mobility through achievements at school. Functionalists generally see the education system as flexible, allowing most students the opportunity to attend university if they apply themselves (Ballantine and Hammack, 2012).

In contrast, most conflict theorists stress that schools are agencies for reproducing the capitalist class system and perpetuating inequality in society. From this perspective, education perpetuates poverty. Parents with a limited income are not able to provide the same educational opportunities for their children as are families with greater financial resources. Today, great disparities exist in the distribution of educational resources. Because funding for education comes primarily from local property taxes, school districts in wealthy suburban areas generally pay higher teachers' salaries, have newer buildings, and provide state-of-the-art equipment. By contrast, schools in poorer areas have a limited funding base. Students in core area schools and poverty-stricken rural areas often attend schools that lack essential equipment and teaching resources.

Poverty exacts such a toll that many young people will not have the opportunity to finish high school, much less enter university, which subsequently affects job prospects, employment patterns, and potential earnings.

CRIME AND LACK OF SAFETY

Along with diminished access to quality healthcare, nutrition, housing, and unequal educational opportunities, crime and lack of safety are other consequences of inequality. As discussed in Chapter 8, although people from all classes commit crimes, they commit different kinds of crimes. Capitalism and the rise of the consumer society may be factors in the criminal behaviour of some upper-middle-class and upper-class people, who may be motivated by greed or the competitive desire to stay ahead of others in their reference group. By contrast, crimes committed by people in the lower classes may be motivated by feelings of anger, frustration, and hopelessness.

According to Marxist criminologists, capitalism produces social inequalities that contribute to criminality among people, particularly those who are outside the economic mainstream. Poverty and violence are also linked. Consequences of inequality include both crime and lack of safety on the streets, particularly for people who feel a profound sense of alienation from mainstream society and its institutions. Those who are able to take care of themselves and protect their loved ones against aggression are accorded deference and regard by others.

POVERTY IN CANADA

When many people think about poverty, they think of people who are unemployed or on welfare; however, many hardworking people with full-time jobs live in poverty. In the United States, the government established an *official poverty line,* which is based on what is considered the minimum amount of money required for living at a subsistence level. The poverty level is computed by determining the cost of a minimally nutritious diet and multiplying this figure by three to allow for nonfood costs. Canada, however, has no official definition of poverty, no official method for measuring poverty, and no official set of poverty lines (Zhang, 2010). As a result, in Canada, there is ongoing and contentious debate with respect to how prevalent and how serious the problem of poverty is.

The most accepted and commonly used definition of poverty is Statistics Canada's before-tax **low-income cutoff**—the income level at which a family may be in "straitened circumstances" because it spends considerably more on the basic necessities of life (food, shelter, and clothing) than does the average family. The low-income cutoff depends on family and community size. According to this measure, any individual or family that spends more than 70 percent of their income on the three essentials of life—food, clothing, and shelter—is considered to be living in poverty. There is no single cutoff line for all of Canada because living costs vary by family size and place of residence. When sociologists define poverty, they distinguish between absolute and relative poverty (Statistics Canada, 2013c).

▶ **low-income cutoff** The income level at which a family may be in "straitened circumstances" because it spends considerably more on the basic necessities of life (food, shelter, and clothing) than does the average family.

Absolute poverty exists when people do not have the means to secure the most basic necessities of life. This definition comes closest to that used by the corporate think tanks, such as the Fraser Institute. Absolute poverty often has life-threatening consequences, such as when a homeless person freezes to death on a park bench. By comparison, **relative poverty** exists when people may be able to afford basic necessities but still are unable to maintain an average standard of living. The relative approach is based on equity—that is, on some acknowledgment of the extent to which society should tolerate or accept inequality in the distribution of income and wealth. This definition recognizes that people who have so little that they stand out in comparison to others in their community will feel deprived. In short, regardless of how it is defined, poverty is primarily about deprivation.

absolute poverty
A level of economic deprivation in which people do not have the means to secure the basic necessities of life.

relative poverty
A level of economic deprivation in which people may be able to afford basic necessities but still are unable to maintain an average standard of living.

With the exception of a small percentage of the Canadian population that is living at a bare subsistence level or below, most of the poor people in our society suffer the effects of a relentless feeling of being boxed in, a feeling that life is dictated by the requirements of simply surviving each day. If something unexpected happens, such as sickness, an accident, a family death, fire, theft, or a rent increase, there is no buffer to deal with the emergency. Life is just today because tomorrow offers no hope (Ross, Scott, and Smith, 2000). As analyst Roger Sauvé sums up after an evaluation of what low-income families do without:

> Money may not buy happiness . . . but the lack of money clearly reduces the probability of accessing many of the goods and services that are enjoyed by others. The poorest fifth of households are short of cash, skip on food, are much less likely to achieve home ownership, less likely to own a car, and give up spending on recreation, camps, investing, dental insurance and many other items. (2008:5)

WHO ARE THE POOR?

Poverty in Canada is not randomly distributed, but rather is highly concentrated among certain groups of people—specifically, women, children, persons with disabilities, and Indigenous peoples. When people belong to more than one of these categories—for example, Indigenous children—their risk of poverty is even greater.

AGE Today, children are at much greater risk of living in poverty than are people over age 65. A generation ago, older persons were at greatest risk of being poor; however, increased government transfer payments and an increase in the number of elderly individuals retiring with private pension plans have led to a decline in poverty among the elderly. Recent statistics indicate that the poverty rate for children under age 18 is approximately 17 percent. More than 1.3 million Canadian children are living in poverty, and a large number of children hover just above the official poverty line (Campaign 2000, 2014).

As shown in Table 9.2, child poverty rates are even higher among vulnerable groups. These include children living in female lone-parent families, children with disabilities, immigrant families, and Indigenous families. Approximately 30 percent of children in all of these groups are living in poverty (Campaign 2000, 2014).

Despite the 1989 promise made by the House of Commons to alleviate child poverty by the year 2000, little progress has been made (see Table 9.2 and Figure 9.8). It makes little difference whether they live in one- or two-parent families: Children as a group are poorer now than they were at the beginning of the 1980s. More than half of all poor children live in two-parent families, while the number of poor children living in single-parent households is increasing. These children are poor because their parents are poor, and one of the main reasons for poverty among adults is a lack of good jobs. Government cuts to employment insurance benefits, employment programs, income supports, and social services for families and children will affect not only those who need these services but also the children of these individuals. (See Box 9.2 for a look at child poverty in Canada and other wealthy nations.)

BOX 9.2 SOCIOLOGY IN GLOBAL PERSPECTIVE

How Does Child Poverty in Canada Compare with Child Poverty in Other Nations?

The UNICEF report *Child Poverty in Rich Nations* ranks Canada a lowly 24 out of the 35 industrialized countries belonging to the Organisation for Economic Co-operation and Development (OECD) (see the figure below).

The international rankings show that a nation's level of wealth does not predetermine its ability to prevent children from falling into poverty. Countries with higher economic growth do not necessarily have a lower poverty ranking. Many of the countries with the lowest poverty rates have relatively lower wealth rankings. The wealthiest nation, the United States, has the highest poverty ranking. The contention that child poverty can be addressed only through increased economic growth is contradicted by the available evidence.

UNICEF states that the reason why countries are able to address child poverty with such varying degrees of success relates to how each chooses to set priorities according to its wealth. Most of the nations that have been more successful than Canada at keeping low levels of child poverty are willing to counterbalance the effects of unemployment and low-paid work with substantial investments in family policies. The comprehensive approach to the well-being of children adopted by many European countries includes generous income security and unemployment benefits and national affordable housing programs, as well as widely accessible early childhood education and care.

The contrast of early childhood education and care services in Canada and in Europe is instructive. A recent OECD review of 12 nations found that early childhood education and care had experienced a "surge of policy attention" in Europe during the past decade. The same has not been true in Canada. While the nations of Western Europe now provide universal full-day early childhood education and care for all three- to five-year-olds, Canada has not even begun to consider this. Yet there is widespread agreement, including among Canadian researchers, that early childhood education and care is a critical component of comprehensive family policy and of an effective antipoverty strategy.

Source: Unicef Innocenti Research Centre, 2000.

FIGURE 9.6 POVERTY RATES FOR CHILDREN AND THE TOTAL POPULATION, 2010

This bar graph shows the percentage of children (0–17 years) in households with an income equivalent to less than 50 percent of the median.

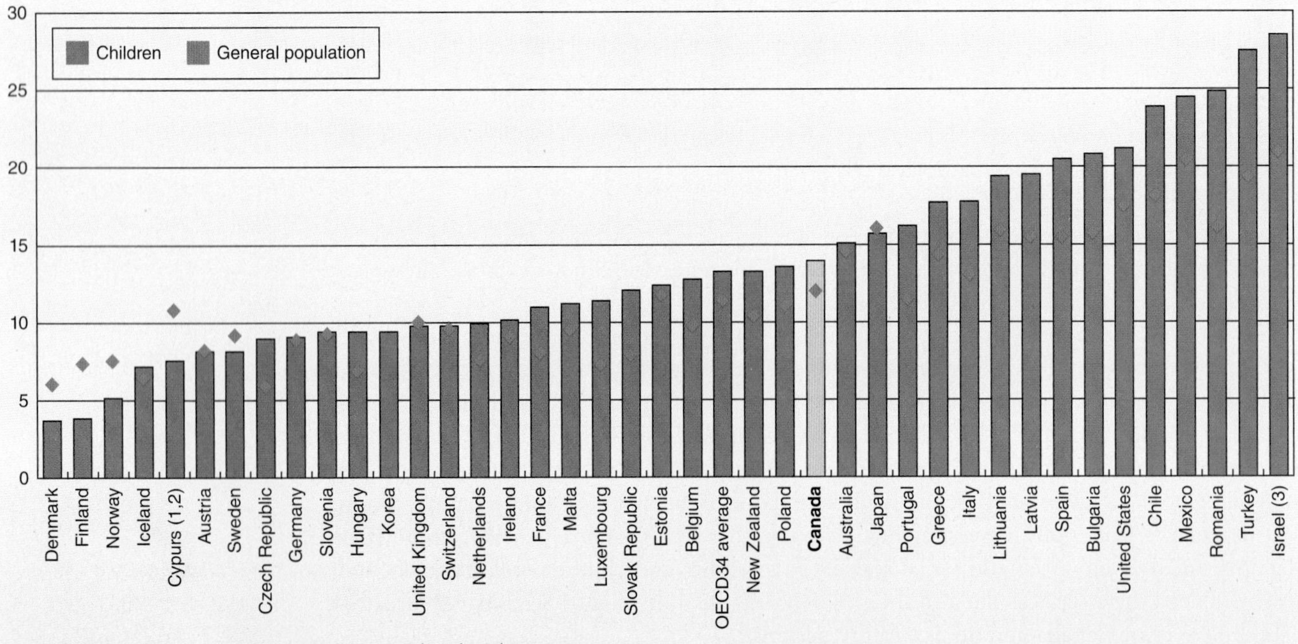

Source: OECD (2015), OECD Family Database, OECD, Paris www.oecd.org/social/family/database.htm

TABLE 9.2 PREVALENCE OF LOW INCOME AMONG CHILDREN IN CANADA, BY SOCIAL GROUP

Group	Prevalence of Low Income (Percent)
All children	8.5%
Children (under 15 years) in recent immigrant families	39.3%
First Nations children (under 15 years)	33.7%
Children in racialized families	27.0%
Children with disabilities	28.0%

Source: Statistics Canada Catalogue no. 97-564-X2006002, 6 December 2008; CANSIM table 202-0802; Canada Census 2001, Catalogue no. 97-564-XCB2006008, 17 December 2008; Employment and Social Development Canada, 2013a, "Snapshot of Racialized Poverty in Canada." Retrieved from: http://www.esdc.gc.ca/eng/communities/reports/poverty_profile/snapshot.shtml

GENDER About two-thirds of all adults living in poverty in Canada are women. Women in all categories are at greater risk for poverty than men are, but the risk is particularly significant in single-parent families headed by women (Naiman, 2008). As Figure 9.6 shows, single-parent families headed by women have a poverty rate of 45 percent, compared with a rate of 10 percent for two-parent families (see Figure 9.7). Furthermore, women are among the poorest of the poor. Poor single mothers with children under 18 are the worst off, struggling on incomes more than $9000 below the low-income cutoff (Campaign 2000, 2011). Sociologist Diana Pearce (1978) coined a term to describe this problem. The **feminization of poverty** refers to the trend in which women are disproportionately represented among individuals living in poverty. According to Pearce, women have a higher risk of being poor because they bear the major economic burden of raising children as single heads of households but earn only 70 cents for every dollar a male worker earns—a figure that has changed little over four decades. More women than men are unable to obtain regular, full-time, year-round employment, and the lack of adequate affordable daycare exacerbates this problem. As we will see in Chapter 11, the feminization of poverty is a global phenomenon.

▶ **feminization of poverty** The trend in which women are disproportionately represented among individuals living in poverty.

Does the feminization of poverty explain poverty in Canada today? Is poverty primarily a women's issue? On one hand, this thesis highlights a genuine problem—the link between gender and poverty. On the other hand, several major problems exist with this argument. First, women's poverty is not a new phenomenon. Women have always been more susceptible to poverty. Second, all women are not equally susceptible to poverty. Many in the upper and upper-middle classes have the financial resources, education, and skills to support themselves regardless of the presence of a man in the household. Some women, however, experience what has been described as *event-driven poverty* as a result of marital separation, divorce, or widowhood (Bane, 1986). Third, event-driven poverty does not explain the realities of poverty for many visible minority women, who instead may experience *reshuffled poverty*—a condition of deprivation that follows them regardless of their marital status or the type of family in which they live. Some women experience *multiple jeopardies,* a term that refers to the even greater risk of poverty experienced by women who are immigrants, members of visible minorities, Indigenous, or who have disabilities (Gerber, 1990).

Finally, poverty is everyone's problem, not just women's. When women are impoverished, so are their children. Moreover, many of the poor people in our society are men, especially the chronically unemployed—older persons, the homeless, persons with disabilities, and members of a visible minority. These men have spent their adult lives without hope of finding work.

The "feminization of poverty" refers to the fact that two out of three impoverished adults in North America are women. Should we assume that poverty is primarily a women's issue? Why, or why not?

FIGURE 9.7 POVERTY RATE BY FAMILY TYPE

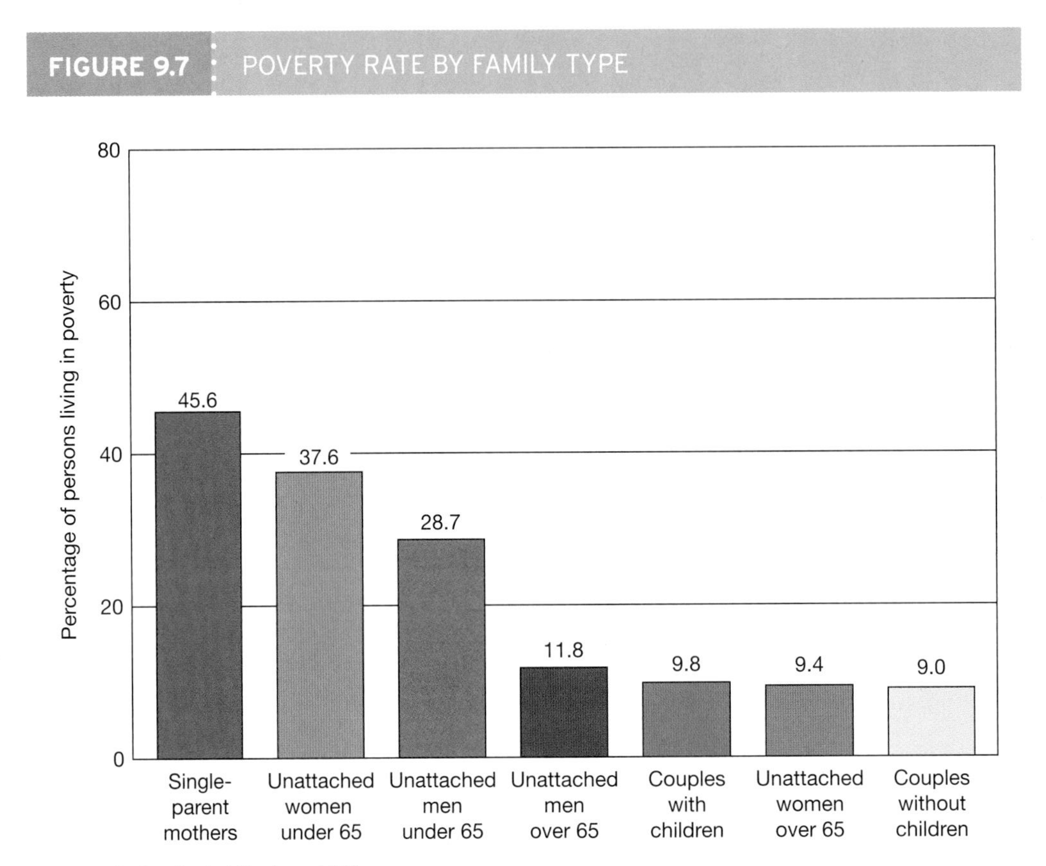

Source: Canadian Council on Social Development, 2009.

RACE/ETHNICITY According to some stereotypes, most of the poor and virtually all welfare recipients are visible minorities. Such stereotypes are perpetuated because a disproportionate percentage of the impoverished in Canada are Indigenous persons and recent immigrants. Members of both of these groups have significantly higher rates of poverty and unemployment and much lower incomes. These families often live in crowded housing, and parents often worry about being able to afford food for their children (Campaign 2000, 2014).

Indigenous people in Canada are among the most severely disadvantaged. About one-third live in poverty, and some live in conditions of extreme poverty. The median income for Indigenous persons is just over $29,000 as compared to the national average income of $75,000 (Statistics Canada, 2011d). A study by the Department of Indian Affairs found that the quality of life for Indigenous persons living on reserves ranks worse than in countries such as Mexico and Thailand. Similarly, the United Nations Human Development Index put their living conditions in line with those in Russia. For Indigenous persons living off reserves, the quality of life is slightly better. The unemployment rate for Indigenous persons in Canada ranges from 13 to 25 percent, while the national average is about 6 percent (Aboriginal Affairs and Northern Development Canada, 2013).

In short, the erosion of Canada's social safety net has had a particularly negative impact on those who have historically experienced exclusion and disadvantage in Canadian society.

PERSONS WITH DISABILITIES Awareness that persons with disabilities are discriminated against in the job market has increased in recent years. As a result, they now constitute one of the recognized "target groups" in efforts to eliminate discrimination in the workplace. People with disabilities have more opportunities to work today than they had a decade ago. Today, although more than 50 percent of people with disabilities are in the labour force, many continue to be excluded from the workplace, not because of the disability itself but because of

FIGURE 9.8 POVERTY RATE FOR CHILDREN IN LOW-INCOME FAMILIES, 1989-2011

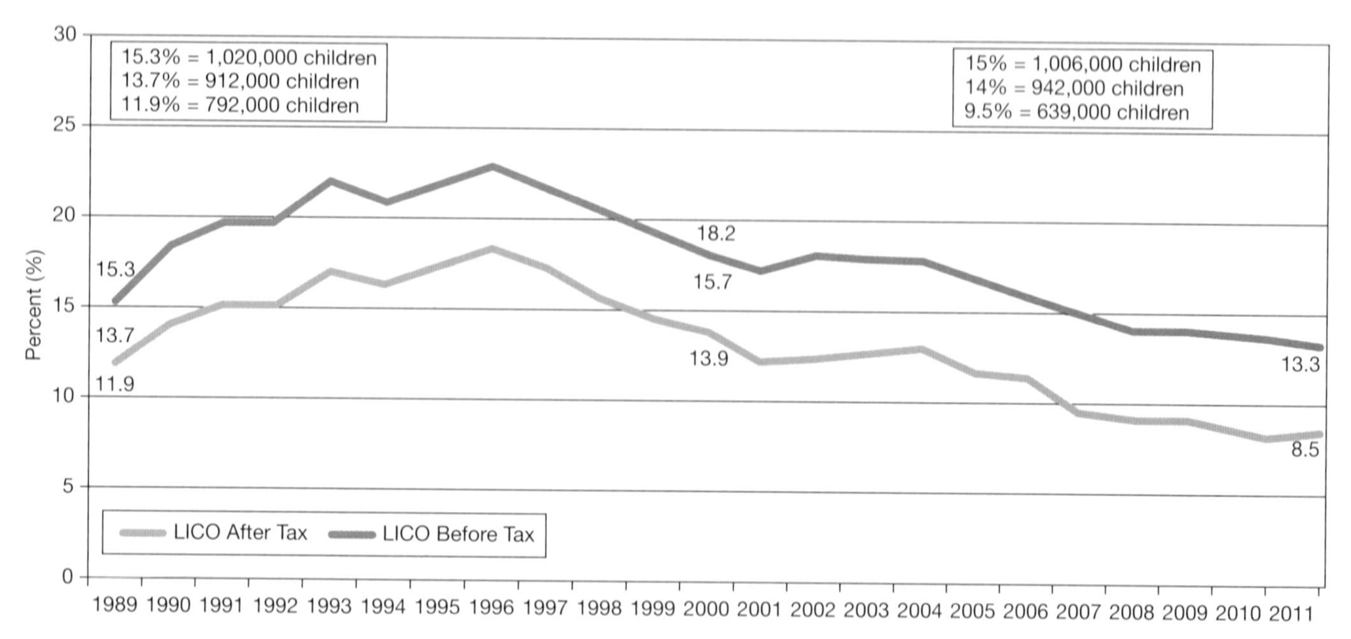

Source: Statistics Canada, Income in Canada, CANSIM Table 202-0802, 2009.

environmental barriers in the workplace (Bonaccio, 2014). The effects of this systemic discrimination continue to be felt by disabled persons, as they are still, as a group, vulnerable to poverty. As discussed in Chapter 15, adults with disabilities have significantly lower incomes than Canadians without disabilities. Recent estimates indicate the self-reported median total income of working-age persons with disabilities was slightly more than $20,000, compared with just over $30,000 for those without disabilities Once again, when gender and disability are combined, we find that women with disabilities are doubly disadvantaged (Galarneau and Radulescu, 2009; Bonaccio, 2014).

ECONOMIC AND STRUCTURAL SOURCES OF POVERTY

Poverty has both economic and structural sources. The low wages paid for many jobs is the major cause: About one in three low-income children have at least one parent who is working full-time but is still in poverty. These parents are not finding jobs with sufficient pay, reasonable amounts of hours of work, and decent benefits that would allow them to move above the poverty line. In 1972, minimum-wage legislation meant that someone who worked 40 hours a week, 52 weeks a year, could earn a yearly income 20 percent over the poverty line. By today's standards, the same worker would have to earn more than $11 per hour simply to reach the poverty line. Minimum wages across Canada range from just over $10 to a high of about $11 per hour. In other words, a person with full-time employment in a minimum-wage job cannot keep a family of four above the official poverty line.

Structural problems contribute to both unemployment and underemployment. The rapid worldwide transformation that is taking place today is changing our societies from industrial-based ones to information-based ones. Automation in the industrial heartland of Quebec and Ontario has made the skills and training of thousands of workers obsolete. Many of these workers have become unemployable and poor. Corporations have been deinvesting in Canada, displacing millions of people from their jobs. Economists refer to this displacement as the *deindustrialization of North America* (Bluestone and Harrison, 1982). Even as they have closed their

job deskilling A reduction in the proficiency needed to perform a specific job, which leads to a corresponding reduction in the wages paid for that job.

Canadian factories and plants, many corporations have opened new facilities in other countries where "cheap labour" exists because people of necessity will work for lower wages. **Job deskilling**—a reduction in the proficiency needed to perform a specific job, which leads to a corresponding reduction in the wages for that job—has resulted from the introduction of computers and other technology. The shift from manufacturing to service occupations has resulted in the loss of higher-paying positions and their replacement with lower-paying and less secure positions that do not offer the wages, job stability, or advancement potential of the disappearing manufacturing jobs. Consequently, there are not enough good jobs available in Canada to enable families to lift themselves out of poverty. In addition, the lack of affordable, high-quality daycare for women who need to earn an income means that many jobs are inaccessible, especially to women who are single parents. The problems of unemployment, underemployment, and poverty-level wages are even greater for members of visible minorities and young people (Canadian Council on Social Development, 2009).

LO-5 SOCIOLOGICAL EXPLANATIONS OF SOCIAL INEQUALITY

Obviously, some people are disadvantaged as a result of social inequality. Is inequality therefore always harmful to a society? Why are all societies stratified?

FUNCTIONALIST PERSPECTIVES

According to sociologists Kingsley Davis and Wilbert Moore (1945), social inequality is not only inevitable but necessary for the smooth functioning of society. The *Davis–Moore thesis,* which has become the definitive functionalist explanation for social inequality, can be summarized as follows:

1. All societies have important tasks that must be accomplished and certain positions that must be filled.
2. Some positions are more important for the survival of society than others.
3. The most important positions must be filled by the most qualified people.
4. The positions that are the most important for society, and that require talent that is scarce and extensive training or both, must be the most highly rewarded.
5. The most highly rewarded positions should be those on which people in other positions rely for expertise, direction, or financing, and that are functionally unique (no one outside the position can perform the same function).

Davis and Moore use the physician as an example of a functionally unique position. Doctors are very important to society and require extensive training, but individuals would not be motivated to go through years of costly and stressful medical training without incentives to do so. The Davis–Moore thesis assumes that social stratification results in **meritocracy**—a hierarchy in which all positions are rewarded based on people's ability and credentials.

meritocracy A hierarchy in which all positions are rewarded based on people's ability and credentials.

A key problem with the Davis–Moore thesis is that it ignores inequalities based on inherited wealth and intergenerational family status. The thesis assumes that economic rewards and prestige are the only effective motivators for people and fails to take into account other intrinsic aspects of work, such as self-fulfillment. It also does not adequately explain how such a reward system guarantees that the most qualified people will gain access to the most highly rewarded positions.

CONFLICT PERSPECTIVES

From a conflict perspective, people with economic and political power are able to shape and distribute the rewards, resources, privileges, and opportunities in society for their own benefit.

Conflict theorists do not believe that inequality serves as a motivating force for people; they argue that powerful individuals and groups use ideology to maintain their favoured positions at the expense of others. A stratified social system is accepted because of the dominant ideology of the society, the set of beliefs that explain and justify the existing social order (Marchak, 1975). Core values in Canada emphasize the importance of material possessions, hard work, and individual initiative to get ahead, as well as behaviour that supports the existing social structure. These same values support the prevailing resource distribution system and contribute to social inequality.

Are wealthy people smarter than others? According to conflict theorists, certain stereotypes suggest that this is the case; however, the wealthy may be "smarter" than others only in the sense of having "chosen" to be born to wealthy parents from whom they could inherit assets. Conflict theorists also note that laws and informal social norms support inequality in Canada. For the first half of the 20th century, both legalized and institutionalized segregation and discrimination reinforced employment discrimination and produced higher levels of economic inequality. Although laws have been passed to make these overt acts of discrimination illegal, many forms of discrimination still exist in educational and employment opportunities.

FEMINIST PERSPECTIVES

According to feminist scholars, the quality of an individual's life experiences is a reflection of both class position and gender. These scholars examine the secondary forms of inequality and oppression occurring *within* each class that have been overlooked by the classical theorists. Feminist theorists focus on the combined effect that gender has on class inequality. Some feminist scholars view class and gender as reinforcing one another and creating groups that are "doubly oppressed." This combined effect of one's class and gender may manifest itself in the workplace or the home or both. Subsequently, feminist authors have identified such terms as the *double ghetto* (Armstrong and Armstrong, 1994) and *second shift* (Hochschild, 1989) to describe women's experiences in the segregated workforce or the home (see Chapters 12 and 6). Rather than male and female spouses maintaining similar class positions within a family unit, women hold a subordinate position. A feminist perspective emphasizes that within any class, women are less advantaged than men in their access to material goods, power, status, and possibilities for self-actualization. For example, upper-class women are wealthy but often remain secondary to their husbands in terms of power. Middle-class women may be financially well off but often lack property or labour force experience and are vulnerable to financial instability in cases of divorce or separation. The position of working-class women varies based on their participation in the paid labour force. Typically, the working-class woman has little income, primary responsibility for the household work, and an inferior position in terms of power and independence to her husband. As a result, the female spouse may become "the slave of a slave" (Mackinnon, 1982:8), allowing the working-class male to compensate for his lower class position in society. The family is viewed as an institution that supports capitalism and encourages or exacerbates the exploitation of women.

SYMBOLIC INTERACTIONIST PERSPECTIVES

Symbolic interactionists focus on microlevel concerns and usually do not analyze larger structural factors that contribute to inequality and poverty. However, many significant insights on the effects of wealth and poverty on people's lives and social interactions can be derived from applying a symbolic interactionist approach. Using qualitative research methods and influenced by a symbolic interactionist approach, researchers have collected the personal narratives of people across all social classes, ranging from the wealthiest to the poorest people.

A few studies provide rare insights into the social interactions between people from vastly divergent class locations. Sociologist Judith Rollins's (1985) study of the relationship between household workers and their employers is one example. Based on in-depth interviews and participant observation, Rollins examined rituals of deference that were often demanded by elite white women of their domestic workers, who were frequently women of colour. According to Erving Goffman (1967), *deference* is a type of ceremonial activity that functions as a symbolic means whereby appreciation is regularly conveyed to a recipient. In fact, deferential behaviour between non-equals (such as employers and employees) confirms the inequality of the relationship and each party's position in the relationship relative to that of the other. Rollins identified three types of linguistic deference between domestic workers and their employers: use of the first names of the workers, contrasted with titles and last names (Mrs. Adams) of the employers; use of the term *girls* to refer to female household workers regardless of their age; and deferential references to employers, such as, "Yes, ma'am." Spatial demeanour, including touching and how close one person stands to another, is an additional factor in deference rituals across class lines. Rollins concludes:

> The employer, in her more powerful position, sets the essential tone of the relationship; and that tone . . . is one that functions to reinforce the inequality of the relationship, to strengthen the employer's belief in the rightness of her advantaged class and racial position, and to provide her with justification for the inegalitarian social system. (1985:232)

Many concepts introduced by the sociologist Erving Goffman (1959, 1967) could be used as springboards for examining microlevel relationships between inequality and people's everyday interactions. What could you learn about class-based inequality in Canada by using a symbolic interactionist approach to examine a setting with which you are familiar?

CONCEPT SNAPSHOT

FUNCTIONALIST PERSPECTIVES	Inequality is necessary for the smooth functioning of society. Social stratification leads to *meritocracy:* a hierarchy in which all positions are rewarded based on ability and credentials.
CONFLICT PERSPECTIVES	Dominant groups maintain and control the distribution of rewards, resources, privileges, and opportunities at the expense of others. Marxist conflict theorists explain that growing inequality is a result of the *surplus value,* profit that is generated when the cost of labour is less than the cost of the goods and services being produced.
FEMINIST PERSPECTIVES	Class and gender reinforce one another and create inequalities and oppressions for women "within" different social classes.
SYMBOLIC INTERACTIONIST PERSPECTIVES	Focus is on the microlevel effects of wealth and poverty on people's social interactions. For example, in its various forms, *deference* confirms inequality between individuals in differing social class positions.

LO-1 Identify three types of stratification systems.

© Jake Norton/Alamy

Slavery is an extreme form of stratification in which some people are owned by others. It is a closed system in which people designated as "slaves" are treated as property and have little or no control over their lives. Like slavery, caste is a closed system of social stratification. A *caste system* is a system of social inequality in which people's status is permanently determined at birth based on their parents' ascribed characteristics. Vestiges of caste systems exist in contemporary India and South Africa. The *class system* is a type of stratification based on the ownership and control of resources and on the type of work people do. At least theoretically, a class system is more open than a caste system because the boundaries between classes are less distinct than the boundaries between castes. In a class system, status comes at least partly through achievement rather than entirely by ascription.

LO-2 Discuss the extent of social inequality in Canada based on measures of income and wealth.

© Shepard Sherbell/CORBIS SABA

Based on measures of income and wealth, there is significant social inequality in Canada. The richest 20 percent of households receive close to 70 percent of the total income "pie," while the poorest receive only 4 percent of all income. Wealth is more unevenly distributed among the Canadian population than is income—the richest 10 percent of Canadian households control close to 60 percent of the country's wealth. The stratification of society into different social groups results in wide discrepancies in income and wealth and in variable access to available goods and services. People with high incomes or wealth have a greater opportunity to control their own lives. People with lower incomes have fewer life chances and must spend their limited resources to acquire basic necessities.

LO-3 Understand the classical analysis of social class by Karl Marx and Max Weber.

Karl Marx and Max Weber acknowledged social class as a key determinant of social inequality and social change. For Marx, people's relationship to the means of production determines their class position. Weber developed a multidimensional concept of stratification that focuses on the interplay of wealth, prestige, and power.

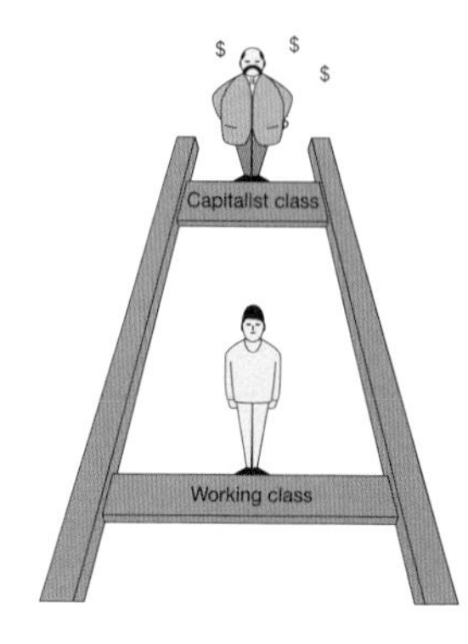

KEY TERMS

absolute poverty A level of economic deprivation in which people do not have the means to secure the basic necessities of life (p. 247).

alienation A feeling of powerlessness from other people and from oneself (p. 236).

capitalist class (bourgeoisie) Karl Marx's term for those who own the means of production (p. 236).

caste system A system of social inequality in which people's status is permanently determined at birth based on their parents' ascribed characteristics (p. 231).

class conflict Karl Marx's term for the struggle between the capitalist class and the working class (p. 237).

class system A type of stratification based on the ownership and control of resources and on the type of work people do (p. 232).

feminization of poverty The trend in which women are disproportionately represented among individuals living in poverty (p. 249).

income The economic gain derived from wages, salaries, income transfers (governmental aid), and ownership of property (p. 233).

intergenerational mobility The social movement (upward or downward) experienced by family members from one generation to the next (p. 230).

intragenerational mobility The social movement (upward or downward) experienced by individuals within their own lifetime (p. 230).

job deskilling A reduction in the proficiency needed to perform a specific job, which leads to a corresponding reduction in the wages paid for that job (p. 252).

life chances Max Weber's term for the extent to which individuals have access to important societal resources, such as food, clothing, shelter, education, and healthcare (p. 230).

low-income cutoff The income level at which a family may be in "straitened circumstances" because it spends considerably more on the basic necessities of life (food, shelter, and clothing) than does the average family (p. 246).

meritocracy A hierarchy in which all positions are rewarded based on people's ability and credentials (p. 252).

pink-collar occupation Relatively low-paying, nonmanual, semiskilled positions primarily held by women (p. 240).

power According to Max Weber, the ability of people or groups to achieve their goals despite opposition from others (p. 238).

prestige The respect or regard with which a person or status position is regarded by others (p. 238).

relative poverty A level of economic deprivation in which people may be able to afford basic necessities but still are unable to maintain an average standard of living (p. 247).

slavery An extreme form of stratification in which some people are owned by others (p. 230).

social mobility The movement of individuals or groups from one level in a stratification system to another (p. 230).

social stratification The hierarchical arrangement of large social groups based on their control over basic resources (p. 230).

LO-4 Provide an overview of poverty and its effects in Canada.

© David Bacon/ The Image Works

Sociologists distinguish between absolute and relative poverty. Absolute poverty exists when people do not have the means to secure the basic necessities of life. Relative poverty exists when people may be able to afford basic necessities but still are unable to maintain an average standard of living. Age, gender, race and ethnicity, and disability tend to be factors in poverty. Children have a greater risk of being poor than do the elderly, while women have a higher rate of poverty than do men. Although white persons account for approximately two-thirds of those below the poverty line, Indigenous peoples and members of visible minorities account for a disproportionate share of the impoverished in Canada. As the gap between rich and poor and between employed and unemployed widens, social inequality will clearly increase in the future.

LO-5 Compare and contrast a functionalist view and a conflict perspective on social inequality.

© Katie Deits/Index Stock Imagery

According to the Davis–Moore thesis, stratification exists in all societies and some inequality is not only inevitable but also necessary for the ongoing functioning of society. The positions that are most important within society and that require the most talent and training must be highly rewarded. Conflict perspectives on inequality are based on the assumption that social stratification is created and maintained by one group to enhance and protect its own economic interests. Conflict theorists measure inequality according to people's relationships with others in the production process.

APPLICATION QUESTIONS

1. Based on the functionalist model of class structure, what is the class location of each of your 10 closest friends or acquaintances? What is their location in relation to yours? What is their location in relation to one another? What does their location tell you about friendship and social class?
2. Should employment be based on merit, need, or affirmative action policies? Discuss.
3. If the gap between rich and poor people continues to widen, what might happen in Canada in the future?

socioeconomic status (SES) A combined measure that attempts to classify individuals, families, or households in terms of indicators, such as income, occupation, and education, to determine class location (p. 238).

wealth The value of all of a person's or family's economic assets, including income and property, such as buildings, land, farms, houses, factories, and cars, as well as other assets, such as money in bank accounts, corporate stocks, bonds, and insurance policies (p. 234).

working class (proleteriat) Karl Marx's term for those who must sell their labour in order to earn enough money to survive (p. 236).

CHAPTER 10

GLOBAL STRATIFICATION

LEARNING OBJECTIVES

AFTER READING THIS CHAPTER, YOU SHOULD BE ABLE TO

- **LO-1** Understand the concept of global stratification and its impact on the lives of billions of people.
- **LO-2** Understand the relationship between global poverty and human development.
- **LO-3** Discuss Rostow's modernization theory and explain the stages that modernization theorists believe all societies must go through.
- **LO-4** Describe how dependency theory differs from modernization theory.
- **LO-5** Understand how world-systems analysis views the global economy.
- **LO-6** Understand the international division of labour theory.

CHAPTER FOCUS QUESTION

What is global stratification, and how does it contribute to economic inequality?

You have learned about the huge differences between rich and poor Canadians. However, the differences between the standard of living of the average Canadian and the world's poorest citizens are far greater.

William Kamkwamba was born in Malawi—one of the world's poorest countries. He loved learning, and going to school was one of his greatest joys. He constantly tinkered with things and, as a child, he taught himself to repair small portable radios. In 2002, the worst famine in 50 years killed hundreds of Malawians. The crop failure made it impossible for William's family to pay school fees, so he was forced to drop out of school and learn from the books in a small library in his village.

Kamkwamba's village did not even have electricity. In fact, only 2 percent of Malawi's population had electricity in their homes. William read about windmills in a library book, and at age 14 he decided to build a windmill that could provide power to light homes and perhaps even improve crops by pumping water for irrigation. Relying on library books and scavenging junk around his community, William was first able to build a crude working model and then a makeshift full-scale windmill. He describes the moment when he first tried the windmill:

Balancing the small reed and wires in my left hand, I used the other to pull myself onto the tower's first rung. The soft wood groaned under my weight, and the compound fell silent. I continued to climb, slowly and assuredly, until I was facing the machine's crude frame. Its plastic arms were burned and blackened, its metal bones bolted and welded into place. I paused and studied the flecks of rust and paint.... Each piece told its own tale of discovery, of being lost and found in a time of hardship and fear....

Reaching over, I removed a bent piece of wire that locked the machine's spinning wheel in place. Once released, the wheel and arms began to turn. They spun slowly at first, then faster and faster, until the force of their motion rocked the tower. My knees buckled, but I held on....

I gripped the reed and wires and waited for the miracle. Finally, it came, at first a tiny light that flickered from my palm, then a surging magnificent glow. The crowd gasped and shuddered. The children pushed for a better look.

"It's true!" someone said.

"Yes," said another. "The boy has done it." (Kamkwamba and Mealer, 2009)

Source: Kamkwamba and Mealer, 2009.

While William's homemade windmill was never powerful enough to bring prosperity to his village, it did generate major changes. A senior government official heard about the young boy's windmill and it turned into a major story in Malawi. This publicity led to William being asked to do a TED talk, which led to William receiving a free education in Malawi. He eventually graduated from Dartmouth College, an elite American university. The TED talk also led to a fundraising initiative that enabled William to start the Moving Windmills Project, which has been bringing power to villages in his district; to begin a one-laptop-per-child project; and to start several other community improvement programs. His book, *The Boy Who Harnessed the Wind*, highlights the incredible distance between life in a poor African village and the lives most of us lead in Canada, and it shows how simple changes such as a windmill can have a profound effect on the world's poorest people.

You can see William's TED talks at

http://www.ted.com/talks/william_kamkwamba_on_building_a_windmill

http://www.ted.com/talks/william_kamkwamba_how_i_harnessed_the_wind.

In this chapter, we examine global stratification. It can be difficult to connect the lives of people living in poverty in distant countries to our lives in one of the world's wealthiest countries. When television shows us thousands of people starving in Somalia or babies being treated for dehydration at aid stations in Darfur, it is sometimes hard to realize that the people we see are individual human beings just like us. The global stratification system determines who will live long, prosperous lives like those of most Canadians, and who will live short, miserable lives, eking out a marginal existence in subsistence agriculture and subject to the vagaries of rain, floods, and political instability like the people in William's village in Malawi.

While statistics may seem boring, they can hold a lot of meaning. Take the time to consider what each of the following points means for the world's poorest people (Oxfam, 2014:2):

- "Almost half of the world's wealth is now owned by just one percent of the population.
- The wealth of the one percent richest people in the world amounts to $100 trillion. That's 65 times the total wealth of the bottom half of the world's population.
- The bottom half of the world's population owns the same [amount of wealth] as the richest 85 people in the world.
- Seven out of ten people live in countries where economic inequality has increased in the last 30 years.
- In the United States, the wealthiest one percent captured 95 percent of post-financial crisis growth since 2009, while the bottom 90 percent became poorer. "

Source: The material on pages 2-3, from *Working for the Few: Political Capture and Economic Inequality*, 2014, is reproduced with the permission of Oxfam GB, Oxfam House, John Smith Drive, Cowley, Oxford OX4 2JY, UK www.oxfam.org.uk. Oxfam GB does not necessarily endorse any text or activities that accompany the materials.

The world has sufficient resources to end poverty, but the wealth is not shared. Why do these inequities persist? In this chapter, you will learn why the life prospects of billions of people remain so dismal. When you read the explanations of global stratification, you should remember that these are not just academic theories. Rather, these are the ideas shaping the way governments and international organizations, including the United Nations, deal with global poverty. As you will see, decisions based on the wrong theories can be disastrous for the poor.

Before reading on, test your knowledge of global wealth and poverty by taking the quiz in Box 10.1.

CRITICAL THINKING QUESTIONS

1. What role do you think higher-income countries play in the continuing poverty of people in lower-income countries?

2. When global corporations such as Apple move their production from North America to countries where wages are low and government regulation is limited, do you think they make life better or worse for workers living in those countries?

3. Educating women is such an important strategy for improving the lives of people living in poor countries that the UN recognized it as a Millennium Development Goal. Why is women's education is so important?

LO-1 WEALTH AND POVERTY IN GLOBAL PERSPECTIVE

Global stratification refers to the unequal distribution of wealth, power, and prestige on a global basis, resulting in people having vastly different lifestyles and life chances. Just as Canada can be divided into classes, the world can be divided into unequal segments characterized by extreme differences in wealth and poverty. *High-income countries* have highly industrialized economies; technologically advanced industrial, administrative, and service occupations; and relatively high levels of income. In contrast, *low-income countries* are undergoing the transformation from agrarian to industrial economies and have lower levels of income.

Where you are born has a huge influence on your life chances. A World Bank report (2006) contrasts the prospects of two children: Nthabiseng, born in a rural area of South Africa, and Sven, born in Sweden. The chance of Sven dying in his first year of life is only 0.3 percent, compared with the 7.2 percent risk faced by Nthabiseng. Sven will likely complete 11.4 years of schooling, while Nthabiseng will have less than one year of formal education. Sven will have access to clean water, good housing, proper food, and excellent medical care, while Nthabiseng may have none of these advantages. As a result, Sven will likely live to the age of 80, while Nthabiseng will likely die before she is 50.

© vietnam/Alamy

Many of the things we buy are produced in low-income countries. Workers in these countries are often poorly paid and work in harsh conditions.

BOX 10.1 SOCIOLOGY AND EVERYDAY LIFE

How Much Do You Know About Global Wealth and Poverty?

True	False	
T	F	1. Because of foreign aid and the globalization of trade, the gap between the incomes of people in the poorest countries and the richest countries has narrowed over the past several decades.
T	F	2. The percentage of people living in extreme poverty in the world has declined since 1990.
T	F	3. The richest one-fifth of the world's population receives about 75 percent of the world's total income.
T	F	4. The political role of governments in policing the activities of transnational corporations has expanded as companies' operations have become more globalized.
T	F	5. In low-income countries, poverty affects women more than men.

For answers to the quiz about global wealth and poverty, go to **www.nelson.com/student.**

CONSUMPTION AND POVERTY

Progress in reducing world poverty has been slow, despite a great deal of talk and billions of dollars in foreign aid flowing from high-income to low-income nations. The notion of "development" has become the primary means of attempting to alleviate global poverty. Often, the nations that have been unable to reduce poverty are blamed for not establishing the necessary social and economic reforms to make change possible. However, some analysts have suggested that the problem of inequality lies not in poverty but in excess:

> "The problem of the world's poor," defined more accurately, turns out to be "the problem of the world's rich." This means that the solution to the problem is not a massive change in the culture of poverty so as to place it on the path of development, but a massive change in the culture of superfluity in order to place it on the path of counter-development. It does not call for a new value system forcing the world's majority to feel shame at their traditionally moderate consumption habits, but for a new value system forcing the world's rich to see the shame and vulgarity of their over-consumption habits, and the double vulgarity of standing on other people's shoulders to achieve those consumption habits. (Lummis, 1992:50)

The United Nations *Human Development Report* (1998) showed how excessive consumption in rich countries threatens the environment, depletes natural resources, and wastes money that might otherwise provide for the needs of the desperately poor in low-income countries. The wealthiest 20 percent of the world's people account for 86 percent of private consumption, while the poorest 20 percent account for only 1.3 percent. See Figure 10.1.

What is overconsumption? Canadians spend about $46 billion a year on clothing (Statistics Canada, 2015a). Much of this is spent keeping up with changes in style. The young Winnipeg woman interviewed for the *Winnipeg Free Press* style section who said her fashion essential was "Shoes, mostly heels. I buy, like, two pairs a week" (2009:F9) was not just trying to protect her feet from the elements. She was motivated by other concerns. Spending even a small portion of our money on disease prevention or education in low-income countries rather than on the latest style of shoes could keep hundreds of thousands of people alive in low-income countries.

But aren't the world's needs so great that channelling some of the money we spend on excessive consumption would make little difference in the lives of the world's poor? Surprisingly, the amount of money required to meet some basic human needs is not that great.

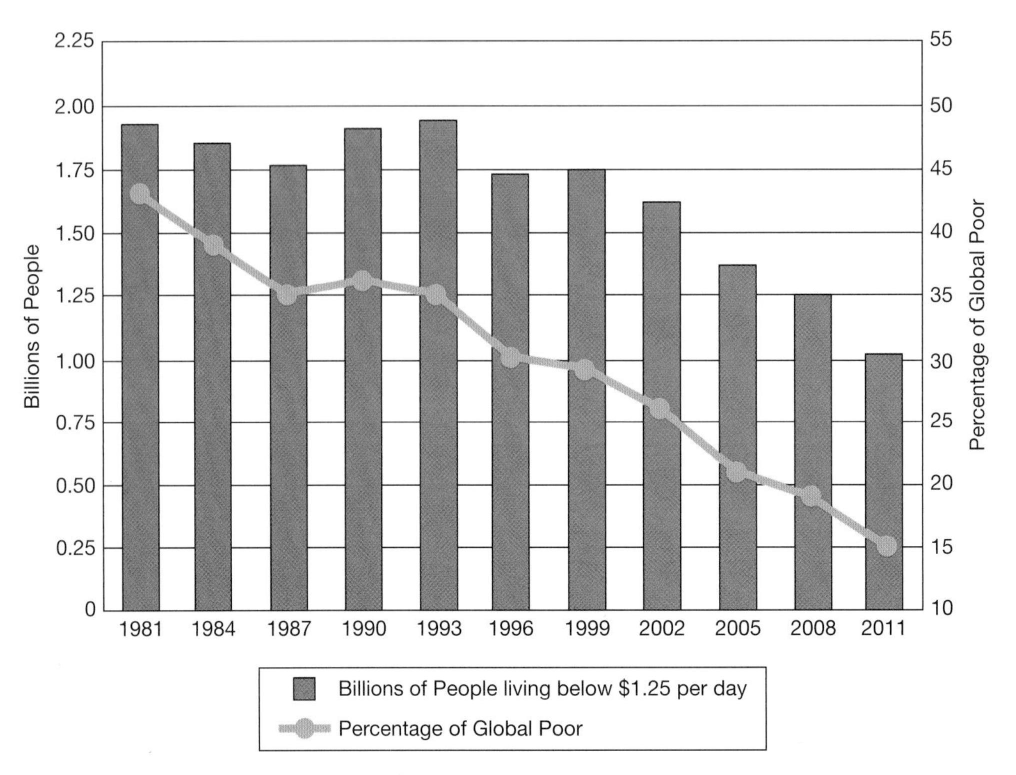

Source: World Bank, Report Card, 2015.

Vast inequalities in income and lifestyle are evident in this photo of slums and nearby higher-priced housing in Mumbai, India. Similar patterns of economic inequality exist in many other cities.

The Hindu, an Indian newspaper, put these needs in perspective. While the specific numbers are from two decades ago, the same message applies today:

> Consumers in Europe and the U.S. spend $17 billion every year on pet foods. Yet the world cannot find the additional $13 billion that is needed every year to provide basic health services to all people in developing countries. Consumers in Europe and the U.S. annually spend $12 billion on perfumes. This is the additional amount needed to meet the basic reproductive health needs of the women in developing countries. Consumers in Europe spend $11 billion every year buying ice cream, which is more than the extra $9 billion required to provide universal access to drinking water and sanitation in the developing countries. (*The Hindu*, 1998:25)

Efforts to alleviate global poverty have been succeeding and the number of people living in extreme poverty has declined significantly. However, progress has been uneven, as some countries lag behind and the disparity between rich and poor has been increasing. Also, the number of people living below the $2 a day poverty line has not decreased nearly as much (World Bank, 2015a).

DEFINING GLOBAL INEQUALITY

There has been a debate over the best way to define global inequality, since these definitions imply judgments about countries, particularly those with low incomes.

THE LEVELS OF DEVELOPMENT APPROACH

There are several ways of describing global inequality. Let's look first at the contemporary origins of the ideas of underdevelopment and underdeveloped nations.

Following World War II, the term *underdevelopment* emerged out of the Marshall Plan, which provided massive sums of money to rebuild the European economic base destroyed during World War II. After the Marshall Plan's success, U.S. leaders decided that Southern Hemisphere nations that had recently been released from European colonialism could also benefit from a massive financial infusion and rapid economic development. Leaders of developed nations argued that problems such as poverty, disease, and famine could be reduced through the transfer of finance, technology, and experience from the developed nations to less developed countries. From this perspective, economic development is the primary way to reduce the problems faced by underdeveloped countries. Hadn't economic growth brought the developed nations to their own high standard of living? Moreover, "self-sustained development" in a nation would require that people in the less developed nations accept the beliefs and values of people in the developed nations, so the development movement had an explicitly political component.

In his 1949 inaugural address, U.S. President Truman stated his view that the nations in the Southern Hemisphere were "underdeveloped areas" because of their low **gross national income (GNI)**—a term that refers to all the goods and services produced in a country in a given year, plus the income earned outside the country by individuals or corporations. If nations could increase their GNI, then social and economic inequality within the country could also be reduced. Accordingly, Truman believed that it was necessary to help the people of economically underdeveloped areas raise their *standard of living*.

▶ **gross national income (GNI)** All the goods and services produced in a country in a given year, plus the income earned outside the country by individuals or corporations.

After several decades of economic development fostered by organizations such as the United Nations and the World Bank, it became apparent by the 1970s that improving a country's GNI did *not* necessarily reduce the poverty of the poorest people in that country. In fact, global poverty and inequality were increasing, and the initial optimism of a speedy end to underdevelopment faded. Even in developing countries that had achieved economic growth, the gains were not shared by everyone. For example, the poorest 20 percent of the Brazilian population received less than 3 percent of the total national income, whereas the richest 20 percent

© Alison Wright/The Image Works

Based on the assumption that economic development is the primary way to reduce poverty in low-income nations, the United Nations has funded projects such as this paper company in Nepal.

of the population received more than 63 percent (World Bank, 2003). Poverty has been reduced in Brazil over the last decade as its economy has improved, but it still has a very high degree of inequality.

Why did inequality increase even with greater economic development? Many attribute this to the impact of actions taken by the industrialized countries. These actions include foreign aid programs and debt-control policies, both of which will be discussed later in this chapter. Some analysts have also linked the growing global inequality to relatively high rates of population growth in the poorest nations.

CLASSIFICATION OF ECONOMIES BY INCOME

An alternative way of describing the global stratification system is simply to measure a country's per capita income. The World Bank (2015a) classifies nations into four economic categories: **low-income economies** (a GNI per capita of $US1045 or less), **lower-middle-income economies** (a GNI per capita between $US1046 and $US4125), **upper-middle-income economies** (a GNI per capita between $US4126 and $US12,746), and **high-income economies** (a GNI per capita of $US12,747 or more).

▶ **low-income economies** Countries with an annual per capita gross national income of $US1045 or less.

▶ **lower-middle-income economies** Countries with an annual per capita gross national income between $US1046 and $US4125.

▶ **upper-middle-income economies** Countries with an annual per capita gross national income between $US4126 and $US12,736.

▶ **high-income economies** Countries with an annual per capita gross national income of $US12,737 or more.

LOW-INCOME ECONOMIES In the world's 34 low-income economies, most people engage in agricultural pursuits, reside in nonurban areas, and are impoverished (World Bank, 2015). As shown on Map 10.1, low-income economies are found primarily in countries in Asia and Africa. Included are such nations as Rwanda, Ethiopia, Cambodia, Afghanistan, and Bangladesh.

Women, children, and Indigenous people are particularly affected by poverty in low-income economies. Many women worldwide are unable to increase their economic power because they do not have the time to invest in additional work that could bring in more income. Also, many have no access to commercial credit and have been trained only in traditionally female skills that are unpaid or produce low wages. These factors have contributed to the *global feminization of poverty,* whereby women tend to be more impoverished than men (Durning, 1993). Despite some gains, the income gap between men and women continues to grow wider in many low-income developing nations. Indigenous people, who comprise 5 percent of the world's population, make up 15 percent of the world's poor and 30 percent of the world's extremely poor (United Nations Human Development Programme, 2014).

MIDDLE-INCOME ECONOMIES One hundred and five nations are classified as middle-income economies (World Bank, 2015). The World Bank divides middle-income economies into lower-middle-income and upper-middle-income. Countries classified as lower-middle-income include the Latin American nations of Bolivia, El Salvador, and Honduras. Even though these countries are referred to as "middle-income," more than half of the people residing in many of them live in poverty, defined as $US1.25 per day in purchasing power (World Bank, 2015).

Compared with lower-middle-income economies, nations with upper-middle-income economies have a somewhat higher standard of living and export diverse goods and services, ranging from manufactured goods to raw materials and fuels. Upper-middle-income economies include Malaysia, Brazil, Hungary, and Mexico.

© Jochen Tack/Glow Images

In developing nations, such as Turkey, many families support themselves by creating handmade products, such as rugs. The loom in this home is a common sight throughout Turkey.

HIGH-INCOME ECONOMIES High-income economies are found in 75 nations, including Canada, the United States, Japan, and Germany. High-income economies dominate the world economy.

MAP 10.1 HIGH-, MIDDLE-, AND LOW-INCOME ECONOMIES IN GLOBAL PERSPECTIVE

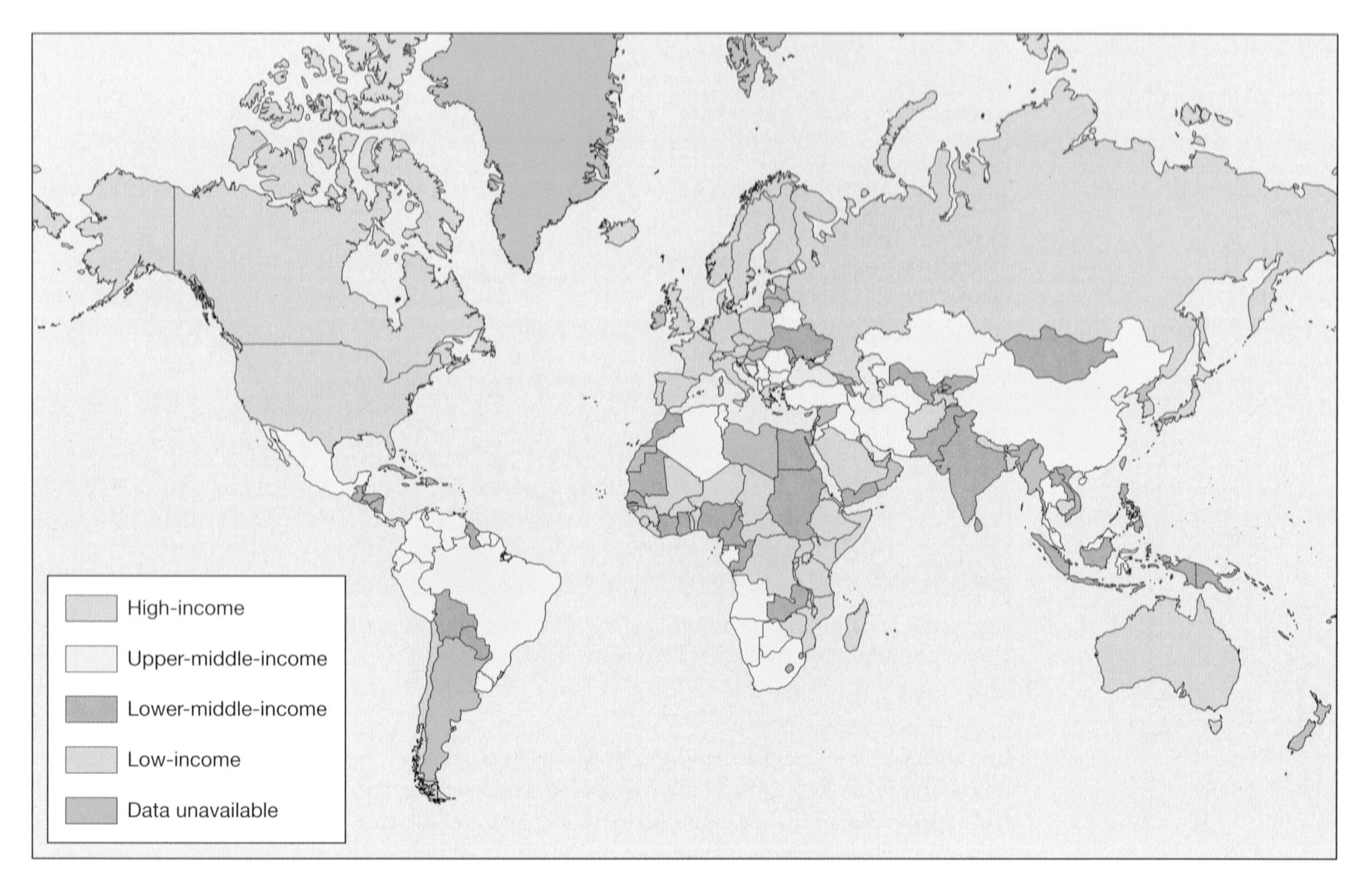

Source: Data from World Bank, 2015. Map from http://chartsbin.com/view/25857

The only significant group of middle- and lower-middle-income economies to close the gap with the high-income industrialized economies over the past few decades has been the nations of East Asia. South Korea has been reclassified from a middle-income to a high-income economy, and several other countries, including China, have grown dramatically over the past two decades.

THE IMPACT OF DEBT AND FOREIGN AID

Why do low-income countries remain poor despite all the efforts that have been made to improve their economic conditions? Some of this is due to the policies of high-income countries.

DEBT AND GLOBAL STRATIFICATION

Massive debt has made it virtually impossible for some countries to move out of poverty. Private banks, governments, and international organizations have lent more money to poor countries than these countries can afford to pay back. Much of the borrowed money was spent on military hardware and other non-productive investments rather than on building the productive capacity that would have allowed the poor countries to develop. Many countries were forced by the International Monetary Fund (IMF) and the World Bank to restructure their economies

arindambanerjee/Shutterstock.com

Many people protest economic summits involving the world's richest countries because they feel these countries promote the interests of capitalism at the expense of the world's poor.

by cutting back on social spending, devaluing their currencies, and reducing the funds spent on economic development. In many respects, these governments lost whatever power they once had to control their own economic destinies because they were forced to follow the dictates of the lenders. This happened to most Latin American economies during the 1980s, to many East Asian and Eastern European economies during the late 1990s, and more recently to European countries, including Greece, which was in a debt crisis in 2014 and 2015.

In many countries, this externally imposed structural adjustment had disastrous consequences. Debt repayment takes money that could otherwise be used to provide social services and expand the country's economic base. Debt repayment and economic restructuring have caused massive unemployment, reduced incomes, and soaring prices that have led to drastically reduced living standards, declines in investment, and political instability. Michel Chossudovsky described the problem:

> The movement of the global economy is "regulated" by "a worldwide process of debt collection" which constricts the institutions of the national state and contributes to destroying employment and economic activity . . . Internal purchasing power has collapsed, famines have erupted, health clinics and schools have been closed down, hundreds of millions of children have been denied the right to a primary education. In several regions of the developing world, the reforms have been conducive to resurgence of infectious diseases including tuberculosis, malaria, and cholera. (1997:33)

Many countries, particularly those in sub-Saharan Africa, have no chance of progressing economically unless the burden of debt repayment is eased by the richer debtholder countries.

In 2005, the wealthy countries agreed to cancel $40 billion in debt—money that would otherwise have gone to interest payments—to allow some of the poor countries to spend on education, agriculture, healthcare, and infrastructure. By 2014, this debt relief had increased to $120 billion and recent research suggests that countries that changed their policies in order to qualify for relief improved conditions within their countries and were developing economically better than countries that had not received relief (Thomas and Giugale, 2014).

FOREIGN AID AND GLOBAL STRATIFICATION

We have all been moved by the plight of starving people in developing countries where droughts or floods have destroyed the annual harvest. Most of us support the emergency exports of food to these countries to prevent famine. However, some analysts have suggested food aid may hurt more than it helps. Consider Somalia, now one of the world's poorest and most politically unstable countries. While Somalia's troubles are commonly blamed on drought and clan rivalries, some analysts feel that economic restructuring and food aid are the real causes (Chossudovsky, 1997). Because of droughts and other internal problems, food aid to Somalia increased dramatically from the mid-1970s to the mid-1980s. This donated food was sold into local markets very cheaply and undercut the price of locally grown food. At the same time, a currency devaluation demanded by the IMF as a condition for restructuring Somalia's foreign debt made the cost of farm equipment and fuel more expensive. The combined result of the financial restructuring and the lower food prices was the virtual destruction of Somalia's agricultural system. While we normally think that a shortage of food is the cause of starvation, the global oversupply of grain may have contributed to famine by destroying the agricultural base of developing countries, making them vulnerable to future food shortages (Chossudovsky, 1997).

Foreign aid can damage low-income countries in other ways. First, aid can be tied to specific projects or objectives that may meet the interests of the donor country more than the interests of the recipients. For example, military aid will do little to help the lives of those who are poor and may do them great harm. Similarly, aid devoted to large infrastructure projects, such as dams, may cause more problems than it solves.

Second, aid may be given to achieve political objectives. The United States provides aid to many countries of strategic interest, including several Latin American countries. Such aid is dependent on the low-income country's continuing political support for the donor country's activities, something that can severely constrain the government's power to make decisions based on its own interests.

Third, even when aid is targeted to individuals, it may not filter down to people who are poor. For example, donor countries may intend food aid to be distributed to those who need it without payment, but elites may simply take the food, sell it, and keep the money.

Finally, there is not enough aid to help low-income countries solve their problems. Globally, foreign aid has remained at a relatively constant level over the past decade and Canada's contributions are very low for a wealthy country. Despite Canada's target of donating 0.7 percent of its gross domestic product (GDP) to foreign aid, the amount donated in 2013 was only 0.3 percent of GDP, and a significant portion of funds allocated to foreign aid were not spent. And not only is there not enough aid, but the aid that is provided is often not spent effectively.

There are signs of change, however. Many aid agencies have changed their focus away from large infrastructure programs, such as dams and railroads, to programs that focus directly on the poor. New ideas include using labour-intensive technologies and other strategies to create employment; providing basic social services, such as healthcare, nutrition, and education; and giving assistance directly to the poorest people in low-income countries (Martinussen, 1997). Above all, aid recipients themselves must have a say in aid programs and should be empowered to make decisions about how the money is spent. Otherwise, foreign aid can be a double-edged sword that creates more problems than it solves.

In addition, high-income countries need to change trade policies that are often devastating to the economies of poor countries and can negate the benefits of the aid provided. For example, in the mid-1990s, Canada gave an average of $44 million per year in aid to Bangladesh.

A young Hong Kong woman wearing a dust mask sews in a garment sweatshop under poor conditions.

However, Canadian quotas (now removed) that restricted the export of textiles from Bangladesh cost the Bangladeshi people $36 for every dollar in aid provided (Oxfam, 2001). These quotas were designed to protect Canadian manufacturers from foreign competition, but they also prevented developing nations from building a viable economic base.

Developed countries often force poor countries to liberalize their own trade laws, and then sell heavily subsidized products into the poor countries' internal markets. The United States devastated Haiti's rice farmers by forcing Haiti to reduce its tariffs on imported rice and selling Haiti large quantities of subsidized American-grown rice. This action depressed local prices and resulted in a 40 percent reduction in Haitian rice production (Oxfam, 2001). More recently, U.S. subsidies to its own farmers have seriously harmed Mexican corn farmers. By 2012, wealthier countries were spending $456 billion to subsidize their farmers, making it impossible for farmers in poor countries to compete, particularly in the export market (Worldwatch Institute, 2014). This far surpassed the amount of aid given by wealthy countries to poor countries. Such practices need to be changed to achieve the goal of alleviating global poverty. However, it has been difficult to convince wealthy countries to lower their agricultural subsidies. Figure 10.2 illustrates the size of agricultural subsidies in the European Union, Japan, and the United States compared to the average income in sub-Saharan Africa and the aid given these countries.

FIGURE 10.2 : COWS AND COTTON RECEIVE MORE AID THAN PEOPLE, 2000

To protect their agricultural industries, high-income countries provide enormous subsidies to farmers and agricultural corporations. In the United States, the subsidy provided to cotton growers is greater than the cash value of the entire cotton crop, amounting to billions of dollars each year. The amount provided in subsidies to farmers in high-income countries far exceeds the foreign aid donated to the low-income countries. These subsidies force down global prices with devastating effects on countries that are heavily dependent upon the export of commodities.

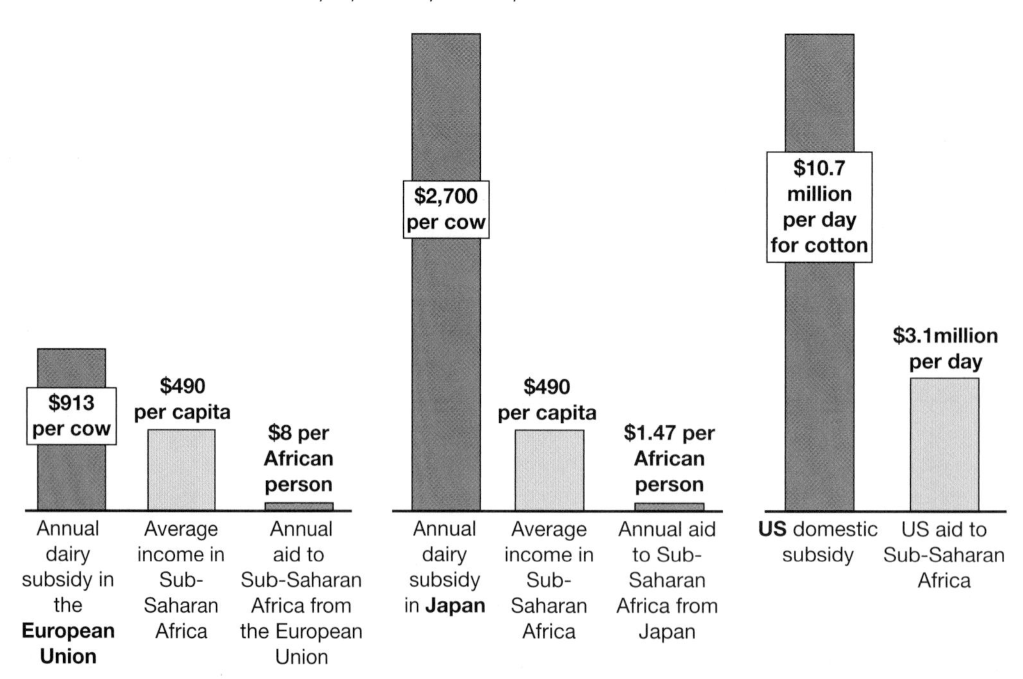

Source: *Human Development Report 2003*, United Nations Development Programme, p. 155.

MEASURING GLOBAL WEALTH AND POVERTY

How is poverty defined? Researchers and policy analysts have developed several ways of measuring global wealth and poverty.

ABSOLUTE, RELATIVE, AND SUBJECTIVE POVERTY

Defining poverty involves more than comparing personal or household income; it also involves social judgments made by researchers. *Absolute poverty*—a condition in which people do not have the means to secure the most basic necessities of life—would be measured by comparing personal or household income or expenses with the cost of buying a given quantity of goods and services. The World Bank (2008c) has defined absolute poverty as living on less than $1.25 a day, and this definition is commonly used by other international organizations. *Relative poverty*—when people may be able to afford basic necessities but are still unable to maintain an average standard of living—is measured by comparing one person's income with the income of others. Finally, *subjective poverty* can be measured by comparing the actual income against the income earner's expectations and perceptions. However, for low-income nations, data on income and levels of consumption are difficult to obtain (World Bank, 2003).

TIME TO REVIEW

- Why does the country you are born in play such an important role in determining your life chances?
- How does excessive consumption in high-income countries affect people living in poverty in low-income countries?
- How have experts conceptualized world poverty and global stratification?
- Discuss the role of debt in determining the economic future of low-income nations.
- Why does foreign aid not always work to the benefit of people in low-income countries?

LO-2 GLOBAL POVERTY AND HUMAN DEVELOPMENT ISSUES

While income disparity is a common measure of well-being, income is not the only factor that defines the impact of poverty. Work by prominent economists, including Nobel Prize winner Amartya Sen, has led to the use of a measure of human welfare as an indicator of development.

In 1990, the United Nations Development Programme introduced the Human Development Index (HDI), establishing three new criteria—in addition to GNI—for measuring the level of development in a country: life expectancy, education, and living standards (see Figure 10.3).

LIFE EXPECTANCY

Although some advances have been made in increasing life expectancy in middle- and low-income countries, major problems still exist. On the positive side, average life expectancy has increased steadily over the past 50 years. However, there are still major income disparities, as the average life expectancy at birth in the countries ranked highest on human development is about 80 years, compared with 59 years in the countries ranked low in human development (United Nations Development Programme, 2013). Especially striking are the declines in life expectancies in sub-Saharan Africa, where estimated life expectancy has dropped significantly in many countries, largely because of HIV/AIDS. While the Safe Motherhood Initiative has had some success, nearly 300,000 women still die each year in pregnancy and childbirth (World Health Organization, 2014).

FIGURE 10.3 : HUMAN DEVELOPMENT INDEX RANKINGS

Human Development Index 2014 Rankings	
Norway	1
Australia	2
Switzerland	3
The Netherlands	4
United States	5
Germany	6
New Zealand	7
Canada	8
Sweden	12
United Kingdom	14
Japan	17
Nigeria	152
Haiti	186
Sierra Leone	183
Chad	184

Source: *Human Development Index and Its Components, Human Development Report 2014*, United Nations Development Programme. http://hdr.undp.org/en/content/table-1-human-development-index-and-its-components

One major cause of shorter life expectancy in low-income nations is the high rate of infant mortality. Comparison of Map 10.1 (in the section "Classification of Economies by Income") with Map 10.2 shows that low-income countries are also those with high rates of infant mortality. The infant mortality rate is more than eight times higher in low-income countries than in high-income countries (World Bank, 2003). In many countries infant mortality is higher among girls than among boys (Box 10.2). Low-income countries typically have higher rates of illness and disease, and they lack adequate healthcare facilities. Malnutrition is a common problem among children, many of whom are underweight, stunted, and have anemia—a nutritional deficiency with serious consequences for child mortality. Consider this journalist's description of a child she saw in Haiti:

> Like any baby, Wisly Dorvil is easy to love. Unlike others, this 13-month-old is hard to hold.
>
> That's because his 10-pound frame is so fragile that even the most minimal of movements can dislocate his shoulders.
>
> As lifeless as a rag doll, Dorvil is starving. He has large, brown eyes and a feeble smile, but a stomach so tender that he suffers from ongoing bouts of vomiting and diarrhea. Fortunately, though, Dorvil recently came to the attention of U.S. aid workers. With round-the-clock feeding, he is expected to survive.
>
> Others are not so lucky. (Emling, 1997:A17)

Over 800 million people suffer from chronic malnutrition, and hunger is a major cause of death (UN Food and Agriculture Organization, 2014). To put this figure in perspective, the number of people worldwide dying from hunger-related causes is the equivalent of more than 60 jumbo-jet crashes a day with no survivors, and half the passengers are children. However, some progress has been made as the number of chronically undernourished people has dropped by 100 million over the past decade.

HEALTH

Health is defined by the World Health Organization as "a state of complete physical, mental and social well-being and not merely the absence of disease or infirmity" (Smyke, 1991:2).

Many people in low-income nations do not have physical, mental, and social well-being—2.5 billion people do not have proper sanitation and one billion do not have safe water. Many do not have adequate housing or access to modern health services (World Bank, 2010). Millions of people die each year from HIV/AIDS, diarrhea, malaria, tuberculosis, and other infectious and parasitic illnesses (World Health Organization, 2012a). Infectious diseases persist in countries with unsanitary and overcrowded living conditions and a lack of basic healthcare. Disease, in turn, can have an impact on economic conditions. For example, the economy in Swaziland stopped growing and began to contract because the AIDS epidemic decimated the workforce and reduced productivity (Nolen, 2007), and the Ebola epidemic in 2014 seriously disrupted the economies of several African nations.

Some middle-income countries are experiencing rapid growth in degenerative diseases, such as cancer and coronary heart disease, and more deaths are expected from smoking-related diseases. Despite the decrease in tobacco smoking in high-income countries, there has been an increase in per capita consumption of tobacco in low- and middle-income countries, many of which have been targeted for free samples and promotional advertising by U.S. and European tobacco companies.

Health is also affected by war and conflict. Countries including Afghanistan, Syria, Iraq, and Somalia have suffered from widespread conflict in recent years. These crises not only contribute

MAP 10.2 INFANT MORTALITY RATES, 2012

Mortality rates for children under one year of age in developing countries have dropped by nearly half since 1990 (World Health Organization, 2015a). Yet many children still die every day, most from preventable causes and almost half of them in sub-Saharan Africa.

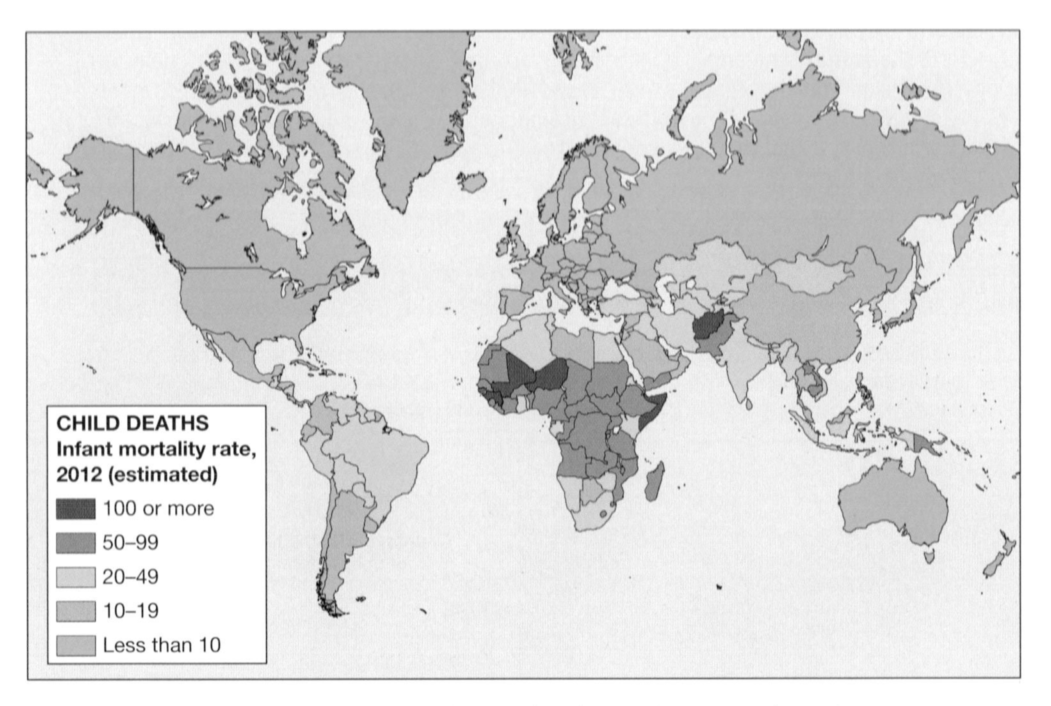

Source: Based on data from *CIA World Factbook*. Retrieved from https://www.cia.gov/library/publications/the-world-factbook/rankorder/2091rank.html

BOX 10.2 SOCIOLOGY IN GLOBAL PERSPECTIVE

The Missing Women

In 1990, economist Amartya Sen brought some very basic demographic data to life in an article called "More Than 100 Million Women are Missing" (1990). Sen's work drew the world's attention to the fact that the ratio of women to men is much lower in many parts of Asia and in North Africa than it is in Europe, North America, and Japan. While boys outnumber girls at birth, women live longer than men, so other things being equal, societies will have more females than males. However, in many countries, there are far fewer females than we would expect (Map 10.3). For example, in some places in Northern India, there are only 830 girls aged 0–6 for every 1000 boys.

The World Bank recently calculated that almost 4 million females are 'missing' each year: "About two-fifths of them are never born, one-fifth goes missing in infancy and childhood, and the remaining two-fifths do so between the ages of 15 and 59" (2012:14). Why do these females disappear?

A cultural preference for boys causes the disparity at birth. Many parents in India and China place a high value on the birth of a son. As family sizes dropped—encouraged by programs such as China's 'one child per family policy' and by higher rates of female education—parents turned to prenatal ultrasound to determine the sex of an unborn child. Some of these parents then aborted female fetuses in order to ensure they would be able to have a son.

The explanation of the higher mortality rates for girls in infancy and early childhood is more complex, but the evidence suggests it can be eliminated by improvements in basic healthcare and nutrition (World Bank, 2012).

(Continued)

The excess mortality in adulthood is mainly due to two factors. The first is that rates of maternal mortality are high in low-income countries. Maternal mortality has all but disappeared in high-income countries, but still takes a toll elsewhere: "One of every 14 women in Somalia and Chad will die from causes related to childbirth" (World Bank, 2012:129). The second factor is that in sub-Saharan Africa, HIV/AIDS disproportionately affects women, particularly between the ages of 15–24. Improved treatment of HIV/AIDS using anti-retroviral therapy means that this cause of death will decline in the future.

The missing females are almost all from low-income nations, illustrating once again the impact of the global stratification system on the world's poorest people. As the World Bank reports, "Poor people everywhere have to choose and make many decisions about many things that richer people take for granted every day. When institutions are bad, so are people's default choices—and 'free to choose' becomes 'forced to choose.' Under these circumstances, many illnesses and many life choices create excess female mortality."

MAP 10.3 : COUNTRIES WITH HIGHEST RATES OF MISSING GIRLS AND WOMEN

In China and India, the number of girls missing at birth remains high, and parts of Africa experienced large increases in excess female mortality during 1990–2008.

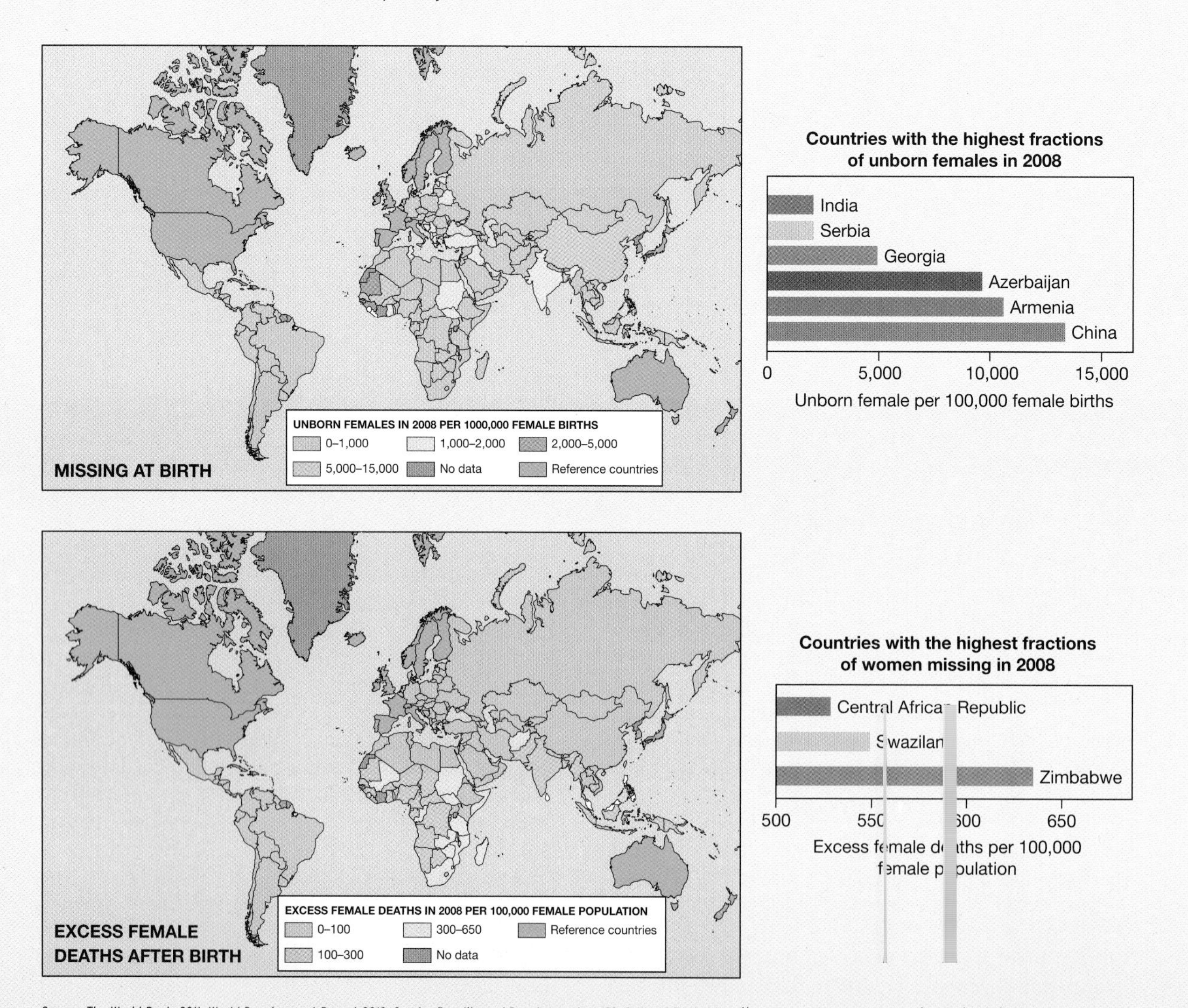

Source: The World Bank. 2011. *World Development Report 2012: Gender Equality and Development*, p. 122. © World Bank. https://openknowledge.worldbank.org/handle/10986/4391 License: Creative Commons Attribution license (CC BY 3.0 IGO). WDR 2012 team estimates based on data from World Health Organization 2010 and United Nations Department of Social and Economic Affairs 2009.

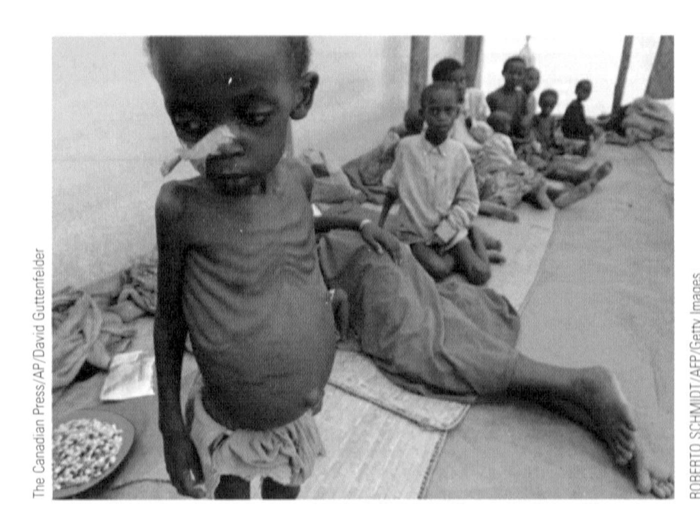

The Canadian Press/AP/David Guttenfelder

ROBERTO SCHMIDT/AFP/Getty Images

Malnutrition is a widespread health problem in many low-income nations. On the left, a Rwandan refugee child waits for help in a field hospital in Zaire; on the right, people in Mogadishu line up to receive food aid.

directly to serious injury and death, but also weaken the effectiveness of a nation's health-related infrastructure—in many cases, health facilities are simply destroyed. As a result, people are more vulnerable to disease and mortality.

EDUCATION AND LITERACY

Education is fundamental to reducing both individual and national poverty. Thus, school enrollment is used as a measure of human development. Literacy is an important result of education, and the adult literacy rate in low-income countries is much lower than that of the high-income countries. For women in these countries, the rate is even lower. However, educational achievements have improved dramatically around the globe. Since 1990, rates of literacy have risen from 73 percent to 84 percent and the average number of years of schooling has gone up by two years. While progress in human development is usually quite uneven, since 1970 no countries have had declines in literacy rates or years of schooling (United Nations Human Development Report, 2010).

GENDER AND EQUALITY

Women suffer more than men from global inequality:

> Women have an enormous impact on the well-being of their families and societies—yet their potential is not realized because of discriminatory social norms, incentives, and legal institutions. And while their status has improved . . . gender inequalities remain pervasive.
>
> Gender inequality starts early and keeps women at a disadvantage throughout their lives (World Bank, 2004:1).

Although more women have paid employment than in the past, women still live in poverty because of increases in single-person and single-parent households headed by women, and because low-wage work is often the only source of livelihood available to them. Despite improvements, in many countries traditional practices still prevent women from achieving their potential:

> Baruani is reflecting on how women's and men's lives have changed over the past decade in . . . a village in Tanzania. "Ten years back was terrible," she recalls. "Women were very behind. They used to be only at home doing housework.

Researchers have found that the education of girls is one of the most important factors in furthering development in low-income countries.

But now, they are in businesses, they are in politics." Others hold similar views. "We do not depend a lot on men as it used to be," says Agnetha. "We have some cash for ourselves, and this assists us in being free from men and to some extent controlling our lives." In addition to managing their businesses, the women now make up half the members of the street committee that runs the village.

Despite these positive changes, many challenges continue to weigh on women's daily lives. Fewer than half the homes in the village have piped water. Even more difficult, Tunginse and other women of the village still fear violence by their partners: "When they are drunk, they can begin beating up women and children in the house. The worst bit of it is forcing sex with you." Although legally women can inherit land or a house, tradition prevails. . . . "In fact, in the will the father is supposed to give each son and daughter something and nowadays the law is strict, equally. But still, men give to their sons and argue that women have the property of where they are married." (quoted in World Bank, 2011:2)*

Women's education is particularly important because it has an impact on many other factors that contribute to human development (United Nations Development Programme, 2003). Figure 10.4 shows that educated females marry later and have smaller families. They do a better job feeding their families and getting medical care, so more of their children survive. The children of illiterate mothers have an under-five mortality rate that is twice as high as the mortality rate for the children of mothers who have a middle school education (United Nations Development Programme, 2005:31). Higher child survival rates lead to a reduction in birth rates, which allows for better child care and less strain on the educational system. In societies where educated women are allowed to work outside the home, they make a significant contribution to family income.

Women's education is improving. According to the World Bank, two-thirds of the world's nations have achieved gender parity in primary education, and female enrolments continue to increase faster than male enrolments. The gap has closed significantly in regions with the greatest disparities: "In 2008, in sub-Saharan Africa, there were about 91 girls for

FIGURE 10.4 EDUCATED WOMEN LEAD DIFFERENT LIVES

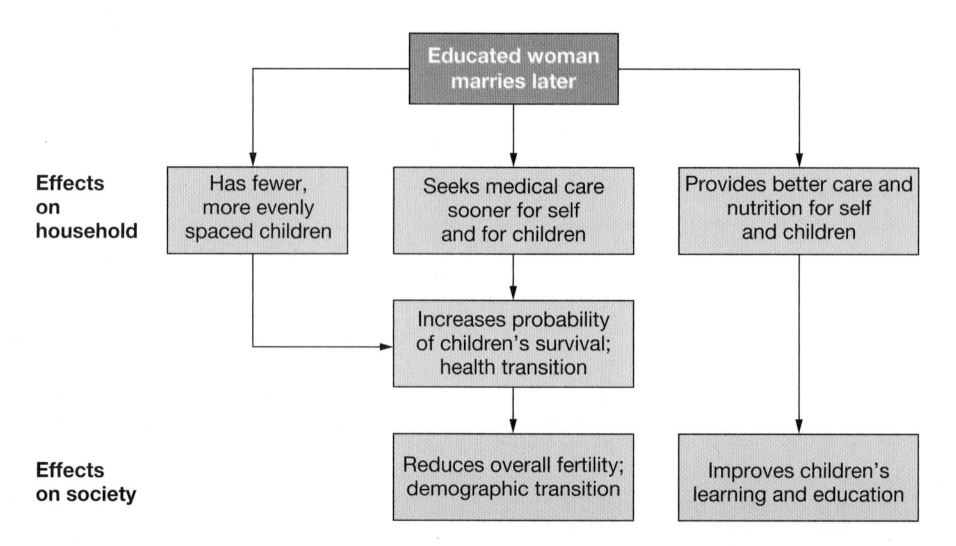

Source: United Nations Development Programme, 2003.

every 100 boys in primary school, up from 85 girls in 1999; in South Asia, the ratio was 95 girls for every boy (World Bank, 2011:61). The figures for secondary education are roughly the same as for primary, while for post-secondary education female enrolment rates were *higher* than males in about two-thirds of developing countries. Girls also tend to outperform boys at all levels. While this picture is very positive, girls have made little progress in some of the very poorest countries and are among the poorest people in many other nations.

TIME TO REVIEW

- In what ways has the Human Development Index helped change people's understanding of global poverty?
- Discuss some of the trends in health and education in low-income nations over the past few decades.
- Why does global inequality affect women more than men?

THEORIES OF GLOBAL INEQUALITY

Social scientists have developed several theories of global stratification that view its causes and consequences somewhat differently. We will examine the development approach and modernization theory, dependency theory, world-systems analysis, and the new international division of labour theory. Modernization theory is part of the functionalist tradition, while the other perspectives are rooted in conflict theory. These approaches are depicted in Figure 10.5.

LO-3 DEVELOPMENT AND MODERNIZATION THEORIES

According to some social scientists, global wealth and poverty are linked to a society's level of industrialization and economic development. These theorists maintain that low-income nations have progressed less than the wealthier industrial countries. They feel that industrialization and economic development are essential steps that nations must go through to reduce poverty and improve the living conditions of their citizens.

The best-known development theory is **modernization theory**—a perspective that links global inequality to different levels of economic development and suggests that low-income economies can move to middle- and high-income economies by achieving self-sustained economic growth. According to Langdon, modernization theory holds that undeveloped countries must follow the road travelled by successful Western capitalist, democratic societies, such as Britain and the United States:

> The usually implicit assumption was that economic development and growth involved a process of becoming like those societies and would be achieved essentially as those societies had achieved it: through economic change focused around industrialization, through social changes that would introduce Western institutions based on universalism and merit/achievement, and through political changes marked by secularization and the bureaucratic efficiency of the state. (1999:41)

▶ **modernization theory** A perspective that links global inequality to different levels of economic development and that suggests that low-income economies can move to middle- and high-income economies by achieving self-sustained economic growth.

Just as Weber had concluded that the adoption of a "spirit of capitalism" facilitated economic development, modernization theorists felt that development would be accompanied by changes in people's beliefs, values, and attitudes toward work. With modernization, the values of people in developing countries should become more similar to those of people in high-income nations.

Walt Rostow (1971, 1978) proposed one version of modernization theory. To Rostow, a major barrier to development in low-income nations was the traditional cultural values people

FIGURE 10.5 APPROACHES TO STUDYING GLOBAL INEQUALITY

Modernization Theory: Low-income, less-developed countries can move to middle- and high-income economies by achieving self-sustained economic growth.

Dependency Theory: Global poverty can at least partially be attributed to the fact that low-income countries have been exploited by high-income economies; the poor nations are trapped in a cycle of dependency on richer nations.

World Systems Analysis: How a country is incorporated into the global capitalist economy (e.g., a core, semiperipheral, or peripheral nation) is the key feature in determining how economic development takes place in that nation.

The New International Division of Labour Theory: Commodity production is split into fragments, each of which can be moved (e.g., by a transnational corporation) to whichever part of the world can provide the best combination of capital and labour.

held, particularly fatalistic beliefs such as viewing extreme hardship and economic deprivation as unavoidable facts of life. Fatalistic people do not see any need to work to improve their lot in life: Their future is predetermined for them, so why bother? According to modernization theory, poverty can be attributed to people's cultural failings, which are reinforced by governmental policies interfering with the smooth operation of the economy.

Rostow suggested that all countries go through four stages of economic development with identical content, regardless of when these nations began to industrialize. He compares the stages of economic development to an airplane flight. The first stage is the *traditional stage,* in which people do not think much about changing their circumstances, hold a fatalistic value system, do not subscribe to the work ethic, and save little money. The second stage is the *take-off stage*—a period of economic growth accompanied by a growing belief in individualism, competition, and achievement. People start to look toward the future, to save and invest money, and to discard traditional values. For Rostow, the development of capitalism is essential for the transformation from a traditional, simple society to a modern, complex one. With the financial help and advice of the high-income countries, low-income countries will be able to "fly" and enter the third stage of *economic development.* The country improves its technology, invests in new industries, and embraces the beliefs, values, and social institutions of the developed nations. In the fourth and final stage, the country reaches the phase of *high mass consumption* and a high standard of living.

According to proponents of modernization theory, studies have supported the proposition that economic development occurs more rapidly in a capitalist economy. The countries that have moved from low- to middle-income status have typically been those most centrally involved in the global capitalist economy. For example, many East Asian countries have successfully made the transition from low-income to higher-income economies through factors such as a high rate of savings, an aggressive work ethic among employers and employees, and the fostering of a market economy.

Critics of modernization theory point out that it is Eurocentric in its analysis of low-income countries, which it implicitly labels as backward (see Evans and Stephens, 1988). In many respects, modernization was equated with Westernization, as modernization theorists assumed that the problems of low-income countries would be alleviated only once they adopted Western values, culture, and economic models. Modernization theory does not take into account the possibility that all nations do not industrialize in the same manner. Allahar (1989) points out that leading industrial nations, such as the United States, Britain, and Japan, followed different paths to industrialization. Critics suggest that the modernization of low-income nations in the early 21st century will require novel policies, sequences, and ideologies that are not accounted for in Rostow's approach (see Gerschenkron, 1962). The theory also does not tell us what causes the move from one stage to another, but simply assumes they are natural stages that must be followed as societies advance economically and socially.

One of the most influential critics of modernization theory was Andre Gunder Frank (1969). Frank's research in Latin America convinced him that modernization theory was badly flawed. While Rostow felt that all societies had to move in a linear fashion from underdevelopment to industrialization, Frank pointed out that underdevelopment was not an original stage but a condition created by the imperial powers that had created dependency through actions, such as the deindustrialization of India, the damage done to African societies during the years of the slave trade, and the destruction of Native civilizations in Central and Latin America (Hettne, 1995). All societies were *un*developed at one time, but not all became *under*developed. While some countries moved from being undeveloped to developed, others moved from being undeveloped to a condition of underdevelopment in which they were dependent on other nations. These dependent countries had structures and institutions that effectively blocked any further development (Allahar, 1989).

Frank's critique of modernization theory was also a critique of the social policies that grew out of the theory. Frank felt that the Western powers, particularly the United States, were imposing their views of development through both political and military means. Modernization theory was linked to the fight against communism during the Cold War. Because communism was an obstacle on the road to modernization, it was necessary to persuade or force countries to adopt alternative forms of government. Many critics felt that modernization theory contributed to underdevelopment by encouraging policies that perpetuated global inequality (Langdon, 1999). The inadequacies of modernization theory and the political injustices that resulted from policies based on it, such as the Vietnam War, moved the next generation of development theorists to *dependency theory,* an approach based on the conflict perspective.

LO-4 DEPENDENCY THEORY

According to dependency theorists, rich countries have an interest in maintaining the dependent status of poor countries, as this ensures them a source of raw materials and a captive market for manufactured goods exported to the dependent nations. Business and political leaders in the poor nations find it in their interests to work with the advanced nations to impose policies that maintain the dependent relationship. Any surpluses created in the dependent country will be taken by the affluent capitalist country rather than being used to build up production infrastructure or raise the standard of living in the dependent nation. Unless this situation changes, poor countries will never match the sustained economic growth patterns of the more advanced capitalist economies.

dependency theory The perspective that global poverty can at least partially be attributed to the fact that low-income countries have been exploited by high-income countries.

Dependency theory states that global poverty can be at least partially attributed to the fact that the low-income countries have been exploited by the high-income countries. Dependency theorists see the greed of the rich countries as a source of increasing impoverishment of the poorer nations and their people. Due to their need for infusions of foreign capital and external markets for their raw materials, poorer nations are trapped in a cycle of structural dependency on the richer nations. This makes it impossible for the poorer nations to pursue their own economic and human development agendas. Frank and other scholars believed that the best way for low-income countries to move ahead was to break their links with the industrialized countries and establish independent socialist governments.

Dependency theory has been most often applied to the newly industrializing countries (NICs) of Latin America, but scholars examining the NICs of East Asia have found that dependency theory has little or no relevance to that part of the world. Therefore, dependency theory has been expanded to encompass transnational economic linkages that affect developing countries, including foreign aid, foreign trade, foreign direct investment, and foreign loans. On the one hand, in Latin America and sub-Saharan Africa, transnational linkages, such as foreign aid, investments by transnational corporations, foreign debt, and export trade, have been significant impediments to development within countries. On the other hand, East Asian countries, such as Hong Kong, Taiwan, South Korea, and Singapore, have also had high rates of dependency on foreign aid, foreign trade, and interdependence with transnational corporations but have still experienced high rates of economic growth.

Dependency theory has contributed to our understanding of global poverty by pointing out that "underdevelopment" is not necessarily the cause of inequality. Rather, this theory points out that exploitation of one country by another, and also of countries by transnational corporations, may limit or slow a country's economic growth and human development.

What remains unexplained is how some East Asian countries have had successful "dependency management," whereas many Latin American countries have not (Gereffi, 1994). In fact, between 1998 and 2005, the annual economic growth in emerging East Asian countries averaged 9 percent, which was much higher than growth in the West (Gill and Kharas, 2007).

NCG/Shutterstock.com

Residential and commercial buildings in Singapore: A variety of factors, including foreign investment and the presence of transnational corporations, has contributed to the economic growth of nations such as Singapore.

Dependency theory has contributed to our understanding of global stratification, but even its proponents no longer believe it is adequate. In addition to the problem of explaining the success of the East Asian economies that were closely linked with global capitalist structures, many dependency theorists have had their faith in development through socialist revolution shaken by the failure of many socialist economies, including that of the former Soviet Union (Frank, 1981). Most have concluded that the global economy is now so pervasive that it is impossible for low-income countries to disconnect themselves from the industrialized world and proceed with their own development (Martinussen, 1997).

LO-5 WORLD-SYSTEMS ANALYSIS

Drawing on Karl Marx's ideas, world-systems analysis suggests that what exists under capitalism is a truly global system held together by economic ties. Global inequality does not emerge solely as a result of the exploitation of one country by another. Instead, economic domination involves a complex world system in which the industrialized, high-income nations benefit from other nations and exploit the citizens of those nations. Wallerstein (1979, 1984) believed that a country's mode of incorporation into the capitalist work economy is the key feature in determining how economic development takes place in that nation. According to **world-systems analysis**, the capitalist world economy is a global system divided into a hierarchy of three major types of nations—core, semiperipheral, and peripheral—in which upward or downward mobility is conditioned by the resources and obstacles that characterize the international system.

Core nations are dominant capitalist centres characterized by high levels of industrialization and urbanization. Core nations, such as the United States, Japan, and Germany, possess most of the world's capital and technology. They can thereby exert massive control over world trade and economic agreements across national boundaries.

Most low-income countries in Africa, South America, and the Caribbean are **peripheral nations**—nations that are dependent on core nations for capital, have little or no industrialization, and have uneven patterns of urbanization. According to Wallerstein, the wealthy in peripheral nations benefit from the labour of poor workers and from their economic relations with core nation capitalists, whom they support in order to maintain their own wealth and position. At a global level, uneven economic growth results from capital investment by core nations. This investment draws poor people from rural areas to cities where they may be able to find work, even though this work may be very poorly paid.

This influx has pushed already overcrowded cities far beyond their capacity. Many people live on the edge of the city in shantytowns made from discarded materials or in low-cost rental housing in central-city slums because their wages are low and affordable housing is nonexistent (Flanagan, 1999). In fact, housing shortages are among the most pressing problems in many peripheral nations. According to most world-systems analysts, it will be very difficult for peripheral countries to change their structural position in the capitalist world economy (Wallerstein, 1979).

Semiperipheral nations are more developed than peripheral nations but less developed than core nations. Nations in this category—Mexico, Brazil, India, and Nigeria are examples—typically provide labour and raw materials to core nations within the world system. These nations constitute a midpoint between the core and peripheral nations that promotes the stability and legitimacy of the three-tiered world economy. According to Wallerstein, semiperipheral nations exploit peripheral nations, just as the core nations exploit both the semiperipheral and the peripheral nations.

Wallerstein (1991) acknowledges that world-systems analysis is an "incomplete, unfinished critique" for long-term, large-scale social change that affects global inequality. However, most scholars acknowledge that nations throughout the world are influenced by a relatively small number of countries and transnational corporations that have prompted a shift from an international to a more global economy (see Knox and Taylor, 1995; Wilson, 1997).

world-systems analysis The perspective that the capitalist world economy is a global system divided into a hierarchy of three major types of nations–core, semiperipheral, and peripheral–in which upward or downward mobility is conditioned by the resources and obstacles that characterize the international system.

core nations According to world-systems analysis, dominant capitalist centres characterized by high levels of industrialization and urbanization, as well as a high degree of control over the world economy.

peripheral nations According to world-systems analysis, nations that are dependent on core nations for capital, have little or no industrialization, and have uneven patterns of urbanization.

semiperipheral nations According to world-systems analysis, nations that are more developed than peripheral nations but less developed than core nations.

LO-6 THE NEW INTERNATIONAL DIVISION OF LABOUR THEORY

According to the **new international division of labour theory,** commodity production is being split into fragments that can be assigned to whichever part of the world can provide the most profitable combination of capital and labour. Consequently, the new international division of labour has changed the pattern of geographic specialization between countries, and high-income countries have become dependent on low-income countries for labour. Low-income countries provide transnational corporations with the ability to pay lower wages and taxes, and face fewer regulations regarding workplace conditions and environmental protection (Waters, 1995).

This new division of labour is part of a global economy based on free trade among countries. Multilateral trade agreements, such as the General Agreement on Tariffs and Trade (GATT) and the North American Free Trade Agreement (NAFTA), have allowed the freer transfer of goods and services among countries, and global corporations now view all the countries of the world both as potential markets and as potential locations for production.

new international division of labour theory The perspective that commodity production is being split into fragments that can be assigned to whichever part of the world can provide the most profitable combination of capital and labour.

Ross Kinnaird/Getty Images

Athletes like Tiger Woods make more money by endorsing products than the combined salaries of thousands of the workers who manufacture the products. According to Nike's chief executive officer, marketing is more important to the company's success than is the manufacturing of their products (Klein, 2000).

These trade liberalization agreements would appear to be beneficial for poor countries, but often they are not. The movement of production into developing countries brings jobs to countries with chronically high unemployment; however, few of the profits stay in these countries. For example, a study of garment manufacturing in Bangladesh found that less than 2 percent of the final value of the product went to production workers and 1 percent went to the local producer. The rest of the money went to profit company owners and to pay expenses, such as shipping costs, and customs duties and sales taxes in high-income countries (Chossudovsky, 1997). Workers are unlikely to get higher wages because the jobs are unskilled and can quickly be moved to another poor country if workers begin to put pressure on the companies (Klein, 2000). In 2015, several large trade agreements were being negotiated, including the massive Trans-Pacific Partnership (TPP). While Mexico and Vietnam are included in the TPP, the vast majority of low-income countries are not being included in TPP or other major trade negotiations, so their interests and those of their workers will not be represented in these agreements. Countries within the agreements will have major incentives to trade with each other, which will leave other countries sitting on the sidelines.

Many large corporations have global operations. Typically, labour-intensive manufacturing operations, from making T-shirts to assembling computers, are established in low-wage countries. Even service industries—completing income tax forms, booking airline flights, and processing insurance claims forms—have now become exportable through the Internet. These activities make up *global commodity chains,* a complex pattern of international labour and production processes that result in a finished commodity ready for sale in the marketplace.

Commodity chains are most common in labour-intensive consumer goods industries, such as toys, garments, and footwear (Gereffi, 1994). Athletic footwear companies such as Nike and Reebok, and clothing companies such as The Gap and Armani, use this model. Manufacturing these products is labour-intensive so the factory system is typically very competitive and globally decentralized. Workers in commodity chains are often exploited by low wages, long hours, and poor working conditions. Most cannot afford the products they make. Tini Heyun Alwi, an assembly line worker at a shoe factory in Indonesia that makes Reebok sneakers, is an example: "I think maybe I could work for a month and still not be able to buy one pair" (quoted in Goodman, 1996:F1). Nike workers in Indonesia make only $2.50 per day, so their monthly salaries would fall short of the retail price of the shoes.

These changes have had a mixed impact on people living in poor countries (see Box 10.3). For example, Indonesia has been able to attract foreign business into the country, but workers remain poor despite working full time in factories making consumer goods such as Nike shoes. As employers feel pressure from workers to raise wages, clashes erupt between the workers and managers or owners. Governments in these countries fear that rising wages and labour strife will drive away the businesses, leaving behind workers who have no other hopes for employment and become more impoverished than they previously were.

THE "FLYING GEESE" MODEL OF DEVELOPMENT The flying geese model was developed by Japanese economist Akamatsu Kaname and is an extension of international division of labour theory. The model tries to explain the rapid development of East Asian economies following World War II (Korhonen, 1994). This process involves one country—in this case, Japan—leading other less-developed countries into more prosperous times. The countries go through the sequential steps of importing goods, manufacturing to serve domestic markets, and, finally, exporting goods. Pressure to increase wages causes production to shift to less

BOX 10.3 POINT/COUNTERPOINT

Ethics and Offshore Production

In May 2010, the Apple corporation became the world's most valuable technology company (by 2015 it was the most valuable business of any kind with a market capitalization of over $700 billion). The same month, a worker at a Foxconn factory in Shenzhen, China, became the ninth worker to jump to his death from one of the company's buildings that year. The connection between these two events is that the Foxconn factory is one of the largest manufacturers of Apple products.

Like many of its competitors, Apple manufactures most of its products outside North America because of lower wages. Apple employs 66,000 people in the United States, but most of its production is done by hundreds of thousands of workers employed by offshore contractors such as Foxconn. The Shenzhen plant, which employs more than 400,000 people, is modern and working conditions are much better than in the sweatshops that exist in many poor countries. However, the workers are poorly paid and are pressured to work long hours of overtime. Many of the workers are rural young people who migrated to Shenzhen to get jobs and who live in company dormitories where they have little contact with their families and friends. The workers who have killed themselves may have seen little in their futures except endless hours on an assembly line.

In response to the suicides, Foxconn significantly increased salaries and placed netting around its buildings to prevent workers from jumping. Apple has responded by opening its offshore production facilities for inspection by the Fair Labor Association and by reaching an agreement with Foxconn to reduce overtime and improve working conditions. However, as late as 2012, suicides continued to occur at Foxconn factories.

Mohammad Asad/LightRocket/Getty Images

The tragic building collapse at Rana Plaza highlighted the dangers faced by workers in many low-income countries. The cheap products demanded by Canadian shoppers can come at a high human cost, particularly if our wholesale and retail merchants don't require that their suppliers meet adequate safety standards.

While workers may labour under harsh conditions at Foxconn's plants, they are much safer than thousands of other manufacturing workers in low-income countries. In 2013, a textile manufacturing facility in Bangladesh collapsed, killing over 1100 workers. The building, Rana Plaza, had been closed because of dangerous structural cracks. However, the owners of a textile factory in the building ordered their workers back to work and many were killed when the building collapsed. The building owners had violated many regulations, including building the top stories without a building permit, and the building had been designed for offices, not for factories. Substandard material had been used in constructing the building. In 2015, murder charges were laid against 41 people, including the owner of the building and a number of government officials, for their role in this tragedy.

The factory owner ordered his workers back because he was concerned that he would lose contracts, as the foreign companies that purchased his clothing typically wanted very quick delivery of new product. In this case, the buyers included several well-known companies, including Benetton and Canada's Loblaws, which owns the Joe Fresh clothing line.

After the building collapsed, these companies faced massive criticism for sourcing their products from manufacturers who endangered their workers. Several of the companies, including Loblaws, signed an agreement binding them to improving their suppliers' worker safety programs and providing compensation for the families of the dead and injured Rana Plaza workers. Walmart, Target, and many other large companies refused to sign these agreements, though several did pay small amounts of compensation.

The Foxconn suicides and the deplorable working conditions in Bangladesh raise an issue we should all consider. Labour costs make up only a small part of the final cost of most high technology and fashion products. Significantly increasing these wages would mean that you would have to pay a few dollars more for these products or that the companies that distribute these products would make slightly less profit.

What are your views about this issue? Would you pay a few dollars more for the technology you use to ensure that the people who manufactured it were being paid a fair wage? Would an additional 25 cents deter you from buying a particular T-shirt? Are companies that source their products offshore contributing to a system that exploits foreign workers, or is sending work offshore an important step in improving the living conditions of people in low-income countries?

advanced economies. This boosts the standard of living of the less advanced economies, raises wages, and, in turn, leads to a shift in production to countries that are even less developed. In the 1960s, Japan was the only developed economy in Asia but was then followed by the "Asian tigers"—Singapore, Hong Kong, Taiwan, and South Korea. Development in these countries was followed by industrialization in Malaysia, Indonesia, the Philippines, and Thailand, and most recently by China and India.

One example of this process has been in textiles, where Japan once had a thriving export business. When wages in Japan became globally uncompetitive, production shifted to the next tier of countries while Japan moved into the production of higher-technology goods such as automobiles and electronics. Production of these goods has also moved to lower-tier countries that have upgraded their technological skills. Companies such as South Korea's Hyundai and Samsung are now producing very high-quality, high technology products that are displacing some Japanese products in global markets.

This perspective provides a dynamic picture of the global division of labour that gives us some reason for optimism as we see technology being transferred from more to less industrialized countries. It is not clear that the flying geese model will apply outside East Asia or whether unique circumstances led to the pattern of development in those countries. However, the fact that a Taiwanese company employs 10,000 people in the African country of Lesotho manufacturing Levi's jeans for sale in North America provides some hope that this process will continue (Kristof, 2012b).

TIME TO REVIEW

- What are the stages that modernization theorists feel all societies must go through before they can become high-income economies?
- According to dependency theorists, in what ways do rich countries exploit the economies of poor countries?
- In world-systems analysis, how do core, semiperipheral, and peripheral countries interact with each other?
- Discuss the ways in which the new international division of labour theory explains current trends in globalization. How does the flying geese hypothesis extend this theory?

GLOBAL INEQUALITY IN THE FUTURE

What are the future prospects for greater equality across and within nations? Social scientists disagree on the answer to this question. Depending on their theoretical framework, they may see either an optimistic or a pessimistic future scenario.

In some regions, economic development has stalled and persistent poverty continues to undermine human development. Most of the countries where human development has worsened are in sub-Saharan Africa and Eastern Europe. Many of these countries have been affected by the HIV/AIDS epidemic or are adjusting to the collapse of the former Soviet Union and then the global recession of 2009. Others have suffered from low prices for the agricultural products that are their main source of income.

The situation of coffee farmers illustrates how low-income producers may fail to benefit from rising prices in high-income countries. Canadians have become used to paying two or three dollars for a cup of coffee at chains like Tim Hortons and Starbucks. Many of us assume that the people who grow the coffee receive a fair share of this price. However, according to the United Nations, between 1990 and 2005, the retail value of coffee sold in high-income countries increased from $30 billion to $80 billion (United Nations Development Programme, 2005). Over the same period, the income received by coffee exporters dropped

While consumers in North America and Europe are content to pay high prices for their double-espresso low-fat lattes, coffee growers are struggling to survive.

from $12 billion to $5.5 billion even though the amount of coffee exported increased. The farmer receives only 1 cent of each dollar you pay for your cup of coffee. The reduction in income has had devastating effects on human development in countries such as Ethiopia, Uganda, and Nicaragua, which rely heavily on coffee exports. At the same time, chains that serve coffee have been enormously profitable because they can charge high prices while paying a very low price for their raw materials. While consumers in North America and Europe happily pay high prices for their morning cup of coffee, coffee growers receive a very small share of this money.

In the future, continued population growth, urbanization, and environmental degradation will threaten even the meagre living conditions of people in low-income nations. Environmental problems will be catastrophic if billions of people in countries such as China and India begin to live like middle-class North Americans, as we are setting an example that is wasteful and resource-intensive.

From this perspective, the future looks dim not only for people in low- and middle-income countries but also for those in high-income countries, who will see their quality of life diminish as natural resources are depleted, the environment is polluted, and high rates of immigration and global political unrest threaten the standard of living.

Others see a more optimistic scenario. Globally, health is improving, a higher proportion of people are educated, incomes have increased, far more females are enrolled in schools, and more countries are holding democratic elections. According to the United Nations, only three countries (the Democratic Republic of the Congo, Zambia, and Zimbabwe) had lower human development scores in 2010 than they did in 1970 (United Nations Human Development Programme, 2010). Most of this improvement has been in East Asia—particularly in China—where the number of people living in absolute poverty was reduced by nearly half during the 1990s (United Nations Development Programme, 2003). It will be challenging to maintain these improvements and to make similar progress in areas, such as sub-Saharan Africa, that have largely been left behind.

These improvements show that it will be possible to alleviate global poverty, but if this positive change is to continue, the practices of global corporations, foreign aid donors, and international lending organizations must begin to focus on the needs of low-income countries rather than solely on the demands of the marketplace. These needs were formally recognized in 2000, when most of the world's heads of state adopted the UN Millennium Declaration that committed countries to make dramatic improvements in the lives of the poor. The declaration contains a series of specific goals and targets that are being monitored to track the progress being made (see Figure 10.6).

FIGURE 10.6 MILLENNIUM DEVELOPMENT GOALS AND TARGETS

Goal 1: Eradicate extreme poverty and hunger.

Target 1: Halve, between 1990 and 2015, the proportion of people whose income is less than $1 a day.

Target 2: Halve, between 1990 and 2015, the proportion of people who suffer from hunger.

Goal 2: Achieve universal primary education.

Goal 3: Promote gender equality and empower women.

Goal 4: Reduce child mortality.

Target 1: Reduce by two-thirds, between 1990 and 2015, the under-five mortality rate.

Goal 5: Improve maternal health.

Target 1: Reduce by two-thirds, between 1990 and 2015, the maternal mortality rate.

Goal 6: Combat HIV/AIDS, malaria, and other diseases.

Goal 7: Ensure environmental sustainability.

Target 1: Integrate the principles of sustainable development into country policies and programs and reverse the loss of environmental resources.

Target 2: Halve, by 2015, the proportion of people without sustainable access to safe drinking water and basic sanitation.

Target 3: Have achieved, by 2020, a significant improvement in the lives of at least 100 million slum dwellers.

Goal 8: Develop a global partnership for development.

Source: © United Nations Development Programme, *Human Development Report, 2003.*

As shown in Figure 10.7, substantial progress has been made toward achieving these goals, but much more remains to be done. The most progress has been made in meeting the educational goals, improving maternal health (attended births), and providing access to safe water. There has been much less success in reducing infant mortality and providing people with proper sanitation. Figure 10.7 also shows that while progress is being made at the global level, many countries lag seriously behind.

The United Nations has suggested six policy changes that will help low-income countries to improve their situations (United Nations Development Programme, 2003:4):

1. Invest in basic education and health and encouraging the equality of women will help to encourage economic growth.
2. Help to improve the productivity of small farmers.
3. Improve roads, ports, communications systems, and the other infrastructure necessary for production and trade.
4. Promote the development of small- and medium-sized businesses to help countries move away from dependence on exporting commodities.
5. Promote democratic governance and human rights, to help ensure that economic growth benefits the poorest people within low-income countries rather than just the elite.
6. Ensure environmental sustainability and urban planning.*

Making these changes is beyond the capability of the poorest countries, so progress will depend on the willingness of the rest of the world to work with them to meet the Millennium Development Goals, as achieving these goals is very important to the world's poorest people.

* United Nations Development Programme, 2003:4

FIGURE 10.7 : PROGRESS TOWARD THE MILLENNIUM DEVELOPMENT GOALS, BY COUNTRY

Goals and Targets	Africa		Asia				Oceania	Latin America and the Caribbean	Caucasus and Central Asia
	Northern	Sub-Saharan	Eastern	South-Eastern	Southern	Western			
GOAL 1 | Eradicate extreme poverty and hunger									
Reduce extreme poverty by half	low poverty	very high poverty	low poverty	moderate poverty	high poverty	low poverty	—	low poverty	low poverty
Productive and decent employment	large deficit	very large deficit	moderate deficit	large deficit	large deficit	large deficit	very large deficit	moderate deficit	small deficit
Reduce hunger by half	low hunger	high hunger	moderate hunger	moderate hunger	high hunger	moderate hunger	moderate hunger	moderate hunger	moderate hunger
GOAL 2 | Achieve universal primary education									
Universal primary schooling	high enrolment	moderate enrolment	high enrolment	high enrolment	high enrolment	high enrolment	high enrolment	high enrolment	high enrolment
GOAL 3 | Promote gender equality and empower women									
Equal girls' enrolment in primary school	close to parity	close to parity	parity	parity	parity	close to parity	close to parity	parity	parity
Women's share of paid employment	low share	medium share	high share	medium share	low share	low share	medium share	high share	high share
Women's equal representation in national parliaments	moderate representation	moderate representation	moderate representation	low representation	low representation	low representation	very low representation	moderate representation	low representation
GOAL 4 | Reduce child mortality									
Reduce mortality of under-five-year-olds by two thirds	low mortality	high mortality	low mortality	low mortality	moderate mortality	low mortality	moderate mortality	low mortality	low mortality
GOAL 5 | Improve maternal health									
Reduce maternal mortality by three quarters	low mortality	high mortality	low mortality	moderate mortality	moderate mortality	low mortality	moderate mortality	low mortality	low mortality
Access to reproductive health	moderate access	low access	high access	moderate access	moderate access	moderate access	low access	high access	moderate access
GOAL 6 | Combat HIV/AIDS, malaria and other diseases									
Halt and begin to reverse the spread of HIV/AIDS	low incidence	high incidence	low incidence	low incidence	low incidence	low incidence	low incidence	low incidence	low incidence
Halt and reverse the spread of tuberculosis	low mortality	high mortality	low mortality	moderate mortality	moderate mortality	low mortality	moderate mortality	low mortality	moderate mortality
GOAL 7 | Ensure environmental sustainability									
Halve proportion of population without improved drinking water	high coverage	low coverage	high coverage	high coverage	high coverage	high coverage	low coverage	high coverage	moderate coverage
Halve proportion of population without sanitation	moderate coverage	very low coverage	moderate coverage	low coverage	very low coverage	high coverage	very low coverage	moderate coverage	high coverage
Improve the lives of slum-dwellers	low proportion of slum-dwellers	very high proportion of slum-dwellers	moderate proportion of slum-dwellers	moderate proportion of slum-dwellers	moderate proportion of slum-dwellers	moderate proportion of slum-dwellers	moderate proportion of slum-dwellers	moderate proportion of slum-dwellers	—
GOAL 8 | Develop a global partnership for development									
Internet users	moderate usage	low usage	high usage	moderate usage	low usage	high usage	low usage	high usage	high usage

The progress chart operates on two levels. The text in each box indicates the present level of development. The colours show progress made towards the target according to the legend below:

- Target met or excellent progress.
- Good progress.
- Fair progress.
- Poor progress or deterioration.
- Missing or insufficient data.

Source: The World Bank, http://data.worldbank.org/news/significant-progress-towards-achieving-MDGs

A recent study reported the results of an experimental study of a program designed to help the world's poorest people. Banerjee and colleagues (2015) conducted a randomized trial of their Graduation Program among the poorest villages in six countries. The program was intended to help people to graduate from poverty. It was a comprehensive initiative that focused on several dimensions: providing families with a productive resource such as goods to sell or livestock; training and support to teach them how to use this resource; weekly coaching in life skills; temporary financial support when needed to help them get established; bank accounts to help them save money; and health information and services. The intervention lasted two years. One year following the end of the program, participants were earning more and consuming more. The program was cost-effective—the best results were in India where every dollar spent on the program led to $4.33 in long-term benefits. Several countries are planning to expand the program, and further evaluations will determine if the program will ultimately contribute to reducing global poverty.

We will continue to focus on issues pertaining to global inequality in subsequent chapters as we discuss such topics as race, gender, education, health and medicine, population, urbanization, social change, and the environment.

LO-1 Understand the concept of global stratification and its impact on the lives of billions of people.

Global stratification refers to the unequal distribution of wealth, power, and prestige on a global basis, which results in people having vastly different lifestyles and life chances both within and among the nations of the world. The income gap between the richest and the poorest people in the world continues to widen and hundreds of millions of people are living in abject poverty.

© Viviane Moos/Corbis

LO-2 Understand the relationship between global poverty and human development.

Income disparity is not the only factor that defines poverty and its effect on people. The United Nations Human Development Index measures the level of development in a country through indicators such as life expectancy, infant mortality rate, proportion of underweight children under age five, and adult literacy rate for low-income, middle-income, and high-income countries.

© Scott Wallace/World Bank

LO-3 Discuss Rostow's modernization theory and explain the stages that modernization theorists believe all societies must go through.

Modernization theory links global inequality to different levels of economic development and suggests that low-income economies can move to middle- and high-income economies by achieving self-sustained economic growth. According to Rostow, all countries go through four stages of economic development: (1) the traditional stage; (2) the take-off stage; (3) the technological maturity stage; and (4) the high mass consumption stage.

© vietnam/Alamy

LO-4 Describe how dependency theory differs from modernization theory.

Dependency theory states that global poverty can be at least partially attributed to the fact that low-income countries have been exploited by high-income countries. Whereas modernization theory focuses on how societies can reduce inequality through industrialization and economic development, dependency theorists see the greed of the rich countries as a source of increasing impoverishment of the people in poorer nations.

© imageBROKER/Alamy Stock Photo

KEY TERMS

core nations According to world-systems analysis, dominant capitalist centres characterized by high levels of industrialization and urbanization, as well as a high degree of control over the world economy (p. 279).

dependency theory The perspective that global poverty can at least partially be attributed to the fact that low-income countries have been exploited by high-income countries (p. 278).

gross national income (GNI) All the goods and services produced in a country in a given year, plus the income earned outside the country by individuals or corporations (p. 263).

high-income economies Countries with an annual per capita gross national income of $US12,737 or more (p. 264).

lower-middle-income economies Countries with an annual per capita gross national income between $US1046 and $US4125 (p. 264).

low-income economies Countries with an annual per capita gross national income of $US1045 or less (p. 264).

modernization theory A perspective that links global inequality to different levels of economic development and that suggests that low-income economies can move to middle- and high-income economies by achieving self-sustained economic growth (p. 275).

new international division of labour theory The perspective that commodity production is being split into fragments that can be assigned to whichever part of the world can provide the most profitable combination of capital and labour (p. 279).

peripheral nations According to world-systems analysis, nations that are dependent on core nations for capital, have little or no industrialization, and have uneven patterns of urbanization (p. 279).

semiperipheral nations According to world-systems analysis, nations that are more developed than peripheral nations but less developed than core nations (p. 279).

upper-middle-income economies Countries with an annual per capita gross national income between $US4126 and $US12,736 (p. 264).

world-systems analysis The perspective that the capitalist world economy is a global system divided into a hierarchy of three major types of nations–core, semiperipheral, and peripheral–in which upward or downward mobility is conditioned by the resources and obstacles that characterize the international system (p. 279).

LO-5 Understand how world-systems analysis views the global economy.

According to world-systems analysis, the capitalist world economy is a global system divided into a hierarchy of three major types of nations: core, peripheral, and semiperipheral. Core nations benefit from their relationships with peripheral and semiperipheral nations.

NCG/Shutterstock.com

LO-6 Understand the international division of labour theory.

The international division of labour theory is based on the assumption that commodity production is split into fragments that can be assigned to whichever part of the world can provide the most profitable combination of capital and labour. This division of labour has changed the pattern of geographic specialization among countries, and high-income countries have become dependent on low-income countries for labour. The low-income countries provide transnational corporations with a situation in which they can pay lower wages and taxes and face fewer regulations regarding workplace conditions and environmental protection. A variation of this theory—the flying geese theory—helps explain how manufacturing technology is successively transferred to lower-income countries.

Ross Kinnaird/Getty Images

KEY FIGURES

Stan Wayman/Time Life Pictures/Getty Images

Walt Whitman Rostow (1916–2003) Rostow was an economist who taught at several universities and worked at high levels of the U.S. government. As a national security advisor to U.S. presidents John F. Kennedy and Lyndon Johnson, he has been criticized for his role in planning Vietnam War strategy. His best-known academic work involved his theory of modernization.

© Rick Maiman/Sygma/Corbis

Amartya Sen (b. 1933) Sen was born in what is now Bangladesh. Raised in India, much of his later work was influenced by sectarian violence in that country in the 1940s and by the Bengal famine that killed two million to three million people in 1943. He became an economist and philosopher and received the Nobel Prize for Economics in 1998. Along with his colleague Mahbub ul Haq, Sen developed the Human Development Index for the United Nations Development Programme.

Used with permission of Immanuel Wallerstein

Immanuel Wallerstein (b. 1930) Wallerstein is a sociologist who taught at McGill University during the 1970s before returning to the United States. His major work involved the development of world-systems analysis. This work has inspired many of those involved in the anti-globalization movement.

APPLICATION QUESTIONS

1. You have decided to study global wealth and poverty. How would you approach your project? Which research methods would provide the best data for analysis? What do you think you might find if you compared your research data with popular presentations—such as films and advertising—of everyday life in low- and middle-income countries?
2. What are some of the positive aspects of globalization? How might the globalization of manufacturing and service industries benefit the world's poorest people?
3. How are people and cultures negatively affected by the practices of global corporations?
4. Using the theories discussed in this chapter, devise a plan to alleviate poverty in low-income countries. Assume that you have the necessary resources, including wealth and political power. Share your plan with others in your class and create a consolidated plan that represents the best ideas and suggestions presented.

CHAPTER 11

ETHNIC RELATIONS AND RACE

Peter Muller/Cultura/Getty Images

CHAPTER FOCUS QUESTION

What is the significance of race in Canadian society?

LEARNING OBJECTIVES

AFTER READING THIS CHAPTER, YOU SHOULD BE ABLE TO

LO-1 Distinguish between race and ethnicity.

LO-2 Define and explain prejudice, discrimination, and racism.

LO-3 Explain the major sociological perspectives on race and ethnic relations.

LO-4 Discuss the unique historical experiences of the racial and ethnic groups in Canada.

LO-5 Describe how Canada's immigration policies have affected the composition of Canada's racial and ethnic population today.

Keisha, a student of South Asian background, tells the story of the racism she experienced when she first went to university in southern Ontario. She describes the following experience as a "big shock":

I came to university with a big mind and an open mind and I was here to learn and it was an environment where my fellow peers, I had hoped, would have the same stand or the same understanding of a lot of things. So walking into lecture one day, I was a little bit late, so I just turned around to one of the girls and asked what was happening, and she turned to me and said something that was very awful and I will quote. She said: "Don't talk to me, filthy Paki." She was pretty loud, and the girls in front of me and behind me kind of heard and there were a couple of guys who heard and they turned around . . . I was so shocked that I couldn't respond because I couldn't fathom that someone my age, someone in the same society that I grew up in or at least at the level of education system that we were in, would not have an open mind and would say some thing like that . . . I couldn't respond because I'm a person who is very naïve . . . I didn't say anything. But I was hoping, I guess, that the people who were sitting in front of me or behind me would've said something. (James, 2010:235)

Source: Carl E. James, *Seeing Ourselves: Exploring Race, Ethnicity and Culture* (4th ed.). Toronto: Thompson Educational Publishing. Inc., 2010.

Contemplation of race and ethnic relations in Canada is a study in contradictions. Canada is a multicultural society composed of racially and ethnically different groups. Since 1990, over 5 million new immigrants have been admitted to Canada, most of whom are visible minorities (Reitz, 2012; Environics, 2013). Our country has a reputation as a tolerant and compassionate country whose success in race and ethnic relations has received international respect and admiration. Our success in integrating immigrants is unparalleled by international standards (Fleras, 2012). Canada is widely renowned for its "cultural democracy" and "harmonious" ethnic diversity (James, 2005). Without question, significant gains have been made in the past 50 years for racial and ethnic minority groups in Canada.

From a distance, Canada maintains its enviable status. However, upon closer examination, we see evidence of cracks in our cultural mosaic. Although Canadians are proud to support the principle of multiculturalism, they also want immigrants to blend in, and they worry whether they will, as evidenced from the results of a recent opinion poll where 70 percent of respondents agreed with the following statement: "There are too many immigrants coming into this country who are not adopting Canadian values" (Environics, 2012).

Furthermore, despite our claims that Canadians are "colour-blind," there is also widespread awareness that ethnic and racial minorities experience discrimination on an ongoing basis. Police-reported hate crimes are on the rise, the majority of which are motivated by hatred toward a race or ethnicity (Allen, 2014). According to a 2014 survey by the Canadian Race Relations Foundation, almost two in three Canadians are "worried" about a rise in racism. Among the university-aged population (18–24 year olds), over 80 percent of respondents indicated they had heard racist comments and 50 percent indicated they had witnessed a racist incident in the past year (Canadian Race Relations Foundation, 2014). The public believes that Muslims experience the most discrimination, followed closely by Indigenous peoples, South Asians, and blacks (see Figure 11.1.) Herein lies the contradiction. It is no secret that racism is a major social problem in Canada, yet when asked, few Canadians openly condone racism and even fewer admit to being racist themselves (Fleras, 2012).

FIGURE 11.1 PERCEIVED DISCRIMINATION IN CANADA, 2011

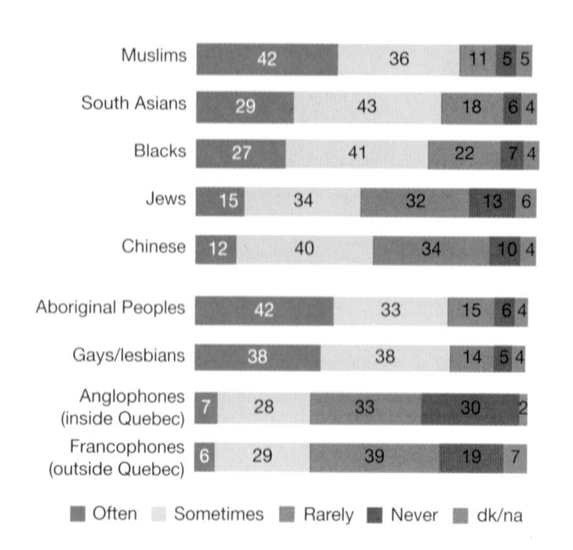

Source: The Environics Institute for Survey Research.

In this chapter, racism will be central to the discussion of race and ethnicity. One of the most important and reliable sources of information on racism is the individual who has experienced it directly. As anti-racism activist Tim Wise argues, "If you want to know if racism is a problem, it would probably do you best to ask the folks who are its targets. They are the ones, after all, who must as a matter of survival know what it is and when it's operating (Wise, 2008a:35). Therefore, we will explore the subjective impact of race and ethnicity on people's lives—and examine whether those effects are changing. Before reading on, test your knowledge about race and ethnic relations in Canada by taking the quiz in Box 11.1.

CRITICAL THINKING QUESTIONS

1. How do you think you would have responded if you had witnessed the racist incident outlined earlier?
2. How might you have responded if you, like Keisha, were the target of such racial hatred?

LO-1 RACE AND ETHNICITY

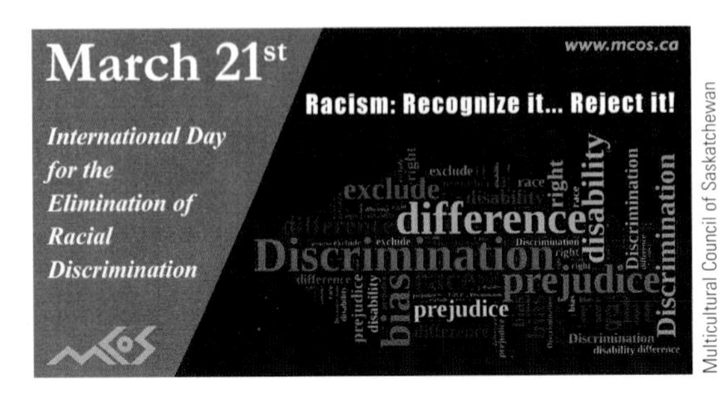

Multicultural Council of Saskatchewan

March 21 is recognized annually as the International Day for the Elimination of Racial Discrimination because on that day, in 1960, police opened fire and killed 69 people at a peaceful demonstration against apartheid in South Africa.

What is "race"? Some people think it refers to skin colour (the Caucasian "race"); others use it to refer to a religion (the Jewish "race"), nationality (the British "race"), or the entire human species (the human "race") (Marger, 2009). Popular usages of *race* have been based on the assumption that a race is a grouping or classification based on *genetic* variations in physical appearance, particularly skin colour. However, social scientists and biologists dispute the idea that biological race is a meaningful concept In fact, the idea of race has little meaning in a biological sense because of the enormous amount of interbreeding that has taken place within the human population. For these reasons, sociologists sometimes place "race" in quotation marks to show that categorizing individuals and population groups on biological characteristics is neither accurate nor based on valid distinctions between the genetic makeup of differently identified "races" (Marshall, 1998).

Race is a *socially constructed reality,* not a biological one. Understanding what we mean when we say that race is a social construct is important to our understanding of how race affects all aspects of social life and society. Race as a *social construct* means that races as such do not actually exist, but some groups are still racially defined because the *idea* persists in many people's minds that races are distinct biological categories with physically distinguishable characteristics and a shared common cultural heritage. However, research on the human genome has been unable to identify any racially based genetic differences in human beings, and fossil and DNA evidence also point to humans all being of one race. Race continues to be an important concern however, not because it is a biological reality but because it takes on a life of its own when it is socially defined and shapes how we see others and ourselves. Race also has significant social consequences, such as which individuals experience prejudice and discrimination and which have the best life chances and opportunities. When we look at race in this way, the *social significance* that people accord to race is more important than any biological differences that might exist among people who are placed in arbitrary racial categories.

▶ **race** A term used by many people to specify groups of people distinguished by physical characteristics, such as skin colour; also, a category of people who have been singled out as inferior or superior, often on the basis of real or alleged physical characteristics, such as skin colour, hair texture, eye shape, or other subjectively selected attributes.

▶ **ethnic group** A collection of people distinguished, by others or by themselves, primarily on the basis of cultural or nationality characteristics.

A **race** is a category of people who have been singled out as inferior or superior, often on the basis of real or alleged physical characteristics, such as skin colour, hair texture, eye shape, or other subjectively selected attributes (Feagin and Feagin, 2011). Categories of people frequently thought of as racial groups include Asian Canadians, African Canadians, and Native or Indigenous peoples.

How do you classify yourself with regard to race? For an increasing number of people, this is a difficult question to answer. What if you were asked about your ethnic origin or your ethnicity? The Canadian census, unlike that of the United States, collects information on ethnic origin rather than race. Whereas race refers only to *physical* characteristics, the concept of ethnicity refers to *cultural* features. An **ethnic group** is a collection of people distinguished, by others or by themselves, primarily on the basis of cultural or nationality characteristics (Feagin and Feagin, 2011). Ethnic groups share five main characteristics:

1. Unique cultural traits, such as language, clothing, holidays, or religious practices.
2. A sense of community.
3. A feeling of ethnocentrism.
4. Ascribed membership from birth.
5. Territoriality, or the tendency to occupy a distinct geographic area.*

Although some people do not identify with any ethnic group, others participate in social interaction with the individuals in their group and feel a sense of common identity based on cultural characteristics, such as language, religion, or politics. Ethnicity provides individuals with a sense

*Feagin and Feagin, 2011

BOX 11.1 SOCIOLOGY AND EVERYDAY LIFE

How Much Do You Know About Racial and Ethnic Relations in Canada?

True	False	
T	F	1. Canadians are significantly less racist than Americans.
T	F	2. Racism occurs only in times of economic decline and recession.
T	F	3. Canada continues to employ racial criteria in the selection of new immigrants.
T	F	4. No civil rights movement has ever existed in Canada.
T	F	5. Slavery has never existed in Canada.

For answers to the quiz about racial and ethnic relations in Canada, go to **www.nelson.com/student.**

of identity and belonging based not only on their perception of being different but also on others' recognition of their uniqueness. Consider the comments from this university student:

> My ethnic identity is Polish. My parents were born in Poland and came to Canada in 1967 . .. I saw my ethnicity as an advantage and disadvantage during my lifetime. When I was younger, I didn't want to admit that I was Polish. Even though I was born here, I felt that admitting my ethnicity would be a barrier to joining the "in crowd" or the "cool group" at school . . . As I became older, I realized I couldn't change my ethnicity. I was who I was. I became more proud of my Polish background. It felt good to be a part of a Polish community where I was able to participate in ceremonies and activities based on my Polish background. It gave me a sense of belonging to a group, a sense of identity, a sense of security. (James, 2010:62)*

THE SOCIAL SIGNIFICANCE OF RACE AND ETHNICITY

Race and ethnicity take on great social significance because how people act in regard to these terms drastically affects other people's lives, including what opportunities they have, how they are treated, and even how long they live. It matters because it provides privilege and power for some. Sociologist Augie Fleras discusses the significance of being white and enjoying what has sometimes been referred to as *white privilege:*

> Being white means one can purchase a home in any part of town without being "blacklisted" or "redlined" by the local real estate market. Being white allows one to go strolling around shopping malls without the embarrassment of being "blackballed" (e.g., followed, frisked, monitored, or fingerprinted). Being white ensures one a freedom of movement without being pulled over by the police for "driving while black" ("DWB") or "flying while Arab" ("FWA"). Being white simplifies identity construction since whiteness is normal, whereas identity construction is an ongoing aspect of minority existence because people must define who they are in relationship to whiteness . . .To put it bluntly, whiteness is a privilege that is largely unearned yet tacitly accepted . . . Whiteness is a kind of "passport" that opens doors and unlocks opportunities (2012:49).

Not only does the privileging of whiteness affect who gets what, but it also contributes to the 'dis-privileging' of Indigenous peoples and racial minorities (Fleras, 2012). (For a more detailed discussion of white privilege and structured inequality based on race, see Box 11.2.

* Carl E. James, *Seeing Ourselves: Exploring Race, Ethnicity and Culture* (4th ed.). Toronto: Thompson Educational Publishing. Inc., 2010.

BOX 11.2 POINT/COUNTERPOINT

Explaining White Privilege to the Deniers and the Haters

Tim Wise is an American anti-racism author, educator and activist. He is the author of numerous books including *White Like Me: Reflections on Race From a Privileged Son*, 2004, and *Colorblind: The Rise of Post-Racial Politics and the Retreat from Racial Equity*, 2010, and *Culture of Cruelty*, 2013. He also starred in the 2013 documentary *White Like Me*. Below is an excerpt from an article he wrote on white privilege and the resistance many majority group members have to the concept.

Explaining White Privilege (Or, Your Defense Mechanism Is Showing)

I would like to address some of the more glaring, and yet reasonable, misunderstandings that many seem to have about the subject of white privilege. That many white folks don't take well to the term is an understatement, and quite understandable. For those of us in the dominant group, the notion that we may receive certain advantages generally not received by others is a jarring, sometimes maddening concept. And if we don't understand what the term means, and what those who use it mean as they deploy it, our misunderstandings can generate anger and heat, where really, none is called for. So let me take this opportunity to explain what I mean by white privilege.

Many take issue with the notion that there was such a thing, arguing, for instance that there are lots of poor white people who have no privilege, and many folks of colour who are wealthy, who do. But what this argument misses is that race and class privilege are not the same thing.

Though we are used to thinking of privilege as a mere monetary issue, it is more than that. Yes, there are rich black and brown folks, but even they are subject to racial profiling and stereotyping (especially because those who encounter them often don't know they're rich and so view them as decidedly not), as well as bias in mortgage lending, and unequal treatment in schools. So, for instance, even the children of well-off black families are more likely to be suspended or expelled from school than the children of poor whites, and this is true despite the fact that there is no statistically significant difference in the rates of serious school rule infractions between white kids or black kids that could justify the disparity (according to fourteen different studies examined by Russ Skiba at Indiana University).

As for poor whites, though they certainly are suffering economically, this doesn't mean they lack racial privilege. I grew up in a very modest apartment, and economically was far from privileged. Yet I received better treatment in school (placement in advanced track classes even when I wasn't a good student), better treatment by law enforcement officers, and indeed more job opportunities because of connections I was able to take advantage of, that were pretty much unavailable to the folks of colour I knew growing up. Likewise, low income whites everywhere are able to clean up, go to a job interview and be seen as just another white person, whereas a person of colour, even who isn't low-income, has to wonder whether or not they might trip some negative stereotype about their group when they go for an interview or sit in the classroom answering questions from the teacher.

The point is, privilege is as much a psychological matter as a material one. Whites have the luxury of not having to worry that our race is going to mark us negatively when looking for work, going to school, shopping, looking for a place to live, or driving for that matter: things that folks of colour can't take for granted.

Let me share an analogy to make the point.

Taking things out of the racial context for a minute: imagine persons who are able bodied, as opposed to those with disabilities. If I were to say that able-bodied persons have certain advantages, certain privileges if you will, which disabled persons do not, who would argue the point? I imagine that no one would. It's too obvious, right? To be disabled is to face numerous obstacles. And although many persons with disabilities overcome those obstacles, this fact doesn't take away from the fact that they exist. Likewise, that persons with disabilities can and do overcome obstacles every day, doesn't deny that those of us who are able-bodied have an edge. We have one less thing to think and worry about as we enter a building, go to a workplace, or just try and navigate the contours of daily life. The fact that there are lots of able-bodied people who are poor, and some disabled folks who are rich, doesn't alter the general rule: on balance, it pays to be able-bodied.

That's all I'm saying about white privilege: on balance, it pays to be a member of the dominant racial group. It doesn't mean that a white person will get everything they want in life, or win every competition, but it does mean that there are general advantages that we receive. So, for instance, studies have found that job applicants with white sounding names are 50% more likely to receive a call-back for a job interview than applicants with black-sounding names, even when all job-related qualifications and credentials are the same. Despite the fact that white men are more likely to be caught with drugs in our car (on those occasions when we are searched), black men remain about four times more likely than white men to be searched in the first place. That's privilege for the dominant group.

That's the point: privilege is the flipside of discrimination. If people of colour face discrimination, in housing, employment and elsewhere, then the rest of us are receiving a de facto subsidy, a privilege, an advantage in those realms of daily life. There can be no down without an up, in other words.

None of this means that white folks don't face challenges. Of course we do, and some of them (based on class, gender, sexual orientation, disability status, or other factors) are systemic and institutionalized. But on balance, we can take for granted that we will receive a leg-up on those persons of colour with whom we share a nation.

And no, affirmative action doesn't change any of this.

Despite white fears to the contrary, even with affirmative action in place (which, contrary to popular belief does not allow quotas except in those rare cases where blatant discrimination has been proven), whites hold about ninety percent of all the management level jobs in this country, receive about ninety-four percent of government contract dollars, and hold ninety percent of tenured faculty positions on college campuses. In other words, despite the notion that somehow we have attained an equal opportunity, or colour-blind society, the fact is, we are far from an equitable nation. People of colour continue to face obstacles based solely on colour, and whites continue to reap benefits from the same. None of this makes whites bad people, and none of it means we should feel guilty or beat ourselves up. But it does mean we need to figure out how we're going to be accountable for our unearned advantages. One way is by fighting for a society in which those privileges will no longer exist, and in which we will be able to stand on our own two feet, without the artificial crutch of racial advantage to prop us up. We need to commit to fighting for racial equity and challenging injustice at every turn, not only because it harms others, but because it diminishes us as well (even as it pays dividends), and because it squanders the promise of fairness and equity to which we claim to adhere to.

It's about responsibility, not guilt. And if one can't see the difference between those two things, there is little that this or any other article can probably do. Perhaps starting with a dictionary would be better.

Source: Tim Wise, "Explaining White Privilege (Or, Your Defense Mechanism Is Showing)", http://www.timwise.org/2008/09/explaining-white-privilege-or-your-defense-mechanism-is-showing/. Reproduced by permission of the author.

Ethnicity, like race, is a basis of hierarchical ranking in society. John Porter (1965) described Canada as a "vertical mosaic," made up of different ethnic groups wielding varying degrees of social and economic power, status, and prestige. Porter's analysis of ethnic groups in Canada revealed a significant degree of ethnic stratification, with some ethnic groups heavily represented in the upper strata, or elite, and other groups heavily represented in the lower strata. The dominant group holds power over other (subordinate) ethnic groups. To what extent does a "vertical mosaic" still exist in Canada? A 2009 study by Philip Oreopoulos found that, despite the fact that immigrants to Canada are selected on the basis of their optimal skills, education, and professional qualifications, immigrants and ethnic minority Canadians still have significantly lower incomes and higher rates of unemployment. Oreopoulos constructed "mock" resumés representative of recent immigrants from the three largest countries of origin (China, India, and Pakistan) and Britain, as well as nonimmigrants with and without ethnic-sounding names. Six thousand resumés were sent out to apply to online job postings in the Toronto area. The findings indicated that applicants with English-sounding names with Canadian education and experience received callbacks 40 percent more often than did applicants with Chinese, Indian, or Pakistani names who had similar Canadian education and experience (Oreopoulos, 2009). This study provides evidence of continued ethnic stratification based on what Oreopoulos described as "substantial discrimination" by employers. Ethnic stratification is one dimension of a larger system of structured social inequality, as examined in Chapter 9.

MAJORITY AND MINORITY GROUPS

majority (dominant) group An advantaged group that has the greatest power and resources in a society.

minority (subordinate) group A group whose members, because of physical or cultural characteristics, are disadvantaged and subjected to unequal treatment and discrimination by the dominant group.

visible minority Government term used to describe those who are nonwhite, non-Indigenous, or non-Caucasian. This term is used interchangeably with "people of colour" and "racialized minorities."

The terms majority group and minority group are widely used, but what do they actually mean? To sociologists, a **majority** (or **dominant) group** is one that has the greatest power and resources in a society (Feagin and Feagin, 2011). In Canada, whites with northern European ancestry (often referred to as Euro-Canadians or white Anglo-Saxon Protestants, or WASPs) are considered the majority group. A **minority** (or **subordinate) group** is one whose members, because of physical or cultural characteristics, are disadvantaged and subjected to unequal treatment and discrimination by the dominant group. All visible minorities and white women are considered minority group members in Canada. The term **visible minority,** first introduced in the 1995 Employment Equity Act, refers to "persons, other than Indigenous peoples, who are non-Caucasian in race or non- white in colour" (Employment Equity Act, 1995). In other words, the term refers to a person who is, because of their race or colour, in a visible minority. Included in this category are Chinese, Japanese, Koreans, Filipinos, Asians, South Asians, Arabs, Southeast Asians, blacks, Latin Americans, and Pacific Islanders (Statistics Canada, 2013a). Indigenous people form a separate category of individuals with minority group status. The term "visible minority" became the subject of controversy in 2007 when a United Nations Committee on the Elimination of Racism suggested that the term might be racist. Although the term was initially created by the government to highlight groups who may be subject to discrimination in employment based on their race or ethnicity, it has become the most widely used term to describe 'people of colour' or 'racialized minorities' in Canada. The United Nations' report suggested that Canada should 'reflect further' on the use of the term 'visible minority' because it may have the unintended result of increasing discrimination by its suggestion that "whiteness was a standard, white people being invisible and others visible" (United Nations Committee on the Elimination of Racial Discrimination, 2007, Para 50). Although the United Nations committee did not expressly prohibit the use of the term, they indicated that they could not condone the use of a term that could be seen as a form of racism. The term has been further criticized because it distorts the fact that "visible minorities" actually comprise the majority of the population in several Canadian cities, including Toronto and Vancouver (Statistics Canada, 2013a). The term racialized minorities is increasingly being used because it better reflects the fact that attaching a race label to minorities is a socially constructed process rather than a description of a reality based on biological traits (Fleras 2012).

Five million Canadians—or close to one in five—identified themselves as members of a visible minority. In the most recent census, South Asians, Chinese, and blacks accounted for close to two-thirds of the visible minority population (Statistics Canada, 2013a).

Although the terms *majority group* and *minority group* are widely used, their actual meanings are not clear. In the sociological sense, *group* is misleading because people who merely share ascribed racial or ethnic characteristics do not constitute a group. Further, *majority* and *minority* have meanings associated with both numbers and domination. Numerically speaking, *minority* means that a group is smaller in number than a dominant group. In countries such as South Africa and India, however, this has not historically been true.

TIME TO REVIEW

- Explain the statement "Race is a social construct."
- How significant do you think this social construct is in the lives of visible minority group members?
- What is the significance of race in the lives of majority group members?

LO-2 PREJUDICE

Prejudice is a negative attitude based on preconceived notions about members of selected groups. The term *prejudice* comes from the Latin words *prae* ("before") and *judicium* ("judgment"), which means that people may be biased either for or against members of other groups before they have had any contact with them. Although prejudice can be either *positive* (bias in favour of a group—often our own) or *negative* (bias against a group—one we deem less worthy than our own), it most often refers to the negative attitudes people may have about members of other racial or ethnic groups. **Racial prejudice** involves beliefs that certain racial groups are innately inferior to others or have a disproportionate number of negative traits.

prejudice A negative attitude based on preconceived notions about members of selected groups.

racial prejudice Beliefs that certain racial groups are innately inferior to others or have a disproportionate number of negative traits.

STEREOTYPES

Prejudice is rooted in stereotypes and ethnocentrism. When used in the context of racial and ethnic relations, ethnocentrism refers to the tendency to regard one's own culture and group as the standard—and thus superior—whereas all other groups are seen as inferior. Ethnocentrism is maintained and perpetuated by **stereotypes**—overgeneralizations about the appearance, behaviour, or other characteristics of members of particular groups. Although all stereotypes are hurtful, negative stereotypes are particularly harmful to members of minority groups. Consider for example, Naomi's experience:

stereotype An overgeneralization about the appearance, behaviour, or other characteristics of members of particular groups.

> People whom I meet frequently ask, "What are you?" as a way of determining my racial background. I then proceed to tell them that I am Canadian. Then they ask me, "Where are your parents from?" I tell them Poland and they then look confused . . . And then when they learn that I am Jewish, their responses always amaze me. People express surprise and say, "You are Jewish!" as if I had a disease or something. And some people think they are paying me a compliment by saying, "We do not think of you as Jewish; you are different than most Jewish people we know." This is an outright insult to my ethnicity, of which I am proud. Another typical comment is that I "do not look Jewish." I do not understand what it means to "look Jewish" considering that there are Jewish people from all over the world. (James, 2010:216)*

How do people develop these stereotypes? Although stereotypes can be either positive or negative, examples of negative stereotyping abound in sports. Think about the Native American names, images, and mascots used by sports teams such as the Chicago Blackhawks, Cleveland Indians, and Washington Redskins. Members of First Nations groups have been actively working to eliminate the use of stereotypic mascots (with feathers, buckskins, beads, spears, and "warpaint"), "Indian chants," and gestures (such as the "tomahawk chop"), which they claim trivialize and exploit Indigenous culture. In 2010, Wisconsin passed a law that allows residents of a school district to lodge complaints regarding race-based names of school sports teams in an effort to eliminate any hostile environment that might be created by the use of such names and imagery (Keen, 2010). College and university sports teams with Native American names and logos also remain the subject of controversy. In 2005, the NCAA prohibited the use of Native American mascots and nicknames in bowl games and other postseason competition (Keen, 2010). According to sociologist Jay Coakley (2009), the use of stereotypes and words such as *redskin* symbolizes a lack of understanding of the culture and heritage of Indigenous peoples and is offensive to many Indigenous groups. Although some people see these names and activities as "innocent fun," others view them as a form of racism.

* Carl E. James, *Seeing Ourselves: Exploring Race, Ethnicity and Culture* (4th ed.). Toronto: Thompson Educational Publishing. Inc., 2010.

THEORIES OF PREJUDICE

Are some people more prejudiced than others? To answer this question, some theories focus on how individuals may transfer their internal psychological problems onto an external object or person. Others look at factors such as social learning and personality types.

The frustration-aggression hypothesis states that people who are frustrated in their efforts to achieve a highly desired goal will respond with a pattern of aggression toward others (Dollard et al., 1939). The object of their aggression becomes the **scapegoat**—a person or group that is incapable of offering resistance to the hostility or aggression of others (Marger, 2012). Scapegoats are often used as substitutes for the actual source of the frustration. For example, members of subordinate racial and ethnic groups are often blamed for societal problems (such as unemployment or an economic recession) over which they have no control.

▶ **scapegoat** A person or group that is incapable of offering resistance to the hostility or aggression of others.

According to some symbolic interactionists, prejudice results from social learning; in other words, it is learned from observing and imitating significant others, such as parents and peers. Initially, children do not have a frame of reference from which to question the prejudices of their relatives and friends. When they are rewarded with smiles or laughs for telling derogatory jokes or making negative comments about outgroup members, children's prejudiced attitudes may be reinforced.

Psychologist Theodor W. Adorno and his colleagues concluded that highly prejudiced individuals tend to have an **authoritarian personality,** which is characterized by excessive conformity, submissiveness to authority, intolerance, insecurity, a high level of superstition, and rigid, stereotypic thinking (Adorno et al., 1950). It is most likely to develop in a family environment in which dominating parents who are anxious about status use physical discipline but show very little love in raising their children (Adorno et al., 1950). Other scholars have linked prejudiced attitudes to traits such as submissiveness to authority, extreme anger toward outgroups, and conservative religious and political beliefs (Altemeyer, 1981, 1988; Weigel and Howes, 1985).

▶ **authoritarian personality** A personality type characterized by excessive conformity, submissiveness to authority, intolerance, insecurity, a high level of superstition, and rigid, stereotypic thinking.

DISCRIMINATION

Whereas prejudice is an attitude, **discrimination** involves actions or practices of dominant group members (or their representatives) that have a harmful impact on members of a subordinate group (Feagin and Feagin, 2011). For example, people who are prejudiced toward South Asian, Jewish, or Indigenous people may refuse to hire them, rent an apartment to them, or allow their children to play with them. In these instances, discrimination involves the differential treatment of minority group members not because of their ability or merit but because of irrelevant characteristics, such as skin colour or language preference. Discriminatory actions vary in severity from the use of derogatory labels to violence against individuals and groups.

▶ **discrimination** Actions or practices of dominant group members (or their representatives) that have a harmful impact on members of a subordinate group.

Discrimination takes two basic forms: *de jure,* or legal discrimination, which is encoded in laws; and *de facto,* or informal discrimination, which is entrenched in social customs and institutions. *De jure* discrimination has been supported with explicitly discriminatory laws, such as the *Chinese Exclusionary Act,* which restricted immigration to Canada on the basis of race, or the Nuremberg laws passed in Nazi Germany, which imposed restrictions on Jews. The *Indian Act* provides other examples of *de jure* discrimination. According to the act, a Native woman who married a non-Native man automatically lost her Indian status rights and was no longer allowed to live on a reserve. Native men had no such problem. The *Indian Act* also specified that Indigenous persons who graduated from university, or who became doctors, lawyers, or ministers before 1920, were forced to give up their status rights. An amendment to the *Indian Act* in 1985 ended this legalized discrimination. The *Charter of Rights and Freedoms* prohibits discrimination on the basis of race, ethnicity, or religion. As a result, many cases of *de jure* discrimination have been eliminated. *De facto* discrimination is more subtle and less visible to public scrutiny and therefore much more difficult to eradicate.

Prejudiced attitudes do not always lead to discriminatory behaviour. Sociologist Robert Merton (1949) identified four combinations of attitudes and responses. *Unprejudiced nondiscriminators* are not personally prejudiced and do not discriminate against others. These are individuals who believe in equality for all. *Unprejudiced discriminators* may have no personal prejudices but still engage in discriminatory behaviour because of peer group pressure or economic, political, or social interests—for example, an employee who has no personal hostility toward members of certain groups but is encouraged by senior management not to hire them. *Prejudiced nondiscriminators* hold personal prejudices but do not discriminate due to peer pressure, legal demands, or a desire for profits. Such individuals are often referred to as "timid bigots" because they are reluctant to translate their attitudes into action (especially when prejudice is considered to be "politically incorrect"). Finally, *prejudiced discriminators* hold personal prejudices and actively discriminate against others—for example, the landlord who refuses to rent an apartment to an Indigenous couple and then readily justifies his actions on the basis of racist stereotypes.

Merton's typology shows that some people may be prejudiced but not discriminate against others. Do you think it is possible for a person to discriminate against some people without holding a prejudiced attitude toward them? Why, or why not?

RACISM

Racism is a set of ideas that implies the superiority of one social group over another on the basis of biological or cultural characteristics, *together* with the power to put these beliefs into practice in a way that controls, excludes, or exploits minority women and men.

Racism involves elements of prejudice, ethnocentrism, stereotyping, and discrimination. For example, racism is present in the belief that some racial or ethnic groups are superior while others are inferior—this belief is a prejudice. Racism may be the basis for unfair treatment toward members of a racial or ethnic group. In this case, the racism involves discrimination.

Fleras (2012) makes distinctions among a number of diverse types of racism (see Table 11.1). **Hate racism** (or **overt racism**) may take the form of deliberate and highly personal attacks, including derogatory slurs and name-calling toward members of a racial or ethnic group who are perceived to be "inferior" (James, 2010). Examples of hate racism, although rare, are available in Canada. In 2009, a Winnipeg case made national headlines when a young girl attended school with white supremist symbols and slogans drawn all over her skin. The girl told social workers that she watched violent racist videos in her home and her parents regularly discussed killing minorities. Hate racism is also demonstrated in the racist violence perpetuated by members of white supremacist groups, including the Heritage Front, White Aryan Nation, and Western Guard, that are active in Canada. These groups are committed to an ideology of racial supremacy in which the white "race" is seen as superior to other races. This type of overt racism is becoming increasingly unacceptable in Canadian society, and few people today will tolerate the open expression of racism. In fact, overt acts of discrimination motivated by hatred are considered hate crimes. *The Criminal Code* includes four specific offences that are classified as hate crimes: advocating genocide, public incitement of hatred, willful promotion of hatred, and mischief. According to the most recent statistics, over half of all police-reported hate crimes were motivated by hatred by race or ethnicity, and another another 28 percent were motivated by hate in relation to religious property. Although *The Criminal Code,* the *Charter of Rights and Freedoms,* and human rights legislation have served to limit the expression of, hate, there were still approximately 1500 incidents of hate crime reported to the police last year (Allen, 2014).

While blatant forms of racism have dissipated to some extent, less obvious expressions of bigotry and stereotyping that allow people to discuss their dislike of certain groups in "coded language" remain in our society. **Polite racism** is an attempt to disguise a dislike of others through behaviour that is outwardly nonprejudicial. Polite racism may consist of subtle remarks or looks that result in members of racialized minorities feeling inferior or out of place.

▶ **racism** A set of ideas that implies the superiority of one social group over another on the basis of biological or cultural characteristics, together with the power to put these beliefs into practice in a way that controls, excludes, or exploits minority women and men.

▶ **hate racism (or overt racism)** Racism that may take the form of deliberate and highly personal attacks, including derogatory slurs and name-calling toward members of a racial or ethnic group who are perceived to be "inferior."

▶ **polite racism** A term used to describe an attempt to disguise a dislike of others through behaviour that is outwardly nonprejudicial.

TABLE 11.1 THE FACTS OF RACISM

	WHAT: CORE SLOGAN	WHY: DEGREE OF INTENT	HOW: STYLE OF EXPRESSION	WHERE: MAGNITUDE AND SCOPE
Hate racism	"X, get out."	Conscious	Personal and explicit	Interpersonal
Polite racism	"Sorry, the job is taken."	Moderate	Discreet and subtle	Interpersonal
Subliminal racism	"I'm not racist, but ..."	Ambivalent	Oblique	Cultural
Institutionalized racism	"We treat everyone the same here."	Unintentional or intentional	Impersonal	Institutional and societal

Sources: Fleras and Elliott, 1996, 2003; Fleras, 2012.

Polite racism is often operating in situations where racialized minorities are turned down for jobs, promotions, or accommodation. For example, "sorry the apartment was just rented" is a polite way of rejecting undesirable tenants (Fleras, 2012). A number of studies have confirmed the extent to which this type of racism continues to manifest itself in the workplace (Henry, 2006; Kunz, Milan, and Schetagne, 2000; Oreopoulos, 2009). Although this form of racism may appear to be less hurtful, it is no less harmful in terms of its damaging effect on the victims.

▶ **subliminal racism** A term used to describe an unconscious racism that occurs when there is a conflict of values.

Subliminal racism is a form of subconscious racism that involves prejudices which individuals are unaware of but that display themselves in discriminatory beliefs and behaviours. Subliminal racism is not directly expressed but is demonstrated in opposition to progressive minority policies (such as Canada's immigration policy) or programs (such as employment equity or affirmative action). For example, after the 9/11 terrorist attacks, there were insinuations that Canada's "weak" immigration policies allowed the terrorists to enter the United States. Subliminal racism allows us to understand how mainstream whites can simultaneously demonstrate nearly universal support for principles of equality and at the same time undermine progressive policies and strategies directed at achieving that equality. As Fleras and Elliott highlight:

AFP/Getty Images

Recent anti-Semitic attacks on Jewish synagogues are an unfortunate indication that some forms of hate racism still exist.

> Refugee claimants are not condemned in blunt racist terminology; rather their landed entry into Canada is criticized on procedural grounds ("jumping the queue"). Or they are belittled for taking unfair advantage of Canada's generosity or ability to shoulder the processing costs . . . Minority peoples have rights, but minority demands that fall outside conventional channels are criticized as a threat to national identity or social harmony . . . Employment equity initiatives are endorsed in principle but rejected in practice as unfair to the majority. (2003:73)

Subliminal racism, more than any other type, demonstrates the ambiguity concerning racism. Values that support racial equality are publicly supported while, at the same time, resentment at the prospect of moving over and making space for newcomers is also present. Subliminal racism enables individuals to maintain two apparently conflicting values—one rooted in the egalitarian virtues of justice and fairness, the other in beliefs that result in resentment and selfishness. Not surprisingly, this form of racism is often found among people who openly detest discriminatory treatment while at the same time make no effort to do something about it (Fleras, 2012).

Institutionalized racism occurs where the established rules, policies, and practices within an institution or organization produce differential treatment of various groups based on race. Although institutions can no longer openly discriminate against minorities without attracting legal sanctions, negative publicity, or consumer resistance, this type of racism nevertheless continues to exist (Fleras, 2012). The practice of word-of-mouth recruitment is an example of an institutional practice that has the result of excluding racial minorities from the hiring selection process.

institutionalized racism A situation where the established rules, policies, and practices within an institution or organization produce differential treatment of various groups based on race.

Institutional racism may also be reflected in organizational practices, rules, and procedures that have the unintended consequence of excluding minority group members. For example, occupations such as police officer and firefighter historically had minimum weight, height, and educational requirements for job applicants. These criteria resulted in discrimination because they favoured white applicants over members of many minority groups, as well as males over females. Other examples of this type of institutional racism include the requirement of a college or university degree for non-specialized jobs, employment regulations that require people to work on their Sabbath, and the lack of recognition of foreign credentials. Institutional racism is normally reflected in statistical underrepresentation of certain groups within an institution or organization. For example, a given group may represent 15 percent of the general population but only 2 percent of those promoted to upper-management positions in a large company.

Efforts to eliminate this kind of disproportionate representation are the focus of employment equity legislation. The target groups for employment equity in Canada are visible minorities, women, persons with disabilities, and Indigenous peoples. Strategies include modified admissions tests and requirements, enhanced recruitment of certain target groups, establishment of hiring quotas for particular minority groups, or specialized training or employment programs for specific target groups. Consideration of affirmative action strategies inevitably leads to claims of reverse discrimination by some individuals who enjoy majority group status. (For a more detailed discussion of reverse discrimination, see Box 11.3.) Consider the comments from this white male student:

> I am a white male and I am discriminated against all the time. Faced with trying to get jobs that have been reserved for minorities and not the best candidate. There is racism in Canada and as a white male I feel lots of it is aimed at myself. There was an article not long ago that Toronto Police want to hire more minority police. I think that comment in itself is racist. We don't want the best person for the job anymore? (James, 2010:243)*

The most recent analysis of employment equity programs indicates that these programs have had the most significant effect on women and Indigenous peoples, while people with disabilities have made the fewest gains. As for members of visible minorities, although they have higher levels of education, on average, than other Canadians and very high labour-force participation rates, they continue to be concentrated in low-status, low-paying occupations (Henry and Tator, 2009, Oreopoulos, 2009).

TIME TO REVIEW

- Former Prime Minister Stephen Harper recently indicated that Muslim women should not be permitted to wear hijabs at their Canadian citizenship swearing-in ceremonies. Is this proposed prohibition a form of racism? What type of racism is it?
- In considering all of the types of racism, which is the most difficult to control? Which type does the most damage?

* Carl E. James, *Seeing Ourselves: Exploring Race, Ethnicity and Culture* (4th ed.). Toronto: Thompson Educational Publishing. Inc., 2010.

BOX 11.3 POINT/COUNTERPOINT

The Myth of Reverse Racism

Is reverse racism possible? According to race relations scholars Augie Fleras and Jean Leonard Elliott, the answer is no. In the following excerpt from *Unequal Relations* (2003), they explain why:

> Are affirmative actions policies that favour visible minority group members and Aboriginal persons racist? Can minority women and men express racism ("reverse racism") against the majority sector? Can ethnic minorities be racist toward other ethnic minorities? Is it racist for Aboriginal peoples to accuse all whites of complicity in the destruction of Indigenous societies? Answers to these questions may never be settled to everyone's satisfaction, given the politics or intellectual dishonesty at play, but their very asking provides a sharper understanding of racism.
>
> Responses depend on how one defines racism—as biology or power. A reading of racism as biology suggests that anyone who approaches, defines, or treats someone else on the basis of race is a racist. Thus, minorities can be racist if they criticize or deny whites because of their whiteness ("reverse racism").
>
> But reference to racism as power points to a different conclusion. Accusations of minority ("reverse") racism must go beyond superficial appearances. There is a world of difference in using race to create equality (employment equity) versus its use to limit opportunity (discrimination), even if the rhetoric sounds the same. Emphasis must be placed instead on the context of the actions and their social consequences. Racism is not about treating others differently because they are different. Rather, it involves different treatment in colour-conscious contexts of power that limits opportunity or privileges (Blauner 1972).
>
> In short, racism is about the politics of difference within the context of power. Statements made by a minority group, however distasteful or bigoted, may not qualify as racist in the conventional sense of outcomes. They are largely preferences or prejudices without the capacity for harm, since minorities lack the institutional power to put bigotry into practice in a way that "stings."
>
> To be sure, minorities are not entirely powerless; after all, there is recourse to alternative sources of power-brokering, such as boycotts, civil disobedience, lobby groups, and moral suasion. And even though they may not have institutional power, minorities may have other ways to put bigotry into practice (e.g., stealing from a store owned by a member of another minority or threatening others on the basis of appearance).
>
> Still, the power that minority individuals wield in certain contexts rarely has the potential to deny or exclude. Those without access to institutionalized power or resources cannot racialize the other in ways that demean, control, or exploit. Minorities do not have the power to dominate and enforce prejudices, oppression, or subdomination. They have neither the resources to topple the dominant sector nor the critical mass to harass, exclude, exploit, persecute, dominate, or undermine the empowered. Conditions of relative powerlessness reduce minority hostility to the level of rhetoric or a protective shell in defence of minority interests. In other words, reverse racism may be a contradiction in terms. Racism is not a two-way street; more accurately, it resembles an expressway with controlled access points for those privileged enough to control the switches.

Source: Fleras and Elliot 2003.

LO-3 SOCIOLOGICAL PERSPECTIVES ON RACE AND ETHNIC RELATIONS

Symbolic interactionist, functionalist, conflict, and feminist perspectives examine race and ethnic relations in different ways. Symbolic interactionists examine how microlevel contacts between people may produce either greater racial tolerance or increased levels of hostility. Functionalists focus on the macrolevel intergroup processes that occur among members of majority and minority groups in society. Conflict theorists analyze power and economic differentials between the dominant group and subordinate groups. Feminists highlight the interactive effects of racism and sexism on the exploitation of women, who are members of a visible minority.

SYMBOLIC INTERACTIONIST PERSPECTIVES

What happens when people from different racial and ethnic groups come into contact with one another? In the *contact hypothesis,* symbolic interactionists point out that contact between people from divergent groups should lead to favourable attitudes and behaviour when certain factors are present. Members of each group must (1) have equal status, (2) pursue the same goals, (3) cooperate with one another to achieve their goals, and (4) receive positive feedback when they interact with one another in positive, nondiscriminatory ways. However, if these factors are not present, intergroup contact may lead to increased stereotyping and prejudice.

What happens when individuals meet someone who does not conform to their existing stereotype? According to symbolic interactionists, they frequently ignore anything that contradicts the stereotype, or they interpret the situation to support their prejudices. For example, a person who does not fit the stereotype may be seen as an exception—"You're not like other [persons of a particular race]." Conversely, when a person is seen as conforming to a stereotype, he or she may be treated simply as one of "you people."

When people from different racial and ethnic groups come into contact with one another, they may treat one another as stereotypes, not as individuals. For example Kai James writes of his experience as a student at a Toronto school: "Gym teachers are perhaps the most overt in their interpretations of stereotypes. I remember the track coach coming into my grade 9 class and asking all the Black students if they would be participating in the track meet" (James, 2010:242).* Symbolic interactionist perspectives make us aware of the importance of intergroup contact and the fact that it may either intensify or reduce racial and ethnic stereotyping and prejudice.

FUNCTIONALIST PERSPECTIVES

How do members of subordinate racial and ethnic groups become part of the dominant group? To answer this question, early functionalists studied immigration and patterns of majority and minority group interaction.

ASSIMILATION **Assimilation** is a process by which members of subordinate racial and ethnic groups become absorbed into the dominant culture. To some analysts, assimilation is functional because it contributes to the stability of society by minimizing group differences that otherwise might result in hostility and violence.

▶ **assimilation** A process by which members of subordinate racial and ethnic groups become absorbed into the dominant culture.

Assimilation occurs at several distinct levels, including the cultural, structural, biological, and psychological stages. *Cultural assimilation,* or *acculturation,* occurs when members of an ethnic group adopt dominant group traits, such as language, dress, values, religion, and food preferences. Cultural assimilation in this country initially followed an "Anglo-conformity" model; members of subordinate ethnic groups were expected to conform to the culture of the dominant white Anglo-Saxon population (Gordon, 1964). However, members of some groups, such as Indigenous peoples and Québécois, refused to be assimilated and sought to maintain their unique cultural identity.

Structural assimilation, or *integration,* occurs when members of subordinate racial or ethnic groups gain acceptance in everyday social interaction with members of the dominant group. This type of assimilation typically starts in large, impersonal settings, such as schools and workplaces, and only later (if at all) results in close friendships and intermarriage.

Biological assimilation, or *amalgamation,* occurs when members of one group marry those of other social or ethnic groups. Biological assimilation has been more complete in some other countries, such as Mexico and Brazil, than in Canada.

Psychological assimilation involves a change in racial or ethnic self-identification on the part of an individual. Rejection by the dominant group may prevent psychological assimilation by members of some subordinate racial and ethnic groups, especially those with visible characteristics, such as skin colour or facial features that differ from those of the dominant group.

* Carl E. James, *Seeing Ourselves: Exploring Race, Ethnicity and Culture* (4th ed.). Toronto: Thompson Educational Publishing. Inc., 2010.

ethnic pluralism The coexistence of a variety of distinct racial and ethnic groups within one society.

ETHNIC PLURALISM Instead of complete assimilation, many groups share elements of the mainstream culture while remaining culturally distinct from both the dominant group and other social and ethnic groups. **Ethnic pluralism** is the coexistence of a variety of distinct racial and ethnic groups within one society.

Ongoing battles over such issues as the proposed prohibition of Muslim women wearing face-covering veils during citizenship ceremonies or while testifying in court suggest Canada has yet to achieve equalitarian pluralism.

teena137/Shutterstock.com

Equalitarian pluralism, or *accommodation,* is a situation in which ethnic groups coexist in equality with one another. Switzerland has been described as a model of equalitarian pluralism; more than six million people with French, German, and Italian cultural heritages peacefully coexist there.

Has Canada achieved equalitarian pluralism? The *Canadian Multiculturalism Act* of 1988 states that "All Canadians are full and equal partners in Canadian society." The Department of Multiculturalism and Citizenship was established in 1991 with the goal of encouraging ethnic minorities to participate fully in all aspects of Canadian life while at the same time maintaining their distinct ethnic identities and cultural practices. The objective of multiculturalism is to "promote unity through diversity." Under multiculturalism, citizens are accepted as racially or ethnically different yet no less Canadian, with a corresponding package of citizen rights and entitlement, regardless of origin, creed, or colour (Fleras and Elliott, 2003:280). Multiculturalism programs provide funding for education, consultative support, and a range of activities, including heritage language training, race relations training, ethnic policing and justice, and ethnic celebrations. In implementing this pluralistic strategy, Canada gained international respect and admiration as a society that is both united and distinct, where citizens are valued as "different" yet recognized as "equal."

In recent years, multiculturalism policies have been under increasing attack. For example, multiculturalism has been described as a policy that creates and maintains an "illusion" of respect for racial and ethnic differences, when in reality the pressures toward conformity and the experiences with exclusion and discrimination for Canadian "multicultural minorities" are impossible to ignore (see Box 3.2 in Chapter 3, Multiculturalism, Reasonable Accommodation, and "Veiled" Hostility).

Neil Bissoondath, author of *Selling Illusions: The Cult of Multiculturalism in Canada* (1994), suggests that multiculturalism does not promote equalitarian pluralism. Rather, he argues, multiculturalism serves to discourage immigrants from thinking of themselves as Canadian; it exaggerates differences, which fosters racial animosity; and it alienates people from the mainstream society, which detracts from national unity. Bissoondath argues:

> Whatever policy follows multiculturalism it should support a new vision of Canadianness. A Canada where no one is alienated with hyphenation. A nation of cultural hybrids, where every individual is unique and every individual is a Canadian, undiluted and undivided. A nation where the following conversation, so familiar—and so enervating—to many of us will no longer take place: "What nationality are you?" "Canadian." "No, I mean, what nationality are you *really?*" (1998:1)

The challenge for a pluralistic society such as Canada lies in attaining some degree of balance between the equally important values of racial and ethnic equality and national unity. To date, any consensus on multiculturalism in terms of definition, policy, or practice remains illusive.

segregation A term used to describe the spatial and social separation of categories of people by race/ethnicity, class, gender, and/or religion.

INEQUALITARIAN PLURALISM, OR SEGREGATION *Inequalitarian pluralism,* or ***segregation***, exists when specific ethnic groups are set apart from the dominant group and have unequal access to power and privilege. Segregation is the spatial and social separation of categories of people by race, ethnicity, class, gender, and/or religion. Segregation may be enforced by law (de jure) or by custom (de facto).

An example of *de jure* segregation was the Jim Crow laws, which legalized the separation of the races in all public accommodations (including hotels, restaurants, transportation, hospitals, jails, schools, churches, and cemeteries) in the Southern United States after the Civil War (Feagin and Feagin, 2011).

De jure segregation of blacks is also part of the history of Canada. Blacks in Canada lived in largely segregated communities in Nova Scotia, New Brunswick, and Ontario, where racial segregation was evident in the schools, government, the workplace, residential housing, and elsewhere. Segregated schools continued in Nova Scotia until the 1960s. Residential segregation was legally enforced through the use of racially restrictive covenants attached to deeds and leases. Separation and refusal of service were common in restaurants, theatres, and recreational facilities (Henry and Tator, 2009). Sociologist Adrienne Shadd describes her experiences growing up in North Buxton, Ontario, in the 1950s and 1960s:

> When we would go into the local ice cream parlour, the man behind the counter would serve us last, after all the Whites had been served, even if they came into the shop after us. Southwestern Ontario may as well have been below the Mason-Dixon line in those days. Dresden, home of the historic Uncle Tom's cabin, made national headlines in 1954 when Blacks tested the local restaurants after the passage of the *Fair Accommodation Practices Act* and found that two openly refused to serve them. This came as no surprise, given that for years certain eateries, hotels, and recreational clubs were restricted to us, and at one time Blacks could only sit in designated sections of movie theatres (usually the balcony) if admitted at all. (1991:11)

One of the most blatant examples of segregation in Canada is the federal government's reserve system for status Indians. Canadian Indians were placed on Reserves in the late 18th century in order to clear land for newly arrived European immigrants and settlers from the United States. Although these Reserves were originally located within the areas which various tribes had long occupied, the actual size of the enclosures was greatly reduced from their previous territories, which resulted in segregation of Indigenous peoples on reserves in remote areas across the country.

Segregation laws existed and were enforced with signs such as these in both Canada and the United States until the 1960s.

Although functionalist explanations provide a description of how some early white ethnic immigrants assimilated into the cultural mainstream, they do not adequately account for the persistent racial segregation and economic inequality experienced by some minority group members.

TIME TO REVIEW

- Compare and contrast assimilation, ethnic pluralism, and segregation.

CONFLICT PERSPECTIVES

Why do some ethnic groups continue to experience subjugation after many years? Conflict theorists focus on economic stratification and access to power in their analysis of race and ethnic relations.

internal colonialism According to conflict theorists, a situation in which members of a racial or ethnic group are conquered or colonized and forcibly placed under the economic and political control of the dominant group.

INTERNAL COLONIALISM Conflict theorists use the term **internal colonialism** to refer to a situation in which members of a racial or ethnic group are conquered or colonized and forcibly placed under the economic and political control of the dominant group. Groups that have been subjected to internal colonialism often remain in subordinate positions longer than groups that voluntarily migrated to North America.

Indigenous peoples in Canada were colonized by Europeans and others who invaded their lands and conquered them. In the process, Indigenous peoples lost property, political rights, aspects of their culture, and often their lives (Frideres and Gadacz, 2011). The capitalist class acquired cheap labour and land through this government-sanctioned racial exploitation. The effects of past internal colonialism are reflected today in the number of Indigenous people who live in extreme poverty on government reserves (Frideres and Gadacz, 2011).

Charla Jones/GetStock.com

The effects of past colonialism are reflected in what has been described by the United Nations as the sub-humane housing conditions of many Indigenous persons on some reserves in Canada.

The internal colonialism model is rooted in historical foundations of racial and ethnic inequality in North America. However, it tends to view all voluntary immigrants as having many more opportunities than do members of colonized groups. Thus, this model does not explain the continued exploitation of some immigrant groups, such as Chinese, Filipinos, and Vietnamese, and the greater acceptance of others, primarily those from Northern Europe.

THE SPLIT LABOUR MARKET THEORY Who benefits from the exploitation of racialized (visible minorities)? The **split labour market** theory states that both white workers and members of the capitalist class benefit from the exploitation of visible minorities. Split labour market refers to the division of the economy into two areas of employment: a primary sector, or upper tier, composed of higher-paid (usually dominant group) workers in more secure jobs, and a secondary sector, or lower tier, made up of lower-paid (often subordinate group) workers in jobs with little security and hazardous working conditions (Bonacich, 1972, 1976). According to this perspective, white workers in the upper tier may use racial discrimination against nonwhites to protect their positions. These actions most often occur when upper-tier workers feel threatened by lower-tier workers hired by capitalists to reduce labour costs and maximize corporate profits. In the past, immigrants were a source of cheap labour that employers could use to break strikes and keep wages down. Agnes Calliste (1987) applied the split labour market theory in her study of sleeping-car porters in Canada. Calliste found a split labour market with three levels of stratification in this area of employment. While "white" trade unions were unable to restrict access to porter positions on the basis of race, they were able to impose differential pay scales. Consequently, black porters received less pay than white porters, even though they were doing the same work. Furthermore, the labour market was doubly submerged because black immigrant workers from the United States received even less pay than both black and white Canadian porters. Throughout history, higher-paid workers have responded with racial hostility and joined movements to curtail immigration and thus do away with the source of cheap labour (Marger, 2012).

▶ **split labour market**
A term used to describe the division of the economy into two areas of employment: a primary sector, or upper tier, composed of higher-paid (usually dominant group) workers in more secure jobs; and a secondary sector, or lower tier, composed of lower-paid (often subordinate group) workers in jobs with little security and hazardous working conditions.

Proponents of the split labour market theory suggest that white workers benefit from racial and ethnic antagonisms. However, these analysts typically do not examine the interactive effects of race, class, and gender in the workplace.

FEMINIST PERSPECTIVES

Minority women (women of colour, immigrant women, and Indigenous women) are doubly disadvantaged as a result of their gender. The term *gendered racism* refers to the interactive effect of racism and sexism in the exploitation of women of colour. According to social psychologist Philomena Essed (1991), women's particular position must be explored within each racial or ethnic group, because their experiences will not have been the same as the men's in each grouping. For example, university-educated immigrant women have a more difficult time finding a job than university-educated male immigrants.

Capitalists do not equally exploit all workers. Gender and race or ethnicity are important in this exploitation. Historically, the high-paying primary labour market has been monopolized by white men. Racialized minorities and most white women more often hold lower-tier jobs. Below that tier is the underground sector of the economy, characterized by illegal or quasi-legal activities, such as drug trafficking, prostitution, and working in sweatshops that do not meet minimum wage and safety standards. Many undocumented workers and some white women and racialized minorities attempt to earn a living in this sector.

POSTMODERN PERSPECTIVES

Conventional theories of race and ethnicity tend to see racial or ethnic identities as organized around social structures that are fixed and closed, such as nations, tribes, bands, and communities.

As such, there is little movement in or out of these groups. Postmodern perspectives, in contrast, view ethnic and racial identities as largely a consequence of personal choice and subjective definition. Ethnic and racial identities are socially constructed and given meaning by our fragmented society. These identities are constantly evolving and subject to the continuous interplay of history, power, and culture.

A postmodernist framework may ask how social actors come to understand who they are in "race" terms. Central to a postmodern perspective on race is the concept of *discourse.* Based on the work of Michael Foucault, *discourse* is used to refer to "different ways of structuring knowledge and social practice" (Fiske, 1994, cited in Henry and Tator, 2009). Postmodernists view reality as constructed through a broad range of discourses, which includes all that is written, spoken, or otherwise represented through language and communication systems (Anderson, 2006:394).

Postmodernist scholars shift the frame of analysis away from race relations to an examination of racist discourse. *Racist discourse,* or *racialized discourse,* is defined as a collection of words, images, and practices through which racial power is directed against ethnic and racial minority groups. An analysis of racist discourse is central to understanding the ways in which a particular society gives a voice to racism and advances the interests of whites.

Frances Henry and Carol Tator (2009) have identified examples of racist discourse that serve to sustain or perpetuate racism in our society. For example, the *discourse of denial* suggests that racism does not exist in our Canadian democratic society. When racism is shown to exist, the discourse of denial will explain it away as an isolated incident rather than an indication of systemic racism. There are numerous examples of the discourse of denial in policing agencies across the country. Despite numerous complaints of racism directed at visible minority groups and Indigenous persons, police agencies continue to respond to allegations with, "We don't have a problem with racism within our organization," or "I have never witnessed a racist incident."

A second, related discourse identified by Henry and Tator is "the discourse of colourblindness," in which white people insist that they do not notice the skin colour of a racial minority. In doing so, white people also fail to "recognize that race is a part of the 'baggage' that people of colour carry with them, and the refusal to recognize racism as part of everyday values, policies, programs, and practices is part of the psychological power of racial constructions" (2009:25). By claiming to be colour-blind, members of the dominant white majority are allowed to ignore the power differentials they experience as a result of their "whiteness," as well as negating the racialized experiences of visible minority persons.

A postmodern perspective not only examines how identities of racial and ethnic minorities are formed, but also asks the same question about white identities. For example:

> [W]hite people are "raced" just as men are "gendered." And in a social context where white people have too often viewed themselves as nonracial or racially neutral, it is crucial to look at the "racialness" of the white experience . . . Whiteness is first a location of structural advantage of race privilege. Second, a "standpoint," a place from which white people look at ourselves, at others, at society. Third, "whiteness" refers to a set of cultural practices that are usually unmarked and unnamed. (Frankenberg, 1993, cited in Gann, 2000)

AN ALTERNATIVE PERSPECTIVE: CRITICAL RACE THEORY

Emerging out of scholarly law studies on racial and ethnic inequality, critical race theory derives its foundation from the U.S. civil rights tradition and the writing of people like Martin Luther King, Jr., W.E.B. Du Bois, Malcolm X, and Cesar Chavez. The growth of critical race theory began in Canada during the 1980s, and it is based on the same theoretical foundation as its American counterpart; that is, a growing dissatisfaction with the failure to acknowledge and recognize the critical roles that race and racism have played in the political and legal structures of Canadian society (Aylward, 1999).

Critical race theory has several major premises, including the belief that racism is such an ingrained feature of North American society that it appears to be ordinary and natural to many people (Delgado, 1995). As a result, civil rights legislation and affirmative action laws (formal equality) may remedy some of the more overt, blatant forms of racial injustice but have little effect on subtle, business-as-usual forms of racism that people of colour experience as they go about their everyday lives. According to this approach, the best way to document racism and ongoing inequality in society is to listen to the lived experiences of people who have experienced such discrimination. In this way, we can learn what actually happens in regard to racial oppression and the many effects it has on people, including alienation, depression, and certain physical illnesses (Razack, 1998).

Central to this argument is the belief that *interest convergence* is a crucial factor in bringing about social change. According to the legal scholar Derrick Bell, white elites tolerate or encourage racial advances for people of colour *only* if the dominant-group members believe that their own self-interest will be served in so doing (cited in Delgado, 1995). From this approach, civil rights laws have typically benefited white North Americans as much as (or more than) people of colour because these laws have been used as mechanisms to ensure that "racial progress occurs at just the right pace: change that is too rapid would be unsettling to society at large; change that is too slow could prove destabilizing" (Delgado, 1995:xiv).

Critical race theory is similar to postmodernist approaches in that it calls our attention to the fact that things are not always as they seem. Formal equality under the law does not necessarily equate to actual equality in society.

CONCEPT SNAPSHOT

SYMBOLIC INTERACTIONIST PERSPECTIVES

Symbolic interactionists examine how microlevel contacts between individuals may produce greater racial tolerance or increase levels of hostility. According to the contact hypothesis, when members of divergent groups have equal status, shared goals, cooperation, and positive feedback, favourable attitudes and behaviour between groups can result.

FUNCTIONALIST PERSPECTIVES

Early functionalists examined immigration and patterns of majority and minority group interaction. Intergroup processes include cultural, biological, structural, and psychological assimilation and ethnic pluralism—equalitarian and inequalitarian pluralism (segregation).

CONFLICT PERSPECTIVES

Conflict theorists focus on power and economic differentials between dominant and subordinate groups. Internal colonialism occurs when members of racial or ethnic groups are conquered or colonized and forcibly controlled by the dominant group. Split labour market theory examines the division of the economy into two unequal areas of employment.

FEMINIST PERSPECTIVES

Feminist perspectives highlight the fact that minority women are doubly disadvantaged as a result of their gender. *Gendered racism* describes the interactive effect of racism and sexism in the exploitation of visible minority women.

POSTMODERN PERSPECTIVES

Postmodern perspectives view racial and ethnic identities as socially constructed through a range of discourses. Postmodern perspectives focus on racist discourse that serves to sustain and reinforce patterns of discrimination against racial and ethnic minorities.

CRITICAL RACE THEORY

Critical race theorists believe that racism is such an ingrained feature of society that it appears to be ordinary and natural to many. According to critical race theorists, human rights legislation and employment equity strategies may remedy overt discrimination but have little effect on subtle racism. Interest convergence is required to effect positive change for visible minority group members.

LO-4 ETHNIC GROUPS IN CANADA

How do racial and ethnic groups come into contact with one another? How do they adjust to one another and to the dominant group over time? Sociologists have explored these questions extensively; however, a detailed historical account of each group is beyond the scope of this chapter. Given the diversity of our population, imposing any kind of conceptual order on a discussion of ethnic groups in Canada is difficult. We will look briefly at some of the predominant ethnic groups in Canada. In the process, we will examine a brief history of racism with respect to each group.

INDIGENOUS PEOPLES

Canada's Indigenous peoples are believed to have migrated to North America from Asia about 14,000 years ago. The term *Indigenous* itself refers to the "first," or indigenous, occupants of this country. Indigenous peoples as a group are striking in their diversity in terms of, among other things, size, language and culture, geographic location (urban, rural), and levels of well-being (Graham & Leveque, 2010). Today, the terms *Native, First Nations,* or *Indigenous* refer to over 600 First Nations across the country with approximately 50 Indigenous languages, including Inuktitut, Cree, Ojibway, Wakashan, and Haida. Other categories of Indigenous peoples are status Indians (those Indians with legal rights under the *Indian Act),* nonstatus Indians (those without legal rights), Métis, and Inuit. Those who settled in the southern part of Canada, Yukon, and the Mackenzie Valley can be termed *North American Indians.* Those located in the eastern Arctic and northern islands, formerly referred to as Eskimos, are now referred to as *Inuit.* A third category, *Métis,* who live mostly on the Prairies, are descendants of mixed European-Indigenous unions. (Fleras, 2012).

When European settlers arrived on this continent, the Indigenous inhabitants' way of life was changed forever. Experts estimate that approximately two million Indigenous people lived in North America at the time of contact; by 1900, however, their numbers had been reduced to under 240,000. What factors led to this drastic depopulation?

GENOCIDE, FORCED MIGRATION, AND FORCED ASSIMILATION Indigenous people have been the victims of genocide and forced migration. Many Native Americans were either massacred or died from European diseases (such as tuberculosis, smallpox, and measles) and starvation (Daschuk, 2013). In battle, Indigenous people often were no match for the Europeans, who had the latest weaponry. Europeans justified their aggression by stereotyping Indigenous as "savages" and "heathens" (Frideres and Gadacz, 2011).

The federal government offered treaties to the Native Americans so that more of their land could be acquired for the growing white population. Scholars note that the government broke treaty after treaty as it engaged in a policy of wholesale removal of indigenous nations in order to clear the land for settlement by white "pioneers." Entire nations were forced to move in order to accommodate the white settlers.

First Nations rights were clearly defined in the *Royal Proclamation* of 1763, which divided up the territory acquired by Britain. In the large area called the Indian Territory, the purchase or settlement of land was forbidden without a treaty (Dyck, 1996:154). The Canadian government then passed the *Indian Act* of 1876, which provided for federal government control of almost every aspect of Indian life. The regulations under the act included prohibitions against owning land, voting, purchasing and consuming alcohol, and leaving reserves without permission and a ticket from the government's agent (Frideres and Gadacz, 2011).

Indigenous children were placed in residential boarding schools to facilitate their assimilation into the dominant culture. The Jesuits and other missionaries who ran these schools believed that Indigenous peoples should not be left in their "inferior" natural state and considered it their mission to replace Indigenous culture with Christian beliefs, values, rituals, and practices (Bolaria and Li, 1988). Many Indigenous children who attended these schools were sexually, physically, and emotionally abused. They were not allowed to speak their language or engage in any of their traditional cultural practices. The coercive and oppressive nature of this educational experience is one of the most horrific examples of institutionalized racism. It was not until 2008 that the Federal government offered a formal apology acknowledging that the policy of forced assimilation "was wrong, has caused great harm, and has no place in our country" and that the treatment of children in residential schools "is a sad chapter in our history" (CBC News, 2008).

INDIGENOUS PEOPLES TODAY According to the 2011 Census, 1.4 million people reported they were Indigenous. Of these, 60 percent (851,560) identified themselves as First Nations, 451,795 Metis, and 59,445 Inuit. These numbers represent approximately 4 percent of Canada's total population (Statistics Canada, 2013d). Figure 11.2 displays the composition of the Indigenous population. Although the majority of registered Indians live on reserves, the majority of all Indigenous people live off reserves. The Indigenous population is unevenly distributed across Canada, with the heaviest concentrations in western and northern Canada.

FIGURE 11.2 ABORIGINAL IDENTITY POPULATION, 2011

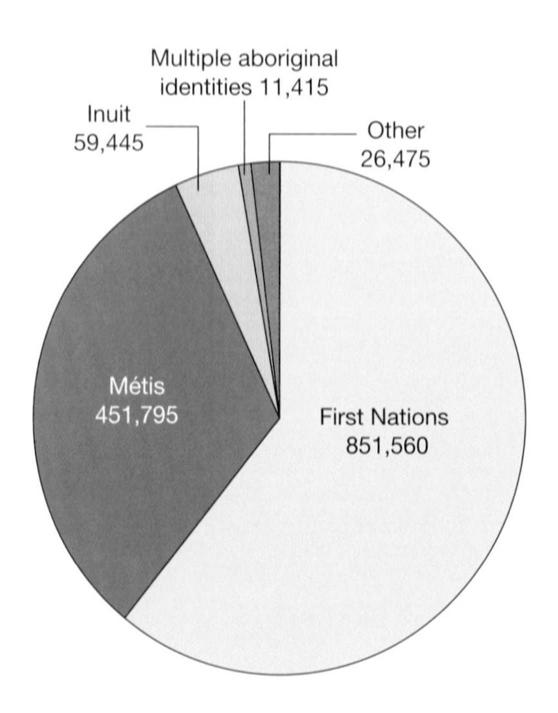

Source: Statistics Canada, 2013d, Aboriginal Peoples in Canada: First Nations People, Métis and Inuit in Canada National Household Survey, 2011, Catalogue no. 99-011-X2011001 http://www12.statcan.gc.ca/nhs-enm/2011/as-sa/99-011-x/99-011-x2011001-eng.pdf

The results of government assimilationist policies, forced segregation, and discrimination continue to be experienced by First Nations children, youth, and families across the country. In terms of income, employment, housing, nutrition, and health, Indigenous peoples are the most disadvantaged racial or ethnic group in Canada. Various human rights organizations have repeatedly criticized Canada's mistreatment of Indigenous peoples, referring to it as shameful, contrary to International law and human rights legislation. For example, a 2012 United Nations report outlined a range of human rights issues facing Indigenous peoples:

> By every measure, be it respect for treaty and land rights, levels of poverty, average lifespans, violence against women and girls, dramatically disproportionate levels of arrest and incarceration, or access to government services such as housing, healthcare, education, water and child protection, Indigenous peoples across Canada continue to face a grave human rights crisis. Despite determined and courageous organizing and legal action nationally and internationally, Indigenous peoples continue to face immense obstacles in ensuring that their rights are even acknowledged, let alone protected. (Amnesty International, 2013:9)

The life chances of Indigenous peoples who live on reserves are especially limited. According to a United Nations report, an alarming number of First Nations children live in Third World conditions, with an estimated 80 percent of urban Indigenous children under the age of six living in poverty. Housing on many reserves is inadequate, overcrowded, and fails to meet minimum standards of amenities and structure, including running water and proper sewage.

Indigenous people living in urban areas were more than twice as likely as non-Indigenous people to live in poverty. And the number of Indigenous children involved with the child welfare system across Canada continues to grow. In some provinces, over 95 percent of children involved with family services are Indigenous. Indigenous peoples have the highest rates of infant mortality and death by exposure and malnutrition, as well as high rates of tuberculosis, alcoholism, and suicide (Frideres & Gadacz, 2011). The overall life expectancy of Indigenous people in Canada is five years less than that of non-Indigenouss, largely due to poor health services and inadequate housing on reserves. Indigenous peoples also have had limited

The Canadian Press/Geoff Robins

Idle No More is a grassroots protest movement started in 2012 by four women, three of whom are First Nations, to promote environmental protection and Indigenous sovereignty.

educational opportunities (the functional illiteracy rate for Indigenous peoples is 45 percent, compared with the overall Canadian rate of 17 percent).

Economic disadvantage is reflected in both employment and income inequality among First Nations populations. Their rate of unemployment is twice that for non-Indigenous Canadians. On reserves, the unemployment rate is about 29 percent, nearly three times the Canadian rate (Frideres & Gadacz, 2011). Finally, incomes for Indigenous persons are about two-thirds the level of non-Indigenous. For First Nations people on reserves, average incomes are less than half the rest of the population's incomes (Pendakur & Pendakur, 2013).

Despite the state's efforts to assimilate Indigenous peoples into Canadian culture and society, many Indigenous people have been successful in resisting oppression. National organizations like the Assembly of First Nations, Sisters in Spirit, the Native Council of Canada, and the Métis National Council have been instrumental in bringing the demands of those they represent into the political and constitutional arenas. Of these demands, the major ones have been—and still are—self-government, Indigenous rights, and the resolution of land claims (see Frideres and Gadacz, 2011).

Indigenous peoples today are in a period of transition from a long history marked by racism, exploitation, and domination to a contemporary life in which they are regaining control of programs directed at protecting their children, delivering education, and promoting and restoring health and commerce in Indigenous communities. Many see the challenge for Indigenous peoples today as being to erase negative stereotypes while maintaining their heritage and obtaining recognition for their contribution to this country's development and growth. A report released in 2012 indicated that Indigenous Canadians are making some measurable progress toward improving their economic outcomes.

THE QUÉBÉCOIS

The French were the first Europeans to immigrate to Canada in large numbers, establishing settlements in what was then known as Acadia and along the St. Lawrence River. In 1608, the first permanent settlement in New France was established at Quebec City. At this time, France's North American empire extended from Hudson Bay to Louisiana.

Following the British conquest of the French in Canada in the Seven Years' War (1756–1763), Canada became a British dominion. However, given the numerical dominance of the French, their links to the fur trade, and the fact that the French colony shared a border with the United States, the British felt it advantageous to accommodate the French. What emerged was a plural, or segmented, society in which the French were able to maintain French civil law, language, and religion; however, the overall economic, social, and political power passed to English Canada (Breton, 1988).

The British North America Act (1867) formally acknowledged the rights and privileges of the French and British as the founding, or charter, groups of Canadian society. French was recognized as an official language in Quebec and the provincial government in Quebec was able to maintain significant authority over culture and education. The Catholic Church was also able to retain control over educational and religious matters and many other aspects of Quebec society. During this time, it was assumed that in the future, French- and English-speaking groups would coexist and complement one another. However, during the period between Confederation and World War II, the French struggled for cultural survival because English-speaking Canadians controlled the major economic institutions in both English Canada and Quebec. During the early 1960s, the Catholic Church's control began to erode and the old elite consensus began to break down, laying the foundation for what was described as the Quiet Revolution (Satzewich and Liodakis, 2010).

During the Quiet Revolution (1960–1966), Quebec nationalism grew sharply. Under the leadership of Premier Jean Lesage, Quebec began undergoing a rapid process of modernization. During this time, the authority of the Catholic Church over the educational system was reduced as the Quebec government established a department of education. More French Canadians began pursuing higher education, particularly in business and science. The Church also lost some of its influence over moral issues, which was reflected in a declining birth rate and an increase in common-law marriages. Finally, nonfrancophone immigrants were challenging French culture by choosing to learn English and having their children learn English rather than French. Francophones came to view their language and culture as endangered, and as a result, many rejected their Canadian identity in favour of a distinctly Québécois identity.

In the 1970s, the separatist Parti Québécois was elected under the leadership of Premier René Lévesque. During this time, the controversial Bill 101 was introduced and established French as the sole official language in Quebec. In 1980, the Parti Québécois held a referendum on the question of pursuing a more independent relationship with Canada called *sovereignty association.* The proposal was narrowly rejected, but the matter was once again addressed in a second referendum in 1995 (Dyck, 2000). Although Quebeckers once again rejected sovereignty (this time by a narrow margin of 1 percent) the issue remains controversial (see Chapter 22).

FRENCH CANADIANS TODAY Today, approximately 20 percent of the Canadian population is francophone, of which 85 percent is located in Quebec; 80 percent of Quebeckers are French-speaking and over 90 percent of French-speaking Canadians live in Quebec (Statistics Canada, 2012h). Many Quebec nationalists now see independence or separation as the ultimate protection against cultural and linguistic assimilation, as well as the route to economic power. As political scientist Rand Dyck comments:

> Given their historic constitutional rights, given their geographic concentration in Quebec and majority control of such a large province, and given their modern-day self-consciousness and self-confidence, the French fact in Canada cannot be ignored. If English Canada wants Quebec to remain a part of the country, it cannot go back to the easy days of pre-1960 unilingualism and federal government centralization. (2008:116)

© Al Harvey

As more Chinese Canadians have made gains in education and employment, many have also made a conscious effort to increase awareness of Chinese culture and develop a sense of unity and cooperation. This Chinese dragon parade exemplifies the desire to maintain traditional celebrations.

French Canadians have at least forced Canada to take its second language and culture seriously, which is an important step toward attaining cultural pluralism.

CANADA'S MULTICULTURAL MINORITIES

Home to close to seven million foreign-born immigrants, Canada is well described as a land of immigrants. Approximately 80 percent of immigrants arriving in Canada today are members of a visible minority group (Statistics Canada, 2013a). But Canada's policies toward some of these groups have been far from exemplary. In fact, initial Canadian immigration policies have been described as essentially racist in orientation, assimilationist in intent, and exclusionary in outcome. For example, the *Immigration Act* of 1869 excluded certain types of undesirables, such as criminals and the diseased, and imposed strict limitations on the Japanese, Chinese, and East Asians. A "racial pecking order" was established to select potential immigrants on the basis of race and perceived ability for assimilation (Lupul, 1988; Walker, 1997, cited in Fleras and Elliott, 2003). As much energy was expended in keeping out certain "types" as was put into encouraging others to settle.

A preferred category was that of *white ethnics*—a term coined to identify immigrants who came from European countries other than England, such as Scotland, Ireland, Poland, Italy, Greece, Germany, Yugoslavia, and Russia and other former Soviet republics. Immigration from "white" countries was encouraged to ensure the British character of Canada. With the exception of visa formalities, this category of "preferred" immigrants was virtually exempt from entry restrictions (Fleras and Elliott, 2003:253). On the other hand, Jews and other Mediterranean populations required special permits for entry, and Asian populations were admitted only because they could serve as cheap labour for Canadian capitalist expansion. The restrictions regarding the Chinese, Japanese, and Jews highlighted the racist dimension of Canada's early immigration policies.

CHINESE CANADIANS The initial wave of Chinese migrants came to Canada in the 1850s, when Chinese men were attracted to emigrate by the British Columbia gold rush and by employment opportunities created by the expansion of a national railroad. Nearly 17,000 Chinese were brought to Canada at this time to lay track for the Canadian Pacific Railway. The work was brutally hard and dangerous, living conditions were appalling, food and shelter were insufficient, and due to scurvy and smallpox the fatality rate was high. These immigrants were "welcomed" only as long as there was a shortage of white workers. However, they were not permitted to bring their wives and children with them or to have sexual relations with white women, because of the fear they would spread the "yellow menace" (Henry and Tator, 2009).

The Chinese were subjected to extreme prejudice and were referred to by derogatory terms, such as *coolies, heathens,* and *Chinks.* Some were attacked by working-class whites who feared they would lose their jobs to Chinese immigrants. In 1885, the federal government passed its first anti-Chinese bill, the purpose of which was to limit Chinese immigration, and a $50 head tax was imposed on all Chinese males arriving in Canada. In 1903, the tax was raised to $500 in a further attempt to restrict entry to Canada (Satzewich and Liodakis, 2010). Other hostile legislation included a range of racist exclusionary policies, such as prohibiting the Chinese from voting, serving in public office, serving on juries, participating in white labour unions, and working in the professions of law and pharmacy. Not until after World War II were these discriminatory policies removed from the *Immigration Act.* After immigration laws were further relaxed in the 1960s, the second and largest wave of Chinese immigration occurred, with immigrants coming primarily from Hong Kong and Taiwan (Henry and Tator, 2009).

JAPANESE CANADIANS When Japanese Canadians first arrived in British Columbia in the 1870s, they experienced similar discriminatory policies and practices. Like Chinese immigrants two decades earlier, the Japanese were viewed as a threat by white workers and became victims of racism and discrimination. They were paid lower wages than white labourers, had restrictions placed on their fishing licences, and were segregated in schools and public places.

In 1907, an organization known as the Asiatic Exclusion League was formed with the goal of restricting admission of Asians to Canada. Following the arrival of a ship carrying more than a thousand Japanese and a few hundred Sikhs, the league carried out a demonstration that precipitated a race riot. After the riot, the Canadian government negotiated a "gentlemen's agreement" that permitted entry only of certain categories of Japanese persons. In this agreement, the government further allowed only 400 Japanese to immigrate to Canada in a given year (Henry and Tator, 2009).

Japanese Canadians also experienced one of the most vicious forms of discrimination ever sanctioned by Canadian law. During World War II, when Canada was at war with Japan, nearly 23,000 people of Japanese ancestry—13,300 of whom were Canadian-born—were placed in jails and internment camps, forced to work, and had their property confiscated. Those interned in camp were not released until two years after the war was over (Miki and Kobayashi, 1991). German immigrants avoided this fate even though Canada was at war with both Japan and Germany. Four decades after these events, the Canadian government issued an apology for its actions and agreed to pay $20,000 to each person who had been placed in an internment camp (Henry and Tator, 2009).

SOUTH ASIANS Immigrants from India were also subjected to widespread anti-immigration sentiments in the early 20th century. One of the first discriminatory immigration laws was the "continuous passage" rule of 1908, which specified that South Asians could immigrate only if they came directly from India and did not stop at any ports on the way. This law made it almost impossible for them to enter the country, since no ships made direct journeys from India. For example, in 1914, a Sikh businessman chartered a ship in Hong Kong to transport more than 300 Indian passengers to Canada. On arrival in Vancouver, the passengers were refused entry. After a two-month standoff, the ship was forced to return to India (Satzewich and Liodakis, 2010).

South Asians who did manage to immigrate to Canada were subject to ongoing exclusion and hostility. Their property and businesses were frequently attacked, and they were denied citizenship and the right to vote in British Columbia until 1947 (Henry and Tator, 2009). Because they were denied their political rights, they were also precluded from entering the more prestigious professions of law, medicine, education, and pharmacy.

JEWISH CANADIANS Between 1933 and 1945, many Jews sought refuge from the persecution of the Nazis. During this time, Canada admitted fewer Jewish refugees as a percentage of its population than any other Western country. In 1942, a ship carrying Jewish refugees from Europe attempted to land in Halifax and was denied entrance. Jews who did immigrate experienced widespread discrimination in employment, business, and education. Other indicators of anti-Semitism included restrictions on where Jews could live, buy property, and attend university. Signs posted along Toronto's beaches warned, "No dogs or Jews allowed." Many hotels and resorts had policies prohibiting Jews as guests (Abella and Troper, 1982, quoted in Henry and Tator, 2000:80).

LO-5 IMMIGRATION TRENDS POST WORLD WAR II TO THE PRESENT

Although the more blatantly racist aspects of immigration policy were moderated after World War II, the underlying philosophy behind immigration to Canada retained its discriminatory agenda—immigration needed to be carefully controlled, the encouragement of nonwhite immigration was not in the best interests of the country, and any economic benefits of immigration needed to be measured against the potential "social costs" of unrestrained immigration of visible minority immigrants. This philosophy was clearly articulated by Prime Minister William Lyon Mackenzie King in a 1947 speech to the House of Commons:

> The people of Canada do not wish, as a result of mass immigration, to make a fundamental alteration in the character of our population. Large scale immigration from the Orient would change the fundamental composition of the Canadian population. Any considerable oriental immigration would, moreover, be certain to give rise to social and economic problems. (Canada, 1947: Debates of the House of Commons, cited in Satzewich and Liodakis, 2010)

After the war, some of the more overtly racist immigration legislation, such as the *Chinese Immigration Act* and the continuous journey stipulations, were repealed and immigration from India was permitted on a fixed quota basis of 300 persons per year. However, the focus of postwar immigration continued to be on the "preferred" white immigrants from Europe and the United States (Satzewich and Liodakis, 2010).

Changes to the *Immigration Act* in 1962 opened the door to immigration on a nonracial basis. Canada became one of the first countries in the world to announce that "any suitable qualified person from anywhere in the world" would be considered for immigration, based solely on the criteria of personal merit. Education, occupation, and language skills replaced ethnicity and nationality as criteria for admission (Fleras, 2012) when a *points system* was introduced. All applicants, regardless of origin or colour, were rated according to the total of points given for the following: job training, experience, skills, level of education, knowledge of English or French, degree of demand for the applicant's occupation, and job offers (Henry and Tator, 2009). Although, as shown in Figure 11.3, this act opened the doors to those from previously excluded countries, critics have suggested that it maintained some of the same racist policies. In 2002, in response to the numerous concerns of continued exclusionary and racist immigration practices, the *Immigration and Refugee Protection Act* was implemented. This act recognizes three classes of immigrants—economic, family class, and refugee—and reflects a more open policy with selection criteria based on language skills, education, age, employment experience, and a category called "adaptability" (Henry and Tator, 2006:78).

GROWING RACIAL AND ETHNIC DIVERSITY IN CANADA

Racial and ethnic diversity is increasing in Canada. This changing demographic pattern is largely the result of the elimination of overtly racist immigration policies and the opening up of immigration to low-income countries. Canada has evolved from a country largely inhabited by whites and Indigenous peoples to a country made up of people from more than 70 countries.

FIGURE 11.3 REGION OF BIRTH OF RECENT IMMIGRANTS TO CANADA, 1971 TO 2011

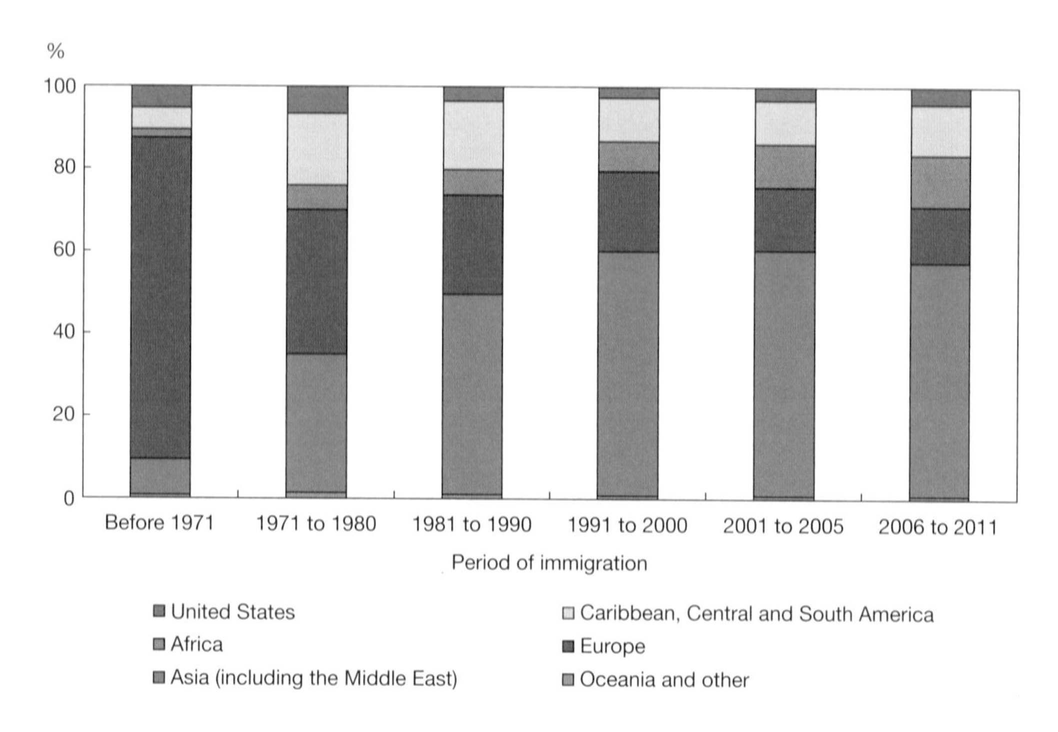

Note: "Oceania and other" includes immigrants born in Oceania, in Canada, in Saint Pierre and Miquelon and responses not included elsewhere, such as "born at sea."

Source: Statistics Canada, 2013a, Immigration and Ethnocultural Diversity in Canada: National Household Survey, 2011, p. 9. Catalogue no. 99-010-X2011001. http://www12.statcan.gc.ca/nhs-enm/2011/as-sa/99-010-x/99-010-x2011001-eng.pdf

Today, people born outside of Canada make up more than 20 percent of the total population of Canada (see Figure 11.4). Newcomers from Asia make up the largest proportion of immigrants, followed by newcomers from Europe (Statistics Canada, 2013a).

Almost all immigrants to Canada live in cities. Recent immigrants are especially attracted to Canada's three largest cities. The majority of recent immigrants have chosen to live in Toronto, Montreal, or Vancouver. Today, nearly half of the population of Toronto and 40 percent of the population of Vancouver is composed of immigrants.

What effect will these changes have on racial and ethnic relations? Several possibilities exist. On the one hand, conflict between whites and racial and ethnic minorities may become more overt and confrontational. Certainly, the concentration of visible minorities will mean that these groups will become more visible than ever in some Canadian cities. Increasing contact may lead to increased intergroup cohesion and understanding, or it may bring on racism or prejudice. Rapid political changes and the global economy have made people fearful about their future and may cause some to blame "foreigners" for their problems. People may continue to use *discourses of denial*—personal beliefs that reflect larger societal mythologies, such as "I am

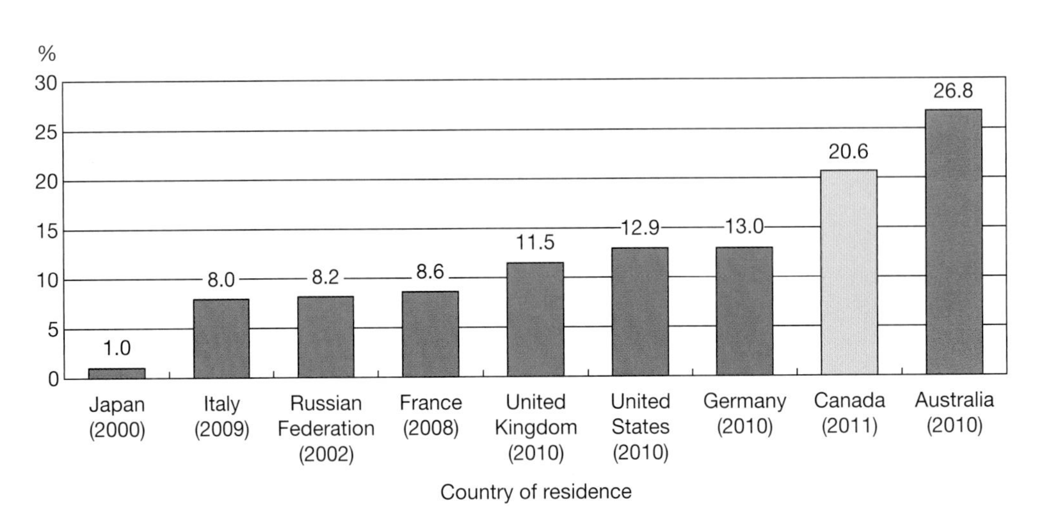

Source: Statistics Canada, 2013a, Immigration and Ethnocultural Diversity in Canada: National Household Survey, 2011, p. 7. Catalogue no. 99-010-X2011001. http://www12.statcan.gc.ca/nhsenm/ 2011/as-sa/99-010-x/99-010-x2011001-eng.pdf

not racist" or "I have never discriminated against anyone"—even when these are inaccurate perceptions (Henry and Tator, 2009).

On the other hand, there is reason for cautious optimism. Throughout Canadian history, subordinate racial and ethnic groups have struggled to gain the freedom and rights that were previously withheld from them. Today, employment equity programs are alleviating some of the effects of past discrimination against minority groups, as well as addressing systemic and institutional forms of racism that exist in employment.

LO-1 Distinguish between race and ethnicity.

A race is a category of people who have been singled out as inferior or superior, often on the basis of real or alleged physical characteristics, such as skin colour, hair texture, eye shape, or other subjectively selected characteristics. An ethnic group is a collection of people distinguished by others or by themselves, primarily on the basis of culture or nationality.

Peter Muller/Cultura/Getty Images

LO-2 Define and explain prejudice, discrimination, and racism.

Prejudice involves attitudes, but discrimination involves actions or practices of dominant group members that have a harmful impact on members of a subordinate group. Discriminatory actions range from name-calling to violent actions and can be either *de jure* (encoded in law) or *de facto* (informal). *Racism* refers to an organized set of beliefs about the innate inferiority of some racial groups, combined with the power to discriminate on the basis of race. There are many different ways in which racism may manifest itself, including hate racism, polite racism, subliminal racism, and institutionalized racism.

AFP/Getty Images

LO-3 Explain the major sociological perspectives on race and ethnic relations.

Interactionists suggest that increased contact between people from divergent groups should lead to favourable attitudes and behaviour when members of each group (1) have equal status, (2) pursue the same goals, (3) cooperate with one another to achieve goals, and (4) receive positive feedback when they interact with one another. Functionalists stress that members of subordinate groups become absorbed into the dominant culture. Conflict theorists focus on economic stratification and access to power in race and ethnic relations. Feminist analysts highlight the fact that women who are members of racial and ethnic minorities are doubly disadvantaged as a result of their gender. There is an interactive effect of racism and sexism on the exploitation of women of colour. Postmodern theorists view racial and ethnic identities as fluid and examine how these concepts are socially constructed. Critical race theorists emphasize the significant role that race and racism have played in legal and political structures in society.

teena137/Shutterstock.com

KEY TERMS

assimilation A process by which members of subordinate racial and ethnic groups become absorbed into the dominant culture (p. 303).

authoritarian personality A personality type characterized by excessive conformity, submissiveness to authority, intolerance, insecurity, a high level of superstition, and rigid, stereotypic thinking (p. 298).

discrimination Actions or practices of dominant group members (or their representatives) that have a harmful impact on members of a subordinate group (p. 298).

ethnic group A collection of people distinguished, by others or by themselves, primarily on the basis of cultural or nationality characteristics (p. 292).

ethnic pluralism The coexistence of a variety of distinct racial and ethnic groups within one society (p. 304).

hate racism (or overt racism) Racism that may take the form of deliberate and highly personal attacks, including derogatory slurs and name-calling toward members of a racial or ethnic group who are perceived to be "inferior" (p. 299).

institutionalized racism A situation where the established rules, policies, and practices within an institution or organization produce differential treatment of various groups based on race (p. 301).

internal colonialism According to conflict theorists, a situation in which members of a racial or ethnic group are conquered or colonized and forcibly placed under the economic and political control of the dominant group (p. 306).

majority (dominant) group An advantaged group that has the greatest power and resources in a society (p. 296).

minority (subordinate) group A group whose members, because of physical or cultural characteristics, are disadvantaged and subjected to unequal treatment and discrimination by the dominant group (p. 296).

polite racism A term used to describe an attempt to disguise a dislike of others through behaviour that is outwardly nonprejudicial (p. 299).

prejudice A negative attitude based on preconceived notions about members of selected groups (p. 297).

race A term used by many people to specify groups of people distinguished by physical characteristics, such as skin colour; also, a category of people who have been singled out as inferior or superior, often on the basis of real or alleged physical characteristics, such as skin colour, hair texture, eye shape, or other subjectively selected attributes (p. 292).

racial prejudice Beliefs that certain racial groups are innately inferior to others or have a disproportionate number of negative traits (p. 297).

LO-4 Discuss the unique historical experiences of the racial and ethnic groups in Canada.

Charla Jones/GetStock.com

When European settlers arrived on this continent, the Indigenous inhabitants were the victims of genocide and forced migration. Indigenous children were placed in residential boarding schools to facilitate their assimilation into the dominant culture. Although the French were the first Europeans to immigrate to Canada in large numbers after being defeated by the English in the Seven Years War, they found themselves in an inferior position. Even though the French were able to maintain French civil law, language, and religion, the overall economic, social, and political power passed to English Canada. Nonwhite immigrants (including Chinese, Japanese, and South Asians) were only welcomed into Canada as a source of cheap labour and were subject to racist laws and immigration policies.

LO-5 Describe how Canada's immigration policies have affected the composition of Canada's racial and ethnic population today.

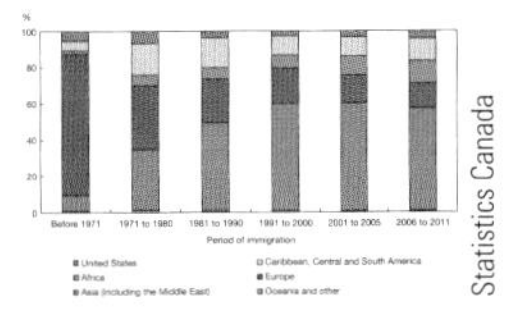

Statistics Canada

Canada's early immigration policies were described as racist and included exclusionary policies directed at Asian populations, including Chinese, Japanese, and South Asians, as well as Jews. "White ethnics" who came from European countries comprised the preferred category of immigrants. Changes to the *Immigration Act* in 1962 involving the implementation of a points system opened the door to immigration on a nonracial basis. In 2002, the *Immigration and Refugee Protection Act* was implemented, with selection criteria based on attributes of human capital and the skills of potential immigrants.

KEY FIGURES

Archive Photos/Getty Images

W.E.B. Du Bois (1868–1963) Born in Massachusetts, Du Bois was both a race relations scholar and a civil rights activist. Du Bois was the first African American to earn a Ph.D. at Harvard University. Over the years, he became frustrated with the lack of progress in race relations and became a co-founder of the National Association for the Advancement of Colored People (NAACP).

Courtesy of Frances Henry

Frances Henry Now retired as a professor emerita from York University, Henry is one of Canada's leading experts in the study of racism and antiracism. Since the mid-1970s, when she published the first study of attitudes toward people of colour, she has consistently pioneered research in this field. She is co-author with Carol Tator of *The Colour of Democracy: Racism in Canadian Society* and more recently *Racism in the Canadian University*.

Courtesy of Carol Tator

Carol Tator Tator is the author of numerous books on racism in Canada, including Racism in the Canadian University and Racial Profiling in Canada: Challenging the Myth of "A Few Bad Apples." For more than three decades, she has worked on the front lines of the antiracism and equity movement in the areas of the development and implementation of antiracism policies and programs, strategic planning, training, and research.

APPLICATION QUESTIONS

1. Do you consider yourself defined more by your race, your ethnicity, or neither of these concepts? Explain.
2. Given that minority groups have some common experiences, why is there such deep conflict between certain minority groups?
3. What would need to happen in Canada, both individually and institutionally, for a positive form of ethnic pluralism to flourish?
4. Is it possible for members of racial minorities to be racist? Discuss.

racism A set of ideas that implies the superiority of one social group over another on the basis of biological or cultural characteristics, together with the power to put these beliefs into practice in a way that controls, excludes, or exploits minority women and men (p. 299).

scapegoat A person or group that is incapable of offering resistance to the hostility or aggression of others (p. 298).

segregation A term used to describe the spatial and social separation of categories of people by race/ethnicity, class, gender, and/or religion (p. 304).

split labour market A term used to describe the division of the economy into two areas of employment: a primary sector, or upper tier, composed of higher-paid (usually dominant group) workers in more secure jobs; and a secondary sector, or lower tier, composed of lower-paid (often subordinate group) workers in jobs with little security and hazardous working conditions (p. 307).

stereotype An overgeneralization about the appearance, behaviour, or other characteristics of members of particular groups (p. 297).

subliminal racism A term used to describe an unconscious racism that occurs when there is a conflict of values (p. 300).

visible minority Government term used to describe those who are nonwhite, non-Indigenous, or non-Caucasian. This term is used interchangeably with "people of colour" and "racialized minorities." (p. 296).

CHAPTER 12

GENDER

LEARNING OBJECTIVES
AFTER READING THIS CHAPTER, YOU SHOULD BE ABLE TO

LO-1 Understand how gender is defined and how it differs from "sex."

LO-2 Explain the significance of gender in our everyday lives.

LO-3 Discuss how the nature of work affects gender equality in societies.

LO-4 Identify and discuss the primary agents of gender socialization.

LO-5 Explain the causes of gender inequality in Canada.

LO-6 Understand how functionalist, conflict, feminist, and interactionist perspectives on gender stratification differ.

CHAPTER FOCUS QUESTION

How have traditional constructions of gender changed, and what impact have these changes had in terms of how individuals live their "gendered" lives?

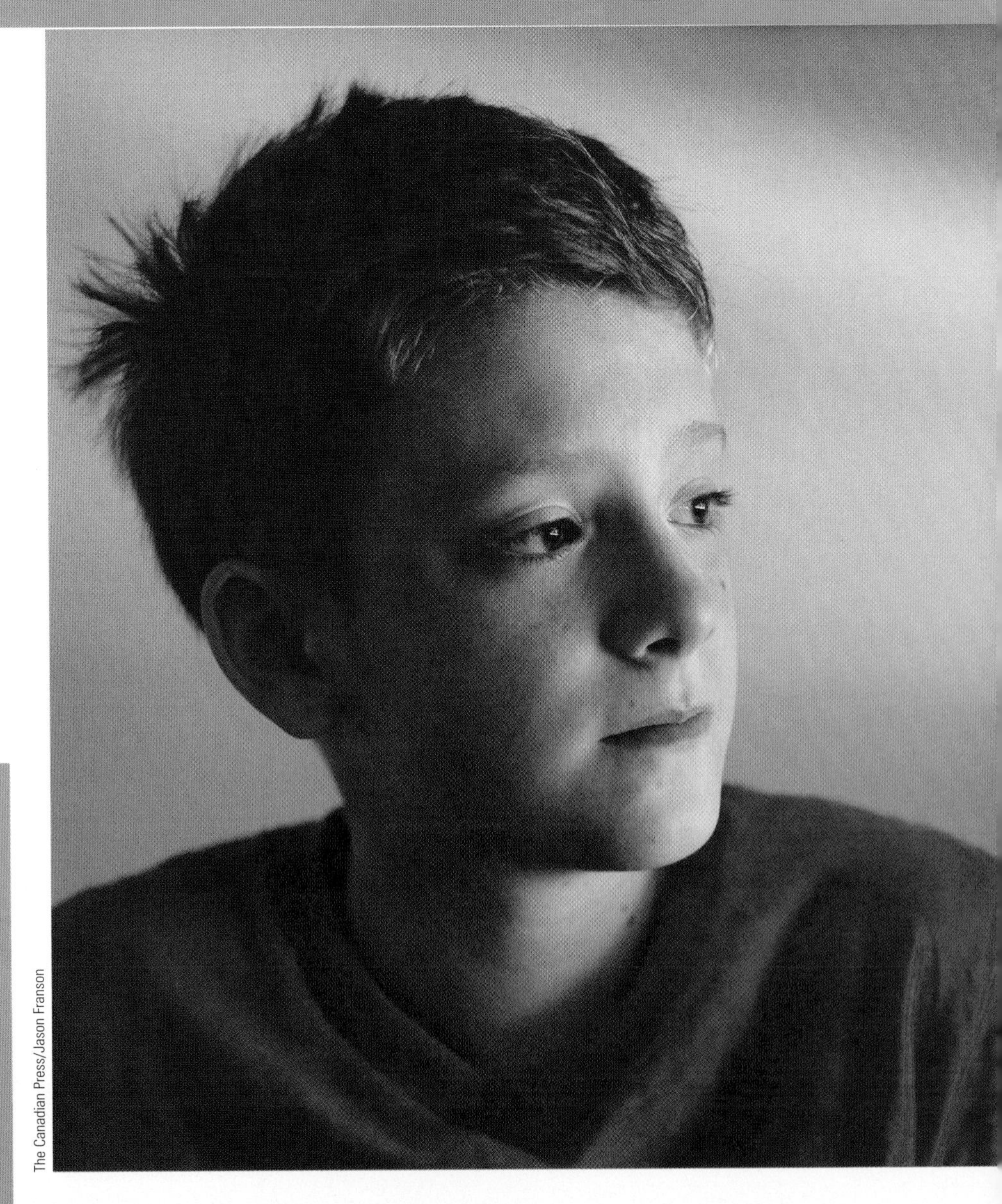
The Canadian Press/Jason Franson

When 11-year-old Wren goes back to school this week, he won't be hiding the fact that he's actually a girl.

Teachers, friends and other students at his Edmonton school know the truth—that he's a girl on the outside but feels like a boy on the inside. And that's why, even at such a young age, he has chosen to live in the world as the opposite sex, and not keep it a secret.

"If you're not yourself, then it kind of gets sad and depressing," says the freckle-faced kid with short-cropped hair. "I'm glad that I told everybody."

Wren, born Wrenna, says he doesn't remember a time when he didn't feel like a boy.

Growing up, he hated wearing dresses. He liked Spiderman and dressed up as comic book hero The Thing one Halloween. When he was five, he had his mom take him to a hairdresser to cut off his long, brown locks. He wanted to look like Zac Efron from the movie "High School Musical."

Wendy Kauffman says she and her husband, Greg, knew their daughter was different. She would often ask: "When do I get to be a boy?" And she pleaded to be born again in order to come out right.

They thought it was a phase. Then they thought their child might be gay.

But as Wren got bigger, so did the sadness and frustration.

Kauffman says it finally hit home when Wren was about nine and Kauffman was tucking her six-year-old child, Avy, into bed one night. "She said to me, 'You know, Mom, Wren is a boy and he told me to tell you.'"

Kauffman says she got a bit defensive. "I said, 'Well, I know Wren wants to be a boy.'"

"Avy said, 'No, Mom, he REALLY wants to be a boy.'"

Kauffman, tears welling up in her eyes, says it was a pivotal moment. Her youngest child had seen it all so clearly and, now, she did too.

Kauffman later told Wren: "I love you whether you're a boy or a girl and I understand now. And we'll figure out how we can help you. And we'll do it together."

Wren was in Grade 5 at Belgravia School, where students occasionally gathered in sharing circles to talk about life events such as the separation of parents or a family death. He took his turn to tell his classmates that he was now living his life as a boy.

Some kids had questions, but they were all supportive, Kauffman says.

The following year, Wren transferred to Victoria School of the Arts. At first, he was private about his actual sex, but after a few months he told friends and shared his story with his class.

There are a couple of older transgender students at the school, but Wren is by far the youngest.

He says it hasn't been a big deal. He uses the boys' washroom "which, by the way, is much grosser than the girls' bathroom." He also changes in a stall in the boys' gym locker room. . . .

His school is part of the Edmonton Public School Board, which, in 2011, became the first in the province to develop a policy to protect gay, lesbian, and transgender students and staff from discrimination based on sexual orientation.

Wren says he hasn't been subjected to any harassment. Just some teasing when he first came out—one student winked and called him "Mr. Kauffman." . . .

Wren says he knows it won't always be this easy and he's prepared for the possibility that he may be bullied later in life.

"People tease me right now and I can handle it. The way that I like to look at it is that they're just practice for the real jerks in life.

"And, besides, if they say something to me, then they don't have to be part of my life . . . I don't think I need people who don't like me." (National Post, 2013)

Source: Chris Purdy,"When do I get to be a boy," *The Canadian Press*, September 2, 2013.

Wren's story is one of many that challenges us to think about how we think about, define, and understand gender. In June 2014, Laverne Cox, star of the HBO series *Orange is the New Black*, graced the cover of *Time* with the sub-headline claiming that trans issues will be North America's next human rights frontier. The same year, *Facebook* recognized the diversity of gender identities by providing over 50 options for users to describe their gender identity to their Facebook friends. In April 2015, former Olympic athlete Bruce Jenner revealed he was transgender in a televised interview watched by a record 17 million viewers. School boards across Canada are creating controversy and challenging parents as they struggle to create inclusive sexual orientation and gender identity policies for a diverse student population (see Figure 12.1).

Although many of you may have never thought about or questioned your gender, it is a central part of our identity. How do you describe yourself to others? Where does your gender enter into the description? Or alternatively, ask yourself a question posed by Professor Ted Cohen in his Gender Studies class to demonstrate the significance of gender: *'Imagine and describe how your life would differ if you had been born and were living as the opposite sex.'* (Cohen, 2001:2–3)

Although significant changes have occurred in the last half of the 20th century in terms of work activities, family roles and responsibilities, and what it means to be "masculine" or "feminine," and recognition that "gender" is no longer comprised of simply "men" and "women," notions of gender continue to play a significant role in social institutions and relationships. As such, gender continues to both constrain and enable us. Men and women continue to live "gendered lives" with societal "scripts" that impose accepted parameters of being male or female. Specific ideas of femininity and masculinity are inescapable products of the society in which we are socialized. For example, children as young as three years old are able to describe "girl" and "boy" toys and "girl" and "boy" clothing. Gender is so much a part of who we are that it often goes unexamined. We are going to do just that in this chapter. We will examine the issue of gender: what it is, how it is changing, and how it affects us. Before reading on, test your knowledge about gender inequality by taking the quiz in Box 12.1.

CRITICAL THINKING QUESTIONS

1. What would you identify as the best and worst aspects of being female or male in today's society?
2. What do your answers reveal about the constraining and enabling effects of gender?

BOX 12.1 SOCIOLOGY AND EVERYDAY LIFE

How Much Do You Know About Gender Inequality?

True	False	
T	F	1. The average earnings of employed women are still substantially lower than those of men, even when they are employed full time.
T	F	2. The percentage of women who are politicians, senior officials, and managers increased significantly in the past 20 years.
T	F	3. Most Canadians living in poverty are female.
T	F	4. It is estimated that the gender wage gap will no longer exist in 30 years.
T	F	5. Married couples today who both work full time tend to share the "unpaid labour" fairly equally.

For answers to the quiz about gender inequality, go to **www.nelson.com/student.**

LO-1 UNDERSTANDING GENDER

▶ **gender** The culturally and socially constructed differences between females and males found in the meanings, beliefs, and practices associated with "femininity" and "masculinity."

▶ **gender role** Attitudes, behaviour, and activities that are socially defined as appropriate for each sex and are learned through the socialization process.

▶ **gender identity** A person's perception of the self as female or male or other.

Gender refers to the culturally and socially constructed differences between females and males found in the meanings, beliefs, and practices associated with "femininity" and "masculinity." In contrast, as we will see in Chapter 13, *sex* refers to the biological and anatomical differences between females and males. Although biological differences between women and men are important, most "sex differences" are socially constructed "gender differences." According to sociologists, social and cultural processes—not biological "givens"—are most important in defining what females and males are, what they should do, and what sorts of relations do or should exist between them.

Virtually everything social in our lives is *gendered:* People continually distinguish between males and females and evaluate them differentially. Gender is an integral part of the daily experiences of both women and men.

A microlevel analysis of gender focuses on how individuals learn gender roles and acquire a gender identity. **Gender role** refers to the attitudes, behaviour, and activities that are socially defined as appropriate for each sex and are learned through the socialization process. For example, in Canadian society, males traditionally are expected to demonstrate aggressiveness and toughness, while females are expected to be passive and nurturing. **Gender identity** is a person's perception of the self as female or male or other. In the past, gender was described in binary terms, such as masculine or feminine, to describe gender in the same way that sex was considered to be comprised of only two categories. However, today we recognize that gender is more complex and encompasses more than just two possibilities. Included in the Other category are transgender, androgynous, gender nonconforming, two-spirit (see Box 12.2). Typically established between 18 months and three years of age, gender identity is a powerful aspect of our self-concept. Although this identity is an individual perception, it is developed through interaction with others. As a result, most people form a gender identity that matches their biological sex: Most biological females think of themselves as female and most biological males think of themselves as male.

A macrolevel analysis of gender examines structural features, external to the individual, that perpetuate gender inequality. These structures have been referred to as *gendered institutions,* meaning that gender is one of the major ways by which social life is organized in all sectors of society. Gender is embedded in the images, ideas, and language of a society and is used as a means to divide up work, allocate resources, and distribute power. For example, every society uses gender to assign certain tasks—ranging from child rearing to warfare—to females and males, and differentially rewards those who perform these duties.

FIGURE 12.1 GENDER GUIDE

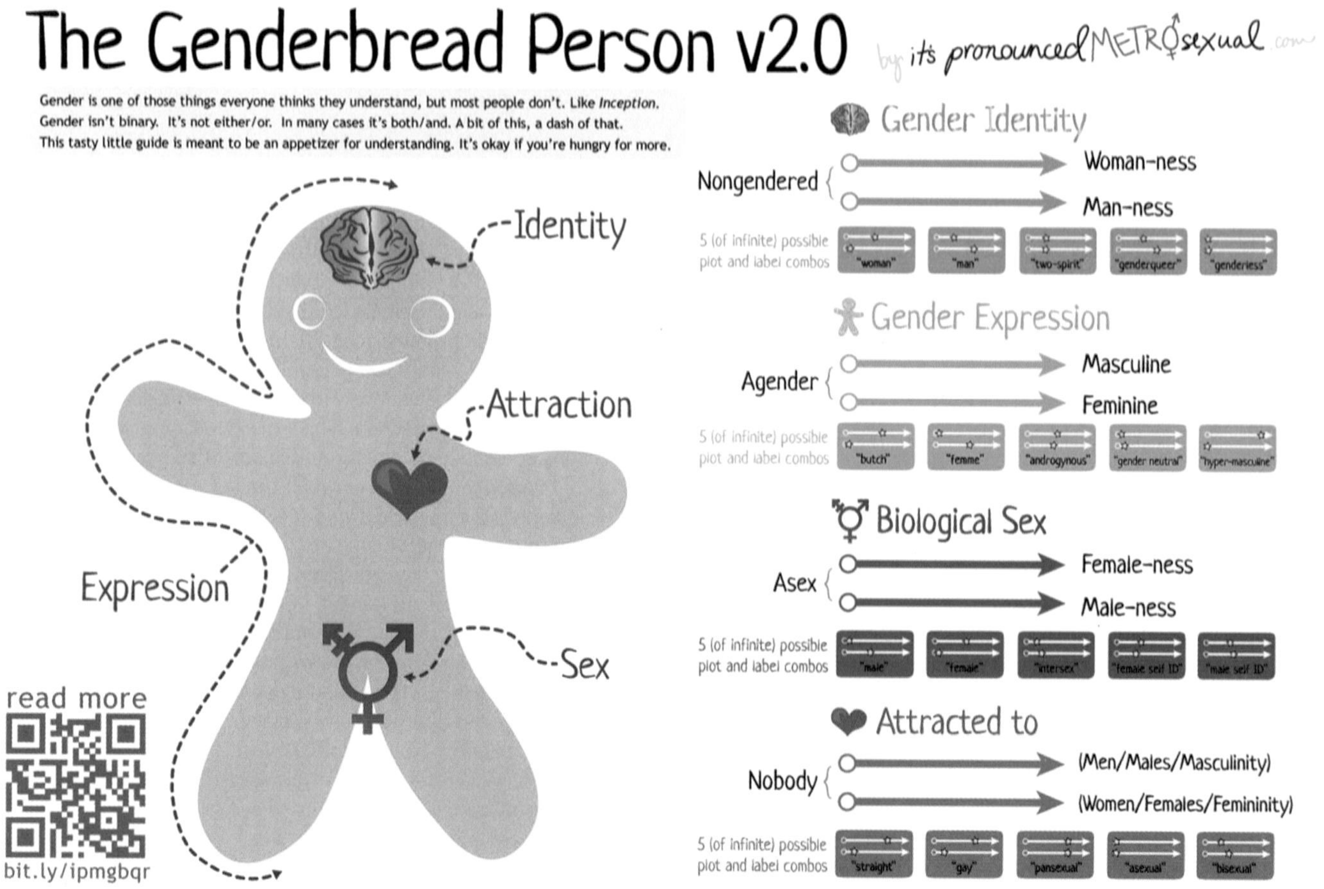

The Genderbread Person v2.0

by it's pronounced METROsexual .com

Gender is one of those things everyone thinks they understand, but most people don't. Like *Inception*. Gender isn't binary. It's not either/or. In many cases it's both/and. A bit of this, a dash of that. This tasty little guide is meant to be an appetizer for understanding. It's okay if you're hungry for more.

Identity
Attraction
Expression
Sex

read more
bit.ly/ipmgbqr

Gender Identity
Nongendered { Woman-ness / Man-ness
5 (of infinite) possible plot and label combos: "woman" "man" "two-spirit" "genderqueer" "genderless"

Gender Expression
Agender { Masculine / Feminine
5 (of infinite) possible plot and label combos: "butch" "femme" "androgynous" "gender neutral" "hyper-masculine"

Biological Sex
Asex { Female-ness / Male-ness
5 (of infinite) possible plot and label combos: "male" "female" "intersex" "female self ID" "male self ID"

Attracted to
Nobody { (Men/Males/Masculinity) / (Women/Females/Femininity)
5 (of infinite) possible plot and label combos: "straight" "gay" "pansexual" "asexual" "bisexual"

Source: Sam Killermann, "The Genderbread Person v2.0," itspronouncedmetrosexual.com, http://itspronouncedmetrosexual.com/2012/03/the-genderbread-person-v2-0/ (updated March 16, 2015).

BOX 12.2 POINT/COUNTERPOINT

How Many Genders: 56 or 2?

In 2014 the world's largest social media network-Facebook, made it possible for its users to self-identify their gender as other than male or female. After consultation with a group of leading LGBTQ advocacy groups, over 50 gender categories were identified and made available via a drop-down menu. Facebook users are also now able to choose the pronoun they would like to be referred to: he/his. She/her, or the gender neutral they/their.

Sexual health educator Dr. Debby Herbenick strongly supports this change: "While Facebook has always used **sex terms** (male and female) to describe users' genders, their recognition of the diversity of **gender identities** and presentations is a welcome one. She explains:

> Most of us never question or think much about our gender, but it's an essential part of our identity. And given the endlessly diverse ways people experience their gender, their bodies, and their masculinity or femininity, it's a wonder there are so few words to describe it. Except there are actually (at least) dozens of gender terms, and Facebook is now offering its

(Continued)

FIGURE 12.2 FACEBOOK GENDER IDENTIFIERS

Facebook now offers more than 50 custom gender identifiers, but the only way to see the options is via a drop-down autocomplete menu.

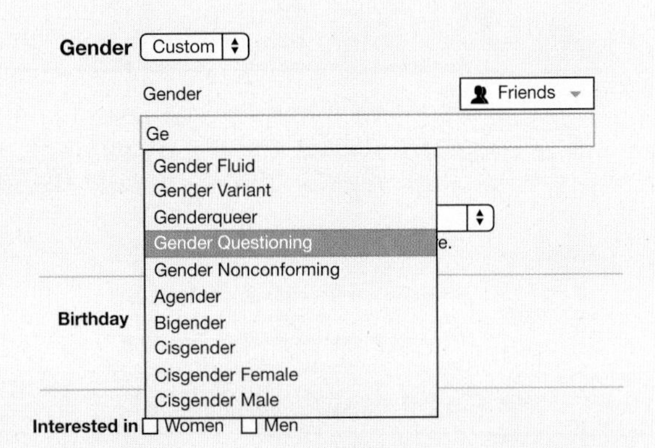

Source: Facebook.

users numerous options to present their gender identity to their Facebook friends in the same way they do in the real world. . . . Gender identity can be a sensitive issue and it's best to let other people tell you about their gender (if they want to) rather than make assumptions. Facebook's new gender options give people a chance to do just that and, we feel, are a good step toward expanding conversations about gender (Herbenick and Baldwin, 2014).

Although 56 gender categories may seem like a lot, many of the categories overlap or may be variations on a similar theme. For example, the terms Cisgender Male, Cis Male, Cis Man, and Cis Gender all refer to a person who is not trans or does not have a gender diverse identity or presentation. Following are definitions of some of the more commonly used terms to describe gender identity:

Agender – Refers to people who do not identify with any gender identity or someone who intentionally has no recognizable gender presentation.

Androgyne/Androgynous – People who do not identify with, or present as, a man or woman. This term can also refer to having both masculine and feminine qualities.

Bigender – People who identify as male and female at different times.

Cis/Cisgender – Cisgender is the opposite of. People who identify as cisgender are males or females whose gender aligns with their birth sex.

Gender Questioning – People who may be questioning their gender or gender identity, or are considering other ways of experiencing or expressing their gender or gender presentation.

Intersex – A person who was born with sexual anatomy, organs, or chromosomes that do not conform to the expected presentations of *either* male typical **or** female typical bodies.

Transsexual – Refers to transgender people who have made lasting changes to their physical body through surgery to align with their gender. Many transsexuals are transitioning (or have transitioned) from male to female or female to male through hormone therapy and/or gender reassignment surgery. The term is problematic and is rarely used by trans individuals because of its association with the psychiatric diagnosis 'transexualism,' which historically was required of trans persons to obtain sex reassignment surgery.

Trans/Transgender – An umbrella term that encompasses all people who feel their gender is different from their birth sex. They may or may not choose to physically transition from their birth sex to their experienced gender. People who identify as transgender *may or may not* have altered their bodies through surgery and/or hormones.

Two-Spirit – A term used by some Indigenous people to identify themselves, rather than as lesbian, gay, bisexual, or transgender. Historically, Two-Spirit persons were often leaders who were given special status based upon their ability to see the world from both male and female perspectives.

Sources: Herbenick, Debby 2014 "IT'S COMPLICATED What Each of Facebook's 51 New Gender Options Means" *The Daily Beast*, Feb 14, 2014 http://www.thedailybeast.com/articles/2014/02/15/the-complete-glossary-of-facebook-s-51-gender-options.html; Peter Weber, "Confused By All the Facebook Genders? Here's What They Mean" *The Week*, Feb 21, 2014 http://www.slate.com/blogs/lexicon_valley/2014/02/21/gender_facebook_now_has_56_categories_to_choose_from_including_cisgender.html

What do you think about all of these gender categories? Do they eliminate the constraints of the binary categories of male and female or masculine and feminine, or do they serve to confuse us even further? Opponents of these changes argue that the 'expansion of gender' does more harm than good. For example, the President of Canada Family Action sees gender as "factual." If you have a penis at birth, you are male. A vagina? Female. There is nothing in between, nor is it transferable.

"As soon as we move away from the fact that there's two genders, as soon as we start undermining or overriding that and say 'No there's three or four or five,' then we have no standard based on fact anymore. The standard becomes whatever's the flavour of the day" (Boesveld, 2014).

These institutions are reinforced by a *gender belief system* that includes all of the ideas regarding masculine and feminine attributes that are held to be valid in a society. This belief system is legitimized by religion, science, law, and other societal values. For example, gendered belief systems may change over time as gender roles change. Many fathers take care of young children today, and there is a much greater acceptance of this change in roles. However, popular stereotypes about men and women, as well as cultural norms about gender-appropriate appearance and behaviour, serve to reinforce gendered institutions in society.

LO-2 THE SOCIAL SIGNIFICANCE OF GENDER

Gender is a social construction with important consequences in everyday life. Just as stereotypes regarding race and ethnicity have built-in notions of superiority and inferiority, gender stereotypes hold that men and women are inherently different in attributes, behaviour, and aspirations. Stereotypes define men as strong, rational, dominant, independent, and less concerned with their appearance. Women are stereotyped as weak, emotional, nurturing, dependent, and anxious about their appearance.

The social significance of gender stereotypes is illustrated by eating problems. The three most common eating problems are anorexia, bulimia, and obesity. With *anorexia,* a person has an overriding obsession with food and thinness that constantly controls her or his activities and eating patterns, resulting in a body weight of less than 85 percent of the average weight for a person of that individual's age and height group. With *bulimia,* a person binges by consuming large quantities of food and then purges the food by induced vomiting, excessive exercise, laxatives, or subsequent fasting. In the past, *obesity* was defined as being 20 percent or more above a person's desirable weight, as established by the medical profession. Today, however, medical professionals use the BMI (body mass index) to define obesity. To determine this index, a person's weight in kilograms is divided by his or her height in metres and squared to yield the BMI. Obesity is defined as a BMI of 30 and above (about 30 pounds overweight for the average person).

In the past, it was assumed that the individuals most likely to have eating disorders were white, middle-class, heterosexual women; however, such problems also exist among women of colour, working-class women, lesbians, and some men. Explanations regarding the relationship between gender and eating disorders must take into account a complex array of social factors, including gender socialization and women's responses to problems such as racism and emotional, physical, and sexual abuse.

Bodybuilding is another gendered experience. *Bodybuilding* is the process of deliberately cultivating an increase in the mass and strength of the skeletal muscles by means of lifting and pushing weights. In the past, bodybuilding was predominantly a male activity; musculature connoted power, domination, and virility. Today, however, an increasing number of women engage in this activity. As gendered experiences, eating problems and bodybuilding have more in common than we might think. As some women's studies scholars have pointed out, the anorexic body and the muscled body are not opposites: Both are united against the common enemy of soft, flabby flesh. In other words, the *body* may be objectified both through compulsive dieting and compulsive bodybuilding.

For males, objectification and gender stereotyping may result in excessive bodybuilding.

SEXISM

Sexism is the subordination of one sex, usually female, based on the assumed superiority of the other sex. Sexism directed at women has three components: (1) negative attitudes toward women; (2) stereotypical beliefs that reinforce, complement, or justify the prejudice; and (3) discrimination—acts that exclude, distance, or keep women separate.

sexism The subordination of one sex, usually female, based on the assumed superiority of the other sex.

Can men be victims of sexism? Although women are more often the target of sexist remarks and practices, men can be victims of sexist assumptions. Examples of sexism directed against men are the assumption that men should not be employed in certain female-dominated occupations, such as nurse or elementary school teacher, and the belief that it is somehow more harmful for families when female soldiers are killed in battle than male soldiers.

Like racism, sexism is used to justify discriminatory treatment. Obvious manifestations of sexism are found in the undervaluing of women's work, and in hiring and promotion practices that effectively exclude women from an organization or confine them to the bottom of the organizational hierarchy. Even today, some women who enter nontraditional occupations (such as firefighting and welding) or professions (such as dentistry and architecture) encounter hurdles that men do not face.

▶ **patriarchy** A hierarchical system of social organization in which cultural, political, and economic structures are controlled by men.

▶ **matriarchy** A hierarchical system of social organization in which cultural, political, and economic structures are controlled by women.

Sexism is interwoven with **patriarchy**—a hierarchical system of social organization in which cultural, political, and economic structures are controlled by men. By contrast, **matriarchy** is a hierarchical system of social organization in which cultural, political, and economic structures are controlled by women; however, few societies have been organized in this manner. Patriarchy is reflected in the way men may think of their position as men as a given, while women may deliberate on what their position in society should be (see Box 12.3 for an example).

BOX 12.3 SOCIOLOGY IN GLOBAL PERSPECTIVE

The Rise of Islamic Feminism in the Middle East?

> I would like for all of the young Muslim girls to be able to relate to Iman, whether they wear the hijab [head scarf] or not. Boys will also enjoy Iman's adventures because she is one tough, smart girl! Iman gets her super powers from having very strong faith in Allah, or God. She solves many of the problems by explaining certain parts of the Koran that relate to the story.
>
> —Rima Khoreibi, an author from Dubai (United Arab Emirates), explaining that she has written a book about a female Islamic superhero because she would like to dispel a widely held belief that sexism in her culture is deeply rooted in Islam (see theadventuresofiman.com, 2007; Kristof, 2006)

Although Rima Khoreibi and many others who have written fictional and nonfictional accounts of girls and women living in the Middle East typically do not deny that sexism exists in their region or that sexism is deeply interwoven with patriarchy around the world, they dispute the perception that Islam is inherently misogynistic (possessing hatred or strong prejudice toward women). As defined in this chapter, patriarchy is a hierarchical system of social organization in which cultural, political, and economic structures are controlled by men. The influence of religion on patriarchy is a topic of great interest to contemporary scholars, particularly those applying a feminist approach to their explanations of why persistent social inequalities exist between women and men, and how these inequalities are greater in some regions of the world than in others.

According to some gender studies specialists, a newer form of feminist thinking is emerging among Muslim women. Often referred to as "feminist Islam" or "Islamic feminism," this approach is based on the belief that greater gender equality may be possible in the Muslim world if the teachings of Islam, as set forth in the Qur'an, the Islamic holy book, are followed more closely. Islamic feminism is based on the principle that Muslim women should retain their allegiance to Islam as an essential part of their self-determination and identity, but that they should also work to change patriarchal control over the basic Islamic worldview (Wadud, 2002). According to journalist Nicholas D. Kristof, both Islam and evangelical Christianity have been on the rise in recent years because both religions provide "a firm moral code, spiritual reassurance and orderliness to people vexed by chaos and immorality around them, and they offer dignity to the poor" (2006: A22).

Islamic feminists believe that the rise of Islam might contribute to greater, rather than less, equality for women. From this perspective, stories about characters such as Iman may help girls and young women realize that they can maintain their deep religious convictions and their head scarf (hijab) while working for greater equality for women and more opportunities for themselves. In *The Adventures of Iman,* the female hero always wears a pink scarf around her neck and she uses the scarf to cover her hair when she is praying to Allah. Iman quotes the Qur'an when she is explaining to others that Muslims are expected to be tolerant, kind, and righteous. For Iman, religion is a form of empowerment, not an extension of patriarchy.

The focus of Islamic feminism is quite different from what most people view as Western feminism. For example, Islamic feminism puts less emphasis than might be expected on issues such as the wearing of the hijab or the fact that in Saudi Arabia, a woman may own a motor vehicle but cannot legally drive it. As rapid economic development and urbanization affect the lives of many people, however, change is clearly underway in many regions of the Middle East and in other areas of the world.

In light of such differences, consider the following. Why is women's inequality a complex issue to study across nations? What part does culture play in defining the roles of women and men in various societies? How do religious beliefs influence what we think of as "appropriate" or "inappropriate" behaviours for men, women, and children? What do you think?

Sources: Kristof, 2006; theadventuresofiman.com, 2007; Wadud, 2002.

LO-3 WORK AND GENDER INEQUALITY

How do tasks in a society come to be defined as "men's work" or "women's work"? Three factors are important in determining the gendered division of labour in a society: (1) the type of subsistence base, (2) the supply of and demand for labour, and (3) the extent to which women's child-rearing activities are compatible with certain types of work. *Subsistence* refers to the means by which a society gains the basic necessities of life, including food, shelter, and clothing. Based on subsistence, societies are classified as hunting and gathering, horticultural and pastoral, agrarian, industrial, or post-industrial.

HUNTING AND GATHERING SOCIETIES

The earliest known division of labour between women and men is in hunting and gathering societies. While the men hunt for wild game, women gather roots, nuts, seeds and berries. A relatively equitable relationship exists because neither sex has the ability to provide all the food necessary for survival. When wild game is nearby, both men and women may hunt. When it is far away, hunting becomes incompatible with child rearing (which women tend to do because they breastfeed their young), and women are placed at a disadvantage in terms of contributing to the food supply (Lorber, 1994). In most hunting and gathering societies, women are full economic partners with men; relations between them tend to be cooperative and relatively egalitarian (Bonvillain, 2001). Little social stratification of any kind is found because people do not acquire a food surplus.

HORTICULTURAL AND PASTORAL SOCIETIES

In horticultural societies, which first developed 10,000 to 12,000 years ago, a steady source of food becomes available. People are able to grow their own food because of hand tools, such as the digging stick and the hoe. Women make an important contribution to food production because cultivation with hoes is compatible with child care. A fairly high degree of gender equality exists because neither sex controls the food supply.

When inadequate moisture in an area makes planting crops impossible, *pastoralism*—the domestication of large animals to provide food—develops. Herding is done primarily by men, and women contribute relatively little to subsistence production in such societies. In some herding societies, women have relatively low status; their primary value is their ability to produce male offspring so that the family lineage can be preserved and enough males will exist to protect the group against attack.

In contemporary horticultural societies, women do most of the farming while men hunt game, clear land, work with arts and crafts, make tools, participate in religious and ceremonial activities, and engage in war. A combination of horticultural and pastoral activities is found in some contemporary societies in Asia, Africa, the Middle East, and South America. These societies are characterized by more gender inequality than in hunting and gathering societies but less gender inequality than in agrarian societies (Bonvillain, 2001).

AGRARIAN SOCIETIES

In agrarian societies, which first developed about eight to ten thousand years ago, gender inequality and male dominance become institutionalized. The most extreme form of gender inequality developed about five thousand years ago in societies in the fertile crescent around the Mediterranean Sea. Agrarian societies rely on agriculture—farming done by animal-drawn or mechanically powered plows and equipment. Because agrarian tasks require more labour and greater physical strength than horticultural ones, men become more involved in food production. It has been suggested that women are excluded from these tasks because they are viewed as too weak for the work and because child-care responsibilities are considered incompatible with the full-time labour that the tasks require.

Male dominance is very strong in agrarian societies. Women are secluded, subordinated, and mutilated as a means of regulating their sexuality and protecting paternity. Most of the world's population currently lives in agrarian societies in various stages of industrialization.

INDUSTRIAL SOCIETIES

An *industrial society* is one in which factory or mechanized production has replaced agriculture as the major form of economic activity. As societies industrialize, the status of women tends to decline further. Industrialization in North America created a gap between the nonpaid work performed by women at home and the paid work that was increasingly performed by men and unmarried young women (Krahn and Lowe, 2007).

In Canada, the division of labour between men and women in the middle and upper classes became much more distinct with industrialization. The men were viewed as "breadwinners"; the women were seen as "homemakers." In this new "cult of domesticity" (also referred to as the "cult of true womanhood"), the home became a private, personal sphere in which women created a haven for the family. Those who supported the cult of domesticity argued that women were the natural keepers of the domestic sphere and that children were the mother's responsibility. Meanwhile, the "breadwinner" role placed enormous pressures on men to support their families—providing for them well was considered a sign of manhood. This gendered division of labour increased the economic and political subordination of women.

The cult of true womanhood not only increased white women's dependence on men but also became a source of discrimination against women of colour, based on both their race and the fact that many of them had to work to survive. Employed, working-class white women were similarly stereotyped; they became more economically dependent on their husbands because their wages were so much lower.

POST-INDUSTRIAL SOCIETIES

Chapter 5 defines *post-industrial societies* as societies in which technology supports a service- and information-based economy. In such societies, the division of labour in paid employment is increasingly based on whether people provide or apply information or are employed in service jobs, such as fast-food restaurant counter help or healthcare workers. For both women and men in the labour force, formal education is increasingly crucial for economic and social success. However, even as some women have moved into entrepreneurial, managerial, and professional occupations, many others have remained in the low-paying service sector, which affords few opportunities for upward advancement.

Will technology change the gendered division of labour in post-industrial societies? Scholars do not agree on the effects of computers, the Internet, cellphones, tablets, and many other newer forms of communications technology on the role of women in society. Although some analysts presumed that technological developments would reduce the boundaries between women's and men's work, researchers have found that the gender stereotyping associated with specific jobs has remained remarkably stable even when the nature of work and the skills required to perform it have been radically transformed. Today, men and women continue to be segregated into different occupations, and this segregation is particularly visible within individual workplaces (as discussed later in the chapter).

How does the division of labour change in families in post-industrial societies? For a variety of reasons, more households are headed by women with no adult male present. This means that women in these households truly have a double burden, both from family responsibilities and from the necessity of holding gainful employment in the labour force. In post-industrial societies such as Canada, close to 80 percent of adult women are in the labour force. This reality means that despite living in an information- and service-oriented economy, women will continue to bear the heavy burden of finding time to care for children, help aging parents, and meet the demands of the workplace (Marshall, 2011).

How people accept new technologies and the effect these technologies have on gender stratification are related to how people are socialized into gender roles. However, gender-based stratification remains rooted in the larger social structures of society, which individuals have little ability to control.

TIME TO REVIEW

- How do new technologies influence gender relations in the workplace and the division of labour in the home?
- Is it likely that technology will increase or decrease the divisions between men and women at work and at home?

LO-4 GENDER AND SOCIALIZATION

We learn gender-appropriate behaviour through the socialization process. Our parents, teachers, friends, and the media all serve as gendered institutions that communicate to us our earliest, and often most lasting, beliefs about the social meanings of being male or female, and thinking and behaving in masculine or feminine ways. Some gender roles have changed dramatically in recent years; others remain largely unchanged over time.

Many parents prefer boys to girls because of stereotypical ideas about the relative importance of males and females to the future of the family and society. Although some parents prefer boys to girls because they believe old myths about the biological inferiority of females, research suggests that social expectations also play a major role in this preference. We are socialized to believe that it is important to have a son, especially as a first or only child. For many years, it was assumed that a male child could support his parents in their later years and carry on the family name.

Across cultures, boys are preferred to girls, especially when the number of children that parents can have is limited by law or economic conditions. In China and India, fewer girls are born each year than boys because a disproportionate number of female fetuses are aborted. In China, a one-child government policy favours males over females; in India, a cultural belief that boys are an asset to families while girls are a liability contributes to selective abortions of female fetuses (Flintoff, 2011). However, now these nations are faced with a shortage of marriageable young women and many other problems that result from an imbalance in the sex ratio. Perhaps seeing the consequences of favouring one sex over the other will produce new ideas among parents regarding sex and gender socialization.

Howard Sayer/Shutterstock.com

Alexey Losevich/Shutterstock.com

Are children's toys a reflection of their own preferences and choices? How do toys reflect gender socialization by parents and other adults?

PARENTS AND GENDER SOCIALIZATION

From birth, parents treat children differently on the basis of the child's sex. Baby boys are perceived to be less fragile than girls and tend to be treated more roughly by their parents. Girl babies are thought to be "cute, sweet, and cuddly" and receive more gentle treatment. Parents strongly influence the gender-role development of children by passing on—both overtly and covertly—their own beliefs about gender. Although contemporary parents tend to play more similarly with their male and female children than their own parents or grandparents might have played with them as they were growing up, there remains a difference in how they respond toward their children based on gender even when "roughhousing" with them or engaging in sports events or other activities.

Children's toys reflect their parents' gender expectations. Gender-appropriate toys for boys include video games, trucks and other vehicles, sports equipment, and war toys such as guns and soldiers. Girls' toys include stuffed animals and dolls, makeup and dress-up clothing, and home-making items. Ads for children's toys appeal to boys and girls differently. Most girl and boy characters are shown in gender-specific toy commercials that target either females or males. These commercials typically show boys playing outdoors and engaging in competitive activities. Girls are more often engaged in cooperative play in the ads, and this is in keeping with gender expectations about their behaviour (Kaklenberg and Hein, 2010).

When children are old enough to help with household chores, boys and girls are often assigned different tasks. Maintenance chores (such as mowing the lawn) are assigned to boys, while domestic chores (such as shopping, cooking, and cleaning the table) are assigned to girls. Chores may also become linked with future occupational choices and personal characteristics.

Many parents are aware of the effect that gender socialization has on their children and make a conscientious effort to provide gender neutral experiences for them.

PEERS AND GENDER SOCIALIZATION

Peers help children learn prevailing gender role stereotypes, as well as gender-appropriate—and inappropriate—behaviour. During the school years, same-sex peers have a powerful effect on how children see their gender roles; during adolescence, they are often more influential agents of gender socialization than adults.

Children, especially boys, are more socially acceptable to their peers when they conform to implicit societal norms governing the "appropriate" ways that girls and boys should act in social situations. Male peer groups place more pressure on boys to do "masculine" things than female peer groups place on girls to do "feminine" things. For example, girls wear jeans and other "boy" clothes, play soccer and softball, and engage in other activities traditionally associated with males. But if a boy wears a dress, plays hopscotch with girls, and engages in other activities associated with being female, he will be ridiculed by his peers. This distinction between the relative value of boys' and girls' behaviours strengthens the cultural message that masculine activities and behaviour are more important and more acceptable.

Peers are thought to be especially important in boys' development of gender identity. Male bonding that occurs during adolescence is believed to reinforce masculine identity and to encourage gender-stereotypical attitudes and behaviour. For example, male peers have a tendency to ridicule and bully others about their appearance, size, and weight. Because peer acceptance is so important, such actions can have very harmful consequences.

TEACHERS, SCHOOLS, AND GENDER SOCIALIZATION

From kindergarten through university, schools operate as gendered institutions. Teachers provide important messages about gender through both the formal content of classroom assignments and informal interaction with students. Sometimes, gender-related messages from teachers and other students reinforce gender roles that have been taught at home; however, teachers may also contradict parental socialization. During the early years of a child's schooling, the teacher's influence is very powerful; many children spend more hours per day with their teachers than they do with their parents.

According to some researchers, the quantity and quality of teacher–student interactions often vary between the education of girls and that of boys (Sadker and Zittleman, 2009). One of the messages that teachers may communicate to students is that boys are more

Teachers often use competition between boys and girls because they hope to make a learning activity more interesting. Here, a middle school girl leads other girls against boys in a Spanish translation contest. What are the advantages and disadvantages of gender-based competition in classroom settings?

gender bias
Behaviour that shows favouritism toward one gender over the other.

important than girls. Research spanning the past 30 years shows that unintentional gender bias occurs in virtually all educational settings. **Gender bias** consists of showing favouritism toward one gender over the other. Researchers consistently find that teachers devote more time, effort, and attention to boys than to girls (Sadker and Zittleman, 2009). Males receive more praise for their contributions and are called on more frequently in class, even when they do not volunteer.

Teacher–student interactions influence not only students' learning but also their self-esteem (Sadker and Zittleman, 2009). A comprehensive study of gender bias in schools suggested that girls' self-esteem is undermined in school through such experiences as (1) a relative lack of attention from teachers; (2) sexual harassment by male peers; (3) the stereotyping and invisibility of females in textbooks, especially in science and math texts; and (4) test bias based on assumptions about the relative importance of quantitative and visual–spatial ability, as compared with verbal ability, that restricts some girls' chances of being admitted to the most prestigious colleges and being awarded scholarships.

Teachers also influence how students treat one another during school hours. Many teachers use sex segregation as a way to organize students, resulting in unnecessary competition between females and males. In addition, teachers may take a "boys will be boys" attitude when girls complain of sexual harassment. Even though sexual harassment is prohibited by law, and teachers and administrators are obligated to investigate such incidents, the complaints may be dealt with superficially. If that happens, the school setting can become a hostile environment rather than a site for learning.

MASS MEDIA AND GENDER SOCIALIZATION

The media, including newspapers, magazines, television, and movies, are powerful sources of gender stereotyping. Although some critics argue that the media simply reflect existing gender roles in society, others point out that the media have a unique ability to shape ideas. Think of the impact that television might have on children if they spend one-third of their waking time watching it, as has been estimated. From children's cartoons to adult shows, television programs are sex-typed, and many are male oriented. More male than female roles are shown, and male characters act strikingly different from female ones. Typically, males are more aggressive, constructive, and direct, and are rewarded for their actions. By contrast, females are depicted as acting deferential toward other people or as manipulating them through helplessness or seductiveness to get their way.

In prime-time television, a number of significant changes in the past three decades have reduced gender stereotyping; however, men still outnumber women as leading characters, and they are often "in charge" in any setting where both men's and women's roles are portrayed. Recently, retro-series on network and cable television have brought back an earlier era when men were dominant in public and family life and women played a subordinate role to them. For example, the award-winning series *Mad Men* (on AMC) was set in a 1960s New York advertising agency, where secretaries were expected to wear tight sweaters and skirts and bring men hot coffee throughout the day, while the men's wives were supposed to be the perfect companions and hostesses at home. Although many other TV series, such as *Modern Family,* have changed traditional norms, offering a wide diversity of families, including gay dads with a child, the shift to retro-gender roles in some television programming and films in the second decade of the 21st century has raised questions about the extent to which change actually occurs in the portrayal of women and men in the media.

Advertising—whether on television and billboards or in magazines and newspapers—can be very persuasive. The intended message is clear to many people: If they embrace traditional notions of masculinity and femininity, their personal and social success is assured; if they purchase the right products and services, they can enhance their appearance and gain power over other people.

TIME TO REVIEW

- Consider all of the primary socialization agents considered in this section. Which of these socialization agents have had the greatest impact on your gender identity?

LO-5 CONTEMPORARY GENDER INEQUALITY

According to feminist scholars, women experience gender inequality as a result of economic, political, and educational discrimination (See Figure 12.3). Women's position in the Canadian workforce reflects the years of subordination that women have experienced in society.

GENDERED DIVISION OF PAID WORK

The workplace is another example of a gendered institution, and where people are located in the occupational structure of the labour market has a major effect on their earnings. In industrialized countries, most jobs are segregated by gender and by race and ethnicity. In most workplaces, employees are either gender-segregated or all of the same gender. *Gender-segregated work* refers to the concentration of women and men in different occupations, jobs, and places of work. Despite some progress, the majority of employed women continued to work in occupations in which they have been traditionally concentrated. The most recent statistics indicate that almost 70 percent of all employed women were working in teaching, nursing and related health occupations, clerical or other administrative positions, or sales and service occupations (Ferrao, 2010).

To eliminate gender-segregated jobs in North America, more than half of all men or all women workers would have to change occupations. Moreover, women are severely underrepresented at the top Canadian corporations, at only about 17 percent of the corporate officers in the *Financial Post 500* list (comprising the 500 largest companies in Canada). Of these, only about 6 percent hold the highest corporate officer titles, and only 23 women serve as chief executive officer (Catalyst Canada, 2014). Based on current rates of change, the number of women reaching the top ranks of corporate Canada will not reach an acceptable level of 25 percent until the year 2025.

Although the degree of gender segregation in parts of the professional labour market has declined since the 1970s, racial–ethnic segregation has remained deeply embedded in the social structure. The relationship between visible minority status and occupational status is complex and varies by gender, however. Visible minority males are overrepresented in both lower- and higher-status occupations; nonwhite women are heavily overrepresented in lower-paying, low-skilled jobs (Ferrao, 2010).

Labour market segmentation—the division of jobs into categories with distinct working conditions—results in women having separate and unequal jobs. Why does gender-segregated work matter? Although we look more closely at the issue of the pay gap in the following section, it is important to note here that the pay gap between men and women is the best-documented consequence of gender-segregated work. Most women work in lower-paying, less-prestigious jobs, with less opportunity for advancement than their male counterparts.

Gender-segregated work affects both men and women. Men are often kept out of certain types of jobs. Those who enter female-dominated occupations often have to justify themselves and prove that they are "real men." Even if these assumptions do not push men out of female-dominated occupations, they affect how the men manage their gender identity at work. For example, men in occupations such as nursing emphasize their masculinity, attempt to distance themselves from female colleagues, and try to move quickly into management and supervisory positions.

Occupational gender segregation contributes to stratification in society. Job segregation is structural; it does not occur simply because individual workers have different abilities, motivations, and material needs. As a result of gender and racial segregation, employers are able to pay many men of colour and all women less money, promote them less often, and provide fewer benefits.

THE GENDER WAGE GAP

▶ **wage gap** A term used to describe the disparity between women's and men's earnings.

Occupational segregation contributes to a second form of discrimination—the **wage gap**, a term used to describe the disparity between women's and men's earnings. It is calculated by dividing women's earnings by men's to yield a percentage, also known as the *earnings ratio*. Today, women who work full-time for the whole year still earn just over 70 cents for each dollar earned by their male counterparts (Williams, 2011). See Figure 12.4. One study found that the majority of this wage gap can be explained by fields of study chosen by men and women, the continued overrepresentation of women in low-paying sectors of the economy, and gender differences in the division of time between caregiving and paid employment (Drolet, 2011). Once again, this gap is partially attributable to occupational segregation. The majority of female university students enrol in degree programs in education, health professions, fine arts, and the humanities, while males continue to dominate in the fields of science and engineering. Even within occupations that require specialized educational credentials, the wage gap does not disappear—for every dollar earned by men, women earned 65 cents as dentists, 68 cents as lawyers, and 77 cents as university professors (Cool, 2010). See Figure 12.3.

FIGURE 12.3 GENDER WAGE GAP IN EDUCATION

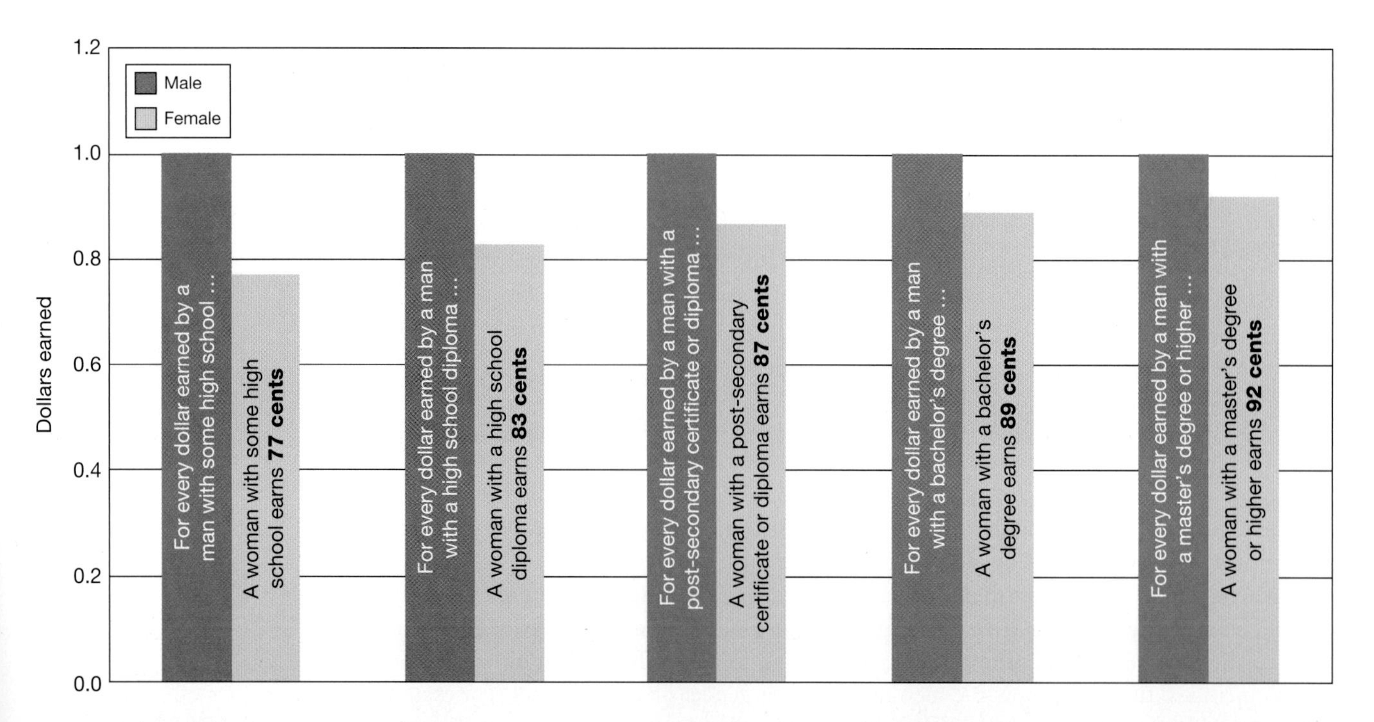

Source: Morissette, René, Garnett Picot, and Yuqian Lu, 2013, The Evolution of Canadian Wages over the Last Three Decades, Analytical Studies Branch Research Paper Series, Statistics Canada, Catalogue no. 11F0019M - No. 347. http://www.statcan.gc.ca/pub/11f0019m/11f0019m2013347-eng.pdf

FIGURE 12.4 : AVERAGE HOURLY WAGES, BY OCCUPATION AND SEX, 2014

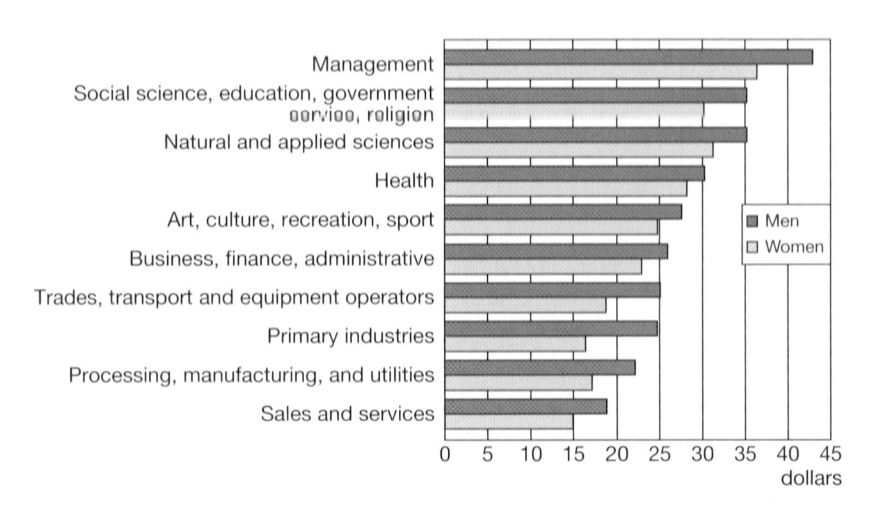

Sources: Adapted from Statistics Canada, The Canadian Market at a Glance, Cat. No 71-22-X, 2009; CANSIM Table 282-0069, "Labour force survey estimates (LFS) wages of employees by type of work, National Occupational Classification for Statistics (NOC-S), sex and age group, 2015, http://www5.statcan.gc.ca/cansim/a26 http://www5.statcan.gc.ca/cansim/a26

pay equity (comparable worth) The belief that wages ought to reflect the worth of a job, not the gender or race of the worker.

PAY EQUITY AND EMPLOYMENT EQUITY

A number of strategies have been implemented in an attempt to achieve greater gender equality in the labour market. *Pay equity* attempts to raise the value of the work traditionally performed by women. *Employment equity* strategies focus on ways to move women into higher-paying jobs traditionally held by men. Since the 1980s, the federal government, some provincial governments, and several private companies have implemented pay equity and employment equity policies.

Pay equity (or, as it is sometimes called, **comparable worth**) reflects the belief that wages ought to reflect the worth of a job, not the gender or race of the worker. How can the comparable worth of different kinds of jobs be determined? One way is to compare the work involved in women's and men's jobs and see if there is a disparity in the salaries paid for each. To do this, analysts break a job into components—such as the education, training, and skills required, the extent of responsibility for others' work, and the working conditions—and then allocate points for each. For pay equity to exist, men and women in occupations that receive the same number of points should be paid the same. In short, pay equity promotes the principle of equal pay for work of equal value.

FIGURE 12.5 : EVOLUTION OF THE GENDER WAGE GAP

Although incomes for both men and women have grown over the past 20 years, the gender wage gap has remained virtually unchanged, going from .728 to .788.

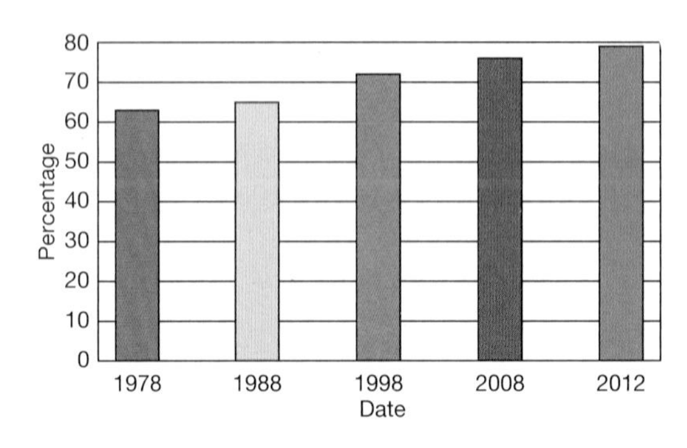

Sources: Cool, Julie. The Wage Gap Between Women and Men. Library of Parliament. http://www.parl.gc.ca/Content/LOP/ResearchPublications/2010-30-e.pdf; Table prepared using data obtained from Statistics Canada, "Female-to-male earnings ratios, by selected characteristics, 2008 constant dollars, annual (percent)," Table 202-0104, CANSIM (database), Using E-STAT (distributor), 13 September 2010; McInturff, Kate, 2013 Closing Canada's Gender Gap: Year 2240 Here We Come, Canadian Centre for Policy Alternatives http://www.policyalternatives.ca/sites/default/files/uploads/publications/National%20Office/2013/04/Closing_Canadas_Gender_Gap_0.pdf

A second strategy for addressing inequality in the workplace is **employment equity**—a strategy to eliminate the effects of discrimination and to make employment opportunities available to groups who have been excluded (Krahn, Lowe, and Hughes, 2007). The target groups for employment equity are visible minorities, persons with disabilities, Indigenous peoples, and women. In comparison with pay equity, which addresses wage issues only, employment equity covers a range of employment issues, such as recruitment, selection, training, development, and promotion. Employment equity also addresses issues pertaining to conditions of employment, such as compensation, lay-offs, and disciplinary action. Critics of employment equity policies have pointed out that the *Employment Equity Act* of 1996 has jurisdiction over a tiny percentage of the population; it covers only federal government employers or companies that have contracts with the federal government.

employment equity A strategy to eliminate the effects of discrimination and to make employment opportunities available to groups who have been excluded.

PAID WORK AND FAMILY WORK

As previously discussed, the first big change in the relationship between family and work occurred with the Industrial Revolution and the rise of capitalism. The cult of domesticity kept many middle- and upper-class women out of the workforce during this period. Primarily working-class and poor women had to deal with the work/family conflict. Today, however, the issue spans the entire economic spectrum. The typical married woman in Canada combines paid work in the labour force and family work as a homemaker. Although this change has occurred at the societal level, individual women bear the brunt of the problem.

While dramatic changes have occurred in women's participation in the workforce, men's entry into housework has been gradual, prompting some to refer to the latter as a stalled revolution. Recent time-use surveys have confirmed that the burden of unpaid work continues to rest disproportionately on women. Even when women work full-time, most maintain primary responsibility for child care, elder care, housework, shopping, and meal preparation. Among couples, the woman does close to two hours per day more housework than her male partner. Consequently, many women have a "double day" or "second shift" because of their dual responsibilities for paid and unpaid work (Hochschild, 2003; see also Chapter 6). Working women have less time to spend on housework; if husbands do not help do routine domestic chores, some chores do not get done or get done less often. Although the income that many women earn is essential for the economic survival of their families, they must spend part of their earnings on family maintenance, such as daycare, fast food, and laundry and housecleaning, in an attempt to keep up with their obligations.

© iStockphoto.com/jonas unruh

© Spencer Grant/PhotoEdit

What stereotypes are associated with men in female-oriented occupations? with women in male-oriented occupations? Do you think such stereotypes will change in the near future?

Especially in families with young children, domestic responsibilities consume much time and energy. Although some kinds of housework can be put off, the needs of children often cannot be ignored or delayed. When children are ill or school events cannot be scheduled around work, parents (especially mothers) may experience stressful role conflicts. ("Shall I be a good employee or a good mother?")

Many working women care not only for themselves, their husbands, and their children but also for elderly parents or in-laws. Some analysts refer to these women as "the sandwich generation"—caught between the needs of their young children and elderly relatives. Many women try to solve their time crunch by forgoing leisure time and sleep. Some married women with children found more fulfillment at work and worked longer hours because they liked work better than facing the pressures of home (see Chapter 6).

Although the transition into housework has been slow for men, there is room for optimism as the household–work gender gap slowly narrows. As shown in Figure 12.6, gender differences in the division of labour remain, but they are slowly diminishing. Since 1986, as women have increased their participation in paid work, men have increased their time spent on housework. In particular, men have made significant changes in their participation in core housework, such as meal preparation and cleanup, cleaning, and laundry. For couples with children, there have also been noticeable changes in men's participation in meeting child-care responsibilities and duties (Marshall, 2011).

TIME TO REVIEW

- Based on your understanding of the gender wage gap, do you think the gap will get larger, smaller, or disappear completely as younger men and women enter the paid workforce? Why, or why not?
- Analysts believe that the burden of the "double day" or "second shift" will likely preserve women's inequality at home and in the workplace for another generation. If this is the case, what needs to be done to address the gendered division of unpaid work?

FIGURE 12.6 : PAID AND UNPAID WORK

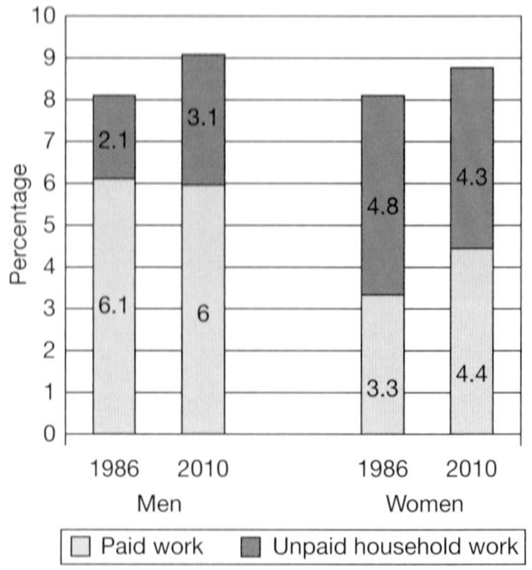

Source: Katherine Marshall, "Generational Change in Paid and Unpaid Work," Canadian Social Trends no. 82, Statistics Canada catalogue no. 11-008-x, 2011, Chart 1, p. 18. http://www.statcan.gc.ca/pub/11-008-x/2011002/article/11520-eng.pdf

LO-6 PERSPECTIVES ON GENDER STRATIFICATION

Sociological perspectives on gender stratification vary in their approach to examining gender roles and power relationships in society. Some focus on the roles of women and men in the domestic sphere; others note the inequalities arising from a gendered division of labour in the workplace. Still others attempt to integrate both the public and private spheres into their analyses. The Concept Snapshot in this chapter outlines the key aspects of each sociological perspective on gender stratification.

FUNCTIONALIST AND NEOCLASSICAL ECONOMIC PERSPECTIVES

As seen earlier, functionalist theory views men and women as having distinct roles that are important for the survival of the family and society. The most basic division of labour is biological: Men are physically stronger, while women are the only ones able to bear and nurse children. Gendered belief systems foster assumptions about appropriate behaviour for men and women and may have an impact on the types of work that women and men perform.

THE IMPORTANCE OF TRADITIONAL GENDER ROLES According to functional analysts such as Talcott Parsons (1955), women's roles as nurturers and caregivers are even more pronounced in contemporary industrialized societies. While the husband performs the instrumental tasks of providing economic support and making decisions, the wife assumes the expressive tasks of providing affection and emotional support for the family. This division of family labour ensures that important societal tasks will be fulfilled; it also provides stability for family members.

This view has been adopted by a number of conservative analysts who assert that relationships between men and women are damaged when changes in gender roles occur and that family life suffers as a consequence. From this perspective, the traditional division of labour between men and women is the natural order of the universe.

THE HUMAN CAPITAL MODEL Functionalist explanations of occupational gender segregation are similar to neoclassical economic perspectives, such as the human capital model. According to this model, individuals vary widely in the amount of human capital they bring to the labour market. Human capital is acquired by education and job training; it is the source of

Getty Images

According to the human capital model, women earn less in the labour market because of their child-rearing responsibilities. What other explanations are offered for the lower wages that women receive?

a person's productivity and can be measured in terms of the return on the investment (wages) and the cost (schooling or training).

From this perspective, what individuals earn is the result of their own choices (the kinds of training, education, and experience they accumulate, for example) and of the labour-market need (demand) for and availability (supply) of certain kinds of workers at specific points in time. For example, human capital analysts might argue that women diminish their human capital when they leave the labour force to engage in childbearing and child-care activities. While women are out of the labour force, their human capital deteriorates from non-use. When they return to work, women earn lower wages than men because they have fewer years of work experience and have "atrophied human capital" because their education and training may have become obsolete. One study found that over a 15-year period, women compared to men worked fewer years and fewer hours when the women were married and had dependent children. As a result, the women were more likely to work fewer hours in the labour market and be low earners.

EVALUATION OF FUNCTIONALIST AND NEOCLASSICAL ECONOMIC PERSPECTIVES Although Parsons and other functionalists did not specifically endorse the gendered division of labour, their analysis views it as natural and perhaps inevitable. Critics argue that problems inherent in traditional gender roles, including the strains placed by these roles on both men and women, and the social costs to society, are minimized by this approach. For example, men are assumed to be "money machines" for their families when they might prefer to spend more time in child-rearing activities. Also, the woman's place is assumed to be in the home, an idea that ignores the fact that many women hold jobs out of economic necessity.

Another limitation of the functionalist approach is that it does not take a critical look at the structure of society—especially the economic inequalities—that make educational and occupational opportunities more available to some than to others. Furthermore, it fails to examine the underlying power relations between men and women or to consider that the tasks assigned to women and men are unequally valued by society. Similarly, the human capital model is rooted in the premise that individuals are evaluated based on their human capital in an open, competitive market where education, training, and other job-enhancing characteristics are taken into account. From this perspective, those who make less money (often men from visible minority groups and all women) have no one to blame but themselves.

CONFLICT PERSPECTIVES

According to many conflict analysts, the gendered division of labour within families and in the workplace results from male control of and dominance over women and resources. Differentials between men and women may exist in terms of economic, political, physical, and/or interpersonal power. The importance of a male monopoly in any of these arenas depends on the significance of that type of power in a society (Richardson, 1993). In hunting and gathering and horticultural societies, male dominance over women is limited because all members of the society must work to survive (Collins, 1971; Nielsen, 1990). In agrarian societies, however, male sexual dominance is at its peak. Male heads of household gain a monopoly not only on physical power but also on economic power, and women become sexual property.

Although men's ability to use physical power to control women diminishes in industrial societies, men still remain the household heads and control the property. Men also gain more power through their predominance in the most highly paid and prestigious occupations and the highest elected offices. In contrast, women have the ability to trade their sexual resources, companionship, and emotional support in the marriage market for men's financial support and social status; as a result, however, women as a group remain subordinate to men (Collins, 1971; Nielsen, 1990).

All men are not equally privileged, though. Some analysts argue that women and men in the upper classes are more privileged, because of their economic power, than men in lower-class positions and members of some minority groups (Lorber, 1994). In industrialized societies, persons who occupy elite positions in corporations, universities, the mass media, and government or who have great wealth have the most power (Richardson, 1993). Most, however, are men.

Conflict theorists in the Marxist tradition assert that gender stratification results from private ownership of the means of production; some men not only gain control over property and the distribution of goods but also gain power over women. According to Friedrich Engels and Karl Marx, marriage serves to enforce male dominance. Men of the capitalist class instituted monogamous marriage (a gendered institution) so that they could be certain of the paternity of their offspring, especially sons, whom they wanted to inherit their wealth. Feminist analysts have examined this theory, among others, as they have sought to explain male domination and gender stratification.

FEMINIST PERSPECTIVES

Feminism—the belief that women and men are equal and that they should be valued equally and have equal rights—is embraced by many men as well as women. Gender is viewed as a socially constructed concept that has important consequences in the lives of all people (Craig, 1992). According to sociologist Ben Agger (1993), men can be feminists and propose feminist theories; both women and men have much in common as they seek to gain a better understanding of the causes and consequences of gender inequality.

feminism The belief that women and men are equal and that they should be valued equally and have equal rights.

Although all feminist perspectives begin with the assumption that the majority of women occupy a subordinate position to men, they often diverge in terms of their explanations of how and why women are subordinated, and the best strategies for achieving true equality for women (Chunn, 2000). Feminist perspectives vary in their analyses of the ways in which norms, roles, institutions, and internalized expectations limit women's behaviour. Taken together, they all seek to demonstrate how women's personal control operates even within the constraints of a relative lack of power (Stewart, 1994).

LIBERAL FEMINISM In liberal feminism, gender equality is equated with equality of opportunity. Liberal feminists assume that women's inequality stems from the denial to them of equal rights (Mandell, 2001). Liberal feminism strives for sex equality through the elimination of laws that differentiate people by gender. Only when these constraints on women's participation are removed will women have the same chance of success as men. This approach notes the importance of gender-role socialization and suggests that changes need to be made in what children learn from their families, teachers, and the media about appropriate masculine and feminine attitudes and behaviour. Liberal feminists fight for better child-care options, a woman's right to choose an abortion, and elimination of sex discrimination in the workplace.

RADICAL FEMINISM According to radical feminists, male domination causes all forms of human oppression, including racism and classism (Tong, 1989). Radical feminists often trace the roots of patriarchy to women's childbearing and child-rearing responsibilities, which make them dependent on men (Chafetz, 1984; Firestone, 1970). In the radical feminist view, men's oppression of women is deliberate, and other institutions, such as the media and religion, provide ideological justification for this subordination. For women's condition to improve, radical feminists claim, patriarchy must be abolished. If institutions currently are gendered, then alternative institutions—such as women's organizations seeking better healthcare, daycare, and shelters for victims of domestic violence and sexual assault—should be developed to meet women's needs.

SOCIALIST FEMINISM Socialist feminists suggest that the oppression of women results from their dual roles as paid and unpaid workers in a capitalist economy. In the workplace, women are exploited by capitalism; at home, they are exploited by patriarchy (Kemp, 1994). Women are easily exploited in both sectors; they are paid low wages and have few economic resources. Gendered job segregation is "the primary mechanism in capitalist society that maintains the superiority of men over women, because it enforces lower wages for women in the labour market" (Hartmann, 1976:139). As a result, women must do domestic labour either to gain a better-paid man's economic support or to stretch their own wages (Lorber, 1994). According to socialist feminists, the only way to achieve gender equality is to eliminate capitalism and develop a socialist economy that would bring equal pay and rights to women.

MULTICULTURAL FEMINISM During the "second wave" of feminism (1970–1990), the mainstream feminist movement was criticized for ignoring the experiences of poor women, women of colour, and women with disabilities. Feminism in its various forms described middle-class white women's experiences as the norm, and other women's experiences were treated as "different" (Cassidy, Lord, and Mandell, 2001). Recently, academics and activists have been attempting to address these criticisms and working to include the experiences of women of colour and Indigenous women. Antiracist feminist perspectives are based on the belief that women of colour experience a different world than do middle-class white women because of multilayered oppression based on race and ethnicity, gender, and class (Khayatt, 1994). Building on the civil rights and feminist movements of the late 1960s and early 1970s, contemporary feminists have focused on the cultural experiences of marginalized women, such as women of colour, immigrant women, and Indigenous women. A central assumption of this analysis is that race, class, and gender are forces that simultaneously oppress some women (Hull, Bell-Scott, and Smith, 1982). The effects of these statuses cannot be adequately explained as "double" or "triple" jeopardy (class plus race plus gender) because these ascribed characteristics are not simply added to one another. Instead, they are multiplicative (race times class times gender); different characteristics may be more significant in one situation than another. For example, a wealthy white woman (class) may be in a position of privilege as compared with people of colour (race) and men from lower socioeconomic positions (class), yet be in a subordinate position as compared to a white man (gender) from the capitalist class (Andersen and Collins, 1998). To analyze the complex relationship among these characteristics, the lived experiences of women of colour and other previously "silenced" people must be heard and examined within the context of particular historical and social conditions.

Feminists who analyze race, class, and gender suggest that equality will occur only when all women, regardless of race and ethnicity, class, age, religion, sexual orientation, or ability (or disability), are treated more equitably (Cassidy, Lord, and Mandell, 2001).

POSTMODERNIST FEMINISM One of the more recent feminist perspectives to emerge is postmodernist feminism. Postmodernist feminists argue that the various feminist theories—liberal, Marxist, radical, and socialist among them—that advocate a single or limited number of causes for women's inequality and oppression are flawed, inadequate, and typically based on suppression of female experiences. In keeping with the assumptions of postmodernist theory, postmodernist feminists resist making generalizations about "all women." Rather, they attempt to acknowledge the individual experiences and perspectives of women of all classes, races, ethnicities, abilities, sexualities, and ages. To postmodernist feminists, a singular feminist theory is impossible because there is no essential "woman." The category woman is seen as a social construct that is "a fiction, a non-determinable identity" (Cain, 1993, cited in Nelson, 2006:94).

Given that the category *woman* is regarded as socially constructed, the challenge of postmodernist feminism is to "deconstruct" these notions of the natural or essential woman. For example, the traditional sciences, in particular medicine, have viewed reproduction as a central construct of "woman." As Phoenix and Woolett have argued, "Women continue to be defined in terms of their biological functions" such that "motherhood and particularly childbearing continues to be defined as the supreme route to physical and emotional fulfillment and as essential for all women" (1991:7). Postmodernist feminists challenge the concept of the reproductive woman as essential and highlight the oppressive nature of such so-called scientific knowledge.

Postmodernist feminists strive to deconstruct our traditional understanding of what constitutes being female or male in society today. They argue that nothing is essentially male or female. In fact, they go so far as to challenge the idea of any real biological categories of male or female—suggesting, rather, that our understanding of biological differences between the sexes is of socially constructed categories that have emerged from specific cultural and historical contexts. Some scholars view the distinction between sex and gender as false because it is based on the assumption of biological differences as real. In sum, the categories of male and female, and man and woman, are viewed by postmodernist feminists as fluid, artificial, and malleable (Andersen, 2006:395).

Critics have suggested that this understanding of gender contradicts the fundamental principle of other feminist perspectives—that is, a central focus on women. As one critic asks, "How can it ascribe to be feminist, since feminism is a theory that focuses on the unitary category 'woman'?" (Cain, 1993:76).

EVALUATION OF CONFLICT AND FEMINIST PERSPECTIVES Conflict and feminist perspectives provide insights into the structural aspects of gender inequality in society. These approaches emphasize factors external to individuals that contribute to the oppression of women; however, they have been criticized for emphasizing the differences between men and women without taking into account their commonalities. Feminist approaches have also been criticized for their emphasis on male dominance without a corresponding analysis of the ways in which some men also may be oppressed by patriarchy and capitalism.

SYMBOLIC INTERACTIONIST PERSPECTIVES

In contrast to functionalist, conflict, and feminist theorists, who focus primarily on macrolevel analysis of structural and systemic sources of gender differences and inequities, symbolic interactionists focus on a microlevel analysis that views a person's identity as a product of social interactions. From this perspective, people create, maintain, and modify gender as they go about their everyday lives. Candace West and Don Zimmerman utilized a symbolic interactionist perspective to explain what they refer to as "doing gender." An individual is "doing gender" whenever he or she interacts with another in a way that displays characteristics of a particular gender. This perspective views gender not as fixed in biology or social roles, but rather as something that is "accomplished" through interactions with others. They explain:

> Gender is not a set of traits, nor a variable, nor a role, but the product of social doings of some sort. What then is the social doing of gender? It is more than the continuous creation of the meaning of gender through human actions. We claim that gender itself is constituted through interaction. (1991:16)

In illustrating the concept of "doing gender," West and Zimmerman refer to a case study of Agnes, a transgender raised as a boy until she adopted a female identity at age 17. Although Agnes underwent a sex reassignment operation several years later, she had the challenging task of displaying herself as female even though she had never experienced the everyday interactions that women use to attach meaning to the concept of being female. Agnes had to display herself as a woman while simultaneously learning what it was to be a woman. To make matters more difficult, she was attempting to do so when most people at that age "do gender" virtually without thinking. As West and Zimmerman explain, this does not make Agnes's gender artificial:

> She was not faking what real women do naturally. She was obliged to analyze and figure out how to act within socially constructed circumstances and conceptions of femininity that women born with the appropriate biological credentials take for granted early on . . . As with others who must "pass" . . . Agnes's case makes visible what culture has made invisible—the accomplishment of gender. (1991:18)

Can you think of ways in which you "do gender" in your daily interactions? Using a symbolic interactionist perspective helps us to understand how we create, sustain, or change the gender categories that constitute being a man or a woman in our society. Analysts emphasize that socialization into gender roles is not simply a passive process whereby people internalize others' expectations, but rather people can choose to "do gender" (Messner, 2000, cited in Andersen, 2006). The interactionist perspective has been criticized for failing to address the power differences between men and women, as well as the significant economic and political advantages that exist in the larger social structure (Andersen, 2006).

TIME TO REVIEW

- Do any of these socialization agents discussed previously provide you with conflicting messages about how you "do gender"?

CONCEPT SNAPSHOT

FUNCTIONALIST PERSPECTIVES

Key thinker: Talcott Parsons

According to functionalists, the division of labour into instrumental tasks for men and expressive tasks for women ensures stability in society.

NEOCLASSICAL ECONOMIC PERSPECTIVE

Human capital analysts argue that women create "atrophied human capital" when they leave the labour force to engage in childbearing and child-care activities. While women are out of the labour force, their human capital deteriorates from non-use. When they return to work, women earn lower wages than men because they have fewer years of work experience and because their education and training may have become obsolete.

CONFLICT PERSPECTIVES

Key thinkers: Friedrich Engels, Karl Marx

According to conflict theorists Engels and Marx, marriage serves to enforce male dominance. Men of the capitalist class instituted monogamous marriage (a gendered institution) so that they could be certain of the paternity of their offspring, especially sons, whom they wanted to inherit their wealth.

FEMINIST PERSPECTIVES

Feminist perspectives vary in their analyses of the ways in which norms, roles, institutions, and internalized expectations limit women's behaviour. Taken together, they all seek to demonstrate how women's personal control operates even within the constraints of a relative lack of power.

SYMBOLIC INTERACTIONIST PERSPECTIVES

Key thinkers: Candace West, Don Zimmerman

From this perspective, people create, maintain, and modify gender as they go about their everyday lives. Candace West and Don Zimmerman utilized a symbolic interactionist perspective to explain what they refer to as "doing gender" whenever he or she interacts with another in a way that displays characteristics of a particular gender.

LO-1 Understand how gender is defined and how it differs from "sex."

Sex refers to the biological categories and manifestations of femaleness and maleness; *gender* refers to the socially constructed differences between females and males. In short, sex is what we (generally) are born with; gender is what we acquire through socialization.

The Canadian Press/Jason Franson

LO-2 Explain the significance of gender in our everyday lives.

Gender role encompasses the attitudes, behaviours, and activities that are socially assigned to each sex and that are learned through socialization. Gender identity is an individual's perception of self as either female or male. Gendered institutions are those structural features that perpetuate gender inequality.

Alexey Losevich/Shutterstock.com

LO-3 Discuss how the nature of work affects gender equality in societies.

In most hunting and gathering societies, fairly equitable relationships exist because neither sex has the ability to provide all of the food necessary for survival. In horticultural societies, cultivation with hoes is compatible with child care, and a fair degree of gender equality exists because neither sex controls the food supply. In agrarian societies, male dominance is apparent; agrarian tasks require more labour and physical strength, and women often are excluded from these tasks because they are viewed as too weak or too tied to child-rearing activities. In industrialized societies, a gap exists between nonpaid work performed by women at home and paid work performed by men and women. A wage gap also exists between men and women in the marketplace. In post-industrial societies, the division of labour in paid employment is increasingly based on whether people provide or apply information or are employed in service jobs.

Getty Images

KEY TERMS

employment equity A strategy to eliminate the effects of discrimination and to make employment opportunities available to groups who have been excluded. (p. 337).

feminism The belief that women and men are equal and that they should be valued equally and have equal rights (p. 341).

gender The culturally and socially constructed differences between females and males found in the meanings, beliefs, and practices associated with "femininity" and "masculinity" (p. 324).

gender bias Behaviour that shows favouritism toward one gender over the other (p. 334).

gender identity A person's perception of the self as female or male (p. 324).

gender role Attitudes, behaviour, and activities that are socially defined as appropriate for each sex and are learned through the socialization process (p. 324).

matriarchy A hierarchical system of social organization in which cultural, political, and economic structures are controlled by women (p. 328).

patriarchy A hierarchical system of social organization in which cultural, political, and economic structures are controlled by men (p. 328).

pay equity (comparable worth) The belief that wages ought to reflect the worth of a job, not the gender or race of the worker (p. 337).

sexism The subordination of one sex, usually female, based on the assumed superiority of the other sex (p. 327).

wage gap A term used to describe the disparity between women's and men's earnings (p. 336).

LO-4 Identify and discuss the primary agents of gender socialization.

Parents, peers, teachers and schools, sports, and the media are agents of socialization that tend to reinforce stereotypes of gender-appropriate behaviour.

© Mary Kate Denny/PhotoEdit

LO-5 Explain the causes of gender inequality in Canada.

Gender inequality results from economic, political, and educational discrimination I against women. In most workplaces, jobs are either gender-segregated or the majority of employees are of the same gender. Although the degree of gender segregation in the professional workplace has declined since the 1970s, racial and ethnic segregation remains deeply embedded.

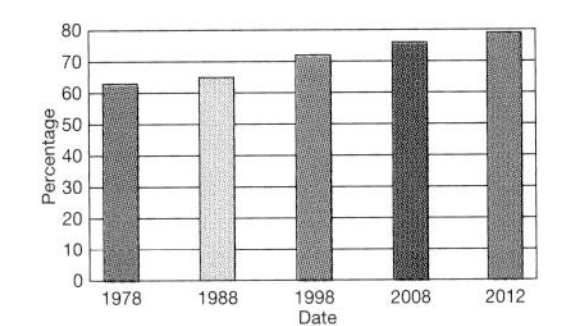

LO-6 Understand how functionalist, conflict, feminist, and interactionist perspectives on gender stratification differ.

According to functional analysts, women's roles as caregivers in contemporary industrialized societies are crucial in ensuring that key societal tasks are fulfilled. Whereas the husband performs the instrumental tasks of economic support and decision making, the wife assumes the expressive tasks of providing affection and emotional support to the family. According to conflict analysis, the gendered division of labour within families and the workplace—particularly in agrarian and industrial societies—results from male control and dominance over women and resources.

Howard Sayer/Shutterstock.com

Although feminist perspectives vary in their analyses of women's subordination, they all advocate social change to eradicate gender inequality. In liberal feminism, gender equality is connected to equality of opportunity. In radical feminism, male dominance is seen as the cause of oppression. According to socialist feminists, women's oppression results from their dual roles as paid and unpaid workers. Antiracist feminists focus on including knowledge and awareness of the lives of marginalized women in the struggle for equality. Postmodernist feminists focus on deconstructing what they see as fluid, artificial notions of the category *woman*. Symbolic interactionists view gender not as fixed in biology or social roles, but rather as something that is "accomplished" through interactions with others. An individual is viewed as "doing gender" whenever he or she interacts with another in a way that displays characteristics of a particular gender.

APPLICATION QUESTIONS

1. As discussed throughout this chapter, gender may be viewed as a social construction. "Doing gender," whether you are male or female, is something you have learned through a process of socialization. What changes would you have to make in your "gender performance" if you were to wake up one morning as the opposite gender?

2. Do the media reflect societal attitudes on gender, or do the media determine and teach gender behaviour? (As a related activity, watch television for several hours and list the roles women and men play in the shows and the advertisements.)

3. Examine the various academic departments at your university. What is the gender breakdown of the faculty in selected departments? What is the gender breakdown of undergraduates and graduate students in those departments? Are there major differences among various academic areas of teaching and study? What hypotheses can you come up with to explain your observations?

CHAPTER 13

SEX, SEXUALITIES, AND INTIMATE RELATIONSHIPS

With contributions by Caitlin Forsey

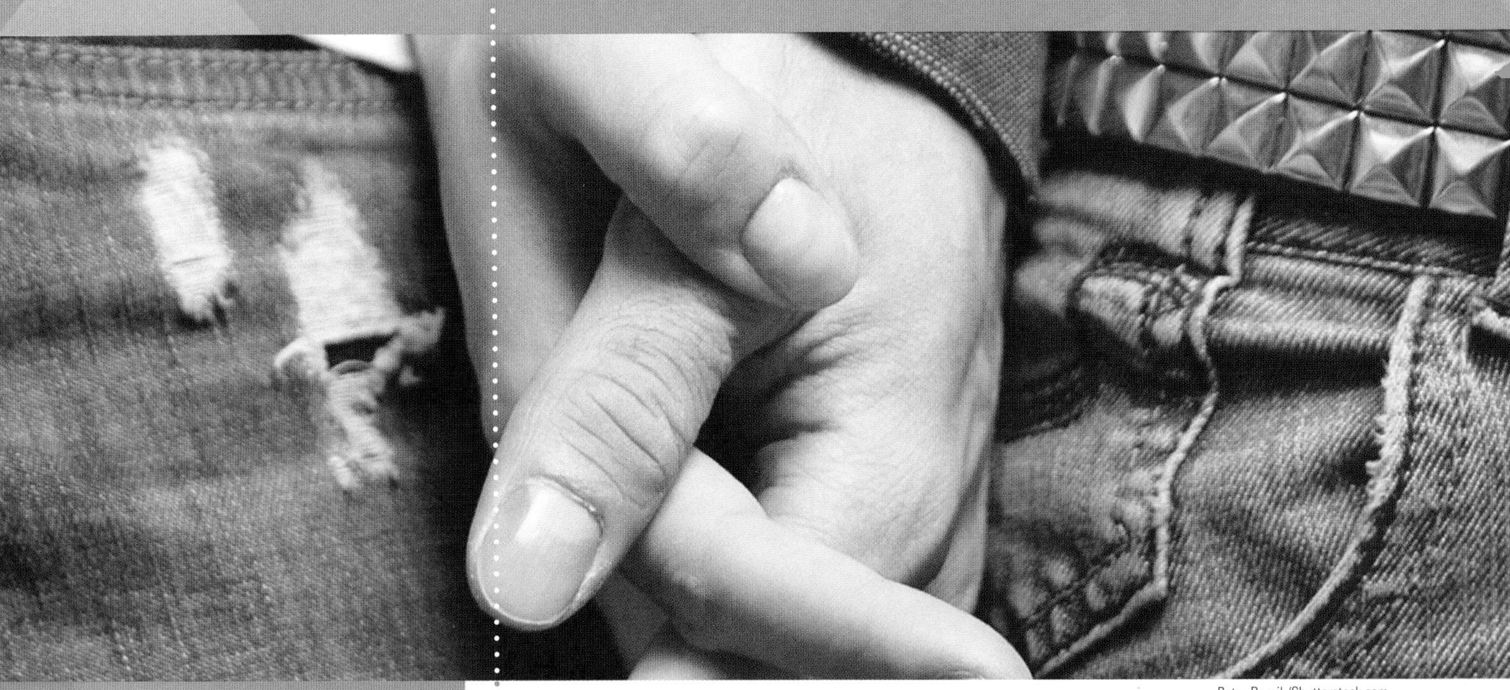

Peter Bernik/Shutterstock.com

CHAPTER FOCUS QUESTION

What are some of the current attitudes toward sexuality, and how have these changed in recent decades?

LEARNING OBJECTIVES

AFTER READING THIS CHAPTER, YOU SHOULD BE ABLE TO

LO-1 Explain how the terms *sex*, *gender*, and *sexuality* relate to one another and how they are different.

LO-2 Understand how various social scientists have classified sexual orientation.

LO-3 Compare and contrast monogamous and polyamorous intimate relationships.

LO-4 Explain a functionalist, conflict, symbolic interactionist, and postmodern analysis of sexuality.

LO-5 Evaluate the sexual health of Canadian youth and young adults.

LO-6 Identify the significant issues and controversies regarding sexuality in Canadian society.

Hookups, friends with benefits, $%!$ buddies, booty call,* and a variety of other terms are used to describe casual sexual relationships or sexual encounters outside of committed relationships. Although it is unlikely that these phenomena are new, it appears that less committed relationships may be more common among today's youth and young adult population (Weaver, MacKeigan, and MacDonald, 2011; Olmstead, Pasley, and Fincham, 2013).

Generally speaking, hooking up includes a range of behaviours, from kissing to intercourse. Most research on the hookup culture to date has focused primarily on heterosexual college and university students to the exclusion of other sexual orientations. Despite a significant amount of inconsistency in how the term *hooking up* is used, it is often the way sexual and romantic relationships among heterosexual young adults are initiated and maintained.

In their classic sociological analysis, Gagnon and Simon (1973) argue that sexuality, like other forms of human behaviour, is socially learned and that we internalize "sexual scripts" in order to learn how to interact with those we are sexually attracted to. **Sexual scripts** are culturally created guidelines that define how, where, with whom, and under what conditions a person is to behave as a sexual being (Nelson, 2006).

Sociologist Kathleen Bogle's (2008) qualitative study of college students and young alumni from two universities indicated that hooking up has its own sexual script, with its own norms on how to meet, get together, become sexually intimate, and manage the potential formation of relationships. Students from universities and colleges across North America shared what they had learned about the "hookup script." Many of these students explained that hookups allowed them to be sexual without the demands of a relationship. For example, Emily, a second-year university student, explains:

More often than not it happens once and it doesn't happen again. I think that's the accepted way that it is and I think that people drink and then they hook up and maybe there's attraction there and then it's not there anymore or maybe it's awkward or maybe you hook up with someone you don't really know and then you don't really take the time to get their number. Like sometimes when I hook up with people like I might not have any interest in them but it just happens to happen and you don't expect any more from it. (Bogle, 2008:40)

Source: Bogle, Katheleen. *Hooking Up: Sex, Dating, and Relationships on Campus.* © 2008 New York University Press, p. 40.

As the following comments from a female student demonstrate, hooking up helped them clarify their values and embrace their own sexuality:

Every experience that I have had with these guys has taught me more about who I am and what type of role sex has in my life. I am no longer afraid to give into my desires, but I am also aware of what I feel is right and wrong. (Wade, 2011:10)

Because hookup scripts are less defined and more varied than other sexual scripts (e.g., dating scripts), the "rules" of the hookup culture are often difficult to describe. As third-year university student Tony explains, even the definition of hookup is difficult to articulate:

If you take someone home and hook up, then that's hooking up . . . it depends on who the person is, like I can read my friends really, really easily. Like if one of my roommates says he "hooked up," that means he brought a girl home and this, that and the other thing. . . . But, like if other kids tell me they hooked up, you got to ask, not to pry into their life, but it could mean a lot of things.

Source: Bogle, Katheleen. *Hooking Up: Sex, Dating, and Relationships on Campus.* © 2008 New York University Press, p. 24.

It becomes apparent in listening to these students that sexuality can often be confusing, ambiguous, and full of contradictions. Unquestionably, sexual scripts are changing, as are sexual values and practices.

For example, in a recent Manitoba case, the judge determined that a man convicted of sexual assault would not go to jail because the victim sent signals that "sex was in the air" through her suggestive attire and flirtatious conduct (McIntyre, 2011). This judgment, which created public outrage and debate, highlights the ambiguity that continues to revolve around issues of sex and sexuality.

In this chapter, we will examine sex and sexuality, what it is, and how it affects us. We will also consider issues pertaining to sex and sexuality such as sexual orientation, intimate relationships, sexual attitudes and practices, and sexual health. It is also important to examine some of the more controversial issues related to sexuality, such as intimate partner violence and the sexual double standard. The voices of individuals will be utilized to assist in understanding the complexities and challenges faced by young adults in developing and managing their sexuality. Before reading on, test your knowledge of sexual attitudes and practises by taking the quiz in Box 13.1.

CRITICAL THINKING QUESTIONS

1. How do you think sexual scripts, sexual values, and sexual practices have changed since your parents were your age? Do you agree that we live in more permissive times?
2. What effects will changes in sexual scripts have on young Canadians as they are struggling to establish their sexuality?

sexual scripts
Culturally created guidelines that define how, where, with whom, and under what conditions a person is to behave as a sexual being.

BOX 13.1 SOCIOLOGY AND EVERYDAY LIFE

How Much Do You Know About Sexual Attitudes and Practices?

True	False	
T	F	1. The age of first intercourse has remained relatively stable among Canadian youth since the 1970s.
T	F	2. Youth today are more inclined than previous generations to have multiple sexual partners.
T	F	3. Today's youth are experiencing better sexual health and are more proactive about protecting their sexual health than were previous generations.
T	F	4. "Hooking up" has largely replaced committed relationships among university and college students.
T	F	5. Young people are having more sex at an earlier age and in a more casual context than their baby boomer parents did.

For answers to the quiz about sexuality, go to **www.nelson.com/student.**

LO-1 UNDERSTANDING SEXUALITY

Sexuality includes our sexual identity, sexual orientation, sexual acts, and intimate relationships. Specific ideas of sex, sexuality, and intimate relationships are the products of the society in which we live. As such, sexuality is a social construction that has significant consequences in our everyday lives.

SEX AND GENDER

▶ **sex** A term used to describe the biological and anatomical differences between females and males.

Sex refers to the biological and anatomical differences between females and males. *Gender,* as discussed in Chapter 12, refers to the distinctive culturally and socially created qualities associated with being a man or a woman (masculinity and femininity). These two concepts are often used interchangeably. This is because *sex*—the biological and physiological aspects of the body—interacts with *gender*—the roles and responsibilities associated with the socially constructed concepts of "masculinity" and "femininity." Given the interdependence of these two concepts, it is virtually impossible to think about sex without thinking about gender at the same time.

At the core of sex differences is the chromosomal information transmitted at the moment a child is conceived. The mother contributes an X chromosome and the father either an X chromosome (which produces a female embryo) or a Y chromosome (which produces a male embryo). At birth, male and female infants are distinguished by **primary sex characteristics**—the genitalia used in the reproductive process. At puberty, an increased production of hormones results in the development of **secondary sex characteristics**—the physical traits (other than reproductive organs) that identify an individual's sex. For women, these include larger breasts, wider hips, and narrower shoulders, a layer of fatty tissue covering the body, and menstruation. For men, they include development of enlarged genitals, a deeper voice, greater height, a more muscular build, and more body and facial hair.

▶ **primary sex characteristics** The genitalia used in the reproductive process.

▶ **secondary sex characteristics** The physical traits (other than reproductive organs) that identify an individual's sex.

SEX AND SEXUALITY

The terms sex and sexuality are also used interchangeably in our everyday speech. For example, although sex may refer to the biological determination of being male or female, it may also be used to describe a range of behaviours or feelings associated with human sexuality (e.g., "sex was in the air," he had "sex" with his girlfriend, she is "sexy"). The fact that the terms sex and sexuality are used interchangeably may contribute to many common misconceptions about the biological determination of human sexuality, such as uncontrollable sexual urges. For our purposes, **sexuality** and **sexual** will refer to the range of human activities designed to produce erotic response and pleasure. Sexuality, like other forms of human behaviour, is shaped by the society and culture we live in.

sexuality (sexual) The range of human activities designed to produce erotic response and pleasure.

INTERSEX INDIVIDUALS

Sex is not always clear-cut. An **intersex person** is an individual who is born with a reproductive or sexual anatomy that does not correspond to typical definitions of male or female; in other words, the person's sexual differentiation is ambiguous. Formerly referred to as hermaphrodites by some in the medical community, intersex persons may appear to be female on the outside at birth but have mostly male-type anatomy on the inside, or they may be born with genitals that appear to be in between the usual male and female types. For example, a chromosomally normal (XY) male may be born with a penis just one centimeter long and a urinary opening similar to that of a female. However, although intersexuality is considered to be an inborn condition, intersex anatomy is not always known or visible at birth. In fact, intersexual anatomy sometimes does not become apparent until puberty, or when an adult is found to be infertile, or when an autopsy is performed at death. It is possible for some intersexual people to live and die with intersex anatomy but never know that the condition exists.

intersex person An individual who is born with a reproductive or sexual anatomy that does not correspond to typical definitions of male or female; in other words, the person's sexual differentiation is ambiguous.

Guided largely by the pioneering work of Money and Ehrhardt (1972), physicians attempted to relieve the distress of parents of intersex infants by assuring them that early surgical "correction" would allow intersex children to perceive themselves as clearly one sex or another—allowing them to grow up as "whole men" or "whole women." More recently, these medical practices have been criticized and many of the incorrect assumptions made about sex, gender, and gender identity have been challenged. For example, gender reassignment surgeries are based on the assumption that unaltered intersex persons have no clear identities. Furthermore, medical practitioners assume that "typical" women and men feel completely at home in their bodies and gender identities, and do not experience crises in relating their identity and appearance (Holmes, 2002:162). Finally, the voices of intersex individuals tell a story of feeling deeply ashamed and abused by their medical treatments. Kiira Triea was reassigned a male in the 1960s and treated by John Money and his associates at John Hopkins University throughout the next decade. She describes her experience:

> I was an interesting lab rat. I call myself a lab rat because that is how intersexed kids are treated. Tested, photographed, tested again, photographed some more. . . . it was precisely my treatment and how it was inflicted on my being which really damaged me more than anything else and prevented me from having what I think of as 'normal' happiness. (Triea, 1997 cf. Holmes, 2002:169)

According to the Intersex Society of North America (2011), intersex is a socially constructed category that includes numerous biological variations:

> Nature presents us with sex anatomy spectrums [, but] nature doesn't decide where the category of "male" ends and the category of "intersex" begins, or where the category of "intersex" ends and the category of "female" begins. Humans decide. (Intersex Society of North America, 2011)

As they explain, it is typically doctors who decide whether a person with XXY chromosomes and XY chromosomes and androgen insensitivity will count as intersex.

San Jose Mercury News/MCT/LANDOV

The complexities of sex and gender came to light in the case of Olympic runner Caster Semenya. After her gold-medal performance at the 2009 World Athletics Championship, Semenya was asked to undergo medical testing to verify that she was in fact "female." The South African sports ministry refused to make the results public and Semenya was able to keep her gold medal.

Brad Barket/Getty Images for Kuato Studios

In 2014, Laverne Cox, star of *Orange Is the New Black*, graced the cover of *Time* magazine. She is one of the strongest public faces of the transgender equality movement.

Western societies acknowledge the existence of only two sexes; some other societies recognize three—men, women, and berdaches (or hijras or xaniths), biological males who behave, dress, and work and are treated in most respects as women. Although outside observers viewed the berdache as shameful or deviant, multiple sex and/or gender roles were common in many Native American societies. In many Indigenous cultures, stories of creation involved themes of sex/gender ambiguity (Nanda, 2000:20). More recently, the term berdache has been rejected as insulting and inaccurate in its presentation of diversity in Indigenous cultures. A more widely accepted term is *two-spirited*, which emphasizes that in Indigenous societies, gender variance is respected and valued and seen as a source of spiritual power (Brotman et al., 2002). Researchers and analysts continue to engage in dialogue about the correct terminology to use when referring to persons in the diverse sexual minority groups.

TRANSGENDER PERSONS

Some people may be genetically of one sex but have a gender identity of the other. That is true for a **transgender person**—an individual whose gender identity (self-identification as woman, man, neither, or both) does not match the person's assigned sex (identification by others as male, female, or intersex based on physical/genetic sex). Consequently, transgender persons may believe that they have the opposite gender identity from that of their sex organs and may be aware of this conflict between gender identity and physical sex as early as the preschool years. Some transgender individuals choose to take hormone treatments or have a sex-change operation to alter their genitalia so that they can have a body congruent with their sense of gender identity. Many then go on to lead lives that they view as being compatible with their true gender identity. But the issue of hormonal and surgical sex reassignment remains highly politicized.

▶ **transgender person** An individual whose gender identity (self-identification as woman, man, neither, or both) does not match the person's assigned sex (identification by others as male, female, or intersex based on physical/genetic sex).

LO-2 SEXUAL ORIENTATION

Sexual orientation refers to an individual's preference for emotional–sexual relationships with members of the opposite sex (heterosexuality), the same sex (homosexuality), both (bisexuality), or neither (asexuality) Some scholars believe that sexual orientation is rooted in biological factors that are present at birth; others believe that sexuality has both biological and social components and is not preordained at birth.

▶ **sexual orientation** A person's preference for emotional–sexual relationships with members of the opposite sex (heterosexuality), the same sex (homosexuality), both (bisexuality), or neither (asexuality).

The terms *homosexual* and *gay* are most often used in association with males who prefer same-sex relationships; the term *lesbian* is used in association with females who prefer same-sex relationships. Heterosexual individuals, who prefer opposite-sex relationships, are sometimes referred to as *straight*. Heterosexual people are much less likely to be labelled by their sexual orientation than are people who are gay, lesbian, or bisexual. This is because we live in a *heteronormative* society where heterosexuality is considered the 'normal' or preferred sexual orientation. This is demonstrated with the Heterosexual Questionnaire (see Figure 13.1) that highlights many of the questions commonly asked of people who are not heterosexual.

What criteria do social scientists use to classify individuals as heterosexual, lesbian, bisexual, homosexual, or asexual? For more than 40 years, the work of biologist Alfred C. Kinsey was considered the definitive research on human sexuality. It was Kinsey's research that first challenged the idea that human beings could be categorized simply as either "heterosexual" or "homosexual." As he explained in his first book, *Sexual Behaviour in the Human Male:*

> Males do not represent two discrete populations, heterosexual and homosexual. The world is not divided into sheep and goats . . . Only the human mind invents categories and tries to force fit individuals into separated pigeon holes. The living world is a continuum in each and every one of its aspects. The sooner we learn this concerning human sexual behaviour the sooner we will reach a sound understanding of the realities of sex. (Kinsey et al., 1948:639, cited in Nelson, 2006)

Instead of three categories (heterosexual, homosexual, bisexual), Kinsey constructed a seven-point scale of sexual behaviour (see Figure 13.1). This scale was constructed to measure the fluidity of human sexual experience. An eighth category, for asexuals, was later added by Kinsey's associates. The term **asexual** refers to an absence of sexual desire toward either sex.

FIGURE 13.1 : KINSEY'S HETEROSEXUAL-HOMOSEXUAL RATING SCALE

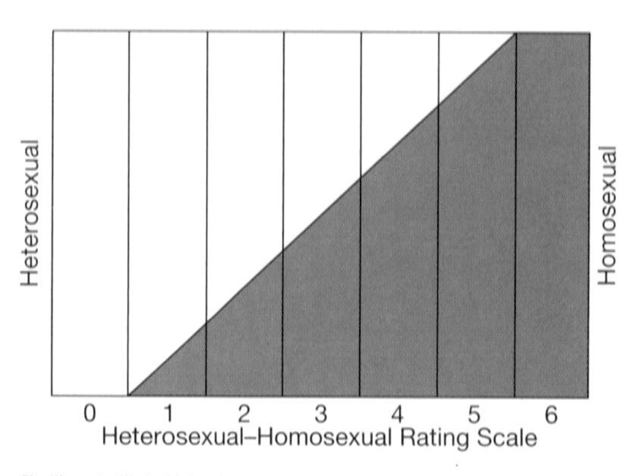

Source: The Kinsey Institute, 2012. "Kinsey's Homosexual-Heterosexual Rating Scale" retrieved August 1, 2012 http://www.iub.edu/~kinsey/research/ak-hhscale.html. Courtesy of The Kinsey Institute for Research in Sex, Gender, and Reproduction.

Findings from Kinsey's research indicated that while 60 percent of his nonrandom sample of white males reported engaging in some same-sex experiences before adulthood, only 4 percent developed exclusively homosexual patterns of sexuality as adults. He further found that although 13 percent of his sample of women reported experiencing at least one same-sex experience, only 3 percent reported being "mostly" or "exclusively" lesbian as adult women. As a result of Kinsey's research, it became widely known that bisexuality was much more common than exclusive homosexuality, and human sexuality was much less definitive and much more fluid than previously believed (Kinsey, 1948, 1953).

asexual An absence of sexual desire toward either sex.

Recently, the work of Kinsey and his associates has been superseded by the National Health and Social Life Survey conducted by the National Opinion Research Center at the University of Chicago. The researchers established three criteria for identifying people as homosexual or bisexual: (1) *sexual attraction* to persons of one's own gender, (2) *sexual involvement* with one or more persons of one's own gender, and (3) *self-identification* as gay, lesbian, or bisexual (Michael et al., 1994). According to these criteria, having engaged in a homosexual act does not necessarily classify a person as homosexual. Many respondents in the University of Chicago study reinforced this idea by indicating that, although they may have had at least one homosexual encounter when they were younger, they had never identified themselves as gay, lesbian, or bisexual. Most respondents reported that they engaged in heterosexual relationships, although 9 percent of the men said they had had at least one homosexual encounter. While approximately 6 percent of men and 4 percent of women said that they were at least somewhat attracted to others of the same gender, only 3 percent of men and just over 1 percent of women identified themselves as gay or lesbian.

TIME TO REVIEW

- How has research on intersex and transgender individuals changed our perception of sex and gender identity?
- Discuss how the research on homosexuality has changed our understanding of sexuality since Kinsey's pioneering work.

LO-3 LOVE AND INTIMACY

It has been said that North Americans are "in love with love." Why is this so? Perhaps the answer lies in the fact that our ideal culture emphasizes *romantic love,* which refers to a deep emotion, the satisfaction of significant needs, a caring for and acceptance of the person we love, and involvement in an intimate relationship (Lamanna and Riedmann, 2011). Although the methods we employ to pursue romance may have changed, we are certainly no less enthralled with the idea.

How have Canadians viewed love and intimacy in the past? During the Industrial Revolution in the late 19th century, people came to view work and home as separate spheres in which different feelings and emotions were appropriate (Coontz, 1992). The public sphere of work—men's sphere—emphasized self-reliance and independence. In contrast, the private sphere of the home—women's sphere—emphasized the giving of services, the exchange of gifts, and love. Accordingly, love and emotions became the domain of women, and work and rationality the domain of men (Lamanna and Riedmann, 2011).

Although the roles of women and men have changed dramatically over the past century, men and women still do not always share the same perspectives about romantic love. According to sociologist Francesca Cancian (1990), women tend to express their feelings verbally, whereas men tend to express their love through nonverbal actions, such as running an errand for someone or repairing a child's broken toy.

Love, intimacy, and sexuality are closely intertwined. Intimacy may be psychic ("the sharing of minds") or sexual, or both. Although sexuality is an integral part of many intimate relationships, perceptions about sexual activities vary from one culture to the next and from one era to another. For example, depending on the time and place, a kiss may (or may not) be regarded as a sexual act, a health threat, a ceremonial celebration, or a disgusting behaviour. Psychologist Leonore Tiefer explains that while sexual kissing is customary in Western cultures, many African and Asian cultures view kissing negatively. When the Thonga of Africa first saw Europeans kissing, they laughed, remarking, "Look at them—they eat each other's saliva and dirt" (2005:24).

INTIMATE RELATIONSHIPS

Across cultures, intimate relationships are characterized by different forms. All sexual or intimate relationships, whether they are monogamous or nonmonogamous, are a social product.

MONOGAMY In Canada, marriage continues to be the most common form of committed intimate relationship among adults. **Marriage** is a legally recognized and/or socially approved arrangement between two or more individuals that carries certain rights and obligations and usually involves sexual activity. In Canada, the only legally sanctioned form of marriage is **monogamy**—an intimate relationship with only one person at a time. For some people, this takes the form of marriage as a lifelong commitment that ends only with the death of a partner.

▶ **marriage** A legally recognized and/or socially approved arrangement between two or more individuals that carries certain rights and obligations and usually involves sexual activity.

▶ **monogamy** An intimate relationship with one person at a time.

Members of some religious groups believe that marriage is "forever"; if one spouse dies, the surviving spouse is precluded from marrying anyone else. For others, marriage is a commitment of indefinite duration. Through a pattern of marriage, divorce, and remarriage, some people practise *serial monogamy*—a succession of relationships in which a person has several partners over a lifetime but is legally married to only one person at a time.

Extramarital or, more generally, sexual infidelity in monogamous relationships is common. Infidelity can involve a wide range of behaviours occurring outside of a committed relationship. Most research on heterosexual relationships focuses on vaginal sex occurring outside of a marital relationship. However, behaviours can range from intense emotional relationships or close friendships to kissing, oral sex, or other sexual behaviours, and the primary dyad need not be married (Mark, Janssen, and Milhausen, 2011). And the Internet has created a new forum for infidelity that provides individuals with an opportunity to engage in secretive, anonymous, often sexually charged interactions with others online. Consider, for example, the website AshleyMadison.com, which has an estimated 30 million users looking for extramarital intimacy.

Research on infidelity in heterosexual relationships suggests that approximately one-third of men and one-quarter of women may engage in sexual relationships outside of their committed relationship at least once in their lives (Kinsey, Pomeroy, and Martin, 1948; Kinsey et al., 1953; Mark, Janssen, and Milhausen, 2011). According to a University of Chicago study, persons who engaged in extramarital sex found their activities to be more thrilling than those with their marital partner, but they also felt guilty. Persons in sustained relationships, such as marriage or cohabitation, found sexual activity to be the most satisfying emotionally and physically (Milhausen and Mark, 2009).

Research on infidelity in same-sex relationships is less prevalent and fraught with the same measurement issues faced by researchers studying infidelity in heterosexual unions. However, one study on infidelity in same-sex relationships found that 45.2 percent of gay men who indicated they were in committed, monogamous relationships reported that they or their partner had had sex with a third party since the beginning of their relationship (Milhausen and Mark, 2009).

According to sociologist Eric Anderson, author of *The Monogamy Gap: Men, Love and the Reality of Cheating* (2012), monogamy does not necessarily provide a lifetime of sexual contentment. Anderson interviewed 40 heterosexual male undergraduate university students in England about their experience with monogamy and cheating. The participants very clearly described competing social scripts of emotional desire for monogamy and sexual desire for cheating. For example, Tony says, "I struggle all the time. I get mad at her. I want sex with other women, and I know she'd never let me, so sometimes I just feel like cheating because I'm not supposed to." Similarly, another participant, James, indicates he desperately wants other women: "I can't stop thinking about other women, I'm sure I'll cheat. I mean, I don't want to but I will. It sucks, really it sucks. I don't want to cheat, but I really want sex [with someone other than his girlfriend]" (Anderson, 2010:859).

Anderson concludes that monogamy is a cultural illusion imposed on young men and women that falsely promises that once they find true love, they will no longer experience sexual boredom and the desire to cheat. Although society cherishes monogamy for most couples, the expectation of exclusive sexual activity may be unsustainable. The solution, according to Anderson, is to explore alternatives to monogamy. He suggests:

> We may need to investigate other relationship models: open arrangements or . . . "monogamish" relationships in which couples have flings, affairs or threesomes. These ways of loving, along with polyamorous relationships and even singlehood, should be as equally culturally valued as monogamy. Only when men and women are able to make sexual choices free of stigma will people be honest with their partners about their desires (2012b).

One alternative to monogamy is a form of intimate relationship that is increasingly presenting itself as an option in North American society—polyamory.

NONMONOGAMY **Polyamory** involves mutually acknowledged emotional, sexual, or romantic relationships with multiple partners. The Latin part of polyamory means "love" and the Greek part means "many," so the word translates into "many loves" or "more than one love" (Klesse, 2006). In the following narrative, Angi talks about her polyamorous family and some of the challenges she faces:

polyamory Intimate relationships that involve mutually acknowledged emotional, sexual, or romantic relationships with multiple partners.

> It is continually surprising to me that people, in general, seem far more able to accept the possibility of sharing a partner sexually than they are able to accept the possibility of having multiple loves. It says a great deal about our society's rigid definition of romantic love that people are able to somewhat easily accept the concept of sexually open relationships—and even dishonest infidelity—while insisting that it cannot be possible to actually love multiple partners simultaneously. Frustratingly, I have been told on more than one occasion that what I share with my partners cannot, by definition, be love, as if anyone can define for others what love is and what it is not. These attitudes strike me as incredibly reminiscent of a society that—30 years ago—viewed same-sex relationships only as a deviant sexual behaviour. (Angi Becker Stevens, 2012.)

© Catchlight Visual Services/Alamy

Unlike other sexual relationships involving multiple partners, love is central to polyamorous relationships.

Polyamory includes a wide range of relationship and sexual practices, including open or closed group marriages, triads (three partners), and quads (four partners). Polyamorous partners may be primary, secondary, or tertiary, depending on the level of commitment and amount of time they spend with the partner(s). Unlike other sexual relationships involving multiple partners, such as swinging or casual sex, love is central to polyamorous relationships. Christian Klesse conducted interviews with 44 men and women from the United Kingdom who were in polyamorous relationships. His subjects emphasized the importance of love, friendship, and commitment in these relationships and de-emphasized the importance of sexuality. As Marianne explained:

> I mean sometimes when I try to explain polyamory to my friends, they don't get it, because they say . . . I mean especially, if it's relationships that aren't sexual, they say, "Oh but they're just close friendships." And I'm trying to explain that "No, they're not just close friendships—they're closer than a close friendship—they're people that I love." And you know, some of my friends that probably are truly monogamous, they just don't get it. Erm, it's funny really. (Klesse, 2006:568)

The blurring of the boundaries between friend, partner, and lover is an important aspect of polyamory. Consequently, having sex with friends is not unusual and may work in a number of different directions: Sometimes, long-standing friendships can turn into more sexual relationships; sometimes, sexual attraction marks the beginning of a later nonsexual relationship. Pal describes his approach to polyamory as a fluid process of moving from friendship to partnership:

> I am very much coming from becoming very, very close friends and then having those friends being very intimate, much more intimate and sort of sharing a life with them, but I really don't like the idea of having a mould for a relationship in the way of doing it. (Klesse, 2006:570)

Individuals in polyamorous unions describe their intimate relationships as "responsible nonmonogamy" because all partners are aware of and share a consensus on the nonmonogamous aspect of their relationship. As such, honesty is central to the success of these relationships (Jamieson, 2004).

Despite media portrayals of patriarchal and nonegalitarian forms of polyamory in *Sister Wives* and *Big Love,* the voices of individuals in committed polyamorous relationships tell a different story. For example, Angi says:

> I don't know that there's any way to offer a universal definition of romantic love. I certainly don't claim to be up to the task. What I do know is that my partners, while vastly different from one another, share several important qualities. They are my two best friends in the world. They are both men I can sit up all night talking with. . . . They both make me feel loved, respected, and desired. And I can say with as much certainty as it is ever possible to have about such things that I am madly in love with them both, and want to live the rest of my life with them both by my side. (Angi Becker Stevens, 2012.)

▶ **polygamy** The concurrent marriage of a person of one sex with two or more members of the opposite sex.

▶ **polygyny** The concurrent marriage of one man with two or more women.

CROSS-CULTURAL VARIATIONS IN INTIMATE RELATIONSHIPS **Polygamy** is the concurrent marriage of a person of one sex with two or more members of the opposite sex (Ward and Belanger, 2011). The most prevalent form of polygamy is **polygyny**—the concurrent marriage of one man with two or more women. Polygyny has been practised, for example, in a number of Islamic societies, including in some regions of contemporary Africa and southern Russia. The reality television show Sister Wives, which documents the life of a polygamist family that includes a husband with four wives and 17 children, certainly challenges traditional notions of family.

The second type of polygamy is **polyandry**—the concurrent marriage of one woman with two or more men. Polyandry is rare; when it does occur, it is typically found in societies where men greatly outnumber women because of high rates of female infanticide or where marriages are arranged between two brothers and one woman (fraternal polyandry). According to recent research, polyandry is never the only form of marriage in a society: Whenever polyandry occurs, polygyny co-occurs (Trevithick, 1997). Although Tibetans are the most frequently studied population where polyandry exists, anthropologists have also identified the Sherpas, Paharis, Sinhalese, and various African groups as sometimes practising polyandry (Trevithick, 1997). An anthropological study of Nyinba, an ethnically Tibetan population living in northwestern Nepal, found that fraternal polyandry (two brothers sharing the same wife) is the normative form of marriage and that the practice continues to be highly valued culturally (Levine and Silk, 1997).

© Dick Hemingway

In Canada, the notion of romantic love is deeply intertwined with our beliefs about how and why people develop intimate relationships and establish families. Not all societies share this concern with romantic love.

polyandry The concurrent marriage of one woman with two or more men.

TIME TO REVIEW

- Discuss the alternatives to monogamy.
- How pervasive is infidelity? What are some of the reasons why people may have relationships outside of their committed relationships?
- What are some of the different types of polyamorous relationships?

LO-4 THEORETICAL PERSPECTIVES ON SEX, SEXUALITY, AND INTIMATE RELATIONS

Sociological perspectives on sex, sexuality, and intimate relations vary in their emphasis on the function of sexuality, the relationship between sexuality and social inequality, the meaning people attach to sexual scripts, and the degree to which sexual identity is understood as either fixed or fluid.

FUNCTIONALIST PERSPECTIVES

Functionalists use the terms *functional* and *dysfunctional* to describe how certain practices and norms contribute to the stability of social institutions and society more generally. On a biological level, sexual reproduction is functional because it ensures the continuation of the human species. For this reason, regulating norms and values surrounding sexual reproduction is an important aspect of maintaining social stability. Most members of society share a common set of values, beliefs, and behavioural expectations regarding legitimate and illegitimate sexual reproduction. The incest taboo is one of the oldest and most universally recognized norms that forbids marriage and sexual relations between certain relatives. In Canada, Europe, Mexico, and some U.S. states, marriage and sexual relations between brothers and sisters and between parents and their children is prohibited. Part of the reason is biological, since reproduction between close relatives increases the odds of genetic abnormalities. However, functionalists also recognize that regulating sexual contact between close relatives integrates people into society, limits sexual competition within the family, and protects children.

Although sexual relations between close relatives are considered dysfunctional because they threaten the stability of the family, functionalists argue that a certain amount of deviance is necessary for the overall stability of society because deviance clarifies social norms and helps maintain social control. For example, although sex is widely viewed as a legitimate form of intimacy and recreation outside of marriage, North American culture still has very strict attitudes concerning extramarital affairs. In Canada, extramarital sex is widely condemned with more than 85 percent of adults reporting that adultery is "almost always wrong" or "always wrong." Functionalists would highlight the fact that these negative attitudes to sexual infidelity within marriage serve an important function in ensuring the stability of the family.

From a functionalist perspective, some sexual practices are both functional *and* dysfunctional for society. Although prostitution is often considered harmful to society because it exploits women and spreads disease, Kingsley Davis (1937) argued that prostitution will always serve an important function in societies where restrictive norms govern sexual conduct. From a functionalist perspective, prostitution offers quick, impersonal sexual gratification for people who are not looking for emotional attachment or long-term relationships. It also provides a sexual outlet for people who may not have ready access to sexual relationships or those who may engage in sexual practices that regular sex partners may consider distasteful or immoral.

In viewing society as a set of interrelated parts, functionalists recognize that solutions to a social problem such as prostitution can also lead to undesirable social consequences. As sociologist Becki Ross (2010) explains, the expulsion of sex trade workers from Vancouver's West End neighbourhood to the uninhabited industrial tracts of the city's Downtown Eastside had lethal consequences for outdoor sex workers in the city. Community residents joined forces with business owners, politicians, lobbyists, journalists, urban planners, and police to strengthen prostitution laws and implement traffic controls that limited the outdoor street sex trade As a result, many sex workers were forced to work in isolated, poorly lit industrial zones where they began to go "missing" (Jiwani and Young, 2006).

CONFLICT PERSPECTIVES

Whereas the functionalist perspective views society as comprised of different parts that work together, the conflict perspective explores issues of power, inequality, and competition over scarce resources. From this perspective, the relationship between the economic system and sexuality is largely a question of exploitation. Conflict theorists focus on unequal relations of power and ask questions such as who benefits and who loses when sex is bought and sold.

In terms of prostitution laws, enforcement is largely unequal and women tend to face more serious consequences than men. While two (or more) people may be involved in the purchase of sexual services, research shows that police are far more likely to arrest and convict less powerful female sex trade workers than more powerful male clients.

Similarly, of all women engaged in sex work, street prostitutes—women with the least income and most likely to be visible minorities—face the highest risk of arrest (Ross, 2010). Street prostitutes also occupy the lowest stratum and receive the strongest dose of stigma from the general public and other sex workers in the industry (Weitzer, 2005). Control over working conditions is lowest among street prostitutes, many of whom do not have access to resources for protection, do not have the freedom to refuse abusive clients or particular sex acts, and are dependent on pimps or other third parties (Chapkis, 2000). Disparities in social and economic status place street prostitutes at risk for victimization, especially when compared to indoor prostitutes who work in brothels, massage parlours, and escort agencies.

FEMINIST PERSPECTIVES

The Western feminist movement of the 1960s and 1970s was instrumental in bringing the issue of sexuality out of the private realm and into the public arena. During this time, feminists challenged the sexual double standard, raised questions about definitions of sex and gender, and advocated for protection for women from sexual violence and coercion.

During the late 19th and early 20th centuries, the connection between gender and sexuality was often described in biological, medical, and psychological terms. Central to these early accounts was the assumption that the relationship between gender and sexuality expressed a natural order, one that relied on universal and fixed dualisms that were presumed to complement each other (male/female, heterosexual/homosexual, masculine/feminine). From this perspective, sex/gender/sexuality relate in a hierarchical, coherent manner.

Feminists put forward new ideas about the relationship between gender and sexuality that challenged earlier perspectives that viewed sex as a product of biological forces and sexual identities as fixed and stable. Feminist analysis paved the way for thinking about the *social* relationship between sex and gender. From this perspective, the relationship between gender and sexuality is one of the key mechanisms by which gender inequalities are maintained. Marxist feminists, and others not clearly aligned with this perspective, view gender as the outcome of a hierarchy in which one class of people (usually men) have systematic power and privilege over another class of people (usually women).

According to some feminist writers, sexuality is central to the maintenance of patriarchal domination where institutionalized heterosexuality is enforced through sexual coercion and violence. According to Kathleen Gough (1975), some of these methods include denying women their own sexuality (e.g., clitoridectomy, chastity belts, capital punishment for female adultery), forcing male sexuality upon women (e.g., rape, arranged marriage, idealization of heterosexual romance), exploiting the reproductive labour of women (e.g., male control of abortion, contraception, sterilization, and childbirth), limiting visibility and movement (e.g., sexual harassment, purdah, high heels, and other "feminine" modes of dress), and using women as objects of exchange (e.g., geisha girls, pimping, bride price).

Adrienne Rich (1980) uses the concept of "compulsory heterosexuality" to draw links between the social ordering of gender and the regulation of sexuality. Specifically, Rich questions the assumption that women are naturally heterosexual and argues that heterosexuality is imposed upon women and reinforced by a variety of social constraints. Cannon's (1998) analysis of the *Indian Act* builds on the concept of compulsory heterosexuality to explore how sexual relations between Indigenous men and women are regulated in Canada. While the literature suggests that a broad range of erotic relationships existed prior to European settlement, compulsory heterosexuality was instituted in Indigenous communities through the *Indian Act* of 1876, in which heterosexual marriage became the only possible avenue through which to convey Indian status and rights. Elements of sexism also appear in the status and citizenship sections of the *Indian Act* that denied Indian status to Indigenous women—along with their children—who marry non-Indigenous men. Some of these inequalities in law were addressed in 1985 when the *Indian Act* was amended. For recent developments in sexuality and the law, see Box 13.2.

BOX 13.2 POINT/COUNTERPOINT

Sexuality and the Law

Canada has a long history of legal regulation of sexuality. The prohibition, condemnation, and regulation of unconventional sexual practices became a priority for criminal justice as a result of the efforts of the "moral entrepreneur" of the 19th century. For example, "wickedly, devilishly, feloniously, and against the order of nature [committing that] sodomitical, detestable, and abominable sin called buggery" was an offence in Canadian common law punishable by execution (Young, 2008:207). Young girls committing the act of "sexual immorality," which was defined as "having sexual intercourse with a man who was not their husband," warranted detention in an industrial school for

(Continued)

an indeterminate period (Lothian, 1990). It was not until the "sexual revolution" of the 1960s that there was a significant liberalization of law pertaining to sexuality. In 1967, Pierre Elliott Trudeau suggested that "the state has no business in the bedrooms of the nation." However, according to legal scholar Alan Young, 40 years after Trudeau's famous declaration, there has yet to be a fundamental shift in criminal justice policy that reflects the significant changes in sexuality emerging from the sexual revolution. According to Young, regardless of whether the law has led, followed, or even been responsive to the sexual morality of the masses, it continues to maintain the ability, capacity, and desire to monitor and control private sexual choices of individuals (Young, 2008:206). This becomes evident when we consider recent changes to laws pertaining to age of consent, obscenity and indecency, and prostitution.

Age of Consent

The age of consent to sexual activity has always been a contentious issue. In the past decade, numerous unsuccessful attempts to raise the age of consent have been put forth in the Canadian Parliament. Both opponents and proponents of raising the age of consent emphasize the need to protect children and youth from harm and exploitation (Wong, 2006:163). Until recently, Canada had one of the youngest ages of consent among developed countries. In May 2008, the age of consent for sexual touching was finally raised from 14 to 16 years of age.

The primary motivation for this change was to provide vulnerable youth some protection in law from sexual exploitation at the hands of adults. However, the legislation does create a "close in age" exemption for sexual partners who are up to five years older for anyone 14 or older, or for partners already married under pre-existing laws at the time of the new legislation. There is also a close-in-age exception for 12 and 13 year olds: They can consent to sexual activity with another person who is *less than two years older* and with whom there is no relationship of trust, authority, or dependency, or other exploitation. In addition, the age of consent has been raised to 18 years of age where the sexual activity "exploits" a young person—for example, through prostitution or pornography, or when touching occurs in a relationship of authority, trust, or dependency (e.g., a teacher).

Obscenity and Indecency Law

The Supreme Court of Canada, in its landmark case *R. v. Labaye* (2005), changed the law governing indecency and obscenity in Canada. Jean-Paul Labaye of Montreal had been charged with keeping a common bawdy house for the practice of acts of indecency under section 210(1) of the *Criminal Code*. The accused operated a club in Montreal, the purpose of which was to permit couples and single people to meet each other for group sex. Only members and their guests were admitted to the club. Prospective members were interviewed to ensure that they were aware of the nature of the club's activities. A doorman manned the main door of the club to ensure that only members and their guests entered. All of these activities were consensual, and while members paid the club membership fees, the members did not pay each other in exchange for sex.

Based on what is referred to as a "harm-based" test, the Supreme Court of Canada determined that Mr. Labaye's club was not harmful and was therefore not illegal. What do you think? Does operating a swingers club constitute indecency? Is it harmful?

Prior to this decision, in assessing harm, the determination of indecency and obscenity had included a consideration of community standards of tolerance. Conduct or expression was considered harmful (and illegal) by considering what other Canadians would or would not tolerate.

The Supreme Court was unable to find that the swinger's club did any harm to participants or passersby, and the club was thus found to have been operating legally. The use of harm as the test for obscenity and indecency was seen as a positive outcome by activists. This legal test was the terrain of much struggle in the 1990s, while the test in *Labaye* has been viewed as something of a victory for marginalized sexual communities and minority groups in that some scholars have described the decision as a measure of tolerance for unruly sexual subjects (Craig, 2008, 2009).

Importantly, censoring sexually explicit materials as a means of promoting women's equality also came into direct conflict with the value of sexual freedom, including the freedom to consume all sorts of pornography. This tension between feminist scholars has been termed by some a Canadian "sex war," where a division was evident between those who "framed sexuality primarily as a site of danger and oppression for women and those who saw sexuality more ambivalently, as also a site of pleasure and liberation" (Cossman, 2004:851; Jochelson, 2009:742).

Recent Developments in Prostitution Law

The arguments in favour of and against the criminalization of prostitution have been extensively covered in both popular media and academic research (see The Challenge of Change, 2006; The Fraser Report, 1985; Gorkoff and Runner, 2003; van der Meulen and Durisin, 2008). This is an issue about which many reasonable Canadians disagree and scores of scholarly articles and other media representations debate. Recently, there have been certain provisions of the *Criminal Code* that have been challenged in order to improve the working conditions of sex trade workers (see *Canada v. Bedford*, 2010, 2012, 2013).

In Canada, prior to 2014, there was no law that determined that the act of selling sex (i.e., prostitution) was illegal. Certain aspects of the sex trade had been criminalized. For example, keeping a brothel, soliciting for the purposes of prostitution, and living off the avails of prostitution all constitute criminal activity in Canada. That legal void was used as a forum for lobbying for decriminalization of prostitution-related conduct. In *Bedford*, for instance, the applicants argued that the criminalization of solicitation for the purposes of prostitution, combined with the criminalization of prostitution in common bawdy houses, caused harm to women sex trade workers. Because the law did not technically criminalize the act of selling sex for money alone, it drove prostitution into a dangerous workspace. The applicants in *Bedford*, sex trade workers represented by prominent lawyer Alan Young, successfully challenged the constitutionality of bawdy house and solicitation laws. The new challenges to these laws stemmed from the concerns inherent in the anti-criminalization movement: that these laws created unsafe working conditions for sex trade workers, limited the economic liberty of these workers, violated their freedom of expression, and drove the sex trade onto the streets instead of in a workplace where better safety screening could occur. In short, the advocates claimed that the laws denied sex workers their right to life, liberty, and security of the person under section 7 of the *Charter*. In part, the impetus for these challenges stemmed from more recent parliamentary studies that confirmed the acute dangers associated with the sex trade. Using over 50 affidavits from sex trade workers, the applicants attempted to show that Canada's sex trade laws contributed to the violence experienced by women, who disproportionately make up the population of sex trade workers in Canada.

Ultimately, in 2013, the Supreme Court of Canada agreed and struck down the anti-prostitution regime as unconstitutional, giving Parliament one year to redraft the laws. Parliament passed new laws noting that the purpose of anti-prostitution laws had moved away from nuisance prevention and "toward treatment of prostitution as a form of sexual exploitation that disproportionately and negatively impacts on women and girls" (Department of Justice 2014, 3).

The new *Criminal Code* Amendments include provisions that prohibit the purchase of sex, place restrictions on public communications for the purposes of prostitution near playgrounds or daycare centres, limit the advertising for prostitution, and limit the ability of people to make money off of or live off of sex workers. Activists have reasons to be concerned that these newest amendments may recreate the harms that sex workers experienced everyday prior to the *Bedford* decision—insecure work that is driven underground into unsafe and under-policed spaces. It is a matter of time before, in the name of harms suffered, these new laws are challenged in the courts.

Sources: Dr. Richard Jochelson, University of Winnipeg. Used with permission; Department of Justice Canada, "Technical Paper: Bill C-36, Protection of Communities and Exploited Persons Act" (2014); *Canada (Attorney General) v. Bedford*, 2012 ONCA 186; *Canada (Attorney General) v. Bedford*, 2013 SCC 72, [2013] 3 S.C.R. 1101.

SYMBOLIC INTERACTIONIST PERSPECTIVES

Symbolic interactionist perspectives on sexuality emerged in the 1960s alongside theories of deviance that stressed the importance of social definition, or "labelling," rather than the features of particular acts and actors. One of the earliest studies that draws on this view is Mary McIntosh's (1968) analysis of the "homosexual role." Rather than viewing homosexuality as an innate or acquired condition, McIntosh suggests that expectations regarding homosexuality vary according to culture and history. Thus, while in Western societies there is an expectation that homosexual men are effeminate and exclusively homosexual, other societies in Australia, Greece, and Northern Africa accept homosexual liaisons between men as part of a varied heterosexual pattern.

William Simon and John Gagnon (1986) also played an important role in the development of interactionist perspectives on sexuality. Simon and Gagnon argue that feelings, practices, and body parts are not inherently sexual. Instead, they derive their sexual significance through the application of *sexual scripts*—cultural guidelines that prescribe "with whom one should have sexual activity, when and where sexual activity should occur, what types of activities are appropriate, and acceptable reasons for participating (or not) in sexual activity" (Baber and Murray, 2001:25).

There are normative sexual scripts that guide how men and women think about themselves as sexual beings and make choices about partners and sexual practices (Brown, 2012). For instance, some of the messages women and girls receive about sex include: Say no to sex (or be swept away); pursue love, security, and romance; "get a man" (aim to be attractive but passive); and don't be too knowledgeable about sex or ask for what you want (Crane and Crane-Seeber, 2003). In contrast, men and boys receive messages about sex that include: Get as much sex as you can; love is a trap/responsibility; "be a man" (get a woman); don't give up, she'll give in; and act like you know all about sex (Crane and Crane-Seeber, 2003).

Sexual scripts may also vary in response to the social messages received by people with different cultural backgrounds. Shirpak, Maticka-Tyndale, and Chinichiani (2007) analyze the social meaning Iranian immigrants attribute to sexuality in Canada and draw attention to the tensions that exist between Canadian and Iranian sexual scripts. In Iran, for instance, women who publicly expose body parts other than their hands, feet, and face are considered immodest because they are believed to inspire enticing thoughts and feelings of sexual arousal among men other than their spouse. Although negative stereotypes are sometimes ascribed to women who wear revealing clothing, the sexual scripts regarding appropriate dress for women are more permissive in Canada. The cultural tension between these sexual scripts is evident among Iranian immigrants who view the dress of Canadian women as symbolizing sexual availability and seduction. As one study participant put it, "Here in Canada women dress in a revealing way and their body is visible. Some Iranian men's minds might become preoccupied with that. The way men dress [in Canada] is not different from Iran. Women are different" (2007:118).

POSTMODERN PERSPECTIVES

Postmodern thinkers reject claims that sexuality is biologically based. Within this broad tradition, *queer theory* emerged in the late 1980's as an approach to understanding gender and sexuality that emphasizes that sexuality is fluid rather than fixed. The name *queer theory* reflects the desire to 'take back' and redefine the term 'queer,' which had always been a derogatory term directed at gays and lesbians. Broadly speaking, the term *queer* describes any mismatch between sex, gender, and sexual desire that disrupts *heteronormativity*— cultural bias toward heterosexuality that privileges sexual relations between men and women.

This area of scholarship is heavily influenced by the work of Judith Butler and Michel Foucault, both of whom use the term *discourse* to describe ways of speaking, thinking and writing about the social world. In *The History of Sexuality, Volume 1* (1978), Foucault challenges the idea that sexuality was silenced and repressed during the 18th century. Foucault (1978:11) is not investigating sexuality; instead, he is exploring "the discourse of sexuality." Specifically, Foucault examined the Christian practice of confessing sinful desires and the 19th-century science of sex as evidence of how discourse surrounding sexuality grew during the Victorian era. The operation of power through these sexual discourses was not repressive; rather, it created new sexual subjects and new modes of being. From this perspective, homosexuality is not a pre-existing form of sexuality but a product of the discourses that construct *the* homosexual as a particular type of person that makes it possible to *be* a homosexual.

The relationship between sex and gender is often made complicated by the body, especially when primary and secondary sex characteristics appear in combinations that deviate from norms established by the medical community. Intersex bodies create problems for any sex/gender/sexuality system that insists bodies are limited to only two categories: male and female. Because intersexuals disrupt the presumed connection between sex, gender, and sexuality, they do not fit within the boundaries of what Judith Butler (1999) calls a "heterosexual matrix." The cultural anxiety about bodies that do not fit this scheme, and responses to that expression launched by the medical community, have made intersex persons targets for neutralization of abnormalities (Fausto-Sterling 2002). This was achieved through hormonal therapy and surgical management, both of which render intersex bodies "intelligible" or "recognizable" in a matrix that uses heterosexuality to define how sex, gender, and sexuality should correspond. In other words, an individual whose sex assignment is male is expected to have a masculine gender identity, act in a masculine way, and be sexually attracted (only) to women.

One of the principal tasks of queer theory is to challenge and question the norms that constrain sex and sexualities. In her first paper, "The Five Sexes," Anne Fausto-Sterling (1993) suggested that rather than thinking about sex as consisting of one of two categories, female and male, sex should be viewed on a continuum comprised of five categories. In addition to male and female, she proposed three new sex categories to account for the huge variability among individuals who present aspects of male and female genitalia: Intersex persons or hermaphrodites (herms) possess one testis and one ovary; male pseudohermaphrodites (merms) have testes, some female genitalia, but no ovaries; and female pseudohermaphrodites (ferms) have ovaries, some male genitalia, but no testes. Intersex individuals and advocates for sexual minorities were offended by both Fausto-Sterling's attempt to further categorize and label intersex persons and focus on genitalia in creating difference. In her later paper, "The Five Sexes Revisited," she retracted her earlier proposal, clarifying her position that the two-sex system was not adequate to reflect the range of diversity in human sexuality and "sex and gender are best conceptualized as points in a multidimensional space" (2000:22).

CURRENT ISSUES IN SEXUALITY

Given that sex and issues of sexuality permeate most aspects of our daily lives, it is no surprise that sex is often a contributing factor to significant social problems and issues. The sexual health of youth, including birth control, teen pregnancy, sexually transmitted infections, and sexual violence, are all issues of significant concern in today's society.

LO-5 THE SEXUAL HEALTH OF YOUTH

Despite media claims of an unprecedented increase in sexual behaviour among youth today—including dramatic increases in the transmission of sexually transmitted infections, unplanned pregnancies, masturbatory displays via Facebook, YouTube, and smartphones, and oral sex games involving multiple partners—there is strong evidence that young people are experiencing better sexual health than ever before.

Sexual health is multidimensional and includes such factors as mutually rewarding intimate relationships, positive attitudes toward sexuality, and avoidance of negative outcomes such as sexually transmitted infections and, in the case of heterosexual health, unplanned pregnancies. Promoting sexual health involves equipping young people with the knowledge, motivation, and behaviours to enhance their sexual health and avoid sexual health-related problems. Trends in teen pregnancy, sexually transmitted infections, age of first intercourse, and condom use are often used to evaluate the status of the sexual health of Canadian youth (McKay, 2009). Sociologist Eleanor Maticka-Tyndale (2008) compared the sexual health and behaviours of Canadian youth today to that of adolescents from previous generations and found that in terms of a number of criteria, today's youth are experiencing better sexual health.

TEEN PREGNANCY The overall teen pregnancy rate has declined in the past 30 years as a result of an increase in the availability of contraception, an increase in the awareness of the dangers of unprotected sex brought on by the HIV/AIDs epidemic, and legal access to abortion. More recently, youth also now have legal and medical access to emergency contraception, commonly referred to as the morning-after pill (Maticka-Tyndale, 2008). In 2011, an estimated 14,000 women aged 15 to 19 gave birth (see Figure 13.2). Over the past decades, the teen pregnancy rate in Canada has dropped significantly, with an almost 40 percent drop between 1996 and 2011. According to Alexander McKay, author of a recent report for the Sex and Education Council of Canada:

> This is a good news story. It's important to look at teen pregnancy rates because they're a basic fundamental indicator of young women's sexual and reproductive health.

FIGURE 13.2 : TEEN BIRTH RATES PER 1000 15- TO 19-YEAR-OLDS, CANADA, 1974-2010

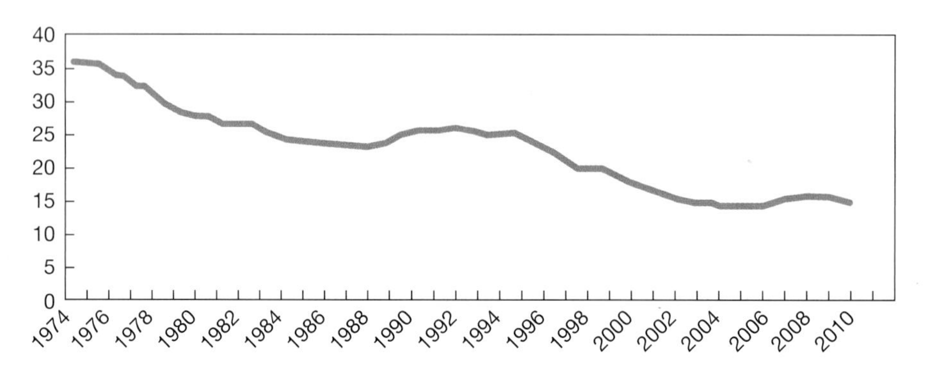

Source: Statistics Canada, 2013, "Live births by age of mother, Canada, provinces and territories," Table 102-4503, http://www5.statcan.gc.ca/cansim/a26?lang=eng&id=1024503&p2=46

> While not all teen pregnancies are a bad thing, when we see [rates] dropping, it's a fairly clear indicator that young women are doing increasingly well in terms of controlling and protecting their reproductive health." (Bielski, 2010)

This decline is not unique to Canada; the United States, England, Wales, and Sweden have witnessed decreases in teenage pregnancies in recent years (McKay, 2010). However, Canada's teenage pregnancy birth rate is less than half that of the United States. This is largely attributable to cultural differences in approaches to teenage sexuality, with Canadians taking a more open and accepting approach to sexuality while certain regions in the United States continue to advocate an abstinence approach to sexuality, which contributes to lower rates of contraceptive use and higher rates of unplanned pregnancies (McKay, 2009).

SEXUALLY TRANSMITTED INFECTIONS In comparison to the successes displayed by Canadian youth in pregnancy prevention, the picture in terms of sexually transmitted infections is not as positive. Sexually transmitted infections (STIs) can be transmitted from one person to another during unprotected oral, anal, or vaginal sex. STIs pose a significant threat to the health and well-being of Canadian youth, and the prevalence of common STIs, such as chlamydia and human papillomavirus (HPV), is highest among youth and young adults. Chlamydia is of particular concern because, if left untreated, it can have serious long-term consequences for the reproductive health of women. Following a steady decline in reported rates of chlamydia (the number of positive test reports made to public health agencies) in the 1990s, rates have been increasing steadily (Public Health Agency of Canada, 2015). It is important to recognize, however, that reported rates are not a measure of prevalence (the percentage of the population that is infected), and that much of the increase in the reported rate of chlamydia is likely due to the increasing use of more sensitive testing technologies and the greater number of young people being tested (McKay and Barrett, 2008). Rates of chlamydia infection range from approximately 3 percent among young women tested at family physicians' offices to almost 11 percent among female street youth.

By Public Health Agency of Canada's standards, the prevalence of chlamydia infection among youth and young adult Canadians is unacceptably high. In fact, a range of STIs—including HPV and herpes simplex virus (HSV)—are common in youth and require that both general health campaigns and school-based sex education programs continue to educate youth around using effective methods to protect themselves from STIs (Maticka-Tyndale, 2008). It has been suggested that today's youth are less sensitive to the potential risks of sexual activity than those who grew up in the fear-filled era of the HIV/AIDS epidemic of the 1980s and early 1990s.

People who became sexually active in the 1980s did so with a clear understanding that sexual choices were life and death choices. In contrast, today's youth may believe, incorrectly, that sexual risk is easily managed through condom use during intercourse. Sex education programs have reinforced this idea by focusing safe sex practices on vaginal-penile intercourse. Many youth are not aware that oral sex carries a significant risk of infection—incorrectly equating safe sex with oral sex. Researchers suggest that this misconception may contribute to their willingness to engage in the sexually risky behaviour of engaging in unprotected oral sex with multiple partners (Heldman and Wade, 2010:329).

AGE OF FIRST INTERCOURSE, MULITIPLE PARTNERS, AND CONDOM USE For a majority of Canadians, their first sexual intercourse occurs during the teenage years (Maticka-Tyndale, 2008, Rotermann, 2008). Overall, the percentage of Canadian youth who report ever having had sexual intercourse has declined since the mid-1990s. The data from the Canadian Community Health Survey indicate that the percentage of 18- to 19-year-olds who had ever had intercourse declined between 1996–1997 and 2005 (Rotermann, 2008). Research from both Canada and the United States indicates that oral sex is about as common as intercourse and typically occurs at about the same time as intercourse, although up to a quarter of teens may begin having oral sex before starting to have intercourse. As Eleanor Maticka-Tyndale explains, "With respect to oral sex, it is important to remember that over the last 30 to 40 years oral sex has become a normative aspect of the adult sexual script and this trend has been followed by youth" (2008:86). As shown in Figure 13.3, the percentage of youth who became sexually active before 15 also declined in the period between 1996 and 2005. The average age of first intercourse among Canadian youth aged 15 to 24 is 16.5 years for both male and females (Rotermann, 2008).

Multiple sexual partners is an important indicator of sexual risk behaviour, especially in terms of the risk of contracting an STI. As shown in Figure 13.4, approximately one-third of youth reported having sex in the previous year with more than one partner. Also indicated in Figure 13.4 is the fact that males are slightly more likely to have multiple sexual partners than females—a reflection, perhaps, that a sexual double standard still exists.

FIGURE 13.3 PERCENTAGE OF 15- TO 16-YEAR-OLDS WHO HAD SEXUAL INTERCOURSE BEFORE AGE 15 OR AT AGE 15 AND 16, BY GENDER AND AGE GROUP, CANADA

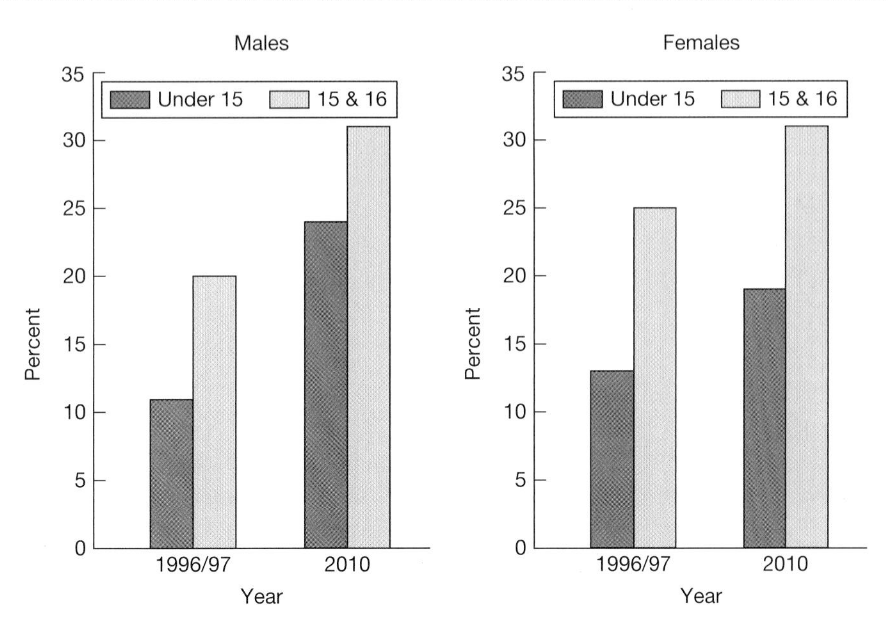

Source: Freeman et al. (2011). *The Health of Canada's Young People: A Mental Health Focus*. Ottawa, ON: Public Health Agency of Canada.

FIGURE 13.4 PERCENTAGE OF SEXUALLY ACTIVE 15- TO 19-YEAR-OLDS WHO REPORTED HAVING MULTIPLE PARTNERS IN THE PAST YEAR, BY GENDER AND AGE GROUP, CANADA

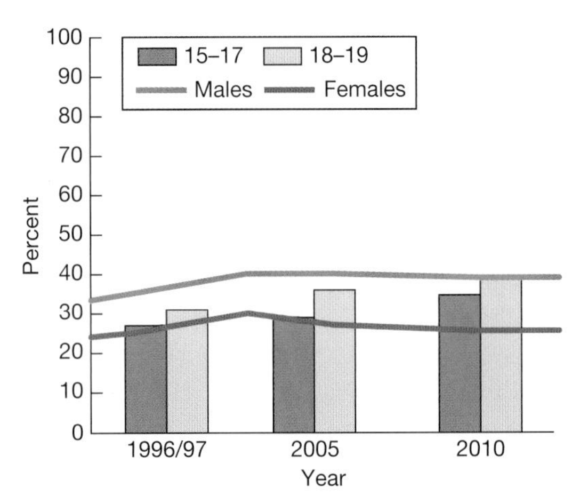

Source: Rotermann, Michelle. 2008. "Trends in Teen Sexual Behaviour and Condom Use," Statistics Canada Cat. no. 82-003-X. Retrieved April 21, 2012. Available: http://www.statcan.gc.ca/pub/82-003-x/2008003/article/10664-eng.pdf

Condom use is an effective means of both preventing unplanned pregnancy and providing protection against sexually transmitted infections. As shown in Figure 13.5, the majority of sexually active youth are effectively protecting their sexual health by using condoms. However, a large proportion of young people move from one sexually active dating relationship to another over the course of their teen and young adult years. A common behavioural pattern is for dating couples to use condoms the first time they have sex, but discontinue condom use as the relationship progresses and increase contraceptive pill use. A study in Toronto in 2009 involved interviews with young women in dating relationships, aged 18 to 24, about their use and discontinuation of condom use. Most of the participants simply assumed that they were in a monogamous relationship without explicitly discussing sexual exclusivity with their partners. As one participant explained:

> I did ask him before we had sex. I said "Look, do I need to be worried about things like sexually transmitted diseases cause if so, put that back in your pants." And he was like, "No, no, no." So I was like, "Okay, you don't need to worry about that with me either." So this is okay. So we have had that discussion without going into detail. (Bolton, McKay, and Schneider, 2010:97)

A repeated pattern of discontinuing condom use in a series of dating relationships over time places individuals at an increased risk of an STI. Sexual health educators need to caution young adults that assumed monogamy is, for a number of reasons, a poor indicator of low STI risk (Bolton, McKay, and Schneider, 2010).

FIGURE 13.5 PERCENTAGE OF SEXUALLY ACTIVE YOUTH WHO USED A CONDOM THE LAST TIME THEY HAD INTERCOURSE, BY GENDER AND AGE GROUP, CANADA

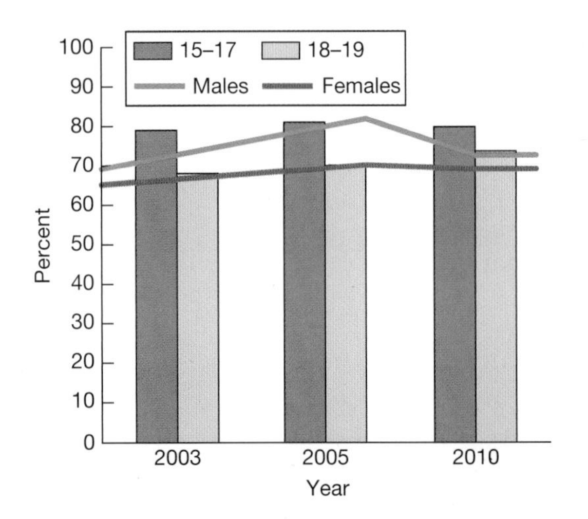

Source: Rotermann, Michelle. 2008. "Trends in Teen Sexual Behaviour and Condom Use," Statistics Canada Cat. no. 82-003-X. Retrieved April 21, 2012. Available: http://www.statcan.gc.ca/pub/82-003-x/2008003/article/10664-eng.pdf

VULNERABLE GROUPS The improvements in sexual health experienced by the majority of Canadian youth are not evenly distributed. In particular, youth living in low-income families, in isolated rural areas, or in regions with greater concentrations of rural and Indigenous populations are more likely to be sexually active in their early teens, more susceptible to STIs, and more likely to experience unplanned pregnancies and sexual abuse (Yee, 2010). Indigenous populations are particularly vulnerable to poor sexual health. For example, in First Nations, 20 percent of births involve a teenage mother, and reported cases of chlamydia are seven times higher than the general population (Health Canada, 2009).

Another group of youth who are particularly at risk for poor sexual health are gay, lesbian, bisexual, transgendered, or questioning (GLBTQ) teens. The First National Climate Survey on Homophobia in Canadian High Schools found that GLBTQ youth face frequent discrimination in their school lives. One of the nearly 4000 students surveyed across Canada made the following comment:

> Most of the gay community in my school are bullied, we all stick together, but that doesn't always help. Many gays are depressed because of this, and teachers and adults need to help and stand up for our community. We are not aliens, we're people, and we have rights. (Taylor and Peter, 2011:307)

Respondents also indicated that they received insufficient sexual health information relevant to their needs. GLBTQ youth also experience higher levels of distrust of health and social services providers that leads them to feel they must hide their sexual identities (Maticka-Tyndale, 2008).

SEX EDUCATION School-based programs are one of the primary avenues for educating young people about sexual health. The primary objective of sexual health education is to provide young people with the necessary knowledge, motivation, and strategies to make informed choices that promote their sexual health and overall well-being (McKay, 2009). Most sex education programs use a biomedical approach, focusing on anatomy and biology rather than on more complex issues related to human sexuality and intimacy, such as developing a positive sexual identity or equality in sexual relationships. Sociologists emphasize that sex education programs could be improved by incorporating these social issues related to sexual health rather than focusing almost exclusively on the biological and medical measures of sexual health. When schools avoid presenting alternative perspectives, including those that incorporate a discussion of eroticism and pleasure, they may perpetuate existing class, race, gender, and sexuality hierarchies (Schneider and Jenness, 2005:396).

One GLBTQ student in the First National Climate Survey on Homophobia in Canadian High Schools wrote:

> Teachers must be educated. They discuss other teachers' sexuality, other students' sexuality, they tell homophobic jokes and do not realize. They also must implement resources into the curriculum, or LGBT individuals will always be viewed as one-dimensional abstractions. I am not satisfied with my rights in the school curriculum. (Taylor and Peter, 2011:207)

An effective sexual health education curriculum needs to "recognize that responsible individuals may choose a variety of paths to achieve sexual health" (Public Health Agency of Canada, 2007:25). This would include information on a variety of options for youth, including the necessary information for youth who choose not to engage in sexual activity, as well as information on sexual diversity issues. To ensure the sexual health and overall well-being of our youth, Canadian families, schools, healthcare providers, public health agencies, and communities must share in the responsibility of providing effective and inclusive sexual health education and services (McKay, 2009).

LO-6 VIOLENCE IN INTIMATE RELATIONSHIPS

It has been suggested that "intimacy can be hazardous to one's health." Unfortunately, this statement reflects the sad reality that we often suffer the most damage within our intimate relationships. The term **intimate partner violence** refers to the physical and nonphysical violence experienced by women and men at the hands of current or former partners. Across cultures and over time, women have been particularly vulnerable to frequent, severe, and sometimes lethal intimate partner violence. Although women can be violent in relationships with men, and violence is sometimes found in same-sex partnerships, the overwhelming burden of partner violence is borne by women at the hands of men. Physical assault, emotional abuse, sexual assault, sexual harassment, and stalking occur at disturbingly high rates within intimate relationships in various forms—dating partners, common-law partners, married partners (Sinha, 2013).

intimate partner violence The physical and nonphysical violence experienced by women and men at the hands of current or former partners.

The extent of violence in relationships was first discovered in the 1980s largely as a result of the efforts of the feminist movement. What had been considered a "private" matter became a "public" issue, precipitating a dramatic increase in services and organizations to serve victims of domestic violence, as well as significant reforms in social policy, law, and police practices. Although significant gains have been made over the past 30 years in reducing various forms of intimate partner violence against women, the numbers remain shocking. In 2013, approximately 70,000 women in

Canada were physically assaulted or sexually assaulted by their intimate partners. In the same year, there were approximately 80 spousal homicides in Canada, and 83 percent of the victims were women (Statistics Canada, 2015b). In a survey of over 3000 students in colleges and universities across the country, 22 percent of female students reported that they had been physically abused, and 29 percent reported being sexually abused by a boyfriend or male acquaintance in the previous year (Johnson & Dawson, 2011). It is evident from these statistics that violence within intimate relationships continues to be a reality for far too many women today. Given that sexual assault and domestic violence are discussed elsewhere in the text (see Chapters 8 and 6), we will focus here on the most recent form of intimate violence—dating violence.

DATING VIOLENCE Consider this story told to sociologist Kathleen Bogle when she was interviewing university students about their experiences with hooking up:

> I'll tell you this story . . . It was final week my junior year and I was done finals . . . so I went out for some drinks . . . There were these two girls there and the bartender started feeding her shots and next thing you know I started talking to her. The bar wasn't crowded at all. Next thing you know we are back at my frat house, she's like, she can't even walk, she is really messed up. So we start hooking up, nothing major. She's coherent, she knows what is going on, but she is really drunk. So we are hooking up and we are sleeping together and she gets sick on me. She's on top of me and throws up on me . . . There were two guys out there watching TV, I'm like, "One of you guys has to help me". . . We put her in the shower and she is like all out of it. My friend that was helping me out tried to help me clean her up a bit, I put her up on his bed . . . I go back in and she's totally fine, she is totally coherent. So we start "going at it" again [laughs] and then she starts calling me by the wrong name . . . I can't believe I am telling you this. (2008:91)*

How do you respond to this story? How would you define the actions of the young man telling the story? Is this a story of a "bad date," or is it a story of sexual assault? Scenarios like this one inevitably come up in discussions about dating violence. These debates emphasize once again that definitions of intimate violence remain unclear for many.

The term **dating violence** refers to various forms of sexually and nonsexually assaultive behaviours that occur within dating relationships. According to the Canadian Violence Against Women Survey, an estimated 1.7 million Canadian women have experienced at least one incident of physical or sexual assault by a date or boyfriend since the age of 16. Included in the category of "dates or boyfriends" were "one-night stands," first dates, long-term boyfriends, and committed relationships (Johnson, 1996). Overall, women reported higher rates of sexual assaults than physical assaults. Consequently, most discussions of dating violence have focused on the subject of date rape or acquaintance rape. **Date rape** is defined as acts in which a date or boyfriend forced or attempted to force any type of sexual activity through threats or use of violence. Date rape is especially common on college and university campuses, where it is estimated that approximately one in four, or 25 percent, of female university students will be date-raped (Brownridge, 2006a; Schwartz et al., 2001).

▶ **dating violence** A term used for various forms of sexually and nonsexually assaultive behaviours that occur within dating relationships.

▶ **date rape** Acts in which a date or boyfriend forced or attempted to force any type of sexual activity through threats or use of violence.

Numerous explanations have been offered as to why women continue to be viewed by some men as legitimate targets of sexual violence. Sociologist Diana Russell (1975) describes sexual assault as an *overconforming act,* or an exaggerated form of "normal" relations that exist between the sexes. Sexual aggression, even violent assault, represents qualities regarded by some men as "supermasculine"—strength, power, forcefulness, domination, toughness. As Russell explains, "to conquer, to win, to induce respect through force" are all attributes commonly associated with masculinity in our culture. Especially for men who feel powerless or inadequate, the conquest or domination of a woman may become a means through which they maintain their sense of masculinity and self-esteem.

The fact that men can sexually assault women in their dating or intimate relationships is facilitated by a set of highly stereotyped beliefs and values about women. Studies have shown that holding stereotypical beliefs about women is strongly related with abusive sexual relationships, acceptance of rape myths, and tolerance of violence against women (Andersen, 2010).

According to sociologists Walter DeKeseredy and Molly Dragiewicz (2014), some men acquired the justifications and support for sexually aggressive behaviours from male peers. These peers provided abusive men with a *rapist vocabulary* comprised of shared motives and rationalizations that consistently blame the victim. Common messages in rapist vocabulary are "Some women need to be raped," "Some women deserve to be raped," and "Some women want to be raped."

Consider the explanation for date rape provided by the following second-year university student:

> I met a girl at a party and I considered her snobbish and phony. She latched onto me at the party, and we laughed and had a good time and went somewhere else. She was an attractive woman in her thirties, but she irritated me. When I took her home to her apartment she was telling me goodnight and I raped her. I didn't really feel the urge. As a matter of fact, I had a hell of a time getting an erection . . . I'd had quite a bit of booze. But I forced myself to do it to prove a point to her, to prove that she wasn't as big as she thought she was. (Russell, 2003: 261)

It is through interactions with like-minded male peers that these young men develop the justifications for their actions, which allow them to maintain a favourable self-image and define themselves as normal despite their abuse of their dating partners.

Not all women are equally at risk of intimate partner violence. Indigenous women are four times more likely to experience intimate partner violence than non-Indigenous women. In addition, Indigenous women are not only more likely to experience violence at the hands of their current or former spouses, but the violence is more frequent, more serious, more likely to result in injury, and more likely to result in the women fearing for their lives (Brennan, 2011; Brownridge, 2008; Sinha, 2013). It is also estimated that persons with disabilities are between 50 and 100 percent more likely than those without disabilities to have experienced spousal violence (Brownridge, 2006b; Perreault, 2009).

TIME TO REVIEW

- Describe the trends in Canada's teenage pregnancy rates over the past 30 years.
- Discuss the characteristics of youth who are at high risk of poor sexual health. Why are these groups particularly vulnerable? What might be done to reduce their risk of sexually transmitted infections and unwanted pregnancies?
- Discuss the issue of intimate partner violence among young people. Why do you think that some men believe women are legitimate targets for sexual violence?

SEXUALITY AND SOCIAL CHANGE

THE SEXUAL REVOLUTION

The term **sexual revolution** refers to the dramatic changes that occurred regarding sexual attitudes, behaviours, and values during the 1960s. These dramatic social changes were precipitated by a number of factors. The women's liberation movement challenged gender-role stereotyping, advocating that women should be free to be sexual both in and out of marriage. In addition, the availability of the birth control pill, along with the liberalization of attitudes toward sexuality, led to significant changes in sexual norms and values and much greater sexual freedom.

sexual revolution A term used for the dramatic changes that occurred regarding sexual attitudes, behaviours, and values during the 1960s.

Steve Russell/Toronto Star via Getty Images

Gay pride parades originated in the 1970s as a celebration of sexual diversity.

Increasingly, sexual intercourse before marriage was no longer viewed as 'wrong' or 'immoral' (Freedman and D'Emilio, 1998). Expressions such as "If it feels good, do it" and "Make love, not war" were mantras of the baby boomers, who were becoming young adults during this era. The gay and lesbian movement also played an important role during this time by bringing issues of sexual diversity and sexual oppression into the public realm and profoundly changing levels of tolerance and understanding of lesbian and gay sexuality (Andersen, 2010:295).

Women's sexual behaviour has changed more than men's since the 1960s, narrowing the differences and inequities between the sexual experiences of men and women. Today, if one considers several factors, such as age of first intercourse, number of sexual partners, and the variety of sexual behaviours, women and men are more similar than different. Over a decade ago, sociologists suggested the sexual revolution is over. According to Risman and Schwartz, "Premarital, unmarried, and post-divorce sex are now seen as individual choices for both women and men. The revolutionary principle that divorced the right to sexual pleasure from marriage (at least for adults) is no longer controversial; it goes unchallenged by nearly everyone but the most conservative of religious fundamentalists" (2002:21).

THE SEXUAL DOUBLE STANDARD

▶ **sexual double standard** The belief that men and women are expected to conform to different standards of sexual conduct.

The **sexual double standard** refers to the belief that men and women are expected to conform to different standards of sexual conduct. In other words, there are different guidelines for men and women when it comes to what is acceptable sexual behaviour. For example, according to the sexual double standard, boys and men are expected to desire and pursue sexual opportunities, while girls and women are stigmatized and viewed critically for similar behaviour (Bogle, 2008). The sexual double standard is believed to inhibit young women's sexual behaviour by making it socially costly in terms of negative labels, peer group relations, or disrespect (Lyons, et al., 2011:437). To what extent has the double standard changed since the sexual revolution? Is there a contemporary or modern sexual double standard? What does the contemporary form of the sexual double standard look like?

There is considerable debate regarding the extent to which this sexual double standard still exists. For example, is it acceptable in today's society for both women and men to have multiple sexual partners? to engage in casual sex? Are there expectations in terms of who initiates sex? Although numerous studies have examined the sexual double standard in various contexts, the results remain inconclusive (Kreager and Staff, 2009). A 2009 study using the National Longitudinal Study of Adolescent Health found that women were labelled "sluts" for the same sexual behaviour that earned boys the label "stud." Milhausen and Herold (2001) found evidence for the erosion of the sexual double standard, finding that although a small minority of men continued to endorse a sexual double standard, the majority of both men and women judged men and women's sexual behaviour from a single standard. More recently, principle researcher Heidi Lyons (Lyons et al., 2011) asked young women to evaluate the sexual activity of other female peers with multiple sexual partners. Interestingly, although the interviewees recognized the continued existence of a sexual double standard at a societal level, their associations with peers with similar sexual attitudes served as a buffer against negative labels.

Although women and men today are equally likely to be sexually active, a remnant of the sexual double standard appears to be alive and well. According to Risman and Schwartz:

> Young women still report being worried about being labeled a slut. . . . Girls today may be able to have sex without stigma, but only with a steady boyfriend. For girls, love justifies desire. A young woman still cannot be respected if she admits an

> appetite-driven sexuality. If a young woman has sexual liaisons outside of publically acknowledged "coupledom," she is at risk of being defamed. If a girl changes boyfriends too often and too quickly, she risks being labeled a slut. (2002:20)

Similarly, sociologist Kathleen Bogle's qualitative research on the hookup culture on university and college campuses found similar evidence of a modern sexual double standard in the hookup scripts of university students. A senior at a large U.S. university explains how the label of "slut" is assigned:

> The perception is that if a girl sleeps with a lot of guys she's a slut. If a guy sleeps with a lot of girls he's a stud . . . I mean, I see it every day . . . you can ask anyone on campus randomly, and they would say that would be the perception. (2008:104)*

For a more detailed discussion of the sexual double standard in the hookup culture, see Box 13.3.

BOX 13.3 POINT/COUNTERPOINT

Is the Hookup Culture Bad for Girls?

Is the hookup culture an indication of increased sexual freedom for men and women and a demonstration of young women's sense of empowerment and control over their sexuality? Or is the hookup culture another variation on inequalities that continue to exist between men and women's sexuality?

Sociologists Elizabeth Armstrong, Laura Hamilton, and Paula England examine the divergent viewpoints and assumptions underlying this debate.

Is Hooking Up Bad for Young Women?

Sociologists and psychologists have begun to investigate adolescent and young adult hookups more systematically, drawing on systematic data and studies of youth sexual practices over time to counter claims that hooking up represents a sudden and alarming change in youth sexual culture. The research shows that there is some truth to popular claims that hookups are bad for women. However, it also demonstrates that women's hookup experiences are quite varied and far from uniformly negative, and that monogamous, long-term relationships are not an ideal alternative. Scholarship suggests that pop culture feminists have correctly zeroed in on sexual double standards as a key source of gender inequality in sexuality.

Before examining the consequences of hooking up for girls and young women, we need to look more carefully at the facts. Laura Sessions Stepp, author of *Unhooked: How Young Women Pursue Sex, Delay Love, and Lose at Both*, describes girls "stripping in the student center in front of dozens of boys they didn't know." She asserts that "young people have virtually abandoned dating" and that "relationships have been replaced by the casual sexual encounters known as hookups." Her sensationalist tone suggests that young people are having more sex at earlier ages in more casual contexts than their baby boomer parents did.

This characterization is simply not true. Young people today are not having more sex at younger ages than their parents did. The sexual practices of American youth changed in the 20th century, but the big change came with the baby boom cohort, who came of age more than 40 years ago. The pervasiveness of casual sexual activity among today's youth may be at the heart of boomers' concerns. England surveyed more than 14,000 students from 19 universities and colleges about their hookup, dating, and relationship experiences. Seventy-two percent of both men and women participating in the survey reported at least one hookup by their senior year in college. What the boomer panic may gloss over, however, is the fact that college students do not, on average, hook up that much. By senior year, roughly 40 percent of those who ever hooked up had engaged in three or fewer hookups, 40 percent between four and nine hookups, and only 20 percent in 10 or more hookups. About 80 percent of students hook up, on average, less than once per semester over the course of college.

In addition, the sexual activity in hookups is often relatively light. Only about one-third had engaged in intercourse in their most recent hookup. Another third had engaged in oral sex or manual stimulation of the genitals. The other third of hookups only involved kissing and nongenital touching. A full 20 percent of survey respondents in their fourth year of college had never had vaginal intercourse. In addition, hookups between total strangers are relatively uncommon, while hooking up with the same person multiple times is common.

(Continued)

*Source: Bogle, Katheleen. *Hooking Up: Sex, Dating, and Relationships on Campus*. © 2008 New York University Press, p. 104.

Ongoing sexual relationships without commitment are labelled as "repeat," "regular," or "continuing" hookups, and sometimes as "friends with benefits." Often there is friendship or socializing both before and after the hookup.

Hooking up hasn't replaced committed relationships. Students often participate in both at different times during college. By their senior year, 69 percent of heterosexual students had been in a college relationship of at least six months. Hookups sometimes became committed relationships and vice versa; generally, the distinction revolved around the agreed upon level of exclusivity and the willingness to refer to each other as "girlfriend/boyfriend."

And, finally, hooking up isn't radically new. As suggested above, the big change in adolescent and young adult sexual behaviour occurred with the baby boomers. This makes sense, as the forces giving rise to casual sexual activity among the young—the availability of the birth control pill, the women's and sexual liberation movements, and the decline of *in loco parentis* on college campuses—took hold in the 1960s. But changes in youth sexual culture did not stop with the major behavioural changes wrought by the sexual revolution. Contemporary hookup culture among adolescents and young adults may rework aspects of the sexual revolution to get some of its pleasures while reducing its physical and emotional risks. Young people today are expected to delay the commitments of adulthood while they invest in careers. They get the message that sex is okay, as long as it doesn't jeopardize their futures; STDs and early pregnancies are to be avoided. This generates a sort of limited-liability hedonism. For instance, friendship is prioritized a bit more than romance, and oral sex appeals because of its relative safety. Hookups may be the most explicit example of a calculating approach to sexual exploration. They make it possible to be sexually active while avoiding behaviours with the highest physical and emotional risks (e.g., intercourse, intense relationships).

Hookup Problems, Relationship Pleasures

Hookups are problematic for girls and young women for several related reasons. As many observers of North American youth sexual culture have found, a sexual double standard continues to be pervasive. As one woman Hamilton interviewed explained, "Guys can have sex with all the girls and it makes them more of a man, but if a girl does, then all of a sudden she's a 'ho' and she's not as quality of a person." Sexual labelling among adolescents and young adults may only loosely relate to actual sexual behaviour; for example, one woman complained in her interview that she was a virgin the first time she was called a "slut." The lack of clear rules about what is "slutty" and what is not contributes to women's fears of stigma.

The most commonly encountered disadvantage of hookups, though, is that sex in relationships is far better for women. England's survey revealed that women orgasm more often and report higher levels of sexual satisfaction in relationship sex than in hookup sex. This is in part because sex in relationships is more likely to include sexual activities conducive to women's orgasm. In hookups, men are much more likely to receive fellatio than women are to receive cunnilingus. In relationships, oral sex is more likely to be reciprocal. In interviews conducted by England's research team, men report more concern with the sexual pleasure of girlfriends than hookup partners, while women seem equally invested in pleasing hookup partners and boyfriends.

The sexual double standard mars women's hookup experiences. In contrast, relationships provide a context in which sex is viewed as acceptable for women, protecting them from stigma and establishing sexual reciprocity as a basic expectation. In addition, relationships offer love and companionship.

Toward Gender Equality in Sex

Like others, Stepp, the author of *Unhooked*, suggests that restricting sex to relationships is the way to challenge gender inequality in youth sex. Certainly, sex in relationships is better for women than hookup sex. However, research suggests two reasons why Stepp's strategy won't work. First, relationships are also plagued by inequality. Second, valorizing relationships as the ideal context for women's sexual activity reinforces the notion that women shouldn't want sex outside of relationships and stigmatizes women who do. A better approach would challenge gender inequality in both relationships and hookups.

Source: Armstrong, E. A., Hamilton, L., England, P. (2010). Is Hooking Up Bad for Young Women? *Contexts* Vol. 9(3) pp. 22–27.

TIME TO REVIEW

- Describe the factors that led to the sexual revolution. What changes occurred in the lives of men and women during this period?
- What is the sexual double standard? Why do you think this double standard still exists?

LO-1 Explain how the terms *sex, gender*, and *sexuality* relate to one another and how they are different.

The terms *sex* and *gender* are often used interchangeably. *Sex* refers to the biological and anatomical differences between females and males. *Gender* refers to culturally and socially created qualities associated with being a man or a woman (masculinity and femininity). *Sex*—the biological and physiological aspects of the body—interacts with *gender*—the roles and responsibilities associated with the socially constructed concepts of "masculinity" and "femininity." Similarly, the terms *sex* and *sexuality* are used interchangeably. *Sexuality* refers to the range of human activities designed to produce erotic response and pleasure. Sexuality, like other forms of human behaviour, is shaped by the society and culture we live in.

San Jose Mercury News/MCT/LANDOV

LO-2 Understand how various social scientists have classified sexual orientation.

Sexual orientation refers to an individual's preference for emotional–sexual relationships with members of the opposite sex (heterosexuality), the same sex (homosexuality), both sexes (bisexuality), or neither sex (asexuality). Alfred Kinsey constructed a seven-point scale of sexual behaviour to emphasize the fluidity of human sexual experience. Kinsey's scale did not measure sexual identity or unacted-upon sexual feelings. More recently, researchers at the University of Chicago established three criteria for identifying people as homosexual or bisexual: (1) *sexual attraction* to persons of one's own gender, (2) *sexual involvement* with one or more persons of one's own gender, and (3) *self-identification* as gay, lesbian, or bisexual.

Brad Barket/Getty Images for Kuato Studios

LO-3 Compare and contrast monogamous and polyamorous intimate relationships.

Monogamy refers to the practice of having an intimate relationship with only one person at a time. Some people practise serial monogamy—a succession of relationships in which a person has several partners over a lifetime but is legally married to only one person at a time. Polyamory involves mutually acknowledged emotional, sexual, or romantic relationships with multiple partners.

© Catchlight Visual Services/ Alamy

KEY TERMS

asexual An absence of sexual desire toward either sex (p. 353).

date rape Acts in which a date or boyfriend forced or attempted to force any type of sexual activity through threats or use of violence (p. 368).

dating violence A term used for various forms of sexually and nonsexually assaultive behaviours that occur within dating relationships (p. 368).

intersex person An individual who is born with a reproductive or sexual anatomy that does not correspond to typical definitions of male or female; in other words, the person's sexual differentiation is ambiguous (p. 351).

intimate partner violence The physical and nonphysical violence experienced by women and men at the hands of current or former partners (p. 367).

marriage A legally recognized and/or socially approved arrangement between two or more individuals that carries certain rights and obligations and usually involves sexual activity (p. 354).

monogamy An intimate relationship with one person at a time (p. 354).

polyamory Intimate relationships that involve mutually acknowledged emotional, sexual, or romantic relationships with multiple partners (p. 355).

polyandry The concurrent marriage of one woman with two or more men (p. 357).

polygamy The concurrent marriage of a person of one sex with two or more members of the opposite sex (p. 356).

polygyny The concurrent marriage of one man with two or more women (p. 356).

primary sex characteristics The genitalia used in the reproductive process (p. 350).

secondary sex characteristics The physical traits (other than reproductive organs) that identify an individual's sex (p. 350).

sex A term used to describe the biological and anatomical differences between females and males (p. 350).

sexual double standard The belief that men and women are expected to conform to different standards of sexual conduct (p. 370).

sexuality (sexual) The range of human activities designed to produce erotic response and pleasure (p. 351).

sexual orientation A person's preference for emotional-sexual relationships with members of the opposite sex (heterosexuality), the same sex (homosexuality), both (bisexuality), or neither (asexuality) (p. 352).

sexual revolution A term used for the dramatic changes that occurred regarding sexual attitudes, behaviours, and values during the 1960s (p. 369).

LO-4 Explain a functionalist, conflict, symbolic interactionist, and postmodern analysis of sexuality.

© Dick Hemingway

According to functionalists, the regulation of norms and values surrounding sexual reproduction is functional because it maintains social stability. Some sexual practices, such as prostitution, can be both functional and dysfunctional. Conflict theorists draw attention to the exploitative aspects of sexuality, focusing on workers and consumers and unequal power relations in the buying and selling of sex. Symbolic interactionists argue that sexual feelings, practices, and body parts are given meaning through the application of sexual scripts. Postmodern theorists view gender identities and sexual identities as fluid and malleable rather than presocial and biologically based.

LO-5 Evaluate the sexual health of Canadian youth and young adults.

In terms of using birth control, rates of teenage pregnancy, and condom use, today's youth are doing a good job of protecting their sexual health and well-being. The picture in terms of sexually transmitted infections (STIs) is not as positive. A range of STIs, including human papillomavirus (HPV) and herpes simplex virus (HSV), are common in youth and require that both general health campaigns and school-based sex education programs continue to educate youth around using effective methods to protect themselves from STIs.

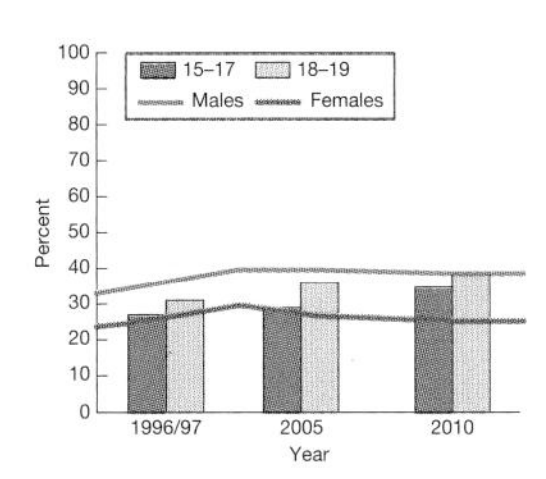

Source: Rotermann, Michelle. 2008. "Trends in Teen Sexual Behaviour and Condom Use," Statistics Canada Cat. no. 82-003-X. Retrieved April 21, 2012. Available: http://www.statcan.gc.ca/pub/82-003-x/2008003/article/10664-eng.pdf.

LO-6 Identify the significant issues and controversies regarding sexuality in Canadian society.

Peter Bernik/Shutterstock.com

Intimate partner violence—the physical and nonphysical violence experienced by women and men at the hands of their current or former intimate partners—remains a significant social problem. Women are particularly vulnerable to frequent, severe, and sometimes lethal intimate partner violence. Dating violence, which includes forms of sexually and nonsexually assaultive behaviours that occur within dating relationships, has been found to be far too common. The sexual double standard—the belief that men and women are expected to conform to different standards of sexual conduct—continues to constrain the sexual behaviour of young men and women today.

KEY FIGURES

Arthur Siegel/Time & Life Pictures/Getty Images

Alfred Kinsey (1894-1956) Kinsey was a biologist and professor of zoology who in 1947 founded the Institute for Sex Research at Indiana University. Kinsey's research on human sexuality provoked controversy in the 1940s and 1950s and profoundly influenced social and cultural values regarding sexuality in North American society.

Jean Pierre FOUCHET/RAPHO/ Gamma-Rapho/Getty

Michel Foucault (1926-1984) Foucault was a French philosopher, social theorist, and historian. In one of his best-known works, the three-volume *History of Sexuality* (1976–1984), Foucault develops an "analytics of power"—the conceptual instruments that make it possible to analyze sex in terms of power. He emphasized that the power mechanisms of sexuality are socially constructed, unstable, and historically situated.

APPLICATION QUESTIONS

1. The media claim that young people engage in far more sexual behaviour than their predecessors did. Assess the validity of this claim.
2. Describe some of the possible consequences if a culture in which people are more likely to hook up than to engage in long-term committed relationships continues to evolve.
3. Discuss cases in which sex and gender do not correspond. What are the consequences for the individuals involved?
4. Our culture believes that marriage should be monogamous. Do you think this is a realistic expectation for marital partners to have for themselves and their partners? Why, or why not?
5. The evidence suggests that sexual health among young people is improving. Why do you think this improvement has occurred?
6. Imagine that your parents have just informed you that you were born intersex and had undergone gender reassignment surgery. How would this affect you? Who do you think should make decisions regarding surgical interventions with intersex newborns? When do you think these decisions should be made?

sexual scripts Culturally created guidelines that define how, where, with whom, and under what conditions a person is to behave as a sexual being (p. 349).

transgender person An individual whose gender identity (self-identification as woman, man, neither, or both) does not match the person's assigned sex (identification by others as male, female, or intersex based on physical/genetic sex) (p. 352).

CHAPTER 14 AGING

LEARNING OBJECTIVES

AFTER READING THIS CHAPTER, YOU SHOULD BE ABLE TO

LO-1 Explain how functional age differs from chronological age.

LO-2 Discuss how views of aging differ in preindustrial and industrialized societies.

LO-3 Explain ageism and elder abuse.

LO-4 Understand the differences between the functionalist, symbolic interactionist, conflict, feminist, and postmodern explanations of aging.

LO-5 Identify the most common living arrangements for older people.

CHAPTER FOCUS QUESTION

Given the fact that aging is an inevitable consequence of living (unless an individual dies young), why do many people in Canada devalue older persons?

Source: The Canadian Press/AP Photo/Rick Rycroft

It all became clear one day, as I stood at the MAC makeup counter of a Toronto department store, trying to catch a salesperson's eye. The black-clad 20-somethings kept avoiding my steely gaze and waited on all of the younger women around me.

I worried that perhaps I was in the middle of a dream, or possibly even invisible. When I finally grabbed a young man at the counter with a couple of piercings to his lower lip, he apologized. 'I thought you were with your daughter,' he said, pointing to one of the Goths beside me. God forbid a woman in her 40s should buy lipstick on her own.

I realized then that part of me was starting to disappear. Unlike Alice, in Woody Allen's movie of the same name, who was given potions to make her invisible, all I had to do was continue to sit around and age and I, too, would start to fade away.

I did what many women do when they first realize they're disappearing—I fought back. We can be perked up or puffed out. I coloured my hair, increased my daily dosage of makeup . . . And started wearing my daughter's clothes . . .

What I didn't realize initially is that invisibility runs deeper than just our skin. (Binks, 2003)

When Morley Callaghan turned 80, his sons gave him a birthday party in a fine restaurant. During a graceful thank-you speech, the novelist said, "Being 80 is like walking through a jungle: you never know what is going to pounce on you next."

I've been 80 for a few months now and I don't see it as a very big deal. My packaging is wearing out: my skin is too big for me and falls in folds, I have less hair, I am shorter and stooped, and my knees hurt on stairs. Plus a bit of cancer. But I'm still me; exactly me, irreducibly, for better and also for worse, me.

When I was young, I expected that age would make me sage and I would become so serene that strangers would stop to take my pulse. I now realize that all age does is make you old.

Almost all the invincible women of the 1950s are now widows. I am almost the only one of our gang with a living husband. After 60 years, our marriage is a mellow one. We seem to be bobbing on a sea of affection, and disagreements are rare. Perhaps we are maturing. Just in time. (Callwood, 2005:56)

We are all affected by **aging**—the physical, psychological, and social processes associated with growing older (Atchley, 1997). In this chapter, you will learn that the way in which societies and individuals respond to aging can be at least as important as the physical processes in determining how we will spend our older years. We will examine how older people live in a society that often devalues people who do not fit the ideal norms of youth, beauty, physical fitness, and self-sufficiency. Before reading on, test your knowledge about aging and age-based discrimination by taking the quiz in Box 14.1.

aging The physical, psychological, and social processes associated with growing older.

CRITICAL THINKING QUESTIONS

1. Why do you think older people are often devalued in our society?
2. Most Canadian jurisdictions have abolished the mandatory retirement age. Who is most likely to benefit and who is most likely to be hurt by this policy change?
3. Older people are often the victims of social stereotypes. What are some of these stereotypes? How do you think these affect the lives of older people?

LO-1 THE SOCIAL SIGNIFICANCE OF AGE

We often hear the question: "How old are you?" Age is socially significant because it defines what is appropriate for or expected of people at various stages. Age is one of the few ascribed statuses that changes over time. Behaviour that is considered appropriate at one stage of a person's life may be considered unusual at another stage. Nobody thinks it is unusual if a person in her 20s goes bungee jumping. However, if her 85-year-old grandfather does the same thing, he may receive some odd looks and even media coverage because he is defying norms regarding age-appropriate behaviour.

▶ **chronological age** A person's age based on date of birth.

▶ **functional age** A term used to describe observable individual attributes–such as physical appearance, mobility, strength, coordination, and mental capacity–that are used to assign people to age categories.

We all have a **chronological age**—a person's age based on date of birth. However, we often assess people on the basis of their **functional age**—observable individual attributes, such as physical appearance, mobility, strength, coordination, and mental capacity, that are used to assign people to age categories (Atchley and Barusch, 2004). Characteristics such as youthful appearance or grey hair and wrinkled skin are common criteria for determining whether someone is "young" or "old." Feminist scholars have noted that functional age works differently for women and men—as they age, men may be viewed as distinguished or powerful, whereas when women grow older, they are thought to be "over the hill" or grandmotherly (Banner, 1993).

TRENDS IN AGING

Just as people get older, societies also age. Today, older Canadians make up about 14 percent of the population, and the number of people over 65 will continue to increase as the baby boomers get older (see Figure 14.1). In 1971, the median age (the age at which half the people are younger and half are older) was 26. By 2011, it was 40. This substantial increase—fourteen years in four decades—is partly the result of the baby boomers (people born between 1946 and 1964) moving into middle and older age and partly the result of more people living longer.

▶ **life expectancy** The average length of time a group of individuals of the same age will live.

▶ **cohort** A category of people born within a specified period of time or who share some specified characteristic.

This *greying of Canada* resulted from an increase in life expectancy combined with a decrease in birth rates following the baby boom. **Life expectancy** is the average length of time a group of individuals of the same age will live. Life expectancy shows the average length of life of a **cohort**—a category of people born within a specified period of time or who share some specified characteristic. In 2011, life expectancy at birth was 84 for females and 79 for males (see Figure 14.2; Statistics Canada, 2013f).

In 1900, about 5 percent of the Canadian population was over age 65; by 1981, that number had risen to approximately 10 percent. By 2011, this had increased to 14 percent and it is predicted to increase to 22 percent in 2031 and 26 percent in 2061 (Figure 14.1).

BOX 14.1 SOCIOLOGY AND EVERYDAY LIFE

How Much Do You Know About Aging?

True	False	
T	F	1. Most older persons have serious physical disabilities.
T	F	2. Women in Canada have a longer life expectancy than men.
T	F	3. Scientific studies have documented the fact that women age faster than men.
T	F	4. The majority of older people have incomes below the poverty line.
T	F	5. Studies show that advertising no longer stereotypes older persons.

For answers to the quiz about aging, go to **www.nelson.com/student.**

FIGURE 14.1 : POPULATION AGED 65 AND OVER, CANADA, 1901-2061

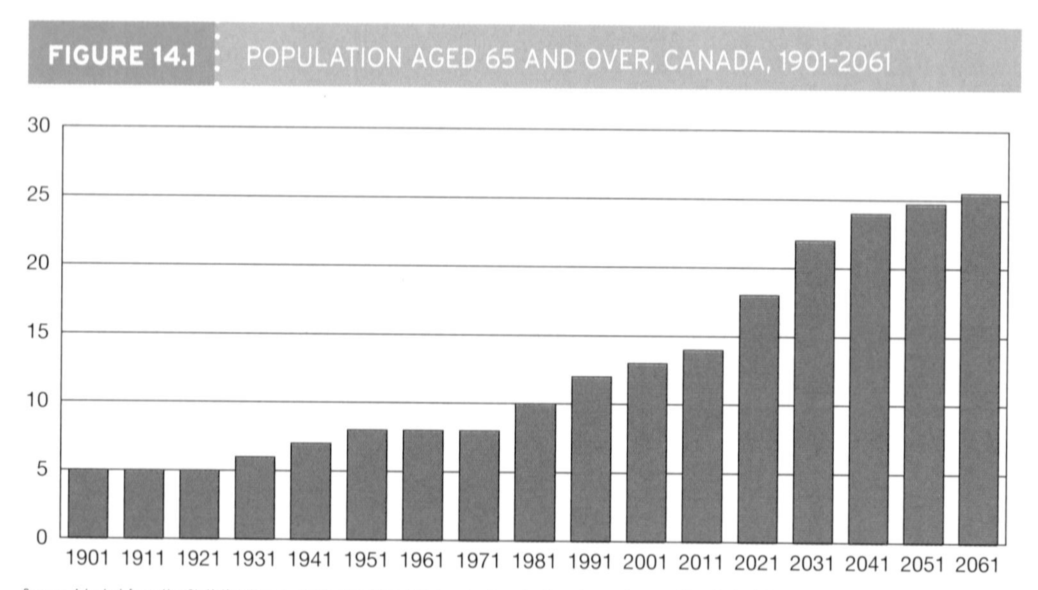

Source: Adapted from the Statistics Canada publication "Population projections for Canada, provinces and territories," Catalogue 91-520, 2005 to 2031. Released December 15, 2005/ Indicators of Well-Being in Canada, Canadians in Context–Aging Population, 2014.

FIGURE 14.2 : EVOLUTION OF LIFE EXPECTANCY AT BIRTH BY AGE AND SEX, CANADA, 1921-2011

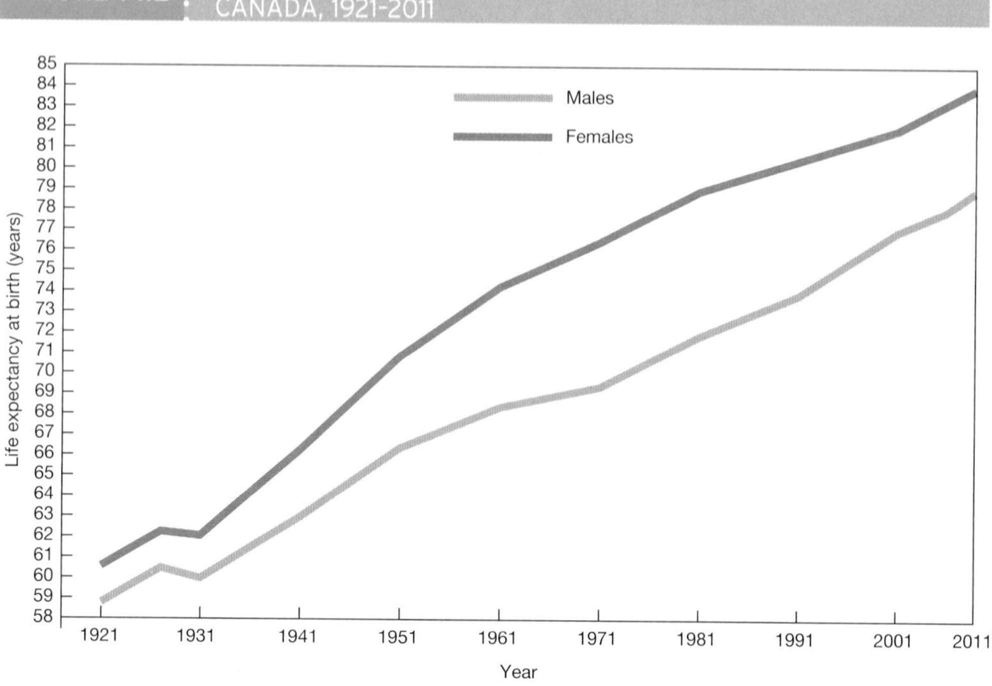

Source: Statistics Canada, CANSIM database table 102-0511 (1990 to 2006), 2011.

Since the beginning of the 20th century, life expectancy has steadily increased as industrialized nations developed better water and sewage systems, improved nutrition, and made tremendous advances in medical science. The economic development that contributed to the lower death rate also helped lower the birth rate. In industrialized nations, children came to be viewed as an economic liability: They could not contribute to the family's financial well-being and had to be supported.

The distribution of the Canadian population is depicted in the age pyramid in Figure 14.3. If, every year, the same number of people are born as in the previous year and a certain number die in each age group, the plot of the population distribution would be pyramid-shaped because

FIGURE 14.3 : DIFFERENT COHORTS AMONG THE AGE PYRAMID OF CANADA, 2011

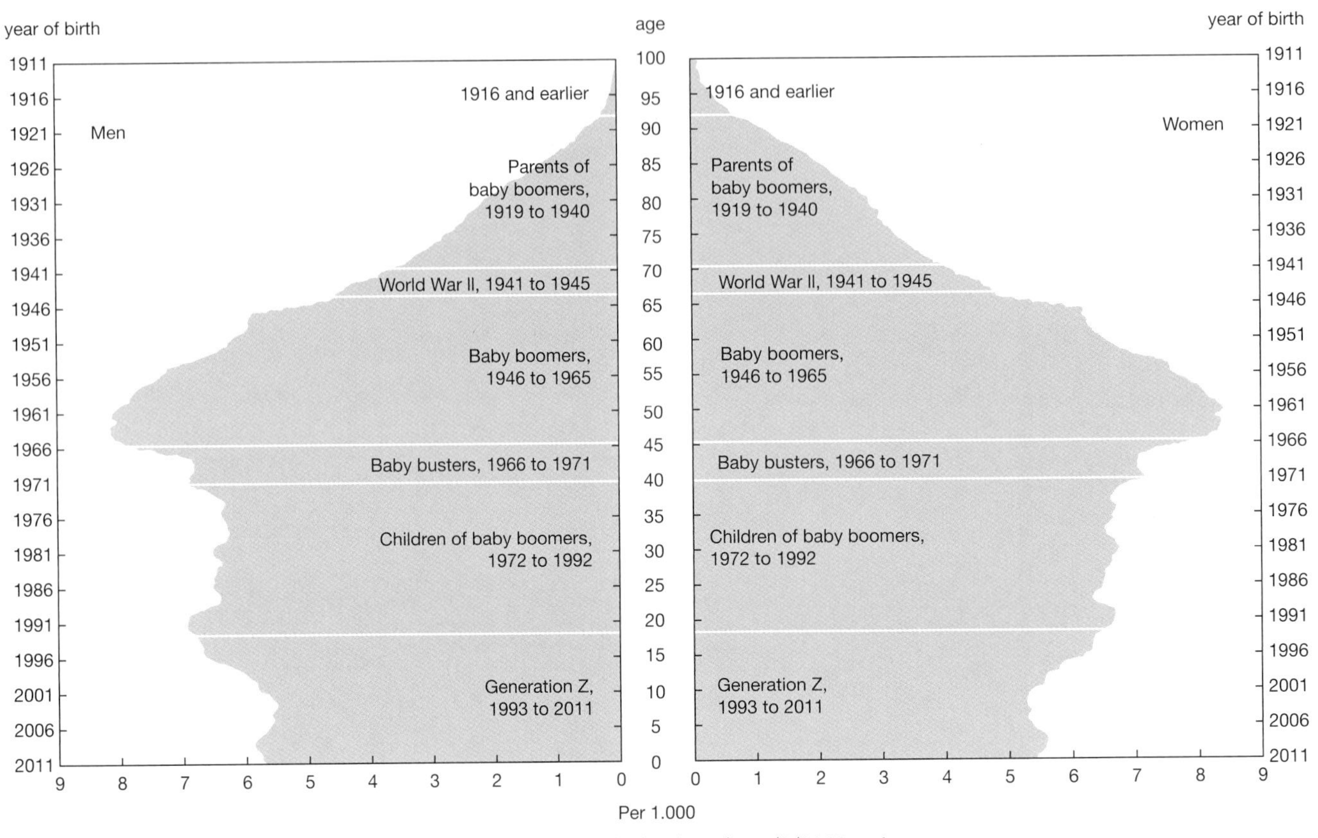

Source: Statistics Canada, Census of Population, 2011, http://www12.statcan.gc.ca/census-recensement/2011/as-sa/98-311-x/2011003/fig/fig3_2-2-eng.cfm

fewer people survive to reach the next age level. Figure 14.3 is not a perfect pyramid because the declining birth rate since the baby boom has resulted in fewer young people at the bottom of the pyramid and a higher average age.

Populations in all parts of the world are aging. By 2050, it is estimated that 30 percent of Canada's population will be over the age of 60 compared with 42 percent in Japan, 39 percent in South Korea, and 38 percent in Spain (World Economic Forum, 2012). Globally, the population is expected to increase by 2 billion people between 2010 and 2050 and almost two-thirds of those (1.3 billion) will be 60-plus.

LO-2 AGE IN HISTORICAL PERSPECTIVE

People are assigned to different roles and positions based on each society's age and role structures. Age structure is the number of persons at each age level within the society; role structure is the number and types of positions available to them (Riley and Riley, 1994). Over the years, the age continuum has been divided into more and more segments. Two hundred years ago, people divided the age spectrum into "babyhood," a very short childhood, and then adulthood. What we would consider "childhood" today was very different 200 years ago, when agricultural societies needed many people to work on the land. When most of the population was involved in food production, categories such as toddlers, preschoolers, preteens, teenagers, young adults, the middle-aged, or older persons did not exist.

If the physical labour of young persons is necessary for society's survival, then they are expected to act like adults and do adult work. Older persons are also expected to continue to be productive for as long as they are physically able. During the 17th and 18th centuries in North America, for example, older individuals helped with the work and were respected because they were needed—and because few people lived that long (Gratton, 1986). The presence of older people in large numbers is a modern phenomenon and their place in society is still evolving.

LATE ADULTHOOD

Late adulthood is generally considered to begin at age 65—the "normal" retirement age. Some gerontologists—researchers who study aging (**gerontology**)—subdivide late adulthood into three categories: (1) the "young-old" (ages 65 to 74), (2) the "old-old" (75 to 85), and (3) the "oldest-old" (85+) (see Moody, 1998).

▶ **gerontology** The study of aging and older people.

One consequence of aging is that older persons have greater chances of serious illness. Some diseases affect virtually only persons in late adulthood. Alzheimer's disease (a progressive and irreversible deterioration of brain tissue) is an example. Persons with this disease have an impaired ability to function. Eventually, they can no longer recognize people they have always known and become totally dependent upon help from others. We do not know the cause of this disease and there is no cure. About half a million Canadians suffer from Alzheimer's disease and related dementias. By the year 2034, this number will likely grow to well over a million (Alzheimer Society of Canada, 2010).

Fortunately, most older people are in relatively good health. Although older people experience some decline in strength, flexibility, stamina, and other physical capabilities, much of that decline does not result simply from the aging process—it is avoidable. With proper exercise, some of it is even reversible.

Along with physical changes come changes in roles, such as retirement and becoming a grandparent. An extended family is a great source of pleasure for many older Canadians. Grandparenting is an interesting role that gives members of younger and older generations the opportunity to build mutually beneficial relationships. Some of the benefits of having grandparents are described in the following comments by Grade 3 students about the role of a grandmother:

> When they read to us they don't skip words and they don't mind if it is the same story.
>
> They don't have to be smart, only answer questions like why dogs hate cats, and how come God isn't married.
>
> Grandmas are the only grownups who have got time—so everybody should have a grandmother, especially if you don't have television. (Huyck, cited in McPherson, 1998:213)

Some of the physical and psychological changes that come with increasing age can cause stress. According to Erik Erikson (1963), older people must accept that the life cycle is inevitable

and that their lives are nearing an end. Mark Novak interviewed several older people about what he termed "successful aging." One respondent, Joanne, commented:

> For me getting older was very painful at first because I resisted change. Now I'm changed, and it's okay. I would say I have a new freedom . . . I thought I had no limits, but for me a great learning [experience] was recognizing my limits. It was a complete turnover, almost like a rebirth. I guess I've learned we're all weak really. At least we should accept that—being weak—and realize, "Hey, I'm only a fragile human being." (1995:125)

Despite its negative aspects, aging has many positive dimensions and most older people are quite content with their lives. Many are financially secure, with home mortgages paid off and no children remaining at home. They thereby have a great deal of personal freedom. Most report that they are in good or excellent health (Novak, Campbell, and Northcott, 2013). A national study (Health and Welfare Canada, 1998) found that older people were much less likely than younger people to report that their lives were stressful, and the vast majority (92 percent) reported that they were quite happy or very happy. Novak and colleagues (2013) report that older couples have a higher level of satisfaction with their marriages than couples in their child-rearing years.

RETIREMENT

Retirement means the end of a status that has been a source of both income and personal identity. Perhaps the loss of a valued status explains why many retired persons introduce themselves by saying, "I'm retired now, but I was a [banker, lawyer, plumber, supervisor, and so on] for 40 years."

Retirement is a relatively recent phenomenon. The first national pension system was established in Germany in 1889. In Canada, the *Old Age Pension Act* was introduced in 1927 to provide a basic income to needy retired people. Initially, pensions were given only to needy people over the age of 70 and payments were minimal. Older people were still expected to look after themselves or to receive help from their families. Retirement was not common until the amount paid to retirees by public pension plans increased. Pension coverage became universal in 1951, and benefits were improved between 1951 and 1975 (Northcott, 1982).

All working people in Canada are covered by the Canada or Quebec pension plans, and all those over age 65 receive the Old Age Security pension, though this is taxed back from higher-income recipients. About 1.7 million lower-income Canadians also receive the benefits of the Guaranteed Income Supplement program (Figure 14.4 shows low income by age). These programs are responsible for the reduction in poverty among older people (see Figure 14.5, which shows how the overall situation for people with low incomes has improved over time). Many people have company pension plans and others have invested in registered retirement savings plans (RRSPs).

Retirees in many other countries are not as fortunate as Canadians. Many Western industrial countries, including the United States, have underfunded their government pension plans and will be forced to raise taxes or reduce benefits. In other parts of the world, there are no pension plans because families have always been expected to care for their own members.

Retirement plays an increasingly important role in the lives of Canadians. When the first retirement laws were passed, the retirement age (usually 70) was much higher than the average life expectancy. As a result, most people never retired, and for those who did, retirement was typically short. Because the average retirement age has declined and life expectancy is increasing, the retirement years are making up an increasing proportion of people's lives. On the other hand, many people prefer to work past the normal retirement age of 65. Most provinces no longer allow organizations to force people who wish to keep working to retire at age 65.

Retirement is a major transition, as retirees lose a source of income, identity, lifestyle, and friends they may have had for most of their adult lives. This is especially true of workers who achieve a great deal of satisfaction from their jobs. Jack Culberg explains the difficulties of his transition:

> When you suddenly leave [the corporate jungle], life is pretty empty. I was sixty-five, the age people are supposed to retire. I started to miss it quite a bit. The phone

FIGURE 14.4 LOW-INCOME RATE BY AGE, CANADA, 1976-2011

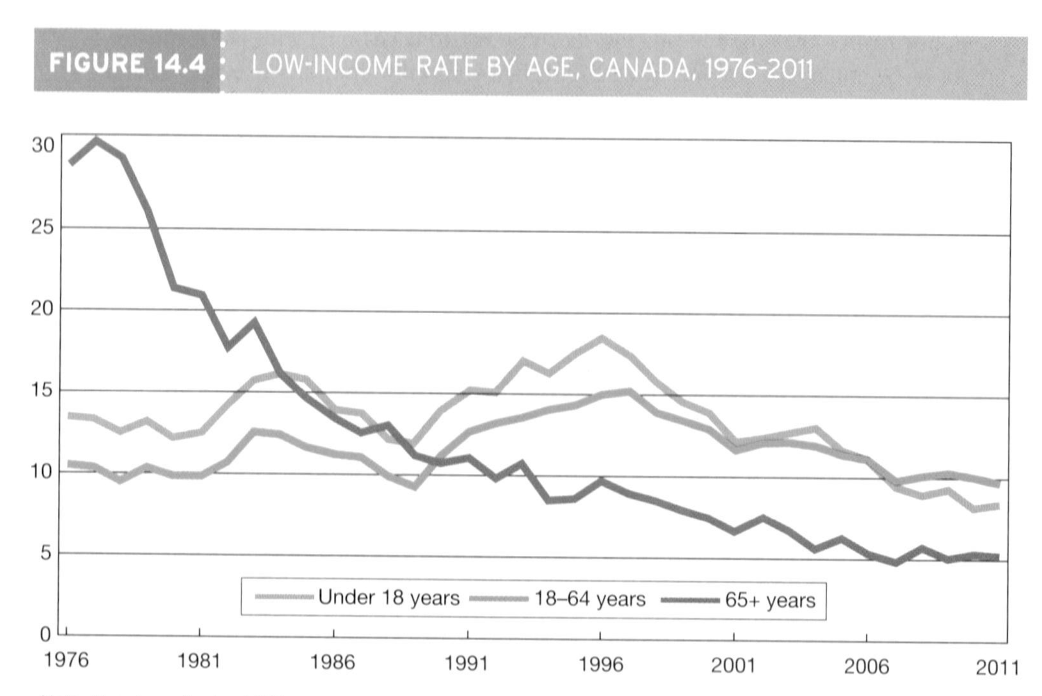

Note: Based on after-tax LICOs.

Source: Employment and Social Development Canada. 2014. Indicators of Well-Being in Canada, Financial Security - Low Income Incidence. Available at: http://www4.hrsdc.gc.ca/.3ndic.1t.4r@-eng.jsp?iid=23. Data source: Statistics Canada. *Table 202-0802 - Persons in low income families*, annual, CANSIM (database).

FIGURE 14.5 LOW-INCOME RATE, PEOPLE 65 AND OLDER, 1980-2008

Poverty rates have steadily declined for older people. However, there are still large income differences between couples and unattached people, particularly females who are much more likely to be poor.

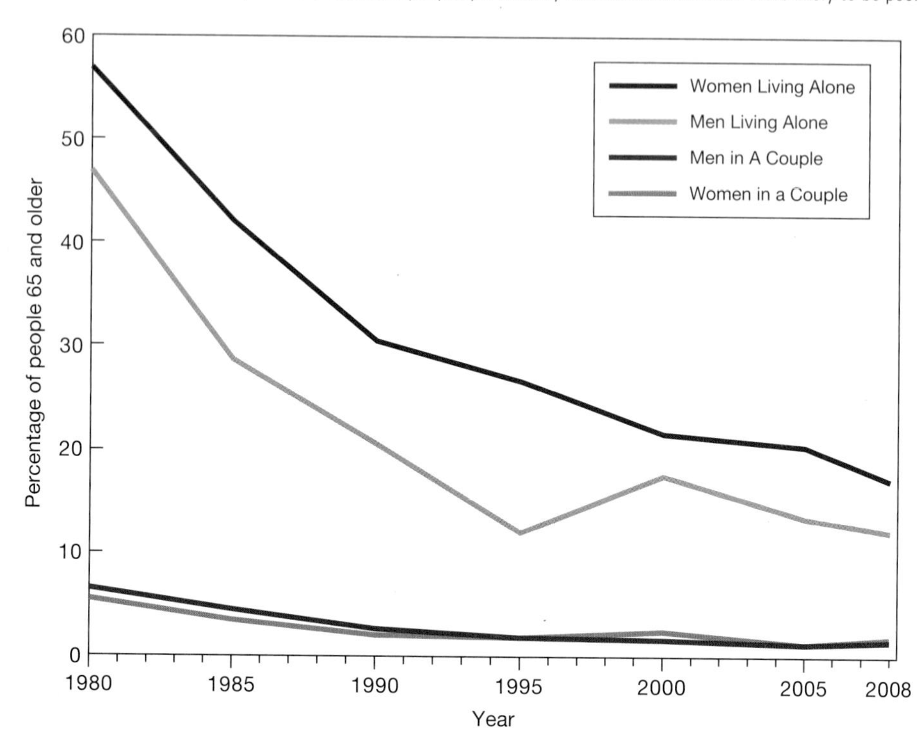

Sources: Statistics Canada, E-Stat CANSIM table 2002-0802; Lindsay, 2002: 108-110; Statistics Canada, 2011: *Women in Canada: A Gender-based Statistical Report*, Table 8, http://www.statcan.gc.ca/pub/89-503-x/89-503-x2010001-eng.htm

> stops ringing. The king is dead. You start wanting to have lunch with old friends. At the beginning, they're nice to you, but then you realize that they're busy, they're working. They've got a job to do and just don't have the time to talk to anybody where it doesn't involve their business . . . You hesitate to call them. (quoted in Terkel, 1996:9–10)

As Culberg describes, retirement moves people out of the mainstream. To ease this transition, researchers suggest that people plan ways of remaining occupied and engaged. Those who are involved with activities, such as volunteer work and hobbies that continue after retirement, are happier and healthier than those who have few retirement interests and withdraw from social life. Many organizations are encouraging older workers to make gradual transitions to retirement by working part time. This allows the organization to retain the skills of senior employees and enables the individual to make a comfortable transition to retirement.

A Statistics Canada survey found that more than half of retirees would have liked to continue working, particularly if they were allowed more flexible working hours and more time off (Schellenberg and Silver, 2004). While the vast majority of seniors do retire by 65, the percentage of seniors who were working has increased significantly since 2000.

People have different reasons for retirement (Park, 2011). Many retire because they have adequate pensions and want to stop working, but a significant proportion retire under less favourable circumstances: some for medical reasons, others because they lose their jobs or because of caregiving obligations. Some of these involuntary retirees have low incomes and are not well prepared for retirement. Many workers return to the labour force after retirement, either because the workers' pensions are not sufficient for a comfortable retirement or to help ease the transition between work and retirement for those nearing the end of their working lives (Hébert and Luong, 2008).

The costs of pensions and healthcare have led to concerns about whether Canada can afford the increasing numbers of retired people. Consider the arguments presented in Box 14.2.

BOX 14.2 POINT/COUNTERPOINT

Will There Be a Generational War Between the Old and the Young?

Seniors have long been considered society's most vulnerable citizens, fragile pensioners on fixed incomes in need of a financial helping hand from both government and agile younger workers. That was true decades ago, but not anymore. Thanks to stock market booms, economic growth, a soaring real estate market, and major expansion in both private and government pension plans, today's seniors are arguably the wealthiest generation in history. The changing fortunes of the elderly have been both swift and profound. In the 1970s, nearly 40 percent of Canadian seniors lived in poverty. Today it's 5 percent, half the poverty rate of the working-age population and one-third the rate of poverty among children.

. . . .[This] dramatic change . . . is one of the greatest policy success stories in Canadian history. Yet there's a dark side to the success, one that threatens to spark an ugly generational crisis, in large part because governments continue to focus so much of their resources on supporting the plight of economically fragile seniors at the expense of their far more fragile children and grandchildren." (McMahon, 2014.)

Is McMahon correct? Will young people be impoverished by their parents and grandparents? Are the fears that the aging baby boomers will reduce the quality of the healthcare system realistic? Will our government pension plans go bankrupt? Let us look at the likely future.

Much of Canada's social welfare system supports the young and the old. An informal social contract between generations obliges the working-age population to support dependent children and older adults. In exchange, these workers received support during their own childhoods and could expect to receive support in their old age.

After World War II, governments in most industrial countries began providing pensions for seniors. However, in the mid-1970s, people began to worry that increasing numbers of elderly and increasing entitlements would cause the collapse of the Old Age Security program (Myles, 1999).

These worries led people, such as McMahon, to suggest that the social contract between generations has been broken and that young people will have to pay far more to support older people than they will ever receive in return. The huge numbers of aging baby boomers, who themselves supported a relatively small number of older people, will have to be supported in their old age by a much smaller number of workers.

The tension between the generations is increased by views, such as that expressed by McMahon at the beginning of this box, that suggest older people are living the good life while younger people struggle economically.

However, you should not be too concerned about the gloomy future predicted by McMahon. Ellen Gee (2000) has described the alarmist view of the costs of aging as "voodoo demography." And several factors should help to alleviate the problem before it turns into a crisis. First, because of declining fertility rates, fewer resources will be required to look after children and it will be easier to direct resources to older people. Second, the government has acted to reduce the future burden of pensions. The 1999 increase in Canada Pension Plan contributions ensures that government pensions will be available for all Canadians in the future. Finally, many older persons are well off and will be paying high levels of tax on their retirement income. Many seniors give money and other support to younger family members and contribute significant amounts of volunteer labour. Older people do not just take from their children and grandchildren.

While medical costs do rise as we get older (Figure 14.6), it is unlikely that the aging baby boomers will cause a crisis in healthcare—the "silver tsunami" predicted by the Canadian Medical Association (2010). A British Columbia study found that aging added only about 1 percent per year to the cost of healthcare—less than other cost drivers such as the increasing prices of pharmaceutical drugs (Morgan and Cunningham, 2011). Community support programs (discussed later in this chapter) can reduce costs. Older people have responded to health promotion efforts focused on diet, exercise, and lifestyle, so their use of the healthcare system should lessen. Finally, research has shown that most of the healthcare costs associated with aging occur in the final six months of life. These costs include keeping the dying patient in an acute-care facility, ordering tests, and using life support to prolong life even when it is clear that the patient is dying. Much of this money is spent without any benefit to the patients (Palangkaraya and Yong, 2009). Many dying patients could be cared for in hospices or in chronic-care hospitals, rather than in more expensive acute-care hospitals.

What do you think about this issue? Are you prepared to pay to support your parents and grandparents? Do you think older people should be viewed as mentors who have valuable experience to pass along or as consumers living the good life on the backs of younger people? How can people be persuaded to act in ways that will reduce their need for healthcare when they are older, for example, by exercising and not smoking?

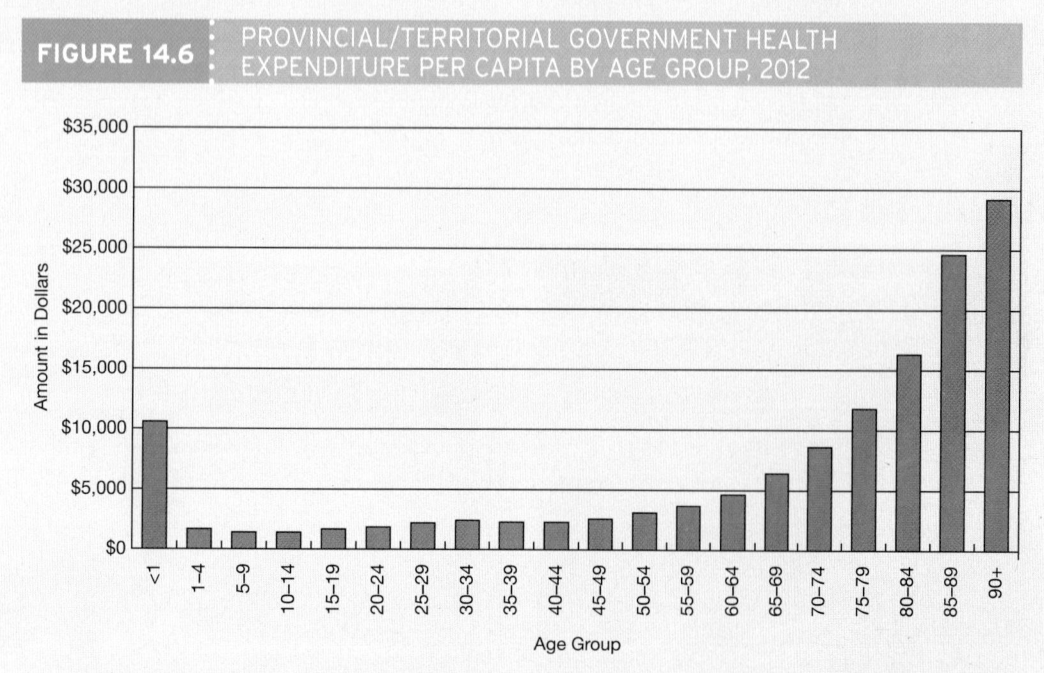

FIGURE 14.6 PROVINCIAL/TERRITORIAL GOVERNMENT HEALTH EXPENDITURE PER CAPITA BY AGE GROUP, 2012

Source: Canadian Institute for Health Information, *Spending and Health Workforce, 2014, National Health Expenditure Trends, 1975 to 2014: Report*, p. 76, http://www.cihi.ca/web/resource/en/nhex_2014_report_en.pdf. Used with permission.

LO-3 INEQUALITIES RELATED TO AGING

You have learned how prejudice and discrimination may be directed toward individuals based on ascribed characteristics—such as ethnicity or gender—over which they have no control. The same holds true for age.

In some respects, older people are treated very well in Canada. Most have adequate incomes—though some do not—and all have access to publicly funded medical care. Many older people feel their biggest problem is that other people have negative views of aging and of the capabilities of older people.

AGEISM

▶ **ageism** Prejudice and discrimination against people on the basis of age.

Stereotypes of older persons reinforce **ageism**—prejudice and discrimination against people on the basis of age (Butler, 1975). Ageism is rooted in the assumption that older people become unattractive, unintelligent, asexual, unemployable, and mentally incompetent as they grow older (Comfort, 1976).

Negative images of older people are common—they are often stereotyped as thinking and moving slowly and as living in the past. They are viewed as cranky, sickly, absent-minded, and lacking in social value (See Figure 14.7).

The media contribute to these negative images of older persons. Older people are underrepresented in the media, and when they are present, they are often shown in a negative light. Advertising often portrays older people as unattractive and incompetent. Stereotypes contribute to the view that women are "old" 10 or 15 years sooner than men. The multibillion-dollar cosmetics industry helps perpetuate the myth that age reduces the "sexual value" of women but increases it for men. Men's sexual value is defined more in terms of personality, intelligence, and earning power than physical appearance. For women, however, sexual attractiveness is based on youthful appearance. By idealizing this "youthful" image of women and playing up the fear of growing older, sponsors sell thousands of products that claim to prevent the "ravages" of aging. Movies and television provide examples of how older male actors represent wealth and virility, while their female co-stars are often half their age. For example, at 37 years of age, Maggie Gyllenhaal was told she was too old to be the romantic partner of a 55 year-old male actor. With notable exceptions such as Meryl Streep and Helen Mirren, fewer female than male actors keep working into their older years.

© Tony Freeman/PhotoEdit

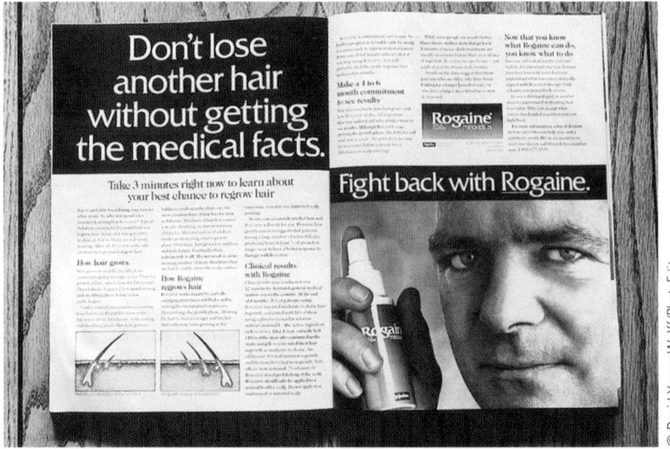

© David Young-Wolff/PhotoEdit

For many years, advertisers have bombarded women with messages about the importance of a youthful appearance. Increasingly, men, too, are being targeted by advertising campaigns that play on fears about the "ravages" of aging.

However, as the population has aged, there has been a change in the media coverage of older persons. Analysis of books, magazines, advertisements, and television shows has shown that media images of older people have begun to become more positive. Features such as a segment on the CBC show *The Mercer Report* showing Mississauga mayor Hazel McCallion, then 88, rushing through her day and bowling, playing hockey, and exercising with host Rick Mercer draw attention to the contributions, talents, and stamina of older persons rather than offering stereotypical and negative portrayals. Because of the increasing consumer power of older persons, this trend will likely continue. Older people will not want to buy from companies that demean them in their advertising.

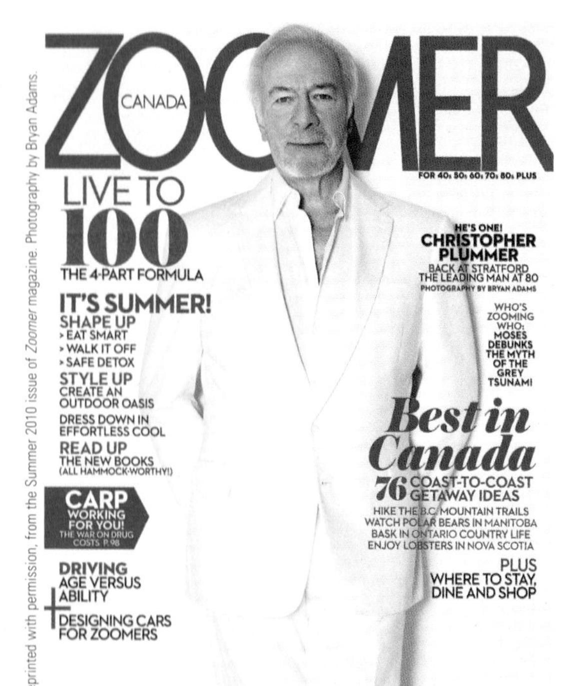

Reprinted with permission, from the Summer 2010 issue of *Zoomer* magazine. Photography by Bryan Adams.

However, stereotyping does continue. TD Canada Trust has run a long series of television commercials featuring two "Grumpy Old Men" who complain about modern conveniences such as bank employees who come to your home to arrange mortgage financing (Laird, 2009). They also appeared in a commercial expressing shock that banks would be open Sunday.

Research has also documented the negative stereotypes many of us hold of "the elderly." Levin (1988) showed photographs of the same man (disguised to appear as ages 25, 52, and 73 in various photos) to a group of college students and asked them to evaluate these apparently different men for employment purposes. Based purely on the photographs, the "73-year-old" was viewed by many of the students as being less competent, less intelligent, and less reliable than the "25-year-old" and the "52-year-old."

At age 27, Patricia Moore, an industrial designer disguised herself as an 85-year-old woman by donning age-appropriate clothing and placing baby oil in her eyes to create the appearance of cataracts. She supplemented the "aging process" with latex wrinkles, stained teeth, and a grey wig. For three years, "Old Pat Moore" went to various locations to see how people responded to her:

> When I did my grocery shopping while in character, I learned quickly that the Old Pat Moore behaved—and was treated—differently from the Young Pat Moore. When I was 85, people were more likely to jockey ahead of me in the checkout line.

FIGURE 14.7 : PERCEPTIONS OF MEDIA PORTRAYALS OF OLDER PERSONS

Seniors in Australia were asked about their perceptions of how older people were portrayed in the media. Most of what they saw was very negative.

Source: Australian Human Rights Commission, 2013, *Fact or Fiction? Stereotypes of Older Australians Research Report 2013.* © Australian Human Rights Commission.

Peter Yang/AUGUST

Stereotypes about appropriate behaviour for older people are increasingly being resisted—though not always as dramatically as demonstrated by this skateboarder.

> And even more interesting, I found that when it happened, I didn't say anything to the offender, as I certainly would at age 27. It seemed somehow, even to me, that it was okay for them to do this to the Old Pat Moore, since they were undoubtedly busier than I was anyway. And further, they apparently thought it was okay, too! After all, little old ladies have plenty of time, don't they? . . . I think perhaps the worst thing about aging may be the overwhelming sense that everything around you is letting you know that you are not terribly important anymore. (Moore with Conn, 1985:75–76)

If we apply our sociological imagination, we find that "Old Pat Moore's" experiences reflect what many older persons already know—it is not their age, but other people's *reactions* to their age that place them at a disadvantage.

These attitudes have consequences. Many older people, particularly women, have negative views about the way others see them and distinguish their outward selves from who they "really" are. A woman of 71 interviewed by Laura Hurd Clarke reported:

> I don't really think about my body much. I know it's there, but I'm still me inside. The outside is sort of a shell. I know some people would look at me and think, "Oh, well, that's an old lady walking her dog." . . . But that isn't me really. I always have that idea that I'm still inside. I'm me inside. And I've always been me. That isn't really me—that's me inside. (2001:448)

Researchers have discovered that physical abilities deteriorated among older people who were exposed to negative images of aging. In the longer term, older persons who held negative images of aging had greater memory loss, worse health, and earlier deaths than comparable people who had more positive views about aging (Kolata, 2006).

Stereotypes of aging can also have an impact on older people who find themselves unemployed. Because of their age, they find it more difficult to find new jobs than their younger colleagues.

WEALTH, POVERTY, AND AGING

Aging is easier for those with money. For those who do not, aging becomes a matter of simply surviving in a society where they do not have sufficient financial resources and where they are devalued because of their age. The consequences of a weak social safety network are shown in Box 14.3.

How have older Canadians as a group fared economically in recent decades? The elderly comprise an extremely diverse group. Some of Canada's wealthiest individuals are old. At the same time, a significant number of this country's older citizens are poor.

To assess the economic situation of older people, it is necessary to address two questions. First, has the economic situation of older Canadians improved? The answer is yes—rather dramatically. The income of people over the age of 65 has improved much more than the income of younger people in the past three decades. According to Lindsay and Almey (2006), the incomes of males over 65 increased by 24 percent and the income of females over 65 increased by 32 percent between 1981 and 2003. This compares with an increase of 2 percent in the incomes of males under 65 years of age.

The second question is whether older Canadians are able to maintain a satisfactory standard of living. The answer to this question is more complex. It is important to remember that

BOX 14.3 SOCIOLOGY IN GLOBAL PERSPECTIVE

Aging in Russia

After the 1991 breakup of the Soviet Union, Russia struggled in its transition from communism to capitalism. Faced with a crumbling economy and an ineffective government, as well as being plagued by corruption and organized crime, Russia struggled to provide for the needs of its citizens. Under the communist regime, the state provided for older persons. While they were not well off by Canadian standards, they had basic medical care and a state pension. Today, the healthcare system has problems and pensions are inadequate.

Political change had a dramatic impact on the structure of the Russian population (see Figure 14.8). As the economic and political structure changed, birth rates dropped and death rates rose (the figure in this box means that for every 120 children born in 2014, there were 140 deaths). Death rates are particularly high for working-age males, many of whom die prematurely because of accidents, alcohol consumption, and coronary problems. This means that males have a very low life expectancy–about 64 years in 2014 compared with 76 years for women.

The difference between birth rates and death rates means that the population of Russia is declining. The population may fall from 140 million to 105 million or less between 2010 and 2050 (Center for Strategic and International Studies). This decline of 25 percent will drop Russia from the fourth most populous country in the world to the 16th largest and will have profound social and economic consequences.

The increasing ratio of old to young persons is an issue in Canada, but is the cause of far greater difficulties for Russia as a declining number of working-age people are supporting a larger proportion of older persons. With a weak economy, it will be difficult to maintain services for older people. Because of oil revenues, rates of poverty have declined over the past decade (although this may change if the dramatic oil price drop in 2015 is sustained), but for many older Russians the situation remains bleak. Russia is 78th of 91 countries ranked in terms of treatment of the elderly by Global AgeWatch (HelpAge International, 2013). By comparison, Canada ranks fifth.

FIGURE 14.8 BIRTH AND DEATH RATES IN RUSSIA, 1979-2014

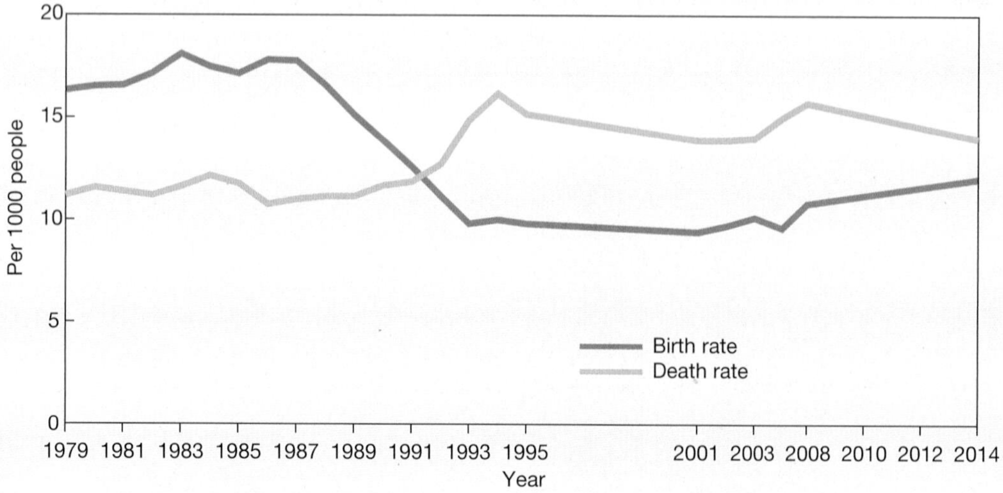

Sources: Index Mundi, 2015; CIA, 2001, 2004, 2008, 2014.

although the economic situation of seniors has improved, many still have low incomes. While income has risen in the past few years for older people in general, certain groups still have incomes below the poverty line in old age. Older people from lower-income backgrounds, people who cannot speak English or French, people with limited education, and people in small towns tend to have low incomes. Very old people and unattached individuals—especially women (see Figure 14.4)—often live at or below the poverty line. Older Indigenous people are more likely than other older Canadians to be poor. Many older Indigenous people also suffer from chronic diseases such as diabetes and live in substandard housing in communities that lack healthcare services.

ELDER ABUSE

▶ **elder abuse** A term used to describe physical abuse, psychological abuse, financial exploitation, and medical abuse or neglect of people age 65 or older.

Abuse and neglect of older persons has received increasing attention due to the rising number of older people and the establishment of more vocal groups to represent their concerns. **Elder abuse** refers to physical abuse, psychological abuse, financial exploitation, and medical abuse or neglect of people age 65 or older (Patterson and Podnieks, 1995).

It is difficult to determine the extent of elder abuse. Many victims are understandably reluctant to talk about it. The 1999 General Social Survey interviewed more than 4000 seniors who were living in private households and asked if they had experienced emotional, financial, physical, or sexual abuse in the previous five years (Dauvergne, 2003). Seven percent of seniors reported having experienced one of these forms of abuse from an adult child, caregiver, or spouse. Emotional abuse was the most common form reported, followed by financial abuse and physical and sexual abuse. Police records showed that those over 65 were more likely to be victimized by a family member than were younger victims. The perpetrators were most likely to be adult children, followed by spouses (Ogrodnik, 2007).

While rates of crime victimization of seniors are lower than for the rest of the population (Perreault and Brennan, 2010), the consequences of victimization may be more serious. For example, seniors may be dependent upon their caregivers and cannot readily escape a dangerous or threatening situation. Also, a young person who is knocked down by a stranger may get up unharmed, while in a similar incident an older person may be seriously hurt.

TIME TO REVIEW

- Why is age such a socially significant characteristic?
- What will the age structure of Canada look like in 2031?
- Describe some of the changes that happen to people as they age.
- How has retirement changed over the past hundred years?
- What are some of the inequities associated with aging?

LO-4 SOCIOLOGICAL PERSPECTIVES ON AGING

age stratification The inequalities, differences, segregation, or conflict between age groups.

Some of the early sociological theories of aging were based on a microlevel analysis of how individuals adapt to changing social roles. More recent theories have used a macrolevel approach to examine the inequalities produced by **age stratification**—the inequalities, differences, segregation, or conflict between age groups—at the societal level.

FUNCTIONALIST PERSPECTIVES ON AGING

Functionalist explanations have examined how older persons adjust to their changing roles. According to Talcott Parsons (1960), the roles of older persons need to be redefined by society. Parsons suggested that devaluing the contributions of older persons is dysfunctional for society; older persons often have knowledge and wisdom to share with younger people.

How does society cope with the disruptions resulting from its members growing older and dying? According to **disengagement theory**, older persons make a normal and healthy adjustment to aging when they detach themselves from their social roles and prepare for their eventual death (Cumming and Henry, 1961). Cumming and Henry (1961) concluded that disengagement can be functional for both the individual and society. Disengagement facilitates a gradual and orderly transfer of statuses and roles from one generation to the next, and the withdrawal of older persons from the workforce provides employment opportunities for younger people. Retirement, then, can be thought of as recognition for years of service and acknowledgment that the person no longer fits into the world of paid work (Williamson, Duffy Rinehart, and Blank, 1992). The younger workers who move into the vacated positions will bring current training and new ideas to their jobs, which should be functional for their organizations.

disengagement theory The proposition that older persons make a normal and healthy adjustment to aging when they detach themselves from their social roles and prepare for their eventual death.

Critics of this perspective disagree with the assumption that all older persons want to disengage while they are still productive and getting satisfaction from their work. Disengagement may be functional for organizations but not for individuals. A corporation that has compulsory retirement may be able to replace higher-paid older workers with lower-paid younger workers, but retirement may hurt some older workers. Contrary to disengagement theory, a number of studies have found that social activity is *more* important as people get older.

Another functionalist theory, the **aging and society paradigm**, says that our lives are shaped by the social institutions through which we pass during our lives (Riley, Foner, and Riley, 1999). These social institutions help define how age is defined: "These broad empirical changes in the place of age in society are reflected in the meanings attached to age . . . In the earlier era [the 1960s] the meaning of age was being redefined and aging became recognized as not entirely fixed or immutable. The theoretical emphasis was shifting from older people as 'disengaged,' a passive burden on society, to older people as active contributors *to* society. Now as the oncoming generation enters a new era [the 2000s] the meanings of age will predictably change again. For the future, a key question is, what will the new meanings be" (1999:328).

aging and society paradigm A functionalist theory that says that our lives are shaped by the social institutions through which we pass during our lives.

These institutional changes mean that those born into different age cohorts will age in different ways—successive cohorts have different opportunities, constraints, and socialization experiences. For example, the generation that was born just after World War II had very different life experiences than those born a generation later. People with high school educations entering the labour force in the 1960s and 1970s had good opportunities for secure, well-paying careers through unionized manufacturing jobs. However, many of these good jobs have moved offshore (Chapter 19), so similarly educated young people in today's work environment will experience aging in very different ways than their predecessors. These age cohorts also influence their social institutions as they age—the size of the baby boom generation meant that it had an enormous impact on Canadian society.

At the time this theory was developed, it had an impact in moving attention away from a strictly biological view of aging. It also showed that historical changes in social institutions were intimately linked with the aging process.

The major weakness of the theory is that it suggests that aging is a homogeneous process. In fact, people experience their societies in very different ways (Novak, Campbell, and Northcott, 2013). People who differ in race, class, and gender will experience aging very differently no matter when they were born.

SYMBOLIC INTERACTIONIST PERSPECTIVES ON AGING

What does growing old mean to an individual? Do all people experience aging in the same way? Are there cultural differences in the aging experience? Symbolic interactionists examine how people deal with aging and at how this experience can vary. For example, a person's culture provides broad ideas of what constitutes aging. Symbolic interactionists show us that the experience of older people in a culture that respects them as sources of wisdom and stability is dramatically different from the experience in a culture in which they are seen as a drain on the resources of younger people. Contrast the earlier discussion of Patricia Moore, who felt diminished and irrelevant when she posed as "Old Pat Moore," with those of older people in Korea whose 60th birthday is celebrated for several days or a Tibetan group whose elders have such status that people exaggerate their ages to "age" more quickly (Tirrito, 2003). Tirrito reports that attitudes toward older people became more negative as societies moved from agrarian to industrial. One reason for this was that people on farms could adjust their workload to match their physical abilities, while owners of factories and large businesses were less likely to make accommodations for older workers (Synge, 1980).

Within their broader cultural context, individuals find different ways of dealing with aging. **Activity theory** states that people tend to shift gears in late middle age and find substitutes for previous statuses, roles, and activities (Havighurst, Neugarten, and Tobin, 1968).

▶ **activity theory** The proposition that people tend to shift gears in late middle age and find substitutes for previous statuses, roles, and activities.

Whether they invest their energies in grandchildren, travelling, hobbies, or new work roles, social activity among retired persons is directly related to longevity, happiness, and health (Palmore, 1981). Consider these differences in the perceptions of people who do and do not remain active:

> The Richardsons came for lunch: friends we hadn't seen for twenty years . . . Helen and Martin had owned and worked together in a very fine women's clothing shop . . . Having some mistaken notion they were getting too old and should retire and "enjoy themselves," they sold the business ten years ago.
>
> During lunch, Larry and I realized we were dealing with two seriously depressed people, in excellent health but with no place to go. When Larry asked Helen what she'd been doing, she replied bitterly, "Who has anything to do?" Martin said sadly he was sorry he gave up tennis ten years ago; if he'd kept it up he could still play . . .
>
> We were embarrassed to indicate we were still so busy that we couldn't see straight. They seemed genuinely shocked that we had no plans to retire at seventy-one and seventy-four. (LeShan, 1994:221–222)

Studies have confirmed that healthy people who remain active have a higher level of life satisfaction than do those who are inactive or in ill health (Havighurst, Neugarten, and Tobin, 1968). Among those whose mental capacities decline later in life, deterioration is most rapid in people who withdraw from social relationships and activities.

Other symbolic interactionist perspectives focus on role and exchange theories. Role theory poses the question: What roles are available for older people? Industrialized, urbanized societies typically do not have roles for older people (Cowgill, 1986). Analysts examining the relationship between ethnicity and aging have found that many older persons are able to find active roles within their own ethnic group. Elders may be esteemed within their racial or ethnic subculture because they provide a rich source of knowledge of traditional lore and history. For example, Ruth Kirk has described the role of Elders in West Coast First Nations culture:

> Native elders are living links to the past. Their vivid memories have the vitality, immediacy and authenticity of those who have experienced the transition from traditional ways to the new. In the short space of two generations, they have gone from travelling the coast in canoes to flying in floatplanes. . . . Not even the social upheaval of losing nine out of every ten people to raging epidemics in the nineteenth century, not even the disorientation of changing to the new cash economy with a more complex technological base, not even the acceptance of a new cosmology and religion, none of these broke native pride in the past or native ties to ancestral lands and waters. This is what elders are about. (1986:8, quoted in Stiegelbauer, 1996)

This important cultural role provides meaning to the lives of older people and helps to link them to younger generations as they carry out their leadership, teaching and mentoring roles.

CONFLICT PERSPECTIVES ON AGING

Conflict theorists view aging as especially difficult in capitalist societies. As people grow older, their power diminishes unless they are able to maintain wealth. Consequently, those who have been disadvantaged in their younger years become even more so in late adulthood. Women's incomes decline significantly following the deaths of their husbands (Statistics Canada, 2006d), and women 75 and over are among the most disadvantaged because, after outliving their spouses, they must often rely solely on government support payments.

Members of racial and ethnic minority groups may also suffer disadvantages. Some do not speak English, and because of racism and geographic isolation, many have been excluded from the broader society. This exclusion can be harmful to older people, as social isolation has been linked with higher mortality among older persons (Steptoe, Shankar, Demakakos, and Wardle, 2013).

Indigenous people who live on reserves often face poverty, poor housing, second-class medical care, and geographic isolation. These living conditions lead to higher rates of disability and more health problems among older Indigenous people than among the rest of the population (Frideres, 1994).

Conflict analysis highlights the diversity in the older population. Differences in social class, gender, and ethnicity divide older people just as they do everyone else. Wealth cannot prevent aging indefinitely, but it can minimize economic hardships. The conflict perspective adds to our understanding of aging by focusing on how capitalism devalues older people, especially women.

The insights of different conflict perspectives have contributed to **cumulative advantage theory**—the notion that the advantages and disadvantages of gender, race, and class accumulate over the life course and have a major impact on aging (Kail, Quadagno, and Keen, 2009). According to O'Rand and Henretta, "Variations in educational achievement, labor force participation and dissolution patterns, pension and asset accumulation, and health trajectories—all of which are cumulative and interrelated over time—produce diverse life chances with unequal outcomes in middle and late life" (1999:6). These accumulated advantages and disadvantages are even transmitted across generations as the rich pass on more wealth to their children and also give them better educations.

cumulative advantage theory
The notion that the advantages and disadvantages of gender, race, and class accumulate over the life course and have a major impact on aging.

Critics of the conflict approach feel that that this approach ignores the fact that industrialization and capitalism have greatly enhanced the longevity and quality of life for many older persons.

FEMINIST PERSPECTIVES ON AGING

Feminist scholars analyze the gender-based inequalities that affect older women:

> One conclusion stands out . . . Poverty in old age is largely a woman's problem, and is becoming more so every year. (National Council of Welfare, cited in Novak, 1993:239)

The poverty rate for elderly women is greater than that for elderly men. *Unattached* elderly women are at the greatest risk of poverty, with a rate that is 42 percent higher than that of unattached elderly men (Milan and Vezina, 2011).

Why do unattached women have such low incomes in old age? Many women who are now 65 or older spent their early adult lives as financial dependants of husbands or as working women

© Strauss/Curtis/CORBIS

© Mark Richard/PhotoEdit

What happens as we grow older? Activity theory assumes that we will find substitutes for our previous roles and activities. Disengagement theory assumes that we will detach ourselves from social roles and prepare for death. Which scenario do you prefer for your future?

trying to support themselves in a culture that did not see women as the heads of households or as sole providers of family income. According to Novak and colleagues (2013), women currently have lower pensions because they were more likely to have left the labour market to raise children, they were more likely to have had part-time jobs, and because they were paid less than men. They also were less likely to have accumulated financial assets during their working lives than their male counterparts had.

As a result of decisions to pay women less and to restrict their eligibility to government pensions, many older women have been forced to rely on inadequate income replacement programs originally designed to treat them as dependants. Furthermore, women tend to marry men who are older than they are, and women live longer than men. Consequently, nearly half of all women over age 65 are widowed and living alone on fixed incomes.

Young women today will be much better equipped to deal with the financial pressures of old age as a result of a number of structural changes in Canadian society. The majority of women are working in the paid labour force, they have begun to enter formerly male-dominated professions, and more of them have private pension plans. At retirement, they will be in much better financial shape than their mothers and grandmothers. This change is already having an impact. Milan and Vezina (2011) report that in 1976, only 15 percent of the income of women over 65 came from employer pension plans. By 2008, this had increased to 54 percent, reflecting changes in patterns of women's employment over that period.

POSTMODERN PERSPECTIVES ON AGING

A central theme of postmodern theory is the notion that identity is a social construct. This implies that we can alter the way our culture thinks about aging—what theorists call our *cultural narratives.* Steven Katz has characterized the postmodern approach to aging as one in which the rigid stages of "childhood, middle age and old age are eroding under pressure from cultural directions that have accompanied profound changes in labor, retirement and the welfare state, and the globalization of Western consumer economies and lifestyles" (1999:3). Katz's analysis emphasizes the role played by marketers of consumer products in portraying an ageless image of older people. Consumers do not want to buy products that are targeted to "old" people, so marketers have developed strategies that mask the dimension of age. At a time when the number of older persons is increasing dramatically, the physical aspects of aging are being concealed (Woodward, 1991). In the postmodern world, nobody should look or act as if they are old.

Technology allows us to change our appearance. No longer must we allow our bodies to visibly age if we choose to surgically mask the aging process. A Toronto plastic surgeon commented on the emerging trend for men to get cosmetic surgery:

> This is a very youth-oriented culture . . . so a lot of men feel they have to look fresh, young and vigorous to get the next promotion at work. And many of these

> same men are divorced or separated and back in the dating game, often competing for much younger women with guys who are frequently 10, 15 or 20 years younger. (Collison, 2006)

Meredith Jones (2004) has compared cosmetic surgery to postmodern architecture. In a world in which anti-aging technologies, such as Botox injections, breast implants, Viagra, and hormone replacement therapy, have been normalized and in which tattoos and piercings are common, the postmodern notion that we can simply choose from a variety of ages and alter that choice at will has currency. Jones illustrates this by quoting Kathryn Morgan's idea of celebrating old age by:

> Bleaching one's hair white and applying wrinkle-inducing "wrinkle-creams," having one's face and breasts surgically pulled down (rather than lifted), and having wrinkles sewn and carved into one's skin. (Morgan, 1991:46, cited in Jones, 2004:99)

In traditional societies, well-defined roles and statuses are attached to older persons, but contemporary society has less certainty. We are not surprised to hear of a 72-year-old person being in a nursing home, nor are we surprised to hear of another 72-year-old running a large company and spending vacations downhill skiing. Retirement at age 65 is no longer mandatory in most provinces, so we no longer have clear signals about when someone should move into a new role as a retired person. This blurring of identity can make it difficult for older people to define themselves.

Negotiating an identity can be particularly hard for lower-income older people who cannot afford to define themselves by their purchases and their travel destinations, and who cannot afford cosmetic surgery and other anti-aging technologies (Polivka, 2000). As the media increasingly portray older people as active and youthful, those who cannot live up to this standard may feel they are to blame for their condition (Hodgetts, Chamberlain, and Bassett, 2003) and feel marginalized. Rozanova analyzed stories about successful aging in the *Globe and Mail* and found that these stories carried consistent messages that "successful aging is an individual choice, and consequently those who do not exercise it are responsible for aging unsuccessfully" (2010:217). As McDaniel has stated, "successful aging is really *not* aging" (quoted in Rozanova, 2010:217). This carries a negative message that those who, by virtue of income, health, or disability, do not fit the criteria of 'successful' aging are to blame for their plight.

More positively, postmodern culture provides older people with the freedom to experiment. Older people no longer have to fit into stereotyped roles, so they can shape their own identities. Through the technology of in vitro fertilization, a Calgary woman of 60 gave birth to twins to fulfill her lifelong dream of motherhood. Affluent seniors can resist traditional aging through pharmaceuticals, cosmetic surgery, technology, and participation in the consumer culture. Older people can keep working but with flexible work times and new paid or volunteer occupations. Some older people now take up activities such as surfing at age 80 or playing tennis at 90. Adopting attitudes such as these may allow seniors to extend active living for many years before biology finally takes its toll—older people can remain creative and engaged with life.

This perspective is very different than the functionalist view of old age as a time for gradual disengagement and preparation for death.

What does the concept of nursing homes imply about the ability of residents to live out their lives with dignity and respect?

CONCEPT SNAPSHOT

FUNCTIONALIST PERSPECTIVES

Key thinkers: Talcott Parsons, Elaine Cumming, William Henry, Matilda Riley

Functionalists look at how people adjust to their changing roles as they grow older. Disengagement theory argues that it is good for society if people detach themselves from their social roles in order to transfer statuses and roles to the next generation. Other functionalists point out that different cohorts experience aging in different ways because of changes in social institutions. In turn, they also affect these social institutions.

SYMBOLIC INTERACTIONIST PERSPECTIVES

Key thinkers: Robert Havighurst, Donald Cowgill

Symbolic interactionists look at how people deal with aging and at how this experience can vary under different circumstances. Older people who live in a culture that values and respects them have a different experience than those who live in a culture in which they are seen as a drain on society's resources.

CONFLICT PERSPECTIVES

As people grow older, their power tends to diminish unless they are able to maintain their health. Those who have been disadvantaged in their younger years become even more disadvantaged in late adulthood. Many conflict theorists feel that aging is particularly problematic in capitalist societies where the people's status and power depends on their wealth. Disadvantages of gender, race, and class accumulate over the life course.

FEMINIST PERSPECTIVES

Key thinkers: Melissa Hardy, Lawrence Hazelrigg

Feminists argue that poverty has become feminized. Because many older women grew up in an era in which women were not treated as men's financial equals, they are much more likely than men to be poor in old age. These women stayed home to raise families or were restricted to low-wage jobs and so they do not have adequate pensions or savings. Because of changes in the role of women, the current generation of women will be much better prepared for retirement than their predecessors.

POSTMODERN PERSPECTIVES

Key thinkers: Stephen Katz, Larry Polivka

A central theme of postmodern theory is the notion that identity is a social construct. Postmodern society has eroded the rigid states of the life course, and older people are now able to experiment with new identities with the assistance of anti-aging technologies, including surgery, hormone replacement therapy, Botox, and fitness programs.

TIME TO REVIEW

- Discuss how disengagement theory explains the aging process.
- According to interactionist theorists, how do different individuals deal with aging?
- How do conflict theorists analyze the aging process?
- According to feminist theorists, what types of discrimination face women as they age?
- How do people construct their identities as they get older?

LO-5 LIVING ARRANGEMENTS FOR OLDER ADULTS

Many older people live alone or in a family setting where care is provided informally by family or friends. Relatives, usually women, provide most of the care. Many women caregivers are employed outside the home; some are still raising a family. The responsibilities of informal caregivers have become more complex. For frail, older persons, for example, family members are often involved in nursing regimes—such as tube feeding—that were previously performed in hospitals.

Only a small percentage of older persons live in nursing homes or other special-care facilities. About 10 percent of older women and 5 percent of older men spend part of their lives in these facilities. As one would expect, the older people are, the more likely it is they will live in institutions. Few people under age 75 are in nursing homes, compared with 35 percent of women and 23 percent of men age 85 and over (Milan, Wong, and Vezina, 2014). Increased social supports and community care have led to a significant decline in rates of institutionalization over the past three decades.

Novak and colleagues (2013) report that living arrangements vary among Canada's different cultural groups. The majority of Chinese Canadians over the age of 55 lived with their children while very few Canadians of European backgrounds chose those living arrangements. Novak and colleagues cite several studies of South Asian immigrant seniors which show that those who immigrate late in life are more likely to be living with their children than those who had arrived in Canada when they were younger. They attribute this both to economic factors—recently arrived seniors often do not have pensions to live on—and to cultural traditions.

Having those older people in the community (referred to as aging in place), where they are cared for by relatives or friends, can pose problems. Most caregivers find this task rewarding, but it does negatively impact many of them (Cranswick, 2003). If seniors are to remain in the community, mechanisms must be found to support the caregivers, many of whom are themselves seniors—often spouses of those needing care (Stobert and Cranswick, 2004).

Caregiving issues exist in many different countries. Between 1980 and 2005, the elderly population of Japan has more than doubled, and people age 65 and over are now 23 percent of the population. The working age population (15–64) has been declining since 1997 and will continue to fall in the future. Because of Japan's declining fertility rate, the total population is projected to fall from 128 million in 2011 to 111 million by 2035, so the proportion of older people will continue to rise.

In the past, older people in Japan were respected and revered; however, recent studies suggest that sociocultural changes and population shifts are producing a gradual change in the social role of the elderly. For example, the proportion of older persons living with a child decreased from about 70 percent in 1980 to 42 percent in 2010. This is high compared to other nations, but is a distinct change for Japanese society. This change can be attributed to contemporary demographic patterns, but it is also related to younger and middle-aged couples working many hours per day and feeling that they have little time to take care of their parents.

Cultural changes are also occurring in China, another country where elders have long been venerated and cared for by their children. As in Japan, several factors are threatening this way of looking at older people. China's elderly population is growing rapidly because of high birth rates in the past. The number of seniors will double between 2015 and 2035, so the country's age distribution will change very rapidly (Wei and Jinju, 2009). China has also rapidly industrialized and urbanized over the past two decades. The new urban factory workers are poorly paid (Chapter 10), and many live in apartments rather than rural houses. These apartments are too small for multi-generational families. Finally, the country's one child per family policy means that many married couples are responsible for both sets of parents. These factors make it difficult to maintain traditional ways of looking after older people, but China has not yet been able to put into place alternative ways of caring for them, such as pension plans. The one child policy was ended in 2015, but its impact will continue to be felt for decades.

SUPPORT SERVICES

Declining family sizes in Canada mean that the baby boomers will likely require higher levels of non-family care than the preceding generation of seniors because there will be fewer children to do this work for their parents. Support services help older individuals cope with the problems in their day-to-day care. For older persons, homemaker services perform basic chores, such as light housecleaning and laundry; other services deliver meals to homes.

Support services and daycare for older persons can be costly, but they are far less expensive than institutional care. Even intensive services that provide the support to allow older persons with serious physical problems to live in their homes are much less expensive than a nursing home. More important, these services allow older persons to live in a familiar environment and retain much of their independence.

A British Columbia program illustrates how careful planning can save money and enhance the quality of life of older persons (Novak, 1997). The Quick Response Team (QRT) was designed to deal with patients who had been admitted to hospital because of medical emergencies, such as broken bones and strokes. After the emergency has been dealt with, the acute-care hospital can do little more for the patient. However, many patients are unable to return home because they cannot care for themselves. The QRT arranges for community support for these patients. These supports can include live-in homemaker services, transportation, home nursing, Meals on Wheels, physiotherapy, and household equipment, such as walkers and bath seats. Because it can cost up to $2000 per day to keep a patient in an acute-care hospital, programs such as the QRT can be a cost-effective way of providing high-quality services to older persons.

NURSING HOMES

nursing home Any institution that offers medical care for chronically ill older people but is not a hospital.

A **nursing home** is defined as any institution that offers medical care for chronically ill older people but is not a hospital (Novak et al., 2013:172). It is the most restrictive environment for older persons.

Why do people live in nursing homes? Many residents have major physical and/or cognitive problems that prevent them from living elsewhere or do not have available caregivers in their family. Women are more likely to enter nursing homes because of their greater life expectancy, higher rates of chronic illness, and higher rates of widowhood. Lower-income seniors are also more likely to stay in institutions because they do not have the financial resources to live on their own (Trottier et al., 2000). Some of the facilities available to the poor do not offer high-quality care.

Some people adjust well to life in a nursing home. For others, the transition to an institutional setting can be stressful. Mortality rates are higher after admission to nursing homes. In part, this is due to the fact that the sickest older people enter these institutions. However, researchers have concluded that institutionalization itself can lower levels of well-being and accelerate mortality (Novak, 1997). Cases of neglect, excessive use of physical restraints, overmedication of patients, and other complaints have occurred in some poorly run homes.

One solution to this problem is to provide higher levels of support so people can remain in the community. However, there will always be people whose physical or mental condition makes institutional care necessary. Gerontologists have found that nursing homes can do many things to reduce the negative effects of institutionalization. They can make life in the institution as much like life outside as possible. They can allow patients as much freedom of choice as possible. They should have programs that allow patients to remain active and maintain their daily, monthly, and yearly rhythms of life (Novak, 1997). It is important for patients to have a normal social life and to keep enough of their own possessions that the institution becomes their home and not just a place where they are forced to stay. The idea of ethnic-focused nursing homes has been very successful, as many older people prefer to live in a facility that serves familiar food and allows them to share lifelong cultural practices with other seniors. People wait as long as a decade to get into facilities such as Mississauga's Yee Hong Centre for Geriatric Care (Bascaramurty, 2012b).

Beverly Suek (2009:n.p.) describes some of the problems with nursing homes and presents an alternative approach based on a community of women who would care for each other as they aged:

> . . . . my mother went into long-term care. She remarried at 80 and had six glorious years with her new partner—then she had a stroke. She needed physical care. The choice was between a private facility where both could live, for $6,000 a month, and subsidized care, where she would be separated from the love of her life. They didn't have $6,000 a month, so he lived in one place and she lived in another.

While we're told that we live in a society that supports families, we divide them in the way we set up long-term care.

> Not only were they separated, but my mother was locked in the building. Many of the home's residents were living with dementia, so none were permitted to leave unless they were accompanied by an adult—which, it was assumed, they weren't. This is what is available in the real world, I realized. And if heterosexual couples are having trouble staying together, I wondered what it would be like for women in lesbian relationships. So, about three years ago, a group of us got together to discuss making things different. Our group is diverse: Native women, immigrant women, women with family members with disabilities, lesbians, and heterosexual women. We call ourselves the Committee for Retirement Alternatives for Women. . . .
>
> As senior housing currently exists, many retirement complexes and long-term care facilities are in isolated locations, far from the communities where people have connections. In the last years of life, you get moved as soon as you lose bladder control or can't feed yourself. Additionally, particularly in long-term care, residents have little control over their environment. Imagine: You're 86 and have been independent all your life, but you can't sleep in till noon and stay up till midnight playing cards with friends. You might even have to sneak that glass of wine into your room.
>
> With these concerns in mind, our committee decided to explore alternatives for affordable housing sensitive to an active, diverse female community. We agreed that the housing would be built on the principles of respect, diversity, and community. We see ourselves helping one another, creating a community where we care for and about one another to the best of our abilities, where we don't ship off some women because they need greater care.*

DEATH AND DYING

Historically, death has been a common occurrence at all stages of the life course. Until the 20th century, infant and child mortality rates were very high, and poor nutrition, infectious diseases, and accidents took the lives of men and women of all ages. In contemporary industrial societies, however, death is unfamiliar to many of us because it has largely been removed from everyday life. Most deaths now occur among older persons and in institutional settings. In the past, explanations for death and dying were rooted in custom or religious beliefs; today, they have been replaced by medical and legal explanations and definitions, and ongoing medical and legal battles.

Several controversial issues relate to the right to die with dignity. Should guardians of persons in permanent vegetative states have the legal right to refuse medical treatment? Should family members be able to overrule doctors' decisions to cease treatment for these patients? Should individuals suffering from an incurable terminal illness have the right to decide when their life should end?

In 1993, the Supreme Court of Canada considered the question of the right to die in the case of Sue Rodriguez, who suffered from Lou Gehrig's disease. The Court decided that the right to life, liberty, and security of the person (as outlined in the *Charter of Rights and Freedoms*) does not include the right to take action to end one's life. Furthermore, the Court decided that prohibition of physician-assisted suicide did not constitute cruel and unusual treatment. However, four of the nine justices dissented from the decision. Justice Beverley McLachlin wrote:

> The denial to Sue Rodriguez of a choice available to others cannot be justified. Such a denial deprived Sue Rodriguez of her security of the person (the right to make decisions concerning her own body which affect only her own body) in a way that offended the principles of fundamental justice. (quoted in Bolton, 1995:391)

* © Beverly Suek, "Raising the Roof on Senior Women's Housing," *Herizons* magazine Summer 2009. www.herizons.ca.

After the Supreme Court declined her challenge for a legal physician-assisted suicide, Rodriguez took her own life with the help of an anonymous doctor. The *Criminal Code* specifies that anyone who counsels or "aids and abets" a suicide is guilty of an indictable offence and subject to 14 years in prison (McGovern, 1995). The patient's consent cannot be used as a defence.

In 2012, the British Columbia Supreme Court found this law unconstitutional on the grounds that it discriminated against people with a physical disability. The ruling was appealed by the federal government, but in 2015 the law was unanimously struck down by the Supreme Court of Canada. The Court gave the government one year to draft new legislation.

A related issue concerns the legality of a doctor's hastening the death of a terminally ill patient without the consent of the family. Ethical guidelines allow doctors to remove life support with the consent of the patient or the patient's family if the patient is incapable of responding. After life support is removed, doctors may administer painkilling drugs to make the patient comfortable even if these drugs may hasten death. In 2013, the Supreme Court of Canada decided the case of Hassan Rasouli. Rasouli was placed on life support after suffering an infection following a brain operation in 2010. Doctors sought to remove Rasouli's life support despite his family's desire to keep him alive, and his family challenged this decision. In 2013, the Supreme Court of Canada ruled that life support could not be withdrawn without the consent of the family or of the Ontario Consent and Capacity Board, which has the power to act in the best interests of incapacitated patients.

Many people have chosen to have a say in how their own lives might end by signing a *living will*—a document stating their wishes about the medical circumstances under which their life should be terminated. Most provinces recognize living wills.

Many issues pertaining to the quality of life and to death with dignity may remain unresolved, but the debate about the right to die will continue. The number of deaths in Canada will increase from about 252,000 per year in 2011 to 500,000 per year in 2039 as the lives of the baby boomers end (Kettle, 1998; Statistics Canada, 2011e).

AGING IN THE FUTURE

How long will you live? Will average lifespans ever extend to 100 years or more? While life expectancy has increased dramatically over the past century, the maximum age has not increased. However, innovative, well-financed organizations such as Google have set up companies whose goal is extending life expectancy. Craig Venter, one of the pioneers of genetic sequencing, has set up a company called Human Longevity, which is targeting aging through genetic research. While there are currently no signs of dramatic breakthroughs in maximum ages, there is at least a possibility that this may occur during the lifespans of most current students.

What would be the consequences of radical life extension? If people routinely lived until age 100, how could society finance 35 years of retirement for everyone? If companies were able to develop biomedical interventions that would extend lifespans, who would pay the cost of these interventions? Would only the rich have access to them?

Dominic Bracco II/The Washington Post/Getty Images

Some older people may find that even a robot companion like Paro can ease their loneliness.

Whether or not scientists ever achieve radical life extension, the number of older people in Canada will continue to increase for the foreseeable future. One trend that will benefit older people is that advances in technology and design will alleviate some of the difficulties that come with aging. Just as eyeglasses, heart pacemakers, and accessible buses have improved life for older persons in the past, computers, robots, and medical advancements will help in the future.

Even simple changes, such as buttons that can be done up with arthritic fingers, can improve the ability of older people to care for themselves, but much more is being done. Robots have begun to take over some elder care duties. Homes can be equipped with sensors that will monitor the

health and activity of older persons. Household robots, such as the vacuum-cleaning Roomba, are taking over some household chores. Medical visits and even some medical tests can be conducted remotely through telemedicine technology. Seniors with disabilities such as visual impairment or mobility problems may benefit from self-driving cars, if this technology proves to be viable. Many older people are lonely and some nursing homes are providing robots as companions, particularly for patients with dementia. One robot companion that has become popular is an electronic seal named Paro, which was designed in Japan but which is used in many different countries.

One of the most positive consequences of population aging is that there will be considerably less age segregation. When the elderly comprise one-quarter of the population, there will be much more interaction between individuals of all age groups. The end of mandatory retirement means that workplaces will have more older people in both full-time and part-time roles. This mixing of generations should help lead to decreases in both ageism and negative stereotypes of old age.

TIME TO REVIEW

- Discuss the living arrangements of most older people.
- Why are community support services so important for older people?
- What are some of the ways in which nursing homes can improve the quality of life for their residents?
- How is the process of dying changing? What are some of the legal issues involved in the dying process?
- How will aging change in the future?

VISUAL SUMMARY

14

KEY TERMS

activity theory The proposition that people tend to shift gears in late middle age and find substitutes for previous statuses, roles, and activities (p. 392).

ageism Prejudice and discrimination against people on the basis of age, (p. 386).

age stratification The inequalities, differences, segregation, or conflict between age groups (p. 390).

aging The physical, psychological, and social processes associated with growing older (p. 377).

aging and society paradigm A functionalist theory that says that our lives are shaped by the social institutions through which we pass during our lives (p. 391).

chronological age A person's age based on date of birth (p. 378).

cohort A category of people born within a specified period in time or who share some specified characteristic (p. 378).

cumulative advantage theory The notion that the advantages and disadvantages of gender, race, and class accumulate over the life course and have a major impact on aging (p. 393).

disengagement theory The proposition that older persons make a normal and healthy adjustment to aging when they detach themselves from their social roles and prepare for their eventual death (p. 391).

elder abuse A term used to describe physical abuse, psychological abuse, financial exploitation, and medical abuse or neglect of people age 65 or older (p. 390).

LO-1 Explain how functional age differs from chronological age.

Chronological age is a person's age based on their date of birth. Functional age is the age a person appears to be based on observable individual attributes, such as physical appearance, mobility, strength, coordination, and mental capacity, that are used to assign people to age categories.

The Canadian Press/AP Photo/Rick Rycroft

LO-2 Discuss how views of aging differ in preindustrial and industrialized societies.

In preindustrial societies, people of all ages are expected to share the work, and the contributions of older people are valued. In industrialized societies, however, older people are often expected to retire so that younger people can take their place.

Peter Yang/AUGUST

LO-3 Explain ageism and elder abuse.

Ageism is prejudice and discrimination against people on the basis of age, particularly against older persons. Elder abuse includes physical abuse, psychological abuse, financial exploitation, and medical abuse or neglect of people aged 65 or older. Passive neglect is the most common form of abuse.

© Mark Richard/PhotoEdit

LO-4 Understand the differences between the functionalist, symbolic interactionist, conflict, feminist, and postmodern explanations of aging.

Functionalist explanations focus on how older persons adjust to their changing roles in society; gradual transfer of statuses and roles from one generation to the next is necessary for the functioning of society. Interactionist theorists show us that the experience of older people in a culture that respects them as sources of wisdom and stability is dramatically different from the experience in a culture in which they are seen as a drain on the resources of younger people. Conflict theorists link the loss of status and power experienced by many older

© Strauss/Curtis/CORBIS

persons to their lack of ability to produce and maintain wealth in a capitalist economy. Feminist scholars have analyzed the gender-based inequalities that affect older women, who are much more likely than men to be poor in old age and who also face other forms of discrimination based on their physical appearance. Postmodern theorists have shown that because identity is a social construct, people are actively trying to avoid being categorized as "old" through means such as plastic surgery, exercise, and staying active in the labour market.

LO-5 Identify the most common living arrangements for older people.

Many older persons live alone or in an informal family setting. Support services and daycare help older individuals who are frail or disabled cope with their day-to-day needs, although many older people do not have the financial means to pay for these services. Nursing homes are the most restrictive environment for older persons. Many nursing home residents have major physical and/or cognitive problems that prevent them from living in any other setting, or they do not have caregivers available in their family.

© Chuck Savage/Corbis

functional age A term used to describe observable individual attributes, such as physical appearance, mobility, strength, coordination, and mental capacity, that are used to assign people to age categories (p. 378).

gerontology The study of aging and older people (p. 381).

life expectancy The average length of time a group of individuals of the same age will live (p. 378).

nursing home Any institution that offers medical care for chronically ill older people but is not a hospital (p. 398).

APPLICATION QUESTIONS

1. Consider your grandparents (or other older persons you know well, or yourself if you are older) in terms of disengagement theory and activity theory. Which theory seems to provide the most insight? Why?
2. Pay close attention to the commercials the next time you spend an evening watching television. How do these commercials portray older persons? What types of products target their advertisements toward older people?
3. Many people are trying to achieve radical life extension. What would the consequences be of achieving this goal if the incidence of dementia was not also changed?

CHAPTER 15

HEALTH, HEALTHCARE, AND DISABILITY

De Visu/Shutterstock.com

CHAPTER FOCUS QUESTION

What effect has HIV/AIDS had on the health of the global population?

LEARNING OBJECTIVES

AFER READING THIS CHAPTER, YOU SHOULD BE ABLE TO

LO-1 Understand the ways in which sociological factors influence health and disease.

LO-2 Compare how functionalist, symbolic interactionist, conflict, feminist, and postmodern theories differ in their analyses of health.

LO-3 Understand how social determinants such as social class, sex, race, and age affect health and healthcare.

LO-4 Identify the consequences of disability.

LO-5 Discuss the state of the healthcare system in Canada today and explain how it could be improved.

Rae Lewis-Thornton describes her experience with HIV/AIDS in the following way:

The day I found out [I was HIV-positive] I was so calm . . . I walked out of the . . . Red Cross office and into the . . . sunshine, flagged a cab and went back to work. I worked late that night . . . I was 24. I'd just been given a death sentence . . . I'm young . . . Well educated. Professional. Attractive. Smart. I've been drug and alcohol-free all my life. I'm a Christian. I've never been promiscuous. Never had a one-night stand. And I am dying of AIDS.

I've been living with the disease for nine years, and people still tell me that I am too pretty and intelligent to have AIDS. But I do. I discovered I was HIV-positive when I tried to give blood at the office. I have no idea who infected me or when it happened. Still, there is one thing I am absolutely certain of; I am dying now because I had one sexual partner too many. And I'm here to tell you one is all it takes. (Lewis-Thornton, 1994:63)

Source: Lewis-Thornton, 1994:63.

AIDS has taken a huge toll on individuals, families, cities, and nations. The disease known as AIDS (acquired immune deficiency syndrome) is caused by HIV, the human immunodeficiency virus, which gradually destroys the immune system by attacking the white blood cells, making the person with HIV more vulnerable to other types of illnesses. The United Nations estimated that 35 million people were infected with HIV/AIDS and 1.5 million people died of AIDS in 2013 (UNAIDS, 2014). In Canada, 71,000 people were living with HIV in 2011 (Public Health Agency of Canada, 2014). Map 15.1 outlines the global distribution of the virus.

After a massive global effort, progress has been made in the fight against AIDS. Because of antiretroviral therapy, the number of AIDS deaths has declined from 2.4 million in 2005 to 1.5 million in 2013, and the number of new infections declined from 3.4 million in 2001 to 2.1 million in 2013 (UNAIDS, 2014).

HIV/AIDS has nonetheless had a devastating impact in sub-Saharan Africa, which has 71 percent of all AIDS cases and 74 percent of AIDS deaths (UNAIDS, 2014). Many of the new infections in Africa are among people aged 15 to 24, and many newborns are infected by their mothers, although better medical care has reduced this number significantly in the past decade. Thus the disease is destroying much of Africa's future. The average life expectancy in some countries there has dropped by as much as 17 years, and the cost of providing even minimal treatment for the disease is taking away many of the hard-won economic gains of some countries.

The problem of AIDS illustrates how sociology can be applied to what, at first glance, appears to be a purely medical matter:

AIDS demonstrates that disease not only affects health, but one's definition of self, relations with others, and behaviours. As well, AIDS has had a significant impact on social institutions. The healthcare system has been most directly affected, requiring assessments of the adequacy of research, treatment modalities, and health care facilities. Legislators have wrestled with issues of privacy and human rights protections for people with AIDS. AIDS has resulted in social and sexual mores and lifestyles being reassessed. (Grant, 1993:395)

Social factors are not just important to an understanding of HIV/AIDS. More broadly, health is strongly linked to social factors. Your age, gender, race and ethnicity, and risk-taking behaviour all are *social determinants* that affect your health. And population health has an impact on society, including the economic impact of healthcare and the ability of healthy populations to build productive societies.

Before reading on, test your knowledge about HIV/AIDS by taking the quiz in Box 15.1.

CRITICAL THINKING QUESTIONS

1. Look at the global distribution of HIV-positive people in Map 15.1. What does this map tell you about the social and cultural factors involved in this disease?
2. What is the role of alternative therapies such as naturopathy and acupuncture in healthcare? Have you or your friends or relatives made use of alternative treatments?
3. Lifelong health is partly determined by habits in diet, exercise, smoking, and drug and alcohol use developed by young people. Are you aware of any efforts made by governments and medical practitioners to encourage you to develop healthy habits? Have these efforts led you to change your behaviour? If not, how could this be done more effectively?

MAP 15.1 NUMBER OF HIV-POSITIVE PEOPLE AROUND THE WORLD IN 2013

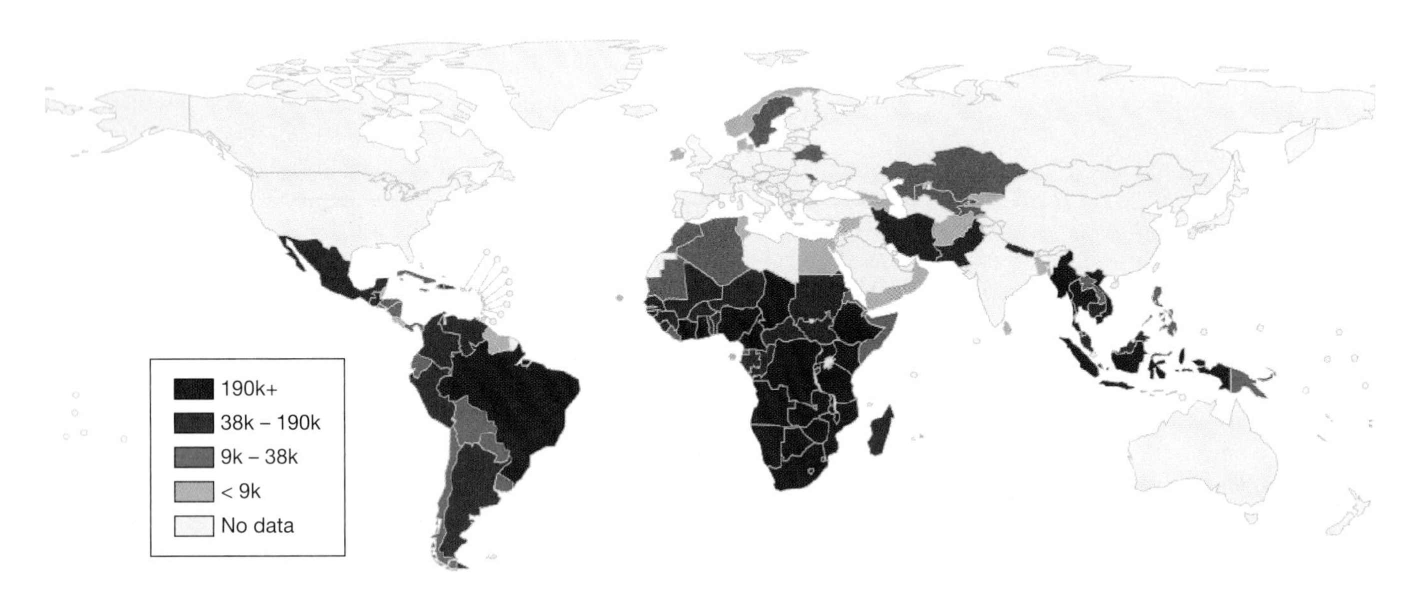

Source: UNAIDS, 2015. http://aidsinfo.unaids.org/

BOX 15.1 SOCIOLOGY AND EVERYDAY LIFE

How Much Do You Know About HIV/AIDS?

True	False	
T	F	1. Worldwide, most people with AIDS are gay men.
T	F	2. In Canada, you can be sent to prison if you knowingly transmit HIV.
T	F	3. HIV is spreading rapidly among women in some nations.
T	F	4. Young people are particularly vulnerable to HIV.
T	F	5. One of the major concerns of AIDS activists is reducing the stigmatization of HIV/AIDS victims.

For answers to the quiz about HIV/AIDS, go to **www.nelson.com/student.**

LO-1 HEALTH AND MEDICINE

► **health** The state of complete physical, mental, and social well-being.

What does 'health' mean to you? If you were asked if you are healthy, how would you respond? Health was once considered to be simply the absence of disease. The World Health Organization (WHO) provides a more inclusive definition of **health**, calling it the state of complete physical, mental, and social well-being. This definition of health has several dimensions; physical, social, and psychological factors are all important. Health does not depend solely on the absence of disease or sickness. Health is socially defined and therefore varies over time and between cultures (Farley, 1992). For example, in our society, obesity is viewed as unhealthy, while in other times and places it has signalled prosperity and good health.

Medicine is an institutionalized system for the scientific diagnosis, treatment, and prevention of illness. Medicine is a vital part of the broader concept of **healthcare**, which is any activity intended to improve health. In North America, medicine is typically used when there is a failure in health. When people get sick, they seek medical attention to make them healthy again. More emphasis is now being placed on the field of **preventive medicine**—medicine that emphasizes a healthy lifestyle in order to prevent poor health before it occurs.

medicine An institutionalized system for the scientific diagnosis, treatment, and prevention of illness.

healthcare Any activity intended to improve health.

preventive medicine Medicine that emphasizes a healthy lifestyle in order to prevent poor health before it occurs.

LO-2 SOCIOLOGICAL PERSPECTIVES ON HEALTH AND MEDICINE

FUNCTIONALIST PERSPECTIVES ON HEALTH: THE SICK ROLE

For functionalists, people are normally healthy and contribute to their society. Talcott Parsons (1951) viewed illness as dysfunctional both for the individual who is sick and for the larger society. Sick people may be unable to fulfill their social roles, such as working or looking after children. Thus illness can disrupt the social system. Societies must therefore establish definitions of who is legitimately sick.

According to Parsons, all societies have a **sick role**—patterns of behaviour defined as appropriate for people who are sick. The characteristics of the sick role are:

sick role Patterns of behaviour defined as appropriate for people who are sick.

1. The sick person is temporarily exempt from normal social responsibilities. When you are sick, you are not expected to go to work or school.
2. The sick person is not responsible for his or her condition, so individuals should not be blamed or punished.
3. The sick person must want to get well. The person who does not do everything possible to return to a healthy state is no longer a legitimately sick person and may be considered a hypochondriac or a malingerer.
4. The sick person should seek competent help and cooperate with healthcare practitioners to hasten his or her recovery.
5. Physicians are the "gatekeepers" who maintain society's control over people who enter the sick role—that is why you need a note from a doctor if you miss an exam or a shift at work because of illness.

Italian School/Getty Images

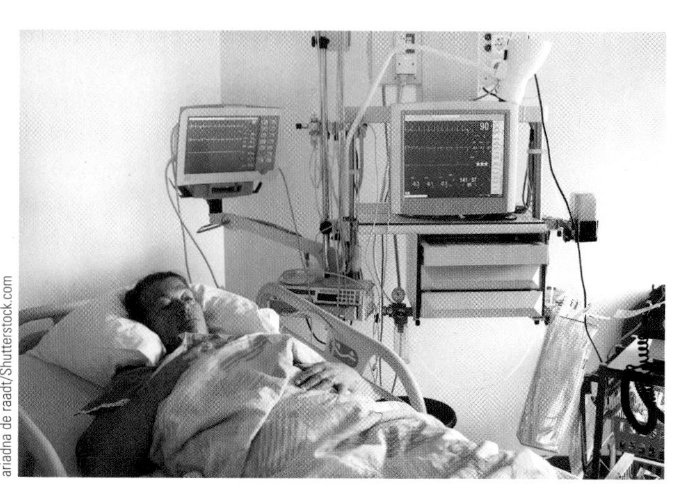

ariadna de raadt/Shutterstock.com

The treatment of the sick, and indeed the way in which we think about sickness, depends upon time and place. Contrast the illustration of the medieval patient surrounded by family and various healers with the modern patient, who is constantly monitored by technology.

Critics of this view of health and illness argue that it places too much responsibility for illness upon the sick people, neglecting the fact that the actions of other people may be the cause of someone's illness. For example, a child may be born with fetal alcohol spectrum disorder (FASD) because the mother consumed alcohol while pregnant. Many people living in poverty become sick because of inadequate food and shelter. Also, contrary to the functionalist view, individuals may be blamed for their illness, as people who contract HIV or lung cancer often are.

CONFLICT THEORIES: INEQUALITIES IN HEALTH AND HEALTHCARE

The conflict approach emphasizes the political, economic, and social forces that affect health and the healthcare system and the inequities that result from these forces. Among the issues of concern for conflict theorists are the ability of all citizens to obtain healthcare; the impact of race, class, and gender on health and healthcare; the relative power of doctors compared with other health workers; and the dominance of the medical model of healthcare.

We will consider several of these issues later in this chapter because they are important enough to consider in depth. For now, the role of conflict in the health field can be illustrated by the debate over the allocation of money for research and treatment for different diseases. Advocates for different diseases compete for funds, as money spent doing research on cancer cannot be spent on heart disease. There is also conflict among those who take different approaches to research and treatment of a particular disease. Should funds be spent on treatment or prevention? Should nontraditional treatment methods be studied, or is the medical model the only legitimate way of responding to disease?

Groups representing victims of particular diseases have lobbied governments to give their problem more funding. Thus the priority given to research, prevention, and medical care for particular types of diseases may reflect the power of lobby groups as well as the seriousness of the problem. AIDS activists have been particularly successful in having their concerns reflected in policy. Breast cancer advocates have also been highly successful, although some of their funding methods have been controversial (see Box 15.2).

FEMINIST PERSPECTIVES ON HEALTH AND ILLNESS

Feminist scholars have studied many different aspects of health and healthcare. One of the first problems they identified was that most medical research was centred on males and ignored diseases that primarily affected females. Other feminist researchers have studied the discrimination against women working in a healthcare system that has traditionally been dominated by male doctors. Women were relegated to subservient roles and have lacked access to leadership roles within the system. Some of the most important work done by feminist scholars has looked at the ways in which medicalization has affected women.

▶ **medicalization** The process whereby an object or a condition becomes defined by society as a physical or psychological illness.

THE MEDICALIZATION OF WOMEN'S LIVES **Medicalization** refers to the process whereby an object or a condition becomes defined by society as a physical or psychological illness. Medicalization has been a focus of feminist researchers because women's health issues, including those pertaining to childbirth, menopause, PMS, and contraception, have been particularly susceptible to medicalization, and this has not necessarily served the interests of women.

Historically, women's health needs, including pregnancy and childbirth, were looked after by other women in their communities (Findlay and Miller, 2002). The era of women looking after women ended when the male profession of medicine successfully challenged midwives and other traditional health practitioners and claimed exclusive jurisdiction over conditions such as childbirth and menopause, which were redefined as medical problems. Medical doctors won this struggle despite the fact that traditional practices often had more favourable outcomes than those of the new profession of medicine. Healing became "men's work." While this development

BOX 15.2 POINT/COUNTERPOINT

Pink Ribbons, Inc.: The Corporatization of Medical Charities

Medical charities compete for the billions of dollars Canadians donate each year. The breast cancer movement has been particularly successful in raising funds. Most of us know someone who has walked, run, or paddled a dragon boat to raise funds for breast cancer, and we have all seen the pink ribbon used in campaigns run by the Canadian Breast Cancer Foundation (Kedrowski, 2010).

The breast cancer movement has been very successful in breaking down the negative stereotypes and shame that used to be attached to breast cancer. They have also provided a way for family and friends to show their support for cancer victims—most of the largest breast cancer charities were founded by relatives of women who died of the disease. Thousands of women have found friendship and support through the breast cancer movement and have also received enormous satisfaction from helping other breast cancer victims.

Despite these successes, the movement has its critics. In 2012, Canada's National Film Board produced the documentary *Pink Ribbons, Inc.*, based on a book by Samantha King. According to King, during the early years of the breast cancer movement in the 1990s, it was focused on broader women's health issues as well as on breast cancer. Now, the emphasis has changed to "the individual breast cancer patient and her participation in uplifting and profit-generating activities to fund high-stakes medical research to the virtual exclusion of other considerations" (2006:113).

King is particularly critical of the way in which the breast cancer movement has partnered with corporations, many of which use this sponsorship as a marketing tool. Rather than using philanthropy as a way of benefiting society, some corporations use it to increase profits through 'cause marketing.' This means that corporations tie themselves to a specific cause in order to "build the reputation of a brand, increase profit, develop employee loyalty to the company, and add to their reputation as good corporate citizens" (King, 2006:9). Many of these sponsorships last for years, so the corporate brand becomes strongly linked with the charity. Some corporations choose to provide support by donating a portion of the money received for each product sold. However, in some cases the amount of money has been very small compared to the cost of the product ($1 for each $200 Eureka vacuum cleaner sold), and in others the product was inappropriate (pink Smith and Wesson handguns and pink shotgun ammunition in the United States).

Tim Boyle/Bloomberg/Getty Images

Some women with life-threatening breast cancer have complained about the upbeat message that everyone can beat cancer if they have the proper attitude and the will to beat the disease. They argue that cancer is not feminine and normal, as many of the campaigns imply, but a terrible disease that kills 5000 Canadian women each year (Canadian Cancer Society, 2015). Not everyone does survive breast cancer, no matter how positive their attitudes might be. Other critics believe that the increase in breast cancer rates over the past few decades is caused by environmental factors such as pollution and the use of carcinogenic chemicals in products, including cosmetics. If cosmetic companies such as Avon and Revlon are important sponsors, projects that ask questions about these environmental issues will not be funded.

The breast cancer movement has made some changes in response to these criticisms. Recently, the Susan G. Komen Foundation, the largest breast cancer organization in the United States, has sponsored a major study on environmental factors and may be moving toward a more preventive approach.

This debate leaves many questions for you to consider. Should we rely on mainstream medical science to find a cure for cancer, or should we look more broadly at environmental and social factors? Do 'awareness' campaigns actually help to reduce disease and illness? What role do you think corporations should play in funding health-related charities? Should corporations use health-funding campaigns to market their products? Does dependence upon corporate donors restrict the questions that foundation-funded researchers are allowed to ask? Should people who wish to fight against disease focus on specific campaigns against problems such as heart disease or breast cancer, or work more broadly for better health policy?

helped to raise the status of the medical profession, it reduced women's control over their own bodies. Medicalization had a profound effect on childbirth, child rearing, and mothering.

Feminist researchers have also questioned the role of medicine in shaping the ways in which women view their physical appearance: "Medical rhetoric itself acts to exacerbate the already powerful cultural demands on women to overemphasize their bodily appearance" (Findlay and Miller, 2002:197).

A paper by the American Society for Plastic and Reconstructive Surgery offers a rather extreme example of this medical rhetoric. This society, the major professional organization representing plastic surgeons, wanted the U.S. government to loosen its restrictions on the use of breast implants. The society based its case on the view that having small breasts constituted a disease. It alleged that this disease (called *micromastia*) resulted in "feelings of inadequacy, lack of self-confidence, distortion of body image, and a total lack of well-being due to a lack of self-perceived femininity" (cited in Weitz, 1996:123). Of course, this "disease" could be cured if the victims received expensive, potentially dangerous breast implants. It is not difficult to imagine the harm that the plastic surgeons' lobbying effort encouraging women to think of their biologically normal bodies as "diseased" might have on some women's self-images. For example, reports of a Penticton, B.C., contest in which 36 women competed for a chance to win breast implants said that many of the losers felt they had "lost a chance at gaining self-confidence" (Carmichael, 2005). The fact that breast size was still a measure of these young women's sense of self-worth, and that many other aspects of physical appearance have come under medical control, is a clear indication of the degree to which women's appearance has become medicalized.

Findlay and Miller feel that the medicalization of women's lives individualizes and depoliticizes their problems. Medicalization also forces women to conform to traditional social norms and "limits women's options—in behaviour, in appearance, and in relationships" (2002:201). There have been incremental changes that may reduce some of the negative effects of medicalization while retaining the benefits of the modern medical system. Women have made headway in reestablishing midwifery and natural childbirth methods that return some of the control over the birthing process to the mother. They have also forced the medical profession to share more information and to empower clients in other ways.

SYMBOLIC INTERACTIONIST THEORIES: THE SOCIAL CONSTRUCTION OF ILLNESS

Interactionists try to understand the meanings and causes that we attribute to particular events. They focus on how the meaning that social actors give their illness or disease affects their self-concept and their relationships with others. The interactionist approach is illustrated by society's response to AIDS.

We often try to explain disease by blaming it on those who are ill. This practice reduces the uncertainty of those of us who fear the disease. Non-smokers who learn that a cancer victim had a two-pack-a-day smoking habit feel comforted that the guilty have been punished and that the same fate is unlikely to befall them. Because of the association of their disease with promiscuous homosexuality and intravenous drug use, victims of AIDS have particularly suffered from blame. How is a person's self-concept affected when he or she is diagnosed with AIDS? How does this diagnosis affect how the person relates with others in his or her social world?

In the case of AIDS, the social definition of the illness can have as profound an impact on the AIDS patient as the medical symptoms. AIDS is an example of illness as stigma (Giddens, 1996). A *stigma* is any physical or social attribute or sign that so devalues a person's social identity that it disqualifies that person from social acceptance. While other illnesses may provoke sympathy or compassion, an illness such as AIDS is perceived by some people as dishonourable or shameful, and sufferers are rejected by the healthy population. Children with AIDS have been driven from their schools, homes of people with AIDS have been burned, employees have been fired, and medical professionals have refused treatment to AIDS patients. These events have happened despite the fact that AIDS cannot be transmitted by casual everyday contact. However, the social definition of an illness is not always based on medical fact. The incidents of hostility and discrimination directed at individuals with AIDS have a profound impact on their self-concept, social relationships, and ability to cope with the illness.

While the stigma of AIDS has lessened, the 2014 Ebola outbreak once again illustrated how peoples' fears can lead to irrational actions and discrimination. In the United States (which had four cases of Ebola despite media predictions of 10,000 new cases a week), medical personnel returning from working in Africa were ordered into quarantine. A bridal shop in Ohio went bankrupt after potential customers learned that a nurse who had recovered from Ebola had visited the store before she had Ebola symptoms. In Canada, the federal government refused admission to visitors from Ebola-affected countries despite the advice of health officials that such bans were unnecessary (Renzetti, 2015).

MARK RALSTON/AFP/Getty Images

While the stigma of AIDS has lessened, the 2014 Ebola outbreak once again illustrated how people's fears can lead to irrational actions and discrimination.

THE SOCIAL DEFINITION OF HEALTH AND ILLNESS: THE PROCESS OF MEDICALIZATION Feminist scholars are not the only ones who have studied medicalization. Interactionists have also focused on the subjective component in the way illness is defined. This subjective component is important when we look at conditions that are more ambiguous than cancer or a broken bone. For example, a child who has difficulty learning may be diagnosed as having attention deficit hyperactivity disorder (ADHD), a man who occasionally behaves strangely may be called mentally ill, and a woman experiencing menopause may be defined as having a hormonal deficiency disease. Alternatively, we could view these conditions as part of the range of normal human behaviour. The child might be seen as a student who needs extra support, the man as a bit eccentric, and the woman as a person going through the normal aging process. The way we view these individuals will depend on our cultural perspectives, which can change over time.

Conrad and Schneider (1992) found that medicalization is typically the result of a lengthy promotional campaign conducted by interest groups, often culminating in legislative or other official changes that institutionalize a medical treatment for the new "disease." The interest groups may include scientists acting on the results of their research, those who have the disease and may be seeking either a cure or a socially acceptable excuse for their behaviour, and members of the medical industry interested in increasing their profits.

Conrad and Schneider (1980) also emphasize that many behaviours that were at one time defined as "badness" have been redefined as "sicknesses" or "illnesses." Until the medical condition "attention deficit hyperactivity disorder" (ADHD) was established, children who had difficulty sitting still and concentrating or were impulsive and full of energy were labelled "active" or "energetic"—or called "problem children" (Conrad, 1975). In the early 1970s, the medical profession began to treat such children as deviant. The "discovery" of ADHD coincided with the development of Ritalin, a drug that suppresses hyperactive behaviours, and medication became the accepted treatment for this condition. For schools, the social construction of this illness results in fewer disruptive students and more manageable classrooms—the illness also creates a huge new patient population for doctors and a profitable new market for the pharmaceutical industry. Ritalin can help children whose problem behaviour is organically based by enabling them to concentrate and function better in the classroom. However, for children whose disruptive behaviour is a reflection of their acting "like children" rather than symptomatic of ADHD, it results in unnecessary medication.

In 2012, a controversy emerged over the American Psychiatric Association's attempt to classify grief after the death of a loved one as a mental illness. The eminent British medical journal *The Lancet* opposed this medicalization:

> Medicalizing grief, so that treatment is legitimized routinely with antidepressants . . . is not only dangerously simplistic, but also flawed . . . Grief is not an illness; it is more usefully thought of as part of being human and a normal response to death of

> a loved one . . . For those who are grieving, doctors would do better to offer time, compassion, remembrance, and empathy, than pills. (2012:589)

Behaviours can also be *demedicalized.* For many years, homosexuality was defined as a mental illness, and gays and lesbians were urged to seek psychiatric treatment. Gay activists fought for years to convince the American Psychiatric Association to remove homosexuality from the association's psychiatric diagnostic manual. Women's groups have been trying to demedicalize childbirth and menopause, and to redefine them as natural processes rather than as illnesses.

POSTMODERN PERSPECTIVES ON HEALTH: THE CROSSROADS OF BIOLOGY AND CULTURE

David Morris has proposed a postmodern perspective on health that understands disease and illness "whatever [their] particular causes, as created in the convergences between biology and culture" (1998:76). While most of us are aware of the biological dimensions of disease and illness, culture is a factor in health in many ways. Human activities, such as coal mining and the pollution caused by the burning of fossil fuels, have health implications for workers and for the general population. Culture also affects how we experience illness. The experience of having cancer—part of which involves the "sick role" described by Parsons—is very different for someone in a poor village in Uganda than it is for someone in a Canadian city. While the biological factors may be identical, the understanding of the illness, the treatment available, the suffering experienced by the patient, and the likelihood of survival differ greatly between the two cultural contexts. According to Morris, another characteristic of illness, viewed through a postmodern lens, is an ambiguity about the nature of some disorders. Patients and doctors have contested the existence of ailments such as chronic fatigue syndrome, post-traumatic stress disorder, and even some types of addictions. Our understanding of, and experience with, health and illness is socially constructed—it is not simply a matter of biology.

Postmodern culture has a fixation on health. Health-related products are heavily advertised, new developments in health research are widely reported in the media, and healthcare is an important political issue. While 16th-century explorers sailed the world searching for the Fountain of Youth, people in postmodern societies search for immortality through medical research, plastic surgery, fitness programs, and miracle cures. In this search for perfection, "our culture has declared war on biology" (Morris, 1998:2). Alexander Segall and Christopher Fries (2011) have noted that the way people look after their bodies has become a measure of their self-worth.

One aspect of this war on biology is the belief in the perfectibility of the body. People now have "the option of transforming bodies into a facsimile of their own ideal vision" (Morris, 1998:138). The desire for bodily perfection is manifested in many ways, including cosmetic surgery, legal and illegal performance-enhancing drugs used by athletes, the very different body types attained by competitive bodybuilders and by anorexics, and the growing occupation of personal trainer. We can alter our bodies for aesthetic purposes through devices such as breast implants or make them function more effectively by replacing defective body parts with artificial hips, knees, and heart valves.

While these measures may change the look of the body, the search for perfection may also be harmful. The premature deaths caused by anorexia and steroid use, and the psychological consequences of realizing that no matter what one does, perfection is not attainable, show the futility of pursuing a vision of bodily perfection. Also, this utopian vision is ultimately contradicted by the biology of aging and the inevitability of death. When old age is seen as just another stage in an active life rather than as a time to prepare for the end, physical decline is viewed as an embarrassment and there is often a denial of death.

Even when death is imminent, postmodern patients are trapped between two conflicting realities. The first is the ability of biomedicine to keep failing bodies alive almost indefinitely; the second is the public pressure in many cultures to allow doctor-assisted suicide to alleviate suffering and provide a dignified death at a time of the patient's choosing. In 2015, the Supreme

Court of Canada determined that Canadians with severe illnesses should have the right to assisted suicide, so society's views on this issue are continuing to evolve.

While Morris feels that modernist Western biomedicine is being challenged by the postmodern emphasis on culture, the future of our culture's understanding of health and illness is not yet clear:

> It is an untold, unnoticed story in which the cultural fantasy of living forever—or at least pushing back death through an unending series of medical purchases—creates sickly lives obsessed with heartburn, bowels, megavitamins, and miracle cures. This new postmodern narrative, in short, represents for us the confusing historical moment we are living through when the biomedical model has begun to reveal its inherent limitations but when a biocultural model . . . has not yet proven its power to constitute a satisfying and coherent replacement. (1998:278)

TIME TO REVIEW

- How do functionalists define the sick role? What are some of the criticisms of this view of health and illness?
- What are some of the questions conflict theorists ask about health and healthcare?
- Describe how the process of medicalization has affected women's lives.
- What do interactionists mean when they say that illness is socially constructed? Is this a complete explanation of illness?
- How do people's ideas of the perfectibility of the body affect the way they think about health and healthcare?

CONCEPT SNAPSHOT

FUNCTIONALIST PERSPECTIVES **Key thinker:** Talcott Parsons	Illness is dysfunctional both for the individual who is sick and for the larger society. All societies have a sick role–patterns of behaviour defined as appropriate for people who are sick.
CONFLICT PERSPECTIVES	Among the issues for conflict theorists are the ability of all citizens to obtain healthcare; the impact of race, class, and gender on healthcare; the relative power of doctors and the medical model in the healthcare system; and the role of profit in the healthcare system.
INTERACTIONIST PERSPECTIVES **Key thinkers:** Peter Conrad and Joseph Schneider	Interactionists examine how the meaning that people give their illness or disease affects their self-concept and their relationships with others and at how the social definition of disease affects those who are ill. A related subjective component of illness is medicalization–the process whereby an object or a condition becomes defined by society as a physical or psychological illness.
FEMINIST PERSPECTIVES **Key thinkers:** Deborah Findlay and Leslie Miller	Feminist scholars have been critical of the male-centred focus of medical research and of discrimination against women working in the healthcare system. They have also studied the way that medicalization blames women for their physical condition and forces women to conform to traditional role expectations, limiting their freedom of behaviour, appearance, and relationships.
POSTMODERN PERSPECTIVES **Key thinker:** David Morris	Some postmodern theorists who study illness and disease focus on the interaction between biology and culture. Culture is a factor in health in many ways–human activities contribute to disease and culture affects how we experience illness. Our culture also emphasizes the perfectibility of the body, and many people seek to alter their bodies through surgery, exercise, and drugs.

The Canadian Press/Wayne Glowacki

Brian Sinclair died after waiting in an emergency room waiting room for 34 hours without receiving medical attention. Mr. Sinclair was a homeless Indigenous man—a double amputee who waited in his wheelchair for care that he never received. His case illustrates how social factors can affect health and healthcare.

LO-3 SOCIAL DETERMINANTS OF HEALTH

The strong correlations between health and social variables such as class, race, and sex have led researchers to consider the social determinants of health. According to the World Health Organization, "The **social determinants of health** are the conditions in which people are born, grow, live, work and age, including the health system. These circumstances are shaped by the distribution of money, power and resources at global, national and local levels, which are themselves influenced by policy choices. The social determinants of health are mostly responsible for health inequities—the unfair and avoidable differences in health status seen within and between countries" (2012b).

▶ **social determinants of health** The conditions in which people are born, grow, live, work and age, including the health system.

Rather than placing responsibility for illness on individuals ("Don't smoke," "Eat a proper diet," "Stay fit"), this perspective focuses on factors—race, class, social inequality—that cannot be controlled by individuals. The high incidence of diabetes among Canada's Indigenous people—who have rates of diabetes three to five times higher than the rest of the population (Canadian Diabetes Association, 2015)—illustrates the social determinants approach to health. The social factors involved in this serious medical condition have been summarized by Canada's Auditor General: "In addition to the common risk factors of unhealthy eating, physical inactivity, smoking, and alcohol misuse, the higher rates of adverse health outcomes in Aboriginal peoples are associated with factors such as low income, lack of education, high unemployment, poor living conditions, and poor access to health services" (2013, n.p.). Indigenous people do not just randomly 'catch' this disease at higher rates than other Canadians. Rather, the intersection of issues related to race and to class has led to higher rates of diabetes and other illnesses among Indigenous people, particularly those living on isolated reserves where employment is limited, food is expensive, housing is inadequate, and medical care is difficult to access.

We often think of health in only physical terms. However, the health of any group is a product of the interaction of a wide range of physiological, psychological, spiritual, historical, sociological, cultural, economic, and environmental factors (Waldram, Herring, and Young, 1995). In this section, we will examine how age, gender, and class affect the health of Canadians, and how these factors operate on a global level.

The Canadian Press/Edmonton Journal/Bruce Edwards

Alzheimer's disease is a tragedy for the afflicted individuals and for their families. As our population ages, such debilitating conditions as Alzheimer's will also increasingly place a burden on our healthcare system and on relatives of those afflicted with the disease.

AGE

Rates of illness and death are highest among the old and the very young. Mortality rates drop shortly after birth and begin to rise significantly during the middle years. After age 65, rates of chronic illness and mortality increase rapidly. This has obvious implications for individuals and their families, but also has an impact on society.

Canada is an aging society. About 12 percent of the population is 65 or over; by 2036, this will double to 25 percent. Because healthcare costs are high for some older people, these costs will rise as the baby boomers age. However, population aging only accounted for 1 percent of the increase in healthcare costs, and the share of health dollars spent on seniors has been stable over the past decade (Canadian Institute for Health Information, 2014), so we should not overestimate the impact of aging on these costs (see Chapter 14).

SEX

Prior to the 20th century, women had shorter lives than men because of high mortality rates during pregnancy and childbirth. Childbirth is now much safer and women now live longer than men. Females born in Canada can expect to live about 84 years, compared with 79 years for males (Statistics Canada, 2013f). Three factors contribute to this sex difference in life expectancy (Waldron, 1994). First, differences in gender roles mean that females are less likely than males to engage in risky behaviour, such as using alcohol and drugs, driving dangerously, and fighting. Males are also more likely to work in dangerous occupations, such as commercial fishing and mining. Second, females are more likely to seek medical care and so may have problems identified at an earlier, more treatable stage than men, who are more reluctant to consult doctors. Third, there may be biological differences, as females have higher survival rates than males at every stage, from fetus to old age.

As the social roles played by females have changed, the mortality gap has narrowed. Women are moving into traditionally male-dominated occupations, such as farming and policing, where they face the same risks as males. As female rates of participation in risky behaviour, such as smoking and illicit drug use, approach those of males, females are paying the price in illness and early death.

Women live longer than men, but they also have higher rates of disease and disability. While men at every age have higher rates of fatal diseases, women have higher rates of nonfatal chronic conditions (Turcotte, 2011; Crompton, 2011).

One important women's health issue is the lack of medical research on women. Many major studies of diseases such as heart disease have excluded women. These studies, however, are the basis for the diagnosis and treatment of both sexes, even though there appear to be sex differences in the diseases. Most funding agencies now require researchers to include both men and women subjects unless there are clear reasons for limiting the study to one sex. However, much of our existing medical knowledge is based on the earlier male-centred research.

SOCIAL CLASS

People who are poor have worse health and die earlier than the rich. This observation is also true of poor and rich countries, as illness and mortality rates are far higher in low-income countries than in high-income countries.

A recent study examined the association between mortality rates and income (Tjepkema et al., 2013). Canadians in the lowest income quintile had mortality rates that were 52 percent higher for females and 67 percent higher for males than those in the highest income quintile. In Winnipeg, this meant that life expectancy was 16 years higher in the highest income neighbourhood than in the lowest income neighbourhood (Welch, 2015). Tjepkema and colleagues concluded that if the mortality rates for all Canadians were the same as the rates of the highest income group, there would be 40,000 fewer deaths each year. The increased mortality risk for the poor was highest for causes of death associated with risky behaviours such as drinking and smoking, and lowest for causes such as breast cancer and prostate cancer that are not associated with these behaviours. However, these high-risk behaviours were not the only factors related to higher mortality among the poor—the data also suggested that there may also have been "differences in the accessibility, use, or quality of medical care" (2013:18).

Good healthcare policy can help reduce the impact of poverty on health. Providing the poor with access to medical advice and treatment through universal medicare is one way of doing this. A study comparing cancer survival rates for the poorest one-third of Toronto residents, who all had government-funded healthcare, with their counterparts in Detroit, who typically had little or no health insurance, showed the benefits of providing adequate healthcare for the poor (Gorey et al., 1997). Survival rates were higher in Toronto for 12 of the 15 most common types of cancer. For many of these cancers, survival rates after five years were 50 percent higher among the poor in Toronto than in Detroit. The benefits of government-funded care go particularly to the poor, as this study found no differences among middle- or high-income patients in the two countries.

If access to medical care improves the health of the poor, why are Canada's poor less healthy than its middle and upper classes despite our universal healthcare? The answer is that medical care cannot compensate for the other disadvantages of poverty, such as poor housing, hazardous employment, inadequate diet, greater exposure to disease, and the psychological stresses of poverty. People who are poor are more likely to engage in risky behaviours, such as smoking and excessive drinking. One study found that class differences in smoking accounted for over half the difference in death rates between upper and lower classes (Jha et al., 2006).

Poor people may also lack knowledge of preventive strategies and services. For example, college-educated women are twice as likely as women who have not graduated from high school to have mammograms, which means that less educated women are at higher risk of dying of breast cancer. Their health is worse despite the availability of care once the medical problem has occurred or been identified. Finally, when they are ill, the poor are less likely to visit doctors than are wealthier people (Roos et al., 2004).

While differences remain between rich and poor, it is encouraging to note that the gap in life expectancy between people living in the highest- and lowest-income neighbourhoods has declined substantially (Statistics Canada, 2002b). One reason for this was that the difference in infant mortality rates between high- and low-income neighbourhoods declined from 9.8 deaths per 1000 births in 1971 to 2.4 deaths per 1000 births in 1996. Despite this decline, class differences in infant mortality still persist (Gilbert et al., 2013).

RACE, CLASS, AND HEALTH: CANADA'S INDIGENOUS PEOPLES

The experience of Canada's Indigenous peoples illustrates how the disadvantages of race can interact with those of class to cause health problems.

HEALTH PROBLEMS AMONG INDIGENOUS PEOPLES IN CANADA

▶ **epidemics** Sudden, significant increases in the numbers of people contracting a disease.

Indigenous people have a history of serious health problems that begins with their early contact with Europeans. **Epidemics**—sudden, significant increases in the numbers of people contracting a disease—of contagious diseases such as tuberculosis, measles, smallpox, and influenza broke out in the early years of this contact. These epidemics occurred partly because Indigenous people had no immunity to these European diseases. Another factor was new patterns of trade that led to contact with more diverse groups of people after European settlement. Trade also led to higher population densities around trading posts, and this density helped to sustain epidemics. Tuberculosis epidemics were particularly devastating in the late 19th century, when Indigenous people were moved to reserves. Crowded and lacking proper sanitation facilities, the reserves were ideal settings for the spread of disease, and mortality rates for tuberculosis remained high until the 1950s.

While their mortality rates have improved significantly, Indigenous people still have shorter lives than other Canadians. Infant mortality rates among Indigenous people have declined significantly since the 1970s but are still well above the Canadian average. The infant mortality rate for First Nations people is about twice the Canadian average, and for the Inuit it is four times the national average (Smylie and Adomako, 2009). Life expectancy is seven years less than average for First Nations men and five years less for First Nations women (Health Canada, 2005). While most infectious diseases have been brought under control among Indigenous people (though their rates remain higher than those of other Canadians), their health problems are now chronic diseases, such as heart disease and diabetes. HIV/AIDS is also having a serious impact on Indigenous people, whose rates of new HIV infection are 3.6 times higher than those of other Canadians. The majority of Indigenous people contracted HIV through intravenous drug use (Public Health Agency of Canada, 2010). A Statistics Canada study (2015c) found that First Nations people were more than twice as likely to die of avoidable causes than were non-Indigenous people.

What are the reasons for the poorer health of Indigenous people? The major factor is poverty. Indigenous people are among the poorest in Canada and suffer from the poor nutrition and other social conditions that go with poverty. Because of the high costs of transporting food to remote reserves, it is difficult for some families to afford a well-balanced diet, so health problems due to nutritional deficiencies are common.

Many of the diseases that affect Indigenous people can also be traced to the inadequate housing, crowding, and poor sanitary conditions common on reserves and in other communities where they live. In 2015, for example, 120 reserves were under orders to boil their water because it was not safe to drink (First Nations Health Authority, 2015, Health Canada, 2015). The isolation of many Indigenous communities is also a factor; an illness that could be easily treated in a city hospital can be fatal in a community 800 kilometres from the nearest doctor.

Indigenous people also have high rates of violent death, with rates of death by murder and suicide higher than those of other Canadians. Rates of adolescent suicide are particularly high, especially in Nunavut. Accidental deaths, particularly by motor vehicle accidents and drowning, are also higher for Indigenous people. George and colleagues (2015) report that potential years of life lost to suicide, homicide, and vehicle accidents were two to three times higher for Status Indians in British Columbia than for other residents of the province, even though differences in rates of injury hospitalizations had declined over the past two decades. In Saskatchewan, rates of HIV/AIDS are 11 times higher than the national average, and the disease is typically spread through intravenous drug use (Leo, 2015).

Finally, the legacy of colonialism still affects Indigenous people's health. Anastasia Shkilnyk (1985), who studied the Ojibwa community of Grassy Narrows in northwestern Ontario, attributed the high rates of suicide and violent death, as well as health problems on the reserve, to colonial actions such as the destruction of Indigenous language and religion, the family breakdown caused by enforced attendance at residential schools, and the forced relocation of the community by the Department of Indian Affairs. Environmental destruction by local industries that dumped methyl mercury into the lakes and rivers around the reserve was another contributor. This toxic substance had a direct impact on the health of Grassy Narrows residents, and also an indirect impact because it destroyed the fishery that was the foundation of the community's way of life.

INDIGENOUS HEALING METHODS

Indigenous healing traditions are holistic and deal with the interactions between spirit, mind, emotions, and body. However, the Western model of medicine has been as dominant in Indigenous communities as in the rest of Canada. Traditional healing practices fell into disuse for many years but are becoming popular again. Medical authorities have responded to Indigenous demands that culturally appropriate healing methods should be available, though the acceptance of these methods has been mixed. Some hospitals and clinics now have Indigenous healers, and combine traditional and Western treatment methods—a plaque in a Kenora, Ontario, hospital reads: "We believe traditional Native healing and culture have a place in our provision of healthcare services to the Native people" (Waldram Herring, and Young, 1995). Indigenous people are also gaining greater control over the delivery of medical services in their communities. These changes mean that in the future, Indigenous people will be more involved in their healthcare, and traditional and Western medical traditions will be better integrated.

The Canadian Press/Francis Vachon

Many residents of First Nations communities are forced to live in deplorable housing conditions, which have a negative impact on their health.

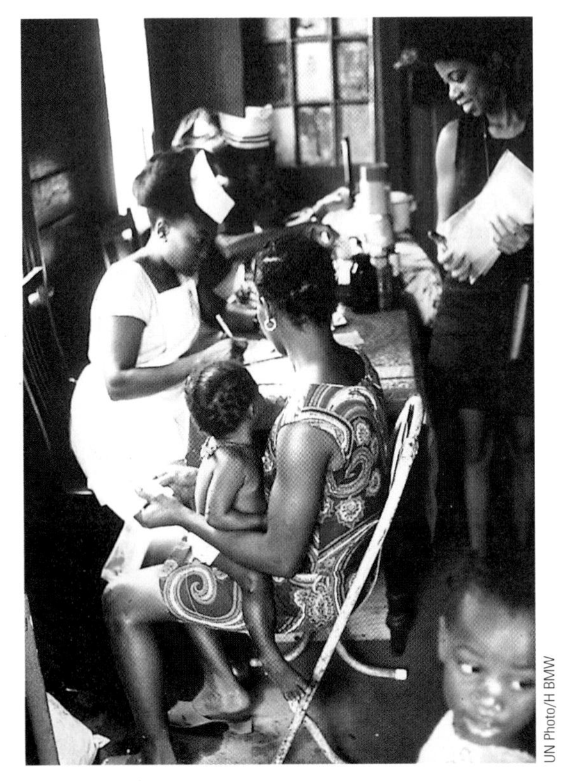

UN Photo/H BMW

A nurse interviews a mother at a rural health clinic in Sierra Leone. With the support of the World Health Organization, these clinics were established to reduce infant mortality and improve the health of mothers and their children.

The Canadian Press/AP Photo/Tsvangirayi Mukwazhi

The availability of safe water has been a major reason why life expectancy has risen dramatically since the 1950s.

Evidence supports the idea that restoring Native control over their healthcare will lead to improved health among Indigenous people (Canadian Institute for Health Information, 2004). Moreover, researchers in many parts of the world have found that many traditional medicines are effective. Acting on this, pharmaceutical companies now market many products (including Aspirin) with the same chemical composition as traditional remedies. However, it will take much more than better healthcare for Indigenous people to become as healthy as the rest of Canadian society.

SOCIAL DEVELOPMENT AND HEALTH: A GLOBAL PERSPECTIVE

The poverty and colonialism that have affected the health of Canada's Indigenous people also operate on a global scale. For example, Hunt (1989) attributes the rapid spread of diseases, such as HIV/AIDS in Africa (see Box 15.3), to the underdevelopment and dependency that are the legacy of colonialism. The underlying roots of this health problem lie in the economic and social marginalization of most African people.

The difference between rich and poor countries is dramatically reflected in infant mortality rates. While five of every 1000 infants in Canada die before their first birthday, infant mortality in the world's poorest countries is far higher. Afghanistan, Somalia, and Nigeria, for instance, have infant mortality rates of 117, 100, and 74 per 1000 live births respectively (CIA, 2015). Life expectancy is correspondingly low. For persons born in Canada in 2011, life expectancy at birth was about 82 years (Statistics Canada, 2013g), compared with about 50 years in many poor African nations. Most deaths in less developed countries are caused by diseases now rare in the industrialized world.

A UNICEF (2015) report on the health of the world's children puts the situation in stark terms. Over 6 million children under five years of age die each year—and most of these deaths could be prevented easily. Most child deaths are caused by malnutrition and by preventable diseases, including measles, diarrhea, malaria, and pneumonia. As Sharma and Tulloch tell us, "Children in rich countries do not die from the common, preventable diseases of childhood. Children in poor countries do" (1997:1).

Tremendous progress has nevertheless been made in saving the lives of children. The number of child deaths is now less than one-third the 20 million who died in 1960 (UNICEF, 2010), and the decline is continuing. Measures such as immunization, rehydration therapy for diarrhea, and child mosquito nets save the lives of millions of children each year. Millions more deaths could be prevented through simple measures such as improved sanitation, clean water, better immunization programs, and the provision of better local health services.

Of course, not only children are dying in poor countries. Each year, nearly 300,000 women die of complications arising from pregnancy and childbirth (World Health Organization, 2014). Virtually all of these deaths take place in poor countries. Many other diseases, such as malaria, take millions of lives, but drug companies are not interested

BOX 15.3 SOCIOLOGY IN GLOBAL PERSPECTIVE

The AIDS Epidemic in Africa

> Rakai, Uganda: From the shadows of his hut, the gaunt and weary young man stares outside at the pigs playing in the dust under the banana palms. His chest is covered with open sores; skin rashes have left his ebony arms looking as if they are covered in chalk; his army fatigues hang loosely around his waist.
>
> Outside, Charles Lawanga glances toward his ailing second son and lowers his voice. Last year, when the Ugandan army gave him his medical furlough, his son was sick, but at least he could walk, says Lawanga.
>
> Lawanga's brows are furrowed; he has the face of a man who is watching his son die. His eyes sharpen when he hears that an American journalist knows many of the Western doctors working on the disease. He knows that the United States is a country of immense wealth, and that the medicine that will save his country and his son will probably come from there. Tears gather in his brown eyes, and he asks, "When will it come? When will there be the cure?" (Shilts, 1988:621)

In the mid-1990s, Uganda had the highest number of recorded HIV cases in Africa—around 1.5 million—and AIDS has touched virtually all families in this country. Uganda was, however, the first African country to make major gains in the fight against HIV/AIDS. One key indicator was the infection rate of pregnant women: Between 1992 and 1998, this rate dropped from 31 percent to 14 percent in the capital city of Kampala, and from 21 percent to 8 percent in the rest of the country. The rate among men attending clinics for sexually transmitted diseases dropped from 46 percent to 20 percent over the same period (Global Health Council, 2002).

The fight against HIV/AIDS was personally led by Uganda's President Museveni. The government worked with community partners to implement school sex education programs that encouraged abstinence but also promoted condom use for those who were sexually active; quick treatment of other sexually transmitted diseases; and same-day results for HIV tests, and immediate counselling for those who were tested. Uganda developed the ABC approach to AIDS prevention: Abstinence, Be faithful to one partner, and use Condoms. This program was implemented through grassroots organizations throughout the country and received funding from the World Bank and from the United States (Avert, 2005). Successful programs in Thailand and Senegal have also been based on encouraging condom use and discouraging risky sexual behaviour, such as casual sex and sex with prostitutes (Global Health Council, 2002).

Unfortunately, few other countries have followed Uganda's lead. Many countries deny that they have a problem, and others reject modern treatment methods. South Africa's former president, Thabo Mbeki, denied that HIV caused AIDS, and his government refused to provide treatment to pregnant women to prevent the transmission of HIV to their children. His health minister promoted garlic and beets as remedies for AIDS. These policies resulted in hundreds of thousands of needless deaths.

In other countries, public discussion of sexual behaviour is so taboo that governments and other leaders refuse to address rising infection rates. Even in Uganda, progress was slowed because under former president George W. Bush, U.S.-based funding organizations discouraged the promotion of condoms and advocated "abstinence-only" programs. In response, President Museveni condemned condom use as immoral, and the safe-sex advertising campaign and free distribution of condoms were drastically curtailed. As evidence clearly shows that condom use is a vital component of successful AIDS prevention programs and that abstinence-only programs do not work, the ideology of the faith-based organizations that helped deliver U.S. funding jeopardized the lives of millions of Ugandans (Avert, 2005; New York Times, 2005). These restrictions were modified under President Barack Obama.

in developing cures because the residents of poor countries where these diseases are epidemic cannot afford to pay high prices for such drugs.

Despite these problems, tremendous progress has been made in improving world health. Since 1955, the average life expectancy in the world has increased from 46 years to 71 years (World Health Organization, 2015b). This increase has been attributed to a number of factors, one of the most important being the development of a safe water supply. The percentage of the world's population with access to safe water nearly doubled between 1990 and 2000 (United Nations Development Programme, 2003).

TIME TO REVIEW

- Life expectancy is far lower in poor countries than in wealthy ones. What could the global community do to increase the life expectancy of the poorest people?

LO-4 DISABILITY

What is a disability? There are many different definitions. In business and government, it is often defined in terms of work—for instance, "an inability to engage in gainful employment." Medical professionals tend to define it in terms of organically based impairments—the problem being entirely within the body (Albrecht, 1992). However, this definition is too narrow. Many people have vision problems that would make it difficult for them to read this textbook. Do these people consider themselves disabled? The answer is no, because eyeglasses or contact lenses can make it possible for most people with vision problems to see normally. Thus the definition of disability must have a social component. We can define a **disability** as a physical or health condition that reduces a person's ability to perform tasks he or she would normally do at a given stage of life and that may result in stigmatization or discrimination against the person. This definition of disability is based not only on physical conditions but also on social attitudes and the social and physical environments in which people live.

▶ **disability** A physical or health condition that reduces a person's ability to perform tasks he or she would normally do at a given stage of life and that may result in stigmatization or discrimination against the person.

Society has not provided the universal access that would allow people with disabilities to participate fully in all aspects of life (Blackford, 1996). For example, many buildings are still not accessible to persons using a wheelchair. In this context, disability derives from a lack of accommodation rather than from someone's physical condition. Oliver (1990) used the term *disability oppression* to describe the barriers that exist for disabled persons in Canadian society. These include economic hardship (from the additional costs of accessibility devices, transportation, and attendant care, or from employment discrimination), inadequate government assistance programs, and negative social attitudes toward disabled persons. According to disability rights advocates, disability must be thought of in terms of how society causes or contributes to the problem—not in terms of what is "wrong" with the person with a disability.

SOCIOLOGICAL PERSPECTIVES ON DISABILITY

How do sociologists view disability? Functionalists often apply Parsons's sick role model, which is referred to as the *medical model* of disability. According to the medical model, disability is deviance. The deviance framework is also apparent in some symbolic interactionist perspectives. According to symbolic interactionists, people with a disability experience *role ambiguity* because many people equate disability with deviance (Murphy et al., 1988). By labelling individuals with a disability as "deviant," other people can avoid them or treat them as outsiders. Society marginalizes people with a disability because they have lost old roles and statuses and are labelled as "disabled" persons.

While he was still in college, British theoretical physicist Stephen Hawking learned he had Lou Gehrig's disease (amyotrophic lateral sclerosis). Hawking nevertheless went on to develop a quantum theory of gravity that forever changed our view of the universe. He is considered one of the leading figures in modern cosmology.

According to Freidson (1965), people are labelled based on (1) their degree of responsibility for their impairment, (2) the apparent seriousness of their condition, and (3) the perceived legitimacy of the condition. Freidson concluded that the definition of and expectations for people with a disability are socially constructed factors.

Titchkosky also feels that disability can be understood only within its social and cultural context. A main feature of this context is "the fact that disability is necessarily an experience of marginality" (2003:232).

Titchkosky describes how the arrival of a guide dog affected her partner, a professor whose sight had been deteriorating for years but who was usually able to pass as a sighted person:

> I had not anticipated that acquiring a guide dog would also mean acquiring a new identity. Rod arrived home with a beautiful new dog and an expert guide. But there was more. With Smokie, Rod was seen as blind. Staring, grabbing, helping, offers of prayers or medical advice, groping for words or even a voice are some of the many ways in which sighted people show that they are seeing a blind person. Through these interactions, Rod was given the identity—blind person. (2003:82)

Finally, from a conflict perspective, persons with a disability are members of a subordinate group in conflict with persons in positions of power in the government, in the healthcare industry, and in the rehabilitation business, all of whom are trying to control their destinies (Albrecht, 1992). Those with power have created policies and artificial barriers that keep people with disabilities in a subservient position (Asch, 1986; Hahn, 1987). In a capitalist economy, disabilities are big business. Persons with a disability have an economic value as consumers of goods and services that will allegedly make them "better" people. Many persons with a disability endure the same struggle for resources faced by people of colour, women, and older persons. Individuals who hold more than one of these ascribed statuses, combined with experiencing disability, are doubly or triply oppressed by capitalism.

DISABILITY IN CONTEMPORARY SOCIETY

An estimated 3.8 million people aged 15 and over, representing 14 percent of the adult population in Canada, report having a disability that limited their daily activities (Statistics Canada, 2015d). This number is increasing for several reasons. First, with advances in medical technology, many people who formerly would have died from an accident or illness now survive, although with an impairment. Second, as people live longer, they are more likely to experience disabling diseases such as arthritis (Albrecht, 1992). Only 4 percent of people aged 15–24 reported disabilities compared with 33 percent of those over 65. Third, persons born with serious disabilities are more likely to survive infancy because of medical technology. Finally, there is some indication that people are now more willing to report disabilities (Statistics Canada, 2007b).

Some people are more likely to become disabled than others. Indigenous people and persons with lower incomes have higher rates of disability. Environment, lifestyle, and working conditions all contribute to disability. Air pollution and smoking lead to a higher incidence of chronic respiratory disease and lung damage. In industrial societies, workers in many types of low-status jobs are at greatest risk for certain health hazards and disabilities. Employees in data processing and service-oriented jobs may also be affected by work-related disabilities, such as arthritis and carpal tunnel syndrome.

People with disabilities have been kept out of the mainstream. They have been denied equal opportunities in education by being consigned to special education classes or special schools. Snowdon (2012) has described the social isolation of children with disabilities. The majority of parents reported that their disabled child had "no close friends" or "only one close friend."

Many people with disabilities have been restricted from entry into schools and the workforce, not due to their own limitations, but by societal barriers. Why are disabled persons excluded? Susan Wendell offers an explanation:

> In a society that idealizes the body, the physically disabled are often marginalized. People learn to identify with their own strengths (by cultural standards) and to hate, fear, and neglect their own weaknesses. The disabled are not only de-valued for their de-valued bodies; they are constant reminders to the able-bodied of the negative body—of what the able-bodied are trying to avoid, forget, and ignore . . . In a culture which loves the idea that the body can be controlled, those who cannot control their bodies are seen (and may see themselves) as failures. (1995:458)

The Canadian Press/Tom Hanson

Steven Fletcher is the first quadriplegic to be elected to Canada's House of Commons, and in Cabinet. After he was injured in an automobile accident, Fletcher served two terms as president of the University of Manitoba's Students' Union while earning his M.B.A. That was the beginning of his political career.

The combination of a disability and society's reaction to the disability has an impact on the lives of many people. The disabled often suffer from stereotyping. Movies often depict villains as individuals with disabilities (think of the villains in the Batman movies, such as the *Dark Knight,* in which both the Joker and former District Attorney Harvey Dent were driven to crime by serious disfigurement). Charitable fundraising campaigns may contribute to the perception of the disabled as persons who are to be pitied. Prejudice against persons with disabilities may result in either subtle or overt discrimination. According to Asch, "Many commentators note that people with disabilities are expected to play no adult social role whatsoever; to be perceived as always, in every social interaction, a recipient of help and never a provider of assistance" (2004:11). This attitude is one of the reasons why people with disabilities have difficulty finding employment.

While the role of persons with disabilities has expanded in the Canadian labour force in recent years, compared to persons without disabilities, a much smaller proportion of the disabled population is employed—47 percent of persons with disabilities aged 15–64 were employed compared with 74 percent of those without disabilities (Statistics Canada, 2015d). As a result, people with activity limitations had incomes that were only two-thirds of those of people without disabilities. Not surprisingly, many persons with disabilities feel that they are disadvantaged in terms of employment and that they have been discriminated against in the workplace because of their condition. Of men aged 25–34 who reported having a severe or very severe disability and who were not employed, 62 percent said they had been turned down for a job because of their disability (Statistics Canada, 2014a).

Ensuring equality for people with disabilities is not just a matter for governments. Legislation is important but social attitudes must also change. Adrienne Asch, a university professor who is blind, says that even her close friends do not treat her the same way they treat others. For example, they "do not feel comfortable accepting my offers to pick up food as part of a dinner we plan to have . . . or who would prefer that a high-school-age stranger take care of their six-year-old son for an evening than have me do it, even though I have known their son and their home ever since his birth" (2004:11). Thus equality isn't just a matter of ending discrimination against people with disabilities, but rather of ensuring full integration into mainstream society.

TIME TO REVIEW

- What is the relationship between health and age, sex, and social class? How do these relationships affect the healthcare system?
- Why are Canada's Indigenous people much less healthy than the rest of the population?
- What is the role of traditional practices in Indigenous healthcare?
- Describe the state of health and healthcare in the world's poorest countries.
- Discuss the degree to which disabilities are a function of the social and physical environments in which people live.

HEALTHCARE IN CANADA

Healthcare is a major social and political issue in Canada. Healthcare costs have steadily increased, while governments have made cuts in other areas such as support for college and university students. However, attempts by governments to make changes in the way healthcare is provided often meet with great resistance from the public and from those who work in the healthcare system.

UNIVERSAL HEALTHCARE

Canadians have a **universal healthcare system** in which all citizens receive medical services paid for through tax revenues. Until the early 1960s, Canadians had a user-pay system, in which people paid for healthcare directly out of their pockets. Individuals who did not have health insurance and who required expensive medical procedures or long-term care or who developed a chronic illness often suffered severe financial losses. Under our universal system, if you are sick, you have the right to receive medical care regardless of your ability to pay. Individuals do not pay doctor or hospital costs directly, but they are responsible for at least part of the costs of other medical services, such as prescription drugs and ambulances.

▶ **universal healthcare system** System in which all citizens receive medical services paid for through tax revenues.

Healthcare is a provincial and territorial responsibility, and each province and territory has its own healthcare system. However, the federal government contributes billions of dollars to the provinces for healthcare and enforces basic standards that each province must follow. Provincial plans must meet the following requirements:

1. *Universality.* All Canadians should be covered on uniform terms and conditions.
2. *Comprehensiveness.* All necessary medical services should be guaranteed, without dollar limit, and should be available solely on the basis of medical need.
3. *Accessibility.* Reasonable access should be guaranteed to all Canadians.
4. *Portability.* Benefits should be transferable from province to province.
5. *Public administration.* The system should be operated on a nonprofit basis by a public agency or commission. (Grant, 1993:401)

Canada spends about 11 percent of its gross domestic product on healthcare. While this figure is low in comparison with the United States (see Table 15.1), it is higher than in most other industrialized countries and represents a huge expenditure. Canada's 2014 healthcare expenditures were $215 billion (Canadian Institute for Health Information, 2014). Of this amount, $151 billion came from government funding, while the remainder came from individuals and medical insurance companies for services not covered by medicare. Healthcare costs have increased faster than government revenues (though the increases slowed dramatically between 2012 and 2014), and factors such as greater use of technology, more expensive drugs, and the aging baby boomers will place further demands on the system.

One cause of increasing costs is overutilization of healthcare services. Some of this is a result, for example, of people using emergency rooms for routine care instead of visiting a doctor's office.

TABLE 15.1 LIFE SPAN AND HEALTHCARE EXPENDITURE, 2011

COUNTRIES IN ORDER OF LIFE EXPECTANCY	LIFE EXPECTANCY AT BIRTH, IN YEARS	TOTAL EXPENDITURE ON HEALTH, % OF GDP	EXPENDITURE ON HEALTH, PER CAPITA[1]
Japan	83	9.6%	$3213
Canada	81	11.2%	$4522
France	82	11.6%	$4118
United Kingdom	81	9.4%	$3405
Germany	81	11.3%	$4495
United States	79	17.7%	$8508

[1]In U.S. dollars adjusted for purchasing power parities.

Source: OECD, 2013.

However, while members of the public may not always make the most economical choices, many of the costs of our system are controlled by doctors, who prescribe drugs, admit patients to hospitals, determine patients' lengths of stay in hospital, order tests and examinations, determine the course of treatment that will be used, and recommend follow-up visits. Since patients will do almost anything to ensure their health and since they do not pay directly, they have no incentive to question doctors' recommendations. On the other hand, doctors have a financial interest in providing more treatment. Ensuring that doctors make decisions about treatment only on medical grounds is a major challenge faced by taxpayer-funded healthcare systems.

Another reason for the high costs of the healthcare system is our focus on hospitals and doctors. Our healthcare system overemphasizes acute care and underemphasizes community care and disease prevention. Cheaper forms of noninstitutional healthcare, such as home-care services, are not subject to national standards, so these services vary widely from province to province and may not be available even when they are the most cost-effective type of care. Thus, people who need minimal care may be taking up expensive acute-care hospital beds because community alternatives are not available. Use of the Internet, however, may enable people to have greater control over their own healthcare (see Box 15.4).

HEALTHCARE IN THE UNITED STATES

The United States is the only industrialized country that does not provide universal health coverage to all its citizens. While Canada and Western European countries treat healthcare as a basic human right, the United States sees it as a market commodity. Most Americans receive healthcare coverage through private insurance programs that are sometimes paid for or subsidized by their employers. Government-funded Medicare and Medicaid programs are available for seniors, people with disabilities, and some people with low incomes. Approximately 13 percent of the U.S. population—42 million people—have no medical coverage (U.S. Census Bureau, 2014). The Affordable Care Act, passed by President Obama in 2010, has improved the situation for millions of Americans, but there are still major gaps in coverage. Many Americans are still at risk of incurring major medical expenses that can lead to financial ruin. Medical expenses are the leading cause of personal bankruptcy in the United States (Mangan, 2013).

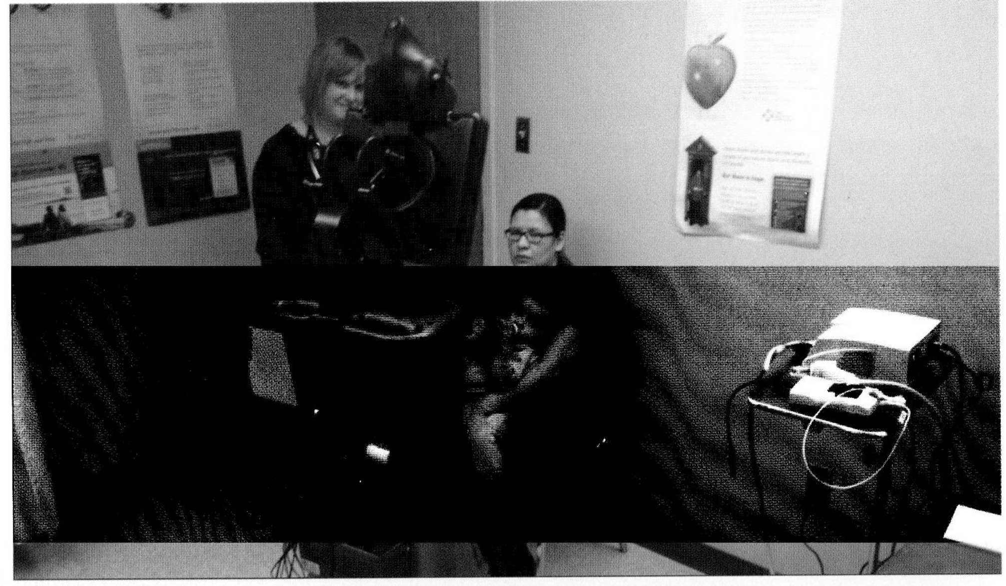

Ryan Pilon/CBC News

A wide variety of diagnostic tools that can be used remotely are now being developed, which will reduce the number of visits needed to doctor's offices and emergency wards, and which will be invaluable to people living in isolated communities. This device provides doctors with diagnostic information from patients in remote communities.

BOX 15.4 SOCIOLOGY AND NEW MEDIA

Dr. Google: Health on the Web

What is the first thing that people do when they experience a sudden chest pain or other medical symptom? For millions of Canadians, the first step is to consult with Dr. Google. The Web is a valuable resource. People can check symptoms and get an idea of when they need to go to a doctor. They can also quickly get information that might help provide emergency care at home. In the past, patients have been passive recipients of healthcare, but this is changing. People are increasingly arriving at doctor's offices or hospital emergency rooms with printouts of material from the Web (Meisel, 2011). In the United States, 80 percent of all Internet users have looked online for health information (Fox, 2011), and the percentage of Canadians who do this is likely very similar.

The Web also has great potential as a collaborative tool where people could access their healthcare records and interact with healthcare providers online. People in remote areas are consulting doctors using their phones and computers, and electronic consultations may also save time and money in urban areas once people become more comfortable with telemedicine. A wide variety of diagnostic tools that can be used remotely are now being developed, which will reduce the number of visits needed to doctor's offices and emergency wards, and which will be invaluable to people living in isolated communities. Phone fitness-tracking apps can be linked to medical databases to monitor patients and warn of potential problems.

People with serious illnesses can also get support by joining online communities of people with similar problems. The www.patientslikeme.com site allows people to learn from others who share their illnesses, to read the latest research on their condition, and to access forms they can use to track their symptoms. Not surprisingly, people who reported having chronic health problems were more likely to have read other people's health commentaries online, watched health videos, and received email updates about their health issues (Fox, 2011).

The Internet can also aid patient care. For example, adolescent diabetics can use an app called Bant, which links an iPhone with the glucometer used to measure blood sugar levels. The app tracks these levels over time and will prompt the user to manage any negative changes. The app also encourages regular testing by awarding users points that can be converted to iTunes gifts. The data can also be accessed by healthcare providers to help them oversee the young person's care. In the future, this technology might be expanded to include phones that could monitor many different facets of our health, including heart rate and blood pressure (Aw, 2011). Physicians are now beginning to use apps to help them access patient records, prescribe medications and determine dosage, and perform countless other tasks.

One interesting use of the Web is to track potential epidemics. Many people respond to early symptoms of illness by entering these symptoms into a Google search. Using these search terms, Google has put together flu trend data that are available one or two weeks earlier than medical diagnostic data, which start only when patients decide they are sick enough to visit a doctor (Ginsberg et al., 2009). Along with search terms, mentions of illnesses on Twitter and Facebook could also be used to give public health officials advance notice.

Unfortunately, there is also a great deal of incorrect information online. Some of the misinformation is deliberate. There has always been medical fraud, but the Internet has made it easier. People promoting miracle "cures" attract the gullible with false arguments and misleading testimonials. Sites such as http://www.hoxseybiomedical.com promise alternative cures for cancer, but the Hoxsey treatments have been called a scam by the U.S. Food and Drug Administration because there is no proper research to back up these claims (FDA, 2011).

Others put false and misleading information online for ideological reasons. For example, many websites oppose the use of vaccines. One early study found that searching for the key term *vaccination* resulted in anti-vaccination sites 60 percent of the time (Wolfe and Sharp, 2005). Many of those who refuse to vaccinate their children have been influenced by these sites (Kata, 2011).

Another negative aspect of using the Internet for medical advice is a lack of privacy. While your medical records are protected by privacy laws, your online information is much more accessible. Search engines such as Google harvest massive amounts of information and sell it to a wide range of clients. In 2014, Canada's Privacy Commissioner forced Google to changes its practices following a complaint from a man who had done a Google search for ways of treating sleep apnea and subsequently received online advertisements from companies offering devices used to treat this condition. Clearly, Google's practices were violating peoples' privacy.

The Web's influence on health and healthcare is just beginning and will move in directions that can barely be predicted now, but it has enabled many individuals to take greater responsibility for their own health and healthcare.

Despite the lack of universal coverage, per capita healthcare costs in the United States are much higher than in Canada and are growing more rapidly. Table 15.1 shows that per capita healthcare costs in the United States of $8508 were almost double Canada's cost of $4522. Much of the difference in costs between Canada and the United States reflects the efficiency of Canada's government insurance system compared with the fragmented U.S. system, with its large number of different healthcare insurers and providers, each eager to maximize profits and each adding its costs to the final bill. The salaries of healthcare workers, particularly doctors, are lower in Canada than in the United States. A comparison of the costs of heart bypass surgery in Canada and the United States (Eisenberg et al., 2005) found that the same operation cost twice as much in the United States ($20,673 as opposed to $10,373 in Canada), even though the success rate for the procedure was the same in both countries.

In the debates over healthcare reform, many U.S. politicians and lobbyists are critical of Canada's "socialized" healthcare. They claim that the Canadian system is beset with delays and that we get care that is inferior to that provided in the United States. Are these critics correct? Do Canadians have an inferior system that forces people to travel to the United States to get proper treatment?

The answer to these questions is no. Despite the higher costs of U.S. healthcare, Canadians are healthier than Americans and have better access to healthcare. Canada has lower infant mortality and longer life expectancy than the United States; these two indicators are often used as broad measures of the quality of healthcare. Many studies demonstrate the superiority of our system to that of the United States (and it should be noted that many European healthcare systems are ranked higher than Canada's). A study cited earlier in this chapter showed that poor Canadians had much higher cancer survival rates than poor Americans. A more recent review (Guyatt et al., 2007) examined 38 studies comparing outcomes of medical treatments in the United States and Canada. Half the studies showed no differences in outcomes, and 14 of the remaining 19 studies (74 percent) showed better results in Canada. These differences between the countries are recognized by the public: 27 percent of Americans say that their healthcare system needs to be completely rebuilt, compared with 8 percent of Canadians (Commonwealth Fund, 2015). Figure 15.1 shows how Canadians' satisfaction with their healthcare system has improved since a serious drop in the 1990s.

FIGURE 15.1 CANADA: SATISFACTION WITH THE HEALTHCARE SYSTEM, 1988, 1998, 2001, 2007, AND 2014

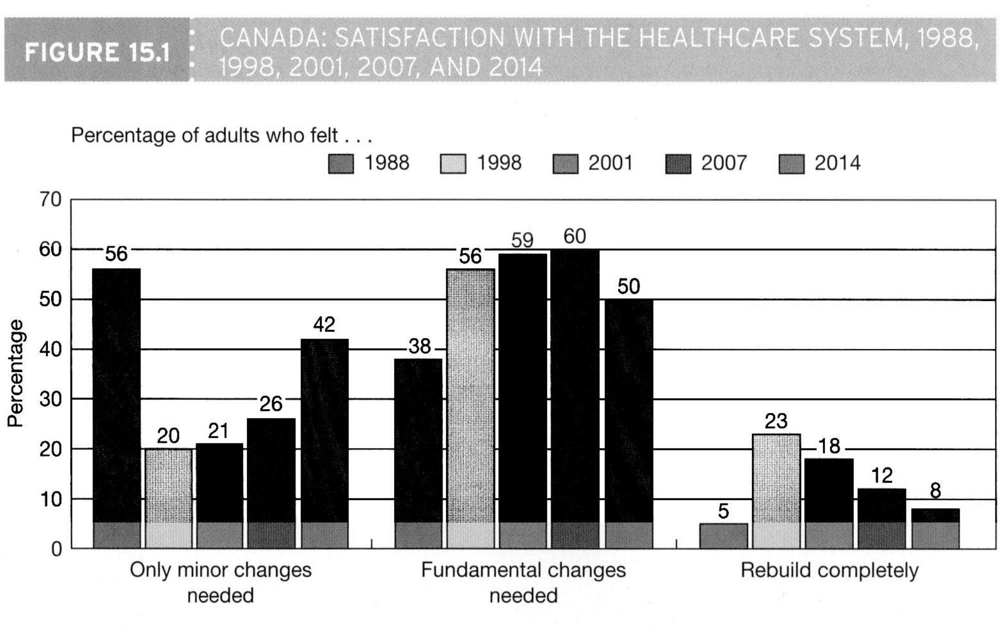

Sources: Commonwealth Fund, 2002, 2007.

LO-5 APPROACHES TO HEALTHCARE

THE MEDICAL MODEL OF ILLNESS

The medical model has been the predominant way of thinking about illness in industrialized societies for many years. This model has five basic assumptions: that illness is "(1) deviation from normal, (2) specific and universal, (3) caused by unique biological forces, (4) similar to the breakdown of a machine whose parts can be repaired, and (5) defined and treated through a neutral scientific process" (Weitz, 1996:129). Acceptance of this model has given great power to doctors, who are viewed as the experts in diagnosing and treating illness. Doctors have gone to great lengths to protect their role at the centre of the healthcare system. For example, they have actively resisted those with conflicting views, such as midwives, advocates of natural healing methods, and those more concerned with preventing disease than with treating it.

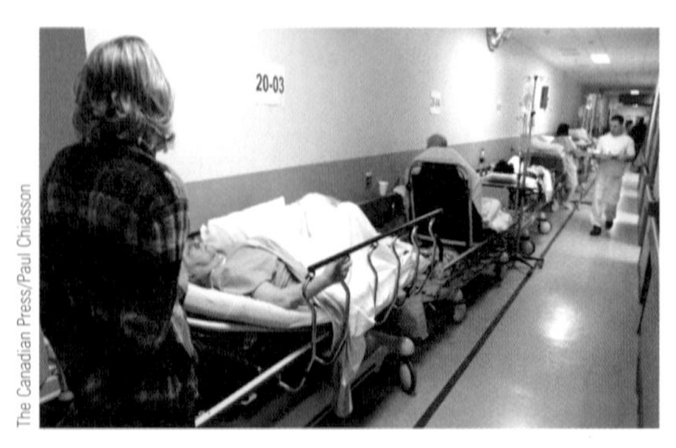

The Canadian Press/Paul Chiasson

One of the consequences of the demands on Canada's healthcare system has been hospital overcrowding. In some hospitals, patients are forced to spend time in emergency room hallways because no rooms are available.

ALTERNATIVE APPROACHES

Despite the successes of modern medicine, the medical model of illness is losing some of its dominance as Canadians are recognizing that their health needs cannot be met by medical services alone and that more medical care does not necessarily lead to better health. Rising medical costs have led governments to implement programs emphasizing environmental and lifestyle factors in health promotion. For example, education about the hazards of smoking, combined with more effective legislation against the use and advertising of tobacco products, has improved public health and saved the money spent treating smoking-related diseases such as emphysema and lung cancer. Responsibility for healthcare is shifting away from the government and the healthcare system and toward the individual and the family.

The popularity of holistic healthcare is a further indication of the move toward a new definition of health. Holism reflects the orientations of many ancient therapeutic systems, including those of Canadian Indigenous peoples. Modern scientific medicine has been widely criticized for focusing on diseases and injuries rather than on preventing illness and promoting overall well-being. Critics say the medical model looks at problems in a mechanical fashion without considering their context, while the holistic approach emphasizes the interdependence of body, mind, and environment.

Holism is adaptable to traditional medical practice and is being adopted by some medical doctors and nurses as well as by practitioners of alternative healthcare, including chiropractors, osteopaths, acupuncturists, and naturopathic doctors. Supporters of the holistic health movement encourage people to take greater individual responsibility for their health and healthcare, especially with regard to lifestyle-related diseases and disabilities.

The holistic approach's emphasis on the role of social factors in illness is now receiving support in research done by traditional practitioners. For example, authors of a *Journal of the American Heart Association* article found that middle-aged men with high levels of despair had a 20 percent greater chance of developing atherosclerosis—narrowing of the arteries—than more optimistic men with similar physiological risk factors. This difference in the risks for heart disease was as great as that between a non-smoker and a pack-a-day smoker (Cable News Network, 1997). Research on the health of older adults has also shown that nonmedical factors, such as isolation, the death of a family member or a friend, and the loss of status after retirement, have a major influence on health.

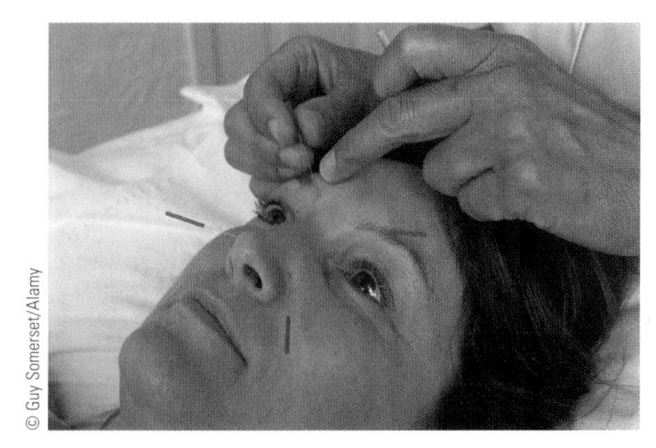

© Guy Somerset/Alamy

Canadians are increasingly using alternative healthcare methods, such as acupuncture.

Accordingly, programs for the elderly must deal with these issues as well as with physical problems (Crichton et al., 1997).

The National Population Health Survey found that many Canadians were using alternative medicine. While about 20 percent of those surveyed had consulted an alternative practitioner within the previous year, few of them had relied exclusively on alternative medicine (Statistics Canada, 2005). Thus it seems that alternative medicine is being used as a complement to traditional medicine rather than as a replacement. Christopher Fries (2008) has described some of the factors that have facilitated the integration of traditional medicine and alternative techniques. He reports that people no longer accept the dominance of traditional medicine and shop around for whatever combination of treatment they feel meets their physical and psychological needs. Their goal is not just to avoid sickness but to achieve a state of well-being. This individualized view of consuming healthcare is reinforced by the government's attempt to encourage people to take responsibility for their own health rather than relying on the healthcare system.

While its use is growing, some types of alternative medicine have been criticized. The main criticism is that some of the claims of alternative medical practice have not been empirically verified. Beyerstein blames the acceptance of such claims on the fact that most people know little about science:

> Even an elementary understanding of chemistry should raise strong doubts about the legitimacy of homeopathy; a passing familiarity with human anatomy would suggest that "subluxations" of the vertebrae cannot cause all the diseases that chiropractors believe they do; and a quite modest grasp of physiology should make it apparent that a coffee enema is unlikely to cure cancer. But when consumers have not the foggiest idea of how bacteria, viruses, carcinogens, oncogenes, and toxins wreak havoc on bodily tissues, then shark cartilage, healing crystals, and pulverized tiger penis seem no more magical than the latest breakthrough from the biochemistry laboratory. (1997:150)

Beyerstein does see some benefits in alternative medicine. It has, he says, added a comforting human component to a medical world that has become increasingly impersonal and technological. Many alternative healers offer sound advice about prevention and a healthy lifestyle, and some alternative practices do have strong scientific backing. However, he fears that some alternatives can divert sick people from more effective treatment. People will need to be sufficiently well informed about the variety and nature of the options available to make sound treatment choices in the future. These options will certainly grow in number as alternative therapies become more widely accepted and as more become integrated with conventional medicine.

TIME TO REVIEW

- What requirements must provinces meet in Canada's universal healthcare system?
- How has the Internet enabled people to take more control over their own health and healthcare?
- How does the U.S. healthcare system differ from that of Canada and most other industrialized countries? Discuss how these differences affect health outcomes.
- Discuss the strengths and weaknesses of alternative approaches to healthcare.

15

LO-1 Understand the ways in which sociological factors influence health and disease.

The health of any group is a product of the interaction of a wide range of factors. Sociological variables such as age, race, sex, and social class have a major impact on the health of every individual.

The Canadian Press/Francis Vachon

LO-2 Compare how functionalist, symbolic interactionist, conflict, feminist, and postmodern theories differ in their analyses of health.

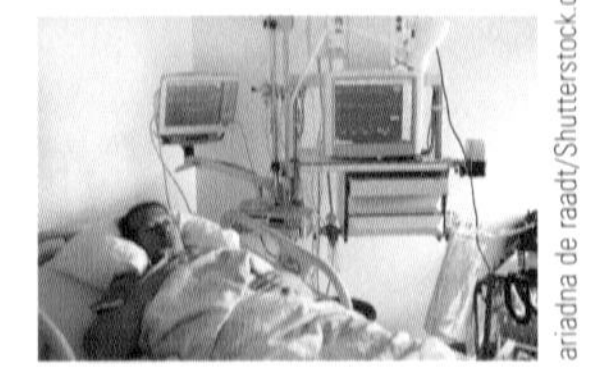

ariadna de raadt/Shutterstock.com

Functionalists see illness as dysfunctional for both the individual who is sick and for society. Sickness may result in an inability on the part of the sick person to fulfill his or her necessary social roles. The conflict approach to health and illness considers the political and social forces that affect health and the healthcare system, as well as the inequities that result from these forces. Feminist scholars have studied several issues, including the manner in which medicalization has affected the lives of women, the male-centred focus of medical research, and discrimination against women working in the healthcare system. Symbolic interactionists attempt to understand the specific meanings and causes that we attribute to particular events. In studying health, interactionists focus on the fact that the meaning that people give their illness or disease will affect their self-concept and their relationships with others. Finally, postmodern theorists have examined the relationship between biology and culture.

LO-3 Understand how social determinants such as social class, sex, race, and age affect health and healthcare.

People with low incomes have worse health and die earlier than the rich. Illness and mortality rates are far higher for low-income than high-income countries. Within the industrialized world, citizens of those countries with the most equal distribution of income have the best health as measured by life expectancy.

The Canadian Press/AP Photo/Tsvangirayi Mukwazhi

KEY TERMS

disability A physical or health condition that reduces a person's ability to perform tasks he or she would normally do at a given stage of life and that may result in stigmatization or discrimination against the person (p. 420).

epidemics Sudden, significant increases in the numbers of people contracting a disease (p. 416).

health The state of complete physical, mental, and social well-being (p. 406).

healthcare Any activity intended to improve health (p. 407).

medicalization The process whereby an object or a condition becomes defined by society as a physical or psychological illness (p. 408).

medicine An institutionalized system for the scientific diagnosis, treatment, and prevention of illness (p. 407).

preventive medicine Medicine that emphasizes a healthy lifestyle in order to prevent poor health before it occurs (p. 407).

sick role Patterns of behaviour defined as appropriate for people who are sick (p. 407).

social determinants of health The conditions in which people are born, grow, live, work and age, including the health system (p. 414).

universal healthcare system System in which all citizens receive medical services paid for through tax revenues (p. 423).

LO-4 Identify the consequences of disability.

Disability may result in stigmatization or discrimination. Persons with disabilities are more likely to be unemployed or underemployed and poor.

The Canadian Press/Tom Hanson

LO-5 Discuss the state of the healthcare system in Canada today and explain how it could be improved.

The costs of the healthcare system are increasing faster than those of almost any other social institution. Those responsible for the system need to reduce cost increases and provide better service to Canadians.

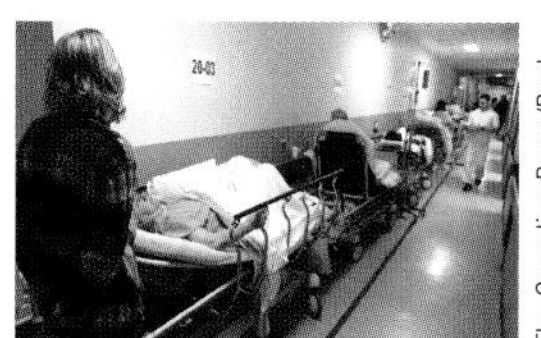
The Canadian Press/Paul Chiasson

APPLICATION QUESTIONS

1. Some doctors are refusing to treat people who smoke, and increasing numbers of pediatricians have begun to remove families from their practices if the parents refuse to have their children vaccinated. Do you think that medical practitioners should have the right to "fire" patients who refuse to look after their own health, or should everyone have the right to be treated?
2. Find some sites on the Web that promise cures for cancer and other serious diseases—you can start with some of the treatments at www.cancure.org. How do these sites try to convince people to buy their products? What are some of the negative consequences that could happen to people who follow these treatments?
3. In 2012, the Northern Ontario community of Attawapiskat received national media attention because of its desperately inadequate housing, which was overcrowded, physically deteriorated, mould-ridden, and lacked running water. Consider how these conditions explain the higher mortality rates and comparatively bad health of First Nations people.
4. Find examples of the ways the media encourage people to look for perfection in their bodies. What techniques do advertisers, such as those for plastic surgery or drugs that grow hair, use to try to encourage people to buy these products or services?

CHAPTER 16 EDUCATION

LEARNING OBJECTIVES

AFTER READING THIS CHAPTER, YOU SHOULD BE ABLE TO

LO-1 Define education and trace how the social institution of education has changed throughout history.

LO-2 Identify the key assumptions of functionalist, conflict, feminist, symbolic interactionist, and postmodern perspectives on education.

LO-3 Describe the major problems in elementary and secondary schools in Canada today.

LO-4 Identify the major challenges facing students in higher education institutions.

CHAPTER FOCUS QUESTION

How do race, class, and gender affect people's access to and opportunities in education?

otnaydur/Shutterstock.com

Journalist Ashley Redmond discussed the challenges faced by young people making difficult decisions around post secondary education:

When students graduate high school they are generally naive and inexperienced—yes they can drive, and some can even drink legally, but they haven't seen the world or likely been left to fend for themselves. It seems comical, then, to ask a 17- or 18-year-old to decide on a post-secondary institution, as well as a career path that will withstand the next 40 to 50 years of their life.

When I was 18, I was worried about whether my fake ID would fool anyone, since 19 is the legal drinking age in Ontario. I was also concerned about the correct rules of beer pong—specifically, the elbow/wrist rule where players MUST keep their elbows and wrists behind the edge of the table or risk disqualification; this tends to be the most argued rule at college parties.

Was I equipped to make major decisions that would likely affect the rest of my life? Across Canada this has been the expectation, now and in the previous century. However, millennials have thrown a wrench in that plan because many of them are choosing to embark on a gap year instead. A gap year is when high school graduates take time off to work, travel or volunteer before they go to college or university. Gap years are common in other parts of the world like the U.K. or Australia, where they are treated like a rite of passage. They've only recently picked up steam in Canada. According to a Statistics Canada survey of approximately 8,500 high school graduates, from 2000 to 2008, only half of students had started college or university within the usual three months, whereas 30 per cent had taken a gap year.

There are no earlier statistics available regarding gap years, but Michelle Dittmer, director of leadership and outreach at My Gap Year says it is an increasingly popular trend for a few reasons. "Teenagers are more connected globally because of the internet and they want to fit into a local and global community. They also want to make sure they are on the right path because post-secondary is such a huge investment." . . .

Source: Redmond, Ashley, 2014, "Gap Years Aren't Just for Brits Anymore," *Huffington Post*. http://www.huffingtonpost.ca/ashley-redmond/gap-year-canada_b_5948708.html

There are two central themes in Redmond's article. Firstly, more than ever before, education—in particular, higher education—is regarded as the key to success. The question for most is not *if* they will get some form of higher education but *when*. The second theme relates to the experience of most post-secondary students—rising tuition costs and increasing debt. University education is a huge expense. This is why more and more Canadian students are taking time off to ensure they make the right choice.

Most young Canadians today will attend university, correct? Although you may be one of the fortunate ones, the answer to this question is no. Increasingly, a university education is becoming a valuable life asset that only a few can afford to obtain. The most recent statistics indicate that just over 25 percent (or one in four) of Canadians aged 25 to 64 have a university degree (OECD, 2013). Given the costs associated with attaining this higher level of education, these numbers should come as no surprise. Students from families in the highest income group are twice as likely to attend university than students whose families come from the lowest income group (Belley, Frenette & Lochner, 2011). It is not only in the system of higher education that we are witnessing unequal access to "intellectual capital." Parents can contribute directly to a young child's educational success by providing a supportive environment for learning or indirectly by paving the way for a higher level of educational attainment. Increasingly, we hear of parents opting out of the public school system, placing their children in private schools, charter schools, homeschooling, or "supplementing" their education with specialized extracurricular programming—computer camps, mini-universities, or private tutoring—in an effort to make sure their child "makes it." It is apparent that only parents with the financial resources (that is, middle- or upper-income families) can afford these programs.

Does this mean that education is stratified by social class? What effect will this have on students from low-income families? Education is one of the most significant social institutions in Canada and other high-income nations. Although most social scientists agree that schools are supposed to be places where people acquire knowledge and skills, not all of them agree on how a wide array of factors—including class, race, gender, age, religion, and family background—affect individuals' access to educational achievement or to the differential rewards that accrue at various levels of academic achievement. Canada has become a "schooled society," and the education system has become a forum for competition (Davies & Guppy, 2014).

In this chapter, we will explore the issue of educational inequality in Canada, as well as look at other problems facing contemporary elementary, secondary, and higher education. Before reading on, test your knowledge about education in Canada by taking the quiz in Box 16.1.

CRITICAL THINKING QUESTIONS

1. All children in Canada have an equal opportunity to participate in elementary and secondary public schools. Does this mean that all students have an equal opportunity to succeed in school?
2. Is education in Canada stratified by social class?
3. What effects do you think this is likely to have on students from low-income families?

BOX 16.1 SOCIOLOGY AND EVERYDAY LIFE

How Much Do You Know About Education in Canada?

True	False	
T	F	1. Canada has the largest population with a university education among developed countries.
T	F	2. Children of parents with high levels of education are more likely to pursue a university education.
T	F	3. Students from families of low socioeconomic status are more likely to not attend college or university.
T	F	4. Finances are the number one source of stress for university and college students.
T	F	5. Tuition fees for postsecondary education have doubled in the past 25 years.

For answers to the quiz about education in Canada, go to **www.nelson.com/student.**

LO-1 AN OVERVIEW OF EDUCATION

Education is the social institution responsible for the systematic transmission of knowledge, skills, and cultural values within a formally organized structure. As a social institution, education imparts values, beliefs, and knowledge considered essential to the social reproduction of individual personalities and entire cultures (Bourdieu and Passeron, 1990). Education grapples with issues of societal stability and social change, reflecting society even as it attempts to shape it. Education serves an important purpose in all societies. At the microlevel, people must acquire the basic knowledge and skills they need to survive in society. At the macrolevel, the social institution of education is an essential component in maintaining and perpetuating the culture of a society across generations. **Cultural transmission**—the process by which children and recent immigrants become acquainted with the dominant cultural beliefs, values, norms, and accumulated knowledge of a society—occurs through informal and formal education. However, the process of cultural transmission differs in preliterate, preindustrial, and industrial nations.

The earliest education in *preliterate societies,* which existed before the invention of reading and writing, was informal in nature. People acquired knowledge and skills through **informal education**—learning that occurs in a spontaneous, unplanned way—from parents and other group members who provided information on survival skills, such as how to gather food, find shelter, make weapons and tools, and get along with others. Formal education for elites first came into being in *preindustrial societies,* where few people knew how to read and write. **Formal education** is learning that takes place within an academic setting, such as a school, that has a planned instructional process and teachers who convey specific knowledge, skills, and thinking processes to students. Perhaps the earliest formal education occurred in ancient Greece and Rome, where philosophers such as Socrates, Plato, and Aristotle taught elite males the skills required to become thinkers and orators who could engage in the art of persuasion (Ballantine and Hammack, 2012). During the Middle Ages, the first colleges and universities were developed under the auspices of the Catholic Church. In the Renaissance era, the focus of education shifted to the importance of developing well-rounded and liberally educated people. With the rapid growth of industrial capitalism and factories during the Industrial Revolution, it became necessary for workers to have basic skills in reading, writing, and arithmetic, and pressure to provide formal education for the masses increased significantly.

In Canada, the school reformers of the late 1800s began to view education as essential to economic growth. Ontario school reformer Egerton Ryerson promoted free schooling for all children, arguing that sending rich and poor children to the same schools would promote social harmony (Wotherspoon, 2014). By the early 1900s, **mass education**—providing free, public schooling for wide segments of a nation's population—had taken hold in Canada as the

▶ **education** The social institution responsible for the systematic transmission of knowledge, skills, and cultural values within a formally organized structure.

▶ **cultural transmission** The process by which children and recent immigrants become acquainted with the dominant cultural beliefs, values, norms, and accumulated knowledge of a society.

▶ **informal education** Learning that occurs in a spontaneous, unplanned way.

▶ **formal education** Learning that takes place within an academic setting, such as a school, that has a planned instructional process and teachers who convey specific knowledge, skills, and thinking processes to students.

▶ **mass education** Free, public schooling for wide segments of a nation's population.

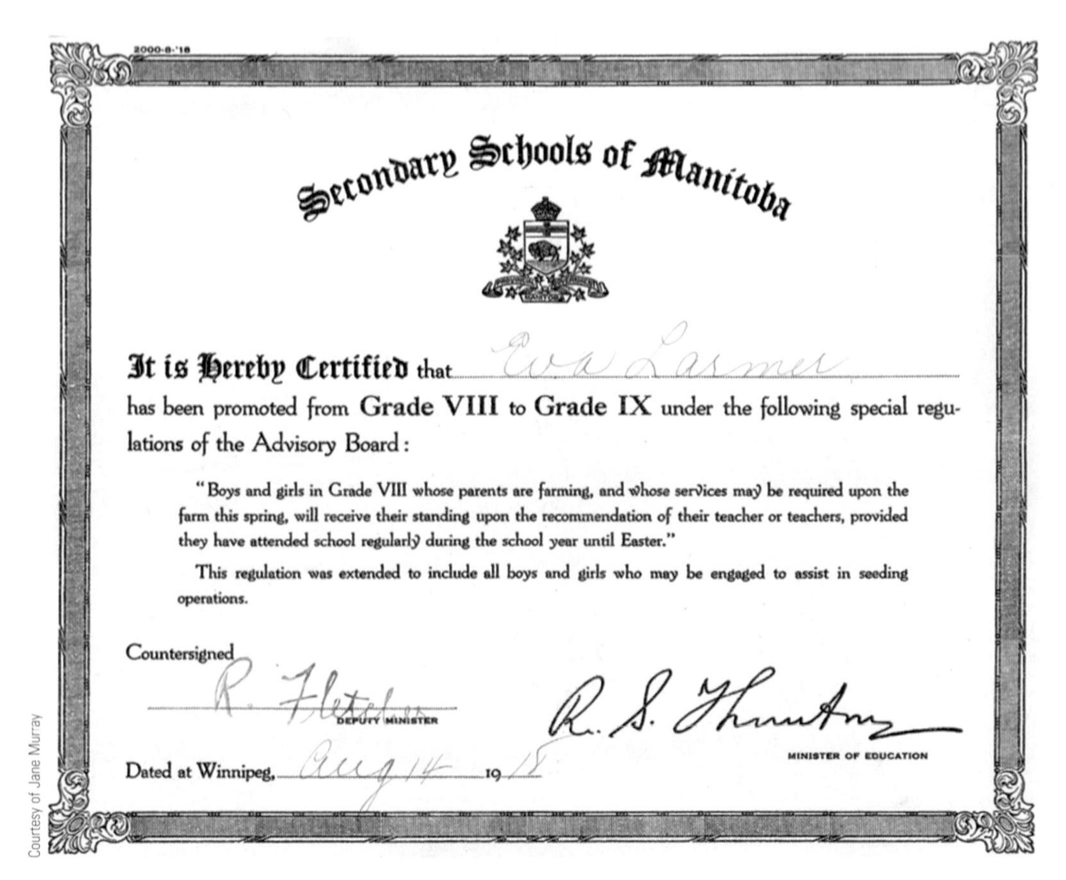
2000-8-'18

Secondary Schools of Manitoba

It is Hereby Certified that Eva Larmer
has been promoted from Grade VIII to Grade IX under the following special regulations of the Advisory Board:

"Boys and girls in Grade VIII whose parents are farming, and whose services may be required upon the farm this spring, will receive their standing upon the recommendation of their teacher or teachers, provided they have attended school regularly during the school year until Easter."

This regulation was extended to include all boys and girls who may be engaged to assist in seeding operations.

Countersigned
R. Fletcher
DEPUTY MINISTER

R. S. Thornton
MINISTER OF EDUCATION

Dated at Winnipeg, Aug 14 1918

The first schools in Canada, typically one-room schoolhouses, combined children of all ages. Attendance was sparse, as other priorities, such as working on the farm, took precedence.

provinces established free, tax-supported elementary schools available to all children. As industrialization and bureaucratization intensified, managers and business owners demanded that schools educate students beyond the third or fourth grade so that well-qualified workers would be available for rapidly emerging "white-collar" jobs in management and clerical work.

Today, schools attempt to meet the needs of industrial and post-industrial society by teaching a wide diversity of students a myriad of topics, including history and science, computer skills, how to balance a chequebook, and how to avoid contracting sexually transmitted infections (STIs). According to sociologists, many functions performed by other social institutions in the past are now under the auspices of the public schools.

LO-2 SOCIOLOGICAL PERSPECTIVES ON EDUCATION

Sociologists have divergent perspectives on education in contemporary society. Here, we examine functionalist, conflict, feminist, symbolic interactionist, and postmodernist approaches to analyzing schooling.

FUNCTIONALIST PERSPECTIVES

Functionalists view education as one of the most important components of society. According to Émile Durkheim, education is crucial for promoting social solidarity and stability in society: Education is the "influence exercised by adult generations on those that are not yet ready for

© Dick Hemingway

What values are these children being taught? Is there a consensus about what today's schools should teach? Why, or why not?

social life" (1956:28) and helps young people travel the great distance that it has taken people many centuries to cover. In other words, we can learn from what others have already experienced. Durkheim also asserted the importance of *moral education* because it conveys moral values—the foundation of a cohesive social order. He believed that schools are responsible for teaching a commitment to the common morality.

From this perspective, students must be taught to put the group's needs ahead of their individual desires and aspirations. Contemporary functionalists suggest that education is responsible for teaching social values. The 1994 Royal Commission on Learning outlined three purposes of schooling: first, to ensure for all students high levels of literacy by building on basic reading, writing, and problem-solving skills; second, to develop an appreciation of learning, the wish to continue learning, and the ability and commitment to do so; and, finally, to prepare students for responsible citizenship, including developing "basic moral values, such as a sense of caring and compassion, respect for the human person and anti-racism, a commitment to peace and non-violence, honesty and justice" (Osborne, 1999:4).

Functionalists emphasize that "shared" values should be transmitted by schools from kindergarten through university. However, not all analysts agree on what those shared values should be or what functions education should serve in contemporary societies. In analyzing the values and functions of education, sociologists using a functionalist framework distinguish between manifest and latent functions. Manifest and latent functions are compared in Figure 16.1.

MANIFEST FUNCTIONS OF EDUCATION Some functions of education are manifest functions—previously defined as open, stated, and intended goals or consequences of activities within an organization or institution. Education serves six major manifest functions in society:

1. *Socialization.* From kindergarten through university, schools teach students the student role, specific academic subjects, and political socialization.

FIGURE 16.1 MANIFEST AND LATENT FUNCTIONS OF EDUCATION

Manifest functions—open, stated, and intended goals or consequences of activities within an organization or institution. In education, these are

- socialization
- transmission of culture
- social control
- social placement
- change and innovation

Latent functions—hidden, unstated, and sometimes unintended consequences of activities within an organization. In education, these include

- restricting some activities
- matchmaking and production of social networks
- creation of a generation gap

2. *Transmission of culture.* Schools transmit cultural norms and values to each new generation. However, questions remain as to *whose* culture is being transmitted.
3. *Multiculturalism.* Schools promote awareness of and appreciation for cultural differences so that students can work and compete successfully in a diverse society and a global economy.
4. *Social control.* Schools are responsible for teaching values, such as discipline, respect, obedience, punctuality, and perseverance.
5. *Social placement.* Schools are responsible for identifying the most qualified people to fill the positions available in society. As a result, students are channelled into programs based on individual ability and academic achievement.
6. *Change and innovation.* Schools are a source of change and innovation to meet societal needs. Faculty members are responsible for engaging in research and passing on their findings to students, colleagues, and the general public.

LATENT FUNCTIONS OF EDUCATION Education serves at least three latent functions, which we have previously defined as hidden, unstated, and sometimes unintended consequences of activities within an organization or institution:

1. *Restricting some activities.* In Canada, *mandatory education laws* require that children attend school until they reach a specified age (usually 16) or complete a minimum level of formal education (generally the eighth grade). Out of these laws grew one latent function of education: keeping students off the streets and out of the full-time job market until they are older.

2. *Matchmaking and production of social networks.* Because schools bring together people of similar ages, social class, and race and ethnicity, young people often meet future marriage partners and develop lasting social networks.
3. *Creation of a generation gap.* Students learn information and develop technological skills that may create a generation gap between them and their parents, particularly as the students come to embrace a newly acquired perspective.

Functionalists acknowledge that education has certain dysfunctions. Some analysts argue that education systems in Canada are not promoting the high-level skills in reading, writing, science, and mathematics that are needed in the workplace and the global economy. However, when it comes to reading, mathematics, and science, a new international report that assesses the skill levels of students nearing the end of their compulsory education ranked Canadian students among the best in the world. Among the students of 32 participating nations, 15-year-old Canadians ranked sixth in science and reading, and fourth in mathematics (CMEC, 2014).

TIME TO REVIEW

- To what extent do you think schools today are fulfilling the manifest and latent functions outlined by functionalists?

CONFLICT PERSPECTIVES

Conflict theorists emphasize that schools solidify the privileged position of some groups at the expense of others by perpetuating class, racial–ethnic, and gender inequalities (Ballantine and Hammack, 2012). Contemporary conflict theorists also focus on how politics and corporate interests dominate schools, particularly higher education.

CULTURAL CAPITAL AND CLASS REPRODUCTION Although many factors—including intelligence, family income, motivation, and previous achievement—are important in determining how much education a person will attain, conflict theorists argue that access to high-quality education is closely related to social class. From this approach, education is a vehicle for reproducing existing class relationships. According to French sociologist Pierre Bourdieu, the school legitimates and reinforces the social elites by engaging in specific practices that uphold the patterns of behaviour and the attitudes of the dominant class. Bourdieu asserts that students from diverse class backgrounds come to school with differing amounts of **cultural capital**—social assets that include values, beliefs, attitudes, and competencies in language and culture (Bourdieu and Passeron, 1990). Cultural capital involves "proper" attitudes toward education, socially approved dress and manners, and knowledge about books, art, music, and other forms of high and popular culture. Middle- and upper-income parents endow their children with more cultural capital than do working-class and poverty-level parents. And because cultural capital is essential for acquiring an education, children with less cultural capital have fewer opportunities to succeed in school. For example, standardized tests that are used to group students by ability and assign them to classes often measure students' cultural capital rather than their "natural" intelligence or aptitude. Thus, a circular effect occurs: Students with dominant cultural values are more highly rewarded by the educational system; in turn, the educational system teaches and reinforces those values that sustain the elite's position in society.

▶ **cultural capital** Pierre Bourdieu's term for people's social assets, including their values, beliefs, attitudes, and competencies in language and culture.

TRACKING AND DETRACKING Closely linked to the issue of cultural capital is how tracking in schools is related to social inequality. **Tracking** refers to the practice of assigning students to specific curriculum groups and courses on the basis of their test scores, previous grades, or other criteria. Conflict theorists believe that tracking seriously affects many students' educational

▶ **tracking** The assignment of students to specific courses and educational programs based on their test scores, previous grades, or both.

performance and their overall academic accomplishments. Tracking first came into practice in the early 20th century, when a large influx of immigrant children entered schools for the first time and were sorted by ability and past performance. In elementary schools, tracking is often referred to as ability grouping and is based on the assumption that it is easier to teach a group of students who have similar abilities. However, class-based factors also affect which children are most likely to be placed in "high," "middle," or "low" groups, often referred to by such innocuous terms as "Blue Birds," "Red Birds." This practice is described by well-known journalist Ruben Navarrette Jr. (1997:274–275), who tells us about his own childhood experience with tracking:

> One fateful day, in the second grade, my teacher decided to teach her class more efficiently by dividing it into six groups of five students each. Each group was assigned a geometric symbol to differentiate it from the others. There were the Circles. There were the Squares. There were the Triangles and Rectangles.
>
> I remember something else, an odd coincidence. The Hexagons were the smartest kids in the class. These distinctions are not lost on a child of seven. Even in the second grade, my classmates and I knew who was smarter than whom. And on the day on which we were assigned our respective shapes, we knew that our teacher knew, too. . . .
>
> We knew also that, along with our geometric shapes, our books were different and that each group had different amounts of work to do. The Circles had the easiest books and were assigned to read only a few pages at a time. Not surprisingly, the Hexagons had the most difficult books of all, those with the biggest words and the fewest pictures, and we were expected to read the most pages.
>
> The result of all of this education by separation was exactly what the teacher had imagined that it would be: Students could, and did, learn at their own pace without being encumbered by one another. Some learned faster than others. Some, I realized only [later], did not learn at all. (1997:274–75)*

As Navarrette suggests, tracking does make it possible for students to work together based on their perceived abilities and at their own pace; however, it also extracts a serious toll from students who are labelled as "underachievers" or "slow learners." Race, class, language, gender, and many other social categories may determine the placement of children in elementary tracking systems, as much as or more than their actual academic abilities and interests.

Until the late 1990s, tracking involved grouping students by ability within subjects so that they are assigned to advanced, regular, or basic courses depending on their past performance. Although this practice was 'officially' eliminated, in reality streaming has been replicated in high schools through the placement of students in Academic, Applied, or Essentials courses (Livingston, 2014).

Overall, many scholars have documented in their research that tracking does not improve student achievement but does intensify educational inequality, particularly along racial–ethnic and class-based lines.

The detracking movement, stressing that students should be deliberately placed in classes of mixed ability, has influenced some schools and teachers. An important benefit of detracking is closing the achievement gap among students based on class or race; however, detracking is a major concern to parents of high-achieving students because they believe their children should have classes that maximize their potential, rather than holding them back with less able or less talented students. According to sociologist Maureen Hallinan (2005), rather than tracking students, schools should provide more engaging lessons for all students, alter teachers' assumptions about students, and raise students' performance requirements. Eventually, independent studies and technologies such as online learning may eliminate the "tracking" debate.

hidden curriculum The transmission of cultural values and attitudes, such as conformity and obedience to authority, through implied demands found in the rules, routines, and regulations of schools.

THE HIDDEN CURRICULUM According to conflict theorists, the **hidden curriculum** is the transmission of cultural values and attitudes, such as conformity and obedience to authority, through implied demands found in the rules, routines, and regulations of schools. Although students from all social classes are subjected to the hidden curriculum, working-class and poverty-level students may be affected the most adversely (Ballantine & Hammack, 2012; Davies and Guppy, 2014). When teachers are from a higher-class background than their students, they tend to use more structure in the classroom and to have lower expectations for students' academic achievement. Schools for working-class students emphasize procedures and rote memorization without much decision making, choice, and explanation of why something is done a particular way. Schools for middle-class students stress the processes (such as calculating and decision making) involved in getting the right answer. Schools for affluent students focus on creative activities in which students express their own ideas and apply them to the subject under consideration. Schools for students from elite families work to develop students' analytical powers and critical thinking skills, applying abstract principles to problem solving.

Over time, low-income students become frustrated with the educational system and drop out or become very marginal students, making it even more difficult for them to attend college and gain the appropriate credentials for gaining better-paying jobs. Educational credentials are extremely important in a nation such as ours that emphasizes **credentialism**—a process of social selection in which class advantage and social status are linked to the possession of academic qualifications. Credentialism is closely related to meritocracy—previously defined as a social system in which status is assumed to be acquired through individual ability and effort. Persons who acquire the appropriate credentials for a job are assumed to have gained the position through what they know, not who they are or whom they know. According to conflict theorists, the hidden curriculum determines in advance that the most valued credentials will primarily stay in the hands of the elites. Therefore, Canada is not as meritocratic as some might claim.

credentialism A process of social selection in which class advantage and social status are linked to the possession of academic qualifications.

FEMINIST PERSPECTIVES

As mentioned in previous chapters, feminism represents not only a theoretical perspective but also a broad movement for social change. One social institution on which feminism has had a significant impact is the educational system. As feminist scholar Jane Gaskell explains, "When feminists demanded equal opportunity for women, education was one of the first areas targeted for reform and rethinking" (2009:17).

GENDER BIAS AND GENDER STEREOTYPING Early feminist analysis of schooling focused on questions about sexism in the classroom and curricula, and unequal distributions of male and female educators in the system (Wotherspoon, 2014). In Canada, in the late 1960s, the Royal Commission on the Status of Women examined the relationship between education and patterns of gender inequality. It identified gender bias and gender stereotyping in the educational curriculum as a significant issue that had to be addressed. For example, girls were underrepresented in the school books, and when they were included, they appeared in rigid sex-typed roles. Boys and girls were also segregated in the playgrounds and in sporting activities. Furthermore, boys received more attention from teachers than girls, and teachers displayed stereotypical expectations of male and female students' aptitudes and interests.

In response to this report, ministries of education across the country appointed advisory groups on sexism and established guidelines to eliminate gender bias and stereotyping in school curriculum and classroom practices. According to Gaskell, "These changes had an effect . . . the critique of stereotyping had caught on. The idea that biology did not mean destiny, that equality meant open access and equal treatment, was increasingly accepted. The numbers of women in science and math, in universities, and in leadership positions in the teaching profession increased" (2009:21).

These strategies are characteristic of liberal feminism (see Chapter 12). Reforms focus on creating change within the existing social, educational, and economic systems—eliminating gender bias is an example (Wotherspoon, 2014).

More than 40 years later, women now surpass men on a number of educational indicators. Ironically, after many years of discussion about how the hidden curriculum and other problems in schools served to disadvantage female students, the emphasis has now shifted to the question of whether girls' increasing accomplishments from elementary school to college and beyond have come at the expense of boys and young men.

The issue of the "boy gap" or "boy crisis" in education has been the subject of increasing attention as discussed in a recent *Globe and Mail* series on 'failing boys': "data suggests that boys, as a group, rank behind girls by nearly every measure of scholastic achievement. They earn lower grades overall in elementary school and high school. They trail in reading and writing, and 30 per cent of them land in the bottom quarter of standardized tests, compared with 19 per cent of girls. Boys are also more likely to be picked out for behavioural problems, more likely to repeat a grade and to drop out of school altogether" (Abraham, 2010; Cappon, 2011:1).

At a postsecondary level, university and college enrolments for women now surpass those of their male counterparts. Today, over 60 percent of university graduates are female and women outnumber men in the faculties of law and medicine (OECD, 2013; Statistics Canada, 2013h). However, differences still remain. Females are highly overrepresented in nursing and teaching, while men are overrepresented in engineering, computer science, and applied mathematics (Statistics Canada, 2013h).

Rather than focusing on gender bias and stereotyping, some feminist scholars have directed our attention to the economic consequences of educational attainment, arguing that gender inequalities are not only built into the structure of schooling but also its links to the labour market. Radical feminism points out that patriarchy, or the systematic oppression of women, is reflected in the relationship between education and work. Although one might assume that more education will lead to higher income and a better job, that assumption is not borne out for many women. As Gaskell explains:

> Instead of arguing that equality will be achieved when there are as many girls as boys in mathematics and physics classes, radical feminism critiques the wages and prestige associated with the jobs that women have traditionally done. Equal-pay legislation has forced employers to recognize that the work women have done is underpaid in relation to the skills, education, and responsibility it entails. Day care workers have been paid less than dog catchers; secretaries are paid less than male technicians with equal levels of education. (2009:24)

Feminist scholars have also challenged common assumptions about learning and traditional teaching methods. They argue that men and women learn in different ways and that formal educational institutions may not adequately attend to women's "ways of knowing" (Gaskell, 2009:23). Language, science, politics, and the economy, for example, are organized around ways of knowing and doing that take the male experience as the norm. In contrast, women are frequently absent as subjects and objects of study, and women's experiences are undermined and relegated to the margins of what is considered socially important (Wotherspoon, 2009:43). Rather than denying difference, some feminists emphasize the need to recognize and incorporate female strategies of learning as well as value the knowledge constructed by women.

TIME TO REVIEW

- What are the indicators of greater gender equity in elementary schools? in universities and colleges?

SYMBOLIC INTERACTIONIST PERSPECTIVES

Unlike functionalist analysts, who focus on the functions and dysfunctions of education, and conflict theorists, who focus on the relationship between education and inequality, symbolic interactionists focus on classroom communication patterns and educational practices that affect students' self-concept and aspirations. Labelling is one such educational practice.

LABELLING AND THE SELF-FULFILLING PROPHECY According to symbolic interactionists, the process of labelling is directly related to the power and status of those persons who do the labelling and those who are being labelled. Chapter 8 explains that labelling is the process whereby others identify a person as possessing a specific characteristic or exhibiting a certain pattern of behaviour, such as being deviant. In schools, teachers and administrators are empowered to label children in various ways, including grades, written comments on classroom behaviour, and placement in classes. For some students, labelling amounts to a self-fulfilling prophecy—an unsubstantiated belief or prediction resulting in behaviour that makes the originally false belief come true (Merton, 1968).

A classic form of labelling and the self-fulfilling prophecy has occurred for many years through the use of various IQ (intelligence quotient) tests, which claim to measure a person's inherent intelligence apart from any family or school influences on the individual. Schools have used IQ tests as one criterion in determining student placement in classes and ability groups (see Figure 16.2). The way in which IQ test scores may become a self-fulfilling prophecy was revealed in the 1960s when two social scientists conducted an experiment in an elementary school during which they intentionally misinformed teachers about the scores of students in their classes (Rosenthal and Jacobson, 1968). Although the students had no measurable differences in intelligence, the researchers informed the teachers that some of the students had extremely high IQ test scores whereas others had average to below-average scores. As the researchers observed, the teachers began to teach "exceptional" students in a different manner from other students. In turn, the "exceptional" students began to outperform their "average" peers and to excel in their classwork. This study called attention to the labelling effect of IQ scores. Although the findings from this study are more than four decades old, evidence still supports what has come to be known as the Pygmalion effect in the classroom, which means that if teachers are led to expect enhanced performance from students, those students are more likely to show improvement or have higher scores than they otherwise might have (Jussim, 2013).

Today, so-called IQ fundamentalists continue to label students and others on the basis of IQ tests, claiming that these tests measure some identifiable trait that predicts the quality of

FIGURE 16.2 IQ TEST SAMPLE QUESTION

Question 2: Consider the following two statements: all farmers who are also ranchers cannot come near town; and most of the ranchers who are also farmers cannot surf. Which of the following statements MUST be true?

- Most of the farmers who cannot come near town can surf.
- Only some farmers who ranch can surf near town.
- A surfer who ranches and farms cannot surf near town.
- Some ranchers who farm can come to town to learn to surf.
- Any farmer who cannot surf also ranches.

people's thinking and their ability to perform. Critics of IQ tests continue to argue that these exams measure a number of factors—including motivation, home environment, type of socialization at home, and quality of schooling—not intelligence alone (Yong, 2011).

POSTMODERN PERSPECTIVES

How might a postmodern approach describe higher education? Postmodern theorist Jean-Francois Lyotard (1984) described how knowledge has become a commodity that is exchanged between producers and consumers. "Knowledge" is now an automated database, and teaching and learning are primarily about data presentation, stripped of their former humanistic and spiritual associations.

In the postmodern era, an emphasis in higher education is on how to make colleges and universities more efficient and how to bring these institutions into the service of business and industry. A major objective is looking for the best way to transform these schools into corporate entities such as the "McUniversity," which refers to a means of educational consumption that allows students to consume educational services and eventually obtain "goods" such as degrees and credentials, and to think of themselves and their parents as consumers. The rapidly increasing cost of higher education has contributed to the perception of "McUniversity" and to the idea of students as consumers.

Savvy college and university administrators are aware of the permeability of higher education and the "students-as-consumers" model. To attract new students and enhance current students' opportunities for consumption, most campuses have amenities such as spacious food courts with many franchise choices, ATMs, video games and gigantic HDTV screens, Olympic-sized swimming pools, and massive rock-climbing walls. Wi-Fi-enabled campuses are also a major attraction for student consumers, and virtual classrooms make it possible for some students to earn university credits without having to look for a parking place at the traditional brick-and-mortar campus.

Based on a postmodern approach, what do you believe will be the predominant means by which future students will consume educational services and goods at your college or university? For many in the second decade of the 21st century, the answer becomes that the digital age will continue to rapidly transform what we think of as knowledge and the social institution of education in which people consume new information.

The Concept Snapshot summarizes the major theoretical perspectives on education.

CONCEPT SNAPSHOT

FUNCTIONALIST PERSPECTIVES	Education is one of the most important components of society: Schools teach students not only content but also to put group needs ahead of the individual's.
CONFLICT PERSPECTIVES	Schools perpetuate class, racial-ethnic, and gender inequalities through what they teach to whom.
FEMINIST PERSPECTIVES	Schools perpetuate gender bias and stereotyping. Educational institutions also fail to recognize that men and women have different ways of learning and knowing.
SYMBOLIC INTERACTIONIST PERSPECTIVES	Labelling and the self-fulfilling prophecy are an example of how students and teachers affect one another as they interpret their interactions.
POSTMODERNIST PERSPECTIVES	In contemporary schools, educators attempt to become substitute parents and promulgators of self-esteem in students; students and their parents become the consumers of education

LO-3 CURRENT ISSUES IN ELEMENTARY AND SECONDARY SCHOOLS

Public schools in Canada today are a microcosm of many of the issues and problems facing the country. Canada is the only advanced industrialized country without a federal educational system—a fact that has made it difficult to coordinate national educational and teaching standards. Each province enacts its own laws and regulations, with local school boards frequently making the final determinations on the curriculum. Accordingly, no general standards exist as to what is to be taught to students or how, although many provinces have now adopted standards for what (at a minimum) must be learned in order to graduate from high school.

COMPETITION FOR PUBLIC SCHOOLS

Public schools do not have a monopoly on K–12 education. Today, parents have more choices of where to send their children than in the past.

PUBLIC SCHOOLS VERSUS PRIVATE SCHOOLS Often, there is a perceived conflict between public and private schools for students and financial resources. However, far more students and their parents are dependent on public schools than on private ones for providing a high-quality education. Enrolment in Canadian elementary and secondary education (kindergarten through Grade 12) totals approximately 5 million students. Almost 95 percent of elementary and secondary students are educated in public schools. The remaining 5 percent of students are educated in private schools or home schooled, and approximately 1 percent of all students attend private schools (Statistics Canada, 2013i).

Private secondary boarding schools tend to be reserved for students from high-income families and for a few lower-income or minority students who are able to acquire academic or athletic scholarships that cover their tuition, room and board, and other expenses. The average cost for seven-day tuition and room and board at secondary boarding schools is approximately $40,000 a year, whereas day-school tuition can be as high as $20,000 (Waine, 2014).

An important factor for many parents whose children attend private secondary schools is the emphasis on academics that they believe exists in private as opposed to public schools. Another is the moral and ethical standards that they believe private secondary schools instill in students. Overall, many families believe that private schools are a better choice for their children because they are more academically demanding, more motivating, more focused on discipline, and without many of the inadequacies found in public schools. However, according to some social analysts, there is little to substantiate the claim that private schools—other than elite academies attended by the children of the wealthiest and most influential families—are inherently better than public schools (Zehr, 2006).

Universal Images Group/Getty Images

Home-schooling has grown in popularity in recent decades as parents have sought to have more control over their children's education. Although some home-school settings may resemble a regular classroom, other children learn in more informal settings, such as the family kitchen.

HOME-SCHOOLING

Another alternative, home-schooling, has been chosen by some parents who hope to avoid the problems of public schools while providing a quality education for their children. Home-schooling is legal in all provinces and territories. In the past, school boards made demands of home-schooling parents, such as detailed documentation of the curriculum used, activities engaged in, hours of instruction, and methods of assessment. Today, school boards are required to accept a family's letter of intent to home-school as sufficient evidence that the parents are providing satisfactory education.

In Canada, it is estimated that about 60,000 children are home-schooled in Grades K through 12 (Ontario Federation of Teaching Parents, 2014).

The primary reasons that parents indicated for preferring to home-school their children are (1) concern about the school environment, (2) the desire to provide religious or moral instruction, and (3) dissatisfaction with the academic instruction available at traditional schools. Typically, the parents of home-schoolers are better educated, on average, than other parents; however, their income is about the same.

Parents who educate their children at home believe that their children are receiving a better education at home because instruction can be individualized to the needs and interests of their children. Some parents also indicate religious reasons for their decision to home-school their children. The Internet has made it possible for home-schoolers to gain information and communicate with one another. In some provinces, parents organize athletic leagues, proms, and other social events so that their children will have an active social life without being part of a highly structured school setting. According to advocates, home-schooled students typically have high academic achievements and high rates of employment.

Critics of home-schooling question how much parents know about school curricula and how competent they are to educate their own children at home, particularly in rapidly changing subjects such as science and computer technology.

SCHOOL DROPOUTS

Although the overall school dropout (or school leaver) rate has significantly decreased in recent decades, less than 8 percent of people under the age of 20 still leave school before earning a high school diploma. Ethnic and class differences are important factors in the data on dropout rates. For example, Indigenous students have a dropout rate of slightly less than 24 percent (Statistics Canada, 2013j). Recent estimates indicate that in Toronto, over 30 percent of black students had failed to complete high school (Livingston, 2014).

In response to these disturbingly high dropout rates, alternative specialized Indigenous and Afrocentric schools have been established across the country. As the Africentric Alternative School Support Committee explains:

> Students who have been failed by the current system will have the opportunity to learn the importance and value of their own histories and community. By learning from a perspective that cherishes the learner and his/her own history, students will be motivated to succeed. (2009:7)

Despite claims by critics that these schools are exclusionary and promote segregation, students and teachers from alternative schools report successes in terms of higher grades, lower drop out rates, and greater student motivation (Allen, 2010).

The dropout rate also varies by region—Manitoba, Alberta, and Quebec had the highest proportion of school dropouts in 2012, while British Columbia had the lowest dropout rate of just under 6 percent (Employment and Social Development Canada, 2013b).

Why do students drop out of school? Some students believe that their classes are boring; others are skeptical about the value of schooling and think that completing high school will not increase their job opportunities. Upon leaving school, many dropouts have high hopes of making money and enjoying newfound freedom; however, many find that few jobs are available and they do not have the minimum education required for any "good" jobs that exist.

Although critics of the public education system point to dropout rates as proof of failure in the public education system, these rates have steadily declined in Canada since the 1950s, when more than 70 percent of students did not complete high school (Employment and Social Development Canada, 2013b).

EQUALIZING OPPORTUNITIES FOR STUDENTS WITH DISABILITIES

Another relatively recent concern in education has been how to provide better educational opportunities for students with disabilities. See Figure 16.3.

As shown in Chapter 15, the term *disability* has a wide range of definitions (see Shapiro, 1993). For the purposes of this chapter, disability is regarded as any physical and/or mental condition that limits students' access to, or full involvement in, school life.

The barriers facing students with disabilities are slowly being removed or surmounted by new legislation. Today, most people with disabilities are no longer prevented from experiencing the full range of academic opportunities. Under various provincial human rights guidelines and the *Charter of Rights and Freedoms,* all children with disabilities are guaranteed a free and appropriate public education. This guarantee means that local school boards must make the necessary efforts and expenditures to accommodate students with special needs (Uppal, Kohen, and Khan, 2007).

A recent case before the Supreme Court reaffirmed the rights of disabled students to participate fully in public education and the mandatory obligation of school boards to provide the services necessary to accommodate the 'special needs' of these students. The case involved Jeffrey Moore, a severely dyslexic student who had lost access to special education services when the North Vancouver school district closed its diagnostic centre for budgetary reasons. The Supreme Court was unanimous in its decision that this was discrimination. They wrote:

> Adequate special education . . . is not a dispensable luxury. For those with severe learning disabilities, it is the ramp that provides access to the statutory commitment to education made to all children in British Columbia. (*Moore v. British Columbia,* 2012 SCC 61 cited in KOMLJENOVIC, 2013:21)

Many schools have attempted to *mainstream* children with disabilities by *inclusion programs,* under which the special education curriculum is integrated with the regular education program and each child receives an *individualized education plan* that provides annual education goals. (Inclusion means that children with disabilities work with a wide variety of people; over the course of a day, children may interact with their regular education teacher, the special education teacher, a speech therapist, an occupational therapist, a physical therapist, and a resource teacher, depending on the child's individual needs.) Today, more than 85 percent of children with disabilities are integrated into mainstream schools. Twenty years ago, more than 80 percent

FIGURE 16.3 : PERSONS WITH AND WITHOUT DISABILITIES BY EDUCATIONAL ATTAINMENT

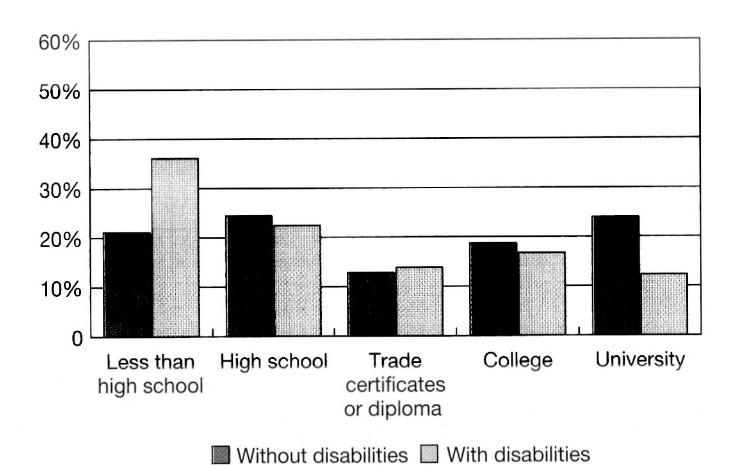

Source: Human Resources and Social Development, *Advancing the Inclusion of People with Disabilities 2006.*

of these children were placed in segregated schools. This dramatic change reflects growing acceptance of the fact that children with a range of disabilities often thrive in an integrated learning environment.

Despite progress, access to education for persons with disabilities remains challenging. What the parents of Jeffrey Moore had to endure is not isolated as parents of children with special needs report often facing roadblocks to support, unnecessary bureaucratic requirements, budgetary constraints, and a lack of involvement and knowledge in the resources available for their children (Komljenovic, 2013).

Courtesy of Jane Murray

This is author Jane Lothian Murray's son. As Drew demonstrated, children with a range of disabilities thrive in an integrated learning environment as do the other students who learn so much from children with challenges.

SCHOOL SAFETY AND VIOLENCE AT ALL LEVELS

Although incidents of extreme violence in school are rare in Canada compared to other countries such as the United States, we are not immune to these tragedies. In 1989, 14 women were shot and killed at Montreal's École Polytechnique. In 1999, a 14-year-old boy walked into his high school in Taber, Alberta, and began shooting, killing one student and wounding another. In September 2006, 18 students were shot and one student was killed at Dawson College in Montreal. The aftermath of each of these tragedies saw a massive outpouring of public sympathy and a call for greater campus security. Gun-control advocates called for greater control over the licensing and ownership of firearms and for heightened police security on college campuses.

Today, officials in schools from the elementary level to two-year colleges and four-year universities are focusing on how to reduce or eliminate violence. In many schools, teachers and counsellors are instructed in anger management and peer mediation, and they are encouraged to develop classroom instruction that teaches values such as respect and responsibility (Canadian Safe Schools Network, 2012). Some schools create partnerships with local law enforcement agencies and social service organizations to link issues of school safety to larger concerns about safety in the community and the nation.

Clearly, some efforts to make schools a safe haven for students and teachers are paying off. Statistics related to school safety continue to show that Canadian schools are among the safest places for young people (Canadian Safe Schools Network, 2012). However, these statistics do not keep many people from believing that schools are becoming more dangerous with each passing year and that all schools should have high-tech surveillance equipment to help maintain a safe environment. And despite the public fear and concern precipitated by highly publicized and often sensationalized events, recent school safety efforts are focused on a more pervasive form of school violence that occurs on a daily basis in schools across the country—bullying.

Bullying affects up to 60 percent of students, depending on grade level and gender. Over 90 percent of students have witnessed bullying in their school environment (Craig, Peplar, and Haner, 2012). This type of school violence takes many forms and includes physical, psychological, and emotional abuse, threats, and intimidation. Victims are often targeted on the basis of race or ethnic origin, sexual orientation, or disability. The effects of bullying can include learning problems, low self-esteem, mental health problems, such as anxiety and depression, substance abuse, and, in far too many cases, suicide.

A more recent form of bullying—cyberbullying—has become a common method of accessing both students and teachers with relative anonymity. Several tragic incidents involving cyberbullying have highlighted the serious consequences of this form of bullying. In 2013, 17-year-old Nova Scotia student Rehtaeh Parsons committed suicide after enduring several months of cyberbullying. A year and a half before she took her life, Parsons was sexually assaulted by four boys in her school. One of the boys involved in the incident distributed photos of the rape on a social media site, which led to months of unrelenting cyberbullying. Rehtaeh's tragic death has raised public awareness of cyberbullying and sexual assault, and has led to anti-bullying initiatives across Canada, including amending the *Criminal Code* to include "bullying" and/or "cyberbullying," legislative changes directed at defining bullying and its aggressive and harmful forms, providing support to victims, and strengthening the responsibilities of schools to investigate and impose sanctions on individuals involved in cyberbullying (Mian, 2013).

TIME TO REVIEW

- What strategies are being used in Canadian schools to address the problem of school violence and bullying?

LO-4 ISSUES IN POSTSECONDARY EDUCATION

INCREASING COSTS OF POSTSECONDARY EDUCATION

Who attends college or university? What sort of college or university do they attend? Even for students who complete high school, access to colleges and universities is determined not only by prior academic record but also by the ability to pay.

Postsecondary education has been described as the dividing line of the modern labour market. Today more than ever before, employers want employees with a university degree, college diploma, or some other form of postsecondary educational certificate. As shown in Figure 16.4, for most Canadians, higher education will result in higher earnings. To obtain a university education, however, students must have the necessary financial resources. What does a university education cost?

In Canada, postsecondary education is funded by the federal and provincial governments, and by parents and students through personal savings. As governments cut their funding to higher education, an increasing financial burden is falling on the shoulders of students and their parents. To make matters worse, the cost of attending university has increased dramatically over the past 25 years.

FIGURE 16.4 LEVELS OF EARNING AND EDUCATION, 2006

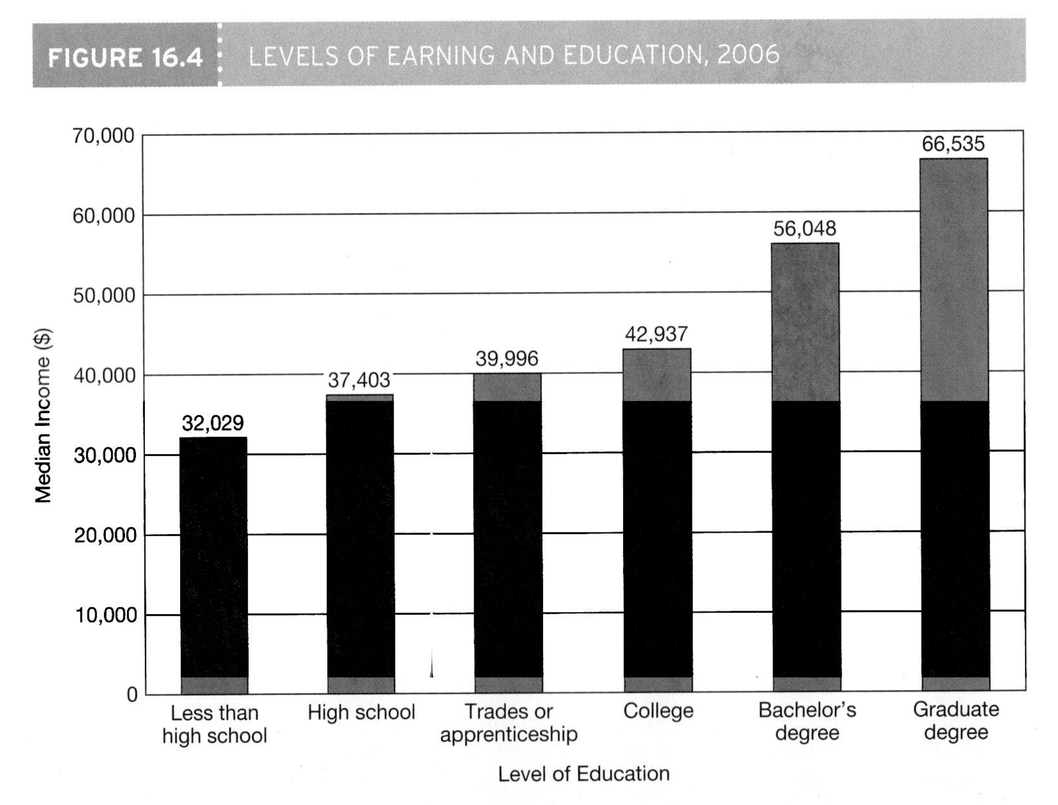

Source: Adapted from Statistics Canada, Census of Canada 2006, 2008 Income and Earnings Highlight Tables.

Tuition fees have been rising faster than the rate of inflation (see Figure 16.5). For example, between 1991 and 2014, average tuition fees for an undergraduate degree went from approximately $1500 to more than $6000. College fees, although lower than the price of university, have also increased (Canadian Federation of Students, 2013). A four-year degree costs on average approximately $60,000. Despite the soaring cost of postsecondary education, the percentage of young people attending university continued to rise in the first years of the 1990s; however, undergraduate enrolments have started to level off in the past five years, an indication that for some students, the cost of a university education has become prohibitive (Statistics Canada, 2013k).

How do students afford this increasingly costly education? In a survey that explored this question, both college and university graduates identified employment earnings and student loan programs as their primary sources of funding. Parents ranked a close third for university graduates. Scholarships, fellowships, grants, and bursaries were rarely identified as a significant source of funding. Approximately 60 percent of college and university students indicated that they relied on student loans to finance their education. Students surveyed anticipated debt of close to $30,000 once their education was completed (MacDonald & Shaker, 2012).

A substantial proportion of postsecondary students choose community college because of the lower costs. However, the overall enrolment of low-income students in community colleges has dropped as a result of increasing costs and also because many students must work full-time or part-time to pay for their education. Many Canadian colleges have implemented three- and four-year degree-granting programs that are an excellent option but may be cost and time prohibitive for low-income students. In contrast, students from more affluent families are more likely to attend prestigious public universities or private colleges outside of Canada, where tuition fees alone may be more than $20,000 per year.

According to some social analysts, a university education is a bargain and a means of upward mobility. However, other analysts believe that the high cost of a university education reproduces the existing class system: Students who lack money may be denied access to higher education, and those who are able to attend college or university tend to receive different types of education based on their ability to pay. For example, a community college student who receives an associate's degree or completes a certificate program may be prepared for a position in the middle of the occupational status range, such as a dental assistant, computer programmer, or auto mechanic.

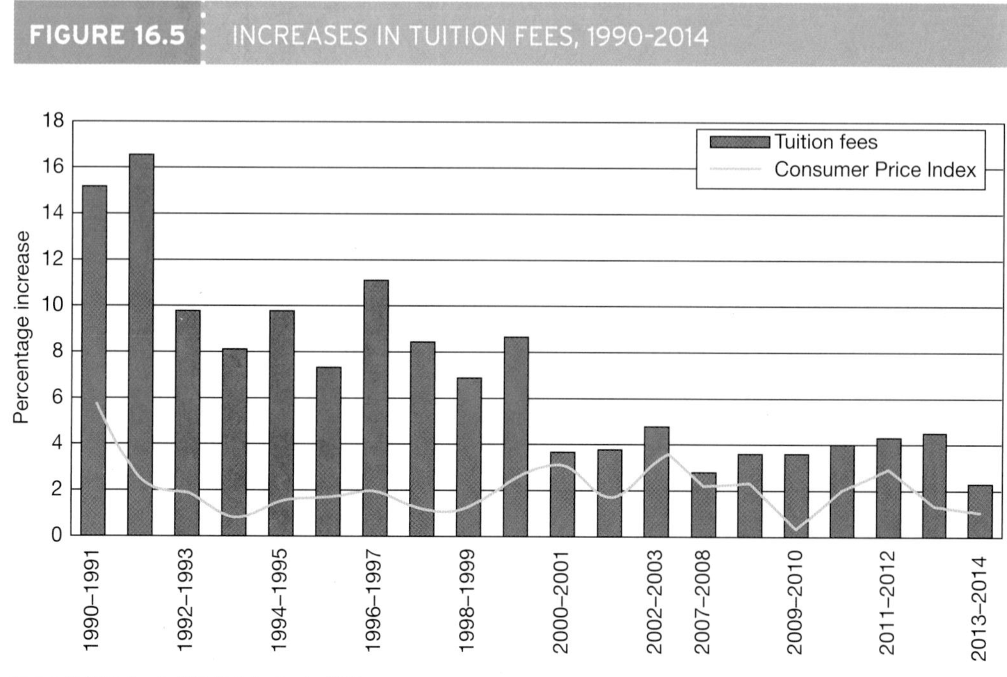

FIGURE 16.5 : INCREASES IN TUITION FEES, 1990-2014

Sources: Statistics Canada, 2004; Statistics Canada, 2012d.

ROGERIO BARBOSA/AFP/GettyImages

In the spring of 2012, tens of thousands of students took to the streets of Montreal to protest tuition increases.

In contrast, university graduates with four-year degrees are more likely to find initial employment with firms where they stand a chance of being promoted to high-paying management and executive positions. Although higher education may be a source of upward mobility for talented young people from poor families, the Canadian system of higher education is sufficiently stratified that it may also reproduce the existing class structure (Barlow and Robertson, 1994; Davies and Guppy, 2014; MacDonald & Shaker, 2012).

DECREASING GOVERNMENT FUNDING

Thirty years ago, 80 percent of the operating costs of universities and colleges were covered by government grants. Today, as a result of federal government funding cuts, government grants now cover only 50 percent of operational costs. Tuition fees and other compulsory fees paid by students make up the shortfall in government funding (Canadian Federation of Students, 2014). Although some provincial governments have directed money to offset tuition fee increases, these tend to be in the form of loans rather than grants and tax credits that are only available long after tuition is paid. Among those most harmed by spiralling tuition costs, rising student debt, and less availability of grant money that does not have to be repaid are lower- and middle-income students. Collectively, the changes in higher education have led to a system that is much less progressive and accessible for students. As education scholar Erika Shaker comments:

> It would be much more efficient—and much less divisive—to apply government transfers up-front in the form of immediate tuition fee reductions . . . or even (gasp!) eliminating fees altogether. This would benefit all students regardless of their home province, or their income before or after graduation. It would also demonstrate a universal commitment to affordability—ensuring all students regardless of individual circumstance have the opportunity to broaden their educational horizons without being saddled with debt upon graduation—and to the recognition that the very high social and economic returns from higher education are collectively shared and well worth the investment. (Shaker & Macdonald, 2013:133)

LO-1 Define education and trace how the social institution of education has changed throughout history.

Education is the social institution responsible for the systematic transmission of knowledge, skills, and cultural values within a formally organized structure. In preliterate societies, people acquired knowledge and skills through informal education from parents and other group members. Formal education—learning that takes place within an academic setting such as a school, that has a planned instructional process and teachers who convey specific knowledge—first became available to members of the elite class in preindustrial societies. In industrial and post-industrial societies, education is provided through mass education, where all members of society have access to publicly funded education.

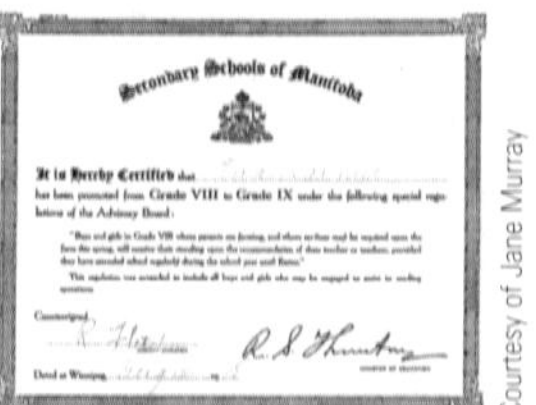

Courtesy of Jane Murray

LO-2 Identify the key assumptions of functionalist, conflict, feminist, symbolic interactionist, and postmodern perspectives on education.

According to functionalists, education has both manifest functions (socialization, transmission of culture, multiculturalism, social control, social placement, and change and innovation) and latent functions (keeping young people off the streets and out of the job market, matchmaking and producing social networks, and creating a generation gap). From a conflict perspective, education is used to perpetuate class, racial–ethnic, and gender inequalities through tracking, ability grouping, and a hidden curriculum that teaches subordinate groups conformity and obedience. Feminist scholars direct our attention to gender bias and stereotyping, the economic consequences of educational attainment. They also argue that men and women learn in different ways and that formal educational institutions may not adequately attend to women's "ways of knowing." Interactionists examine classroom dynamics and study ways in which a practice, such as labelling, may become a self-fulfilling prophecy for some students, such that these students come to perform up—or down—to the expectations held for them by teachers. Some postmodernists suggest that in a consumer culture, education becomes a commodity that is bought by students and their parents. Colleges and universities function as "McUniversities" that allow students to consume educational services and obtain goods such as degrees and credentials.

© Dick Hemingway

LO-3 Describe the major problems in elementary and secondary schools in Canada today.

Dropping out, unequal educational opportunities for students with disabilities, and school violence, primarily in the form of bullying, are among the most pressing issues in elementary and secondary schools.

Courtesy of Jane Murray

KEY TERMS

credentialism A process of social selection in which class advantage and social status are linked to the possession of academic qualifications (p. 440).

cultural capital Pierre Bourdieu's term for people's social assets, including their values, beliefs, attitudes, and competencies in language and culture (p. 438).

cultural transmission The process by which children and recent immigrants become acquainted with the dominant cultural beliefs, values, norms, and accumulated knowledge of a society (p. 434).

education The social institution responsible for the systematic transmission of knowledge, skills, and cultural values within a formally organized structure (p. 434).

formal education Learning that takes place within an academic setting, such as a school, that has a planned instructional process and teachers who convey specific knowledge, skills, and thinking processes to students (p. 434).

hidden curriculum The transmission of cultural values and attitudes, such as conformity and obedience to authority, through implied demands found in the rules, routines, and regulations of schools (p. 440).

informal education Learning that occurs in a spontaneous, unplanned way (p. 434).

mass education Free, public schooling for wide segments of a nation's population (p. 434).

tracking The assignment of students to specific courses and educational programs based on their test scores, previous grades, or both (p. 438).

LO-4 Identify the major challenges facing students in higher education institutions.

The combined effects of funding cuts and the soaring cost of a college or university education are among the most pressing issues in higher education in Canada today.

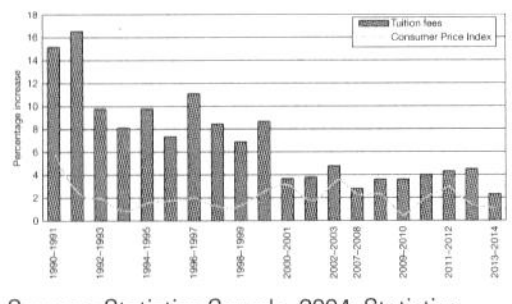

Sources: Statistics Canada, 2004; Statistics Canada, 2012d.

APPLICATION QUESTIONS

1. Why do some theorists believe that education is a vehicle for decreasing social inequality whereas others believe that education reproduces existing class relationships?
2. Why does so much controversy exist over what should be taught in Canadian public schools?
3. How has education shaped your life in both direct and indirect ways?

CHAPTER

17 RELIGION

The Canadian Press/Darryl Dyck

CHAPTER FOCUS QUESTION

What is the relationship between society and religion, and what role does religion play in people's everyday lives?

LEARNING OBJECTIVES

AFER READING THIS CHAPTER, YOU SHOULD BE ABLE TO

LO-1 Describe what religion is and understand its purpose in society.

LO-2 Understand the differences between the functionalist, conflict, interactionist, feminist, and postmodern perspectives on religion.

LO-3 Understand the different types of religious organizations.

LO-4 Discuss the impact of religion on people's attitudes and behaviour.

LO-5 Consider the future role of religion in Canada.

For most of Canada's history, religion has played a major role in the development and maintenance of many of our institutions, including our schools and universities. Religious instruction was an essential component of "becoming educated." The religion taught was Christianity. However, because religion is becoming increasingly diverse and because Canadian society is now very secular, there is no longer any consensus regarding the role religion should play in education. Should students receive religious instruction in the classroom? If so, which religions should be included? Should prayer be offered in schools? Should students be required to follow religious codes of behaviour? Answers to these questions are becoming increasingly controversial.

The debate over the establishment of a law school at Trinity Western University (TWU) highlights some of the issues concerning religious education as well as broader issues of the role of religion in society. TWU is a private Christian university in British Columbia whose staff and students must sign a Christian Covenant that, among other things, states that sexual relations are permissible only within a marital relationship and between a man and a woman.

Many people have objected to the establishment of the law school. Consider the comments of law student Stephane Erickson (2014):

How hurtful, how absolutely senseless—the thought of having a Canadian law school accessible to all, to the exclusion of those who do not, or cannot, adhere to heterosexual marriage. The obvious questions follow: How is a law school, which does not recognize the legitimacy of civil unions, same-sex marriage, and non-traditional family structures, going to ensure an accurate and sincere legal education? How is the Charter going to be taught with respect to women's rights, LGBT rights, and other issues. . . . Moreover, and maybe most importantly, how is the school going to ensure that students feel safe in an environment morally bound by religious doctrine and skewed interpretations of sacred texts?

The legal battle over TWU's law school has centred on whether its graduates will be accredited by provincial law societies. The Canadian Council of Law Deans said that the Covenant was "fundamentally at odds with the core values of all Canadian law schools," and the law societies of British Columbia, Ontario, and Nova Scotia have refused to accredit TWU graduates. On the other hand, the Law Society of Canada, the law societies of Alberta and Saskatchewan, and the B.C. Civil Liberties Association have supported accreditation.

Bob Kuhn, the President of TWU, has argued:

[The] debate has raised the issue of the balance of equal rights of gays and lesbians, and religious freedoms. But in reality, Parliament and the courts have already answered that question. Within the legislation that changed the definition of marriage in 2005, it is clear that religious communities, such as TWU, retain the right to define marriage according to their religious precepts. Further, the Supreme Court of Canada came to the same conclusion when it ruled in favour of Trinity Western University on the same issue [concerning graduates of TWU's Faculty of Education] in 2001. And in the Reference re Same-Sex Marriage in 2005, the court again affirmed the need for protection of religious freedom. (Kuhn, 2014)

In response to the notion that graduates would be biased in their practice of law, President Kuhn said that "Thirty-five years, I've been practising law. . . I signed that covenant very easily because that's what I believe, but it doesn't eliminate my ability to perform my legal practice for any number of clients, some of whom are members of the LGBT [lesbian, gay, bisexual, transgender] community (Patten, 2014)."

In 2015, a Nova Scotia court ruled that the Nova Scotia Law Society could not deny Trinity Western's accreditation, and an Ontario court ruled that the Ontario Law Society was allowed to deny accreditation to the school. Both decisions are being appealed and, ultimately, it is likely that the Supreme Court of Canada will have to decide the matter.

What do you think about this issue? Should private universities be allowed to set policies based on their religious beliefs? Can graduates of schools that have discriminatory beliefs practice professions such as teaching and law in a fair and objective fashion? Should religion ever play a role in education in Canada?

What role does religion play in Canada's educational systems today? Religion plays almost no role in the public school system. Those who wish to combine education with religious instruction must do so through private schooling. Education falls within provincial jurisdiction, however, and some provinces provide public funding to Roman Catholic separate schools (Holmes, 1998). Saskatchewan and Ontario fully fund Roman Catholic schools, but Ontario does not support schools operated by members of other religious denominations, while Saskatchewan does provide funding for private schools, including religious schools. Manitoba and British Columbia provide funding to a wide variety of private schools (many of which are religious) based on academic criteria. Alberta allows school boards to establish faith-based schools. At the post-secondary level, there are over 40 Christian colleges and universities in Canada.

As the issues concerning religious education suggest, religion can be a highly controversial topic. One group's deeply held beliefs or cherished religious practices may be a source of irritation to another group. Religion is a source of both stability and conflict throughout the world. In this chapter, we will examine how religion influences life in Canada and in other countries. Before reading on, test your knowledge about how religion affects public education in this country by taking the quiz in Box 17.1.

CRITICAL THINKING QUESTIONS

1. In 2004, a Canadian university was criticized for not providing prayer space for Muslim students. Do you think that public universities should be required to provide on-campus prayer space for all observant students?

2. What role does religion play in your life? Does it ever come up in everyday conversations? Is it discussed in any of your classes? Do you see any differences between your acquaintances who are religious and those who are not?

3. A Nova Scotia high school student was suspended for a week for wearing a T-shirt saying "Life is wasted without Jesus" (Winnipeg Free Press, 2012). Was the school justified in restricting this student's freedom of expression when other students are allowed to wear religious symbols to school? Would you have a different view if the student had been aggressively trying to get other students to support his Christian beliefs?

LO-1 THE SOCIOLOGICAL STUDY OF RELIGION

▶ **religion** A system of beliefs, symbols, and rituals, based on some sacred or supernatural realm, that guides human behaviour, gives meaning to life, and unites believers into a community.

What is religion? **Religion** is a system of beliefs, symbols, and rituals, based on some sacred or supernatural realm, that guides human behaviour, gives meaning to life, and unites believers into a community (Durkheim, 1995/1912). Religion is one of our most significant social institutions. It consists of several elements, including beliefs about the sacred or supernatural, rituals, and a social organization of believers drawn together by their common religious tradition (Kurtz, 1995). Most religions attempt to answer fundamental questions about the meaning of life and death, and how the world was created. Most religions also provide comfort to persons facing emotional traumas, such as illness, grief, and death.

According to Kurtz, religious beliefs are typically woven into a series of narratives, including stories about how ancestors and other significant figures had meaningful experiences with supernatural powers (1995:9). Religious beliefs help to bind people together and are usually involved in rites of passage, such as birth, marriage, and death. People with similar religious beliefs and practices often come together in a moral community based on a church, mosque, temple, or synagogue.

RELIGIOUS BELIEF AND RITUAL

▶ **faith** Unquestioning belief that does not require proof or scientific evidence.

▶ **sacred** A term used to describe those aspects of life that are extraordinary or supernatural.

▶ **profane** A term used for the everyday, secular, or "worldly" aspects of life.

▶ **rituals** Regularly repeated and carefully prescribed forms of behaviour that symbolize a cherished value or belief.

Religions seek to answer important questions, such as why we exist and what happens after death. Peter Berger (1967) referred to religion as a *sacred canopy*—a sheltering fabric hanging over people that gives them security and provides answers to the questions of life. This sacred canopy requires that people have **faith**—unquestioning belief that does not require proof or scientific evidence. Religious answers point to the **sacred**, those aspects of life that are extraordinary or supernatural—those things that are set apart as "holy" (Durkheim, 1995/1912). People feel a sense of awe, reverence, deep respect, or fear for what is considered sacred. Across cultures and at different times, many things have been considered sacred, including invisible gods, spirits, specific animals or trees, altars, crosses, and holy books (Collins, 1982). The things that people do not consider sacred are referred to as the **profane**—the secular, or "worldly" aspects of life (Collins, 1982). Sacred beliefs are rooted in the holy or supernatural, while secular beliefs have their foundation in scientific knowledge or everyday explanations. For example, in the educational debate over creationism and evolutionism, creationists view their belief as founded in sacred teachings, but evolutionists argue that their beliefs are based on scientific data.

In addition to beliefs, religion also involves symbols and rituals. Religious **rituals** are regularly repeated and carefully prescribed forms of behaviour that symbolize a cherished value or belief (Kurtz, 1995). Rituals range from songs and prayers to offerings and sacrifices that worship or praise a supernatural being or a set of supernatural principles. For example, Muslims bow toward the holy city of Mecca five times a day to pray to God, and Christians participate in Holy Communion to commemorate the life, death, and resurrection of Jesus Christ. The rituals involved in praying or in observing communion are carefully orchestrated and must be followed with precision. According to Collins, "In rituals, it is the forms that count.

This Nova Scotia student was suspended from high school for wearing a T-shirt expressing his religious beliefs.

BOX 17.1 SOCIOLOGY AND EVERYDAY LIFE

How Much Do You Know About the Impact of Religion on Education in Canada?

True	False	
T	F	1. Parents who home-school their children for religious reasons are free to teach the children whatever curriculum they wish.
T	F	2. Enrolment in parochial (religious) schools has decreased in Canada as interest in religion has waned.
T	F	3. In Canada, the public school system recognizes only Christian religious holidays by giving students those days off.
T	F	4. The number of children from religious backgrounds other than Christian and Judaic has grown steadily in schools over the past several decades.
T	F	5. Debates over textbook content focus only on elementary education because of the vulnerability of young children.

For answers to the quiz about the impact of religion on education in Canada, go to **www.nelson.com/student**.

Saying prayers, singing a hymn, performing a primitive sacrifice or a dance, marching in a procession, kneeling before an idol or making the sign of the cross—in these, the action must be done the right way" (1982:34).

Rituals and other religious regulations are important because the purpose of religion is to explain fundamental questions, such as death and the meaning of life. Rodney Stark has pointed out that religions do more for humans than "supply them with answers to questions of ultimate meaning.

The assumption that the supernatural exists raises a new question: *What does the supernatural want or expect from us?*" (1998:386). Thus, religions also provide the faithful with rules about how they must act if they are to please the gods. These rules are justified in religious terms, and following the rules is a sign of faith.

CATEGORIES OF RELIGION

Anthropologists have concluded that all known groups over the past 100,000 years have had some form of religion (Haviland, 1993). Religions have been classified into four main categories based on their dominant belief: simple supernaturalism, animism, theism, and transcendent idealism (McGee, 1975). In simple preindustrial societies, religion often takes the form of **simple supernaturalism**—the belief that supernatural forces affect people's lives. This type of religion does not acknowledge specific gods or spirits but focuses on impersonal forces that may exist in people or natural objects. For example, simple supernaturalism has been used to explain mystifying events of nature, such as sunrises and thunderstorms, and ways that some objects may bring a person good or bad luck. By contrast, **animism** is the belief that plants, animals, or other elements of the natural world are endowed with spirits or life forces that have an impact on events in society. Animism is associated with early hunting and gathering societies in which everyday life is not separated from the elements of the natural world (Albanese, 1992).

The third category of religion is **theism**—a belief in a god or gods. Horticultural societies were among the first to practise **monotheism**—a belief in a single, supreme being or god who is responsible for significant events, such as the creation of the world. Three of the major world religions—Christianity, Judaism, and Islam—are monotheistic. By contrast, Shinto and a number of the Indigenous religions of Africa are forms of **polytheism**—a belief in more than one god. The fourth category of religion, transcendent idealism, is a **nontheistic religion**—a religion based on a belief in divine spiritual forces, such as sacred principles of thought and conduct, rather than a god or gods. Transcendent idealism focuses on principles, such as truth, justice, affirmation of life, and tolerance for others, and its adherents seek an elevated state of consciousness in which they can fulfill their true potential.

▶ **simple supernaturalism** The belief that supernatural forces affect people's lives either positively or negatively.

▶ **animism** The belief that plants, animals, or other elements of the natural world are endowed with spirits or life forces that have an impact on events in society.

▶ **theism** A belief in a god or gods.

▶ **monotheism** A belief in a single, supreme being or god who is responsible for significant events, such as the creation of the world.

▶ **polytheism** A belief in more than one god.

▶ **nontheistic religion** A religion based on a belief in divine spiritual forces, such as sacred principles of thought and conduct, rather than a god or gods.

LO-2 SOCIOLOGICAL PERSPECTIVES ON RELIGION

The major sociological perspectives have very different views about the relationship between religion and society. Functionalists emphasize the ways in which religious beliefs and rituals can bind people together by giving them a sense of purpose and connecting them to their societies. Conflict explanations suggest that religion can be a source of false consciousness in society. Interactionists focus on the meanings that people give to religion in their everyday lives. Feminists look at the ways in which women's religious experiences differ from those of men. Postmodern theorists examine the changing nature and role of religion in the 21st century.

The Concept Snapshot later in this chapter outlines what the proponents of various perspectives say about religion.

FUNCTIONALIST PERSPECTIVES ON RELIGION

DURKHEIM ON RELIGION Durkheim believed that religion is essential for the maintenance of society. Religion was found in all societies because it met basic human needs and served important societal functions.

For Durkheim, the central feature of all religions is the presence of sacred beliefs and rituals that bind people together. In his studies of the religion of the Australian Aborigines, for example, Durkheim found that each clan had established its own sacred totem, including kangaroos, trees, rivers, rock formations, and other animals or natural creations. To clan members, their totem was sacred; it symbolized some unique quality of their clan. People developed a feeling of unity by performing ritual dances around their totem, which caused them to abandon individual self-interest.

FUNCTIONS OF RELIGION Religion has three important functions in any society:

1. *Providing meaning and purpose to life.* Religion offers meaning for the human experience. Some events create a profound sense of loss for both individuals and groups. Individual losses include the death of a loved one; group losses include famines or earthquakes. Inequality may cause people to wonder why their personal situation is no better than it is. Most religions offer explanations for these concerns. Explanations differ among religions, but each tells the individual or group that life is part of a larger system (McGuire, 1997). Some (but not all) religions offer the hope of an afterlife for persons who follow the religion's tenets of morality in this life. Such beliefs help make injustices in this life easier to endure.
2. *Promoting social cohesion and a sense of belonging.* By emphasizing shared symbolism, religion helps promote social cohesion. The Christian ritual of Holy Communion not only commemorates a historical event but also allows followers to participate in their unity ("communion") with other believers (McGuire, 1997). Religion can also help members of subordinate or minority groups develop a sense of social cohesion and belonging. For example, in the late 1980s and early 1990s, Russian Jewish immigrants to Canada found a sense of belonging in their congregations. Even though they did not speak the language of their new country, they had religious rituals and a sense of history in common with others in the congregation. In Calgary, the Baptist minister at a church with a congregation made up of 1500 people of Korean background commented, "The church is more than a Christian institution, it is also a means of cultural fellowship. It is a place to feel comfortable. They are in a strange country and here there is friendship" (Nemeth, Underwood, and Howse, 1993:33).
3. *Providing social control and support for the government.* If individuals consider themselves part of a larger order that holds the ultimate meaning in life, they will feel bound to one another and to past and future generations (McGuire, 1997). Religion also helps maintain social control by conferring supernatural legitimacy on the norms and laws in society. In some societies, social control occurs as a result of direct collusion between the dominant classes and the dominant religious organizations. Absolute monarchs have often claimed that God gave them the right to rule so their citizens could not question their legitimacy to govern.

CONFLICT PERSPECTIVES ON RELIGION

KARL MARX ON RELIGION While most functionalists feel that religion serves a positive role, many conflict theorists view religion negatively. For Marx, ideologies—"systematic views of the way the world ought to be"—are embodied in religious doctrines and political values (Turner, Beeghley, and Powers, 1995:135). These ideologies also justify the status quo and block social change. The capitalist class uses religious ideology to mislead the workers about their true interests. For Marx, religion is the "opium of the people." People become complacent because they have been taught to believe in an afterlife in which they will be rewarded for their suffering and misery in this life. Although these religious teachings soothe the masses, any relief is illusory. Religion unites people under a "false consciousness" that leads them to believe they have common interests with members of the dominant class (Roberts, 1995b).

Rick Diamond/Getty Images

The shared experiences and beliefs associated with religion have helped many groups maintain a sense of social cohesion and a feeling of belonging in the face of prejudice and discrimination.

© Hemis/Alamy

According to Marx and Weber, religion serves to reinforce social stratification in a society. According to Hindu belief, for example, a person's social position in his or her current life is a result of behaviour in a former life.

From a conflict perspective, religion also tends to promote strife between groups and societies. The conflict may be *between* religious groups (religious wars), *within* a religious group (a splinter group leaving an existing denomination), or between a religious group and *the larger society* (conflict over religion in the classroom). Conflict theorists assert that, in attempting to provide meaning and purpose in life while at the same time promoting the status quo, religion is used by the dominant classes to impose their own control over society and its resources (McGuire, 1992).

The experience of Canada's Indigenous people is consistent with the conflict approach to religion. Prior to European settlement, Indigenous spirituality was based on living in harmony with nature and on the interconnectedness of people. This was very different from the Christianity that was promoted by the missionaries who played a major role in European settlement. The Christian religions were much more individualistic and competitive than traditional Indigenous spirituality.

The missionaries saw their role as one of 'civilizing' Indigenous people, and part of this process was converting them to Christianity. William Cockran, one of the Anglican missionaries, described the nature of his mission in Manitoba's Red River settlement:

> I thought of making the red men Christians, & then Christians & Englishmen were so closely united in my imagination, they appeared as one. Consequently I expected that when the red man became a Christian, I should see all the active virtues of English Christians immediately developed in his character. Here I endeavoured to make the red man not only a Christian but an Englishman, pressed the necessity of industry, economy, cleanliness, taste, good order & all the other moral virtues, which make the Christian shine among a perverse generation. (quoted in Peikoff, 2000)

The missionaries used very harsh measures in their attempt to convert Indigenous people to Christianity (Peikoff, 2000), and when these measures failed to produce the desired results, they forced children to attend residential schools in order to 'kill the Indian in the child.' The impact of these schools has been disastrous and they are still having a cross-generational impact on peoples' lives (Truth and Reconciliation Commission, 2012).

WEBER'S RESPONSE TO MARX While Marx believed that religion hindered social change, Weber argued just the opposite: that religion could help produce social change. In The Protestant Ethic and the Spirit of Capitalism (1976/1904–1905), Weber asserted that the religious teachings of John Calvin were directly related to the rise of capitalism. Calvin emphasized the doctrine of predestination—the belief that all people are divided into two groups, the saved and the damned. Only God knows who will go to heaven (the elect) and who will go to hell. Because people cannot know whether they will be saved, they look for earthly signs that they are among the elect. According to the Protestant ethic, those who have faith, perform good works, and achieve economic success are more likely to be among the chosen of God. As a result, people work hard, save their money, and do not spend it on worldly frivolity; instead, they reinvest it in their land, equipment, and labour (Chalfant, Beckley, and Palmer, 1994).

The spirit of capitalism grew under the Protestant ethic. As people worked harder to prove their religious piety, structural conditions in Europe led to the Industrial Revolution, free markets, and the commercialization of the economy—developments that worked hand in hand with Calvinist religious teachings. From this viewpoint, wealth was an unintended consequence of religious piety and hard work.

Is Weber's thesis about the relationship between religion and the economy supported by other researchers? Collins examined Weber's claim that the capitalist breakthrough occurred just in Christian Europe and concluded that it is only partially accurate. According to Collins, the foundations for capitalism in Asia, particularly in Japan, were laid in the Buddhist monastic economy of late medieval Japan: "The temples were the first entrepreneurial organizations in Japan: the first to combine control of the factors of labor, capital, and land so as to allocate them for enhancing production" (1997:855). Due to an ethic of self-discipline and restraint on consumption, high levels of accumulation and investment took place in medieval Japanese Buddhism. Gradually, secular capitalism emerged from temple capitalism as new guilds, independent of the temples, arose, and the gap between the clergy and everyday people narrowed. The capitalist dynamic in the monasteries was eventually transferred to the secular economy, opening the way to the Industrial Revolution in Japan.

From these works, we can conclude that the emergence of capitalism through a religious economy happened in several parts of the world, not just one, and that it occurred in both Christian and Buddhist forms (Collins, 1997).

SYMBOLIC INTERACTIONIST PERSPECTIVES ON RELIGION

Thus far, we have looked at religion primarily from a macrolevel perspective. Symbolic interactionists focus on a microlevel analysis that examines the meanings that people give to religion in their everyday lives.

RELIGION AS A REFERENCE GROUP For many, religion serves as a reference group to help them define themselves. For example, religious symbols have meaning for many people. The Star of David holds special significance for Jews, just as the crescent moon and star do for Muslims and the cross does for Christians.

RELIGIOUS CONVERSION John Lofland and Rodney Stark (1965) wanted to learn why people converted to nontraditional religious movements. What would attract people to a small movement outside the religious mainstream, and why were some movements successful while most others failed? In the early 1960s, they studied a small religious movement that had been brought to the San Francisco area from Korea. They called the movement the Divine Precepts but later revealed that the group was the Unification Church, better known as "the Moonies" after their founder, Sun Myung Moon.

Lofland and Stark observed the group's activities and interviewed members and potential recruits. They proposed a seven-step theory of conversion. The most important personal characteristics that made conversion likely were:

1. Individuals had an important tension or strain in their lives, such as financial problems, marital issues, or sexual identity problems.
2. They had a religious problem-solving perspective. While other people respond to strain in their lives by taking direct action (such as divorce or bankruptcy), going to a psychiatrist, or joining a political movement, these individuals try to solve their problems through spiritual means.
3. They defined themselves as religious seekers trying to resolve their problems through some system of religious meaning. Some had tried several different religious alternatives before encountering the Moonies.

   These background factors were present before the potential converts came to the Divine Precepts. Several situational factors then increased the likelihood of conversion:
4. Individuals had come to a turning point in their lives. Some of the future converts had just dropped out of school, while others had moved, lost a job, or experienced some other major life change. This turning point not only increased the tension experienced but gave the people the opportunity to turn to something new.
5. Potential converts had close personal ties with a Divine Precepts member. Many religious seekers heard the message of the Divine Precepts, but only those who also developed a

personal bond with a member underwent conversion. One of the converts, recently recovered from a serious illness, described the process:

> I felt as if I had come to life from a numb state and there was spiritual liveliness and vitality within me by being among this group. As one feels when he comes from a closed stuffy room into the fresh air, or the goodness and warmth after freezing coldness was how my spirit witnessed its happiness. Although I could not agree with the message intellectually I found myself one with it spiritually. (1965:871)

6. A lack of ties with people outside the group. Few of those who converted had strong ties outside the group.
7. Intensive interaction with Divine Precepts members. The Divine Precepts recognized the importance of this interaction and encouraged those who had verbally converted to move into a shared residence with other group members. This helped secure converts' total commitment to the movement.

FEMINIST PERSPECTIVES ON RELIGION

Like other approaches in the conflict tradition, feminist perspectives focus on the relationship between religion and women's inequality. Some feminist perspectives highlight the patriarchal nature of religion and seek to reform religious language, symbols, and practices to eliminate elements of patriarchy. In virtually all religions, male members predominate in positions of power in the religious hierarchy and women play subordinate roles in religious and public life. For example, an Orthodox Jewish man may focus on his public ritual roles and discussions of sacred texts, while an Orthodox Jewish woman may have few, if any, ritual duties and is more likely to focus on her responsibilities in the home.

According to feminist theorists, religious symbolism and language consistently privilege men over women. Women are also more likely than men to be depicted as negative or evil spiritual forces in religion. For example, Eve in the Book of Genesis is a temptress who contributes to the Fall of Man.

Former U.S. president Jimmy Carter left the Baptist Church because he disagreed with their views on the role of women. However, he believed that the problem was far broader than just among Baptists:

> This view that women are somehow inferior to men is not restricted to one religion or belief. It is widespread. Women are prevented from playing a full and equal role in many faiths. Nor, tragically, does its influence stop at the walls of the church, mosque, synagogue, or temple. This discrimination, unjustifiably attributed to a higher authority, has provided a reason or excuse for the deprivation of women's equal rights across the world for centuries. The male interpretations of religious texts and the way they interact with and reinforce traditional practices justify some of the most pervasive, persistent, flagrant, and damaging examples of human-rights abuses. (Carter, 2009)

As President Carter points out, religious discrimination against women is not just a religious matter, as it is used as justification for discrimination in other aspects of life, including education and marriage. In many societies, social policies that restrict the role of women are reinforced by religious ideology that supports practices restricting women's role in society.

Consider one woman's reaction to the serving of Holy Communion in her Protestant church:

> Following the sermon, the worshippers are invited to participate in the celebration of the Lord's Supper. As the large group of male ushers marches down the aisle to receive the communion elements and distribute them to the congregation, I am suddenly struck with the irony of the situation. The chicken suppers, the ham suppers, the turkey suppers in the church are all prepared and served by the women. But not the Lord's Supper . . . the privilege of serving the Lord's Supper in worship

> is reserved for the men. This particular morning I find it very difficult to swallow the bread and drink the wine, knowing that within the Body of Christ, the Church, the sisters of Christ are not given the same respect and privileges as are his brothers. (Johnstone, 1997:237)

WOMEN IN THE MINISTRY

> A woman can't represent Christ. Men and women are totally different—that's not my fault—and Jesus chose men for his disciples. (MacDonald, 1996:47)

This quotation highlights the absence of women in significant roles within many religious institutions. Women from many religious backgrounds are demanding an end to traditions that do not reflect their religious commitment. Some women are choosing alternative spiritual belief systems, while others are working inside their church to create change. The battles have been intense, and the issue of the role of women has polarized some churches. In 1992, the Church of England allowed the ordination of women priests. In response, a British vicar made a point of telling the media that he would "burn the bloody bitches" (MacDonald, 1996:47).

While the Church of England does have women serving as priests—and in 2014 finally allowed them to become bishops—other major churches do not. The Roman Catholic Church does not allow women to serve as clergy, and women are still struggling to make their voices heard within the church. In 2012, Pope Benedict was critical of the largest association of U.S. nuns, the Leadership Conference of Women Religious (LCWR). The nuns were accused of pursuing a radical feminist agenda of social justice and poverty reduction and of disagreeing with official Catholic Church positions on the ordination of women, ministry to homosexuals, and several other issues. Pope Benedict appointed an archbishop (of course a male) to oversee the LCWR to ensure that they follow Catholic doctrine more faithfully. The Mormon Church also does not allow women to serve as priests and recently excommunicated Kate Kelly, who established an organization lobbying to allow women to serve as priests. Her group is also trying to change the church doctrine that men should preside over their wives and families (Green, 2014).

Despite opposition, there have been some advances. Women make up an increasing proportion of the clergy in some religious denominations. The United and Anglican churches have significant numbers of female clergy. Reform Judaism has ordained women as rabbis since the early 1970s, and Indigenous Canadian religions have traditionally given status to women in spiritual leadership.

While women are a distinct minority among religious leaders, they make up a substantial majority of the faithful. Rodney Stark has concluded that "in every sizable religious group in the Western world, women outnumber men, usually by a considerable margin" (2004:61). Thus, many churches discriminate against their most faithful members. What do you think lies ahead for those churches that do not follow other social institutions in ensuring that women and men are treated equally?

RATIONAL CHOICE PERSPECTIVES ON RELIGION

Rational choice theory of religion is based on the assumption that religion is essentially a rational response to human needs. However, the theory does not claim that any particular religious belief is necessarily true or more rational than another. The rational choice perspective views religion as a competitive marketplace in which religious organizations (suppliers) offer a variety of religions and religious products to potential followers (consumers), who shop around for the religious theologies, practices, and communities that best suit them.

According to this approach, people need to know that life has a beneficial supernatural element, and they seek to find these rewards in various religious organizations. The rewards include explanations of the meaning of life and reassurances about overcoming death. However, because religious organizations cannot offer religious certainties, they instead offer compensators—a body of language and practices that compensates for some physical lack or frustrated goal. According to some sociologists, all religions offer compensators, such as a belief in heaven, personal fulfillment, and control over evil influences in the world, to offset the fact that they cannot offer certainty of an afterlife or other valued resources that potential followers and adherents might desire (Stark, 2007). Rational choice theory focuses on the process by which actors—individuals, groups, and communities—settle on one optimal outcome out of a range of possible choices (a cost–benefit analysis). These compensators provide a range of possible choices for people in the face of a limited (or nonexistent) supply of the choice (certainty, for example) that they truly desire.

Recently, sociologists have applied rational choice theory to an examination of the religious marketplace and found that people are actively shopping around for beliefs, practices, and religious communities that best suit them. Based on the diverse teaching and practices of various religious bodies, adherents and prospective followers move among various religious organizations, with every major religious group simultaneously gaining and losing adherents. Although some find the religious home they seek, others decide to consider themselves unaffiliated with any specific faith.

POSTMODERN PERSPECTIVES ON RELIGION

THE SECULARIZATION DEBATE One of the most important debates within the sociology of religion deals with the question of whether the world is becoming more secular and less religious or whether we are seeing a renewal of religious belief. The view that modern societies are becoming more secular goes back to the work of Weber, Durkheim, and Marx. Weber felt that as societies became modernized, the role of religion as the sole source of authority would inevitably diminish as other social institutions—particularly economic and political ones—became dominant. Hadden has summarized the secularization perspective:

> Once the world was filled with the sacred—in thought, practice, and institutional form. After the Reformation and the Renaissance, the forces of modernization swept across the globe and secularization . . . loosened the dominance of the sacred. In due course, the sacred shall disappear altogether except, possibly, in the private realm. (1987:598)

Proponents of this view link modernization with secularization. They predict that as the world becomes more rational and bureaucratized, and as knowledge becomes more science-based, the influence of religion will decline. According to Fukuyama, Weber's prediction has proven accurate in many ways: "Rational science-based capitalism has spread across the globe, bringing material advancement to large parts of the world and welding it together into the iron cage we call globalization" (2005:2). Religion has become much less important in Canada and in almost all Western industrial countries other than the United States. In Canada, Wilkins-Laflamme has identified three significant social changes since World War II: the rapid decline of the influence of the Catholic Church with the Quiet Revolution in Quebec; the collapse of the number of members in mainline Protestant churches such as the United Church; and the pluralization of religion in many of our largest cities because of immigration from non-Christian countries. While many people still report an interest in spiritual matters, they do not pursue these interests through the organized church.

Critics of secularization theory concede that in most Western industrialized countries, the separation between church and state has increased and church attendance has declined. At the same time, however, religion is becoming more important in other parts of the world. Religion remains strong in Islamic societies, even in countries that have begun to modernize, such as Turkey and Pakistan. In the former Soviet bloc, where religion had been banned, there has been a resurgence in religious participation since the end of the communist era. Pentecostal churches are growing in South America. The United States, perhaps the world's most modernized society, is still a very religious country, showing that modernization does not inevitably cause a decline in religiosity.

The proportion of the world's population that holds religious beliefs is now growing because of high birth rates in religious countries (Norris and Inglehart, 2004). Even in more secular countries, religious people have higher birth rates than those who are not religious. Because of this, some sociologists, including Rodney Stark, disagree with secularization theory: "After nearly three centuries of utterly failed prophecies and misrepresentations of both past and present, it seems time to carry the secularization doctrine to the graveyard of failed theories, and there to whisper ['rest in peace']" (1999:270).

Jeff Haynes (1997) has looked at this situation from the postmodern perspective and concludes that both sides in this debate are partially correct. While secularization continues in much of the industrialized West, the postmodern condition has led people in many low-income countries to turn to religion. The structural conditions of postmodernism, including the negative consequences of globalization and its perceived threats to the moral order, can destabilize local values and traditions. Instead of leading to secularization, these conditions can lead to a strengthening of faith as some people resist these threats by turning to religion. Proponents of secularization theory would not have predicted the role played by religion in global politics in the past four decades. Religion has been critical in major political events as diverse as the ongoing conflicts between India and Pakistan and between Israel and the Palestinians; the violent breakup of the former Yugoslavia; and the ongoing sectarian wars in the Middle East. See Box 17.2.

Haynes explains the coexistence of secularization in some parts of the world and the spread of religion elsewhere by hypothesizing that secularization will continue, except in circumstances where religion "finds or retains work to do other than relating people to the supernatural . . . only when religion does something other than mediate between man and God does it retain a high place in people's attentions and in their politics" (1997:713). Haynes predicts that religion will retain or increase its importance in societies where it helps defend culture against perceived threats from outside or from the threat of internal cultural change. Global capitalism has weakened national sovereignty and carries with it only the values of the marketplace. Many people view their own governments as part of the enemy (Juergensmeyer, 2003). Many political observers have attributed the appeal of right-wing politics in the United States to the fact that many Americans feel threatened by this social change. And Islamic militant Sayyid Qutb, who was tortured and executed by Egyptian police in 1966, stated his disgust with the Westernization of the Arab world, which he felt was destroying his basic Islamic values:

> Humanity today is living in a large brothel! One has only to glance at its press, films, fashion shows, beauty contests, ballrooms, wine bars, and broadcasting stations! Or observe its mad lust for naked flesh, provocative postures, and sick, suggestive statements in the literature, the arts and the mass media! (Ruthven, 2004:37)

One way for people to deal with these threats is to turn to fundamentalist beliefs. These beliefs do more than mediate between people and their god. To people who feel that their values and identities are under threat from globalization, poverty, immorality, religious pluralism, or corrupt government, religious fundamentalism provides certainty in an uncertain world. According to Haynes, "For many people, especially in the Third World, postmodernism is synonymous with poverty, leading the poor especially to be receptive to fundamentalist arguments which supply a mobilising ideology" (1997:719). In large part, these fundamentalist religious institutions are based on strong local community organizations. They not only fill people's spiritual needs and provide a moral code that protects them from the consequences of globalization, but they also provide adherents with the support of a strong moral community that can replace older structures that have been weakened by rapid social change.

Haynes's theory is supported by data from the World Values Survey. Several indicators of religiosity are strongly correlated with a country's level of development (see Figure 17.1). Respondents in agrarian countries (including Nigeria, Tanzania, and Zimbabwe) are twice as likely as those in industrialized, high-income countries to attend religious services at least weekly and to pray daily, and three times as likely to say that religion is "very important" in their lives (Norris and Inglehart, 2004). The major exceptions to this pattern are the United States and Ireland, which are both wealthy and religious countries.

BOX 17.2 SOCIOLOGY IN GLOBAL PERSPECTIVE

Religious Terrorism

Religious terrorism has become a serious threat in postmodern societies. While there is a long history of religious wars among states and many earlier instances of religious terrorism, this type of terrorism has intensified over the past three decades. Following the September 11, 2001, attacks on the United States and subsequent bombings in Madrid, Bali, London, Mumbai, and elsewhere, much of the world's attention is now focused on Islamic terrorists, including members of groups like al-Qaeda and ISIS, and Nigeria's Boko Haram. However, all of the world's major religious traditions—as well as many minor religious movements—have been linked with terrorism. Among the questions that interest sociologists are: What are the causes of religious terrorism? How does it differ from other types of terrorist activities?

Violent extremism is not limited to any one faith. In Northern Ireland, the Catholic Irish Republican Army (IRA) exploded hundreds of bombs and killed hundreds of civilians in an attempt to free Northern Ireland from British rule. In 1994, a Jewish right-wing settler, Dr. Baruch Goldstein, shot and killed more than 30 Palestinians who were praying at the Tomb of the Patriarchs in Hebron. On the other side of the Israeli-Palestinian conflict, hundreds of Israelis have been killed by Palestinian suicide bombers. In Canada and the United States, there have been numerous bombings of abortion clinics, and several doctors who perform abortions have been killed or wounded—some of these attacks were carried out by Christian ministers, and other attacks were supported by militant Christian groups.

The largest domestic terrorism incident in the United States was the bombing of the Oklahoma City federal building, which killed 168 people. The bomber, Timothy McVeigh, was inspired by the white supremacist book *The Turner Diaries*. This book condemns the dictatorial secularism that it alleges has been imposed on the United States by a Jewish and liberal conspiracy. McVeigh's religious group, Christian Identity, shared these values and beliefs (Juergensmeyer, 2003).

Militant Sikhs, fighting for an independent homeland, have committed many acts of terrorism. These include the assassination of Indian prime minister Indira Gandhi by her own bodyguards. Sikh extremists were also responsible for Canada's worst act of terrorism, the Air India bombing that killed 329 people (see Chapter 22).

In 1995, members of an apocalyptic Japanese Buddhist sect released sarin nerve gas into the Tokyo subway system—this was the first attempt by religious terrorists to use a weapon of mass destruction. While the attack killed 12 people, thousands more would have died if the terrorists had been able to find a more effective way of vaporizing the sarin gas.

There are differences in the motivation behind these different attacks. The IRA bombing campaign had a strong political component, while members of the Japanese sect had few identifiable political goals. In each of the instances, however, the religious ideology of the terrorists defines the enemy and provides a justification for killing innocents. According to Bruce Hoffman, there are important differences between religious and secular terrorism:

> For the religious terrorist, violence first and foremost is a sacramental act or divine duty executed in direct response to some theological demand or imperative. Terrorism assumes a transcendental dimension, and its perpetrators are thereby unconstrained by the political, moral, or practical constraints that seem to affect other terrorists . . . Thus, religion serves as a legitimizing force—conveyed by sacred text or imparted via clerical authorities claiming to speak for the divine. (1995:272)

The acts of secular terrorists may be restrained by their fear of alienating potential supporters. Religious terrorists must please only themselves and their god, and can justify attacks against all "nonbelievers." Finally, purely religious terrorists are not trying to change an existing system, such as a particular government. Rather, they wish to transform the social order. For example, according to former al Qaeda leader Osama bin Laden:

> It is no secret that warding off the American enemy is the top duty after faith and that nothing should take priority over it . . . jihad has become [obligatory] upon each and every Muslim . . . The time has come when all the Muslims of the world, especially the youth, should unite and soar . . . and continue jihad till these forces are crushed to naught, all the anti-Islamic forces are wiped off the face of this earth and Islam takes over the whole world and all the other false religions. (quoted in Juergensmeyer, 2003:431)

Because negotiating with religious terrorists is almost impossible and because contemporary terrorist organizations have a loose networked form (see Box 7.5), they are very difficult to control. Religious terrorism is a serious threat, but we should keep religion in perspective by remembering that the mass genocides of the 20th century, including the Holocaust, Stalin's purges in Russia and his genocide against the Ukrainian people (the Holodomor), China's Cultural Revolution, and Pol Pot's massacre in Cambodia, were committed in the name of political ideology, not religion.

FIGURE 17.1 RELIGIOSITY BY TYPE OF SOCIETY

The level of a country's development is strongly related to the religiosity of its citizens.

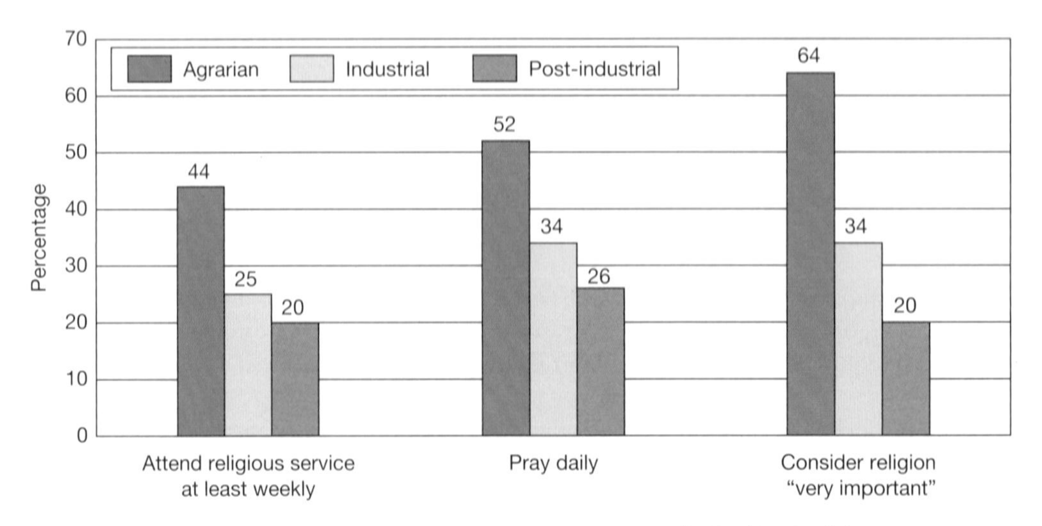

Source: Norris, Pippa and Ronald Inglehart. 2004. *Sacred and Secular: Religion and Politics Worldwide*. Cambridge: Cambridge University Press.

CONCEPT SNAPSHOT

FUNCTIONALIST PERSPECTIVES

Key thinkers: Emile Durkheim, Max Weber

Durkheim considered religion to be essential to the maintenance of society. Religion performs three important functions: (1) providing meaning and purpose to life; (2) promoting social cohesion; (3) providing social control and support for government. Weber believed that religion could lead to positive change in society. His main example was the way in which he believed the religious teachings of Calvinist Protestantism led to the rise of capitalism.

CONFLICT PERSPECTIVES

Key thinker: Karl Marx

Conflict theorists believe that the capitalist class uses religious ideology as a tool of domination to mislead the working class about their true interests. According to Marx, religion is the "opiate of the masses" and unites people under a false consciousness that leads them to believe they have common interests with the capitalists.

INTERACTIONIST PERSPECTIVES

Key thinkers: John Lofland, Rodney Stark

Interactionists look at the role of religion in people's everyday lives. Lofland and Stark studied conversion to a nontraditional religious movement. This conversion was more likely if individuals had certain predisposing background factors; had come to a turning point in their lives; had close personal ties to members of the group; had a lack of ties to people outside the group; and had intensive interaction with group members.

FEMINIST PERSPECTIVES

Key thinker: Meredith McGuire

For many feminist scholars, men and women experience religion in different ways. Religious doctrine and language typically create a social definition of the roles of men and women. Typically, these definitions place women in subordinate roles in the church.

POSTMODERN PERSPECTIVES

Key thinker: Jeff Haynes

The structural conditions of postmodernism, including the negative consequences of globalization and its perceived threats to the moral order, can destabilize local values and traditions. Instead of leading to secularization, these conditions can lead to a strengthening of faith as some people resist these threats by turning to religion.

RATIONAL CHOICE PERSPECTIVES

Key thinker: Rodney Stark

Religious persons and organizations, interacting within a competitive market framework, operate a variety of religions and religious products to consumers, who shop around for religious theologies, practices, and communities that best suit them.

TIME TO REVIEW

- Why is ritual important in the practice of religion?
- Describe the forms that religion can take in different societies.
- How do theorists from each of the functionalist, conflict, interactionist, feminist, postmodern, and rational choice perspectives analyze religion?

LO-3 CHURCHES, SECTS, AND CULTS

church A large, bureaucratically organized religious body that tends to seek accommodation with the larger society in order to maintain some degree of control over it.

sect A relatively small religious group that has broken away from another religious organization to renew what it views as the original version of the faith.

cult A religious group with practices and teachings outside the dominant cultural and religious traditions of a society.

Religious groups vary widely in their organizational structure. While some groups are large and bureaucratically organized, others are small, with a relatively informal authority structure. Some require total commitment from their members; others expect only a partial commitment.

To help explain the different types of religious organizations, Ernst Troeltsch (1960/1931) and his teacher Max Weber (1963/1922) developed a typology that distinguishes between churches and sects. A **church** is a large, bureaucratically organized religious body that tends to seek accommodation with the larger society in order to maintain some degree of control over it. Church membership is largely based on birth; children of church members typically are baptized as infants and become lifelong members of the church, though older people can also join if they go through a training process. Leadership is hierarchical, and clergy often have many years of formal education. Religious services are highly ritualized; they are often led by clergy who wear robes, administer sacraments, and read services from a prayer book or other standardized liturgical format.

By contrast, a **sect** is a relatively small religious group that has broken away from another religious organization to renew what it views as the original version of the faith. Sects offer a more personal religion and an intimate relationship with a supreme being, who is depicted as taking an active interest in the individual's everyday life. Whereas churches use formalized prayers, often from a prayer book, sects often have informal prayers composed at the time they are given. Whereas churches typically appeal to members of the middle and upper classes, sects seek to meet the needs of people who are low in the stratification system (Stark, 1992).

Christians around the world have been drawn to cathedrals such as the Cathedral-Basilica of Notre-Dame de Québec (originally built in 1647) to worship God and celebrate their religious faith.

CULTS/NEW RELIGIOUS MOVEMENTS

While sects represent attempts to renew old religions, cults represent new religious practices. A **cult** is a religious group with practices and teachings outside the dominant cultural and religious traditions of a society. While many of the world's major religions began as cults, the term now has a negative connotation for many people, so the term *new religious movement* is often used to refer to these groups (Barrett, 2001). Cult leadership is based on charismatic characteristics of the individual (Stark, 1992). An example is the religious movement started by Rev. Sun Myung Moon, a Korean engineer who believed that God had revealed to him that Judgment Day was rapidly approaching. Out of this movement, the Unification church, or "Moonies," grew and flourished, recruiting new members through their personal attachments to present members (Stark, 1992). Some cult leaders have not fared well. These include Jim Jones, whose ill-fated cult members ended up committing mass suicide in Guyana, and David Koresh, of the also ill-fated Branch Davidians in Waco, Texas.

Are all cults short-lived? Over time, most cults disappear. However, others undergo transformation into sects or denominations. Some researchers view cults as a means of reviving religious practice when existing churches do not provide satisfaction to those seeking a spiritual home.

In some countries, new religious movements are harshly repressed. This is because they may challenge state religions or, as in the case of the Falun Gong movement in China, are seen by the country's leadership as a potential political threat.

This mass wedding ceremony of thousands of brides and grooms brought widespread media attention to Rev. Sun Myung Moon and the Unification church, which many people view as a religious cult.

TRENDS IN RELIGION IN CANADA

CANADA'S RELIGIOUS MOSAIC

Until the end of the 19th century, Canada's population was made up almost entirely of Protestants and Roman Catholics. The Roman Catholic Church was the dominant religious force during the early settlement of Canada. With the arrival of the United Empire Loyalists from the American colonies in the 1780s, the Protestant population in Canada became larger than the French Catholic population. However, the combined share of the Protestant churches has declined substantially since the 1950s. Changes in the population of the major religious groups in Canada are illustrated in Figure 17.2. In 2011, Roman Catholics, at 39 percent of the population, were the largest religious group in Canada.

Religions other than Christianity were practised in Canada prior to European colonization. Indigenous peoples were excluded from the earliest census collections. Even so, in 1891, almost 2 percent of Canadians reported practising religions other than Christianity. In 2011, more than 11 percent of Canadians were affiliated with "other" religions, including Judaism; Eastern non-Christian religions, such as Islam, Buddhism, Hinduism, and Sikhism; and para-religious

These protesters in Vancouver are drawing attention to the persecution of members of Falun Gong by the Chinese government.

FIGURE 17.2 RELIGIOUS AFFILIATION IN CANADA

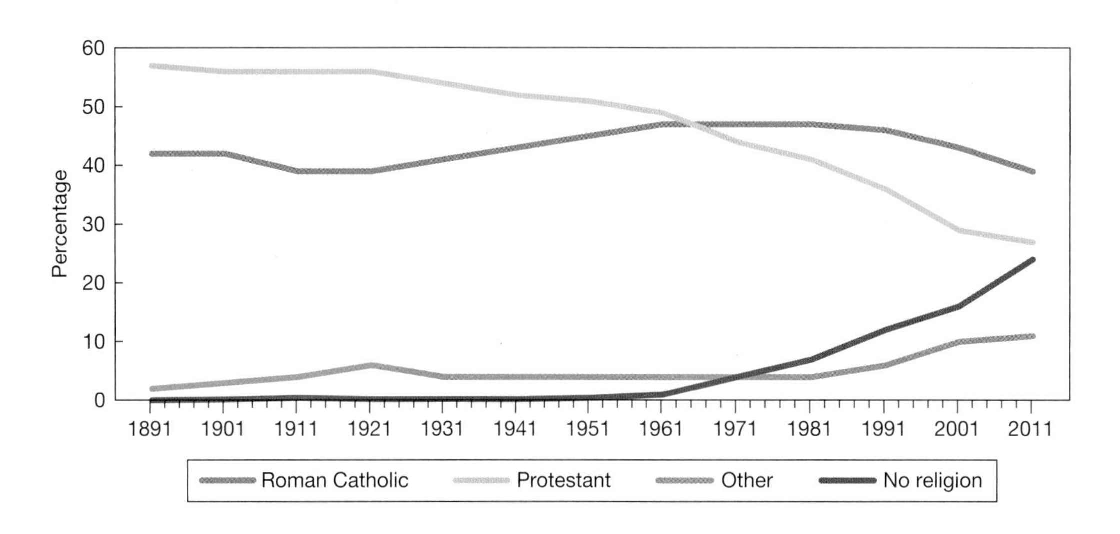

Percentage

	1891	1901	1911	1921	1931	1941	1951	1961	1971	1981	1991	2001	2011
Roman Catholic	42	42	39	39	41	43	45	47	47	47	46	43	39
Protestant	57	56	56	56	54	52	51	49	44	41	36	29	27
Other	2	3	4	6	4	4	4	4	4	4	6	10	11
No religion	N/A	0.1	0.4	0.2	0.2	0.2	0.4	1	4	7	12	16	24

Prepared by the Centre for International Statistics.

Note: The figures for "Other' and "No Religion" were taken from Pew Research Center (2013) based on 2011 National Household Survey (NHS) data. Note that the NHS uses a different methodology than the previous census, so comparisons should be made with caution.

Sources: Vanier Institute of the Family, 1994; Statistics Canada, 2003, 2014.

groups (see Figure 17.3). As a result of changing immigration patterns, Eastern non-Christian religious populations have grown significantly since the 1960s.

The numbers who fall under the category of *No religion* have also increased, going from less than 1 percent in 1951 to 24 percent in 2011. Does this mean that Canadians are rejecting religion? We can answer this question by examining other recent trends in religion in Canada.

RELIGIOSITY

Is Canada a religious society? The answer depends on how you look at things. Nationally, attendance at religious services, public confidence in religious leadership, and religious influence have all gradually declined since the late 1940s.

Over the past 60 years, attendance at religious services has declined precipitously (Clark, 1998). A 1946 Gallup poll reported that 67 percent of Canadian adults reported attending religious services during the previous week. By 2001, the General Social Survey (GSS) found that reported attendance at weekly religious services had declined to only 20 percent (Clark, 2003). A generation ago, most Canadians attended religious services; today only a small minority attend regularly.

While attendance rates have declined for all age groups, the drop has been greatest for younger people: 43 percent of those 65 years of age and over attend religious services at least once a month, compared with only 24 percent of people 15 to 24 years old. Other indicators

FIGURE 17.3 MAJOR RELIGIOUS DENOMINATIONS, CANADA, 2011

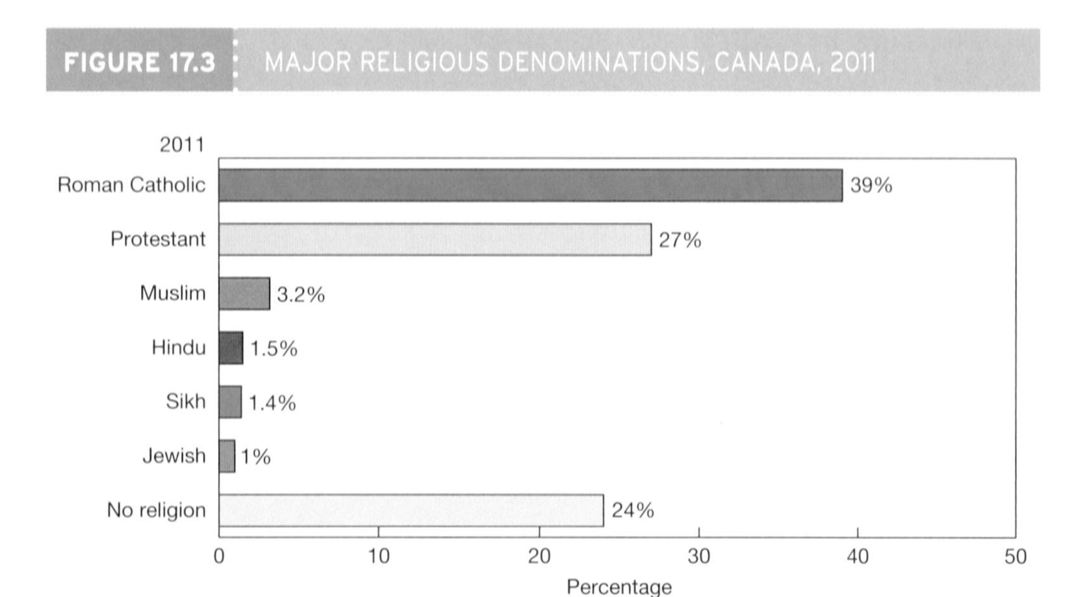

Source: Adapted from Statistics Canada, 2003.

of religiosity were also lower for younger people (Wilkins-Laflamme, 2015). This loss of young members does not bode well for the future of Canadian religions, as the vast majority of adults who attend regularly also attended regularly in childhood. This means that as older members die, fewer and fewer people will take their places in the pews.

Despite the decline in church attendance, the vast majority of Canadians still report a religious affiliation and affirm that they believe in God. However, both of these measures have also declined in recent years. In 1961, only 1 percent of Canadians reported no religious affiliation; by 2011, this number had increased to 24 percent (Statistics Canada, 2003, 2014b). A 2012 survey found that 42 percent of Canadians reported that religion was an important aspect of their lives and that those over 65 were almost twice as likely as those 18 to 24 to report that religion was important (Association for Canadian Studies, 2012).

Wilkins-Laflamme (2015) has found that the 'religious nones'—those who say they have no religious affiliation—fit into several different categories. Some are still believers, but do not belong to a church; others shift back and forth in their affiliations; and others are non-believers (atheists and agnostics). She found that about 20 percent of the unaffiliated attended church at least once a year, so organized religion still played at least some role in the lives of this group.

According to Clark and Schellenberg (2006), although only about one-third of Canadians attended services once or more a month, over half (53 percent) carried out some religious practice, such as praying, on their own. In addition to their religious beliefs, Canadians show an interest in other aspects of spirituality. Table 17.1 shows some data collected by Reg Bibby (2001). The table shows that most Canadians believe that some people have psychic powers; that supernatural and evil forces exist; that there is life after death; and that some people have extrasensory perception. A significant minority believe that astrology has some merit and that it is possible to make contact with the spirit world.

The results of these surveys provide an interesting paradox. Attendance at religious services has dramatically declined despite the fact that most Canadians report some religious affiliation, express a belief in God, and believe in other spiritual aspects of life. People have not rejected religious institutions completely, as most still rely on organized religion for services, such as baptisms, weddings, and funerals (Bibby, 2002). However, they are not regular participants in religious activities, choosing instead to adopt what Bibby calls "religious fragments"—isolated beliefs, isolated practices, and isolated services (1987).

TABLE 17.1 SIPIRITUAL BELIEFS AND INVOLVEMENT OF CANADIANS, 2000

"I BELIEVE ..."	ADULTS	TEENS
Conventional	**Percentage Agreeing**	
God exists	81%	73%
In life after death	68%	78%
Less Conventional		
In ESP	66%	59%
One can have contact with the spirit world	45%	43%
In astrology	35%	57%
"I ... "		
Group Involvement		
Am committed to Christianity or another faith	55%	48%
Attend weekly	21%	22%
Am open to possibility of greater involvement	57%	43%
Spirituality		
Have spiritual needs	73%	48%
Find spirituality very important	34%	30%
Pray privately weekly or more often	47%	33%

Source: Bibby, 2001.

Bruce has pointed out the individualistic nature of these spiritual activities: "In the New Age, the self is the final arbiter of truth and utility. If it works for you, it is true" (Bruce, 2006:42). People choose from a menu of different spiritual practices and beliefs that fit them best as individuals. Spiritual choices such as New Age beliefs and the adoption of elements of Eastern religious practices focus on what the individual can gain from them. These activities include practices such as yoga, spiritual healing (Reiki), meditation, astrology, the use of crystals, and complementary health and healing programs. These practices serve as both adjuncts to and as substitutes for traditional religious practices.

Theologian Tom Harpur sums up this approach to religion:

> There is a huge spiritual quest going on. There's a lot of attempts at quick fixes and spiritual junk food as well. But even the silly fringe is part of it . . . People seem intuitively aware that something is missing in their lives, and there's a reaction against traditional religion. (quoted in MacDonald, 1996:42)

WHY HAVE CANADIANS TURNED AWAY FROM THE CHURCH?

According to Bibby, people have moved from religious commitment to religious consumption. Religious consumers look at the church as just one of many different options for solving their spiritual or worldly problems. Even those with a high religious commitment may not feel church attendance is the best way to express that commitment. As one of Bibby's respondents commented:

> I've been through a great deal in life and my faith is very strong. But I believe that one is closer to God in their own home and garden than a church. I see going to church these days as "keeping up with the Joneses." (1987:83)

These data support rational choice theory. Also, a postmodernist would explain that as our culture has become more individualistic, people are less likely to accept the dictates of an organized church. Religion has now been internalized, and rather than depending on the dictates of organized religion, people can make up their own version of spirituality that meets their needs.

SOME REASONS FOR THE DECLINE OF ORGANIZED RELIGION It is apparent that organized religion no longer seems relevant to the lives of many Canadians. One of Bibby's subjects illustrates this view: "The major issues of the day seem to me to have little to do with religion and morality; economic and political factors are far more important" (1993:59).

When it has tried to address contemporary issues, the church has often had problems. Debates over issues such as the acceptance of homosexuality have led to serious divisions. These moral issues have distracted the churches from other activities. Some churches, notably the United Church, have tried to become more socially relevant by focusing on social justice issues; however, this strategy has not attracted new members and may have driven older members away.

The alienation of women has contributed to the decline in attendance. Many churches have not adapted to the changed role of women. Patriarchal practices can be difficult or impossible to change, as traditional gender roles are part of the core religious ideology of some churches. Many fundamentalist churches believe that a woman's sacred duty is submission to her husband. Some women welcome this role, but others find this subordination unacceptable. Failure to address women's concerns will have serious consequences, as women are more likely to participate in church activities than are men and are also instrumental in ensuring that their children go to church.

The image of the church has also suffered from thousands of charges of child sexual abuse by ministers and priests. The abuse was most pervasive in residential settings, such as the church-run schools that were established for Indigenous children during the first half of the 20th century. More than 150,000 children attended these schools before the system was closed in the 1980s. About 37,000 of the 80,000 survivors filed claims for serious physical and sexual abuse as part of the Indian Residential School Settlement Agreement (Sinclair, 2013). The problem was not limited to the abuse of Indigenous youth, however; the first major residential school scandal in Canada grew out of offences committed by members of the Christian Brothers order at Newfoundland's Mount Cashel orphanage.

The image of the church has been further damaged by the fact that, in many cases, senior church officials knew about these problems, did little or nothing to stop them, and tried to cover them up. These incidents make it more difficult for the churches to speak credibly on moral issues. Some critics have linked abuse within the Roman Catholic Church to its patriarchal structure and its celibate male priesthood, and have called into question these fundamental principles of the church.

Religious diversity has also reduced the influence of the church. If everyone in a society has the same religious beliefs, people are not likely to question those beliefs. However, if there are many different religions, people will be aware that there are many different approaches to religious issues and will be more likely to question their own belief systems.

Finally, we can look at the special case of Quebec. The early development of French and English Canada was strongly influenced by religious principles. In Quebec, most social institutions came under the influence of the Roman Catholic Church. For example, much of the education system was church-run, so it had a daily influence in the lives of most Quebec children. The Roman Catholic Church was politically powerful and dominated the province's social and moral life.

What do you think of the use of this bobblehead Jesus to help advertise the United Church?

Following the Quiet Revolution in the mid-1960s, the church's influence in Quebec dwindled rapidly. Weekly church attendance dropped from a remarkable high of 90 percent in the 1940s (Bibby, 1993) to less than 30 percent by 1990, and it continues to decline (Pew Research Center, 2013). As Quebec became a secular society, the church lost its influence in fields such as education and social services. Church policies, such as the prohibitions on birth control, premarital sex, abortion, and divorce, and the refusal to ordain women priests, also turned people away. The reduced influence of the church is shown by the fact that Quebec has both Canada's highest proportion of Roman Catholics and its highest rate of common-law marriages, a practice that is contrary to Roman Catholic teachings.

TIME TO REVIEW

- What are the key differences between churches, sects, and cults?
- How has church attendance in Canada changed over the past 100 years?
- What are the demographic characteristics of those who are most likely to attend church in Canada?
- Describe the spiritual practices that appear to be replacing formal participation in church services.
- What factors account for the decline in church membership and attendance in Canada?

FUNDAMENTALISM

As many mainline denominations have been losing membership, some fundamentalist churches have steadily grown. The term *religious fundamentalism* refers to a religious doctrine that is conservative, is typically opposed to modernity, and rejects "worldly pleasures" in favour of otherworldly spirituality. Whereas "old" fundamentalism usually appealed to people from lower-income, rural backgrounds, the "new" fundamentalism appears to appeal to persons from all socioeconomic levels, geographical areas, and occupations.

INNER CITY CHRISTIAN OUTREACH CENTER

© Mark Scheuern/Alamy Stock Photo

Storefront missions such as this seek to win religious converts and offer solace to people in low-income, central-city areas.

LO-4 DOES RELIGION MAKE A DIFFERENCE?

What difference does religion actually make in peoples' daily lives? A Statistics Canada study found that people who attended church weekly were much more likely to feel satisfied with their lives and much less likely to feel that their lives were stressful than people who did not attend (Clark, 1998). Lim and Putnam (2010) also found that religious people had higher levels of subjective well-being than those who were not religious. They attribute this both to the social bonds that are part of church attendance and to the parishioners' religious identities.

Religiosity also influences other aspects of behaviour. All religions have ethical codes that govern personal and social behaviour. There is some evidence that religious commitment does influence people's moral conduct. For example, religiosity reduces involvement in delinquent and criminal behaviour (Linden, 2009). This relationship is complex, however. First, it is greatest where there is a strong religious community (Stark, Doyle, and Kent, 1982). Second, it has more impact on behaviour

BOX 17.3 SOCIOLOGY AND NEW MEDIA

Religion and New Media

> In a single telecast, I preach to millions more than Christ did in His entire lifetime.
>
> —Billy Graham (quoted in Roberts, 1995b:360)

> Involvement in the mass media, however, is not meant merely to strengthen the preaching of the Gospel. There is a deeper reality involved here: since the very evangelization of modern culture depends to a great extent on the influence of the media.
>
> —Pope John Paul II (quoted in Catholic News Agency, 2009)

For decades, churches have been using television to broaden their reach. Many religious groups have broadcast their services to get their message to those who are unable or unwilling to attend church. In recent years, the Internet has begun to have an impact on religion. Most established religions have websites. For example, the Catholics on the Net site (www.catholic.net) provides a range of services, including a review of recent church-related news stories, discussions of church teachings, and a discussion of careers with the Church.

While the Internet offers a good way for mainline churches to get their message out, it also provides a means for newer spiritual groups to try to attract new followers. For example, the Falun Gong website both describes the faith and the persecution of believers at the hands of the Chinese government (www.faluninfo.net). The Internet has even been used for religious satire. The very popular website for the Church of the Flying Spaghetti Monster (www.venganza.org) was designed as a protest against the teaching of creationism in Kansas schools and continues to use humour to challenge religious practices.

While many see social media as a way to spread the word, the quotation at the beginning of this box from Pope John Paul II shows that he understood that more is involved and that a new culture has been created.

> It is not enough to use the media simply to spread the Christian message and the Church's authentic teaching. It is also necessary to integrate that message into the "new culture" created by modern communications. (quoted in Catholic News Agency, 2009)

As Pope John Paul II recognized, the interactive nature of social media and the speed at which ideas can be shared may affect religious practices. For example, online religious communities can supplement or even replace local congregations. Religious institutions have become part of the online world, Second Life, so we now see virtual missionaries who try to spread the faith online. According to one religious spokesperson, "This virtual Second Life is becoming populated with churches, mosques, temples, cathedrals, synagogues, places of prayer of all kinds. And behind an avatar, there is a man or woman, perhaps searching for God and faith, perhaps with very strong spiritual needs" (quoted in Rhoten and Lutters, 2009). Cho (2011) has described several virtual communities, including groups of teenage witches and Buddhists, that have a very different organizational form from offline churches.

As people continue to live more of their lives online, the new media will play a more significant role in religious practice. What impact will this have on organized religion? How will new media affect existing churches whose congregations are currently declining? Will the people who worship online develop different religious identities than those who are socialized in real-life churches? How does an online church affect Durkheim's distinction between the sacred and the profane?

While the Internet may be a useful tool, it raises concerns for some religious groups because controlling what young people read on the Internet is almost impossible. Also, many religious organizations are used to top-down control in which church leaders determine religious practices and parishioners follow or risk being expelled from the church. Social media does not necessarily support this top-down model, and it is not clear how religious leaders will be able to ensure that those practicing religion as part of online culture will accept and obey church dogma.

that is not universally condemned by other segments of society than on behaviour that most other social institutions also disapprove of. That means religiosity has more influence on matters such as illegal drug use than it does on theft and assault (Linden and Currie, 1977).

Religiosity is also associated with marital stability. Weekly church attenders place more importance on marriage and children than people who do not attend, although the differences are not large. Church attenders have longer and happier marriages than non-attenders do, and the marriages of church attenders are less than half as likely to break down as the marriages of couples who do not attend (Clark, 1998).

What about the impact of religion on health? In many small-scale societies, the same individual—the healer or *shaman*—was responsible for both physical and spiritual needs. Some people are once more trying to reintegrate medicine and religion. The increasing popularity of alternative medicine has led to an openness to nontraditional approaches, and polls show that many people (including some doctors) believe that religious faith can help cure disease and therefore use prayer as medical therapy (Sloan, Bagiella, and Powell, 1999).

Many researchers have examined the relationship between religion and medical outcomes. In a humorous attempt to test the hypothesis, the eminent British scientist Sir Francis Galton sought to determine whether prayer could increase longevity. He assumed that nobody in England received more prayers for longevity than the British royal family. People sang "God Save the Queen [or King]" and regularly expressed concerns for their rulers in their prayers. Recognizing that the upper-class lifestyle of royalty made them more likely to live longer, Galton knew he had to compare them with other wealthy people. He selected for his comparison group wealthy lawyers, arguing that nobody would pray that lawyers live longer lives. Contrary to his hypothesis, he found that the lawyers lived longer and concluded that prayer had little efficacy in this regard.

Other, more serious studies found that religious people live longer, but failed to account for risk factors. Studies show that priests, monks, and nuns have less illness and live longer than members of the general population. However, the studies have weak validity because they do not control for the lower exposure of those in religious orders to a variety of risk factors. Similarly, studies of Israelis living on secular and religious kibbutzim found that the religious Jews lived longer than those who were nonreligious, but did not control for risk factors, such as smoking, blood cholesterol, and marital status (Sloan, Bagiella, and Powell, 1999).

In their review of research in this area, Sloan and his colleagues (1999) concluded that the evidence of an association between religiosity and health is weak and inconsistent. If it does not affect physical health, however, there is evidence that religion can play a role in comforting the sick. For example, one study found that 40 percent of a group of hospitalized adults reported that their religious faith was the most important factor in their ability to cope with their illness (Johns Hopkins, 1998).

LO-5 RELIGION IN THE FUTURE

What significance will religion have in the future? Religion will continue to be important because it provides answers to basic questions that are important to many people. Moreover, the influence of religion is felt in global politics. In many nations, the rise of *religious nationalism* has led to the blending of strongly held religious and political beliefs. The rise of religious nationalism is especially strong in the Middle East, where Islamic nationalism and sectarian violence are having a major impact on peoples' lives.

In Canada, the influence of religion will be evident in ongoing political battles over social issues, such as school prayer, abortion, and family issues. On the one hand, religion may unify people; on the other, it may result in tensions and confrontations between individuals and groups.

Many religious institutions in Canada face a bleak future as church attendance and membership continue to decline. The Roman Catholic Church faces the additional difficulty that the number of priests and nuns has declined dramatically as few young people are attracted to these occupations.

Some see hope for the future in our aging population. They feel that as the baby boomers age, they are likely to search for spiritual meaning, and some may turn back to the religions of their youth. However, there is little evidence that the baby boomers are returning, and church membership continues to stagnate. Even if the organized church continues to decline, however, spirituality will remain important to many Canadians:

> [People] know that religion, for all its institutional limitations, holds a vision of life's unity and meaningfulness, and for that reason it will continue to have a place

in their narrative. In a very basic sense, religion itself was never the problem, only social forms of religion that stifle the human spirit. The sacred lives on and is real to those who can access it. (Roof, 1993:261)

TIME TO REVIEW

- Discuss the meaning of the term religious fundamentalism.
- What impact does religion have on individuals and on their behaviour?
- What is religious nationalism, and what impact is it having around the world?
- What factors have limited the impact of immigration on religious life in Canada?

VISUAL SUMMARY

17

KEY TERMS

animism The belief that plants, animals, or other elements of the natural world are endowed with spirits or life forces that have an impact on events in society (p. 458).

church A large, bureaucratically organized religious body that tends to seek accommodation with the larger society in order to maintain some degree of control over it (p. 468).

cult A religious group with practices and teachings outside the dominant cultural and religious traditions of a society (p. 468).

faith Unquestioning belief that does not require proof or scientific evidence (p. 456).

monotheism A belief in a single, supreme being or god who is responsible for significant events, such as the creation of the world (p. 458).

nontheistic religion A religion based on a belief in divine spiritual forces, such as sacred principles of thought and conduct, rather than a god or gods (p. 458).

polytheism A belief in more than one god (p. 458).

profane A term used for the everyday, secular, or "worldly" aspects of life (p. 456).

religion A system of beliefs, symbols, and rituals, based on some sacred or supernatural realm, that guides human behaviour, gives meaning to life, and unites believers into a community (p. 456).

rituals Regularly repeated and carefully prescribed forms of behaviour that symbolize a cherished value or belief (p. 456).

LO-1 Describe what religion is and understand its purpose in society.

Religion, based on some sacred or supernatural realm, is a system of beliefs, symbols, and rituals that guides human behaviour, gives meaning to life, and unites believers into a community.

Rick Diamond/Getty Images

LO-2 Understand the differences between the functionalist, conflict, interactionist, feminist, and postmodern perspectives on religion.

According to functionalists, religion has three important functions: (1) providing meaning and purpose to life, (2) promoting social cohesion and a sense of belonging, and (3) providing social control and support for the government. From a conflict perspective, religion can have negative consequences. The capitalist class uses religion as a tool of domination to mislead workers about their true interests. However, Max Weber believed that religion could be a catalyst for social change. Symbolic interactionists focus on a microlevel analysis of religion, examining the meanings people give to religion and the meanings they attach to religious symbols in their everyday life. Feminist theorists have pointed out that even though they may belong to the same religious groups, men and women experience religion in different ways. Religious doctrine and language create a social definition of the roles of men and women that place women in subordinate roles in some religions. Some postmodern theorists have proposed that the structural conditions of postmodernism, including the negative consequences of globalization and its perceived threats to the moral order, can destabilize local values and traditions. Instead of leading to secularization, these conditions can lead to a strengthening of faith as some people resist these threats by turning to religion.

© Julie Thompson Photography/Alamy

LO-3 Understand the different types of religious organizations.

Religious organizations can be categorized as churches, sects, and cults.

© Jack_Art/iStockphoto.com

LO-4 Discuss the impact of religion on people's attitudes and behaviour.

Research looking at the impact of religion on attitudes and behaviour has had mixed results. Religiosity reduces involvement in delinquent and criminal behaviour. Religious people have longer and happier marriages than nonreligious people. Religion and prayer appear to have little impact on health, although they play a strong role in comforting the sick.

© Mark Scheuern/Alamy Stock Photo

LO-5 Consider the future role of religion in Canada.

Religion in Canada is clearly in decline, and the prognosis for the future is not bright. However, Canadians still have a strong interest in spiritual matters and continue to identify with the church. To take advantage of these factors, churches must find new ways to become relevant to the daily lives of Canadians.

© The United Church of Canada, WonderCafe.ca

APPLICATION QUESTIONS

1. Do you think religion will continue as a major social institution in Canada? What factors lead people to turn away from religion? What factors promote a renewed or continued interest in religion?
2. The church is a place where many people mark the milestones in their lives—births, baptisms, marriages, and deaths are all celebrated in church and recorded in church documents. Do you think that social media sites such as Facebook are now replacing churches as the preferred site for recording this information?
3. How does religion contribute to social stability? How is religion a force for social change?
4. How are Canada's religious institutions addressing important social issues such as rising social inequality and women's role in society?
5. What do you think religious leaders could do to attract more young people to their faiths?

sacred A term used to describe those aspects of life that are extraordinary or supernatural (p. 456).

sect A relatively small religious group that has broken away from another religious organization to renew what it views as the original version of the faith (p. 468).

simple supernaturalism The belief that supernatural forces affect people's lives either positively or negatively (p. 458).

theism A belief in a god or gods (p. 458).

KEY FIGURES

© Pictorial Press Ltd./Alamy

Émile Durkheim (1858–1917) Many of the early sociological theorists tried to understand the role of religion in social life. Durkheim's work focused on the role religion plays in contributing to social stability. His book *The Elementary Forms of the Religious Life* is one of the classic works in this field.

Mondadori Portfolio/Getty Images

Max Weber (1864–1920) Weber was another of the early sociologists who studied religion. Weber is best known for his work on the role of Calvinist Protestantism in the development of capitalism. For Weber, the new way of thinking that was part of Protestantism set the stage for the development of the rational, bureaucratized, capitalist economic system.

Courtesy of Rodney Stark

Rodney Stark (b. 1934) Rodney Stark is one of the most prolific sociology of religion scholars. His work has looked at a wide range of religious issues, including examining how religions attract converts, studying the degree to which a rational choice model applies to religion, and debating whether societies have become more secular over time.

Courtesy of Reginald Bibby

Reginald Bibby (b. 1943) Sociologist Reginald Bibby holds the Board of Governors Research Chair at the University of Lethbridge. Much of his work has been based on a series of national surveys he has conducted looking at trends in religious attitudes and behaviour. He has been particularly active in the debate over whether the decline in participation in formal religious activity has ended.

CHAPTER 18 MASS MEDIA

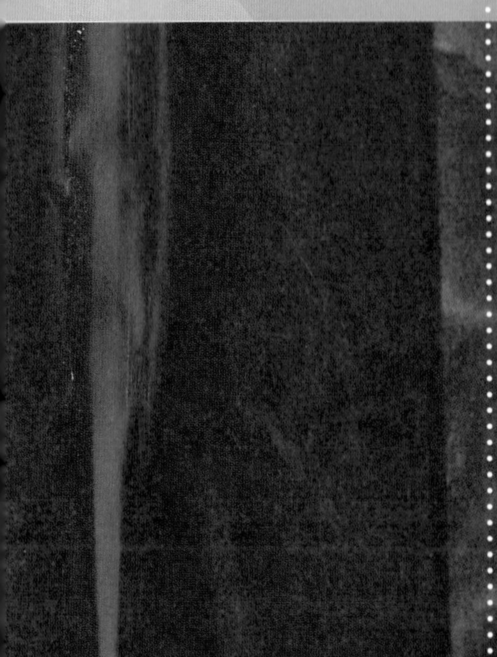

Used with permission of Ayat Mneina

CHAPTER FOCUS QUESTION

What impact are the media having on our culture, our social institutions, and our communities?

LEARNING OBJECTIVES

AFTER READING THIS CHAPTER, YOU SHOULD BE ABLE TO

LO-1 Understand what is meant by the term *mass media*.

LO-2 Consider how the impact of the media is explained by functionalist, conflict, feminist, interactionist, and postmodern theories.

LO-3 Understand how new media both bring people together and keep them apart.

LO-4 Understand how social media sites profit from the information you provide them.

LO-5 Think about the role played by the media in globalization.

SOCIAL MEDIA AND REVOLUTION

In 2011, people living in Tunisia, Egypt, and Libya overthrew their despotic and corrupt governments. These revolutions were unique in their use of social media as an organizing tool. Ayat Mneina, then a Libyan-born University of Manitoba student, describes how she helped in the fight for political freedom in her home country:

The resentment toward the regime and the desire for change are sentiments I share with the youth in Libya. Upon seeing the uprisings take place in neighbouring Tunisia and Egypt in January, a friend (Omar Amer) and I made a conscious decision to get involved if the youth of Libya were ever to take action.

We decided to observe social media outlets used by Libyan youth, predominantly Facebook, to gauge the potential for action. We simultaneously monitored the use of social media, including Twitter, by Tunisian and Egyptian youth in case we had to mobilize.

Within weeks of the Jasmine Revolution in Tunisia, Facebook groups emerged calling for a Day of Rage in Libya, setting the date for February 17, 2011. Although we could not predict whether or not the youth would in fact take to the streets, as the day grew near we equipped ourselves with a Twitter account (called @ShababLibya), a small network of friends and relatives on the ground who we could contact for updates, a Skype account, a YouTube account, a Gmail account, and a Facebook account and page all complementing our effort—which we chose to call the Libyan Youth Movement (ShababLibya).

Because of premature protests occurring on February 15—due to the arrest of a young human rights lawyer named Fathi Terbil and reports of increased security in the resistance stronghold of Benghazi—we grew skeptical about the upcoming Day of Rage. On the evening of the 16th, we received confirmation from several sources that youth were in fact planning to take to the streets the next morning. Fully equipped with caffeine, laptops, and our complete toolbox of social media, we began the task of filling the void that existed because independent and international media were not allowed to operate in Libya.

Overnight, we found ourselves tweeting urgent messages to the world and international media calling for the coverage of what we soon understood was the brutal and indiscriminate murder of innocent, unarmed protesters across the country. The images, reports, and accounts we received were unlike anything we could have prepared for, and in the next week we became the number one source of information on the Libyan uprising. We tweeted and posted reports as we received them, confirmed from the ground based on accounts from at least three people. We also provided media outlets with interviews and contacts with protesters on the ground, and we were even able to deliver coordinates of Gaddafi forces to NATO. We rallied for international support, and effectively became the voice of the Libyan youth during the revolution. I have seen our humble efforts thrive and grow to include more than 49,000 followers on Twitter, nearly 23,000 subscribers to our Facebook page, a website that sees thousands of visitors every week (www.shabablibya.org), and a handful of dedicated volunteers located around the globe who have tirelessly given their time and effort to this cause.

Source: Used with permission of Ayat Mneina.

At the same time, people in Libya were using their mobile phones to record the fighting and to post their videos on YouTube. Outside reporters were not allowed to work freely in the parts of Libya controlled by Gaddafi, so these videos were important in helping to convince other countries to provide support to the revolution.

Social media have played a prominent role in other political events. Repressive governments control their local media and constantly transmit government propaganda. Satellite TV helped to break this control—the Al Jazeera network gave many people in the Middle East their first real look at the outside world and showed the harm being done by their own repressive regimes. The Internet and the World Wide Web further opened up the world and for the first time enabled two-way communication inside and outside these countries. Cellphones enabled people who could not afford computers to access the Internet and to share information through social media. Revolutionaries no longer needed to seize state radio and television stations to get their message out. The transformation of media communication from "one-to-many" to "many-to-many" (Zakaria, 2011) made it possible for citizens to use the media to organize against the state.

People do not need social media to overturn oppressive governments—revolutions occurred long before the Internet. However, social media help people to get their message out and to mobilize others. Social media allow opponents to share information without the knowledge of the government and also to send news about their situation to the rest of the world.

Repressive governments do not just passively allow their opponents to use social media. More than 40 countries, including China, Iran, and Cuba, restrict Internet access (Canada Centre for Global Security Studies, 2011), and during times of crisis, governments frequently try to shut down the Internet and other social media entirely. In 2011, the Egyptian government tried shutting down social media in an unsuccessful attempt to save the regime of Hosni Mubarak, and prior to 2014 elections in Turkey, the government shut down Twitter in order to prevent users from sharing information about government corruption.

While the Internet is a valuable tool, people cannot just dial up a revolution. The efforts to overthrow governments in Egypt, Tunisia, and Libya would not have succeeded without the courage of thousands of local residents who were willing to give their lives to change their government. Gaddafi would likely still be in power in Libya if the rebels had not also received military assistance from NATO. However, this support would probably not have happened without the work of people like Ayat Mneina, who collected information and put together a story that attracted the attention of individuals and governments around the world. The social media message is part of a media ecosystem that can help to amplify a message to enough audiences that a huge undertaking like a revolution becomes possible (Zuckerman, 2011).

The Arab Spring has served as a prototype for protest movements in North America. Both Occupy Wall Street and Idle No More made extensive use of social media. Idle No More began because Indigenous Canadians were unhappy about proposed government legislation they felt would have a negative impact on their rights. After an initial meeting organized by four women in Saskatchewan, the movement quickly spread through Facebook and Twitter. Ultimately, the movement spread across Canada (in January 2013 there were 600,000 tweets using #idlenomore), and ultimately spread throughout the world (see Chapter 22). This wasn't just a social media phenomenon as supporting events were held in Europe, Australia, and the Middle East (Coates, 2015).

Before reading on, test your knowledge of the media by taking the quiz in Box 18.1.

CRITICAL THINKING QUESTIONS

1. Think of a political issue that you and your friends are interested in (tuition fees, student loan policies, environmental issues). How could you use social media to help promote your views on this issue?
2. In the 2012 U.S. presidential election, Barack Obama based much of his campaign strategy on organizing support through social media. Are Canadian politicians doing a good job of using social media? Do you think young voters would be more receptive to political participation if more politicians follow President Obama's example?
3. Do any of your friends or classmates come from countries with repressive governments? If so, are any of them involved in using social media to help people who oppose these governments?

LO-1 WHAT ARE THE MASS MEDIA?

▶ **mass media** Any technologically based means of communicating between large numbers of people distributed widely over space or time.

The media connect those who produce messages with those who receive them. The **mass media** can be defined as any technologically based means of communicating between large numbers of people distributed widely over space or time (Pavlik and McIntosh, 2011:18).

Traditionally, the mass media have involved one-way communication in which a single source sent out a message to a large number of people who passively received that message. In the early days of television, there were only one or two channels and families gathered together in their living rooms to watch their favourite programs. There was little choice of programs and no opportunity for direct interaction between the media and members of the audience.

The nature of the media has changed dramatically. Social media such as blogs, Facebook, and Twitter allow many individuals or groups to send out messages to many others. The monopoly of those who control transmission has been broken. We have access to hundreds of television channels and millions of websites, there are countless YouTube videos to watch, and we can

BOX 18.1 SOCIOLOGY AND EVERYDAY LIFE

How Much Do You Know About the Media?

True	False	
T	F	1. You do not need to be concerned about your privacy when using social media sites such as Facebook.
T	F	2. Canadians spend more time watching television than using the Internet.
T	F	3. Canadian radio stations can broadcast any songs they wish.
T	F	4. When people design their online avatars in virtual worlds such as Second Life, they are not bound by our real-life cultural preferences about body size, hairstyles, and dress.
T	F	5. While Internet dating sites are becoming more common, most people still meet their partners through traditional means such as family, school, and church.

For answers to the quiz about the media, go to **www.nelson.com/student.**

communicate directly with people anywhere in the world, thanks to the **Internet**—the network infrastructure that links together the world's millions of computers.

The development of smartphones, tablets, and other portable devices means that we can stay constantly connected with family and friends and with the vast range of information available on the **World Wide Web**—the computer language that allows us to access information on the Internet. There used to be a clear separation of the different media, but media convergence has blended them together. We can watch movies and television programs, read novels and online newspapers, listen to music, and send text and voice messages on the same device. Anyone can become a publisher or broadcaster through blogs, YouTube videos, or publishing services such as Amazon's CreateSpace—and the traditional media have had to adapt to this new reality.

In this chapter, we will discuss some of the ways in which we are affected by the media and how new media technology is changing many aspects of our lives.

Internet The network infrastructure that links together the world's millions of computers.

World Wide Web The computer language that allows us to access information on the Internet.

LO-2 SOCIOLOGICAL PERSPECTIVES ON THE MASS MEDIA

The media play an important role in our lives. Media theorist Douglas Kellner has set out some of the ways the media affect us:

> Radio, television, film, and the other products of media culture provide materials out of which we forge our very identities; our sense of selfhood; our notion of what it means to be male or female; our sense of class, of ethnicity and race, of nationality, of sexuality; and of "us" and "them." Media images help shape our view of the world and our deepest values: what we consider good or bad, positive or negative, moral or evil. Media stories provide the symbols, myths, and resources through which we constitute a common culture and through the appropriation of which we insert ourselves into this culture. Media spectacles demonstrate who has power and who is powerless, who is allowed to exercise force and violence, and who is not. They dramatize and legitimate the power of the forces that be and show the powerless that they must stay in their places or be oppressed . . .
>
> The media . . . contribute to educating us how to behave and what to think, feel, believe, fear, and desire—and what not to do. The media . . . show us how to dress, look, and consume: how to react to members of different social groups; how to be popular and successful and how to avoid failure; and how to conform to the dominant system of norms, values, practices, and institutions. Consequently, the gaining of critical media literacy is an important resource for individuals and citizens in learning how to cope with a seductive cultural environment. (2011:7)

The Center for Media Literacy defines **media literacy** as the ability to access, analyze, evaluate, and create media in a variety of forms (2011). Sociology provides many of the tools we need to become media literate and to understand the role of the media in our society. Different sociological theories provide us with several perspectives that help us to explain how the media operate in our society and to understand the impact the media have on our attitudes and behaviour.

media literacy The ability to access, analyze, evaluate, and create media in a variety of forms.

FUNCTIONALIST PERSPECTIVES ON THE MEDIA

FUNCTIONS OF THE MEDIA Charles Wright was one of the earliest sociologists to analyze the media from a functionalist perspective. Wright (1959) described four functions of the mass media:

- *Surveillance.* The media tell us what is happening in the world. By warning us about imminent storms, reporting the latest economic news, or telling us about the newest Hollywood movies, the media play an important role in informing the public.

- *Interpretation.* The media interpret what is going on and tell us how we should respond to events. In editorials and in the selection of news stories, the media advance a particular agenda. Many parts of the mass media advocate particular political perspectives and try to influence how citizens vote.
- *Socialization.* The media transmit information, values, and norms from one generation to another and to newcomers. We can learn about society's rules, customs, and ways of behaving from the media.
- *Entertainment.* The media provide us with content that we find interesting and enjoyable. Many of us spend a significant part of our day being entertained by television, video games, radio programs, social media, and recorded music.*

Most would agree that the media do indeed fill these functions. The media sometimes do an excellent job of telling us what is going on—people around the world instantly learned of disasters like the 2011 earthquake and tsunami in Japan and the 2014 economic crisis in Greece. However, media coverage can be selective, so we learn more about some events than others. While a terrorist bombing in a Western industrialized country will receive massive coverage in the media, atrocities elsewhere in the world may remain virtually uncovered—most Canadians are unaware of mass atrocities going on in countries such as the Democratic Republic of the Congo because the media do not want to spend money to cover conflicts in little-known countries. Also, many of us are more concerned with the activities of celebrities such as the Kardashians than with life-and-death global issues.

One important form of socialization is that the media share images and experiences that allow us to imagine ourselves as part of a nation state. The invention of the printing press played an important role in developing a unified culture in France and other European countries (Straw, 2011). For centuries, France consisted of decentralized regional cultures with little sense of a national identity. Print media helped to standardize the French language, and the news and entertainment media contributed to a shared identity throughout the country. Straw also reports that the media, particularly television, drew the regions of Quebec together following the Quiet Revolution of the 1960s. The media often play a similar role everywhere in the world through broadcasting events such as the Olympics, legislative sessions, and national elections.

In many countries, the media have a legislated nation-building mandate. Canada's *Broadcasting Act* specifically gives the Canadian Broadcasting Corporation (CBC) the mandate to be "predominantly and distinctively Canadian, reflect Canada and its regions to national and regional audiences, while serving the special needs of those regions," to "actively contribute to the flow and exchange of cultural expression," to "contribute to shared national consciousness and identity," and to "reflect the multicultural and multiracial nature of Canada." (See Box 18.2.)

These early CBC logos highlight the organization's mandate of reflecting Canada and its regions.

* Wright (1959)

BOX 18.2 POINT/COUNTERPOINT

Should We Regulate the Media?

Many countries strictly regulate their media. China has imprisoned hundreds of journalists over the past decade, and "The Great Firewall of China" (formally named the Golden Shield Project) is one of many ways the Chinese government restricts free access to the Internet. Iran, Cuba, and several other countries are also known for their harsh treatment of journalists who are critical of the government.

Many other governments have also passed laws regulating the media. While Canadian media regulations may not censor free political expression, they do restrict who can own media companies, they regulate the use of the airwaves, and they put restrictions on media content. Children cannot view certain types of movies, and tobacco advertising is prohibited. One of the most interesting laws regulates the music played by Canadian radio stations.

In 1971, the Canadian government passed legislation specifying the amount of Canadian music that must be played on Canadian radio. York has explained why this legislation was passed:

> The average Top Forty station was programmed by an American broadcasting consultant, who told the station what to play, how to play it, and who to hire and fire. The station played either [British] or American records or both; it employed primarily American deejays; and it modeled itself after the best (or worst, whichever way you personally hear it) U.S. stations in Boston, Chicago and New York. The yardstick of success was just how close you could come to sounding exactly like these American stations. (Ranson, 2008)

In the 1950s and 1960s, few Canadian musicians were as popular as those from the United States and England. This situation changed when the Canadian Radio-television and Telecommunications Commission (CRTC) required radio and television stations to promote Canadian culture by broadcasting a regulated amount of content written and performed by Canadians. Television stations must have 50 percent Canadian programming between 6 pm and 11 pm, and radio stations must play music with 35 percent Canadian content. These regulations have had a positive impact. Canadian popular musicians flourished, and many have become international stars (Ranson, 2008), including Arcade Fire, Drake, Michael Bublé, and Justin Bieber.

While this effort to promote Canadian artists was successful, in 1999 the CRTC decided not to regulate the Internet. The CRTC felt that regulation would stifle creativity and innovation, and that most of the content of the Internet at that time was mainly text-based and did not involve broadcasting content (CRTC, 2011). The CRTC also felt that enough Internet content was being produced in Canada, so no regulation was necessary.

The CRTC revisited this issue in 2009, and several groups, including those representing actors, urged the CRTC to regulate online content in the same way as it regulates television and radio because they felt the new media are just providing alternate platforms for viewing and listening to cultural products. While it is easy to control the content produced by radio and television stations operating in Canada, however, Internet content can be transmitted from anywhere in the world. Transmitting radio over the Internet requires few resources other than a computer, a microphone, and some MP3 files—all of which are very portable, so regulation would likely be futile (Ranson, 2008). Thus, Internet content is still not regulated.

What are your views on this issue? Now that we are able to access radio and video broadcasts from anywhere in the world, should the federal government try to regulate content on the Internet to protect Canadian artists?

Building a national consciousness is not always a positive thing. The 1994 genocide in Rwanda was spread by radio stations and newspapers that promoted hatred against the Tutsi minority, and the media were heavily used by the Nazis to promote their vision of German racial superiority prior to World War II (Jackson, Nielsen, and Hsu, 2011).

Bookman (2011) has pointed out that today's interactive media also allow us to become members of very different types of imagined communities. These include online communities such as Second Life, as well as communities organized around interests such as cooking, sports, and environmental issues that can include members from around the globe.

OTHER FUNCTIONS OF THE MEDIA Wright's four categories do not exhaust the functions of the media. For example, new forms of social media give the public a forum to express their views and perspectives. Traditional mass media tightly control what is transmitted, but we are now living in an era when anyone can set up a website, write a blog, or comment on a news story. As you read in Ayat Mneina's story in the chapter introduction, people can be mobilized through these sites. The July 2011 edition of the Vancouver-based anti-consumer magazine *Adbusters* included the message, "What is our one demand? . . . #OCCUPYWALLSTREET, September 17" (Mickleburgh, 2011), which precipitated the Occupy movement in 2011. This movement spread quickly to many other cities throughout North America and Europe, and at least temporarily drew attention to the increasing degree of inequality in Western countries. And some politicians follow public sites and analyze Twitter feeds to help them develop policies that might appeal to voters.

The media also links people through means such as Facebook, dating sites, and sales sites such as eBay. The importance of being able to stay close to people who are important to us is shown by the fact that Skype reports that they host 2 billion minutes of video calls each day (Skype, 2013).

status conferral The process of giving prominence to particular individuals by focusing media attention on them.

The media also have a **status conferral** function—attention by the media can give people high status. This status can apply to people who have made real contributions to society and deserve recognition, but today's mass media also confer "star" status on people who have no particular talent but are simply "famous for being famous."

The celebrity socialite is not a new phenomenon, but reality television and the new social media have magnified the popularity of people such as the Kardashians (Gerds, 2011). Reality television shows enhance celebrity by showing us the details of their homes, cars, parties, holidays, and other personal details that make fans feel they have a close relationship with the celebrity, many of whom use their blogs and Twitter accounts to reinforce these intimate ties with fans. Fans can respond to these social media posts, which helps them believe that they have a mutual relationship with the celebrity. Frequent tweets by the stars give fans the sense that they are part of the celebrity's life. People grieved the deaths of celebrities such as Princess Diana and Michael Jackson as if they were close friends rather than people who were known only through media images (Hodkinson, 2011). The fact that people feel they are friends with people they will never meet blurs the distinction between reality and fiction—an issue that will be discussed later in this chapter when we discuss postmodern theories of media.

The Occupy movement spread to many Canadian cities, including this site in Toronto. Did this movement make you more aware of income inequality in Canada?

DYSFUNCTIONS OF THE MEDIA The media can also be *dysfunctional*. More than 60 years ago, long before the development of the Internet and the 24-hour global news cycle, Paul Lazarsfeld and Robert Merton (1948) coined the term *narcotization* to describe a situation in which people become so overwhelmed by the amount of information they receive that they become numb and do not act on the information. Nonstop broadcasts of scenes from a natural disaster such as the earthquake in Japan in 2011 can mobilize people to contribute to relief efforts, but as the bad news continues, people become desensitized and their attention moves on to other things.

The media can also deliberately distort the news to sway public opinion. For example, Anderson and Robertson have documented the way in which the media have perpetuated stereotypes that have reinforced notions that "degrade, denigrate, and marginalize" Indigenous people (2011:6). For nearly 150 years, the media have contributed to the marginalization of Indigenous people by portraying them as inferior and by failing to present the Indigenous case in discussions of issues such as land claims. Thus Indigenous people have been largely excluded from the nation-building narrative presented by mainstream newspapers.

Other distortions, mainly spread through the Web rather than through traditional media, include claims that the U.S. government was behind the 9/11 attacks on the World Trade Center, that President Obama is not legally the president because he was born outside the United States, and that vaccines cause autism. These stories survive because the new media provide so many possible sources of information that almost every bias and interest can find support online.

The impact of false stories may also be greater because of the new media. A person who watches network news or reads a mainstream newsmagazine will be exposed to a variety of issues and perspectives. However, if news comes through a customized daily news feed—what MIT Media Lab founder Nicholas Negroponte (1995) has called "The Daily Me"—the recipient may be exposed only to points of view he or she already supports. This new information will reinforce these beliefs rather than encourage the recipient to consider alternative views. However, there has not yet been sufficient research to know whether the Internet will lead to selective exposure to a narrow range of opinions or whether people will use its power to find sources that challenge their views.

CONFLICT PERSPECTIVES ON THE MEDIA

Conflict theorists argue that the media help the dominant class control society by reinforcing its capitalist ideology and by encouraging a mass consumer culture that allows those who own the means of production to sell us unnecessary products and services.

MASS MEDIA AND MASS DECEPTION Max Horkheimer and Theodor Adorno (1972) were two of the earliest critics of the mass media. During the 1940s, they asserted that the *culture industry* has turned artistic expression into just another marketable commodity—a commodity that keeps people passively entertained and politically apathetic. Consumption of this mass media culture destroys individual creativity and prevents consumers from taking a critical approach to their life situation. While "true" artistic activities may lead to independent thought and criticism of existing social arrangements, monopoly capitalism has used popular culture to promote its own values and to help preserve the status quo. According to these critics, the culture industry has been responsible for the mass deception and control of the public.

Through advertising and the images created by the popular media, capitalism also creates false needs for consumer products while people's true needs—including freedom and creativity—are ignored. And of course these products are manufactured and sold by the capitalist system that created the need for them. These false needs keep workers motivated to work even harder for their capitalist employers.

Image Courtesy of The Advertising Archives

Conflict theorists believe that capitalism uses the media to create false needs for consumer products. One example of this is advertising for makeup products that try to sell people on the ideal of everlasting youth and beauty.

The mass culture industry also provides workers with a way of escaping the boredom and the routine that define their lives. The workers consume standardized cultural products that follow a consistent formula so the consumer does not have to think much about them or engage with them intellectually. People can just come home after work and absorb media products such as television programs without engaging with others, without thinking about their problems, and without thinking critically about their lives under capitalism.

MEDIA CONCENTRATION Conflict theorists are critical of the fact that members of the dominant class control the media. They feel the media exploit the working class by promoting the cultural values and beliefs of the rich and powerful and by creating a false consciousness among workers. The media in most democratic countries claim to be objective, but conflict theorists reject this claim. For example, Herman and Chomsky (1998) describe the reasons why concentrated capitalist control means that stories critical of capitalism will not appear in the popular media: massive media companies profit from capitalism; their advertisers would not allow them to be critical of corporations; and governments and other organizations, such as pressure groups and corporations, influence media organizations. This means that the public does not receive any information that would seriously threaten the interests of the dominant class.

The Senate has concluded that the ownership of media is more highly concentrated in Canada than in most other countries and has recommended that the government take steps to limit media concentration (Parliament of Canada, 2006). However, since the Senate report, there has been even more concentration: Shaw recently purchased the broadcasting assets of Canwest, and Bell bought the portion of the CTV network that it did not already control. Both of these transactions are examples of media convergence as telecommunications companies (Shaw and Bell) have merged with media companies (Canwest and CTV). All the largest media companies show this convergence. For example, Rogers Communications provides cellular and home phone service, Internet, and cable television. Along with these telecommunications services, Rogers owns 54 radio stations and two television networks; 70 magazines, including *Macleans* and *Chatelaine;* and several television channels, including The Shopping Channel and Rogers Sportsnet. Rogers also owns the Toronto Blue Jays baseball team. In 2011, Rogers and Bell each purchased 37.5 percent of the Toronto Maple Leafs, the Toronto Raptors, and the Toronto FC soccer club.

Thus, Rogers (and its major competitors: Bell, Telus, Shaw, and Quebecor) owns some of the content as well as the means of distributing that content. Content can easily be shifted from one platform to another, and advertising can be targeted to very specific audiences. This gives these corporations a powerful place in the digital world. They have always been hugely profitable, as they spent many years as monopolies—for decades, in many areas, people had only one choice of a cable television provider, so the companies that received these licences were guaranteed large profits. While some of their business has now been deregulated, the massive cost of entering the market helps to limit competition.

Why is media concentration an issue? Those who control the media control the message, and if ownership is concentrated, there will be less likelihood of diverse messages. Conflict theorists such as Hall (1982) believe that the media support the values and interests of the dominant class. For example, while the Fox News Network describes itself as "fair and balanced," it actually has a very deliberate right-wing bias (Martin, 2008) and vigorously supports America's economic elites. While television networks such as Fox (and Sun News, its now-closed Canadian counterpart) do not hide their ideological bias, conflict theorists believe other media that claim to be objective are also biased. For example, Parenti argues that

> [Media bias] moves in more or less consistent directions, favoring management over labor, corporations over corporate critics, affluent whites over low income minorities, officialdom over protestors . . . privatization and free market 'reforms' over public sector development, U.S. dominance of the Third World over revolutionary or populist social change, and conservative commentators and columnists over progressive or radical ones. (2001)

How is media reporting biased? The media report some stories, ignore others that might challenge the dominant view of the world, and discredit stories that do not support the message of the established elites. They may also use loaded words: "freedom fighter" is a positive label, while "terrorist" is a negative label used to describe the same behaviour. Referring to "Third World" countries implicitly adopts the perspective that these countries are inferior to industrialized Western countries. The media also often deal with issues at a superficial level, which has the effect of supporting the status quo. Election campaigns focus on things like the latest poll results or a candidate's bus breaking down rather than on fundamental issues like inequality or the environment. Crime stories provide us with all the gruesome details, and the discussion is often racialized so that particular racial groups are blamed, deflecting attention from the social causes of crime.

Media concentration is an important concern, but the new social media are beginning to change this situation. The chapter introduction showed how social media broke the media monopoly of repressive governments, and anyone can set up a blog to get out information that might not be published or broadcast by media conglomerates. The music industry has much less control over its products since downloading became possible, and individual artists can now sell their work online without going through a record company.

INTERACTIONIST PERSPECTIVES ON THE MEDIA

Symbolic interactionists tell us that individuals continually negotiate their social realities. In his looking-glass-self theory, Charles Horton Cooley (1922/1902) proposed that a person's sense of self is derived from the perceptions of others. According to Cooley, we use our interactions with others as a mirror for our own thoughts and actions; our sense of self depends on how we interpret what they do and say. Consequently, our sense of self is always developing as we interact with others.

Cooley developed his theory more than a century ago, so it is interesting to see how it applies to our new social media. Do social media affect our sense of self? Does interacting with others online have positive or negative effects on our self-concepts? Do our online representations affect our sense of self and our behaviour?

SOCIAL MEDIA AND GENDER IDENTITY Shapiro has studied the impact of the Internet on transgendered people. The Internet has made it easier for everyone to learn more about transgenderism, which makes it easier for the transgendered to work out new identities. It has also helped them to "shape their own gender identity and self-esteem and manage feelings of fear, isolation, and anger" (Shapiro, 2010:107).

Differently gendered people can feel isolated because they are afraid of public stigmatization (Schrock et al., 2004). However, the Internet can help them construct their identities by providing them with information through blogs and information sites. Perhaps more importantly, online discussion forums provide a way of discussing their lives with others who share the same experience. The importance of this contact is apparent in the expression of joy posted by a cross-dresser who had just discovered an online news group:

> just couldn't resist sharing my *elation* at having found you! I stumbled across the group by accident during lunch today and my heart skipped, then skipped again,

> and again. i've been a t-something as long as i can recall, but never had much hope of meeting anyone else. it's so *good* to see you all out there. (Schrock et al., 2004:66)

This contact can help people construct their gender identities. Hill asked members of Toronto's trans community (including cross-dressers, transsexuals, and transgendered persons) how they came to their sense of identity as trans persons. Many reported that communications technologies had played an important role in this process. For example, Miqqi said:

> The biggest impact on the transgender world has been the Internet . . . I cannot overestimate its importance. It is . . . more than anything else how contact has been made for hundreds and hundreds of people who are isolated and otherwise out of contact. (Hill, 2005:39)

Another respondent, Melisa, who had been very isolated in a small northern community, described how she began to develop her trans identity based on her online contacts:

> I got on the Internet and the first thing I did, you know, was talk to anybody I could talk to about it . . . I wanted to talk to anybody or anything I could just to get some kind of rationale behind it, the vocabulary, do something with it . . . To build a story, to build a way to talk about it . . . It was something I could never do before. (Hill, 2005:41)

This research suggests that new social media can affect how people develop their identity, particularly for people who are isolated from other role models.

VIRTUAL IDENTITIES AND REALITY Many people blur the difference between reality and simulation by using the virtual world to experiment with their identity. Audrey, a subject interviewed by Sherry Turkle, told her that creating avatars and Facebook profiles is a "performance of you":

> Making an avatar and texting. Pretty much the same. You're creating your own person; you don't have to think of things on the spot really, which a lot of people can't really do. You're creating your own little ideal person and sending it out. Also on the Internet, with sites like MySpace and Facebook, you put up the things you like about yourself, and you're not going to advertise the bad aspects of you . . . You can write anything about yourself; these people don't know. You can create who you want to be . . . maybe in real life it won't work for you, you can't pull it off. But you can pull it off on the Internet. (2011:191)*

Audrey uses her online virtual life to help construct her real identity. She uses Facebook and Second Life to try different styles—flirting one day and being witty on another—and notes how others respond. If the new style gets a poor response, she changes it, and if her online friends respond positively, she incorporates that into her real-life identity. One of Turkle's respondents came out as gay online to help him deal with this process in real life. Another, who needed a prosthetic leg following a car accident, practised sexual intimacy through her online avatar (which also had a prosthetic leg) before considering a sexual relationship in real life. The Internet's anonymity makes these identity experiments much less risky than if they were done in real life. Just as a pilot can safely learn to fly and to handle aircraft emergencies in a flight simulator, some people can learn to handle real-life problems by practising online.

We know that people create idealized images of themselves when they create avatars on sites such as Second Life (LeBlanc, 2011). Yee and Bailenson (2007) have addressed the interesting question of whether the process works the other way as well—do our avatars also affect our real-life behaviour?

Yee and Bailenson assigned volunteers to either attractive or unattractive avatars and asked the volunteers to interact online with other avatars. They wanted to see whether their subjects

* (Turkle 2011:191)

conformed to the stereotypes of their avatars. They found that those with attractive avatars were more confident and friendlier toward others than were subjects with unattractive avatars. They chatted more often with the other avatars and positioned themselves much more closely to their companions than did the unattractive avatars. In a second experiment, taller avatars did better in a negotiation than shorter avatars. The finding that the nature of our avatars affects the way we behave may help to explain why virtual communities are friendlier and more intimate than real life because people tend to create attractive avatars.

This effect also spills over into our real lives. Yee, Bailenson, and Duchenault (2009) found that the effects of the height of avatars continued when the volunteers later engaged in face-to-face real-life negotiations. Yee and his colleagues conclude that, "Together these studies suggest that neither the virtual nor the physical self can ever be truly liberated from the other. What we learn in one body is shared with other bodies we inhabit, whether virtual or physical" (2009:309). As more people spend more time online with their digital avatars, it will be interesting to see what impact this will have on behaviour and on our interactions with others.

The new media raise some interesting questions concerning the meaning of the "self" in a digital world. Nancy Baym asks where our true selves reside: "[W]hat if the selves enacted through digital media don't line up with those we present face to face, or if they contradict one another? If someone is nurturing face to face, aggressive in one online forum, and needy in another online forum, which is real? Is there such a thing as a true self anymore? Was there ever?" (2010:3). Of course, sociologists such as Goffman (1959) long ago showed us that all of us present ourselves in different ways depending on the circumstances, so this is not just limited to our digital worlds.

FEMINIST PERSPECTIVES ON THE MEDIA

Gender scripts are our "blueprints for behavior, belief, and identity" (Shapiro, 2010:9). These scripts guide how we perform gender roles, and one of the ways we learn about these scripts is through the media. Many feminists are critical of the media's stereotyping of women, and research supports their view that television shows, newspaper stories, advertisements, movies, songs, and video games use sexist imagery to define women. If the media help us define our gender identities, the distorted mirror they hold up to women will have a negative impact.

Some of this bias can be attributed to the fact that the majority of people working in the media are male, and this male overrepresentation is particularly great at the senior management level (Straw, 2011). Advertising has been particularly criticized for being sexist, and most of the people employed in advertising agencies, particularly those who create the ads, are white males (Leiss et al., 2006). Some new media, such as massively multiplayer online games (MMOGs), are dominated by males, and many of their community forums reflect the sexism and misogyny that exists in the broader society (Braithwaite, 2014).

WOMEN IN THE NEWS Since 1995, the Global Media Monitoring Project (2010) has reported on the representation of women in the media. Its first study, which looked at 71 countries (now expanded to 108 countries), found that women were greatly underrepresented in newspaper, television, and radio stories in every country. Between 1995 and 2010, the percentage of stories that mentioned women increased from 17 percent to 24 percent. However, this meant that women were still neglected in media reports. The study also found that media reports stereotyped women's social roles. Women were more likely to appear in stories dealing with science and health than in stories dealing with politics or the economy. Women were underrepresented in all occupational categories except homemakers and students, so their role in other occupations and professions was minimized. Stories about women were more likely to mention their age and their family status than were stories about men. Thus, the media play an important role in reinforcing unequal gender stereotypes around the globe (Global Media Monitoring Project, 2010).

WOMEN IN ADVERTISING Gender roles are an important part of media advertising. While most products are designed to have some practical use, marketers also try to give their products a symbolic value that will encourage consumers to buy (Hodkinson, 2011). Consumers are invited to associate themselves with the image presented in advertising.

The image that many advertisers wish to hold up to women was illustrated by American feminist leader Gloria Steinem. Steinem, the founding editor of *Ms.* magazine, described a conversation she had with Leonard Lauder, president of the Estée Lauder cosmetics company:

> *Ms.* isn't appropriate for his ads anyway, he explains. Why? Because Estée Lauder is selling a "kept-woman mentality."
>
> I can't quite believe this. Sixty percent of the users of his products are salaried, and generally resemble *Ms.* readers. Besides, his company has the appeal of having been started by a creative and hardworking woman, his mother, Estée Lauder.
>
> That doesn't matter, he says. He knows his customers, and they would *like* to be kept women. That's why he will never advertise in *Ms.* (2011:241)

The lack of advertising from Estée Lauder and other companies led to the eventual sale of the magazine.

One of the few companies that did use a feminist message was a tobacco company. During the 1960s and 1970s, advertising for Virginia Slims cigarettes used the slogan "You've Come a Long Way, Baby" to co-opt the progressive image of the feminist movement as a way of attracting women to their product.

Despite decades of criticism, advertisers continue to show women in the role of caregivers or sex objects (Leiss et al., 2006). Current ads for GoDaddy.com—a company that registers Internet domain names—portray several female athletes, including Danica Patrick, in a sexualized manner. When not doing these commercials, Patrick has been breaking down stereotypes

Xiang yang/Imaginechina/AP Images

This beer advertisement shows how advertisers use gender stereotypes to market their products.

in the very dangerous, male-dominated sport of high-performance auto racing, but in the ads she is depicted only as an attractive, scantily clad woman. Other advertisements have depicted women as needing a man's help—sink stains required the help of the Ajax White Knight, and dirty floors and doors led to a call for Mr. Clean to rescue the woman of the house (O'Reilly, 2011). While women prefer advertising that shows them in more egalitarian roles (Leiss et al., 2006), advertisers still persist in showing women in these traditional ways.

Erving Goffman (1979) went beyond the content of the ads to look at other ways in which advertisers engage in gender stereotyping. He looked at the details of print advertisements and found several ways in which women and men were portrayed differently. Women were more often sitting down or even lying on the floor or a bed, while men were more often standing. Men were authorities, while women were more often in subordinate positions—men were shown instructing women or as doctors who had authority over female nurses. The exception to this was ads showing a kitchen, where women were in charge and were typically shown cooking for or serving men. If men were shown doing household chores, they were typically doing an incompetent job, often under the scrutiny of a woman.

The media perpetuate traditional gender roles, despite the dramatic changes in the role of women over the past 40 years. Why should we be concerned about this depiction of women? Media stereotypes may make it harder for women to move into non-traditional roles and to break through the glass ceiling that still exists in many occupations. The way the media portray ideals of personal appearance also influences how people think about themselves and the improvements that they believe their bodies require. Researchers have found, for example, that the media ideal of thin female bodies can lead to body dissatisfaction and depression among those who are most exposed to these idealized images (Mastro and Stern, 2006). If this view is correct, media framing plays an important role in the growing phenomenon of young women opting for cosmetic surgery.

POSTMODERN PERSPECTIVES ON THE MEDIA

The new social media are postmodern developments, so it should not be surprising that we can use the tools of postmodern theory to analyze some of the latest media trends. We can consider several examples of the way postmodern scholars have analyzed new media.

CREATING AVATARS IN SECOND LIFE Cathie LeBlanc (2011) used a postmodern framework to help us understand how participants design their avatars in Second Life. Postmodern theorist Michel Foucault showed that we live in a society where technologies make widespread surveillance possible in many different social settings. Because people feel they are always being watched, they govern their own behaviour through self-surveillance. While some have predicted that virtual environments such as Second Life will enable people to freely choose the bodies of their avatars without concern for physical limitations or cultural constraints, LeBlanc believes that self-surveillance will enforce social constraints on the appearance of avatars.

When people first move into the Second Life environment, they can select a default body, but the default body of a newcomer reduces their status in the virtual world, so most quickly customize their avatar as it becomes clear that the avatar's external appearance matters. The expectation is that Second Life residents will design an avatar to match their idealized view of who they really are and that this avatar will also meet conventional standards of attractiveness.

Avatars are designed by Second Life residents who are sitting outside the online environment looking at their on-screen representations. Like a cosmetic surgeon, residents can change those aspects of their avatar that do not match their view of who they are or would like to be. As residents of Second Life move through their online world, they see both the landscape and the back of their avatar. This creates a feeling that the body is "both a subject and an object" (LeBlanc, 2011:116). This sense of detachment reinforces a self-surveillance that enforces the norms of the virtual community.

These rather normal-looking avatars are the "plus-sized" bodies that generated hostile comments on Second Life. Why do you think people responded so negatively to the appearance of these avatars?

What are these norms? Despite the hope that virtual space would enable people to free themselves from real-life bodily constraints, what actually happens is that external appearance becomes as important online as it is in the real world. Just as people use cosmetic surgery, makeup, tattoos, piercings, and clothing to try to communicate who they are in real life, online residents are bound by cultural constraints. In effect, self-surveillance is even more explicit online than it is in the real world. This explains one researcher's finding that in Second Life, "Almost everyone was thin, beautiful, well dressed, and had chosen features typical of North American societies' Anglo-European ideals including skin tone, facial and body structure" (Shapiro, 2010:89).

Thus, people who could be whatever they wanted chose to follow the familiar cultural norms they had learned in the real-life world. This was likely the result of self-surveillance, but this is reinforced by pressure from others online to use attractive avatars. For example, one researcher found that many fellow Second Lifers were hostile toward the plus-sized avatars that she and several friends used (Ashkenaz, 2008). Typical of the comments were the following:

> Kess also experienced some biting comments from her friends about her new weight gain:
>
> Grow: Hi fatass : P
>
> *Then later on:*
>
> Grow: Kess you put on weight over xmas, hun.
>
> Jad: Eat too much for xmas, kess?

FANTASY FOOTBALL

> For the fantasy football fanatic, the seventh day is hardly an occasion for rest. From the moment he wakes, the clock begins ticking toward kickoff. There are statistical match-ups to analyze; weather and injury reports to consult; imaginary rosters to juggle. He is still a fan, but a very different kind of fan, and while he may not be alone in managing and cheering on his particular fantasy squad, he is not at all alone in his quest for fantasy success. (Serazio, 2008:229)

Fantasy sports have become a significant part of North American culture. Fantasy sports leagues enable people to draft real players onto their fantasy teams. The fantasy "owners" decide who is going to play and can trade players to other teams. The results are based on the performance of each of their chosen players in actual games.

Websites, magazines, talk shows, apps, and statistical services are dedicated to fantasy sports (Harper and Ploeg, 2011). The Internet has had a huge impact on fantasy sports because it enables the real-time tracking of statistics so that participants do not have to spend time making these calculations. In the United States, the number of fantasy sports participants has grown from just over 10 million in 2005 to over 40 million in 2015 (McDuling, 2015). Major media organizations, including NBC and Disney, have invested in fantasy sports companies, and a Canadian company, theScore, has developed a rapidly growing mobile app to help fantasy players track their teams.

The blending of real performance with the fantasy ownership of a team has changed the way people watch and relate to the game (Serazio, 2008). While sports fans normally cheer for a favourite team and often identify strongly with that team, fantasy team owners are concerned about the performance of the individual players who make up their virtual teams. And, of course, these teams are not really "teams" in the sense of being a group of people who work together to attain the common goal of winning but exist only in the simulated realm of fantasy. To the participants, however, the results of their fantasy teams do feel very real and can have real consequences if there are financial prizes for the best teams.

Jean Baudrillard (1995) developed the notion of **hyperreality**—a situation in which the distinction between reality and simulation has become blurred. We are saturated in media that have become a central part of our lives and that help to define our experiences and our understanding of the world. Manufactured images and representations are part of our reality and in turn affect the way we interpret subsequent images.

hyperreality A situation in which the distinction between reality and simulation has become blurred.

Harper and Ploeg argue that fantasy sports illustrate the world of hyperreality: "Reality is no longer diametrically opposed to fantasy; it is indistinguishable from it. Ironically, in *fantasy* sports, fans abandon the dichotomized framework of fantasy/reality and, instead, embrace the paradigm of the hyperreal" (2011:156).

TIME TO REVIEW

- Describe the role played by social media in the 2011/12 revolutions in the Middle East and North Africa.
- What are the functions and dysfunctions of the media?
- Discuss how conflict theorists feel the media support capitalist societies. What role does media concentration play in this support?
- How do interactionists help us to understand the way the new media helps people develop their identities?
- Discuss how stereotyped gender roles are used in advertising to promote products.
- How do online games fit Baudrillard's notion of hyperreality?

CONCEPT SNAPSHOT

FUNCTIONALIST PERSPECTIVES **Key thinker:** Charles Wright	Functionalists have shown that the media help us to learn what is happening in the world; suggest how we should interpret and respond to these events; teach us about our society's rules, customs, and ways of behaving; and keep us entertained. The media also provide us with community forums, link people together, and confer status on certain individuals. The media also have dysfunctions, including misinforming us and "narcotizing" us to the extent that we do not respond to social issues.
CONFLICT PERSPECTIVES **Key thinkers:** Max Horkheimer, Theodor Adorno	Conflict theorists argue that the media help the dominant class control society by reinforcing capitalist ideology and by encouraging a mass consumer culture that allows those who own the means of production to sell us unnecessary products and services. Conflict theorists are critical of the fact that the conventional media are dominated by a few very large corporations that control the message that goes out to media consumers.
INTERACTIONIST PERSPECTIVES **Key thinkers:** Sherry Turkle, Jeremy Bailenson	Interactionists tell us that individuals continually negotiate their social realities. The media, particularly the new social media, play an important role in creating our sense of self. Many people are now experimenting with identities online through the persona they put on their Facebook pages and through avatars in online games and virtual reality communities.

(Continued)

FEMINIST PERSPECTIVES **Key thinkers:** Gloria Steinem, Erving Goffman	Feminist sociologists have been critical of the stereotyped manner in which women are portrayed in the media. Most stories in the media focus on males, and women are depicted in typical gender-linked roles. Many advertisements depict women as traditional homemakers and caregivers or as sex objects. These media stereotypes may make it more difficult for women to move into non-traditional social roles.
POSTMODERN PERSPECTIVES **Key thinkers:** Michel Foucault, Cathie LeBlanc, Jean Baudrillard	Postmodern theorists have approached the media in several different ways. One of the most interesting of these is Baudrillard, whose theory of hyperreality points out how the distinction between reality and simulation has become blurred. Manufactured images and representations are part of our reality and in turn affect the way we interpret subsequent images.

LO-3 THE IMPACT OF SOCIAL MEDIA

In this section, we will look at some of the ways in which new media technologies reshape the ways we interact with each other. Texting is replacing face-to-face contact, and we are increasingly likely to meet our romantic partners online rather than through family or school. It is important to consider the impact of these changes on our culture and our social relationships.

ALONE TOGETHER: DO SOCIAL MEDIA BRING US TOGETHER OR KEEP US APART?

Sherry Turkle was one of the first to study computer culture. When she began her work in the 1980s, phones were mounted on kitchen walls, people played rudimentary video games that plugged into their television sets, and a few hobbyists were building and programming their own home computers. Turkle was interested in learning how these tools were shaping individuals and their cultures.

The relationship between people and their communication devices began to change dramatically in the 1990s and continues to rapidly evolve. Many of us are connected all the time, and communicating electronically is becoming the way we live rather than just something we do. The Internet has dramatically increased the scope of possible relationships, as it is now just as easy to communicate with people halfway around the world as with the people next door. Your parents' old wall-mounted telephones were just a way of talking to other people, but our portable devices have become a "portal that enable[s] people to live parallel lives in online worlds" (Turkle, 2011:xi).

What impact has this had on our lives? Do social media hold us together or keep us apart? Two facets of social media are having a dramatic impact on our social relationships: Mobile devices mean that we can always be contacted, and we can develop online identities in addition to living our "real" lives.

WE ARE ALWAYS AVAILABLE Mobile connections mean that people are no longer tethered to their computers but carry their online lives with them everywhere. We can always be available to others and we may have difficulty escaping this contact even if we wish to do so.

Some people have become dependent on this constant connection. In 2011, when a problem affected many subscribers' use of their BlackBerry smartphones for several days, many called it a "disaster" and filed lawsuits against the manufacturer. Some of Turkle's respondents describe the loss of a cellphone as feeling "like a death" (2011:16).

Communications professor Danna Walker required her students to avoid all electronic media for 24 hours. Many of the students found this painful. One wrote, "I was in shock . . . I honestly did not think I could accomplish this task. The 24 hours I spent in what seemed like complete isolation became known as one of the toughest days I have had to endure" (2007). Walker commented that another "apparently did not see the irony in this statement: 'I felt like I would be wasting my time doing the project. I did not want to give up my daily schedule, which mainly includes lying on my couch, watching television and playing The Sims 2 . . . computer game" (2007). Another said, "There was a moment in my day when I felt homeless . . . I couldn't go home because I knew that would be too tempting . . . I was walking down the street literally with nowhere to go, and I just didn't know what I was going to do" (2007).

Our constant availability might help us to strengthen our relationships with significant others like friends and family. However, Turkle concluded that the new media may actually be keeping us apart: "We are increasingly connected to each other but oddly more alone: in intimacy, new solitudes" (2011:19).

Think of how some people pay more attention to their mobile phones than to the people they are with. Face-to-face conversations are routinely interrupted by the beep or buzz of the mobile device as a text message is given priority over face-to-face communication because cultural norms demand an instant response. The average 18- to 24-year-old sends and receives 3800 text messages each month (MarketingCharts, 2013). While each text message may not take long, dealing with more than a hundred texts a day takes an enormous amount of time. The time spent using social media—whether responding to texts each day, spending hours in virtual reality, or playing an online game—is time that is not spent interacting with other people. Face-to-face relationships are being replaced by networked relationships. This point was highlighted by another of Danna Walker's students who had to give up electronic media for a day: "My mother is thrilled that I'm doing this . . . to her it means I get to spend the day with her. I bite, and we walk into town for some brunch. I draw out the brunch as long as possible" (2007).

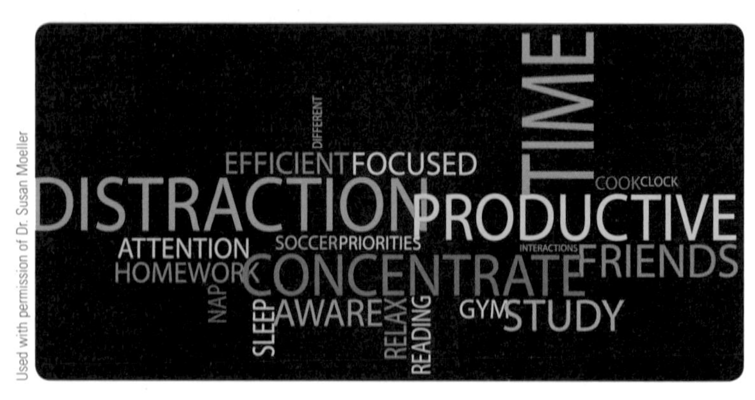

The first of these word clouds shows the feelings of students who had spent 24 hours without media. The second shows their responses to a question about the advantages of unplugging. What lesson can you draw from these two word clouds?

The ability to chat online certainly broadens our range of contacts. People can keep in touch with hundreds of Facebook friends. However, we can ask whether having many brief contacts with a broad range of "friends" is as meaningful as more intimate personal relationships with fewer people.

Others argue that social media do facilitate personal relationships. One of Danah Boyd's respondents told her:

> "I can't really go see people in person. I can barely hang out with my friends on the weekend, let alone people I don't talk to as often. I'm so busy. I've got lots of homework. I'm busy with track, I've got a job. . . . there's some people I've kind of lost contact with and I like keeping connected to them because they're still friends. . . . [and] Facebook makes it a lot easier for me" (2014:20).*

For many, social media sites are the social spaces where they get together with their peers (Boyd, 2014). However, for those who do not wish to have close contacts, technology can be used to keep people at a distance: "Texting puts people not too close, not too far, but at just the right distance" (Turkle, 2011:15).

Turkle also discusses an extreme form of distancing one's self from others when she describes an interview with a reporter from *Scientific American* who accused her of intolerance because she did not think it was appropriate that people might someday have sexual relationships with robots and even marry them. Turkle subsequently interviewed several people who would have welcomed such relationships because they came without the risks inherent in relationships

* Danah Boyd (2014:20)

with other humans. Real-life marital partners can cheat, become alcoholics, and be judgmental about their partner's behaviour. Robotic spouses would be completely predictable and could be programmed to do whatever their partners/owners wished. Many young people have been raised with robotic pets, and some seniors now have robot companions to keep them from getting lonely.

WE CAN LIVE ALTERNATIVE LIVES Social media can provide us with online lives that may be more satisfying to some people than their real lives. Consider the promises made by Second Life (secondlife.com): "Who will you meet in Second Life? Where will you explore? What will you discover? Who will you be? Anything is possible in Second Life. A whole new world is waiting."

To someone who is lonely and unhappy, the promise of virtual reality can be compelling. A new identity, a glamorous appearance, and the chance to live an adventurous lifestyle lie ahead. Your Second Life avatar can help you live your dreams, and you are in control in a way that would never be possible in the real world.

Consider Pete, one of Turkle's interview subjects. Pete was a 46-year-old man who was devoted to Second Life. His online avatar is a handsome young man who had married a young and beautiful avatar, Jade, at a wedding attended by their virtual best friends. While Pete has never met the anonymous woman (or man) who is behind Jade, he shares with Jade a relationship that is both socially and physically intimate. Pete prefers his Second Life to his "real life": "Second Life gives me a better relationship than I have in real life. This is where I feel most myself. Jade accepts who I am. My relationship with Jade makes it possible for me to stay in my marriage, with my family" (2011:159). To Turkle, "The ironies are apparent: an avatar who has never seen or spoken to him in person and to whom he appears in a body nothing like his own seems, to him, most accepting of his truest self" (2011:159). A later study found that people involved in Second Life responded to being cheated on by a virtual partner with the same emotional intensity that they would have to a similar situation in the material world (Gabriels et al., 2013).

While Pete used his online life to help cope with an unhappy family situation, others who are dissatisfied with their real lives may completely withdraw into their online worlds. Adam was another of Turkle's respondents. He got little satisfaction from his life and turned to online games such as Quake and Civilization:

> These games take so long, you can literally play it for days. One time when I played it, I had just got the game and I got so addicted, I stayed home the next day and I played . . . I think it was like noon the next day, or like nine o'clock the next day, I played all night long. And I ended up winning. You get so advanced. You get super-advanced technology. (2011:222)*

Adam felt good only when he was playing games, and he was emotionally attached to the bots he played with. He was letting the rest of his life slip away, and his difficulties dealing with reality turned him even more toward the virtual world.

The next major evolution in new media will likely be virtual reality. As technology improves, the virtual experience will become more like that of the material world. We will be able to visit

* Turkle 2011:222

museums in distant countries in much the same way as we can in our own communities, and we will be able to experience very realistic sensations of flying a plane, driving a race car, or climbing Mount Everest. While escaping into an alternative, virtual world may have unhealthy consequences for some people, virtual reality may also enable us to interact with family, old and new friends, and romantic partners even though they live in other places. The ability to escape into a fantasy world may be a welcome change for many of those whose material lives are not satisfying.

HOW SOCIAL MEDIA AFFECT OUR LIVES

Even those who are not hooked on virtual worlds or online games can be affected by the demands and distractions of the new media. Many of our activities require concentration, but it can be difficult to focus on a task when texts and emails keep arriving and any fleeting thought can lead us to waste minutes or hours tracking down something on the Internet. Many of our students have told us how difficult it can be to study when their friends expect an immediate response to text messages. The Internet has been an invaluable resource that has helped us write this text—we no longer need to walk to the library to find the most recent crime rates or to read an article in an academic journal—but it is also a source of distraction that makes it more difficult to meet deadlines. Sometimes watching cat videos or checking the latest hockey scores is more interesting than writing about how to conduct survey research or why social class affects educational success.

There are other, more serious consequences of distractions. An iPhone app named Type n Walk uses the iPhone camera to show the ground ahead of you so you do not walk into a tree or a lamppost while you are texting on your phone. Hundreds of people are killed each year in North America because drivers continue to text and talk on cellphones while they are driving despite a vast amount of evidence showing that this practice is at least as dangerous as drunk driving. For some people, the need to be in constant contact with others outweighs the need to walk or drive safely.

Some researchers have studied problematic Internet use (Tokunaga and Rains, 2010), which consists of behaviours such as not being able to control the amount of time spent using the Internet, a preference for online interaction over in-person interaction, and withdrawal when the Internet is not available. Some consider the Internet addictive—South Korea has set up Internet addiction recovery programs for young people, and 12-step recovery programs for Internet addiction are even available online (Blascovich and Bailenson, 2011). However, there is some hope for those of you who wish to avoid distraction while studying. A recent study (Patterson, 2015) found that students who used software tools that restricted Internet access were 40 percent more likely to complete courses and received higher grades than students in a control group.

There have always been fears about the impact of media—in the 18th century, the new popularity of novels led to fears of "reading mania" (Korkeila, 2010). However, Elias Aboujaoude (2010) suggests that virtual technology may have a much more profound impact than earlier forms of mass media because the new media have immersive and interactive qualities that the earlier media did not have. Blascovich and Bailenson (2011) speculate that 3-D immersive virtual reality experiences are potentially much more addictive than current online games and activities because the experiences are closer to those of real life. Online avatars will look like humans and will be controlled as easily as we control our own bodies, so the line between real and simulation will continue to fade.

While there may be negative effects, the popularity of social media shows that most people who use it consider it a valuable part of their lives. Families share news and photos on social media, and friends can stay in touch with each other, even if they live in different cities. The earlier discussion of the impact of the Internet on the transgendered and transsexual community shows how important it can be for people to be able to meet like-minded others online.

The ability to have friends and role models from anywhere in the world can be liberating to people whose physical condition, sexual preferences, or other characteristics make them different. The GimpGirl Community, established online by women with disabilities, is another example of how the Internet enables people to set up networks of friendship and support from hundreds of communities to interact with one another.

WILL YOU MEET YOUR PARTNER ON THE INTERNET? THE RISE OF ONLINE DATING The Internet has dramatically changed the way many people meet their partners. In the 1940s and 1950s, most heterosexual couples met through family, friends, and school (Rosenfeld and Thomas, 2012). The family and school have steadily declined as sources of connection since then, but friends are still very important in making introductions, and the proportion of people who meet their future partners in bars and restaurants has doubled since the 1940s. The most significant change has been that in just over a decade, the percentage of people who met their partners on the Internet jumped from zero to 22 percent, and it is now the third most common way of meeting (see Figure 18.1). These people meet through dating sites, online personal ads, online gaming sites, and sites focused on interests such as religion or hobbies.

Same-sex couples were much more likely to meet over the Internet: 61 percent of same-sex couples met online (see Figure 18.2). Rosenfeld and Thomas believe this is because the Internet

FIGURE 18.1 : THE CHANGING WAYS AMERICANS MEET THEIR PARTNERS: HETEROSEXUAL COUPLES

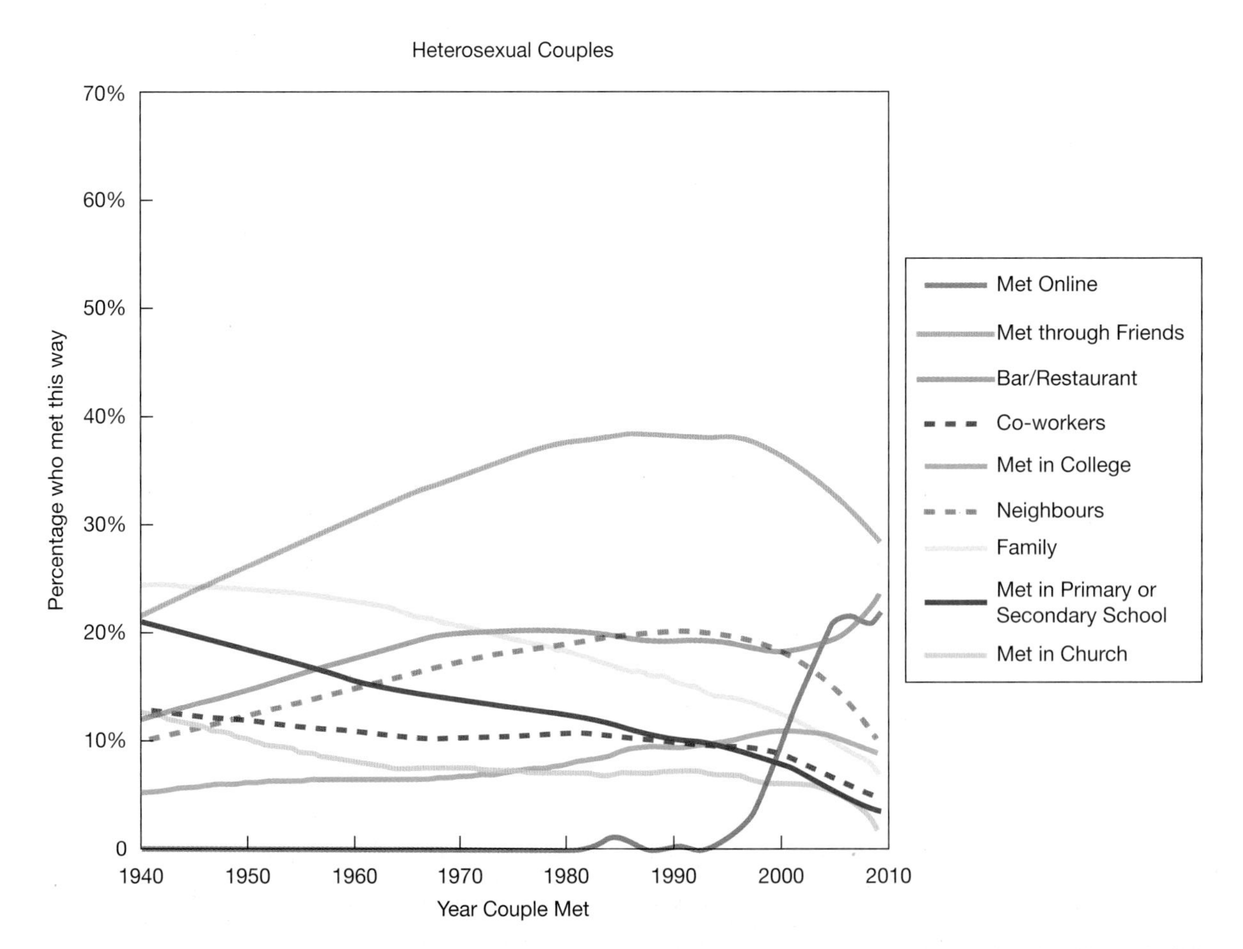

Source: Michael J. Rosenfeld and Reuben J. Thomas, "Searching for a Mate: The Rise of the Internet as a Social Intermediary," *American Sociological Review*, 77 (4) pp. 523-547. Copyright © 2012. Reprinted by Permission of SAGE Publications Inc.

may be particularly useful to people facing thin dating markets—those who are seeking partners who are harder to find and where relationships may meet with social disapproval from some people. Parents will not be helpful in introducing a gay or lesbian child to a future partner if they disapprove of same-sex marriage or if their child has not come out to them. However, the Internet allows for a very efficient search process far beyond local neighbourhoods and friendship groups. Older heterosexuals also have a limited range of potential partners because most people in their age group already have partners and also have much higher than average rates of finding partners online. Rosenfeld and Thomas found that couples who met online were just as satisfied with their relationships as people who met in other ways and were no more likely to break up their relationships.

WILL ONLINE DATING BROADEN OUR SOCIAL TIES? How will digital media transform our social relationships? A Super Bowl ad for telecommunications company MCT told us, "There is no race, there are no genders, there is no age, there are no infirmities, there are only minds. Utopia? No, the Internet" (Baym, 2010:34). However, others feel social media will just reinforce our existing social networks. Those who favour this view cite research on the impact of the telephone, which increased people's ability to communicate but did not change people's social networks because they used the phone to communicate with their existing network of friends and relatives. Rosenfeld and Thomas examined the impact of the Internet on dating and marriage.

FIGURE 18.2 : THE CHANGING WAYS AMERICANS MEET THEIR PARTNERS: SAME-SEX COUPLES

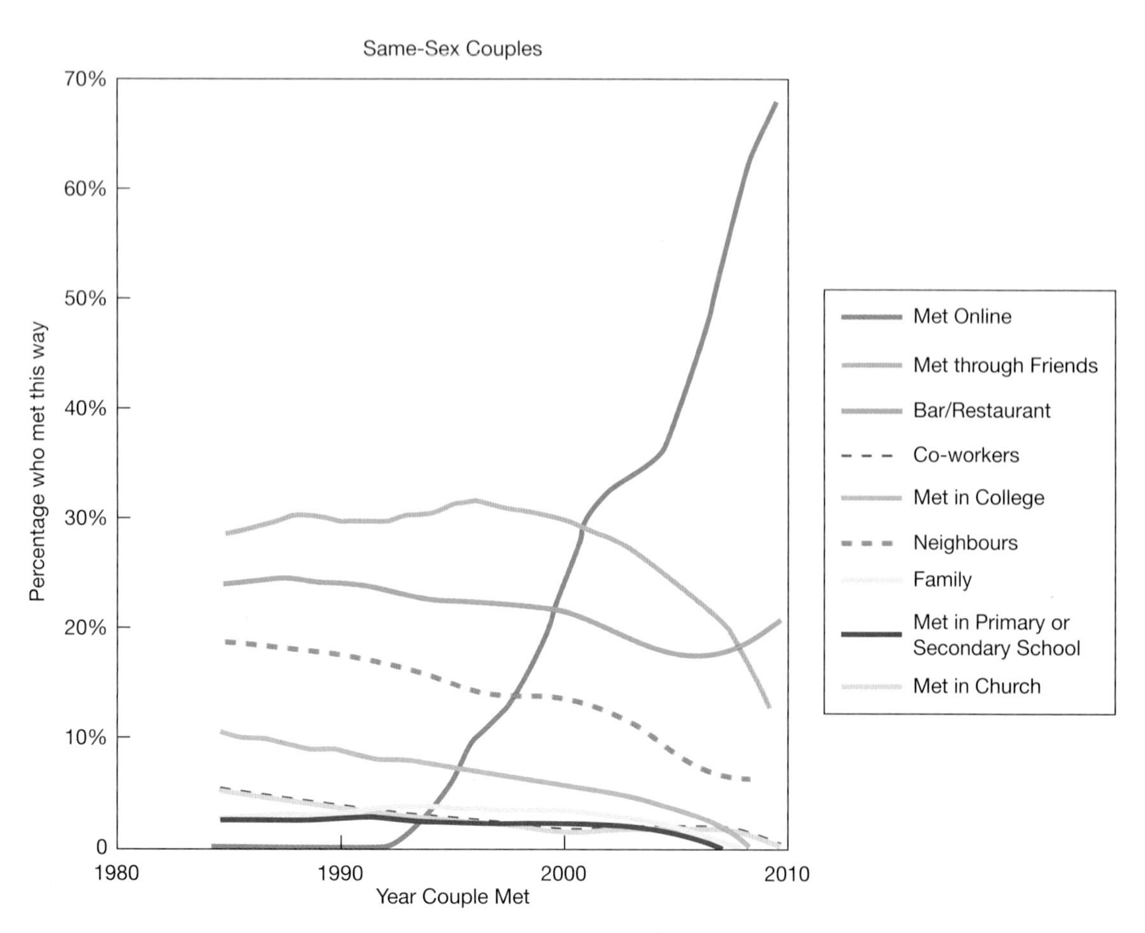

Source: Michael J. Rosenfeld and Reuben J. Thomas, "Searching for a Mate: The Rise of the Internet as a Social Intermediary," *American Sociological Review*, 77 (4) pp. 523–547.

Before World War II, most people chose their mates from their local neighbourhoods, schools, and churches. Because these institutions were homogeneous, people married partners with very similar class and race backgrounds. Does the broader reach of the Internet mean that people will meet potential partners who come from more diverse economic, racial, and religious backgrounds?

The Internet has had a limited effect on broadening social contacts. While couples from different religions were more likely than those from the same religion to have met online, the opposite was true for couples from different racial backgrounds, although the latter difference was not strong (Rosenfeld and Thomas, 2012). A study conducted using data from one of the largest online dating sites found that most people chose to contact and to respond to people whose racial and educational backgrounds were similar to their own. The researchers concluded there was a racial hierarchy in dating preferences: "our results show that although these patterns differ by gender, a common thread is that black women and black men are clearly disadvantaged in the dating market relative to other ethnic groups. Being black on the dating market—particularly being a black female—means that one's invitations are most likely to be ignored" (Lin and Lundquist, 2013).

DO PEOPLE TELL THE TRUTH ABOUT THEMSELVES ONLINE? While people are able to hide their identities in much of the online universe, this does not work in online dating, where the ultimate goal is a real-life romantic relationship. Those seeking a romantic attachment online face a dilemma. To attract potential partners, a person wants to create a positive impression. For online dating, physical attractiveness is particularly important in self-presentation, so people may use their most flattering photographs, photographs taken several years ago, or photographs that have been digitally altered (Toma and Hancock, 2010). However, a person must take care in constructing an online image. If reality is distorted too much, the potential partner may be very disappointed at the first face-to-face meeting—it is immediately obvious if a man is 20 cm shorter or 40 kg heavier than his dating profile had claimed, or if he really does not own a Ferrari and live in a mansion.

Toma and Hancock (2010) assessed the accuracy of daters' online reports of physical attractiveness. Those who were the least attractive were the most likely to enhance their online photographs and to lie about height, weight, and age. They were slightly more likely than more attractive online daters to lie about factors such as income and occupation. On the other hand, attractive daters posted more photographs of themselves, presumably to showcase their desirability.

Given the possibility of distortion, how do those seeking a partner reduce the chance of being disappointed when they actually meet their online contact? In relationships that begin through social contacts, some information about a potential partner can often be obtained from family, friends, and co-workers. Also, when we meet people in person, we know how they look, and we can use nonverbal cues to give us some clues about their personality.

These information sources are not available in online dating, but there are other ways of checking out another person (Gibbs, Ellison, and Lai, 2011). Networking sites such as Facebook and LinkedIn, public records (such as property tax assessments), and Google searches provide information about potential partners. Some online daters also ask direct questions and disclose more of their own personal information in online conversations in the hope of getting their potential partner to disclose more of their information.

Toma and Hancock used computerized linguistic analysis to find clues to the truthfulness of potential partners. These clues can be found in the language people used to describe themselves. The less truthful used fewer first-person pronouns (words such as *I*), used more negatives, and had shorter self-descriptions than those who were being honest (Rosenbloom, 2011).

LO-4 PROFITING FROM SOCIAL MEDIA

In its early days, the Internet was not about money because it was designed to allow scientists to communicate with one another. Businesses soon began to seek ways of profiting from Web activities, however, and the ideologies of "free" versus "profit" have battled throughout the history of the Internet. For example, early in the development of the Web, Microsoft forced computer manufacturers to include Internet Explorer in their computers. Eventually, Explorer captured 97 percent of the browser market and put most of its competitors out of business. This monopoly was broken by the courts, and other browsers, such as Firefox and Chrome, have become popular.

Another economic battle was fought after the invention of Napster, which was the first widely available means of downloading music. This had a huge impact on the music industry as millions of people began to share music online rather than purchasing it from music stores. This permanently changed the music business, although legal music sources such as iTunes have helped the music companies and the artists regain some of their profits.

The new social media are also trying to become profitable. Most users do not understand the role of Facebook (Rushkoff, 2010). They think they are *clients* of Facebook, but they are actually its *products*—because Facebook makes money by selling access to personal information. Because of the information people provide to the site, Facebook can charge a premium for ads that are tailored to the interests of a particular audience (see Box 18.3). Many other social media platforms such as Twitter and LinkedIn have rewarded investors as the value of their stock has increased, but are looking for ways of reaping greater financial rewards from their enormous popularity. Popular games such as *Candy Crush* and *Clash of Clans* can be downloaded free, but have extra features for which users can pay if they choose—and enough choose to pay that the games make millions of dollars a day in revenue (Stevenson, 2014).

BOX 18.3 POINT/COUNTERPOINT

New Media and Privacy

You have just tweeted a friend that you're going for coffee. As you pass a coffee shop, a coupon arrives on your phone offering 50 cents off a large cup of coffee. A friend who is travelling to Paris uses an online site to book a hotel room. Because she is using a Mac computer, the hotels that come up on the booking list are more expensive than if she had used a PC. A new college graduate has submitted a resumé for a job. The potential employer looks at the applicant's Facebook site, finds photos of the applicant using soft drugs at parties, and decides not to hire the person. In each of these cases, information that a person might expect to be private has been used by a third party. In the first two cases, the information was sold to an advertiser.

Online sites such as Facebook, Google, and Twitter provide a useful service for hundreds of millions of users. However, from the perspective of those who own these sites, the service is actually provided to advertisers rather than to users. The commodity they provide to these advertisers is personal information about users that allows advertisers to carefully target their ad campaigns. Advertisers are interested in knowing your location, relationship status, travel plans, musical tastes, occupation, and other interests.

Search engines such as Google make billions of dollars from tracking the key words you use. If you search for terms such as *headache* or *upset stomach*, you may receive ads or coupons for remedies for these maladies. Google also tracks your information across its different products, such as Gmail and YouTube, to develop more complete profiles of users in order to personalize the service.

Facebook and other networking sites frequently change their privacy policies with little notice. In 2009, Facebook suddenly made lists of friends publicly available. This change had serious consequences for many people. For example, the Iranian government detained people who were Facebook friends with people who were critics of the government (Andrews, 2012).

(Continued)

Many of those involved in the Occupy Wall Street movement would not use Facebook to help organize their protests because faces in photographs were automatically tagged so that people could be identified (Taras, 2015). In 2015, shares of Facebook were valued at $81 for each $1 of earnings, which means that investors are counting on rapid increases in the company's profits. Thus shareholders will be putting pressure on Facebook executives to find new ways of selling people's information that will put increasing pressure on privacy. Meanwhile, Google has paid large fines after admitting to using its Street View camera cars to collect people's wireless data, and to violating rules by bypassing privacy settings on the Safari browsers used on iPhones and computers.

While people may feel that receiving targeted advertising is an acceptable price to pay for using the services of a search engine or networking site, many of us are not aware of the other organizations that use these data. Insurance companies may search for evidence of risky activities or chronic illnesses that individuals may not report on their application forms. Law enforcement agencies and tax officials use social media sites to find information about potential wrongdoing and do not need search warrants that would be required if they were accessing the information in another fashion. Law enforcement agencies also track people by using their cellphone records and obtain a great deal of personal information through this tracking:

> The United States Court of Appeals for the District of Columbia Circuit, ruling about the use of tracking devices by the police, noted that GPS data can reveal whether a person "is a weekly church goer, a heavy drinker, a regular at the gym, an unfaithful husband, an outpatient receiving medical treatment, an associate of particular individuals or political groups—and not just one such fact about a person, but all such facts." (Maass and Rajagopalan, 2012)*

Information may last forever on the Internet. Do most people know that personal information they and their friends post online can come back to haunt them in the future when they apply for a job, run for political office, or try to convince their teenage children that they behaved perfectly when they were attending university or college? What about sites such as Banjo that allow people to track you and receive a message if you are near their location? Does it concern you that if you search for information on HIV/AIDS for a term paper, a data aggregator may infer that you have AIDS? Should people have the right to know what data are being collected about them and to prohibit companies from tracking their information without their consent? Would you prefer to pay a fee for Facebook rather than letting them sell your personal information to advertisers? These and other issues need to be worked out as new media technologies become more pervasive in our lives.

Sources: Andrews, 2012; Maass and Rajagopalan, 2012.

While social media are an important part of our economy, we are barely touching the surface of the commercial exploitation of the new media. For example, advertisers have developed a strategy of self-endorsing, which means that an avatar or other representation of the consumer is used in an advertisement. Ahn and Bailenson (2011) found that this was a highly effective technique. If the avatar of the potential customer was shown using the product, favourable brand attitude and purchase intentions increased. For example, if the avatar was shown wearing a particular brand of clothing, the research subjects were more likely to prefer that brand to others representing similar products. The closer avatars were in appearance to the consumer, the more positive the brand attitude. Increasingly, the information we provide about ourselves through social media will be used to personalize the advertising we receive and the impact of that advertising.

LO-5 GLOBALIZATION AND THE MEDIA

MEDIA TECHNOLOGY AND GLOBALIZATION

With its dots and dashes transmitted through a wire, the telegraph seems a very primitive means of communication compared with today's media technologies. After its introduction in the 1840s, however, people predicted it would have a revolutionary impact because, for the first time, people could instantly communicate over a distance:

> Universal peace and harmony seem at this time more possible than ever before, as the telegraph binds together by a vital cord all the nations of the earth.

* Maass and Rajagopalan, 2012

> It is impossible that old prejudices and hostilities should any longer exist, while such an instrument has been created for an exchange of thought between all nations of the earth. (Czitrom, 1982:10)

Canadian media scholar Marshall McLuhan was one of the first to recognize the potential impact of modern media technologies. McLuhan believed these technologies would create a **global village** in which people around the world share information through interactive media: "'Time' has ceased, 'space' has vanished. We now live in a global village . . . a simultaneous happening" (McLuhan, Fiore, and Fairey, 1967:63). McLuhan optimistically believed that the global village would empower people and foster the growth of democracy and equality.

global village A world in which distances have been shrunk by modern communications technology so that everyone is socially and economically interdependent.

Today's media make it possible to form communities of shared interest even though people have no direct contact with each other. For McLuhan, it was important that this communication was not one-way and linear but decentralized and multidirectional:

> In the electric age, when our central nervous system is technologically extended to involve us in the whole of mankind and to incorporate the whole of mankind in us, we necessarily participate, in depth, in the consequences of our every action. It is no longer possible to adopt the aloof and disassociated role of the literate Westerner . . . As electronically contracted, the globe is no more than a village. (McLuhan, 1964:4–5)

McLuhan believed that media technology would change society and shape our social lives. While McLuhan's work predated the Internet by three decades, the impact of today's media has at least partly supported his views. Examples such as the global response to natural disasters such as the 2011 Japanese earthquake and tsunami, and the earthquake in Nepal in 2015 are fostered by the fact that people around the globe could follow the disaster in real time through the mass media. Other examples of positive change are technologies that enable people in isolated rural areas anywhere on the globe to educate themselves, care for their health, and conduct business—a farmer with a smartphone can easily find the most lucrative place to market her products rather than accepting the price offered by local buyers.

Not everyone is as positive about the impact of the media on globalization as McLuhan. Schiller (1992) argued that media technologies would lead to **cultural imperialism**—a process whereby powerful countries use the media to spread values and ideas that dominate and even destroy other cultures, and local cultural values are replaced by the cultural values of the dominant country. The dominant media culture is that of the United States, and the media enable large transnational corporations to profit from their global domination at the expense of cultural diversity. The globalization of culture means that local cultural values are replaced by Western values such as consumerism.

cultural imperialism A process whereby powerful countries use the media to spread values and ideas that dominate and even destroy other cultures, and local cultural values are replaced by the cultural values of the dominant country.

Cultural imperialism has occurred—many people around the world are familiar with Lady Gaga, the Disney characters, and many other products of American culture. However, the flows are becoming multidirectional as there is now a reverse flow of cultural products, such as India's Bollywood movies, which have become popular in many parts of the world, including North America. Countries including China and Brazil have very significant cultural industries that dominate their own local media and are now being exported.

The Qatar-based television news network Al Jazeera has become a significant cultural force since its establishment in 1996. Al Jazeera has transformed news reporting in many Middle Eastern countries, where coverage was often limited because of restrictions imposed by dictatorial governments. It has provided a global alternative to Western media, and its English language services have provided people in Europe and North America with a very different perspective on Middle Eastern issues than they have been given by traditional news sources.

KHALED DESOUKI/AFP/Getty Images

Mohamed Fahmy is a Canadian citizen who worked for Al Jazeera. He was sentenced to three years in an Egyptian prison after his conviction on charges of broadcasting false news. No evidence was presented in court to support these charges and it is likely that the conviction was part of a dispute between the country of Qatar (which owns Al Jazeera) and the current Egyptian government. In late 2015, Fahmy was pardoned after spending a year in jail and another year on bail in Egypt.

THE DIGITAL THIRD WORLD

Information and communications technologies are changing the world, but are these technologies helping people who are poor or just increasing the gap between rich and poor?

Some feel that the ability to share information from around the globe will help low-income countries develop by providing them with the opportunity to become knowledge societies that can compete with industrialized nations. New technologies will help them streamline industry and government, and their competitive advantage in wages will allow them to attract business. Technology can speed up educational reform and help build a more participatory civil society through the sharing of information and ideas. India has generated hundreds of thousands of jobs based on new information and communication technologies (Friedman, 2005). See Figure 18.3.

However, there is a danger that the move to a world linked by new information and communications technology will lead to a greater polarization between the rich, who can exploit the new technologies, and the poor, who do not have access to them. The poor may be excluded from the global information society—and this includes the poor in industrialized countries, as the "digital Third World" does not follow international borders. See Figure 18.4.

The major barrier to the spread of information and communications technology is cost. To become part of the "digital world," low-income countries must build expensive communications infrastructures. Because of these costs, high-income residents of high-income countries are most likely to have access to the Internet. In 2014, 88 percent of North Americans and 73 percent of Europeans had access to the Internet compared with 35 percent of Asians and only 27 percent of residents of Africa. However, there has been significant growth in Internet use in low-income countries. For example, even though it is still low, Internet use in Africa increased by 6500 percent between 2000 and 2014 (Internet World Stats, 2015). Similarly, mobile phone use is much lower in low-income

FIGURE 18.3 WORLD CONNECTIONS

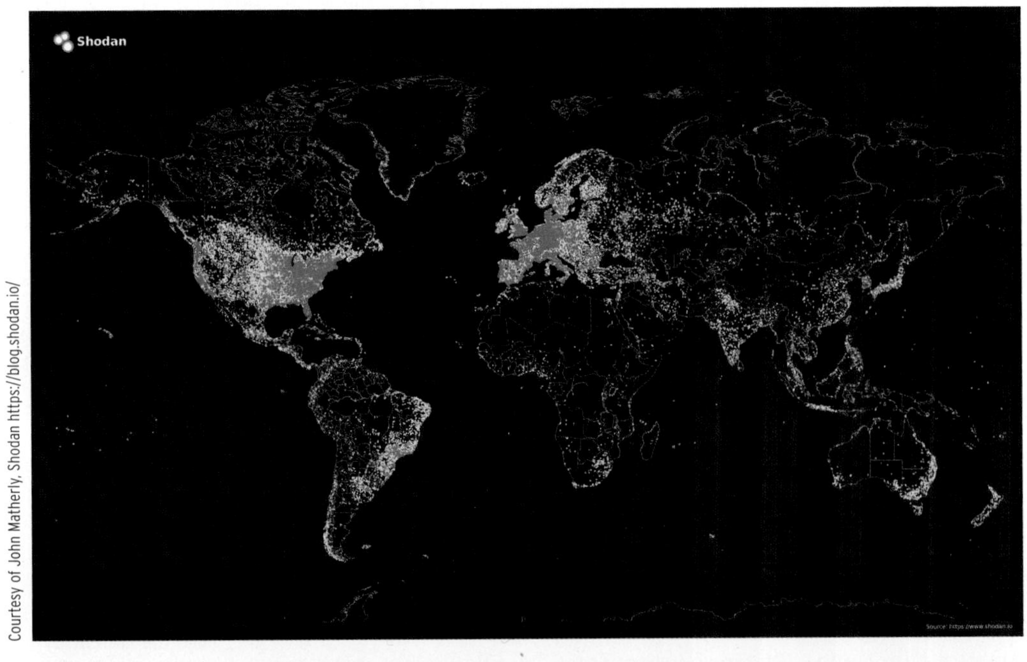

This graphic shows the relative density of Internet connectivity across the globe. You can see how North America, Europe, and Japan are far more connected than other parts of the world.

FIGURE 18.4 : CITY-TO-CITY CONNECTIONS

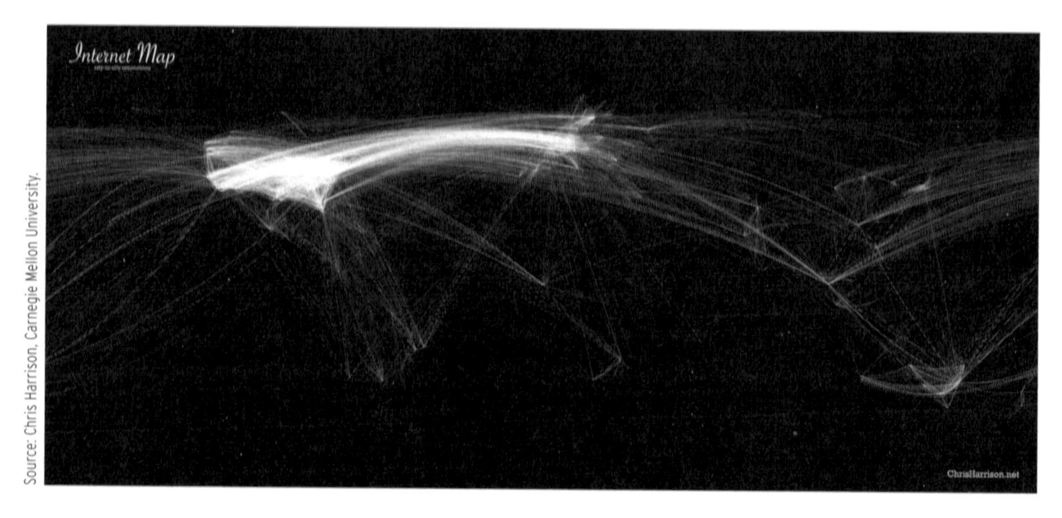

Source: Chris Harrison, Carnegie Mellon University.

This graphic shows the network ties around the globe. North America and Europe are clearly at the core of the digital world, while other countries are on its periphery.

countries but is also growing very rapidly. Despite this growth, it will take many years before access to modern communications approaches that of people in wealthier parts of the world.

Language issues also act as a barrier to Internet access in many parts of the world. About 60 percent of Internet users speak English (29 percent), Chinese (23 percent), or Spanish (8 percent). Without multilingual sites, the name "World Wide Web" will never be accurate. A Web that is dominated linguistically by English, and technologically and culturally by the United States will never reflect the point of view of people in low-income countries, and there is a danger that 'small' languages will not survive. The magnitude of this problem was demonstrated by Charles Kenny (2003), who searched for Web pages in Tgbo, a language spoken by 17 million Nigerians. Kenny found only five sites that used the Tgbo language. Canada's Indigenous languages are similarly under-represented on the Internet.

TIME TO REVIEW

- What impact does the constant connectivity of smartphones have on our lives?
- Discuss the implications of living a life online through networks such as Second Life along with your life in the 'real' world.
- What are some of the positive and negative consequences of the way people use new social media?
- How has the Internet changed the process of selecting a potential spouse?
- Describe how the pressure to make a profit can affect people's privacy rights.
- Do you think that new media will help or harm the world's poor?

VISUAL SUMMARY

18

KEY TERMS

cultural imperialism A process whereby powerful countries use the media to spread values and ideas that dominate and even destroy other cultures, and local cultural values are replaced by the cultural values of the dominant country (p. 507).

global village A world in which distances have been shrunk by modern communications technology so that everyone is socially and economically interdependent (p. 507).

hyperreality A situation in which the distinction between reality and simulation has become blurred (p. 497).

Internet The network infrastructure that links together the world's millions of computers (p. 485).

mass media Any technologically based means of communicating between large numbers of people distributed widely over space or time (p. 484).

media literacy The ability to access, analyze, evaluate, and create media in a variety of forms (p. 485).

status conferral The process of giving prominence to particular individuals by focusing media attention on them (p. 488).

World Wide Web The computer language that allows us to access information on the Internet (p. 485).

LO-1 Understand what is meant by the term *mass media*.

The media connect those who produce messages with those who receive them. The mass media are "any technologically based means of communicating between large numbers of people distributed widely over space or time" (Pavlik and McIntosh 2011:18).

© CBC

LO-2 Consider how the impact of the media is explained by functionalist, conflict, feminist, interactionist, and postmodern theories.

Xiang yang/Imaginechina/AP Images

Functionalist theorists believe the media exist because of the functions they play for society. For example, the media keep us informed, help societies socialize their members, and entertain us. Conflict theorists argue that the media help the dominant class control society by reinforcing capitalist ideology. Interactionist theorists show us how people use the media to help develop their individual identities. Feminist theorists have been critical of the way the media perpetuate gender stereotyping. Postmodern theorists have showed us how the media blur the distinction between reality and simulation, creating a new condition that Baudrillard called "hyperreality."

LO-3 Understand how new media both bring people together and keep them apart.

Social media can enable us to stay connected with friends and family even though we may be living in different parts of the world. However, many people spend so much time interacting with media that they may neglect their personal relationships.

© Philip Street. Used with permission.

LO-4 Understand how social media sites profit from the information you provide them.

Companies like Facebook and Twitter are worth many billions of dollars. They make money by selling information about their subscribers to advertisers or by selling things that people use in their online lives. We need to be aware of how these organizations make their money in order to ensure that we are not exploited through such factors as violations of our privacy.

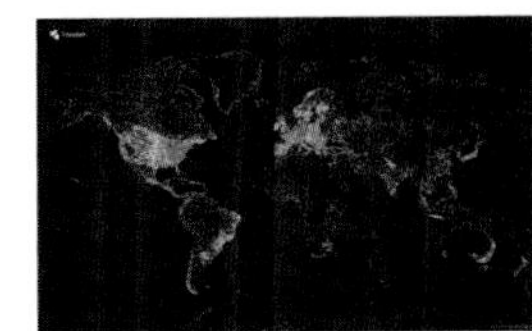
Courtesy of John Matherly, Shodan https://blog.shodan.io/

LO-5 Think about the role played by the media in globalization.

McLuhan believed that media technologies would create a global village in which people around the world shared information through interactive media. This would empower people and foster the growth of democracy and equality. Not everyone agrees with this view. There is a huge digital divide between people living in rich countries and those living in poor countries. Without action to reduce this technological gap, the disparity between rich and poor will get worse, not better.

Used with permission of Ayat Mneina

KEY FIGURES

© INTERFOTO/Alamy

Theodor Adorno (1903-1969) Philosopher Theodor Adorno was one of the earliest critics of the media. He was a member of the Frankfurt School of critical theory, which looked at the role of social factors in the oppression of the lower classes. His work with Max Horkheimer focused on the way in which the culture industry had turned into just another marketable commodity.

© INTERFOTO/Alamy

Jean Baudrillard (1929-2007) Postmodern scholar Jean Baudrillard is best known for his work on the way in which postmodern societies are organized around simulation, which can be more involving than reality. He developed the notion of hyperreality, which refers to a situation in which the distinction between reality and simulation has been blurred.

© Bettmann/Corbis

Marshall McLuhan (1911-1980) McLuhan was Canada's foremost media theorist. One of his best-known assertions was that because of the power of the media, we live in a global village in which people around the world share information through interactive media.

Peter Urban

Sherry Turkle (b. 1948) Psychologist Sherry Turkle was one of the first scholars to study how our lives are affected by the new technology of computers and social media. Her most recent work looks at how social media may cause a flight from conversation to mere connection, thereby reducing the richness of human communication.

APPLICATION QUESTIONS

1. The chapter opener described how social media helped to overthrow Libya's repressive government. Progressive movements are not the only ones to use social media, however, as groups such as ISIS, al-Qaeda, and the Taliban also have a strong online presence. Using information you find on the Internet, examine how these groups make use of social media and consider how successful they have been.
2. In what ways do the media make you feel more Canadian? In what ways do the media make you feel like a citizen of the world (part of the global village) rather than like a Canadian?
3. Some people use virtual reality scenarios, such as those in Second Life, to try out new identities. Which personality trait(s) would you like to "test drive" in virtual reality?
4. People can have strong emotional attachments through their avatars in Second Life and other online worlds. If a married person's avatar has an intimate relationship with another avatar, should that be grounds for divorce in that person's real-life marriage? Why, or why not?
5. Why do you think so many people are meeting their partners online? Do you think that this way of meeting is better than relying on traditional ways, such as meeting people through your family, at church, or in school? Why, or why not?

CHAPTER 19

THE ECONOMY AND WORK

ArTono/Shutterstock.com

CHAPTER FOCUS QUESTION

How is work in Canada affected by changes in the economy?

LEARNING OBJECTIVES

AFER READING THIS CHAPTER, YOU SHOULD BE ABLE TO

LO-1 Understand the primary function of the economy.

LO-2 Describe the differences between the primary, secondary, and tertiary sectors of economic production.

LO-3 Discuss the differences between the three major contemporary economic systems—capitalism, socialism, and mixed economies.

LO-4 Understand the functionalist, conflict, symbolic interactionist, and feminist perspectives on the economy and work.

LO-5 Understand the role of labour unions and identify the challenges currently faced by the union movement.

LO-6 Consider how globalization and new technology have affected Canadian workers.

Wilfred Popoff was the associate editor of Saskatoon's *Star Phoenix* until Conrad Black's Hollinger Corp. purchased the newspaper and immediately downsized its staff. Popoff describes how he, a senior employee with more than 30 years of service, was dismissed:

I can only attribute my sudden firing, within several months of possible retirement, a dignified retirement I had seen so many others receive, to total abandonment of common civility, a phenomenon more and more prevalent today. You see, I was fired not because of anything I did or didn't do, but because of the need to cut costs in the quest for fantastic profits. And how the affair was stage-managed tells more than one wishes to know about the uncivil environment surrounding contemporary capitalism.

On a Friday afternoon all employees, about 300 in all, received a terse letter from the boss commanding attendance at a meeting in a hotel the following morning. The arrangement was reminiscent of military occupations portrayed in countless movies. The vanquished are summoned to the market square where officers of the occupying army register all people and direct them to various camps. In our case the officers were employees of a consulting firm, also strangers, who directed employees to various rooms, separating survivors from those marked for elimination. Of course, I was in the second group, although none of us knew what fate awaited us. Eventually the boss entered, gripped the lectern and read a brief statement: We were all finished, the decision was final.

Not only were we finished, our place of work a few blocks away had been locked up, incapacitating our entry cards, and was under guard. We could never go back except to retrieve our personal belongings, and this under the watchful eye of a senior supervisor and one of the newly retained guards. I felt like a criminal. In my time I had managed large portions of this company, had represented it the world over and, until the previous day, had authority to spend its money. Now I couldn't be trusted not to snitch a pencil or note pad . . . The current phenomenon known as downsizing is threatening to hurt capitalism by depriving it of the very thing it needs most: a market. This, however, speaks to the stupidity of capitalism today, not its abandonment of civility. But perhaps there is a connection.

Source: Popoff, 1996:A22.

Many Canadians face unemployment when their employers restructure or shut down because of business problems. However, Popoff and his colleagues lost their jobs so the new owner could cut costs. The paper had been profitable, but new owner Conrad Black wanted higher profits. Firing staff is the quickest route to short-term profits, so the termination consultants were called. In an interesting postscript to this story, Conrad Black later served a three-and-a-half-year sentence in a Florida prison for misappropriating funds from his publicly owned newspaper chain.

Unemployment shows the linkage between the economy and work. Changes in the economy affect us all, but those who lose their jobs will feel a greater sense of financial and personal loss. For many of us, work helps define who we are. When people meet for the first time, the first question often asked is, Where do you work? or What do you do for a living? Losing your job can effect how others see you and how you see yourself.

In this chapter, we discuss the economy and the world of work—how people feel about their work, how work is changing, what impact these changes will have on your future working lives, and the connections between work and the larger economic structure in Canada and around the globe. Before reading on, test your knowledge about the economy, work, and workers by taking the quiz in Box 19.1.

CRITICAL THINKING QUESTIONS

1. In the years since Wilfred Popoff was fired, the gap between the pay of those running corporations and their workers has become wider. What do you think has led to this increased disparity between the highest and lowest earners?
2. In 2011 and 2012, the Occupy movement drew public attention to the growing power of the "1 percent" who were doing well while the rest of society suffered. Do you think social movements like Occupy can help to redress the imbalance between rich and poor in Western societies?
3. Why is the job a person does so central to that person's identity?

LO-1 THE ECONOMY

▶ **economy** The social institution that ensures the maintenance of society through the production, distribution, and consumption of goods and services.

The **economy** is the social institution that ensures the maintenance of society through the production, distribution, and consumption of goods and services. *Goods* are tangible objects that are necessary or desired. Necessary objects include food, clothing, and shelter; desired objects may include smartphones and stylish clothing. *Services* are activities for which people are willing to pay (a haircut, a movie, or medical care). *Labour* consists of the physical and intellectual services that people contribute to the production process. *Capital* is wealth (money or property) owned or used in business by a person or corporation. Financial capital (money) is needed to invest in the physical capital (machinery, equipment, buildings, warehouses, and factories) used in production.

THE SOCIOLOGY OF ECONOMIC LIFE

Sociologists study the connections among the economy and other social institutions, and the social organization of work. At the macrolevel, sociologists may study the impact of multinational corporations on industrialized and developing nations. At the microlevel, sociologists might study people's job satisfaction.

LO-2 HISTORICAL CHANGES IN ECONOMIC SYSTEMS

The nature of work in Canada has changed dramatically since preindustrial times. This section will illustrate how economic systems have evolved.

▶ **primary sector production** The sector of the economy that extracts raw materials and natural resources from the environment.

PREINDUSTRIAL ECONOMIES Hunting and gathering, horticultural and pastoral, and agrarian societies are all preindustrial economies. Most workers engage in **primary sector production**—extracting raw materials and natural resources from the environment. These materials and resources are typically consumed or used without much processing.

Preindustrial societies show a progression from work by family members to the use of other people to fulfill specialized tasks. In hunting and gathering societies, most goods are produced by family members and the division of labour is by age and gender. The ability to produce surplus goods increases as people learn to domesticate animals and grow their own food.

BOX 19.1 SOCIOLOGY AND EVERYDAY LIFE

How Much Do You Know About the Economy and the World of Work?

True	False	
T	F	1. Women have dramatically increased their representation in the professions of law and medicine.
T	F	2. Sociologists have developed special criteria to distinguish professions from other occupations.
T	F	3. Workers' skills are usually upgraded when new technology is introduced in the workplace.
T	F	4. Many of the new jobs being created in the service sector pay poorly and offer little job security.
T	F	5. Women are more likely than men to hold part-time jobs.

For answers to the quiz about the economy and the world of work, go to **www.nelson.com/student.**

After the Industrial Revolution, many observers were critical of the mechanization of work and its effects on the dignity of workers. Filmmaker Charlie Chaplin satirized the new relationship between workers and machines in the classic *Modern Times*.

In horticultural and pastoral societies, the economy becomes distinct from family life and the accumulation of a *surplus* means some people can do work other than food production. In agrarian societies, production is related primarily to producing food but workers have a greater variety of specialized tasks. For example, warriors are necessary to protect surplus goods from theft by outsiders. There are small-scale commercial enterprises, and most of the population lives in small rural communities.

Most manufactured goods were produced by artisans with the help of family members and apprentices. The individual artisans had control over the work process and kept the profits from their labour. Sjoberg described the flexibility of work schedules:

> Merchants and handicraft workers generally do not adhere to any fixed schedule. Shopkeepers open and close their shops as they see fit. . . . The lunch hour is likely to be longer on some days than on others. (1960:209)

Craftwork was controlled through guilds, made up of artisans such as weavers and stonemasons (Volti, 2008). The guilds set and enforced standards, provided social activities, and gave assistance to members. Guilds also controlled admission to the trades, ensuring that the number of tradespeople did not exceed the demand for the product because this would have led to competition and decreased wages.

INDUSTRIAL ECONOMIES Industrialization dramatically changed the nature of work and life. At the beginning of the 20th century, the majority of workers in Canada were farmers (Drucker, 1994). By the end of the century, however, only 3 percent were agriculture workers, and other primary sector workers were equally rare. Drawing on new forms of

The Granger Collection, New York

Tomasz Wieja/Shutterstock.com

Before assembly lines, goods were manufactured by individual artisans, such as the 15th-century French cabinetmaker shown on the left. These artisans were supported and regulated by guilds made up of their fellow craftsmen. On the right, the elaborate guildhalls in Brussels' town square show the wealth of some of the guilds.

energy (steam, gasoline, and electricity) and technology, factories became the primary means of producing goods, and wage labour became the dominant form of employment. Workers sold their labour to others rather than working independently or with family members. In a capitalist system, this means that the product belongs to the factory owner and not to the workers who create it.

▶ **secondary sector production** The sector of the economy that processes raw materials (from the primary sector) into finished goods.

Most workers engage in **secondary sector production**—processing raw materials into finished goods. For example, steel workers process metal ore and auto workers then convert the steel into automobiles. In industrial economies, work becomes specialized and repetitive, activities are bureaucratically organized, and workers primarily work with machines instead of with one another.

This is very different from craftwork, where individual artisans make and assemble everything. Think of the difference between a skilled artisan, creating an elegant piece of furniture from pieces of wood, and a relatively unskilled assembly line worker stapling the back onto a cabinet as it passes by on an assembly line. This approach was much cheaper than using craftspeople because more goods are produced in a shorter time on an assembly line, and labourers received a lower wage than more skilled workers. However, their work no longer involved using skill and judgment as everything was driven by the assembly line.

Industrialization also affected women's lives. In preindustrial times, much production took place within the household, and men and women often worked together. Factories separated production from the household, causing a gendered division of labour. Men became responsible for the family's income and women for domestic tasks (Cohen, 1993).

Horatio N. Topley/Library and Archives Canada/C-006035

iStockphoto/Thinkstock

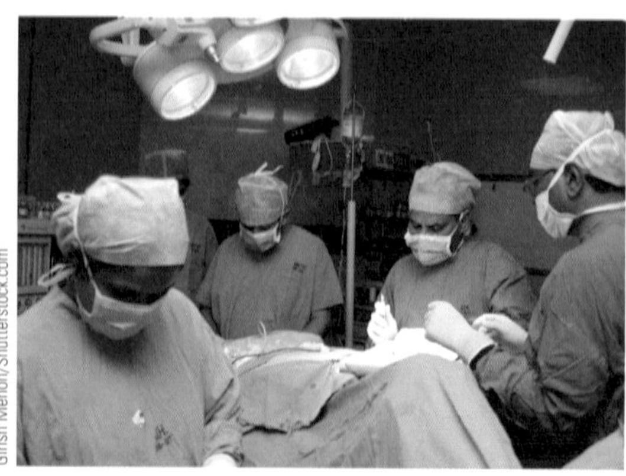

Girish Menon/Shutterstock.com

Work is very different in the three main types of economies. In preindustrial economies, most workers extract or produce raw materials. In industrial economies, production and distribution of goods are more complex and work is specialized and repetitive. In post-industrial economies, most workers provide services, such as healthcare, rather than produce goods.

In the early days of industrialization, work hours were long, wages were low, and workers had no pensions, vacation, or overtime pay. A parliamentary committee in the 1880s documented the exploitation of child labour:

> Many children of tender age, some of them not more than nine years old, were employed in cotton, glass, tobacco and cigar factories . . . Some of them worked from six o'clock in the morning till six in the evening, with less than an hour for dinner, others worked from seven in the evening till six in the morning. (quoted in Rinehart, 1996:xx)

By the 1950s, unionized industrial workers had gained better working conditions and better wages, but many of these gains were lost by the end of the 20th century as manufacturing jobs disappeared in the *post-industrial economy.*

▶ **post-industrial economy** An economy that is based on the provision of services rather than goods.

POST-INDUSTRIAL ECONOMIES Between 1900 and 1950, Canada shifted from a primary sector economy to one focused on manufacturing and service industries. By 1951, 47 percent of Canadian workers were employed in the service sector, 31 percent were employed in manufacturing, and the remaining 22 percent worked in primary industries. More recently, manufacturing jobs have steadily declined as technology has replaced workers and as production has shifted offshore to low-wage countries. By 2008, only 4 percent of workers remained in primary industries, while 20 percent worked in manufacturing and construction. Fully 77 percent worked in service industries (Krahn, Lowe, and Hughes, 2011). We now have a **post-industrial economy** that is based on the provision of services rather than goods. The service sector includes activities such as food services, transportation, communication, healthcare, education, legal services, sports, and entertainment.

In a post-industrial economy:

1. *Information displaces property as the central focus of the economy.* Post-industrial economies are characterized by ideas, and computer and communications technology is the infrastructure of the future. Facebook's 'product' is information about its users that it sells to advertisers.
2. *Workplace culture shifts away from factories and toward a diversity of work settings.* Many people are still employed in traditional

© SimplyCreativePhotography/iStockphoto.com

The BlackBerry, manufactured by Waterloo's Research In Motion company, was one of the first products that made it possible to stay in contact with work day and night and from almost anywhere in the world. However, BlackBerry lost the battle for phone supremacy to Apple and is no longer the dominant manufacturer in this field.

workplaces with set workdays, but these jobs are increasingly being affected by layoffs and outsourcing, and people work in other sites.

3. *The traditional boundaries between work and home are being set aside.* Communications technology allows many workers to work away from the office, and globalization means that some businesses operate 24 hours a day.

Challenging, well-paid jobs in the service sector have grown dramatically, and skilled "knowledge workers" have benefited from the post-industrial economy. However, these benefits are not shared by those who do routine production work, such as data entry, and workers who provide personal services, including restaurant workers and salesclerks. The positions filled by these workers form a poorly paid second tier. Many jobs in the service sector emphasize productivity, often at the expense of job satisfaction. Fast-food restaurants are an example, as the manager of a McDonald's explains:

> As a manager I am judged by the statistical reports which come off the computer. Which basically means my crew's labour productivity. What else can I really distinguish myself by? . . . O.K., it's true, you can overspend your [maintenance and repair] budget; you can have a low fry yield; you can run a dirty store, every Coke spigot is monitored. Every ketchup squirt is measured. My costs for every item are set. So my crew labour productivity is my main flexibility . . . Look, you can't squeeze a McDonald's hamburger any flatter. If you want to improve your productivity there is nothing for a manager to squeeze but the crew. (quoted in Garson, 1989:33–35)

"McDonaldization" is built on many of the ideas and systems of industrial society, including bureaucracy and the assembly line (Ritzer, 1993) (see Box 19.2 and Chapter 7).

Automation and offshore production have eliminated many of the well-paying manufacturing jobs formerly available to young people with low levels of education and training. These workers have now been forced into lower-paying service sector jobs. This change was particularly

BOX 19.2 POINT/COUNTERPOINT

McJobs: Assembling Burgers in the Global Economy

"McJobs" at McDonald's and other fast-food restaurants worldwide have received both praise and criticism from employees and social analysts in recent years. Since the first McDonald's restaurant opened in 1955, the company has grown to more than 33,000 restaurants, serving tens of millions of customers a day in 119 countries and employing 1.7 million people (McDonald's, 2014).

Many argue that McDonald's provides a service by giving job opportunities to people who might otherwise be out of work. Many young people have had their first jobs at McDonald's and were happy to receive some training and a paycheque. However, critics complain that McDonald's and other fast-food restaurants derive huge profits at the expense of workers—often young people and members of minority groups—who are paid low wages, given few breaks, and are required to work shifts that meet the fluctuating demands of customers. According to critics, companies like McDonalds are net destroyers of jobs. They use low wages and the huge size of their businesses to undercut local food outlets and force them out of business (McSpotlight, 2014). McSpotlight (www.mcspotlight.org), a website critical of McDonald's, states that the restaurant chain "feeds" on foreign visitors, women, students, and ethnic minorities who, with few other opportunities, are forced to accept the poor wages and conditions (McSpotlight, 1999). McDonald's spokespersons adamantly disagree. For example, the manager of a McDonald's restaurant in England stated, "We don't look at people's colour or nationality but their availability" (quoted in McSpotlight, 1999).

Could McDonald's workers organize to achieve better working conditions? McDonald's has resisted employees' attempts to unionize in many countries. Employees at a McDonald's in Squamish, B.C., were able to unionize, but McDonald's would not sign a contract with them and the employees eventually voted to decertify the union.

Are corporations such as McDonald's good employers and good corporate citizens in the global community?

hard on young men (18–24) with full-time jobs, whose earnings (in constant dollars) declined by 20 percent between 1977 and 1997 (Gadd, 1998). Over the same period, earnings for young women declined by 9 percent, but their incomes levelled off during the 1990s. Many young men have been stuck in entry-level jobs with few prospects for advancement or for additional training. Some economists believe this wage restructuring was permanent and that many of these young men may never move into a more favourable job situation. More recent data show that the employment picture for young people, particularly males, is still bleak, although in the oil-producing provinces both the employment and wages are high for young people (Statistics Canada, 2013l). This may change with the dramatic decline in oil prices in 2015.

LO-3 CONTEMPORARY ECONOMIC SYSTEMS

Capitalism and socialism have been the principal economic models in industrialized countries. Two criteria—property ownership and market control—enable us to distinguish between these types of economies. Keep in mind, however, that no society has a purely capitalist or socialist economy.

CAPITALISM

Capitalism is characterized by private ownership of the means of production. Personal profits are derived through market competition and without government intervention. Most of us think of ourselves as "owners" of private property because we own a car, an iPhone, or other possessions. However, most of us are not capitalists; we are consumers who *spend money* on the things we own rather than *making money* from them. A few people control the means of production, and the rest are paid to work for these capitalists. "Ideal" capitalism has four distinctive features: (1) private ownership of the means of production, (2) pursuit of profit, (3) competition, and (4) lack of government intervention.

▶ **capitalism** An economic system characterized by private ownership of the means of production, from which personal profits can be derived through market competition and without government intervention.

PRIVATE OWNERSHIP OF THE MEANS OF PRODUCTION Capitalist economies are based on the right of individuals to own income-producing property, such as land and factories, and to "buy" people's labour.

The early Canadian economy was based on *staples*—goods associated with primary industries, including lumber, wheat, and minerals. Harold Innis (1984/1930) showed how the economy was driven by the demand for raw materials by the colonial powers of France and Britain. In 1670, a British royal charter gave the privately held Hudson's Bay Company exclusive control over much of what is now Western Canada, the source of the lucrative fur trade. This was *commercial capitalism,* by which fortunes were made by merchants who controlled the trade in raw materials.

Inventions such as the steam engine led to factory production and the dramatic transformation to *industrial capitalism.* Industrial capitalism changed the very nature of European and North American societies. Urbanization, the growth of the modern nation-state, and the struggle for democracy can all be traced to the growth of industrial capitalism (Krahn, Lowe, and Hughes, 2015). In the early stages of industrial capitalism (1850–1890), virtually all investment capital was individually owned, and a few individuals and families controlled all the major trade and financial organizations in Canada.

Under early monopoly capitalism (1890–1940), ownership shifted from individuals to huge **corporations**—large-scale organizations with legal powers, separate from their individual owners. Major industries came under the control of a few corporations owned by shareholders. For example, the automobile industry in North America was dominated by the "Big Three"—General Motors, Ford, and Chrysler.

▶ **corporations** Large-scale organizations that have legal powers, such as the ability to enter into contracts and buy and sell property, separate from their individual owners.

Industrial development in Canada has lagged behind that of many other countries as Canadian business has focused on exporting raw materials that companies in other countries use in manufacturing. Many of the industries that did establish themselves in Canada were branch plants of large American and British corporations whose profits flowed back to their home countries. By 1983, Canada received more direct foreign investment than any other country (Laxer, 1989). Economist Kari Levitt (1970) was among the first to show how foreign investment threatened Canadian sovereignty, as fundamental economic decisions were made outside the country and did not necessarily reflect Canadian interests. While foreign investment has declined (see Figure 19.1), many of Canada's industries are still controlled by foreign parent corporations, such as General Motors, Toyota, and Walmart. Large corporations, including most of Canada's largest mining companies, have been purchased by foreign owners. Even iconic Canadian companies such as Tim Hortons and Molsons are foreign-owned. However, corporate Canada is no longer being hollowed out. Carroll and Klassen (2010) found that not only is foreign ownership no longer increasing, but that Canadian corporations are increasingly taking over foreign companies, particularly in Europe.

▶ **multinational corporations** Large companies that are headquartered in one country and have subsidiaries or branches in other countries.

Multinational corporations—large companies with headquarters in one country and subsidiaries or branches in others—play a major role in the global economy. The 200 largest global corporations have sales that are greater than the economies of many countries. No more than five corporations control over 50 percent of the global market in several industries, including the automotive, aerospace, and electronics industries (Brownlee, 2005). Such corporations do not depend on the labour, capital, or technology of any one country, and many move their operations to countries with the lowest wages and taxes. For example, multinational automotive manufacturers have moved thousands of jobs to Mexico and to the southern United States where salaries and benefits are much lower than they are in Canada.

PURSUIT OF PROFIT Capitalist doctrine holds that if people are free to maximize their individual gain through personal profit, then the entire society will benefit from their activities (Smith, 1976/1776). Economic development is assumed to benefit both capitalists and workers,

FIGURE 19.1 PERCENTAGE OF CANADIAN NONFINANCIAL INDUSTRIES UNDER FOREIGN CONTROL

Foreign control was at its peak in the early 1970s and has been relatively stable for the past two decades.

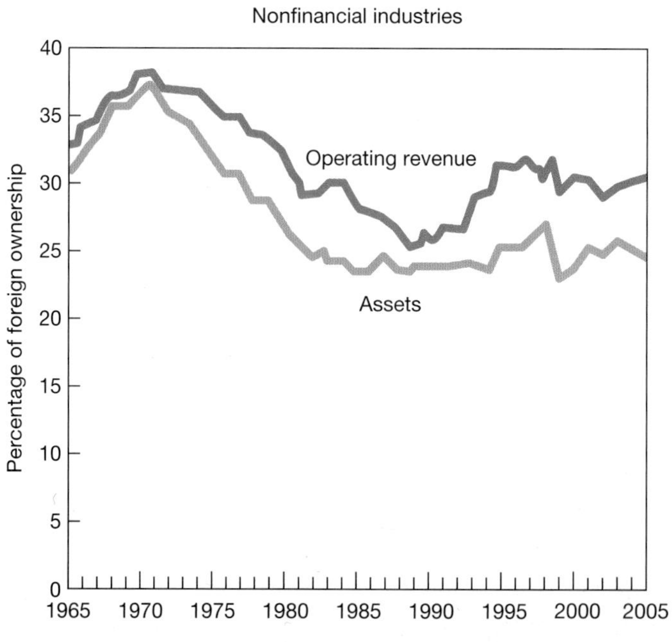

Source: Statistics Canada, *The Daily*, November 2, 2004; "Corporations Returns Act (CRA), major financial variables," CANSIM Table 179-0004.

and the general public also benefits from public expenditures (such as for roads, schools, and parks) made possible through business tax revenues.

During the period of industrial capitalism, however, a small number of individuals and families, not the general public, were the primary recipients of profits. While most corporations are no longer family-controlled, those running these corporations still receive disproportionate benefits—over the past decade, pay received by corporate CEOs has soared while many workers are worse off because of cost-cutting.

Nobel Prize–winning economist Milton Friedman famously stated his belief that the only responsibility of business is to increase its profits. According to this view, corporations should not be concerned about their social responsibilities but only about maximizing the money they make for shareholders. Thus, investing in communities, training workers and paying them a living wage, protecting the environment, and producing safe products should not be of concern unless doing these things will make the corporation more profitable. Businesses that maximize their profits will provide the jobs and investment that are essential to strong communities. To Friedman, the corporate downsizing that cost Wilfred Popoff his job (see the chapter introduction) was positive because of the social benefits Friedman believed would flow from higher profits.

The Canadian Press/Dave Chidley

Caterpillar closed this plant in London, Ontario, when workers refused to agree to a wage cut from $35 to $18 an hour. The jobs were moved to a plant in the United States, where high unemployment rates have forced workers to accept lower wages. At the time the plant was closed, Caterpillar was reporting record revenues and profits.

COMPETITION In theory, and sometimes in practice, competition will ensure that profits are not excessive. When producers compete for customers, they must be able to offer innovative goods and services at competitive prices. However, the trend has been toward less, rather than more, competition among companies because profits are higher when there is less competition.

One way of decreasing competition is by increasing concentration *within* a particular industry. For example, Microsoft so dominates certain segments of the computer software industry that it has virtually no competitors. Large companies may also try to restrict competition by temporarily setting prices so low that weaker competitors are forced out of business. While cutting prices provides a temporary benefit to consumers, it reduces competition by forcing small retailers out of the market. The large companies recoup their losses when the competition has disappeared.

What appears to be competition among producers *within* an industry really may be "competition" among products, all of which are produced and distributed by relatively few corporations. Much of the beer in Canada is produced by Molson Coors and Labatt (both now foreign-owned companies), which use a wide variety of different brand names for their products.

An **oligopoly** exists when several companies control an entire industry. In the music industry, three giant companies control most of the labels and artists. This number has dropped from six companies when the first edition of this text was published about twenty years ago. Another example is the cellphone industry where three companies (Bell, Rogers and Telus) control most of the market—one reason why Canada has high phone rates compared with other countries.

▶ **oligopoly** The situation that exists when several companies overwhelmingly control an entire industry.

LACK OF GOVERNMENT INTERVENTION Proponents believe that capitalism works best without government intervention. This policy was advocated by economist Adam Smith in 1776. Smith argued that when people pursue their own selfish interests, they are guided "as if by an invisible hand" to promote the best interests of society (Smith, 1976/1776). The common terms market economy and free enterprise illustrate the underlying assumption that free market competition, not the government, should regulate prices and wages.

However, the "ideal" of unregulated markets benefiting all citizens has seldom been realized. Individuals and companies have run roughshod

Matthew Staver/Bloomberg/Getty Images

Many of the brands of beer sold in Canada are actually owned by a small number of companies such as Molson Coors.

over weaker competitors in pursuit of higher profits, and small businesses have grown into large, monopolistic corporations. Accordingly, governments implemented regulations in an effort to curb the excesses of the marketplace. Canada has a Competition Bureau to ensure that corporations compete fairly, though its effectiveness can be debated (it hasn't done much to increase competition in the cellphone business),

Ironically, much of government intervention has been in the form of *aid* to business. Canadian governments have always provided financial support to business. To encourage settlement of the West, the government gave subsidies and huge tracts of land to the Canadian Pacific Railway for the construction of a national railway. Many corporations receive government assistance in the form of public subsidies and protection from competition by way of tariffs, patents, and trademarks. Governments provide billions of dollars in tax credits for corporations, large subsidies or loan guarantees to manufacturers, and subsidies and tariff protection for farmers. For example, Chrysler and General Motors received nearly $14 billion in government funds as a bailout during the 2009 economic meltdown. While most corporations have gained much more than they have lost through government involvement in the economy, the free enterprise mythology still persists.

SOCIALISM

socialism An economic system characterized by public ownership of the means of production, the pursuit of collective goals, and centralized decision making.

Socialism is an economic system characterized by public ownership of the means of production, the pursuit of collective goals, and centralized decision making. Marx described socialism as a temporary stage en route to a communist society, but like "pure" capitalism, "pure" socialism does not exist.

Although the terms *socialism* and *communism* are associated with Marx and are often used interchangeably, they are not identical. Marx defined communism as an economic system characterized by common ownership of all economic resources (Marshall, 1998). In *The Communist Manifesto* and *Das Kapital,* he predicted that the working class would become increasingly impoverished and alienated under capitalism. The singular focus on the pursuit of profits meant that capitalists would exploit the labour of the workers, who would not receive a fair return for their labour.

As a result, the workers would become aware of their own class interests, revolt against the capitalists, and overthrow the entire system. After the revolution, private property would be abolished and capital would be controlled by collectives of workers who would own the means of production. The government would no longer be necessary. People would contribute according to their abilities and receive according to their needs (Marx, 1967/1867; Marx and Engels, 1967/1848). Over the years, state control was added as an organizing principle for communist societies. "Ideal" socialism has three distinct features: (1) public ownership of the means of production, (2) pursuit of collective goals, and (3) centralized decision making.

PUBLIC OWNERSHIP OF THE MEANS OF PRODUCTION In a socialist economy, the means of production are owned and controlled by the state, not by private individuals or corporations. Prior to the early 1990s, the state owned all the natural resources and almost all the capital in the Soviet Union. At least in theory, goods were produced to meet the needs of people. Access to housing and medical care was considered a right. In the 1990s, leaders of the former Soviet Union and other Eastern European nations abandoned government ownership and control of the means of production because the system was so inefficient. Shortages and widespread unrest led to the reform movement headed by Soviet president Mikhail Gorbachev in the late 1980s.

China—previously the world's other major communist economy—has privatized many state industries. In *privatization,* resources are converted from state to private ownership and the state protects private property rights. China now has an economy made up of both capitalism and an autocratic form of Communist Party governance. Rapid economic growth has significantly improved the lives of many Chinese, so it is likely that this combination of communism and a modified form of capitalism will remain for the foreseeable future.

PURSUIT OF COLLECTIVE GOALS Ideal socialism is based on the pursuit of collective goals, rather than on individual profits. Equality replaces hierarchy such as that which exists between owners and workers or political leaders and citizens. Everyone shares goods and services—especially necessities, such as food, clothing, shelter, and medical care—based on need, not on ability to pay. In reality, however, few societies pursue purely collective goals.

Andrey Rudakov/Bloomberg/Getty Images

Oil and other natural resources are the most important part of Russia's economy.

CENTRALIZED DECISION MAKING Another tenet of socialism is centralized decision making. In theory, economic decisions are based on the needs of society, and the government is responsible for ensuring the production and distribution of goods and services. Central planners set wages and prices. Problems such as shortages and unemployment are dealt with by the central government.

In the former Soviet Union, the leaders of the Communist Party set economic policy. The production units (factories and farms) at the bottom of the structure had little say in the decision-making process. Wages and prices were based on political priorities and eventually came to be completely unrelated to supply and demand. Thus, while some factories kept producing goods that nobody wanted, there were chronic shortages of other goods. The collapse of state socialism in the former Soviet Union was due partly to the declining ability of the Communist Party to govern and partly to the growing incompatibility of central planning with the requirements of a modern economy (see Misztal, 1993).

While the socialist system of the Soviet Union was not sustainable, privatization has been challenging. More than two decades after the end of socialism, Russians are still faced with high unemployment, and the prices of goods and services have risen. Organized criminal groups have muscled their way into business and trade, and many workers see a dim future. Russia's oil resources will help, but the transition to a capitalist economy will take decades (or longer if President Putin continues policies like the annexation of Crimea from Ukraine that are isolating Russia from Europe and North America).

MIXED ECONOMIES

No economy is truly capitalist or socialist. A **mixed economy** combines elements of a market economy (capitalism) with elements of a command economy (socialism). Sweden is an example of a mixed economy, sometimes referred to as **democratic socialism**—an economic and political system that combines private ownership of some of the means of production with governmental control of some essential goods and services. Government ownership in Sweden is limited primarily to railroads, mineral resources, a public bank, and liquor and tobacco operations (Feagin and Feagin, 1994). Compared with capitalist economies, however, the government in a mixed economy plays a larger role in setting rules, policies, and objectives.

▶ **mixed economy** An economic system that combines elements of a market economy (capitalism) with elements of a command economy (socialism).

▶ **democratic socialism** An economic and political system that combines private ownership of some of the means of production, governmental distribution of some essential goods and services, and free elections.

In Sweden, the government provides all residents with health insurance, housing subsidies, child allowances, paid parental leave, and daycare subsidies. Public funds help subsidize cultural institutions, such as theatres and orchestras ("General Facts on Sweden," 2005; Kelman, 1991). While Sweden has a relatively high degree of government involvement, all industrial countries provide support and services to their citizens. There are, however, significant differences in the degree to which these services are provided. For example, Canada provides medical care to all its citizens, while about 41 million Americans have no health insurance at all, though the number of uninsured has declined since President Obama passed the Affordable Care Act. Canada is much closer to a mixed economy than the United States, but our government provides fewer benefits than do most Western European countries.

TIME TO REVIEW

- What is the economy and what is its role in society?
- How did most people earn their living in preindustrial societies?
- How did the new technology of the industrial era change the nature of work?
- What is the most common form of labour in post-industrial society?
- What are the differences between capitalist, socialist, and mixed economies?

LO-4 PERSPECTIVES ON ECONOMY AND WORK

FUNCTIONALIST PERSPECTIVES ON THE ECONOMIC SYSTEM

Functionalists view the economy as the means by which needed goods and services are produced and distributed. When the economy runs smoothly, other parts of society function more effectively. However, if the system becomes unbalanced, such as when demand does not keep up with production, a maladjustment occurs (in this case, a surplus). Some problems may be fixed by the marketplace (by cutting prices to get rid of the surplus) or through government intervention (buying and storing surpluses). Other problems, such as periodic peaks and troughs in the business cycle, are more difficult to resolve. The *business cycle* is the rise and fall of economic activity relative to long-term growth in the economy (McEachern, 1994).

From this perspective, peaks occur when business has confidence in the country's economic future. During a peak period, the economy thrives. Plants are built, materials are ordered, workers are hired, and production increases. There is upward social mobility for workers and their families.

The possibility of upward mobility is greatest during peaks in the business cycle. Once the peak is reached, however, the economy turns down because too large a surplus of goods has been produced. This downturn is partly due to *inflation*—a continuous increase in prices. Inflation erodes the value of people's money, and they are no longer able to purchase as much. Lower demand means fewer goods are produced, workers are laid off, credit becomes difficult to obtain, and people cut back on their purchases even more, fearing unemployment. Eventually, this produces a *recession*—a decline in an economy's total production that lasts six months or longer. To combat a recession, the government lowers interest rates (to make borrowing easier and to get more money back into circulation), trying to spur the beginning of the next expansion period. A global recession occurred in 2008 and 2009, driven by factors that included a speculative bubble in housing prices and a lack of financial regulation in the United States. In many countries, particularly in Europe, the resulting recession and stagnant growth continued through 2015.

CONFLICT PERSPECTIVES ON THE ECONOMIC SYSTEM

Conflict theorists view the economic system differently. They see business cycles as the result of capitalist greed. To maximize profits, capitalists suppress the wages of workers to the point where workers are not able to purchase the goods that have been produced. The resulting surpluses cause capitalists to reduce production, close factories, and lay off workers, thus contributing to the growth of the reserve army of the unemployed, whose presence helps reduce the wages of the remaining workers. In recent years, many businesses have forced wages down by hiring large numbers of part-time workers and by negotiating pay cuts for full-time employees. Companies justify this on the grounds of meeting the lower wages paid by competitors. The practice of contracting out—governments and corporations hiring outside workers to do some jobs rather than using existing staff—has become a favourite cost-cutting technique. In a slow economy, it is easy to find someone who will do the work more cheaply than existing employees whose seniority and wages have increased over time.

Much of the pressure to reduce costs comes from shareholders, who want their profits to increase. Conflict theorists view the firing or deskilling of workers as class warfare in which the rich benefit at the expense of the poor. The rich have indeed thrived; those with large amounts of capital have seen their fortunes increase dramatically, and the gap between rich and poor has increased over the past two decades.

Marx's critique of capitalism is still valuable. He predicted that the excesses of capitalism, driven by the pursuit of profits, would lead to periodic economic crises. In 2008/09, unregulated financial gambling by some of the world's largest financial institutions almost brought down the global economy and created a crippling recession. Disaster was only averted by massive financial bailout funded by taxpayers. While the profits from the speculation flowed to the wealthy, losses were mainly borne by workers.

Alienation occurs when work is done strictly for material gain, with no accompanying sense of personal satisfaction. According to Marx, workers have little power and no opportunities to make workplace decisions. This lack of control contributes to an ongoing struggle between workers and employers. Contemporary pressures to reduce the labour force and cut payroll costs have likely increased the alienation levels of Canadian workers, as remaining workers often work harder and get fewer rewards from their jobs.

SYMBOLIC INTERACTIONIST PERSPECTIVES

THE MEANING OF WORK Symbolic interactionists have studied the meaning of work and the factors contributing to job satisfaction. Does work play a role in defining us, or is it just something we have to do to put bread on the table? Some critics view work as dehumanizing, oppressive, and alienating. Others take the view that people find both moral meaning and a sense of personal identity through their work. Robert Wuthnow (1996) found that when people were asked about their most important reason for working, the most common response was "the money." However, when asked what they most preferred in a job, 48 percent chose "a feeling of accomplishment" and only 21 percent chose "high income." Even for many lower-level employees, work is not just a source of money, but also a vital part of their identity. Work connects people to their communities, they share personal friendships in their work settings, and work provides a sense of accomplishment.

Murial Johnson, one of Wuthnow's interview subjects, had three jobs but spent most of her work time as a security guard at a city convention centre. She took pride in the job because she was able to do it well:

> Johnson says she has earned the "respect and liking" of other employees to the point that she is often called on to serve in a supervisory capacity. She takes pride in having organizational skills. When a big event takes place, or when there is a special crisis . . . she can be counted on to make things work. (Wuthnow, 1996:219)

William Julius Wilson (1996) studied what happened to inner-city neighbourhoods when work disappeared. The impact was devastating to individuals and to the community. To Wilson, a person without work is incomplete and lacks the concrete goals and expectations, as well as the daily discipline and regularities, that work provides. Without work, it is difficult for the urban poor to take control of their lives.

However, even proponents of the view that work has moral meaning recognize that not all jobs allow the same degree of satisfaction and personal fulfillment. Assembly line work can be particularly alienating:

> Basically, I stand there all day and slash the necks of chickens. You make one slash up on the skin of the neck and then you cut around the base of the neck so the

> next person beside you can crop it . . . The chickens go in front of you on the line and you do every other chicken or whatever. And you stand there for eight hours on one spot and do it. (Armstrong and Armstrong, 1983:128)

JOB SATISFACTION Work is an important source of self-identity for many. It can help people feel positive about themselves or it can cause them to feel alienated. A Canadian study found that men and women who enjoyed their work reported a much higher quality of life than those who disliked their jobs (Frederick and Fast, 2001). Worker satisfaction is highest when employees have some control over their work, when they are part of the decision-making process, when they are not too closely supervised, and when they feel that they play an important part in the outcome (Kohn et al., 1990). The reasons contract administrator Beth McEwen gives for liking her job illustrates this:

> I've worked for employers who couldn't care if you were gone tomorrow—who let you think your job could be done by anyone because 100,000 people out there are looking for work. But here, there's always someone to help you if you need assistance, and they're open to letting you set out your own job plan that suits what they're after and what you're trying to accomplish. They know every person goes about a job in a different way. (quoted in Maynard, 1987:121)

FEMINIST PERSPECTIVES ON WORK AND LABOUR

All societies assign some tasks based on gender and age. In hunting and gathering societies, women typically care for children and men do the hunting. In industrial economies, working-age men are involved in wage labour, while women and males who are too young or too old for paid jobs look after household tasks. This situation has been changing slowly, and "Much political activity of the last 100 years has been aimed at overcoming the gender specificity of these definitions" (Wallerstein, 2004:34).

Feminist researchers have studied the relationship between gender and work. According to Amy Wharton, feminist theory "implies that work and the social practices that compose it are organized in ways that create and reproduce gender distinctions and inequalities" (2000:179). She has identified the three major themes of this research: "(1) characteristics of housework and so-called women's work more generally; (2) economic inequality between men and women; and (3) structural and institutional bases of gender in the workplace" (2000:167).*

WOMEN'S WORK Early feminist scholars drew attention to the fact that the home was a workplace in which women did a great deal of unpaid work. This research was done at a time when many women were full-time housewives and when the prevailing career model was a male who worked steadily from graduation until retirement to support his family. Society assigned women the tasks of staying at home, caring for children, and supporting the husband's career. As a result, while homemakers were full-time workers, they were often in a financially precarious position because their contribution was not recognized when their husbands died or when they were divorced. Programs, such as Employment Insurance and the Canada Pension Plan, do not provide support for women who have not been in the paid labour force. Women who did work outside the home typically worked in stereotypical "women's" fields, such as nursing, and office and clerical work. Women rarely held high-level positions, even in female-dominated occupations. For example, most elementary school teachers were women, but most principals were men. Women were greatly underrepresented in professions such as medicine, law, and dentistry.

When women did work outside the home, they arranged their jobs and work hours to ensure that their work did not interfere with their husbands' careers and that they could care for their children (Nelson, 2006). An important study by Hochschild (1989) showed that working women faced a "second shift" when they got home because they were also primarily responsible

* Amy S. Wharton, "Feminism at Work", *Annals of the American Academy of Political & Social Science*, pp. 167–182.

for household tasks and child rearing. Because of family demands, many women chose to work part-time and even today are much more likely than men to hold part-time jobs (Krahn, Lowe, and Hughes, 2011).

c12/Shutterstock.com

What workplace amenities, such as nursery facilities or diaper-changing stations, may be of value to working mothers?

GENDER INEQUALITY IN WAGES Women in the paid labour force faced discrimination. Men were viewed as the primary earners, so the financial needs of women were not recognized. Women were predominantly in low-wage jobs, they were often paid less than men doing the same work, and a "glass ceiling" limited their access to managerial positions. Structural factors, such as the segregation of women into lower-income segments of the labour force, meant that it was very difficult for them to achieve equality. These issues led to campaigns by feminist scholars and activists to ensure pay equity and affirmative action hiring. Canada's Charter of Rights and Freedoms and provincial pay equity legislation have provided legal support for these actions.

THE STRUCTURAL SOURCES OF GENDER BIAS Rosabeth Kanter (1977) raised concerns about the impact of structural factors on women in the workplace. Kanter found that employees whose gender or race made them a minority in their workplace faced difficulties because of their status as "tokens." They were often singled out as representatives of their gender or race, excluded from formal and informal work groups, and lacked support from their peers. In many occupations, such as police and fire departments, male colleagues felt women lacked basic job-related qualities, such as strength and courage. As a result, many of the women who were among the first to move into formerly "male" jobs had a difficult time. According to Kanter, once the proportion of minorities reached a "tipping point" of 15 or 20 percent, the pressures on the minority would be reduced and the work environment would be normalized.

Subsequent research has supported Kanter's work but has suggested that women suffer from more than just their token status. Williams (1989) found that women in traditionally male jobs faced far more hostility than do men entering female-dominated occupations. A female police officer was more likely to hear derogatory comments from male officers than a male nurse heard from female nurses.

While the situation has improved over the past few decades, male hostility toward women in non-traditional jobs still persists. California's Silicon Valley is the centre of high-tech in North America, and many of these leading-edge companies are particularly difficult for women. Sue Gardner (2014), former CEO of the Wikipedia Foundation, found that women have only 15 percent of technical jobs in the high-tech industry, and many more women than men left technical jobs in mid-career. The women left because of harassment, favouritism to males, lower pay for women, and a glass ceiling that limits promotions. According to Gardner, the CEO of taxi app Uber has been quoted as saying that he calls his company "Boob-er" because women are attracted to his wealth, and the CEO of Snapchat has called women 'frigid' and 'bitches' (2014, n.p.). Men with these misogynistic attitudes are not likely to do a good job of ensuring a work culture that is welcoming to women.

Thus, "gender is embodied in social structures and other forms of social organization" (Wharton, 2000:175). Occupational gender assignments have almost always reflected male superiority. For example, nurses (almost exclusively female until recently) were subordinate to doctors (who were almost always male), and the role of paralegals (female) was to provide support to lawyers (male). Just as in households, women were assigned the caretaker roles in the workplace and men were in charge (Pierce, 1995). These structures and practices had become an integral part of the society, so once an occupation was defined as "male" or "female," it was very difficult to change. Change is resisted by those who benefit from the gender bias. If women had equal access to male-dominated occupations and to high-paying, high-status managerial positions, fewer of these jobs would be available to men. Some men's self-image is also threatened by the presence of women in traditionally male-dominated occupations such as the military and policing.

The area of gender and work is an area where feminist scholarship and advocacy has made a major contribution to social change over the past 40 years. Most occupations and professions are now much more open to women and, while the glass ceiling still persists in some fields, women are now much more likely to hold managerial positions than they were in the past. Progress has been made in equalizing wages, though again some differences still remain.

TIME TO REVIEW

- How do functionalists view the operation of the business cycle?
- Describe how current efforts by business and government to reduce labour costs fit the conflict perspective on the economy.
- According to symbolic interactionists, how does the work we do affect our self-identity?
- How do feminist theorists analyze the labour market?

THE SOCIAL ORGANIZATION OF WORK

OCCUPATIONS

occupations Categories of jobs that involve similar activities at different work sites.

Occupations are categories of jobs that involve similar activities at different work sites (Reskin and Padavic, 1994). Historically, occupations have been classified as blue collar and white collar. Blue-collar workers were primarily factory workers and craftsworkers who did manual labour; white-collar workers were office workers and professionals. However, contemporary workers in the service sector do not easily fit into either of these categories.

CONCEPT SNAPSHOT

FUNCTIONALIST PERSPECTIVES	Functionalists view the economy as the means by which needed goods and services are produced and distributed. When the economy runs smoothly, other parts of society function more effectively. The business cycle involves the rise and fall of economic activity relative to long-term growth in the economy. This system is largely self-correcting, though government intervention may be necessary.
SYMBOLIC INTERACTIONIST PERSPECTIVES Key thinker: Robert Wuthnow	For many people, work is not just a source of money, but is also a vital part of their identity. Work connects people to their communities, they share personal friendships in their work settings, and work gives them a sense of accomplishment. However, not all jobs are equally satisfying.
CONFLICT PERSPECTIVES Key thinker: Karl Marx	Work in capitalist societies is characterized by conflict between workers and employers. Employers seek to maximize their profits by exploiting workers. Work is alienating when workers' needs for self-identity and meaning are not met and when work is done strictly for material gain, with no accompanying sense of personal satisfaction.
FEMINIST PERSPECTIVES Key thinkers: Arlie Hochschild, Rosabeth Kanter, Amy Wharton	Work is gendered, and gender biases persist because they are an integral part of a patriarchal society. Women are not paid as much as men, and women often work a "second shift" because of their extra work at home. Structural factors, such as the "token" status of women, also have a negative impact on women in the workplace.

PROFESSIONS

What occupations are professions? Athletes who are paid for playing sports are referred to as "professional athletes." Dog groomers, automobile mechanics, and manicurists also refer to themselves as professionals. Although sociologists do not always agree on exactly which occupations are professions, most do agree that the term *professional* includes most doctors, scientists, engineers, accountants, economists, lawyers, professors, and some clergy. What criteria do they use to define the term *professional*?

CHARACTERISTICS OF PROFESSIONS **Professions** are high-status, knowledge-based occupations. Sociologists define occupations as professions if they have these characteristics:

professions High-status, knowledge-based occupations.

1. *Abstract, specialized knowledge.* Professionals have abstract, specialized knowledge of their field, based on formal education and interaction with colleagues.
2. *Autonomy.* Professionals are autonomous—they rely on their own judgment in making decisions about their work.
3. *Self-regulation.* In exchange for autonomy, professionals regulate their members. Professions have licensing, accreditation, and regulatory associations that set professional standards and require members to adhere to a code of ethics.
4. *Authority.* Professionals expect compliance with their directions and advice. Their authority is based on mastery of the body of specialized knowledge and on their profession's autonomy.
5. *Altruism.* The term *altruism* implies some degree of self-sacrifice whereby professionals go beyond self-interest so that they can help a patient or client (Hodson and Sullivan, 2002).*

Job satisfaction among professionals has generally been high because of relatively high levels of income, autonomy, and authority.

Legislation gives members of many professions a monopoly on the provision of particular services, and these professions use their power to resist attempts by others to provide these services. Doctors have used their professional associations to protect their right to control the practice of medicine. Chiropractors, midwives, and other alternative healthcare practitioners have been fighting for decades to have governments recognize their right to practise and be paid under provincial healthcare legislation. Similarly, lawyers have actively opposed the attempt by companies of non-lawyers to represent people accused of traffic offences.

Women have made significant gains in most traditionally male-dominated professions in Canada, and continue to increase their representation in the professions of law and medicine. In 1971, women received 9 percent of law degrees and 13 percent of medical degrees. By 2000, more than half the graduates in law and medicine were women. In 2013, 37 percent of all practicing lawyers were female, and 54 percent of all new admissions to the Canadian Bar Association were female (Canadian Bar Association, 2013). In 2014, 57 percent of medical graduates were female (Association of Faculties of Medicine of Canada, 2014), and in 2013, 43 percent of family physicians and 33 percent of specialists were women (Canadian Institute for Health Information, 2013) (see also Figure 19.2).

UPPER-TIER JOBS: MANAGERS AND SUPERVISORS

The term *manager* is used to refer to executives, managers, and administrators. Women are steadily increasing their presence in managerial jobs. In 2009, 37 percent of those working in managerial positions were women, compared with 6 percent in 1971 (Statistics Canada, 1994, 2011f). However, women had less representation at the senior managerial level.

MANAGEMENT IN BUREAUCRACIES Upper-level managers are typically responsible for coordinating the organization's activities and controlling workers. Loss of worker control over the labour process was built into the earliest factory systems through techniques known as scientific management (Taylorism) and mass production (Fordism).

FIGURE 19.2 GENDER DIFFERENCES IN MEDICAL SCHOOL ENROLMENT

In 1993-1994, women outnumbered men in first-year medical school enrolment for the first time. Now, they typically make up about 55 percent.

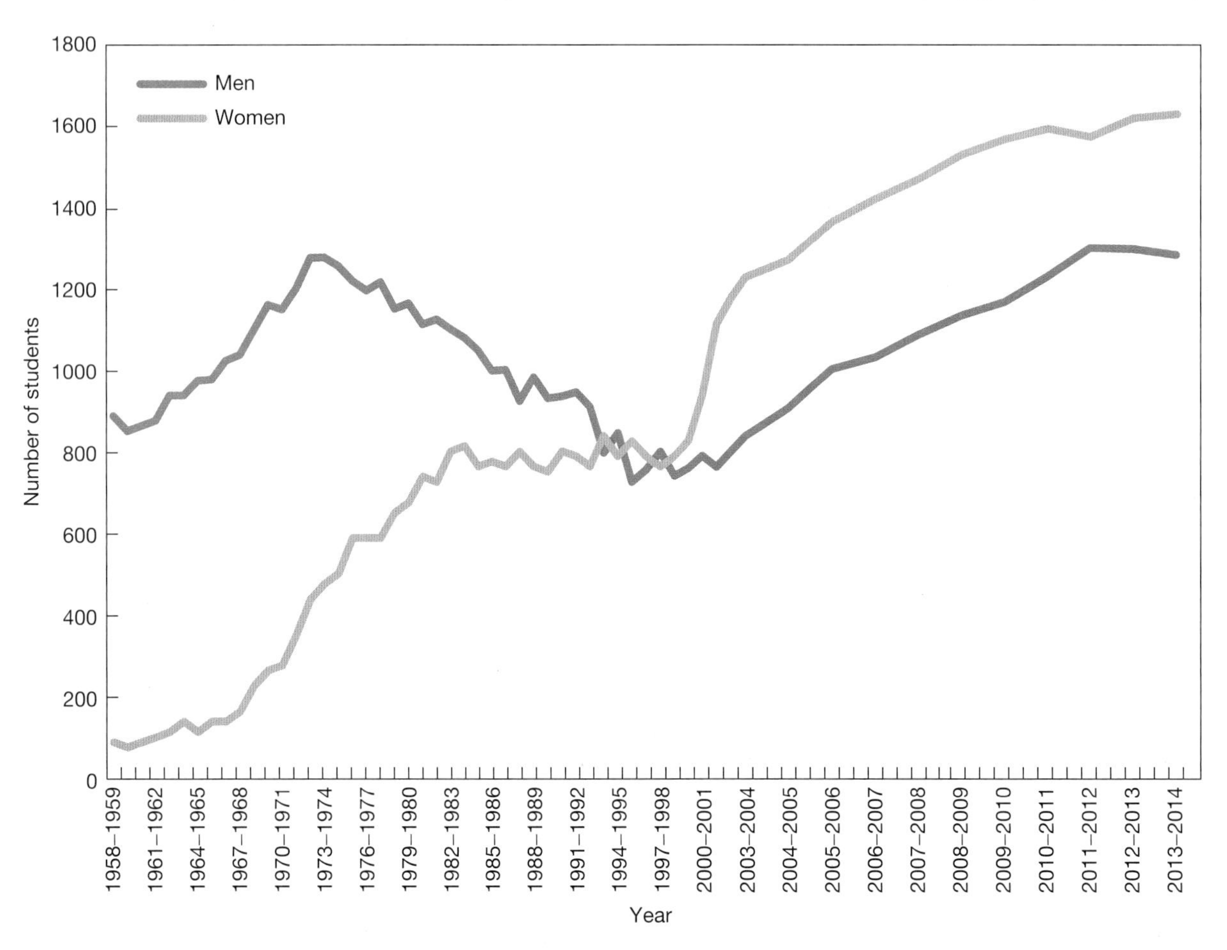

Source: Association of Faculties of Medicine of Canada, Canadian Medical Education Statistics 2004.

SCIENTIFIC MANAGEMENT (TAYLORISM) Early in the 20th century, industrial engineer Frederick Winslow Taylor revolutionized management with a system he called scientific management. To increase productivity, Taylor conducted time-and-motion studies of workers. He then broke down each task into its smallest components to determine the "one best way" of doing that task. Workers were taught to perform the tasks in a concise series of steps. Skills became less essential, since unskilled workers could be trained to follow routinized procedures. The process of breaking up work into small, specialized tasks contributed to the deskilling of work and shifted much of the control of knowledge from workers to management (Braverman, 1974). This greatly reduced the power of workers.

Taylor favoured piece-rate incentives. In this system, workers were paid for the number of units they produced. This often led to estrangement between workers and managers. Workers felt overworked because managers often increased the number of pieces required when workers met their quotas. Overall, management became more removed from workers with the advent of mass production.

MASS PRODUCTION THROUGH AUTOMATION (FORDISM) Fordism, named for Henry Ford, founder of the Ford Motor Company, incorporated hierarchical management structures and scientific management techniques into the manufacturing process (Collier and

Horowitz, 1987). The assembly line, a system in which workers perform a specialized operation on an unfinished product as it is moved by conveyor past their workstation, increased efficiency and productivity. On Ford's assembly line, a Model T automobile could be assembled in one-eighth the time formerly required. Ford broke the production process of the Model T into 7882 specific tasks (Toffler, 1980).

This fragmentation of the labour process meant that individual workers had little to do with the final product. The assembly line allowed managers to control the pace of work by speeding up the line when they wanted to increase productivity. As productivity increased, workers began to grow increasingly alienated as they saw themselves becoming robot-like labourers (Collier and Horowitz, 1987).

© George Steinmetz/Corbis

Robots at this Honda factory exemplify the deskilling of jobs through automation. What are managers' responsibilities in workplaces such as this?

Working conditions on the assembly line were so bad that most of the workers hired by Ford quit within their first week on the job (Volti, 2008). However, the dramatically increased productivity allowed Ford to give pay raises to $5 a day, about double the usual wage in 1914 (Volti, 2008). The wage kept workers relatively content, while Ford's own profits steadily rose. The difficult working conditions on the assembly line were described in a letter written by a worker's wife to Henry Ford:

> The chain system that you have is a *slave driver. My God!,* Mr. Ford. My husband has come home & thrown himself down & won't eat his supper—so done out! Can't it be remedied? That $5 a day is a blessing—a bigger one than you know, but *oh* they earn it. (Hounshell, 1984:259)

Without mass consumers, there could be no mass production; Ford recognized that better wages would allow the workers to buy his products, a lesson that has not been learned by many of today's employers.

Assembly line technology also dominates work settings such as fast-food restaurants (Ritzer, 2000). Some restaurant chains use conveyor belts to cook hamburgers and pizzas, and soft drink dispensers with a sensor that shuts off when the cup is full so that the employee does not have to make this decision (Ritzer, 2000). The task of managers is limited because the restaurant's system is designed to be error free. George Cohon, the former president of McDonald's of Canada, describes how the system works:

> A McDonald's outlet is a machine that produces, with the help of unskilled machine attendants, a highly polished product. Through painstaking attention to total design and facilities planning, everything is built integrally into the technology of the system. The only choice open to the attendant is to operate it exactly as the designers intended. (*Globe and Mail,* 1990:B80)

Of course, managers also hire workers, settle disputes, and take care of other tasks, but in many work settings, automation has also dramatically deskilled managerial jobs.

▶ **marginal job** A position that differs from the employment norms of the society in which it is located.

LOWER-TIER AND MARGINAL JOBS

Typical lower-tier positions include janitor, server, salesclerk, farm labourer, and textile worker. Many of these jobs do not provide adequate pay, benefits, opportunity for advancement, or job security, and are therefore marginal.

Marginal jobs differ from the employment norms of the society in which they are located. Examples are service and household workers. Women are more likely than men to be employed in this sector. Younger

Reza Estakhrian/Stone/Getty

Occupational segregation by race and gender is clearly visible in personal service industries. Women and visible minorities are disproportionately represented in marginal jobs, such as server, fast-food employee, or cleaner—jobs that do not meet societal norms for minimum pay, benefits, or security.

workers are also more likely than older people to work in this sector, as they pay for their studies with part-time work or use these low-level positions as a means of entering the labour force.

Household service work is typical of marginal jobs. Once done by domestics who worked full time with one employer, it is now usually performed by part-time workers who may work several hours a week per household in many different homes. Household work lacks regularity, stability, and adequate pay. The jobs are excluded from most labour legislation, the workers are not unionized, and employers sometimes break the rules and regulations that do apply. The jobs typically have no benefits.

CONTINGENT WORK

▶ **contingent work** Part-time or temporary work.

Contingent work is part-time or temporary work. This type of employment is advantageous for employers but often detrimental to workers. Contingent work is found throughout the workforce, including at colleges and universities, where tenure-track positions are seldom available and a series of one-year temporary appointments as instructors is the only option for many people with Ph.D.s. In the healthcare field, nurses, personal care homeworkers, and others are increasingly employed through temporary agencies as their jobs are contracted out. Supermarkets have replaced full-time workers with part-timers.

Why do employers prefer part-timers? Contingent workers enable them to cut wages and benefits, increase profits, and have workers available only when they need them. Kathy Sayers, the co-owner of a Vancouver technical writing company that relies heavily on temporary employees, told a researcher:

> When the [economic] downturn hits, International Wordsmith will be ready and able to retrench quickly and wait out the storm. It won't have a big payroll to cut, nor high overhead costs. "It takes a load off your mind and lets you sleep easier," says Sayers, "when your company has a plan to deal with a change in the economic weather." When the economy perks up . . . the partners will quickly hire more temporary employees for stints lasting weeks or months. (Zeidenberg, 1990:31)

While Sayers and other employers may sleep easier because of part-timers, think of the workers who earn less than full-time employees, get few benefits, and find themselves quickly unemployed during economic hard times. Consider a single mother in Toronto who "worked on call for several years at a daycare and an afterschool program. Her hours varied between six and 35 hours a week. The variability created havoc at home, where sometimes she'd have to wake her seven-year-old son up at 6 a.m. to get him to daycare before starting her own shift at a daycare 45 minutes away. In months when hours are thin, she's been late paying rent and incurs a $45 penalty. 'It's a stressful life,' she says" (Grant, 2014:n.p.).

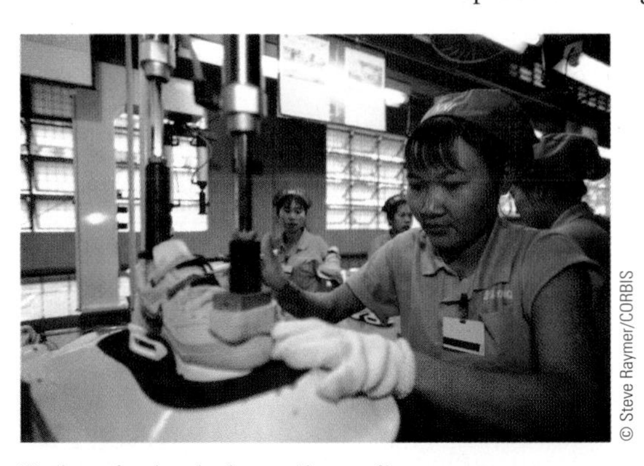
© Steve Raymer/CORBIS

Workers in developing nations—often women or young girls—make or assemble a number of products sold in North America. Workers in Vietnam make many Nike products; in the United States, Nike employees are primarily involved in nonmanufacturing work, including research, design, and retailing.

While some people voluntarily work part time, many people do so because they lack opportunities for full-time employment. Most part-time workers are young people, many of whom work while attending school. Women are much more likely to hold part-time jobs than men are. In 2013, women accounted for about 67 percent of part-time workers (Statistics Canada, 2014c), a figure that has remained relatively constant for many decades.

Crowdsourcing is a new form of contingent work. Online services such as Amazon's Mechanical Turk and Crowdflower enable companies to break large jobs, such as data entry, tagging Twitter messages, or writing programming code, into microtasks and to pay a small amount of money per task to anyone who wants to log on and do the work. Crowdsourcing can be even cheaper for companies than using part-time or casual help (Silverman, 2012), and it typically pays the workers very little. These companies do not have to hire or supervise their crowdsourced workers who sign themselves up online and are supervised electronically. Their work is rated and those who do not meet the standard are automatically kicked off the system.

Encyclopaedias used to be published by large companies who employed hundreds of workers—Wikipedia has been compiled by thousands of unpaid labourers, some of who invest massive amounts of time in this work. Many other websites also depend on volunteer labour to keep them functioning.

Online journalism is a type of contingent employment that has hurt traditional media workers. Contributors to blogs and news sites may be unpaid, or may be paid by the number of times their work is viewed online—in either case the pay is not nearly as good as it was in traditional media jobs such as working as a news reporter. People who now perform these jobs must work for multiple employers and almost always lack any job security or benefits.

The Temporary Foreign Worker (TFW) program is another type of contingent employment. Workers are brought to Canada for limited periods of time to fill jobs that cannot be filled by Canadians. Unfortunately, the program has two major flaws.

The first flaw is that employers have misused the program to exclude Canadians who would actually like to fill these jobs. For example, several McDonald's franchisees have been accused of using the program to deny work to Canadians. The franchisees were alleged to have paid the temporary foreign workers higher wages than Canadians and gave them more hours and better shifts (Tomlinson, 2014a).

Another abuse of the program occurred at a restaurant in a shopping mall serving Alberta's Ermineskin Cree Nation. Despite the community's 70 percent unemployment rate, the non-Indigenous restaurant owner received government approval to hire several temporary foreign workers for the restaurant rather than hiring local people.

The second problem with the program is that the employers can exploit workers because they control whether the workers can stay in Canada. Some McDonald's franchisees were alleged to have required foreign workers to live in crowded apartments and deducted the rent from their wages. Several workers recruited from Belize paid almost half their take-home pay for these accommodations and accused McDonald's of treating them like 'slaves' (Tomlinson, 2014b). Exploiting workers has been a problem in other occupations as well. Domestic workers brought to Canada on a related program have sometimes been badly mistreated by employers who threatened to have them sent home if they did not follow their demands.

In 2014, the federal government responded to criticisms of the program by proposing legislation restricting the use of temporary foreign workers in many parts of the country.

UNEMPLOYMENT

To many Canadians, unemployment rates are just numbers we see in the news every month. However, behind the statistics lie countless individual tragedies. In the chapter introduction, Wilfred Popoff describes the personal devastation of losing his job. When an entire community is affected, the consequences can be far-reaching. A newspaper article on unemployment in Port Hardy, British Columbia, began with the statement, "This picturesque Vancouver Island coastal town is on a deathwatch" (Howard, 1998:A4). The suicide rate increased dramatically in the town following the closing of a copper mine, a reduction in logging activity, and a sharp decline in salmon stocks. With a population of only 5500, Port Hardy had five suicides and 24 attempted suicides in a nine-month period. The victims were young to middle-aged adults—the people whose prospects were most affected by the

© Bill Aron/PhotoEdit

Hiring contingent workers can increase the profitability of many corporations. Other companies make their profits by furnishing these contingent workers to the corporations. What message does this ad convey to corporate employees?

community's poor employment prospects. While suicide is the most serious consequence of a community's loss of jobs, hundreds of communities, particularly in the Atlantic provinces, are in danger of disappearing due to the loss of so many of their young people to other provinces, as traditional resource-related jobs in the fishery and mining industries have disappeared.

▶ **unemployment rate** The percentage of unemployed persons in the labour force actively seeking jobs.

The **unemployment rate** is the percentage of unemployed persons in the labour force actively seeking jobs. The unemployment rate does not include those who have become discouraged and have stopped looking for work, nor does it count students, even if they are looking for jobs. Unemployment rates vary over several dimensions:

- *Yearly variations.* The Canadian unemployment rate reached a post–World War II high in the early 1980s, when it climbed above 11 percent. After a decline to below 6 percent by 2008, unemployment rose to 8.5 percent in 2009, showing the impact of a global recession. By October 2015, the rate had improved to 7 percent.
- *Regional differences.* Canada's regions have widely different rates of unemployment. Rates in the Atlantic provinces and in Quebec are usually higher than the Canadian average, and the Prairie provinces are usually lower. See Figure 19.3.
- *Gender.* In January 2015, the unemployment rate for adult males was 5.8 percent, compared with 5.2 percent for adult women. Females have had lower unemployment rates than men since 1990.
- *Race.* Most immigrants are members of visible minority groups. Despite the fact that immigrants have higher levels of education than nonimmigrants, they are not employed in the same level of jobs. Over one-third of immigrants with university degrees are working in low-skill jobs, compared with 10 percent of native-born Canadians (Krahn, Lowe, and Hughes, 2015). In the 1970s, immigrants' incomes caught up with those of native-born Canadians after 10 years in Canada, but by 2000 this had declined so immigrants made only 80 percent of native-born incomes (Krahn, Lowe, and Hughes, 2015). While immigrants do eventually catch up to native-born Canadians, the rate at which they catch up has slowed down.

There are a number of reasons for this: immigrants have difficulty getting professional associations and employers to recognize their foreign qualifications; language barriers; a lack of contacts in the labour market; and racial discrimination (Houle and Yssaad, 2010; Krahn, Lowe, and Hughes, 2015).

Indigenous people have much higher rates of unemployment than the rest of the population. The 2011 National Household Survey found that the unemployment rate for non-Indigenous people was 7.5 percent while the rates for First Nations and Métis people were 18.3 percent and 10.4 percent respectively. For First Nations people, on-reserve unemployment rates were much higher than for those who lived off-reserve (Krahn, Lowe, and Hughes, 2015). As a result, incomes for Indigenous people are far lower than the national average. There is some good news, however, as employment rates for Indigenous people have been increasing, particularly for those who have postsecondary education (Statistics Canada, 2013m).

- *Age.* Youth have higher rates of unemployment than older persons. Unemployment rates for people aged 15 to 24 are more than double the overall rate (Statistics Canada, 2015e). The gap between youth and other workers has grown since 1990.
- *Presence of a disability.* People with disabilities are much less likely to be employed than other Canadians. Forty-nine percent of Canadians between the ages of 25–64 with disabilities were employed compared with 79 percent of Canadians without disabilities. They are also more likely to hold jobs below their qualifications, although the differences are less for people with university degrees (Turcotte, 2014).
- *Chronic unemployment.* While many people are occasionally out of work, about 10 percent of labour force members are chronically unemployed. These people were more likely to be from the Atlantic provinces, to have less than high school educations, to be recent immigrants, to be male, and to be older (Statistics Canada, 2010b). Long-term unemployment can be difficult to escape, as employers are less willing to hire people who have not had recent work experience.

FIGURE 19.3 UNEMPLOYMENT RATES BY PROVINCE AND TERRITORY

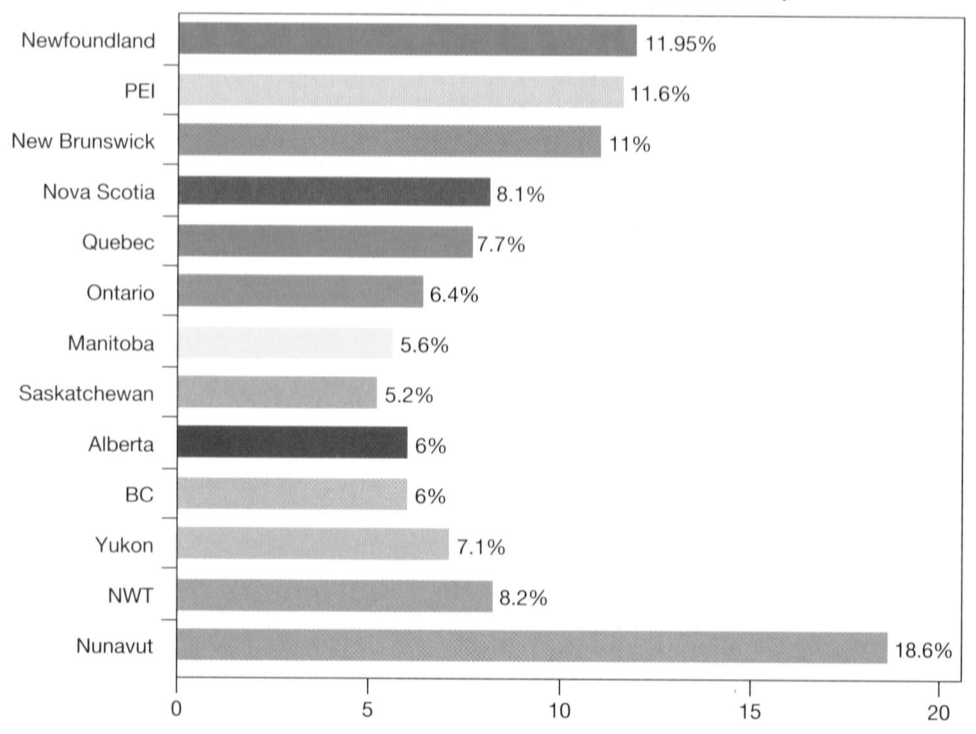

Source: Statistics Canada, "Labour Force Survey, July 2015," *The Daily*, August 7, 2015.

LO-5 LABOUR UNIONS

Workers formed labour unions to improve their work environment and gain some control over their work-related activities. A **labour union** is a group of employees who join together to bargain with an employer over wages, benefits, and working conditions.

labour union A group of employees who join together to bargain with an employer over wages, benefits, and working conditions.

As industries shifted to mass production during the era of monopoly capitalism, workers realized that they needed more power to improve their working conditions. The violent suppression of the Winnipeg General Strike in 1919 and the employment crisis during the Depression of the 1930s had devastated unions. One of the most important events in Canadian labour history was the United Automobile Workers (UAW) strike against General Motors in 1937 in Oshawa, Ontario (Abella, 1974). In 1937, workers suffered their fifth consecutive wage cut while General Motors announced record profits. When the company announced that the assembly line would be speeded up to produce 32 cars per hour instead of the 27 it had been producing, the men stopped work, and organizers started a Canadian chapter of the UAW and went on strike.

An unusual feature of the strike was the involvement of the premier of Ontario, Mitchell Hepburn. He hoped to crush the strike and stop the spread of industrial unions in Ontario. His fear of unions is shown in this statement:

> We now know what these [union] agitators are up to. We are advised only a few hours ago that they are working their way into the lumber camps and pulp mills and our mines. Well, that has got to stop and we are going to stop it! If necessary we'll raise an army to do it. (Abella, 1974:106)

Hepburn did raise an army. He recruited hundreds of men into the provincial police; the newcomers were irreverently called "Hepburn Hussars" and "Sons of Mitches."

Despite the premier's efforts, the union won the strike and received a contract that became a model for unions throughout Canada. This marked the beginning of *industrial unionism* in Canada, in which all workers in a particular industry, covering a wide variety of different trades, belonged to a single union. This gave the workers a great deal of bargaining power, because a strike could shut down an entire industry.

Industrial unions faced a long struggle to organize. Relations between workers and managers were difficult, and violence was often used to fight against unionization. Ultimately, organizers were successful and unions helped to gain an eight-hour workday and a five-day workweek, health and retirement benefits, sick leave and unemployment insurance, and health and safety standards. Collective bargaining also helped to expand the middle class by raising workers' wages.

Union membership grew, particularly during the two world wars, when labour shortages and economic growth made union expansion easy, and during the 1970s, when governments allowed public servants to unionize (Krahn, Lowe, and Hughes, 2015). However, membership declined during the 1980s and 1990s as economic recessions and massive layoffs in government and industry reduced the numbers of unionized workers and severely weakened the bargaining power of unions. Between 1981 and 2012, the percentage of workers who belonged to unions declined from 38 percent to 30 percent (Galarneau and Sohn, 2014), and workers' salaries have stagnated. Membership declined even more rapidly in the United States. The impact of this decline was shown in a U.S. study that found that states with lower rates of unionization have higher rates of working poverty (Brady, Baker and Finnigan, 2013).

Jobs have shifted from manufacturing and manual work to the service sector, where employees have been less able to unionize. While white-collar workers in the public sector, such as teachers and civil servants, have successfully organized, white-collar workers in the private sector remain largely unorganized. Women now make up over half of all union membership and have accounted for most of the growth in union membership over the past two decades. In 2012, 30 percent of all female paid workers were unionized, compared with 29 percent for male paid workers (Uppal, 2011). The rate of union membership among workers in Canada is higher than in the United States, but far lower than in many Western European countries, though rates in most of these countries have declined as well. Thirty years ago, the union membership rate in the United States was the same as in Canada, but at 11 percent, it is now far less than half (Bureau of Labor Statistics, 2014). American industry has been anti-union, and state and federal laws such as "right to work" legislation have made it difficult for workers to unionize. Governments everywhere are contracting out unionized jobs. For example, many cities have replaced unionized garbage collectors with private companies whose employees are not likely to be unionized and who are paid much lower wages.

As our economy becomes increasingly service-based, an important challenge for unions will be organizing the lower tier of service sector workers. McDonald's and Walmart, for example, have strongly resisted attempts to organize unions. Both companies have shut down operations rather than accept unionization. In 2005, Walmart closed a store in Jonquière, Quebec, after staff had obtained union certification. The closure put 176 employees out of work. Ironically, high-status workers have profited most from union activities in recent years. Professional athletes' unions such as the NHL Players' Association have been very successful in ensuring that their members are well paid, and provincial medical associations have also done a good job of increasing doctors' salaries.

The next decade will be critical for the labour movement. The increase in temporary and part-time work, the ease with which jobs can be moved from one country to another, the replacement of jobs with technology, and the increasing popularity of new patterns of work such as telecommuting where workers have little contact with one another are just a few of the future challenges. How do you think union leaders can change their organizations to meet the new realities of the world of work?

TIME TO REVIEW

- How has the gender structure of some occupations and professions affected the way women are treated in those organizations?
- In what ways would you distinguish between occupations and professions? What are the characteristics of a profession?
- How did the development of scientific management and the assembly line change the nature of work?
- How has the increase in contingent work affected workers and their employers?
- Discuss the history of labour unions in Canada and try to predict how unions will change in the future.

LO-6 THE GLOBAL ECONOMY IN THE FUTURE

How will the nature of work change during your lifetime? What are Canada's future economic prospects? What about the global economy? We cannot predict the future, but some general trends can be suggested.

THE END OF WORK?

Corporations around the world eliminated millions of jobs in 2009 during a global recession that dramatically reduced the demand for goods and services. Some employers have cut jobs because of the impact of globalization, and new technology has replaced workers with machines. We live in a post-industrial, information-based economy in which factory work, clerical work, middle management, and other traditional jobs are falling victim to technology.

Think of the staff who now work in a large drugstore and ask how many of them will still be employed when electronic scanners replace cashiers, robots stock the shelves, and robotic dispensing systems fill most of the prescriptions. How many construction jobs will be lost when 3D–printed houses become feasible? In 2014, a Chinese company printed 10 houses in 24 hours at a cost of less than $5000 each and the technology is being adapted to many other purposes. Many mass transit systems such as Vancouver's SkyTrain operate without drivers—what is the future of jobs in the trucking and taxi industries with the disruptive technology of driverless vehicles on the horizon? Suncor has agreed to purchase autonomous-capable vehicles for its oil sands site, and these vehicles may lead to 800 fewer highly paid jobs by the end of this decade. Would you rather spend $20 for the TurboTax computer tax program or pay a tax preparer significantly more for the same service? What is the future of bank tellers as most of us turn to bank machines and online banking? These changes represent a victory for capital over labour. While automation is very costly, it works very reliably for 24 hours a day, and the savings in labour costs can more than make up for its expense. The company's profits will increasingly go to the owners rather than to the workers, which will further increase the extent of social inequality.

The sharing economy—facilitated by smartphones—will also affect future jobs. Car-sharing services such as Zipcar allow people to pay for cars only during the times they need them rather than buying an expensive machine that sits idle for most of the day. Airbnb allows people to use otherwise-vacant private accommodations, and sharing services have also been set up for items such as tools and children's toys. Each of these services affects existing jobs.

Artificial intelligence systems have improved dramatically and will likely affect some well-paid jobs. A lawyer, faced with reviewing 1.3 million documents in preparation for a case, 'taught' a computer rules for selecting relevant documents and ultimately only had to review 2 percent of the documents (Ito, 2014). This significantly reduced the need for a staff of lawyers or paralegals to look through each document. The process took 600 hours with the help of the intelligent computer program compared with 13,000 hours doing it manually. This is just the beginning. University of Toronto law students were given access to 'Watson' (the IBM computer that can beat humans at the game *Jeopardy*) and trained it to do legal research:

> "Here's how Ross's [the computer program's] creators say it works: You ask it a legal question, and it spits out an answer, citing a legal case, providing some relevant readings and a percentage number indicating how confident Ross is he got it right. If a new case that might be relevant to your question comes into the database, Ross knows right away and alerts you on your smartphone, perhaps as you are heading to court" (Gray, 2014, n.p.).

These trends will have greater impact as computers become better at making decisions. Brynjolfsson and McAfee (2014) believe we are in a transformative Second Machine Age because artificial intelligence can now make better decisions in many areas than humans can. And the presence of big data (Chapter 2) means that machines have massive amounts of information to inform their decision making. These changes will radically change many jobs at the same time as they are eliminating many others.

When agriculture mechanized one hundred years ago, displaced workers found jobs in industry, and when industry turned to computers, service jobs were available. However, the information age does not hold the same potential for jobs. Some knowledge workers—the engineers, technicians, and scientists who are leading us into the information age—will gain, but there may be little work for the rest of us. Some predict we will have an elite, not a mass, workforce. The elites will be well paid. Microsoft's Bill Gates is just one of a number of people who have made huge fortunes in the computer industry. However, most people have not received the benefits of the post-industrial society. Many people are unemployed, and those with jobs have seen declines in their salaries.

Without adequate incomes, people will not be able to purchase the goods and services created by our new technology. Henry Ford was no great friend of the worker and strongly opposed unions, but he paid his employees very well. However, service economy employers, such as Walmart and McDonald's, keep costs down and profits up by paying very low wages. While Henry Ford wanted his employees to be able to buy his cars, the Walton family (which runs Walmart and which accounts for four of the 12 richest people in the world with a combined wealth of almost $160 billion) pays wages so low that its employees can shop only at low-cost stores—such as Walmart. Walmart's headquarters are in Arkansas where the minimum wage is the lowest in the United States at $6.25 an hour.

While low-wage workers have not been able to unionize because of employer resistance, a growing social movement called Fight for 15 is using protest marches and walkouts in an attempt to increase the minimum wage in the United States and Canada to $15 an hour, which is much closer to a living wage. In Canada, the NDP unsuccessfully tabled a motion in 2014 to increase the federal minimum wage to $15. It will be interesting to see if the growing movement to increase wages will be more successful in the future.

The Canadian Press/Jacques Boissinot

Immediately after the employees of this Walmart store in Jonquière, Quebec, obtained certification as a union, Walmart officials closed it down. Such actions have discouraged Walmart employees in other communities from attempting to unionize. In 2014, many years after the closure, the Supreme Court of Canada ruled that Walmart had to compensate the terminated employees, because the shutdown had violated the Quebec Labour Code.

THE IMPACT OF COMMUNICATIONS TECHNOLOGY

In a knowledge-based economy, many workers are not tied to a fixed work location, such as a factory or farm. Computer programmers, accountants, writers, and people in many other occupations can do their work from home, a coffee shop, or any other place with Internet access; videoconferencing even allows for "face-to-face" meetings of people scattered around the globe.

Modern communications technology has benefits. People who work from home avoid wasting time commuting and they are able to respond quickly to the needs of their children. Mobile phones enable people to stay in contact with the office if urgent situations arise. However, this also means that some people are never free of their work, as employers or their clients expect an immediate response to their requests. This stress is part of a broader problem of work–life balance for workers.

Technology has hurt many lower-tier workers. Companies focused on minimizing labour costs to maximize profits use technology to micromanage their employees to ensure that they are not wasting any time. Mac McClelland describes her time working at a large warehouse that filled orders for an Internet sales company. The warehouse employed thousands of people, despite low wages ($7.25 an hour) and appalling working conditions. The workers are continually monitored by an electronic scanner that calculates how long each item should take them to retrieve:

> That afternoon, we are turned loose in the warehouse, scanners in hand. And that's when I realize that . . . I will never be able to keep up with the goals I've been given.
>
> The place is immense. Cold, cavernous. Silent, despite thousands of people quietly doing their picking, or standing along the conveyors quietly packing or box-taping, nothing noisy but the occasional whir of a passing forklift. My scanner tells me in what exact section—there are nine merchandise sections, so sprawling that there's a map attached to my ID badge—of vast shelving systems the item I'm supposed to find resides. It also tells me how many seconds it thinks I should take to get there. Dallas sector, section yellow, row H34, bin 22, level D: wearable blanket. Battery-operated flour sifter. Twenty seconds. I count how many steps it takes me to speed-walk to my destination: 20. At 5-foot-9, I've got a decently long stride, and I only cover the 20 steps *and* locate the exact shelving unit in the allotted time if I don't hesitate for one second or get lost or take a drink of water before heading in the right direction as fast as I can walk or even occasionally jog. Olive oil mister . . . Fairy calendar. Neoprene lunch bag. Often as not, I miss my time target. (2012:6)*

These jobs will soon disappear—Amazon owns Kiva Systems, a company that manufactures robots that are now doing much of this retrieval work, and future generations of robots will take it over completely—but McClelland's experience shows the impact that new technology can have on the workforce.

Given the current trends toward globalization of the labour force and displacement by technology, it is difficult to see anything in the future that will create large numbers of well-paying jobs. Large-scale unemployment or underemployment will continue to drive wages down. If these trends continue, society will need to consider options such as shorter workweeks and guaranteed annual incomes to ensure that people are able to live meaningful lives.

THE UNDERGROUND ECONOMY

> The Bougons are a French-Canadian family that has achieved notoriety not only in their home province of Quebec but also across Canada. The father, Paul Bougon, bribed a Canada Post letter carrier to deliver fraudulent welfare cheques to the Bougon house. His wife, Rita, is a self-employed phone sex operator. Their eldest son, Paul Jr., engages in car theft, among other illegal activities, and their daughter, Dolorès, is an exotic dancer/prostitute. . . . Not a penny of the income earned through any of these activities is reported to the appropriate tax collecting agency. (Tedds, 2005:157)

While the Bougons are a fictional family portrayed in a popular Quebec television comedy series, their activities represent the *underground economy.* The underground economy is "legal and illegal market-based transactions not reported to the revenue gathering agency"—that is,

* McClelland, 2012:6

the Canada Revenue Agency (Tedds, 2005:158). Along with criminal activities such as drug sales, much of the underground economy involves people who make cash payments for home renovations and other services so the contractor can avoid paying taxes on the money. While the size of the underground economy can never be known, Statistics Canada has estimated that it amounted to $41 billion in 2011 (Statistics Canada, 2014d). Governments do not collect taxes on this money, so the rest of us pay to make up for this revenue loss.

GLOBAL ECONOMIC INTERDEPENDENCE AND COMPETITION

Global markets and industries do not follow political boundaries. Japanese cars, for example, are assembled in Canada and the United States using components that may be made in many other countries. Capital and jobs can move rapidly, so governments have much less power to intervene in markets.

Multinational corporations are becoming more powerful. As they become global powers, these corporations will become less aligned with the values of any one nation. Those who favour increased globalization typically focus on its potential impact on high-income countries, not on the effect it may have on the 80 percent of the world's population living in middle- and low-income countries. The gap between rich and poor nations may continue to widen as the benefits of global growth are not evenly shared (see Chapter 10).

There is no consensus about the impact of globalization on Canada. Has our economy benefited from agreements, such as the North American Free Trade Agreement (NAFTA), which involves Canada, the United States, and Mexico? We know that the economies of Canada and the United States are becoming more closely linked. The amount of trade has increased and the border has become almost irrelevant to business. Businesses looking for opportunities to expand are increasingly looking south rather than within Canada.

During the first few years following the 1989 implementation of the agreement with the United States that preceded NAFTA, a great number of manufacturing jobs were lost, just as opponents of the deal had predicted. However, these were also years when Canada's economy was in a recession and high interest rates and the high value of the Canadian dollar made our exports uncompetitive. More recently, Canada has usually exported more to the United States than we have imported. We do not know whether the future will bring increased prosperity to all three countries or whether factors such as the low wages paid to workers in Mexico and the Southern United States will result in jobs permanently moving out of Canada.

More broadly, competition with low-wage countries has had an impact on all Canadians. By threatening to close plants and move elsewhere, corporations have forced drastic wage cuts. Governments have also been pressured to adjust taxation and social policies so business remains competitive. Business leaders argue that if Canada's tax rates are higher than those of other countries, particularly the United States, corporations will not locate here and we will lose jobs. This international pressure limits the ability of the Canadian government to significantly raise taxes on business or on wealthy Canadians to redistribute income to the poor; doing so would make us less competitive, though corporate tax rates are already lower than those in the United States.

Governments are steadily losing the ability to set national policies in the face of global market forces and multinational corporations that operate without concern for national boundaries.

LO-1 Understand the primary function of the economy.

The economy is the social institution that ensures the maintenance of society through the production, distribution, and consumption of goods and services.

iStockphoto/Thinkstock

LO-2 Describe the differences between the primary, secondary, and tertiary sectors of economic production.

In primary sector production, workers extract raw materials and natural resources. Industrial societies engage in secondary sector production, based on the processing of raw materials into finished goods. Post-industrial societies provide services rather than goods.

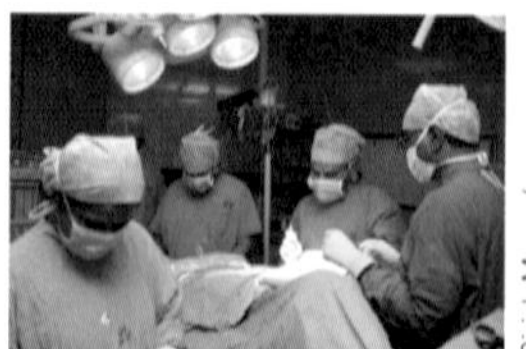

Girish Menon/Shutterstock.com

LO-3 Discuss the differences between the three major contemporary economic systems—capitalism, socialism, and mixed economies.

Capitalism is characterized by ownership of the means of production, pursuit of personal profit, competition, and limited government intervention. Socialism is characterized by public ownership of the means of production, the pursuit of collective goals, and centralized decision making. In mixed economies, elements of a capitalist market economy are combined with elements of a command socialist economy.

Andrey Rudakov/Bloomberg/Getty Images

LO-4 Understand the functionalist, conflict, symbolic interactionist, and feminist perspectives on the economy and work.

Functionalists believe the economy is a vital social institution because it is the means by which needed goods and services are produced and distributed. Business cycles represent the necessary rise and fall of economic activity relative to long-term economic growth. Conflict theorists view business cycles as the result of capitalist greed. To maximize profits, capitalists suppress the wages of workers, who, in turn, cannot purchase products, making it necessary for capitalists to reduce production, close factories, lay off workers, and adopt other remedies that are detrimental to workers and society. Symbolic interactionists focus on issues such as the social organization of work and its effects on workers' attitudes and behaviour. Feminist researchers have documented the ways in which women are discriminated against in the world of work.

c12/Shutterstock.com

KEY TERMS

capitalism An economic system characterized by private ownership of the means of production, from which personal profits can be derived through market competition and without government intervention (p. 521).

contingent work Part-time or temporary work (p. 534).

corporations Large-scale organizations that have legal powers, such as the ability to enter into contracts and buy and sell property, separate from their individual owners (p. 521).

democratic socialism An economic and political system that combines private ownership of some of the means of production, governmental distribution of some essential goods and services, and free elections (p. 525).

economy The social institution that ensures the maintenance of society through the production, distribution, and consumption of goods and services (p. 516).

labour union A group of employees who join together to bargain with an employer over wages, benefits, and working conditions (p. 537).

marginal job A position that differs from the employment norms of the society in which it is located (p. 533).

mixed economy An economic system that combines elements of a market economy (capitalism) with elements of a command economy (socialism) (p. 525).

multinational corporations Large companies that are headquartered in one country and have subsidiaries or branches in other countries (p. 522).

occupations Categories of jobs that involve similar activities at different work sites (p. 530).

oligopoly The situation that exists when several companies overwhelmingly control an entire industry (p. 523).

post-industrial economy An economy that is based on the provision of services rather than goods (p. 519).

primary sector production The sector of the economy that extracts raw materials and natural resources from the environment (p. 516).

professions High-status, knowledge-based occupations (p. 531).

secondary sector production The sector of the economy that processes raw materials (from the primary sector) into finished goods (p. 518).

socialism An economic system characterized by public ownership of the means of production, the pursuit of collective goals, and centralized decision making (p. 524).

unemployment rate The percentage of unemployed persons in the labour force actively seeking jobs (p. 536).

LO-5 Understand the role of labour unions and identify the challenges currently faced by the union movement.

A labour union is a group of employees who join together to bargain with an employer or a group of employers over wages, benefits, and working conditions. Union membership is declining and facing the pressures of globalization and the desire of employers in the private and public sectors to reduce labour costs.

The Canadian Press/Jacques Boissinot

LO-6 Consider how globalization and new technology have affected Canadian workers.

The need to stay competitive with low-wage countries has kept wages down. If workers insist on higher wages, companies can simply close their factories and move production to another country. On the positive side, increased trade has created jobs in Canada. New technology may potentially lead to the loss of jobs in a wide variety of different fields.

The Canadian Press/Dave Chidley

APPLICATION QUESTIONS

1. If you were the manager of a fast-food restaurant, how might you increase job satisfaction and decrease job alienation among your employees? How do you think work differs between a fast-food restaurant and a luxury restaurant? What do these differences mean for employees?
2. What actions could governments and individuals take to ensure that women are fairly treated in the labour market?
3. In what ways do you think new communications technology will affect employment? Do you think it will make work life better or worse?
4. Many occupations will change or disappear in the future. Think of a specific occupation or profession and consider its future. For example, what will be the role of the librarian when books, journals, and abstracts are all instantly accessible on the Internet?

CHAPTER 20

SOCIOLOGY AND THE ENVIRONMENT

IRA MEYER/National Geographic Creative

CHAPTER FOCUS QUESTION

How do the social world and the natural world interact?

LEARNING OBJECTIVES

AFER READING THIS CHAPTER, YOU SHOULD BE ABLE TO

LO-1 Understand what is meant by environmental sociology.

LO-2 Compare how the treadmill theory of production, ecological modernization theory, ecofeminism, and constructionism differ in their analysis of the environment.

LO-3 Understand the primary ways in which the environment affects our health.

LO-4 Explain how human populations are differentially and unequally affected by challenges in the environment.

LO-5 Discuss the emerging field of green criminology.

Climate change is having an impact on people around the world. Canada's North is one of the places which might be the most affected:

My name is Caitlyn Baikie. I am a 20-year-old Inuk from Nain, Nunatsiavut, and have lived there all of my life. Nunatsiavut, in the Inuktitut *language, means "our beautiful land," and what a beautiful land it is. . . .*

My life in Nain is very different than for those who live in the city. Nain is a very isolated community, there are no roads to or from Nain, the only way to get here is by plane or boat, and in the winter by airplane or snowmobile. Though our weekdays are school and work filled as it is in the south, our recreational and sustenance activities are very different. In the winter, we go hunting for small game like partridges, which can be found locally all around the town. On weekends we often go by snowmobile on the ice to our cabins near hunting grounds where we hunt seals, caribou, ukialik *(arctic hare), and migratory birds. This is when we get the bulk of our meat for the year. It is very important to get good sea ice, it provides us with the means to travel and hunt, usually for six months of the year. The people of Nain and the north coast are often referred to as* sikimiut, *people of the sea ice.*

Inuit depend greatly on the sea ice also to travel to other nearby communities. However, the past few seasons have not been good for sea ice—it was not until mid-January that people first went for a snowmobile drive. This is very late considering that we are accustomed to have already travelled these routes and filled our freezers with freshly killed caribou by then. We collectively wondered if these years could just be "off." But how far do you go in saying it's just "off," instead of part of something bigger? We have been faced with precipitation, and week-long periods of fog and cold temperatures during our summer instead of sun and warm temperatures. Summer blends into fall, without the usual chill and freezing ground.

. . . .The elders have noticed this particularly. Their knowledge of our land and climate has sustained them and our people in their survival on the land for decades and it is scary to think that they have not seen a year like 2010 before. They know that our climate is changing because they have lived here for generations, and these generations hold no knowledge of years like this. Our elders know from experience that our climate is changing; scientists have data that says our climate is changing.

. . . .As a collective, the elders have noticed that there is not as much snowfall as there used to be. They say that we don't get near the amount of snow that they did when they were young. The elders have also noted an increased number of polar bear sightings further south. This is very unusual, and likely attributable to the lack of floe ice during the summer months, leaving polar bears no other option than to move to the land where people live. Unusual ice conditions have also brought tragedy to our communities. We have lost lives because of less predictability in ice and snow cover. A few years ago, two experienced Inuit hunters lost their lives while travelling a route that was known to them.

As Inuit, our lives are tied to nature and for that we have a great respect for Mother Nature's strength. The land, the sea, and the climate define us as a culture, and our culture will forever be altered because of the changes we are undergoing today. (Baikie, 2012)

Source: Skeptical Sicence, September 27, 2012, posted by Robert Way, http://skepticalscience.com/Inuit-Climate-Change.html. Reproduced by permission of Skeptical Science.

Caitlyn Baikie's story illustrates the interdependence of nature and society. By burning fossil fuels, human societies have caused global warming. The changes caused by global warming are having a huge impact on society—not just in Canada's Arctic but everywhere on the globe.

The Intergovernmental Panel on Climate Change concluded that, "Warming of the climate system is unequivocal, and since the 1950s, many of the observed changes are unprecedented over decades to millennia. The atmosphere and ocean have warmed, the amounts of snow and ice have diminished, and sea level has risen" (IPCC:2). Climate scientists tell us that we should not allow global average temperatures to rise more than 2 degrees Celsius above preindustrial levels. However, unless governments act soon, temperature may rise four or five degrees by the end of the century. This will almost certainly cause massive changes in the global climate, including severe heat waves, droughts, extreme weather events, and a significant rise in the level of our oceans. According to Pope Francis, in his 2015 papal encyclical on the environment, "Doomsday predictions can no longer be met with irony or disdain. We may well be leaving to coming generations debris, desolation and filth."

Before reading on, test your knowledge of the relationship between the environment and society by taking the quiz in Box 20.1

CRITICAL THINKING QUESTIONS

1. If global warming continues, what impact will it have on the community in which you live?
2. What do you think are the most important things that Canadians can do to help protect the world's environment?
3. Why are the world's most vulnerable citizens the most negatively affected by environmental change?

BOX 20.1 SOCIOLOGY AND EVERYDAY LIFE

How Much Do You Know About Environmental Issues?

True	False	
T	F	1. Scientists are forecasting a global warming of between 2 and 11 degrees Fahrenheit over the next century.
T	F	2. The environmental movement in Canada started in the 1960s.
T	F	3. People who hold strong attitudes regarding the environment are very likely to be involved in social movements to protect the environment.
T	F	4. Environmental groups may engage in civil disobedience or use symbolic gestures to call attention to their issue.
T	F	5. Global warming is not a major concern for most Canadians.

For answers to the quiz about environmental issues, go to **www.nelson.com/student**.

LO-1 WHAT IS ENVIRONMENTAL SOCIOLOGY?

▶ **environmental sociology** The study of the interaction between human society and the physical environment.

In this chapter, we will discuss the interaction between the natural world and the social world. Dunlap and colleagues have defined **environmental sociology** as "the study of the interaction between human society and the physical environment" (2002:331). Humans have always affected the natural world and the environment has always shaped human societies, but environmental sociology began only recently as an offshoot of the environmental movement in the 1960s and 1970s. Rather than focusing on the environment or environmentalism, sociologists examine the way "social systems are organized and change in response to the natural world, just as the changes they produce in the natural world force them [social systems] to further respond and change" (Gould and Lewis, 2009:3). For example, a sociological inquiry might focus on the social factors that cause a particular environmental problem, the social or human costs of the damage to the environment on various populations, and the strategies or policies developed to solve the problem. Turner and Wu (2002, p. 1) described the environment as encompassing "the air people breathe walking down a city or country street, the water drawn from their taps or wells, the chemicals a worker is exposed to in an industrial plant or strawberry field, and the forests people visit to hike, extract mushrooms, and engage in spiritual practice." This conception of the environment links labour and public health, recreation and housing, and culture and history. Furthermore, this understanding of the environment breaks the boundaries between nature and society, work environments and open spaces, and urban and rural places.

For most of its history, the discipline of sociology did not pay much heed to the natural world. According to Bruce Catton and Riley Dunlap (1978), two of the earliest environmental sociologists, sociology was dominated by the Human Exemptionalism Paradigm—the view that humans determine their own fates. Human decisions, along with human constructs such as "economy, politics, language, symbols and cultures" (Young, 2015:7) were the causes of social ills such as inequality and poverty. Many of our sociological theories at least implicitly assume that our social institutions have placed us above the forces of nature and that nature exists primarily for human use.

▶ **New Ecological Paradigm** An alternative sociological approach, which recognizes that the social world does not exist on its own but is embedded in the natural world.

Catton and Dunlap (1978) suggested an alternative approach, which recognizes that human societies are subject to natural forces. They proposed that sociology incorporate what they called a **New Ecological Paradigm** recognizing the constraints the environment imposed on societies. Thus our social world does not exist on its own, but is embedded in the natural world (Buttel, 2010). For example, social inequality is not just a result of economic and political decisions. People living in a country—such as Canada—that has enormous water resources will be much better off economically than the citizens of a country that suffers from chronic drought.

Another simple example illustrates the way in which social institutions interact with the environment. Some of your parents will remember the massive ice storm that affected large parts of southern Ontario and Quebec in 1998. At the time, it was the most expensive natural disaster in Canadian history. Raymond Murphy (2004) has shown how our centralized, technologically dependent society is much more vulnerable to disasters than decentralized pre-modern societies.

Electric power lines are a fundamental part of our infrastructure, and we are highly dependent upon the power that runs through these lines. Many things in our society—heat, light, and the power to run our businesses and factories, bank machines, traffic lights, computers, and phones—depend upon a constant supply of electricity. In January 1998, an intense period of freezing rain brought down many of these power lines in Eastern Ontario and Quebec, and millions of people were left without heat or light. Many residents waited weeks before their power was restored. Factories closed down and business came to a halt during the crisis. The military mobilized thousands of soldiers to help with the cleanup and restoration of power.

Murphy contrasts the massive impact the ice storm had on most residents of the affected areas with its impact on Amish communities in upstate New York, which was also affected by the ice storm. The Amish lived in small communities that rejected modern technological conveniences, and they relied on traditional farming techniques to feed their families. What impact did the storm have on the Amish? "Those families used their labour to work their horses, milk cows by hand, and pump water as in periods of average weather. They heated their homes with wood stoves, which they also used to cook their food, and they ate preserves. By eschewing modern centralized technology, the Amish averted a disaster" (Murphy, 2004: 257).

Thus the disaster of the ice storm was not just caused by an abnormal weather event, but by our dependence upon a centralized infrastructure. Rather than setting us above the forces of nature, the way we have organized our modern societies actually can make us much more vulnerable to natural events. Natural events can also precipitate human-caused disaster. Wars and civil unrest can be caused by conflicts over resources, and if resources run out, these conflicts

MARCOS TOWNSEND/AFP/Getty Images

The 1998 ice storm, which affected millions of people in Ontario and Quebec, was one of the most costly natural disasters in Canadian history. The impact of the ice storm was magnified by our dependence upon centralized sources of electrical power.

will increase. For example, major rivers run through many different countries, and in times of drought, countries that are upstream may take more than their share of the remaining water and those denied water downstream may retaliate. High temperatures and sustained droughts may also lead to famines and mass migrations of starving populations.

Dunlap and Catton (2002) have highlighted three ecosystem functions that are required by human societies. When we overuse these functions, environmental problems result:

1. Nature provides the resources we need to live. It is our 'supply depot.' If we overuse these resources, we may starve. The global decline in the fishery due to overfishing and pollution has had a severe impact on the diets of people in many poor countries.
2. Human societies produce a great deal of waste, and this is deposited into the environment (see Box 20.2). Uncontrolled waste results in pollution—toxic wastes can cause illness and death and can also destroy the natural world.
3. We all need a place to stay—a living space—and this is of course located in the natural environment. In Chapter 23, we discuss some of the consequences of overpopulation, which, in addition to taking up more of Earth's space, also places a huge demand on natural resources and produces mountains of waste.

Dunlap points out that these problems used to be limited to local ecosystems—now, issues like the loss of biodiversity and global warming are global issues. While more serious, these problems are less obvious to us. The loss of biodiversity because of very high levels of species extinction due to human activities (Ceballos, et al., 2015) is much less obvious to us than is the toxic smoke blowing from a local factory.

BOX 20.2 POINT/COUNTERPOINT

Does Your Bottle of Water Harm the Environment?

Dunlap and Catton have highlighted the fact that waste is an environmental problem. Our behaviour as consumers plays a major part in creating waste and pollution.

One example of this is the way in which bottled water has become a consumer staple over the past two decades, despite the fact that most Canadians have access to plentiful amounts of pure, healthy water. While millions of people in low-income nations still die each year of water-borne diseases such as diarrhea and typhoid, Canada's water supply is very safe. With the exception of many First Nations communities and some others who get their water from private wells, nobody in Canada needs to drink bottled water for health reasons.

Despite this, Canadians drink an enormous amount of bottled water—in 2014, we consumed 2.4 billion litres, which is about 68 litres per person (Euromonitor International, 2015). Why do so many of us choose bottled water over the water that flows from our taps at almost no cost? According to Statistics Canada, the reasons include convenience, taste, concerns over the quality of tap water, and the marketing efforts of the bottled water industry (2008).

Marketing of bottled water has been very effective. Buying a single bottle of water can sometimes be more costly than buying a bottle of beer, despite the fact that beer is basically water with the additional costs of barley, hops, and yeast, plus a significant government tax (42 cents on a 355 mL can in Ontario). Water often sells for more than a litre of gasoline, which, like alcohol, is heavily taxed. Thus bottled water is a highly profitable product. The water itself is extremely cheap—British Columbia charges only $2.25 for a million litres of water, about as much as you would pay for 1 litre at a convenience store (Tieleman, 2015). People will pay even more for certain brands of water, and some restaurants even have water sommeliers whose job is to convince customers to order high-priced bottles of water.

As Nathan Young (2015) has noted, the companies profiting from bottled water managed to frame their produce in terms of health (bottled water is pure and good for you), rather than in terms of the environment (it creates waste and uses energy). Their marketing campaigns have been very aggressive. Gleick has reported that high-level executives from PepsiCo (bottlers of Aquafina water) stated that "The biggest enemy is tap water. . . . We're not against water—it just has its place. We think it's good for irrigation and cooking" and "When we're done, tap water will be relegated to showers and washing dishes" (Gleick, 2010:7). Coca Cola, which sells Dasani water, worked with restaurant chains on a program to encourage customers to use bottled water: "Some 20 percent of consumers drink tap water exclusively in Casual Dining restaurants. This trend significantly cuts into retailer profits. . . . This research provides the valuable insight and understanding needed to convert water drinkers to profit-producing beverages" (Gleick, 2010:9). This marketing pitch does not highlight the fact that 40 percent of bottled water sold in Canada—including both Dasani and Aquafina—is tap water (Ellison, 2013).

Our use of bottled water comes with environmental costs. The main issues are the oil used to produce the disposable bottles, most of which are never recycled, and the fuel and other costs of bottling and transporting water by train, truck, and ship rather than just taking it from your local tap. It takes a lot of energy to ship bottles of water from France and Fiji to North America.

While sales are still increasing—at least up to 2014—many interest groups are now mobilizing against bottled water. As a result, several cities have banned the sale of bottled water in municipal buildings, as have many universities and school boards.

These measures to reduce the sale of bottled water represent a victory for environmentalists. However, it is important to recognize that this is only a small victory in the context of our impact on the environment. Some sociologists feel that these small victories may actually do more harm than good because they take our attention away from more serious problems. Environmental problems are so great that major changes in the way our social systems operate are needed to avoid major changes in our environment in the future. If we do not adapt our behaviours to protect our environment, our environment will impose changes upon us, and these changes may not be desirable ones.

Source: Courtesy of Awish Aslam.

Stephane Bidouze/Shutterstock.com

This litter is just one of the environmental impacts of the billions of bottles of water Canadians purchase each year.

LO-2 THEORIES OF ENVIRONMENTAL SOCIOLOGY

At least partly because North American environmental sociology grew out of the environmental movement in the 1970s, people working in this area have tended to focus more on social action than on theory. However, a broad range of theories have been developed and these correspond

The Granger Collection, New York

According to Foster, "The imposition of the treadmill on English workers symbolized for Marx the tendency of the capitalist mode of production to degrade the work and, hence, the worker in mind and body" (2005:9).

to some degree with the theoretical categories we have set out in other chapters of this text. The theories we will discuss here are treadmill of production theory (conflict); ecological modernization theory (functionalist); ecofeminism (feminist), and environmental constructionism (interactionist).

TREADMILL OF PRODUCTION: CONFLICT THEORY

The failure of governments to take actions that will reduce pollution and waste raises the question of why societies continue to allow practices that might ultimately destroy their natural environments. Why isn't the future more important? From a neo-Marxist perspective, the answer to these questions is that economic interests have been able to drive the political agenda to the advantage of capitalism and to the detriment of the environment (Schnaiberg, 1980).

Capitalists and workers benefit from increased economic activity, including resource extraction, factory production, and services. The state also has an economic interest in growth, as increased economic activity increases government tax revenues. Capitalists, workers, and government are all competing for a greater share of revenues. The easiest way to satisfy everyone is to promote continual growth. Growth means that owners can accumulate capital, workers can receive higher wages, and governments can generate more tax revenue. Because of these benefits, politicians like to run on records of economic growth. Growth, then, allows the conflicting needs of the three interests to be satisfied. However, continual growth comes at the expense of the environment because growth requires energy and material to be sustained. As Young has observed, "the contradictions at the heart of capitalism have not been resolved, but rather downloaded to the point of least resistance—the environment" (2015:83). The **treadmill of production** takes resources, including material and energy, out of the environment, and deposits waste and pollution back into it.

▶ **treadmill of production** A cycle in which resources are taken out of the environment to produce goods, and waste and pollution are deposited back into the environment.

▶ **treadmill of accumulation** The never-ending process of accumulating more and more profit.

In addition to the treadmill of production, there is also a **treadmill of accumulation**. Capital accumulation is the primary focus of the capitalist system and is a never-ending process as corporations are valued on increases in profits. The desire is not simply to satisfy human needs, but to keep getting richer. In 2015, Apple was the world's largest company, with a valuation of nearly $1 trillion—a value that had grown by over 20 percent in the previous six months. To sustain this performance, Apple has an aggressive policy of making marginal improvements to technology such as iPhones and iWatches. These improvements eventually make earlier models obsolete. Apple has been highly successful at increasing consumer demand—what other company attracts people to stand in line outside their stores waiting for a small change to one of their phones?

As the treadmill analogy suggests, there is no end to the process. People and societies can never have too much—a standard of living that was quite satisfying to our parents and grandparents would seem far too modest today. However, there is no relationship between happiness and economic growth. Bell (2009) cites surveys showing that the percentage of Americans who reported being 'very happy' peaked in 1957. This was a time in which houses were about half the size of today's new homes and in which most families had only one vehicle—so the growth treadmill has given us material goods that have harmed the environment but which have not made us any happier.

ECOLOGICAL MODERNIZATION THEORY

Given the discipline's roots in the environmental movement, it is not surprising that most theories in the field of environmental sociology have focused on the negative aspects of society's

interaction with the environment. However, not all environmental theories reflect this pessimism. **Ecological modernization theory** has a very different view of capitalism's impact on the environment than do neo-Marxist theories.

▶ **ecological modernization theory** Environmental sociological theory that societies are adapting to the threat of environmental harms by reforming societal practices and policies.

Arthur Mol and Gert Spaargaren (2000) believe that societies are adapting to the threat of environmental harms by reforming societal practices and policies. The treadmill of profit growth can create the wealth needed to protect the environment. New scientific discoveries will lead to new technologies that will allow growth to continue without destroying the environment. Along with these changes, better government will alleviate the danger of environmental harms.

Mol and Spaargaren developed ecological modernization theory in response to the view of many environmentalists that sustainable development was only possible if society's core institutions—including capitalism and the industrial system of production—were fundamentally transformed. Proponents of ecological modernization theory maintain that, while changes are necessary, the environment can be protected within the current system of production. Mol and Spaargaren recognize that capitalism has harmed the environment, but they are confident that capitalism is very flexible in adapting to change and that environmental concerns are one of the factors now pushing these changes. They believe that environmental reform is possible under many different types of economic systems, including capitalism.

Ecological modernization theorists see an important role for the state in promoting sound environmental practices. Governments need to regulate environmental practices and provide incentives encouraging or requiring industry to develop and implement environmentally sound practices and technologies. As the theory's name suggests, the focus is on modernizing the system of environmental regulation rather than on replacing it.

The theory's proponents have cited some successes in protecting the environment. One of the most notable successes was the Montreal Protocol. A buildup of chemicals called chlorofluorocarbons (CFCs) was destroying the world's ozone layer—an atmospheric layer that helps to absorb some of the Sun's harmful ultraviolet radiation. The world's nations signed the Montreal Protocol in 1987 and most stopped production of the chemicals that depleted the ozone. The damage stopped and the ozone layer is slowly being renewed.

Vehicle emissions are one of the major causes of global warming. The U.S. government has imposed some very stringent greenhouse gas emission requirements on the auto industry, and

chris kolaczan/Shutterstock.com

Extracting oil from the oil sands is far more energy-intensive than more conventional methods of oil production.

Canada has agreed to follow the same standards. By 2025, each manufacturers' vehicle fleets will have to average 4.4 litres per 100 kilometres—an 80 percent reduction over the averages in 2012 when the legislation was passed. This standard will be extremely challenging for manufacturers. One of the first steps has been to make vehicles lighter, as weight affects fuel consumption. Ultimately, however, to meet the 2025 standard, manufacturers will have to turn to building affordable hybrid and electric vehicles that the public will want to buy.

Public pressure has placed environmental issues on the corporate radar. Many companies promote 'green' products, have set up internal and external recycling programs, and try to build factories and office buildings that incorporate energy-saving design standards. Despite this progress, environmentally conscious practices have not been implemented nearly as quickly as critics would like. For example, Alberta's oil sands are a major source of Canada's emissions of greenhouse gases. Oil from the oil sands is problematic because it is not simply pumped out of the ground, but is mined in solid form. Intensive processing is required to turn it into usable fuel. Thus not only do greenhouse gases result from burning the fuel in our vehicles and power generators, but significant amounts of energy must be used to turn the oil sands into fuel. Under Prime Minister Harper, the federal government resisted imposing emissions standards on Alberta's oil and gas industries, and taxing carbon—likely the best way of reducing emissions.

However, public pressure may lead to better policies. Environmental groups have delayed the construction of several pipelines that are needed to take oil from Alberta. Realizing that some progress on environmental issues will be the only way to convince environmental groups to allow them to move their products, many of Alberta's largest oil companies have called for governments to tax carbon emissions in order to reduce the consumption of fossil fuels. This pressure may lead to change under the Liberal government elected in 2015.

Despite some progress, the notion that increased growth and prosperity can help provide the economic and technological resources needed to preserve the environment may not be valid. Current evidence tells us that modest improvements will not be sufficient to avoid major environmental problems in the future. As Dunlap concludes, "the evidence suggests that the growing demands of the human population for living space, resources and waste absorption are beginning to exceed long-term global carrying capacity . . . with the result that the current human population is drawing down natural capital and disrupting the functioning of ecosystems from the local to the global level" (2010:18).

ECOFEMINISM

Ecofeminists have linked concerns about gender and the environment with other types of oppression, including race and class (Plumwood, 1993). The perspective is concerned with the way capitalism and patriarchy have combined to dominate and to oppress both humans and the natural world. The basic premise of most **ecofeminism** theories is that "The ideology which authorizes oppressions such as those based on race, class, gender, sexuality, physical abilities, and species is the same ideology which sanctions the oppression of nature (Gaard, 1993:1).

▶ **ecofeminism** Sociological theory that the ideology that authorizes oppressions such as those based on race, class, gender, sexuality, physical abilities, and species is the same ideology that sanctions the oppression of nature.

Nature is feminized—metaphors such as 'Mother Nature,' and 'settle virgin territories' were frequently used. And, of course, males were deemed superior to both nature and women. Edmund Burke, a noted British philosopher, claimed that "a woman is but an animal and an animal not of the highest order" (Bell, 2009:149). Such views are demeaning to both women and to nature.

Early versions of ecofeminism had a spiritual tone—describing both women and nature as possessors of a spiritual life force. Contemporary ecofeminists focus on the cultural patterns of patriarchal domination that pervade our society. In the environmental realm, this domination has had serious negative outcomes—notably a lack of environmental justice for women, the poor, racial minorities, non-humans, and nature itself. The structure and culture of patriarchy has also had negative consequences for the environment. To remedy these impacts, ecofeminists advocate that humans should try to coexist with nature rather than try to dominate it.

Ecofeminists have also pointed out that the consequences of environmental damage are disproportionately borne by women. In many low-income societies, women do most of the work on the land, but men take most of the food they produce. Women have higher rates of poverty than men, so they suffer more from the consequences of environmental damage that limits crops. While they are more affected by climate change and other environmental problems, women are under-represented in the governments and organizations that make decisions that have an impact on the environment.

CONSTRUCTING ENVIRONMENTAL ISSUES: THE SYMBOLIC INTERACTIONIST PERSPECTIVE

The final theoretical perspective to be discussed here is quite different from the others. Rather than focusing on the causes and consequences of environmental problems or on the best ways to improve the planet, another group of sociologists has studied how environmental issues become public concerns. **Constructionism** is concerned with how environmental issues are socially constructed. Constructionists study the "social, political, and cultural processes by which certain environmental conditions are defined as unacceptably risky, and therefore, contributory to the creation of a perceived 'state of crisis' (Hannigan, 2006:29). Constructionism helps us to focus on the often-contentious process involved in defining environmental problems and in taking action to deal with these problems.

constructionism A theory that focuses on the way in which environmental issues are socially constructed.

This perspective is helpful, because it points out that environmental conditions are not automatically defined as social issues. Environmental problems have existed for millennia, but significant social concern with the environment is much more recent. Many attribute the beginning of the environmental movement to the publication of Rachel Carson's *Silent Spring* in 1962. Carson described the damage done by the indiscriminate use of harmful pesticides, particularly DDT, and generated massive public attention. While scientists had known of the effects of these chemicals for years, Carson's book brought the material together in a way the public could understand, and led to a demand for government action to limit pesticide use.

While *Silent Spring* did help to limit the use of insecticides, most notably through a global ban on the use of DDT, Carson faced strong opposition from chemical companies. She was accused of working for the Communists (this was during the Cold War when the Soviet Union was seen as the enemy of western countries) and suffered other attacks on her character. She was also accused of being responsible for the deaths of millions of people because DDT was no longer used to fight malaria-bearing mosquitoes. These attacks continue today—the website *rachelwaswrong.org* is one example. The attacks on Carson were a prelude to those we see today where a broad range of interest groups are trying to convince politicians and the public that the views of scientists who attribute climate change to burning carbon-based fuels are wrong. Much of this work is funded by oil, gas, and coal producers, and also by fiscal conservatives who resist government regulation.

Hannigan (2006) outlines three key tasks that help to successfully construct an environmental problem:

ASSEMBLING AN ENVIRONMENTAL CLAIM Some environmental claims are easy to identify. Everyone can see clear-cut areas on a forested mountainside near a large community, and we can all notice the smell from a malfunctioning sewage treatment facility or toxic waste dump. However, some claims are not as visible because they are based on science that is very difficult for lay people to understand. The ozone layer can only be measured by complex scientific instruments, so the problem of the thinning ozone layer caused by CFCs had to be discovered and proven by scientists. Organizations such as Greenpeace and the World Wildlife Fund have very sophisticated researchers and fundraisers who prepare and present environmental issues to the public and politicians in an easily understandable way.

PRESENTING AN ENVIRONMENTAL CLAIM Once an environmental issue has been identified and documented, the public must be made aware of the problem and convinced that the problem is real. Some problems get massive public attention. Dramatic incidents like major oil spills or nuclear reactor breakdowns quickly draw global attention. The imagery associated with 'mad cow disease'—a disease affecting the brain and nervous system, which can be contracted through eating meat—was very dramatic. Video of helpless cows staggering around before dying and horrible stories of people dying of brain disease caught the public's attention and led to quick action. Despite the fact the disease was not common and affected a very small number of people, many countries banned imports of foreign beef, which harmed many beef producers. On the other hand, the unnecessary use of antibiotics in domestic animals has much more serious consequences because it increases the likelihood that human diseases will become resistant to antibiotics. However, this is a much less dramatic story, and despite a huge amount of scientific evidence, there is no great public concern and politicians have been reluctant to act because of opposition from the farming industry.

CONTESTING AN ENVIRONMENTAL CLAIM Gaining public acknowledgment of environmental problems does not necessarily mean anyone will deal with the problems. For example, Canadian governments have recognized the dangers of global warming, but most have essentially failed to act. The federal government has consistently ignored its commitments under the Kyoto Protocol. When he became prime minister in 2006, one of Stephen Harper's first acts was to withdraw from Kyoto. He believed it was a 'socialist scheme' to take money from wealthy countries (Fanelli, 2014). Prime Minister Harper also reduced funding to federal agencies that were involved in environmental issues. In 2015, he agreed to a meaningless agreement that Canada would achieve a no-carbon economy by the year 2100. Needless to say, this did not unduly upset some of the prime minister's supporters in the oil industry.

As Rachel Carson and many other environmentalists have discovered, claims about environmental issues are often contested. When huge numbers of dollars and jobs are threatened, the reaction can be strong, and many environmental issues involve the interests of many different groups. There are major battles over the recognition of environmental problems. This is particularly the case in the issue of global warming because resource companies will lose massive amounts of money if carbon emissions are limited. While the overwhelming consensus of climate change scientists is that human-made global warming is taking place, a very sophisticated campaign has attempted to convince the public that these conclusions are based on 'junk science.' Environmental skepticism has been promoted by conservative think tanks, including Canada's Fraser Institute, which is part of the 'Cooler Heads Coalition'—a group of lobbying organizations "focused on dispelling the myths of global warming by exposing flawed economic, scientific, and risk analysis" (http://www.globalwarming.org/about/, n.d.). Jacques and colleagues (2008) found that almost all environmentally skeptical books published between 1972 and 2005 were linked to conservative think tanks. They conclude that "skepticism is a tactic of an elite-driven counter-movement designed to combat environmentalism and that the successful use of this tactic has contributed to the weakening of U.S. commitment to environmental protection" (2008:349).

CONCEPT SNAPSHOT

FUNCTIONALIST PERSPECTIVES **Key thinkers:** Arthur Mol, Gert Spaargaren	Societies are adapting to the threat of environmental harms by reforming societal practices and policies. The treadmill of profit growth can create the wealth needed to protect the environment.
CONFLICT PERSPECTIVES **Key thinker:** Allan Schnaiberg	Treadmills of production and accumulation put pressure on societies to emphasize continuous growth. This growth benefits capitalists, workers, and governments, but it places enormous pressure on the environment.
INTERACTIONIST PERSPECTIVES **Key thinker:** John Hannigan	Constructionism is concerned with how environmental issues are socially constructed. If environmental conditions are to be defined as societal problems, people must be persuaded to define them as harmful, overcome the resistance that is often present with environmental problems, and convince politicians and citizens to take action to deal with the problem.
FEMINIST PERSPECTIVES **Key thinker:** Val Plumwood	Ecofeminists have linked concerns about gender and the environment with other types of oppression, including race and class. The perspective is concerned with the way capitalism and patriarchy have combined to dominate and to oppress both humans and the natural world.

TIME TO REVIEW

- What is environmental sociology? How does environmental sociology differ from the traditional approach to sociology?
- How do the treadmills of production and accumulation have a negative impact on the environment?
- Discuss how proponents of the ecological modernization theory believe environmental problems will be solved in the future.
- How do environmental problems have a disproportionate impact on women?
- Discuss how interest groups and individuals go about 'constructing' an environmental problem.

LO-3 THE ENVIRONMENT AND HEALTH

In the following narrative, sociologist Sabrina McCormick talks about the first time she experienced the reality that our environment can make us sick:

> When I was eight, I lost almost everything: my room and all the stuff in it—toys, dance costumes, purple barrettes, green tights, sparkly lip gloss. I had to leave them behind when we abandoned our house on Bishop Lake. That summer, a crew had sprayed our house for powder post beetles. Quickly, my mother got sick and dizzy. She had a piece of furniture tested. The crew had illegally sprayed our house with chlordane, a chemical used as a pesticide. Before that year, 3.6 million pounds of chlordane were applied to corn, fruits, lawns, and houses every year. But the initial testing of chlordane that allowed those agricultural applications was not sufficient. Eventually, the U.S. Environmental Protection Agency (EPA) discovered that the chemical can contaminate air, water, and soil and that it causes damage to the nervous and blood systems, lungs, and kidneys. The year before chlordane was sprayed in my room, 1983, regulations were put in place to ban it. That was when I first learned how the environment affects health. (McCormick, 2015:179)

How does the environment affect health? Environmental change, exposure to environmental toxins, and environmental destruction and overconsumption can have significant and devastating effects on the health of human populations. Rather than focusing on individual causes of disease and health related problems such as diet and exercise, environmental sociologists focus on the social production of disease. How does failing to take care of our environment make us sick? Although there is no simple answer to this question, we are increasingly aware of the fact that many of the most common diseases (such as cancer and heart disease) are caused, at least in part, by environmental factors.

The world's population is just over 7.3 billion. It is predicted that in the next 25 years, the global population will reach almost 10 billion (United Nations, 2015). Will it be possible to sustain an environment with this population? Given that we are already using massive amounts of air, energy, fresh water, and land, while at the same time creating significant damage to our environment, it is unlikely. The term *environmental sustainability* refers to the ability of human society to sustain itself without damaging or destroying the basic ecological support systems. In other words, achieving environmental sustainability means managing the ability of the environment to support human life. The primary question of sustainability is how long can we keep doing what we are doing? If we are failing to maintain the environment by managing oceans, freshwater systems, land, and air, then the health of our populations, both nationally and globally, will suffer. What are the health effects of not managing our atmosphere, land, and water systems?

The following cases are just two of many human tragedies that demonstrate how the environment can make us sick.

BHOPAL TRAGEDY

In 1984 in the small city of Bhopal, India, a pesticide plant owned by Dow Chemical leaked an extremely toxic chemical, *methyl isocyanate,* after an accident at the plant. A cloud of deadly gas was released into the air, and within hours thousands of people from the Jaiprakash Nagar neighborhood, 100 yards from the plant, had died. In the coming hours, days, and years, thousands more would die. Some 5,000 to 15,000 in all would lose their lives, including a quarter of the residents of Jaiprakash Nagar and two other small neighborhoods close to the plant. Some claim that 30,000 died and at lest 500,000 more were injured. Following are some of the descriptions from local newspapers:

> Jaiprakash Nagar, a sleepy locality of Old Bhopal, is today a ghost colony. Every second house in the locality has lost at least one family member in yesterday's night of horror.
>
> This correspondent who went round the locality early morning found more than fifty dead bodies lying unattended and unnoticed. . . . The dead included mostly children below ten years of age.
>
> The scene was so gruesome that it was difficult for survivors to identify their own dead family members. The neighbors were not willing to tell anything to anybody. They just sat glassy eyed, dumb-founded. (Bell, 2012:125)

Dow Chemical denied responsibility for the tragedy, claiming worker error had caused the accident and that only a few people died. After over 20 years of legal battling, some of those who lost loved ones were compensated, although many never received compensation (McCormick, 2015). The tragedy continues among the victims to this day as many continue to experience neurological defects, breathing problems, and mental health problems (Bell, 2012). Many of those who were exposed to the gas have given birth to physically and mentally disabled children.

MERCURY POISONING IN GRASSY NARROWS

In 1970, members of Asupeeschoseewagong First Nation (Grassy Narrows) in Northwestern Ontario learned that Dryden Chemicals, a pulp and paper mill, had dumped approximately

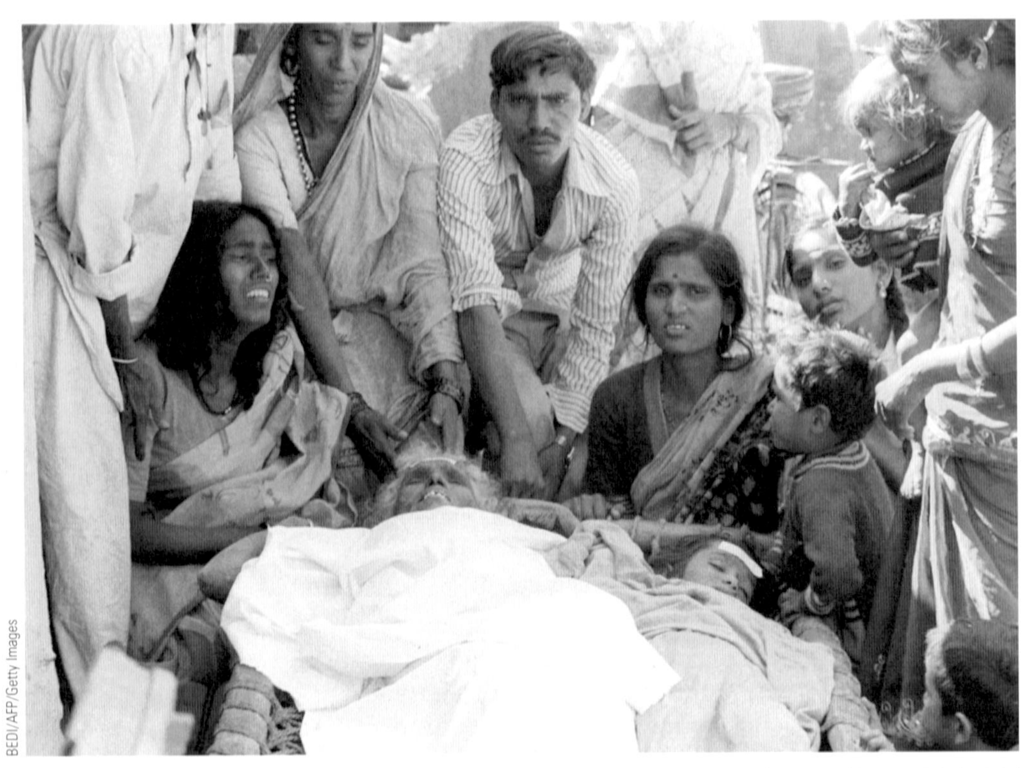

BEDI/AFP/Getty Images

A gas leak in 1984 at a Dow Chemical Plant in Bhopal, India, is considered one of the world's worst industrial accidents. Thousands of residents died and thousands more suffered lasting injuries, including neurological and mental health problems and birth defects in the children of those exposed to the toxic gases.

20,000 pounds of mercury into the Wabigoon River only 100 kilometres upstream from Grassy Narrows. Once in the river, the mercury was converted to methyl mercury, an extremely toxic substance that contaminated the water, soil, fish, and wild game in the area. The main food source for members of Grassy Narrows First Nation was fish. While the community increasingly showed signs of mercury poisoning, the Ontario government insisted that the poisonous fish were safe to eat. Mercury is highly poisonous to humans, with devastating health impacts (Vecsey, 1987). The effects of mercury poisoning include speech, taste, and smell impairment; difficulty swallowing; choking, blurry vision; loss of strength; "tunnel vision; loss of co-ordination; numbness; tremors; loss of balance; and overall impaired motor functioning. In pregnant women, the mercury settles in the fetus, leading to birth defects (Ilyniak, 2014:43).

As community members explained every spring, as the fresh rivers run, mercury is re-released into the waterways. Clear-cut logging also poses a great threat as it leads to erosion, which stirs up more of the mercury.

Mercury levels remain elevated in the animals and fish and community members' health continues to suffer. In 2007, two children were born with brain cancer and many others experienced seizures (Ilyniak, 2014). A report in 2015, 50 years after the mercury contamination took place, again described the toxic environment in Grassy Narrows. In response, Ontario Premier Kathleen Wynne said that more research was needed and that any solution would have to balance "the health of the community, the environment and the economy" (Porter, 2015). This lack of commitment means that another generation of children will grow up suffering from the effects of mercury poisoning.

CLIMATE CHANGE AND HEALTH

Climate change is having a dramatic impact on health around the world. As we continue to fail to 'sustain' our environment, an increasing number of people are expected to die from climate

Brendan Kennedy/GetStock.com

These people are protesting the mercury pollution that continues to affect the community of Grassy Narrows.

related illnesses. One of the world's leading medical journals has predicted that climate change "is the biggest global health threat of the 21st century" (Lancet and University College London Institute for Global Health Commission, 2009:1693). Sociologist Sabrina McCormick identifies five primary health concerns related to climate change: (1) air pollution-related illnesses; (2) temperature-related illnesses; (3) vector-borne disease; (4) water- and food-borne disease; and (5) illness or injury from extreme weather events.

Climate change has both direct and indirect effects. The direct effects include the impact of extreme temperatures (a heat wave in Europe led to about 70,000 deaths in 2003). The effects of weather events such as floods and hurricanes, respiratory problems caused by air pollution, and infectious diseases spread by carriers such as insects and rodents—and some diseases like malaria—may spread more widely as the planet warms. Climate change has already led to the spread of Lyme disease and cryptococcosis (a lung infection that can lead to serious brain damage) into Canada, and many other diseases in other parts of the world.

The indirect effects may be even more severe: the economies of many countries may collapse, causing starvation and mass migration; and major internal and external conflicts over scarce resources may result in the deaths of many people.

One of the most immediate effects of climate change may be massive shortages of food and water. In 2015, California suffered a lengthy drought that seriously affected the state's water supply. This drought significantly reduced the state's agricultural production. Major rivers, including the Ganges, the Colorado, and the Yangtze, have reduced flows that are attributed to climate change, and this reduction will have a major impact on food supplies. Also, the aquifers (underground water supplies) that serve many of the globe's major cities, are drying up. In Mexico City, buildings are sinking and water supplies have dropped by 40 percent because of over-use of ground water (Lancet, 2009). The consequences of this will be less agricultural land that will sustain food production—increased desertification may destroy much of the food production in some countries.

There is no question that climate change will compound problems of starvation, malnutrition, and food insecurity both nationally and globally. While recent estimates indicate that today approximately 800 million people suffer from hunger and food insecurity, it is estimated that half of the world's population could face severe food shortages as a result of climate change (Lancet, 2009:1704).

While some populations will suffer from too little water, others will have too much, as the rise in ocean levels (because of melting icecaps) will force many coastal residents to move, and will

Climate change may transform productive land into desert, as shown in this photo from Mauritania. This desertification may make it impossible for some countries to sustain their food production.

also affect food production in low-lying areas. For example, Bangladesh—home to 160 million people—is currently prone to flooding, and this will only get worse with global warming. Even a one-metre rise in sea level will mean that about 30 million people will be forced from their homes.

Increasing temperatures and acidification in the oceans, along with overfishing, will mean that the world's fisheries will be in danger of collapse. This will have a huge impact on the population.

> Climate change will have the greatest impact on those who have the least access to the world's resources and who have contributed least to its cause. We are experiencing a global health crisis as a result of the damage we have done to the environment. Ten million children die each year; over 200 million children under 5 years of age are not fulfilling their developmental potential; 800 million people go to bed each night hungry; and 1500 million people do not have access to clean drinking water. The inequity of climate change—with the rich causing most of the problem and the poor initially suffering most of the consequences—will prove to be a source of historical shame to our generation if nothing is done to address it. (Lancet, 2009: 1694)*

As with many other societal problems, those in marginalized populations—the poor, racial minorities, and Indigenous peoples—will bear the brunt of harm. "The rich will find their world to be more expensive, inconvenient, uncomfortable, disrupted, and colorless—in general, more unpleasant and unpredictable, perhaps greatly so. The poor will die (Smith, 2008: 11).

ENVIRONMENTAL INEQUALITY AND ENVIRONMENTAL JUSTICE

The fundamental principle of **environmental justice** is the idea that all people and all communities are entitled to equal protection under environmental health laws and regulations (Bullard, 1996). Environmental justice scholars highlight the fact that problems of environmental inequality are fundamentally *social* problems rather than environmental issues. Environmental

environmental justice The idea that all people and all communities are entitled to equal protection under environmental health laws and regulations.

* Reprinted from *The Lancet*, Vol. 373, Lancet and University College London Institute for Global Health Commission, "Managing the health effects of climate change," pp. 1693–1733, Copyright 2009, with permission from Elsevier.

inequality (or environmental injustice), then, refers to a situation in which a specific group is disproportionately affected by negative environmental conditions brought on by unequal laws, regulations, and policies (Pellow & Nyseth Brehm, 2013).

▶ **environmental racism** The targeting of racial minority communities for toxic waste facilities, the official sanctioning of poisons and pollutants in minority communities, and the systematic exclusion of racial minorities from decision making regarding the production of environmental conditions that affect their lives and livelihoods.

A specific form of environmental inequality is the phenomenon of **environmental racism**, which is the targeting of racial minority communities for toxic waste facilities, the official sanctioning of poisons and pollutants in minority communities, and the systematic exclusion of racial minorities from decision making regarding the production of environmental conditions that affect their lives and livelihoods (Mascarenhas, 2015).

The environmental justice movement developed in the United States in the 1980s when it was discovered that environmental hazards disproportionately affected poor and racial minority populations (Pellow and Nyseth, 2013). The movement gained momentum when local residents of the predominately black community of Warren County, North Carolina, took a stand against the construction of a landfill for toxic PCBs. Their efforts to mobilize against the proposed dump led to similar protests in other communities across the country. Shortly after, a national study using U.S census data and the location of hazardous waste facilities, concluded that "Race continues to be an independent predictor of where hazardous wastes are located [that is] stronger than income, education, and other socioeconomic indicators" (Bullard, 2007:xi). The principle researcher of the study, sociologist Robert Bullard, published a national bestseller, *Dumping in Dixie: Race, Class and Environmental Quality* (1990), documenting numerous examples of environmental racism across the United States. In 1994, President Bill Clinton signed an Executive Order requiring federal agencies to develop environmental justice strategies to address the disproportionate human health and environmental effects of their programs on minority and low-income populations. The U.S. Environmental Protection Agency now coordinates environmental justice activities and gives mandatory consideration to issues of racism and justice when evaluating projects and permits and developing new regulations (Mitchell, 2015). Today environmental justice has become a central civil rights issue in the United States.

GLOBAL ENVIRONMENTAL INEQUALITY

A primary focus of environmental sociologists is the highly unequal distribution of environmental harm (referred to as *environmental bads*) and privileges (*environmental goods*) in societies throughout the world (see Box 20.3, Old Environmental Pollution with New Social Pressures in China). As environmental sociologist Michael Bell explains:

> Global warming, sea level rises, ozone depletion, photochemical smog, fine-particle smog, acid rain, soil erosion, salinization, waterlogging, desertification, loss of genetic diversity, loss of farmland to development, water shortages, and water pollution: These have a potential impact on everyone's lives. But the well-to-do and well-connected are generally in a better position to avoid the worst consequences of environmental problems, and *often to avoid consequences entirely*. (2012:25)

As discussed in the previous section, climate change results in pervasive environmental inequality. People in Africa, Asia, and Latin America are the first to experience the 'bad' of climate change in the form of natural disasters (floods, droughts, and storms), food insecurity, malnutrition, and starvation, respiratory illnesses and infectious diseases, health-related illnesses and death, and large decreases in renewable energy sources. They have also suffered significantly more from climate change in terms of loss of life and livelihood than rich nations. For example, an analysis by Roberts (2009) of over 4000 climate-related disasters between 1980 and 2000 found that some poor countries had rates of death and homelessness from climate disasters that were between 200 and 300 times higher than the United States (196). As shown in Figure 20.1, which shows global carbon emissions, the responsibility for environmental damage is even more unequally distributed among rich and poor nations. The United States, Canada, Australia, and Russia are responsible for the vast majority of global carbon emissions. Sub-Saharan Africa (one of the largest populations in the world) contributes only 2 percent of carbon emissions in the world.

FIGURE 20.1 COUNTRIES BY CARBON DIOXIDE EMISSIONS VIA THE BURNING OF FOSSIL FUELS (BLUE THE HIGHEST)

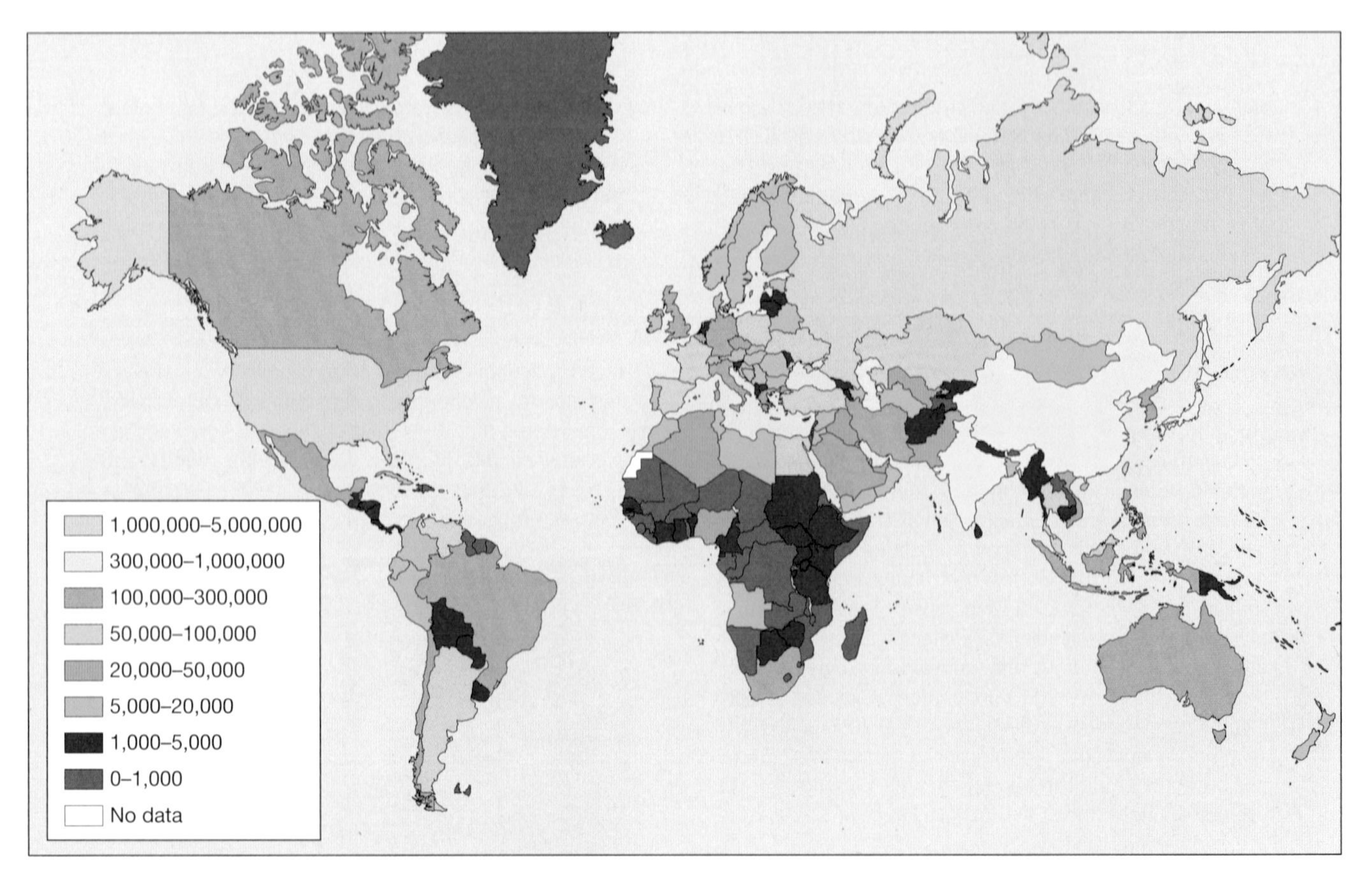

Source: Based on data from Tom Boden and Bob Andres, Carbon Dioxide Information Analysis Center, Oak Ridge National Laboratory; Gregg Marland Research Institute for Environment, Energy and Economics Appalachian State University. Accessed: http://cdiac.ornl.gov/trends/emis/tre_coun.html

How does Canada measure up in terms of responsibility for damage to the environment? Consider these facts:

- 37 million Canadians use more energy than all 760 million people in Africa.
- Canada makes up less than one half of one percent of the world's population, but is the world's eighth largest producer of greenhouse gases. (David Suzuki Foundation, 2015)

ENVIRONMENTAL RACISM IN CANADA

Although the majority of Canadians recognize the need to address problems of the environment generally, few are aware of the numerous cases of environmental racism across Canada. As environmental sociologist Michael Mascarenhas suggests, "because racism is associated with hostile and intentional acts, the majority of whites can exonerate themselves from environmental racism" (2015:176). However, while white Canadians may not *individually* engage in acts of racism, they have been able to benefit *collectively* from social, economic, and environmental privileges at the expense of the health and welfare of First Nations and other racial minority populations. While many Canadians may deny, or reject, the suggestion of "environmental racism," the following examples demonstrate that it is a very real problem.

CHEMICAL VALLEY In Southern Ontario's "Chemical Valley," the petrochemical industry has created what is reported to be Canada's largest concentration of water and air pollution. Between 1974 and 1986, over 300 oils spills led to approximately 10 tons of pollutants entering the St. Clair River. Since then, an average of 100 spills have occurred each year. In addition,

BOX 20.3 SOCIOLOGY IN GLOBAL PERSPECTIVE

Old Environmental Pollution with New Social Pressures in China

- Up to 40 percent of China's rivers are seriously polluted after 75 billion tons of sewage and wastewater were discharged into them.
- Twenty percent of rivers are so polluted that their water quality was rated too toxic for people to come into contact with the rivers' water.
- Nearly 300 million rural residents lack access to drinking water.
- About two-thirds of the cities in China lack sufficient water (*People's Daily Online*, 2012).

For many years, environmental problems in China have been in the news around the world. The Internet and social media have intensified and greatly sped up this coverage, and have made environmental activists worldwide more aware of the problems faced by people in that rapidly growing region of the world. For example, it is well-known that cancer is now the leading cause of death in China, partly because of air pollution, water pollution, and other environmental contaminants, some of which may be attributed to factories related to production of goods for high-income nations such as the United States (Larsen, 2011).

What, if anything, is being done about this? Some social protests and the beginnings of social movements are now being found in various provinces of China. Here are a few examples:

- Approximately 12,000 protesters demonstrated in Dalian, China, in 2011 after a storm damaged a paraxylene (PX) factory. PX is a toxic chemical used to make polyester. The protesters were mobilized by cellphones and the Internet. Officials announced that the plant would be closed, which was no small feat because $1.5 billion had been invested in a joint venture between a state-owned chemical company and a private developer (Economist.com, 2011).
- Apple announced in 2012 that it would allow independent environmental reviews of factories that supply parts for Apple products. Apple has faced rising criticism about toxic pollution and factory injuries of workers in suppliers' factories in China and other countries. Among the environmental problems cited are hazardous-waste leaks and the use of toxic chemicals that might create health risks not only for workers but also for neighboring communities. On its website, Apple stated that it insists that suppliers "provide safe working conditions, treat workers with dignity and respect, and use environmentally responsible manufacturing processes" (quoted in Chu, 2012).

Workers in plants in China and other emerging nations typically do not have the same legal protection as North American workers. If Canadian companies with factories in these nations become an active force in requiring higher environmental standards, will this improve the quality of life in other nations as well?

nearby livestock production facilities produce significant runoff of pesticides and fertilizers that further contaminate the river (Mascarenhas, 2015). Collectively, the facilities in Chemical Valley emit tens of millions of kilograms of air pollution each year. The St. Clair River is the source of water for approximately 800 residents of the Aamijiwnaag First Nation who live next to Chemical Valley and suffer the health consequences of the polluted water and air and its impact on their ability to maintain a livelihood by fishing, hunting and growing food.

LACK OF ACCESS TO WATER AND FOOD Nothing illustrates Canada's startling environmental inequities more clearly than the lack of access to clean drinking water in First Nations communities (Mitchell, 2015). Water is a vital part of our day-to-day lives—we need it for drinking, sanitation, and household uses. Communities need water for economic, social, cultural, and spiritual purposes. Imagine having to boil water rather than just turning on the tap for everything you do in your home: to cook, clean, and wash. Access to safe drinking water is a problem that many families across Canada, predominantly families living in First Nation communities, face—some for years at a time (Lui, 2015:5).

According to a recent United Nations report, First Nations homes are 90 times more likely to be without safe drinking water than other Canadian homes. More than 10,000 homes have no indoor plumbing on First Nations reserves, and one home in four has a substandard water or sewage system. Approximately 55 percent live in communities where half of the houses are inadequate or substandard. Many First Nations have deteriorated homes, toxic mould, lack of heating and insulation, and leaking pipes (United Nations, 2009).

In the absence of a national water law, communities under federal jurisdiction, such as First Nations reservations, have virtually no legal protection of their drinking water. As of January 2015, drinking water advisories were in effect in 126 First Nation communities across Canada. Although most drinking water advisories in Indigenous communities are boil water advisories, there are a handful of communities that are under "do not consume" orders (Lui, 2015).

SHOAL LAKE FIRST NATION In the early 1900s, an aqueduct was constructed on Shoal Lake to carry much-needed clean water to Winnipeg. As a result of the aqueduct, the Shoal Lake First Nations community was cut off and made into an artificial island. Shoal Lake residents were left with water so dirty that weeds sometimes came through the taps. The Anishinaabe people of Shoal Lake First Nation have been under a boil water advisory for over 18 years. In 2015, a state of emergency was declared after the ferry that linked it to other communities was shut down. Without permanent road access, they have no access to food services, no garbage removal, and no emergency or postal services. Not only were people cut off from medical care and grocery stores, they had no way to bring in the bottled water the community has relied on ever since a boil water advisory was put in place. The Province of Manitoba and the City of Winnipeg have each finally agreed to provide one-third of the costs of a new bridge and road to provide access to the mainland. However, prior to the 2015 election, the federal government continued to refuse to put up their share of the funds but the new Liberal government promised during the campaign that it would contribute its share of funds to this project.

Environmental justice advocates argue that Canada needs to follow the United States' example and address issues of environmental justice and environmental racism in Canadian law.

Chief Hector Shorting shows off some of his community's "undrinkable" water outside his mould-infested lakeside home in Little Saskatchewan First Nation. There are 126 First Nation communities in Canada that have drinking water advisories.

The Canadian Press/John Woods

A boy from the Shoal Lake First Nation sits on a bridge over a man-made channel made to support Winnipeg's water system. Residents have been on a boil water advisory for over 18 years.

Nova Scotia began this process in 2015 with the introduction of the proposed "Act to Address Environmental Racism." Although this is a start, environmental justice advocates suggest that as Canadians, we have a long way to go:

> We believe that Canada needs to go further and recognize that everyone in this country—regardless of who they are or where they live—has a Charter right to a healthy environment. An even more basic step would be to admit that we, as a country, have a problem. Turning a blind eye to the links between race, socio-economic status, and environmental risks doesn't make the issue any less real. The fact is, environmentally harmful activities take place in some communities more than others. We have to name that reality before we can begin to address it. (Mitchell, 2015)*

As the previous cases demonstrate, the environmental problems experienced by First Nations communities in Canada are fundamentally human rights issues (Mitchell, 2015). All people and all communities should have equal access to the 'environmental goods' in a society.

CRIME AND THE ENVIRONMENT: GREEN CRIMINOLOGY

Green criminologists believe that criminology should not just study actions that violate the criminal law but also actions that are socially harmful. Some environmental harms are illegal. One of Canada's most serious environmental disasters was a case of water pollution in Walkerton, Ontario, that caused more than 2000 people to fall seriously ill and led to seven deaths. The operators of the water treatment plant who failed to test the water and who falsified test results were successfully prosecuted. However, the real responsibility for this disaster lay with former Ontario premier Mike Harris, whose government put people at risk by cutting 725 positions from the Ministry of the Environment, so enforcement of environmental standards dropped dramatically (Girard, et al., 2010). An inquiry found that these policies contributed in a major way to the tragedy (O'Connor, 2002), but none of the responsible politicians was prosecuted.

* Mitchell, Kaitlyn, 2015, "Environmental Racism Remains a Reality in Canada," *Huffington Post*, May 7, 2015 http://www.huffingtonpost.ca/ecojustice/environmental-racism-canadal_b_7224904.html.

Many of the actions that are the most harmful to the long-term health of the environment, such as the emission of huge quantities of greenhouse gases in the production of oil from Alberta's tar sands (Smandych and Kueneman, 2010) and clear-cutting tropical rainforests, are not against the law. Green criminologists argue that the damage to Earth caused by destructive environmental practices can be far more serious than the illegal acts that have traditionally been the subject matter of criminological study (Lynch and Stretesky, 2007).

Green criminology has its roots in the environmental and animal rights movements, though more green criminologists have focused on environmental issues than on animal rights. The environmental focus of green criminology covers the study of environmental damage, including air and water pollution and harm to natural ecosystems such as oceans and forests. Those interested in animal rights study "individual acts of cruelty to animals and the institutional, socially acceptable human domination of animals in agribusiness, in slaughterhouses and abattoirs, in so-called scientific experimentation and, in less obviously direct ways, in sports, colleges and schools, zoos, aquaria and circuses" (Beirne and South, 2007, xiv) (see Box 20.4). Thus **green criminology** encompasses a broad range of behaviours ranging from acts that are clearly harmful, such as dumping toxic waste into the ocean, to acts that many people consider to be acceptable, such as eating meat or wearing leather shoes.

green criminology A branch of criminology that encompasses a broad range of behaviours ranging from acts that are clearly harmful, such as dumping toxic waste into the ocean, to acts that many people consider to be acceptable, such as eating meat or wearing leather shoes.

Environmental changes, such as climate change, may have a direct impact on other types of crime (Agnew, 2012). Climate change will have a serious negative impact on the global economy. This will likely mean that people will aggressively compete for scarce resources such as food and water, and this competition may well involve criminality. Climate change may force huge numbers of people to migrate from rural areas to urban slums and to refugee camps. Social controls are weak in these areas, so they may have high rates of crime. Climate change will also cause greater inequality between rich and poor, and inequality is associated with higher levels of crime. (Portions of this section are taken from Linden, Rick. *Criminology: A Canadian Perspective*. Toronto: Nelson, 2016:22–24.)

ENVIRONMENTAL ISSUES AND THE LAW

The Canadian government has been very reluctant to pass or enforce laws protecting the environment. For example, despite signing the Kyoto Accord regulating greenhouse gas emissions in 1997, successive Liberal and Conservative governments never put into place regulations that would ensure that emissions targets could be met, in large part because they believed such laws would hurt industrial development and the development of Alberta's oil sands. In 2015, when the Conservative government proposed a new set of targets of reducing emissions by 30 percent by 2030, they excluded the oil sands, the largest source of emissions, from these targets.

Even when public pressure forces governments to pass environmental laws—usually following major environmental disasters caused by negligence or by willful wrongdoing—they do little to enforce these regulations. Girard and colleagues (2010) found that between the 1990s and 2008, the number of environmental prosecutions—and particularly successful prosecutions—dropped steadily, while warnings and 'directives' increased dramatically. The researchers concluded this was because resources were limited, and the corporations that would be most affected by the regulations were given the opportunity to help develop the rules, to self-regulate their own activities, and to act voluntarily to improve their treatment of the environment. Many of Canada's most profitable industries, including oil and gas, mining, and forestry, cause the most environmental damage, so even environmentally conscious governments at all levels are torn between the responsibility to preserve the environment and the desire to maintain jobs and corporate profits. The fact that corporations are far more likely to have governments' attention than environmentalists are means that government policies will more often reflect corporate interests.

The work of green criminologists is grounded in the philosophy of ecological citizenship. This means that notions of morality and rights should be extended to "non-human nature" (White, 2007, 35), and that societies should adopt a notion of ecological citizenship obliging them to recognize that the environment must be protected for future generations. This will require a global perspective because the effects of environmental crimes go far beyond the borders of any single country.

BOX 20.4 POINT/COUNTERPOINT

Animal Cruelty Laws in Canada

In 2015, a New York court considered the case of Leo and Hercules, two chimpanzees who were used in locomotion studies at Stony Brook University. Lawyers for the Nonhuman Rights Project argued that the university did not have the right to confine the animals for research purposes. The government argued that giving the chimps the same rights as people would open the door to granting rights to other animals. The court decided that the chimpanzees did not have these rights but the university agreed to stop using the animals in any future experiments.

There is a wide range of views on this issue. Even among animal rights supporters, there are differences of opinion between those who advocate better treatment for animals to those who feel they should have rights similar to those of humans. One of these groups would still allow the exploitation of animals for food, while the other takes a much more animal-friendly approach.

What do you think about this issue? Do you agree with Angus Taylor who argues that "If we ascribe certain moral rights to humans on the basis of particular qualities such as the ability to suffer or the capacity for self-awareness, then we cannot deny those rights to nonhumans who possess the same qualities" (2015:12). Or do you think that animals are commodities whose only right is not to be subjected to outright cruelty?

Canada has made only minor changes in its animal cruelty legislation since 1892, and there is widespread agreement that stronger laws are needed to prevent animal abuse. Many critics of the current legislation cite an Edmonton case in which two men tied a dog to a tree and beat it to death with a baseball bat. The men were not convicted of animal cruelty because the evidence showed that the dog had died when it was first hit with the bat so it did not suffer cruelty. It is also very difficult to get convictions for people who neglect their animals because the Crown must prove that the neglect is "willful." Thus a farmer whose animals have starved to death will be acquitted unless it can be proved that he acted willfully. As a result, very few people in Canada have been convicted of animal abuse.

New animal cruelty laws have been brought before Parliament since 1999. However, the proposed legislation has been opposed by hunters, trappers, farmers, and medical researchers who experiment on animals and who fear that the laws would affect their livelihoods. The agriculture industry was also concerned about stronger legislation, particularly those involved in industrial food production where animals such as pigs and chickens are raised in small cages with no contact with the outside world. The current law prohibits brutal behaviour toward farm animals. For example, the gratuitous violence inflicted upon cattle at the Chilliwack Cattle Sales—Canada's largest dairy farm—was universally condemned and should be criminal, although a year after the 2014 release of video showing the violence, no charges had been laid against the workers. However, a 1957 court decision means that if the violence is committed for a socially approved purpose, such as slaughtering an animal for food, it is legal (Sykes, 2015). The act in question in this case was hoisting a hog high in the air by means of a shackle on one leg, and slamming it into a wall in order to stun it prior to cutting its throat—a practice that was apparently common at the time.

In 2008, Parliament was faced with two competing bills. The first bill (S-203) involved minimal changes beyond making the penalties tougher for existing offences. The second bill (C-229) would have added significant protection for animals. It would have removed the 'willful neglect' provision and would have made it more difficult to kill stray animals. The second bill was supported by many groups, including the Canadian Veterinary Medical Association and virtually all of Canada's humane societies and animal support organizations. On the other side were groups such as the Canadian Sportfishing Industry Association, which claimed that the proposed legislation would jeopardize the $10 billion-a-year sportfishing industry by making it "possible for a Grandfather to face a federal criminal prosecution for taking his grandchildren fishing" (Canadian Sportfishing Industry Association, 2007). Other opponents claimed that the bill would give animals the same legal standing as humans and would encourage animal rights "terrorists" to keep attacking medical researchers (Senate Committee on Legal and Constitutional Affairs, 2006). Proponents of Bill C-229 argued that the bill excluded harming animals for lawful reasons such as hunting and medical experimentation, but this interpretation was challenged by opponents. The Conservative government was able to pass Bill S-203 in 2008, but the bill that would do much more to protect animals (C-229) has not been passed (Linden, 2016).

Source: Linden, Rick. *Criminology: A Canadian Perspective* (8th ed.). Toronto: Nelson, p. 23.

TIME TO REVIEW

- Why do you think environmental crimes are treated less seriously than other types of crime that cause less harm to society?
- How will changes in the environment affect the nature and amount of crime in the future?

LO-1 Understand what is meant by environmental sociology.

Environmental sociology has been defined as the study of the interaction between human society and the physical environment.

MARCOS TOWNSEND/AFP/ Getty Images

LO-2 Compare how the treadmill theory of production, ecological modernization theory, ecofeminism, and constructionism differ in their analysis of the environment.

The treadmill theory of production looks at the impact of an endless cycle of growth and accumulation on the environment; ecological modernization theory predicts that cultural and technological changes driven by economic growth will mitigate the environmental effects of that growth; ecofeminists are concerned with the way capitalism and patriarchy have combined to dominate and oppress both humans and the natural world; and constructionists examine the way in which environmental issues are socially constructed.

The Granger Collection, New York

LO-3 Understand the primary ways in which the environment affects our health.

Environmental change, exposure to environmental toxins, and environmental destruction and overconsumption can have significant and devastating effects on the health of human populations. Climate change has caused numerous health issues, including illness and injury from extreme weather events, respiratory problems caused by air pollution, an increase in infectious diseases spread by carriers such as insects and rodents and by food insecurity, malnutrition, and starvation.

chris kolaczan/Shutterstock.com

Key Terms

constructionism A theory that focuses on the way in which environmental issues are socially constructed (p. 555).

ecofeminism Sociological theory that the ideology that authorizes oppressions such as those based on race, class, gender, sexuality, physical abilities, and species is the same ideology that sanctions the oppression of nature (p. 554).

ecological modernization theory Environmental sociological theory that societies are adapting to the threat of environmental harms by reforming societal practices and policies (p. 553).

environmental justice The idea that all people and all communities are entitled to equal protection under environmental health laws and regulations (p. 561).

environmental racism The targeting of racial minority communities for toxic waste facilities, the official sanctioning of poisons and pollutants in minority communities, and the systematic exclusion of racial minorities from decision making regarding the production of environmental conditions that affect their lives and livelihoods (p. 562).

environmental sociology The study of the interaction between human society and the physical environment (p. 548).

green criminology A branch of criminology that encompasses a broad range of behaviours ranging from acts that are clearly harmful, such as dumping toxic waste into the ocean, to acts that many people consider to be acceptable, such as eating meat or wearing leather shoes (p. 567).

New Ecological Paradigm An alternative sociological approach, which recognizes that the social world does not exist on its own but is embedded in the natural world (p. 548).

treadmill of accumulation The never-ending process of accumulating more and more profit (p. 552).

treadmill of production A cycle in which resources are taken out of the environment to produce goods, and waste and pollution are deposited back into the environment (p. 552).

LO-4 Explain how human populations are differentially and unequally affected by challenges in the environment.

Environmental changes are having a negative impact on the health and safety of millions of people. These impacts are disproportionately borne by the world's most disadvantaged populations, in particular, the poor, and racial minorities.

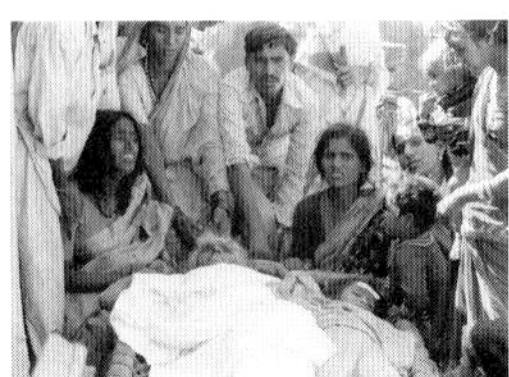

BEDI/AFP/Getty Images

LO-5 Discuss the emerging field of green criminology.

Green criminology encompasses a broad range of behaviours, ranging from acts that are clearly harmful, such as dumping toxic waste into the ocean, to acts that many people consider to be acceptable, such as eating meat or wearing leather shoes.

Brendan Kennedy/GetStock.com

APPLICATION QUESTIONS

1. Technology, government, and consumers all have a role to play in reducing energy consumption. What factors within each of these categories must change in order for solar energy to play a significant role in reducing our reliance on carbon-based energy? What changes will have to occur before electric cars can become more widely used?
2. Discuss some of the factors that helped to construct the environmental problem of global warming; that is, how have some people come to define global warming as a problem that must be solved and how have others opposed this initiative?
3. Both authors of this chapter live with household pets—in our cases, dogs and cats who are treated like important members of the family rather than as potential sources of meat. As we are finishing this chapter, the media are reporting global outrage over the dog meat festival in Yulin, China (note: do not Google this if you are a dog lover). One of the authors—who is not a vegetarian—recently turned down the chance to order rabbit as a dinner item. Why are we able to eat certain types of animals and find the thought of eating others to be offensive? Drawing upon your sociological knowledge, can you explain why we can treat some animals as part of the family and happily put barbecue sauce on others? Do you think that animal lovers should be vegetarians?
4. Major changes will be required if the world is going to avoid the most severe consequences of global warming. The authors argue that these changes will be as much sociological as they are technological. Explain why you agree or disagree with this view.

CHAPTER 21

COLLECTIVE BEHAVIOUR, SOCIAL MOVEMENTS, AND SOCIAL CHANGE

LEARNING OBJECTIVES

AFTER READING THIS CHAPTER, YOU SHOULD BE ABLE TO

LO-1 Understand how collective behaviour is defined and how collective behaviour leads to social change.

LO-2 Identify the common forms of collective behaviour.

LO-3 Describe the primary forms of mass behaviour.

LO-4 Distinguish between different types of social movements.

LO-5 Understand the theoretical explanations of social movements.

CHAPTER FOCUS QUESTION

How might collective behaviour and social movements make people more aware of important social issues such as environmental destruction and global warming?

ANDRE FORGET/QMI AGENCY

In 2011, Senate Page Brigette DePape disrupted the throne speech with a "Stop Harper" sign in protest of the prime minister's policies. In the following narrative, she explains what motivated her to protest. She then addresses the importance of youth involvement in grassroots activism:

You might have seen the photo of the Senate Page holding up the Stop Harper sign that disrupted the Speech from the Throne. That was me! I shifted paths. I was planning on a career as a politician. But I decided to change directions and jump into the hotbed of grassroots activism, and I'm anxious to share what it's been like for me and what I've learnt.

As a Page in the Senate, I saw how people in power were not representing our interests as young people. I saw our government expand the tar sands and fuel climate change with its billions of dollars to an already rich oil industry. It's our future they are gambling with, selling it to the highest bidder. I saw how climate change bills were rejected, and choices made to cater to corporate interests rather than ours, spending money to fuel climate change, rather than on want and need, like alleviating our massive debt loads and providing education for all of us. I realized that it is not by taking power in a system made to maintain the status quo that I'd make the most impact, but by taking part in building the movement we need to shift power back to all of us for a safe, democratic, and equitable present and future.

After the action in the Senate, a lot of people came up to me and said they were happy to see a young person act because youth in Canada are so apathetic. But we are not just about our Macbooks and caramel lattes! The youth I've met across Canada are engaged, motivated, pissed off, and fearless. More than that, I believe they're at the forefront of a really important transformation of our country.

Certainly, many young people in Canada are not engaged. Why aren't youth even voting? Why aren't youth taking the streets in the same numbers of the 60s and 70s? It's easy to get the wrong impression—that we're apathetic. But young people in Canada care deeply. The problem is, we are trapped in a structure that bars us from meaningful engagement. Young people don't have time to fight the system; they are too busy working two jobs to pay for tuition, or working a corporate job they don't want but need to take to get out of the thousands of dollars of debt they have accumulated. Young people are challenging these structural barriers, like the student federation's campaign to drop the debt and achieve education for all, and the Occupy movement working to rethink and change the system.

Historically, youth have been at the forefront of change. Take for example the four young students who staged the first sit-ins in the Southern U.S. to end racial segregation of lunch counters. The student non-violent coordinating committee played an integral role in the Civil Rights movement. And youth continue to be at the forefront of movements across the world, from the millions of students in Chile staging occupations for free and fair education, to the April 6th movement in Egypt that led the mobilizations in Tahrir Square which eventually overthrew a dictator.

People in Canada are part of this incredible movement that is sweeping the globe. We are living at an unprecedented moment in history—from Occupy/de-Occupy to the protests against the Keystone XL and Enbridge pipelines. Our time has come to shake things up. (DePape, 2012:18–20)

Source: Brigette DePape, "Power of Youth: Youth and Community-Led Activism in Canada," *Our Schools / Ourselves*, Spring 2012, pp. 18-20.

Demonstrations and protests are a daily part of life in the 21st century. Almost daily, social media, the Internet, TV, and newspapers inform us of new or unresolved social, political, or environmental problems. As the chapter's opening narrative indicates, the message of social movements is that people should act collectively and immediately to help reduce these problems. In other words, social change is essential. Sociologists define **social change** as the alteration, modification, or transformation of public policy, culture, or social institutions over time. Social change is usually brought about by collective behaviour and social movements. But how much social change can the individual bring about?

In this chapter, we will examine collective behaviour, social movements, and social change from a sociological perspective. We will use numerous examples, including the Occupy movement, Idle No More, and environmental activism (also known as the Green movement), to show how people use social movements as a form of mass mobilization and social transformation. For example, the Occupy movement, which began in 2011 as Occupy Wall Street in Manhattan, referred to itself as a leaderless resistance movement comprising diverse people from all backgrounds and political affiliations. The group claimed on its website to represent 99 percent of the U.S. population while standing up against the wealthiest 1 percent in the nation, which has been labelled as greedy and corrupt and a sure sign that capitalism has run amok. That same year, thousands of Canadians in over 20 cities joined forces in the Occupy movement. Similarly, the "Go-Green" movement has become popular in recent years; however, it should be noted that the environmental movement can be traced back to the 1960s and 1970s, when many people were actively involved in collective behaviour, making demands that efforts be increased to protect the environment and to save our planet. In fact, some ideas expressed by the "Go-Green" movement can be traced back to the 1930s or earlier. Thus, collective behaviour and social movements may come and go over lengthy periods of time, but the themes of concern may remain similar. Before reading on, test your knowledge about collective behaviour and environmental issues by taking the quiz in Box 21.1.

CRITICAL THINKING QUESTIONS

1. What social issues would most motivate you to engage in a protest movement?
2. What results would you most like to see?
3. Would the experiences of protestors in the Occupy movement, the G20 Summit in Toronto, or students in Quebec dissuade you from participating in a non-violent protest? Why, or why not?

social change The alteration, modification, or transformation of public policy, culture, or social institutions over time.

LO-1 COLLECTIVE BEHAVIOUR

collective behaviour Voluntary, often spontaneous activity that is engaged in by a large number of people and typically violates dominant-group norms and values.

Collective behaviour is voluntary, often spontaneous activity that is engaged in by a large number of people and typically violates dominant-group norms and values. Unlike the *organizational behaviour* found in corporations and voluntary associations, such as environmental organizations, collective behaviour lacks an official division of labour, hierarchy of authority, and established rules and procedures. Unlike *institutional behaviour* (in education, religion, or politics, for example), it lacks institutionalized norms to govern behaviour. Collective behaviour can take various forms, including crowds, mobs, riots, panics, fads, fashions, and public opinion.

Early sociologists studied collective behaviour because they lived in a world that was responding to the processes of modernization, including urbanization, industrialization, and proletarianization of workers. Contemporary forms of collective behaviour, such as the Occupy Wall Street movement and similar social protests, are variations on the theme that originated during the transition from feudalism to capitalism and the rise of modernity in Europe. Some forms of collective behaviour and social movements are directed toward public issues, such as air pollution, water pollution, and the exploitation of workers in global sweatshops by transnational corporations.

In 2011, over 300,000 peaceful protesters gathered in Tahrir Square in Egypt.

CONDITIONS FOR COLLECTIVE BEHAVIOUR

Collective behaviour occurs as a result of some common influence or stimulus that produces a response from a collectivity. A *collectivity* is a relatively large number of people who mutually transcend, bypass, or subvert established institutional patterns and structures. Three major factors contribute to the likelihood that collective behaviour will occur: (1) structural factors that increase the chances of people responding in a particular way, (2) timing, and (3) a breakdown in social control mechanisms and a corresponding feeling of normlessness.

A common stimulus is an important factor in collective behaviour. For example, the Occupy movement came at a time when people were becoming more concerned about social issues and beginning to see that they could empower themselves through grassroots activism. Similarly,

BOX 21.1 SOCIOLOGY AND EVERYDAY LIFE

How Much Do You Know About Collective Behaviour and Environmental Issues?

True	False	
T	F	1. The environmental movement in North America started in the 1960s.
T	F	2. A number of social movements in North America are becoming globalized.
T	F	3. Environmental groups may engage in civil disobedience or use symbolic gestures to call attention to their issue.
T	F	4. Influencing public opinion is an important activity for many social movements.
T	F	5. Social movements are more likely to flourish in democratic societies.

For more questions and the answers to the quiz about collective behaviour and environmental issues, go to **www.nelson.com/student**.

protest was inevitable at the G20 Summit in Toronto in 2010. The exclusive focus on economic issues made it an obvious target for protesters increasingly concerned about the human rights and social costs of economic globalization.

Timing and a breakdown in social control mechanisms are also important in collective behaviour. Since the 1960s, most urban riots in Canada and the United States have begun in the evenings or on weekends, when most people are off work. More recently, a report released after the 2011 Vancouver Stanley Cup riot indicated that a major contributor to the riot was that Vancouver police underestimated the number of people who would attend the event in the downtown area and consequently had insufficient police resources in place to manage the crowd. Once the crowd became unruly, the police force's communications systems failed, leading to a complete loss of control of the situation by Vancouver police.

TIME TO REVIEW

- When is collective behaviour likely to occur?

DISTINCTIONS REGARDING COLLECTIVE BEHAVIOUR

People engaging in collective behaviour may be divided into crowds and masses. A **crowd** is a relatively large number of people who are in one another's immediate vicinity (Lofland, 1993). Examples of crowds include the audience at a movie theatre or fans at a sporting event. In contrast, a **mass** is a large number of people who share an interest in a specific idea or issue but are not in one another's immediate vicinity (Lofland, 1993). To further distinguish between crowds and masses, think of the difference between a riot and a rumour: People who participate in a riot must be in the same general location; those who spread a rumour may be thousands of kilometres apart, communicating by telephone or through the Internet.

crowd A relatively large number of people who are in one another's immediate vicinity.

mass A large number of people who share an interest in a specific idea or issue but are not in another's immediate physical vicinity.

Collective behaviour may also be distinguished by the dominant emotion expressed. According to sociologist John Lofland, the *dominant emotion* refers to the "publicly expressed feeling perceived by participants and observers as the most prominent in an episode of collective behaviour" (1993:72). Lofland suggests that fear, hostility, and joy are three fundamental emotions found in collective behaviour; however, grief, disgust, surprise, or shame may also predominate in some forms of collective behaviour.

LO-2 TYPES OF CROWD BEHAVIOUR

When we think of a crowd, many of us think of *aggregates,* previously defined as a collection of people who happen to be in the same place at the same time but have little else in common. The presence of a relatively large number of people in the same location, however, does not necessarily produce collective behaviour. Sociologist Herbert Blumer (1946) developed a typology in which crowds are divided into four categories: casual, conventional, expressive, and acting. Other scholars have added a fifth category: protest crowds.

CASUAL AND CONVENTIONAL CROWDS *Casual crowds* are relatively large gatherings of people who happen to be in the same place at the same time; if they interact at all, it is only briefly. People in a shopping mall or on a bus are examples of casual crowds. Other than sharing a momentary interest, such as watching a busker perform on the street or observing the aftermath of a car accident, a casual crowd has nothing in common. The casual crowd plays no active part in the event—such as the car accident—which would have occurred whether or not the crowd was present; the crowd simply observes.

Conventional crowds are made up of people who specifically come together for a scheduled event and thus share a common focus. Examples include religious services, graduation ceremonies, concerts, and university lectures. Each of these events has established schedules and norms. Because these events occur regularly, interaction among participants is much more likely; in turn, the events would not occur without the crowd, which is essential to the event.

EXPRESSIVE AND ACTING CROWDS *Expressive crowds* provide opportunities for the expression of some strong emotion, such as joy, excitement, or grief. People release their pent-up emotions in conjunction with other persons experiencing similar emotions. Examples include worshippers at religious revival services or mourners lining the streets when a celebrity, public official, or religious leader has died. *Acting crowds* are collectivities so intensely focused on a specific purpose or object that they may erupt into violent or destructive behaviour. Mobs, riots, and panics are examples of acting crowds, but casual and conventional crowds may become acting crowds under some circumstances. A **mob** is a highly emotional crowd whose members engage in, or are ready to engage in, violence against a specific target, which may be a person, a category of people, or physical property. Mob behaviour in this country has included fire bombings, effigy hangings, and hate crimes. Mob violence tends to dissipate relatively quickly once a target has been injured, killed, or destroyed. Sometimes, actions, such as effigy hanging, are used symbolically by groups that otherwise are not violent.

▶ **mob** A highly emotional crowd whose members engage in, or are ready to engage in, violence against a specific target, which may be a person, a category of people, or physical property.

▶ **riot** Violent crowd behaviour that is fuelled by deep-seated emotions but not directed at one specific target.

Compared with mob action, riots may be of somewhat longer duration. A **riot** is violent crowd behaviour that is fuelled by deep-seated emotions but not directed at one specific target. Riots are often triggered by fear, anger, and hostility. For example, the 2011 riot in Vancouver was fuelled by anger and disappointment over the Canucks' loss to the Boston Bruins in game 7

Jason Payne/Vancouver Sun

The Vancouver riots in 2011 resulted in hundreds of people being injured and massive property damage to the downtown Vancouver area.

of the Stanley Cup final. However, not all riots are caused by hostility and hatred. People may be expressing joy and exuberance when rioting occurs.

A **panic** is a form of crowd behaviour that occurs when a large number of people react to a real or perceived threat with strong emotions and self-destructive behaviour. The most common type of panic occurs when people seek to escape from a perceived danger, fearing that few (if any) of them will be able to get away from that danger. Examples include passengers on a sinking cruise ship or persons in a burning nightclub. Panics can also arise in response to larger social, financial, or political conditions that people believe are beyond their control—such as a major disruption in the economy. A "bank run" in which hundreds or thousands of customers seek to take out all of their money at the same time, fearing that the financial institution is becoming insolvent, is an example. Although panics are relatively rare, they receive massive media coverage because they provoke strong feelings of fear in readers and viewers, and the number of casualties may be large. An example was the tragic 2013 garment factory blaze in Bangladesh; more than 100 workers died as fire gutted the large manufacturing warehouse and panic-stricken workers jammed a stairwell trying to escape their workplace.

▶ **panic** A form of crowd behaviour that occurs when a large number of people react to a real or perceived threat with strong emotions and self-destructive behaviour.

PROTEST CROWDS *Protest crowds* engage in activities intended to achieve specific political goals. Examples include sit-ins, marches, boycotts, blockades, and strikes. Some protests sometimes take the form of **civil disobedience**—nonviolent action that seeks to change a policy or law by refusing to comply with it. Acts of civil disobedience may become violent, as in a confrontation between protesters and police officers; in this case, a protest crowd becomes an *acting crowd*.

▶ **civil disobedience** Non-violent action that seeks to change a policy or law by refusing to comply with it.

EXPLANATIONS OF CROWD BEHAVIOUR

What causes people to act collectively? How do they determine what types of action to take? One of the earliest theorists to provide an answer to these questions was Gustave Le Bon, a French scholar who focused on crowd psychology in his contagion theory.

CONTAGION THEORY *Contagion theory* focuses on the social–psychological aspects of collective behaviour; it attempts to explain how moods, attitudes, and behaviour are communicated rapidly and why they are accepted by others. Le Bon (1841–1931) argued that people are more likely to engage in antisocial behaviour in a crowd because they are anonymous and feel invulnerable. Le Bon (1960/1895) suggested that a crowd takes on a life of its own that is larger than the beliefs or actions of any one person. Because of its anonymity, the crowd transforms individuals from rational beings into a single organism with a collective mind. In essence, Le Bon asserted that emotions, such as fear and hate, are contagious in crowds because people experience a decline in personal responsibility; they will do things as a collectivity that they would never do when acting alone.

SOCIAL UNREST AND CIRCULAR REACTION Robert E. Park was the first U.S. sociologist to investigate crowd behaviour. Park believed that Le Bon's analysis of collective behaviour lacked several important elements. Intrigued that people could break away from the powerful hold of culture and their established routines to develop a new social order, Park added the concepts of social unrest and circular reaction to contagion theory. According to Park, social unrest is transmitted by a process

Phillip Chin/WireImage/Getty Images

Convergence theory is based on the assumption that crowd behaviour involves shared emotions, goals, and beliefs. An example is this We Day event, which brings together global leaders, social activists, entertainers, and young people to promote global change.

of *circular reaction*—the interactive communication between persons such that the discontent of one person is communicated to another, who, in turn, reflects the discontent back to the first person (Park and Burgess, 1921).

CONVERGENCE THEORY *Convergence theory* focuses on the shared emotions, goals, and beliefs that many people bring to crowd behaviour. Because of their individual characteristics, many people have a predisposition to participate in certain types of activities (Turner and Killian, 1993). From this perspective, people with similar attributes find a collectivity of like-minded persons with whom they can express their underlying personal tendencies. Although people may reveal their "true selves" in crowds, their behaviour is not irrational; it is highly predictable to those who share similar emotions or beliefs.

Convergence theory has been applied to a wide array of conduct, from lynch mobs to environmental movements. In a study of a lynching in the United States, social psychologist Hadley Cantril (1941) found that the participants shared certain common attributes: They were poor and working-class whites who felt that their own status was threatened by the presence of successful African Americans. Consequently, the characteristics of these individuals made them susceptible to joining a lynch mob even if they did not know the target of the lynching.

Convergence theory adds to our understanding of certain types of collective behaviour by pointing out how individuals may have certain attributes that initially bring them together, such as racial hatred or fear of environmental problems that directly threaten them. However, this perspective does not explain how the attitudes and characteristics of individuals who take some collective action differ from those who do not.

EMERGENT NORM THEORY Unlike contagion and convergence theories, *emergent norm theory* emphasizes the importance of social norms in shaping crowd behaviour. Drawing on the interactionist perspective, sociologists Ralph Turner and Lewis Killian asserted that crowds develop their own definition of a situation and establish norms for behaviour that fit the occasion:

> Some shared redefinition of right and wrong in a situation supplies the justification and coordinates the action in collective behaviour. People do what they would not otherwise have done when they panic collectively, when they riot, when they engage in civil disobedience, or when they launch terrorist campaigns, because they find social support for the view that what they are doing is the right thing to do in the situation. (1993:12)

According to Turner and Killian, emergent norms occur when people define a new situation as highly unusual or see a long-standing situation in a new light (1993:13).

Sociologists use the emergent norm approach to determine how individuals in a given collectivity develop an understanding of what is going on, how they construe these activities, and what types of norms are involved. For example, in a study of audience participation, sociologist Steven E. Clayman (1993) found that members of an audience listening to a speech applaud promptly and independently but wait to coordinate their booing with other people—they do not wish to "boo" alone.

Some emergent norms are permissive—that is, they give people a shared conviction that they may disregard ordinary rules, such as waiting in line, taking turns, or treating a speaker courteously. Collective activity, such as mass looting, may be defined (by participants) as taking what rightfully belongs to them and punishing those who have been exploitative. In the aftermath of the 2010 Haiti earthquake, when relief aid was slow in coming, looting was commonplace, but so too was "mob justice" for those who were caught looting other people's possessions.

Joshua Lott/Getty Images

What type of crowd behaviour is occurring in this photo? Which explanation of crowd behaviour would you use to explain this occurrence?

Emergent norm theory points out that crowds are not irrational. Rather, new norms are developed in a rational way to fit the needs of the immediate situation. Critics, however, note that proponents of this perspective fail to specify exactly what constitutes a norm, how new ones emerge, and how they are so quickly disseminated and accepted by a wide variety of participants. One variation of this theory suggests that no single dominant norm is accepted by everyone in a crowd; instead, norms are specific to the various categories of actors rather than to the collectivity as a whole (Snow, Zurcher, and Peters, 1981).

TIME TO REVIEW

- Name five different types of crowds and provide recent examples other than those already identified.
- Which of these types of crowds occur most often? Why?
- Which of the theoretical explanations of crowd behaviour best apply to the protests that occurred across the United States and Canada when a grand jury decided to not charge the white officer who shot unarmed youth Michael Brown?

LO-3 MASS BEHAVIOUR

mass behaviour Collective behaviour that takes place when people (who are often geographically separated from one another) respond to the same event in much the same way.

Not all collective behaviour takes place in face-to-face collectivities. **Mass behaviour** is collective behaviour that takes place when people (who are often geographically separated from one another) respond to the same event in much the same way. For people to respond in the same way, they typically have common sources of information, and this information provokes their collective behaviour. The most frequent types of mass behaviour are rumours, gossip, mass hysteria, fads, fashions, and public opinion. Under some circumstances, social movements constitute a form of mass behaviour. However, we will examine social movements separately because they differ in some important ways from other types of dispersed collectivities.

rumour An unsubstantiated report on an issue or subject.

RUMOURS AND GOSSIP **Rumours** are unsubstantiated reports on an issue or subject. Rumours may spread through an assembled collectivity, but they may also be transmitted among people who are dispersed geographically, including people spreading rumours on Twitter or posting messages on Facebook or talking by cellphone. Although they may initially contain a kernel of truth, as they spread, rumours may be modified to serve the interests of those repeating them. Rumours thrive when tensions are high and little authentic information is available on an issue of great concern. For example, a week after the terrorist explosion at the 2013 Boston Marathon, a fake Twitter feed of the Associated Press stating that President Barack Obama had been injured in an explosion at the White House caused the stock market to take a sharp decline, wiping out $130 billion in stock value in a matter of seconds. Although the market recovered the value, the power of rumour in an age of rapid communication and social media was proven in a new and frightening manner (Matthews, 2013).

Why do people believe rumours? People are willing to give rumours credence when no offsetting information is available. Once rumours begin to circulate, they seldom stop unless compelling information comes to the forefront that either proves them false or makes them obsolete. In industrialized societies with sophisticated technology, rumours come from a wide variety of sources and may be difficult to trace. Print media (newspapers and magazines) and electronic media (radio and television), fax machines, cellular networks, satellite systems, and the Internet facilitate the rapid movement of rumours around the globe. In addition, modern communications technology makes anonymity much easier. In a split second, messages (both factual and fictitious) can be disseminated to millions of people through various forms of social media.

gossip Rumours about the personal lives of individuals.

Whereas rumours deal with an issue or a subject, **gossip** refers to rumours about the personal lives of individuals. Charles Horton Cooley (1962/1909) viewed gossip as something that spread among a small group of individuals who personally knew the person who was the object of the rumour. Today, this is often not the case; many people enjoy gossiping about people they have never met. Internet sites such as Perez Hilton and Gawker, tabloids and magazines such as the *National Enquirer* and *People,* and television "news" programs, such as *Entertainment Tonight,* that purport to provide "inside" information on the lives of celebrities, are sources of contemporary gossip, much of which has not been checked for authenticity.

mass hysteria A form of dispersed collective behaviour that occurs when a large number of people react with strong emotions and self-destructive behaviour to a real or perceived threat.

MASS HYSTERIA AND PANIC **Mass hysteria** is a form of dispersed collective behaviour that occurs when a large number of people react with strong emotions and self-destructive behaviour to a real or perceived threat. Does mass hysteria really occur? Although the term has been widely used, many sociologists believe this behaviour is best described as panic with a dispersed audience. You will recall that panic is also a form of crowd behaviour that occurs when a large number of people react with strong emotions and self-destructive behaviour to a real or perceived threat.

An example of mass hysteria or panic with a widely dispersed audience occurred on Halloween evening 1938 when actor Orson Welles presented a radio dramatization of the science fiction classic *The War of the Worlds* by H.G. Wells. A dance music program on CBS radio was interrupted suddenly by a news bulletin informing the audience that Martians had landed in New Jersey and were aiming to conquer Earth. Some listeners became extremely frightened, even though an announcer had indicated before, during, and after the performance that the broadcast was a fictitious dramatization. According to some reports, as many as one million of the estimated 10 million listeners believed that this astonishing event had occurred. Thousands were reported to have hidden in their storm cellars or gotten in their cars so they could flee from the Martians (see Brown, 1954). The program probably did not generate mass hysteria, but rather created panic among gullible listeners. Others switched stations to determine if the same "news" was being broadcast elsewhere. When they discovered that it was not, they merely laughed at the joke being played on listeners by CBS. In 1988, on the 50th anniversary of the broadcast, a Portuguese radio station rebroadcast the program, and once again panic ensued.

FADS AND FASHIONS A **fad** is a temporary but widely copied activity, enthusiastically followed by large numbers of people. Fads can be embraced by widely dispersed collectivities: TV, the Internet, and social media bring the latest fads—such as top memes, YouTube videos of cats or dogs doing funny things, zombie films—to the attention of audiences around the world.

fad A temporary but widely copied activity enthusiastically followed by large numbers of people.

fashion A currently valued style of behaviour, thinking, or appearance.

Fashions tend to last longer than fads. **Fashion** may be defined as a currently valued style of behaviour, thinking, or appearance. Fashion also applies to art, drama, music, literature, architecture, interior design, and automobiles, among other things. However, most sociological research on fashion has focused on clothing, especially women's apparel (Davis, 1992).

In preindustrial societies, clothing styles remained relatively unchanged. With the advent of industrialization, however, items of apparel became readily available at low prices because of mass production. Fashion became more important as people embraced the "modern" way of life and advertising encouraged "conspicuous consumption."

Georg Simmel, Thorstein Veblen, and French sociologist Pierre Bourdieu have all viewed fashion as a way to differentiate status among members of different social classes. Simmel (1904) suggested a classic "trickle-down" theory (although he did not use those exact words) to describe the process by which members of the lower classes emulate the fashions of the upper class. As the fashions descend through the status hierarchy, they are watered down and "vulgarized" so that they are no longer recognizable to members of the upper class, who then regard them as unfashionable and in bad taste (Davis, 1992). Veblen (1967/1899) asserted that fashion served mainly to institutionalize conspicuous consumption among the wealthy. Almost 80 years later, Bourdieu (1984) similarly (but most subtly) suggested that "matters of taste," including fashion sensibility, constitute a large share of the "cultural capital" (or social assets) possessed by members of the dominant class.

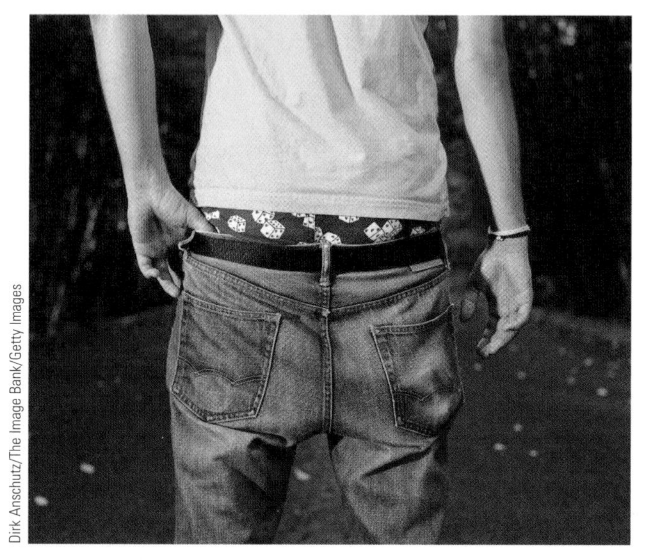

Dirk Anschutz/The Image Bank/Getty Images

javi_indy/Shutterstock.com

Are these fads or fashions?

Herbert Blumer (1969) disagreed with the trickle-down approach, arguing that "collective selection" best explains fashion. Blumer suggested that people in the middle and lower classes follow fashion because it is *fashion,* not because they desire to emulate members of the elite class. Blumer thus shifts the focus on fashion to collective mood, states, and choices: "Tastes are themselves a product of experience . . . They are formed in the context of social interaction, responding to the definitions and affirmation given by others. People thrown into areas of common interaction and having similar runs of experience develop common tastes" (quoted in Davis, 1992:116). Perhaps one of the best refutations of the trickle-down approach is the way in which fashion today often originates among people in the lower social classes and is mimicked by the elites. The mid-1990s grunge look was a prime example of this.

▶ **public opinion** The political attitudes and beliefs communicated by ordinary citizens to decision makers.

PUBLIC OPINION **Public opinion** consists of the political attitudes and beliefs communicated by ordinary citizens to decision makers. It is measured through polls and surveys, which utilize research methods, such as interviews and questionnaires, as described in Chapter 2. Many people are not interested in all aspects of public policy but are concerned about issues they believe are relevant to them. Even on a single topic, public opinion will vary widely based on race and ethnicity, religion, region, social class, education level, gender, age, and so on.

Scholars who examine public opinion are interested in the extent to which the public's attitudes are communicated to decision makers and the effect (if any) that public opinion has on policy making (Turner and Killian, 1993). Some political scientists argue that public opinion has a substantial effect on decisions at all levels of governments.

▶ **propaganda** Information provided by individuals or groups that have a vested interest in furthering their own cause or damaging an opposing one.

As the masses attempt to influence elites and vice versa, a two-way process occurs with the dissemination of **propaganda**—information provided by individuals or groups that have a vested interest in furthering their own cause or damaging an opposing one. Although many of us think of propaganda in negative terms, the information provided can be correct and can have positive effects on decision making.

LO-4 SOCIAL MOVEMENTS

▶ **social movement** An organized group that acts consciously to promote or resist change through collective action.

Although collective behaviour is short-lived and relatively unorganized, social movements are longer lasting and more organized and have specific goals or purposes. A **social movement** is an organized group that acts consciously to promote or resist change through collective action (Goldberg, 1991). Because social movements have not become institutionalized and are outside the political mainstream, they offer "outsiders" an opportunity to have their voices heard.

Social movements are more likely to develop in industrialized societies than in preindustrial societies, where acceptance of traditional beliefs and practices makes such movements unlikely. Diversity and a lack of consensus (hallmarks of industrialized nations) contribute to demands for social change, and people who participate in social movements typically lack power and other resources to bring about change without engaging in collective action. Social movements are most likely to spring up when people come to see their personal troubles as public issues that cannot be solved without a collective response.

Social movements make democracy more available to excluded groups (see Greenberg and Page, 1996). Historically, people in North America have worked at the grassroots level to bring about changes even when elites sought to discourage activism (Adams, 1991). For example, the civil rights movement brought into its ranks African Americans who had never been allowed to participate in politics (see Killian, 1984). The women's suffrage movement gave voice to women, who had been denied the right to vote (Rosenthal et al., 1985).

Social movements provide people who otherwise would not have the resources to enter the game of politics a chance to do so. We are most familiar with those movements that develop around public policy issues considered newsworthy by the media, ranging from abortion and women's rights to gun control and environmental justice. However, other types of social movements exist as well.

TYPES OF SOCIAL MOVEMENTS

Social movements are difficult to classify; however, sociologists distinguish among movements on the basis of their *goals* and the *amount of change* they seek to produce (Aberle, 1966; Blumer, 1974). Some movements seek to change people, while others seek to change society.

REFORM MOVEMENTS Grassroots environmental movements are an example of *reform movements*, which seek to improve society by changing some specific aspect of the social structure. Members of reform movements usually work within the existing system to attempt to change existing public policy so that it more adequately reflects their own value system. Examples of reform movements (in addition to the environmental movement) include labour movements, animal rights movements, antinuclear movements, and the disability rights movement.

Some social movements arise specifically to alter social response to and definitions of stigmatized attributes. From this perspective, social movements may bring about changes in societal attitudes and practices while at the same time causing changes in participants' social emotions. For example, the civil rights, LGBTQ rights, and Indigenous rights movements helped replace shame with pride (Britt, 1993).

REVOLUTIONARY MOVEMENTS Movements seeking to bring about a total change in society are referred to as *revolutionary movements*. These movements usually do not attempt to work within the existing system; rather, they aim to remake the system by replacing existing institutions with new ones. This was apparently the goal of the Occupy Wall Street movement, although it was unclear because it was leaderless and functioned with spokespersons who declined to express specific objectives for their activism. In the 21st century, people in Muslim countries around the world, including Tunisia, Egypt, and Iran, have participated in revolutionary movements and risen up against what they perceived to be tyrannical regimes. For example, people in Tunisia gathered in 2013 to call for unity and support for the resistance movement against tyranny and injustices perpetrated by the former regime of Zine El Abidine Ben Ali, the ousted president of that nation who had banned all public meetings and engaged in other forms of oppression.

Terrorism is acts of serious violence, planned and executed clandestinely and committed to achieve political ends. Movements based on terrorism often use tactics such as bombing, kidnapping, hostage taking, hijacking, and assassination. A number of movements in the United States have engaged in terrorist activities or supported a policy of violence. However, the terrorist attacks in New York City and Washington, D.C., on September 11, 2001, and the events that followed those attacks, demonstrated that terrorism can originate from the activities of extremists from both inside and outside the target.

▶ **terrorism** Acts of serious violence, planned and executed clandestinely and committed to achieve political ends.

Canada is not immune to terrorist activity. In the late 1960s, the Front de Libération du Québec (FLQ), a small group of extremists on the fringe of the separatist movement, carried out 200 bombings. In addition, Sikh separatists are believed to be responsible for the 1985 bombing of an Air India jet that was travelling to India from Canada, killing 329 people.

RELIGIOUS MOVEMENTS Social movements that seek to produce radical change in individuals are typically based on spiritual or supernatural belief systems. Also referred to as expressive movements, religious movements are concerned with renovating or renewing people through "inner change." Fundamentalist religious groups seeking to convert nonbelievers to their belief system are an example of this type of movement. Some religious movements are millenarian—they forecast that "the end is near" and assert that an immediate change in behaviour is imperative. Relatively new religious movements in industrialized Western societies have included the Hare Krishna sect, the Unification Church, Scientology, and the Divine Light Mission, all of which tend to appeal to the psychological and social needs of young people seeking meaning in life that mainstream religions have not provided for them.

ALTERNATIVE MOVEMENTS Movements that seek limited change in some aspect of people's behaviour are referred to as *alternative movements*. For example, in the early 20th century, the Woman's Christian Temperance Union attempted to get people to abstain from drinking alcoholic beverages. Some analysts place "therapeutic social movements," such as Alcoholics Anonymous, in this category; however, others do not, due to their belief that people must change their lives completely to overcome alcohol abuse (see Blumberg, 1977). More recently, a variety of "New Age" movements have directed people's behaviour by emphasizing spiritual consciousness, often combined with a belief in reincarnation and astrology. Such practices as vegetarianism, meditation, and holistic medicine are often included in the self-improvement category. Beginning in the 1990s, some alternative movements have included the practice of yoga (usually without its traditional background in Hindu religion) as a means by which the self can be liberated and union achieved with the supreme spirit or universal soul.

RESISTANCE MOVEMENTS Also referred to as *regressive movements, resistance movements* seek to prevent or undo change that has already occurred. Virtually all the proactive social movements previously discussed face resistance from one or more reactive movements that hold opposing viewpoints and want to foster public policies that reflect their own viewpoints. Examples of resistance movements are groups organized to oppose free trade, gun control, and restrictions on smoking.

STAGES IN SOCIAL MOVEMENTS

Do all social movements go through similar stages? Not necessarily, but there appear to be identifiable stages in virtually all movements that succeed beyond their initial phase of development.

In the *preliminary* (or incipiency) *stage,* widespread unrest is present as people begin to become aware of a problem. At this stage, leaders emerge to agitate others into taking action. In the *coalescence stage,* people begin to organize and to publicize the problem. At this stage, some movements become formally organized at local and regional levels. In the *institutionalization*

Blend Images/Ariel Skelley/Jupiter

Yoga has become an increasingly popular activity in recent years as many people have turned to alternative social movements derived from Asian traditions.

(or bureaucratization) *stage,* an organizational structure develops and a paid staff (rather than volunteers) begins to lead the group. When the movement reaches this stage, the initial zeal and idealism of members may diminish as administrators take over management of the organization. Early grassroots supporters may become disillusioned and drop out; they also may start another movement to address some as yet unsolved aspect of the original problem. For example, some environmental organizations—such as the Sierra Club, the Canadian Nature Federation, and the National Audubon Society—that started as grassroots conservation movements are currently viewed by many people as unresponsive to local environmental problems (Cable and Cable, 1995). As a result, new movements have arisen.

LO-5 SOCIAL MOVEMENT THEORIES

What conditions are most likely to produce social movements? Why are people drawn to these movements? Sociologists have developed several theories to answer these questions.

RELATIVE DEPRIVATION THEORY

According to relative deprivation theory, people who are satisfied with their present condition are less likely to seek social change. Social movements arise as a response to people's perception that they have been deprived of their fair share. Thus, people who suffer relative deprivation are more likely to feel that change is necessary and to join a social movement to bring about that change. *Relative deprivation* refers to the discontent that people may feel when they compare their achievements with those of similarly situated persons and find that they have less than they think they deserve. Karl Marx captured the idea of relative deprivation in this description: "A house may be large or small; as long as the surrounding houses are small it satisfies all social demands for a dwelling. But let a palace arise beside the little house, and it shrinks from a little house to a hut" (quoted in Ladd, 1966:24). Movements based on relative deprivation are most likely to occur when an upswing in the standard of living is followed by a period of decline, such that people have *unfulfilled rising expectations*—newly raised hopes of a better lifestyle that are not fulfilled as rapidly as they expected or not realized at all.

Although most of us can relate to relative deprivation theory, it does not fully account for why people experience social discontent but fail to join a social movement. Even though discontent and feelings of deprivation may be necessary to produce certain types of social movements, they are not sufficient to bring movements into existence.

VALUE-ADDED THEORY

The value-added theory, developed by sociologist Neil Smelser (1963), is based on the assumption that certain conditions are necessary for the development of a social movement. Smelser called his theory the "value-added" approach, based on the concept (borrowed from the field of economics) that each step in the production process adds something to the finished product. For example, in the process of converting iron ore into an automobile, each stage "adds value" to the final product (Smelser, 1963). Similarly, Smelser asserted, the following six conditions are necessary and sufficient to produce social movements when they combine or interact in a particular situation:

1. *Structural conduciveness.* People must become aware of a significant problem and have the opportunity to engage in collective action. According to Smelser, movements are more likely to occur when a person, class, or agency can be singled out as the source of the problem, when channels for expressing grievances either are not available or fail, and when the aggrieved have a chance to communicate among themselves.

2. *Structural strain.* When a society or community is unable to meet people's expectations that something should be done about a problem, strain occurs in the system. The ensuing tension and conflict contributes to the development of a social movement based on people's belief that the problems would not exist if authorities had done what they were supposed to do.
3. *Spread of a generalized belief.* For a movement to develop, there must be a clear statement of the problem and a shared view of its cause, effects, and possible solution.
4. *Precipitating factors.* To reinforce the existing generalized belief, an inciting incident or dramatic event must occur. With regard to technological disasters, some gradually emerge from a long-standing environmental threat, while others involve a suddenly imposed problem.
5. *Mobilization for action.* At this stage, leaders emerge to organize others and give them a sense of direction.
6. *Social control factors.* If there is a high level of social control on the part of law enforcement officials, political leaders, and others, it becomes more difficult to develop a social movement or engage in certain types of collective action.

Value-added theory takes into account the complexity of social movements and makes it possible to use Smelser's assertions to test for the necessary and sufficient conditions that produce such movements.

RESOURCE MOBILIZATION THEORY

Smelser's value-added theory tends to underemphasize the importance of resources in social movements. By contrast, *resource mobilization theory* focuses on the process through which members of a social movement gather resources and mobilize people in order to advance their cause. Resources include money, members' time, access to the media, and material goods, such as property and equipment. Assistance from outsiders is essential for social movements. Reform movements, for example, are more likely to succeed when they gain the support of political and economic elites.

Resource mobilization theory is based on the assumption that participants in social movements are rational people. From this perspective, social movements are formed and dissolved, mobilized and deactivated, based on rational decisions about the goals of the group, available resources, and the cost of mobilization and collective action. Resource mobilization theory also assumes that participants must have some degree of economic and political resources to make the movement a success. In other words, widespread discontent alone cannot produce a social movement; adequate resources and motivated people are essential to any concerted social action.

In the 21st century, scholars continue to modify resource mobilization theory and to develop new approaches for investigating the diversity of movements. Emerging perspectives based on resource mobilization theory emphasize the ideology and legitimacy of movements as well as material resources.

Additional perspectives are also needed on social movements in other nations to determine how activists in those countries acquire resources and mobilize people to advance causes such as environmental protection.

SOCIAL CONSTRUCTIONIST THEORY: FRAME ANALYSIS

Theories based on a symbolic interactionist perspective focus on the importance of the symbolic presentation of a problem to both participants and the general public. *Social constructionist theory* is based on the assumption that a social movement is an interactive, symbolically defined, and negotiated process that involves participants, opponents, and bystanders.

Research based on this perspective often investigates how problems are framed and what names they are given. This approach reflects the influence of the sociologist Erving Goffman's

Frame Analysis (1974), in which he suggests that our interpretation of the particulars of events and activities is dependent on the framework from which we perceive them. According to Goffman (1974:10), the purpose of frame analysis is "to try to isolate some of the basic frameworks of understanding available in our society for making sense out of events and to analyze the special vulnerabilities to which these frames of reference are subject." In other words, various "realities" may be simultaneously occurring among participants engaged in the same set of activities. When people come together in a social movement, they assign meanings to their activities in such a way that they build a framework for interacting and socially constructing their grievances so that they can more effectively voice them and know what resolution they want for these issues.

Sociologists have identified at least three ways in which grievances are framed. First, *diagnostic framing* identifies a problem and attributes blame or causality to some group or entity so that the social movement has a target for its actions. Second, *prognostic framing* pinpoints possible solutions or remedies, based on the target previously identified. Third, *motivational framing* provides a vocabulary of motives that compel people to take action. When successful framing occurs, the individual's vague dissatisfactions are turned into well-defined grievances, and people are compelled to join the movement in an effort to reduce or eliminate those grievances.

Beyond motivational framing, additional frame alignment processes are necessary in order to supply a continuing sense of urgency to the movement. *Frame alignment* is the linking together of interpretive orientations of individuals and social movement organizations so that there is congruence between individuals' interests, beliefs, and values and the movement's ideologies, goals, and activities. Four distinct frame alignment processes occur in social movements: (1) *frame bridging* is the process by which movement organizations reach individuals who already share the same worldview as the organization; (2) *frame amplification* occurs when movements appeal to deeply held values and beliefs in the general population and link those to movement issues so that people's pre-existing value commitments serve as a "hook" that can be used to recruit them; (3) *frame extension* occurs when movements enlarge the boundaries of an initial frame to incorporate other issues that appear to be of importance to potential participants; and (4) *frame transformation* refers to the process whereby the creation and maintenance of new values, beliefs, and meanings induce movement participation by redefining activities and events in such a manner that people believe they must become involved in collective action. Some or all of these frame alignment processes are used by social movements as they seek to define grievances and recruit participants.

Frame analysis provides new insights on how social movements emerge and grow when people are faced with problems such as technological disasters, about which greater ambiguity typically exists, and when people are attempting to "name" the problems associated with things such as nuclear or chemical contamination. However, frame analysis has been criticized for its "ideational biases" (McAdam, 1996). According to the sociologist Doug McAdam (1996), frame analyses of social movements have looked almost exclusively at ideas and their formal expression, whereas little attention has been paid to other significant factors, such as movement tactics, mobilizing structures, and changing political opportunities that influence the signifying work of movements. In this context, *political opportunity* means government structure, public policy, and political conditions that set the boundaries for change and political action. These boundaries are crucial variables in explaining why various social movements have different outcomes.

POLITICAL OPPORTUNITY THEORY

Why do social protests occur? According to political opportunity theorists, the origins of social protests cannot be explained solely by the fact that people possess a variety of grievances or that they have resources available for mobilization. Instead, social protests are directly related to the political opportunities that potential protesters and movement organizers believe exist within the political system at any given point in time. Political opportunity theory is based on the assumption that social protests that take place *outside* of mainstream political institutions are deeply intertwined with more-conventional political activities that take place *inside* these institutions. As used in this context, *opportunity* refers to "options for collective action, with chances and risks attached to them that depend on factors outside the mobilizing group" (Koopmans, 1999:97). *Political opportunity theory* states that people will choose those options for collective action that are most readily available to them and those options that will produce the most favourable outcome for their cause.

What are some specific applications of political action theory? Urban sociologists and social movement analysts have found that those cities that provided opportunities for people's protests to be heard within urban governments were less likely to have extensive protests or riots in their communities because aggrieved people could use more-conventional means to make their claims known. By contrast, urban riots were more likely to occur when activists believed that all conventional routes to protest were blocked. Changes in demography, migration, and the political economy in the United States (factors that were seemingly external to the civil rights movement) all contributed to a belief on the part of African Americans in the late 1960s and early 1970s that they could organize collective action, and that their claims regarding the need for racial justice might be more readily heard by government officials.

Political opportunity theory has grown in popularity among sociologists who study social movements because this approach highlights the interplay of opportunity, mobilization, and political influence in determining when certain types of behaviour may occur. However, like other perspectives, this theory has certain limitations, including the fact that social movement organizations may not always be completely distinct from, or external to, the existing political system. For example, it is difficult to classify the Tea Party movement in the United States, which emerged in the aftermath of the election of President Barack Obama. Some supporters were outside the political mainstream and felt like they had no voice in what was happening in Washington. Keli Carender, who is credited with being one of the first Tea Party campaigners, complained that she tried to call her senators to urge them to vote against the $787 billion stimulus bill but constantly found that their mailboxes were full. As a result, she decided to protest against "porkulus"; in her words, "I basically thought to myself: 'I have two courses. I can give up, go home, crawl into bed and be really depressed and let it happen, or I can do something different, and I can find a new avenue to have my voice get out'" (quoted in Zernike, 2010:A1).

By contrast, other active supporters of the Tea Party movement are players in the mainstream political process. An example is Sarah Palin, the former Alaska governor and Republican vice presidential candidate, who was a frequent spokesperson at Tea Party rallies across the United States. In political movements, social activists typically *create* their own opportunities rather than wait for them to emerge, and activists are often political entrepreneurs in their own right, much like the state and federal legislators and other governmental officials whom they seek to influence on behalf of their social cause. Political opportunity theory calls our attention to how important the degree of openness of a political system is to the goals and tactics of persons who organize social movements.

Sociologist Steven M. Buechler has argued that theories pertaining to 21st-century social movements should be oriented toward the structural, macrolevel contexts in which movements arise. These theories should incorporate both political and cultural dimensions of social activism:

> Social movements are historical products of the age of modernity. They arose as part of a sweeping social, political, and intellectual change that led a significant number of people to view society as a social construction that was susceptible to social reconstruction through concerted collective effort. Thus, from their inception, social movements have had a dual focus. Reflecting the political, they have always involved some form of challenge to prevailing forms of authority. Reflecting the cultural, they have always operated as symbolic laboratories in which reflexive actors pose questions of meaning, purpose, identity, and change. (2000:211)

The Concept Snapshot below summarizes the main theories of social movements.

CONCEPT SNAPSHOT

RELATIVE DEPRIVATION THEORY	People who are discontented when they compare their achievements with those of others consider themselves relatively deprived and join social movements in order to get what they view as their "fair share," especially when there is an upswing in the economy followed by a decline.
VALUE-ADDED THEORY **Key thinker:** Neil Smelser	Certain conditions are necessary for a social movement to develop: (1) structural conduciveness, such that people are aware of a problem and have the opportunity to engage in collective action; (2) structural strain, such that society or the community cannot meet people's expectations for taking care of the problem; (3) growth and spread of a generalized belief as to causes and effects of and possible solutions to the problem; (4) precipitating factors or events that reinforce the beliefs; (5) mobilization of participants for action; and (6) social control factors, such that society comes to allow the movement to take action.
RESOURCE MOBILIZATION THEORY **Key thinker:** Charles Tilly	A variety of resources (money, members, access to media, and material goods, such as equipment) are necessary for a social movement; people participate only when they feel the movement has access to these resources.
SOCIAL CONSTRUCTIONIST (FRAME ANALYSIS) THEORY **Key thinker:** Erving Goffman	Based on the assumption that social movements are an interactive, symbolically defined, and negotiated process involving participants, opponents, and bystanders, frame analysis is used to determine how people assign meaning to activities and processes in social movements.
POLITICAL OPPORTUNITY THEORY	People will choose the options for collective action (i.e., "opportunities") that are most readily available to them and those options that will produce the most favourable outcome for their cause.

As we have seen, social movements may be an important source of social change. Throughout this text, we have examined a variety of social problems that have been the focus of one or more social movements. For this reason, many groups focus on preserving their gains while simultaneously fighting for changes that they believe are still necessary.

TIME TO REVIEW

- What is the primary focus of research based on frame analysis and new social movement theory?

SOCIAL CHANGE IN THE FUTURE

In this chapter, we have focused on collective behaviour and social movements as potential forces for social change in contemporary societies. A number of other factors also contribute to social change, including the physical environment, population trends, technological development, and social institutions.

THE PHYSICAL ENVIRONMENT AND CHANGE

Changes in the physical environment often produce changes in the lives of people; in turn, people can make dramatic changes in the physical environment, over which we have only limited control. Throughout history, natural disasters have taken their toll on individuals and societies. Major natural disasters—including hurricanes, ice storms, floods, and tornadoes—can devastate an entire population. In the 21st century, earthquakes have affected India, Pakistan, El Salvador, Iran, China, Italy, Haiti, Chile, New Zealand, Japan, and the United States, among other nations. Hurricanes, tsunamis, and floods have devastated portions of Pakistan, Australia, and the United States. Even comparatively "small" natural disasters change the lives of many people. As sociologist Kai Erikson (1976, 1994) has suggested, the trauma that people experience from disasters may outweigh the loss of physical property—memories of such events can haunt people for many years.

Some natural disasters are exacerbated by human decisions. For example, floods are viewed as natural disasters, but excessive development may contribute to a flood's severity. As office buildings, shopping malls, industrial plants, residential areas, and highways are developed, less land remains to absorb rainfall. When heavier-than-usual rains and snowfall occur, flooding becomes inevitable; in recent years, some regions in Canada have remained under water for days and even weeks. Clearly, humans cannot control the rain, but human decisions can worsen the consequences.

People also contribute to changes in Earth's physical condition. Through soil erosion and other degradation of grazing land, often at the hands of people, an estimated 24 billion tons of Earth's topsoil is lost annually. As people clear forests to create farmland and pastures and acquire lumber and firewood, Earth's tree cover continues to diminish. As hundreds of millions of people drive motor vehicles, the amount of carbon dioxide in the environment continues to rise each year, resulting in climate change.

Just as people contribute to change in the physical environment, human activities also must be adapted to changes in the environment. For example, we are being warned to stay out of sunlight because of increases in ultraviolet rays, a cause of skin cancer, which have resulted from the increasing depletion of the ozone layer. If this prediction is accurate, the change in the physical environment will dramatically affect those who work or spend their leisure time outside.

POPULATION AND CHANGE

Changes in population size, distribution, and composition affect the culture and social structure of a society and change the relationships among nations. As discussed in Chapter 23, the countries experiencing the most rapid increases in population have a less developed infrastructure to deal with those changes. How will the nations of the world deal with population growth as the global population continues to move toward 7 billion? Only time will provide a response to this question.

Immigration to Canada has created an increasingly multi-ethnic population. The changing makeup of the Canadian population has resulted in children from more diverse cultural backgrounds entering school, producing a demand for new programs and changes in curriculums. An increase in the number of single mothers and of women employed outside the household has created a need for more child care, while an increase in the older population has created a need for services such as home care, and has placed increasing demands on our healthcare systems and programs such as the Canada Pension Plan.

As we have seen in previous chapters, population growth and the movement of people to urban areas have brought changes to many regions and intensified existing social problems. Among other factors, growth in the global population is one of the most significant driving forces behind environmental concerns, such as the availability and use of natural resources.

TECHNOLOGY AND CHANGE

Technology is an important force for change, and in some ways, technological development has made our lives much easier. Advances in communication and transportation have made instantaneous worldwide communication possible but have also brought old belief systems and the status quo into question as never before. Today, we are increasingly moving information instead of people—and doing it almost instantly. Advances in science and medicine have made significant changes in people's lives.

Ranging from impacts on the foods we eat to our reproductive capabilities, scientific advances will continue to affect our lives. Genetically engineered plants have been developed and marketed in recent years, and biochemists are creating potatoes, rice, and cassava with the same protein value as meat (Petersen, 1994). Advances in medicine have made it possible for those formerly unable to have children to procreate; women well beyond menopause are now able to become pregnant with the assistance of medical technology. Advances in medicine have also increased the human lifespan, especially for white and middle- or upper-class individuals in high-income nations; they have also contributed to the declining death rate in low-income nations, where birth rates have not yet been curbed.

Just as technology has brought about improvements in the quality and length of life for many, it has created the potential for new disasters, ranging from global warfare to localized technological disasters at toxic waste sites. As sociologist William Ogburn (1966) suggested, when a change in the material culture occurs in a society, a period of *cultural lag* follows in which the nonmaterial (ideological) culture has not caught up with material development. The rate of technological advancement at the level of material culture today is mind-boggling. Many of us can never hope to understand technological advances in the areas of artificial intelligence, holography, virtual reality, biotechnology, and robotics. One of the ironies of 21st-century high technology, however, is the increased vulnerability that results from the increasing complexity of such systems. We have already seen this in situations ranging from jetliners that are used as terrorist weapons to identity theft and fraud on the Internet.

SOCIAL INSTITUTIONS AND CHANGE

Many changes have occurred in the family, religion, education, the economy, and the political system over the past century. As we saw in Chapter 6, the size and composition of families in Canada changed with the dramatic increase in the number of single-person and single-parent households. Changes in families produced changes in the socialization of children, many of whom now spend much of their time playing video games, texting friends, posting their daily activities on Facebook or Twitter, or spending time in a child-care facility outside their own home.

Public education has changed dramatically in Canada over the past century. This country was one of the first to provide universal education for students regardless of their ability to pay. As a result, Canada has one of the most highly educated populations and one of the best public education systems in the world. However, the education system does not seem to be meeting the needs of some students—namely, those who are failing to learn to read and write or those who are dropping out. As the nature of the economy changes, schools will almost inevitably have to change, if for no other reason than the demands from leaders in business and industry for an educated workforce that allows Canadian companies to compete in a global economic environment.

Although we have examined changes in the physical environment, population, technology, and social institutions separately, they all operate together in a complex relationship, sometimes producing large, unanticipated consequences. We need new ways of conceptualizing social life at both the macrolevels and microlevels. The sociological imagination helps us think about how personal troubles—regardless of our race or ethnicity, class, gender, age, sexual orientation, or physical abilities and disabilities—are intertwined with the public issues of our society and the global community of which we are a part. Using our sociological imaginations also encourages us to think creatively about ways we can contribute to the social issues identified in this text.

A FEW FINAL THOUGHTS

In this text, we have covered a substantial amount of material, examined different perspectives on a wide variety of social issues, and suggested different methods by which to deal with them. The purpose of this text is not to encourage you to take any particular point of view; rather, it is to allow you to understand different viewpoints and ways in which they may be helpful to you and to society in dealing with the issues of the 21st century. Possessing that understanding, we can hope that the future will be something we can all look forward to—producing a better way of life, not only in this country but worldwide.

LO-1 Understand how collective behaviour is defined and how collective behaviour leads to social change.

Collective behaviour is voluntary, often spontaneous activity that is engaged in by a large number of people and typically violates dominant-group norms and values. Social change is the alteration, modification, or transformation of public policy, culture, or social institutions over time; it is usually brought about by collective behaviour, which is relatively spontaneous, unstructured activity that typically violates established social norms.

Phillip Chin/WireImage/Getty Images

LO-2 Identify the common forms of collective behaviour.

People engaging in collective behaviour can be divided into crowds and masses. A crowd is a relatively large number of people who are in one another's immediate vicinity. In contrast, a mass is a number of people who share an interest in a specific idea or issue but are not in one another's immediate vicinity.

Joshua Lott/Getty Images

LO-3 Describe the primary forms of mass behaviour.

Mass behaviour is collective behaviour that occurs when people respond to the same event in the same way, even if they are not geographically close to one another. Rumours, gossip, mass hysteria, fads and fashions, and public opinion are forms of mass behaviour.

Dirk Anschutz/The Image Bank/Getty Images

LO-4 Distinguish between different types of social movements.

A social movement is an organized group that acts consciously to promote or resist change through collective action; such movements are most likely to be formed when people see their personal troubles as public issues that cannot be resolved without a collective response. Reform movements seek to improve society by changing some specific aspect of the social structure. Revolutionary movements seek to bring about a total change in society—sometimes by the use of terrorism. Religious movements seek to produce radical change in individuals based on spiritual or supernatural belief systems. Alternative movements seek limited change to some aspect of people's behaviour. Resistance movements seek to prevent or undo change that has already occurred.

Blend Images/Ariel Skelley/Jupiter

KEY TERMS

civil disobedience Nonviolent action that seeks to change a policy or law by refusing to comply with it (p. 577).

collective behaviour Voluntary, often spontaneous activity that is engaged in by a large number of people and typically violates dominant-group norms and values (p. 574).

crowd A relatively large number of people who are in one another's immediate vicinity (p. 575).

fad A temporary but widely copied activity enthusiastically followed by large numbers of people (p. 581).

fashion A currently valued style of behaviour, thinking, or appearance (p. 581).

gossip Rumours about the personal lives of individuals (p. 580).

mass A large number of people who share an interest in a specific idea or issue but are not in another's immediate physical vicinity (p. 575).

mass behaviour Collective behaviour that takes place when people (who are often geographically separated from one another) respond to the same event in much the same way (p. 580).

mass hysteria A form of dispersed collective behaviour that occurs when a large number of people react with strong emotions and self-destructive behaviour to a real or perceived threat (p. 580).

mob A highly emotional crowd whose members engage in, or are ready to engage in, violence against a specific target, which may be a person, a category of people, or physical property (p. 576).

panic A form of crowd behaviour that occurs when a large number of people react to a real or perceived threat with strong emotions and self-destructive behaviour (p. 577).

propaganda Information provided by individuals or groups that have a vested interest in furthering their own cause or damaging an opposing one (p. 582).

public opinion The political attitudes and beliefs communicated by ordinary citizens to decision makers (p. 582).

riot Violent crowd behaviour that is fuelled by deep-seated emotions but not directed at one specific target (p. 576).

rumour An unsubstantiated report on an issue or subject (p. 580).

social change The alteration, modification, or transformation of public policy, culture, or social institutions over time (p. 573).

social movement An organized group that acts consciously to promote or resist change through collective action (p. 582).

terrorism Acts of serious violence, planned and executed clandestinely and committed to achieve political ends (p. 583).

LO-5 Understand the theoretical explanations of social movements.

© ZUMA Wire Service/Alamy

Social scientists have developed several theories to explain crowd behaviour. Contagion theory asserts that a crowd takes on a life of its own as people are transformed from rational beings into part of an organism that acts on its own. A variation on this is circular reaction—people express their discontent to others, who communicate back similar feelings, resulting in a conscious effort to engage in the crowd's behaviour. Convergence theory asserts that people with similar attributes find other like-minded persons with whom they can release underlying personal tendencies. Emergent norm theory asserts that, as a crowd develops, it comes up with its own norms that replace more conventional norms of behaviour. Relative deprivation theory asserts that, if people are discontented when they compare their accomplishments with those of others similarly situated, they are more likely to join a social movement than are people who are relatively content with their status. Value-added theory asserts that six conditions must exist in order to produce social movements: (1) a perceived source of a problem, (2) a perception that the authorities are not resolving the problem, (3) a spread of the belief to an adequate number of people, (4) a precipitating incident, (5) mobilization of other people by leaders, and (6) a lack of social control. Resource mobilization theory asserts that successful social movements can occur only when they gain the support of political and economic elites, without whom they do not have access to the resources necessary to maintain the movement. Frame analysis often highlights the social construction of grievances through the process of social interaction. Various types of framing occur as problems are identified, remedies are sought, and people feel compelled to take action. Like frame analysis, new social movement theory has been used in research that looked at technological disasters and cases of environmental racism.

KEY FIGURES

Morphart Creation/Shutterstock.com

Gustave Le Bon (1841-1931) Le Bon was a French scholar who focused on crowd psychology in his contagion theory. Le Bon argued that people are more likely to engage in antisocial behaviour in a crowd because they are anonymous and feel invulnerable.

American Sociological Association, asanet.org

Robert E. Park (1864-1944) The first U.S. sociologist to investigate crowd behaviour, Park developed the concepts of social unrest and circular reaction.

American Sociological Association, asanet.org

Herbert Blumer (1900-1987) Blumer developed a typology that divides crowds into four categories: casual, conventional, expressive, and acting.

APPLICATION QUESTIONS

1. What types of collective behaviour in Canada do you believe are influenced by inequalities based on race or ethnicity, class, gender, age, or disabilities? Why?
2. In what ways have your actions or behaviours changed when you have been part of a crowd? Why do you think this has happened? In what ways, if any, would you like to alter these actions?
3. Analyze the environmental movement in terms of the value-added theory. Next, try using the relative deprivation and resource mobilization theories.
4. Using the sociological imagination you have gained during this course, determine some positive steps you believe might be taken in Canada to make our society a better place for everyone in this century. What types of collective behaviour and/or social movements might be required to take those steps?

Note: Page numbers that begin with "21" and "22" (e.g. 21-11, 22-15–22-16) refer to online chapters 22 and 23, found at www.nelson.com/student

absolute poverty A level of economic deprivation in which people do not have the means to secure the basic necessities of life. *(p. 247)*

achieved status A social position a person assumes voluntarily as a result of personal choice, merit, or direct effort. *(p. 116)*

activity theory The proposition that tend to shift gears in late middle age and find substitutes for previous statuses, roles, and activities. *(p. 392)*

ageism Prejudice and discrimination against people on the basis of age, particularly against older persons. *(pp. 105, 386)*

age stratification The inequalities, differences, segregation, or conflict between age groups. *(p. 390)*

agents of socialization The persons, groups, or institutions that teach us what we need to know to participate in society. *(p. 92)*

aggregate A collection of people who happen to be in the same place at the same time but have little else in common. *(p. 166)*

aging The physical, psychological, and social processes associated with growing older. *(p. 337)*

aging and society paradigm A functionalist theory that says that our lives are shaped by the social institutions through which we pass during our lives. *(p. 391)*

alienation A feeling of powerlessness and estrangement from other people and from oneself. *(p. 15)*

altruism Behaviour intended to help others and done without any expectation of personal benefit. *(p. 29)*

analysis The process through which data are organized so that comparisons can be made and conclusions drawn. *(p. 34)*

animism The belief that plants, animals, or other elements of the natural world are endowed with spirits or life forces that have an impact on events in society. *(p. 458)*

anomie Émile Durkheim's term for a condition in which social control becomes ineffective as a result of the loss of shared values and a sense of purpose in society. *(p. 14)*

anticipatory socialization The process by which knowledge and skills are learned for future roles. *(p. 103)*

ascribed status A social position conferred on a person at birth or received involuntarily later in life. *(p. 116)*

asexual An absence of sexual desire toward either sex. *(p. 353)*

assimilation A process by which members of subordinate racial and ethnic groups become absorbed into the dominant culture. *(p. 303)*

authoritarian personality A personality type characterized by excessive conformity, submissiveness to authority, intolerance, insecurity, a high level of superstition, and rigid, stereotypic thinking. *(p. 298)*

authoritarian political system A political system controlled by rulers who deny popular participation in government. *(p. 22-8)*

authority Power that people accept as legitimate rather than coercive. *(p. 22-4)*

big data Very large datasets that can be accessed in digital form and that can be linked with other large datasets. *(p. 47)*

bilateral descent A system of tracing descent through both the mother's and father's sides of the family. *(p. 141)*

bourgeoisie Karl Marx's term for the class comprised of those who own and control the means of production. *(p. 14)*

bureaucracy An organizational model characterized by a hierarchy of authority, a clear division of labour, explicit rules and procedures, and impersonality in personnel matters. *(p. 176)*

bureaucratic personality A psychological construct that describes those workers who are more concerned with following correct procedures than they are with doing the job correctly. *(p. 181)*

capitalism An economic system characterized by private ownership of the means of production, from which personal profits can be derived through market competition and without government intervention. *(p. 521)*

capitalist class (bourgeoisie) Karl Marx's term for those who own the means of production. *(p. 236)*

caste system A system of social inequality in which people's status is permanently determined at birth based on their parents' ascribed characteristics. *(p. 231)*

category A number of people who may never have met one another but who share a similar characteristic. *(p. 166)*

central city The densely populated centre of a metropolis. *(p. 23–20)*

charismatic authority Power legitimized on the basis of a leader's exceptional personal qualities or accomplishments. *(p. 22-5)*

chronological age A person's age based on date of birth. *(p. 378)*

church A large, bureaucratically organized religious body that tends to seek accommodation with the larger society in order to maintain some degree of control over it. *(p. 468)*

civil disobedience Nonviolent action that seeks to change a policy or law by refusing to comply with it. *(p. 577)*

class conflict Karl Marx's term for the struggle between the capitalist class and the working class. *(pp. 14, 237)*

class system A type of stratification based on the ownership and control of resources and on the type of work people do. *(p. 232)*

cohabitation The sharing of a household by a couple who live together without being legally married. *(p. 146)*

cohort A category of people born within a specified period in time or who share some specified characteristic. *(p. 378)*

collective behaviour Voluntary, often spontaneous activity that is engaged in by a large number of people and typically violates dominant-group norms and values. *(p. 574)*

commonsense knowledge A form of knowing that guides ordinary conduct in everyday life. *(p. 5)*

complete observation Research in which the investigator systematically observes a social process, but does not take part in it. *(p. 42)*

conflict perspectives The sociological approach that views groups in society as engaged in a continuous power struggle for control of scarce resources. *(p. 19)*

conformity The process of maintaining or changing behaviour to comply with the norms established by a society, subculture, or other group. *(p. 171)*

constructionism A theory that focuses on the way in which environmental issues are socially constructed. *(p. 555)*

contingent work Part-time or temporary work. *(p. 534)*

control group Subjects in an experiment who are not exposed to the independent variable, but later are compared to subjects in the experimental group. *(p. 36)*

core nations According to world-systems analysis, dominant capitalist centres characterized by high levels of industrialization and urbanization, as well as a high degree of control over the world economy. *(p. 279)*

corporate crime An illegal act committed by corporate employees on behalf of the corporation and with its support. *(p. 207)*

corporations Large-scale organizations that have legal powers, such as the ability to enter into contracts and buy and sell property, separate from their individual owners. *(p. 521)*

correlation Exists when two variables are associated more frequently than could be expected by chance. (p. 38)

counterculture A group that strongly rejects dominant societal values and norms and seeks alternative lifestyles. *(p. 76)*

credentialism A process of social selection in which class advantage and social status are linked to the possession of academic qualifications. *(p. 440)*

crime An act that violates criminal law and is punishable by fines, jail terms, and other sanctions. *(p. 195)*

crowd A relatively large number of people who are in one another's immediate vicinity. *(p. 575)*

crude birth rate The number of live births per 1000 people in a population in a given year. *(p. 23-5)*

crude death rate The number of deaths per 1000 people in a population in a given year. *(p. 23-5)*

cult A religious group with practices and teachings outside the dominant cultural and religious traditions of a society. *(p. 468)*

cultural capital Pierre Bourdieu's term for people's social assets, including their values, beliefs, attitudes, and competencies in language and culture. *(p. 438)*

cultural imperialism A process whereby powerful countries use the media to spread values and ideas that dominate and even destroy other cultures, and local cultural values are replaced by the cultural values of the dominant country. *(pp. 78, 507)*

cultural lag William Ogburn's term for a gap between the technical development of a society (material culture) and its moral and legal institutions (nonmaterial culture). *(p. 71)*

cultural relativism The belief that the behaviours and customs of any culture must be viewed and analyzed by the culture's own standards. *(p. 76)*

cultural transmission The process by which children and recent immigrants become acquainted with the dominant cultural beliefs, values, norms, and accumulated knowledge of a society. *(p. 434)*

cultural universals Customs and practices that occur across all societies. *(p. 62)*

culture The knowledge, language, values, customs, and material objects that are passed from person to person and from one generation to the next in a human group or society. *(p. 59)*

cumulative advantage theory The notion that the advantages and disadvantages of gender, race, and class accumulate over the life course and have a major impact on aging. *(p. 393)*

cybercrime Offences where a computer is the object of a crime or the tool used to commit a crime. *(p. 207)*

date rape Acts in which a date or boyfriend forced or attempted to force any type of sexual activity through threats or use of violence. *(p. 368)*

dating violence A term used for various forms of sexually and nonsexually assaultive behaviours that occur within dating relationships. *(p. 368)*

deductive approach Research in which the investigator begins with a theory and then collects information and data to test the theory. *(p. 32)*

democracy A political system in which the people hold the ruling power, either directly or through elected representatives. *(p. 22-9)*

democratic socialism An economic and political system that combines private ownership of some of the means of production, governmental distribution of some essential goods and services, and free elections. *(p. 525)*

demographic transition The process by which some societies have moved from high birth and death rates to relatively low birth and death rates as a result of technological development. *(p. 23–17)*

demography The subfield of sociology that examines population size, composition, and distribution. *(p. 23–4)*

dependency theory The perspective that global poverty can at least partially be attributed to the fact that low-income countries have been exploited by high-income countries. *(p. 278)*

dependent variable A variable that is assumed to depend on or be caused by one or more other (independent) variables. *(p. 33)*

descriptive study Research that attempts to describe social reality or provide facts about some group, practice, or event. *(p. 31)*

deviance Any behaviour, belief, or condition that violates cultural norms in the society or group in which it occurs. *(p. 194)*

differential association theory The proposition that individuals have a greater tendency to deviate from societal norms when they frequently associate with persons who favour deviance over conformity. *(p. 197)*

diffusion The transmission of cultural items or social practices from one group or society to another. *(p. 71)*

disability A physical or health condition that reduces a person's ability to perform tasks he or she would normally do at a given stage of life and that may result in stigmatization or discrimination against the person. *(p. 420)*

discovery The process of learning about something previously unknown or unrecognized. *(p. 71)*

discrimination Actions or practices of dominant group members (or their representatives) that have a harmful impact on members of a subordinate group. *(p. 298)*

disengagement theory The proposition that older persons make a normal and healthy adjustment to aging when they detach themselves from their social roles and prepare for their eventual death. *(p. 391)*

dramaturgical analysis The study of social interaction that compares everyday life to a theatrical presentation. *(p. 126)*

dual-earner families Families in which both partners are in the labour force. *(p. 152)*

dyad A group consisting of two members. *(p. 170)*

dysfunctions A term referring to the undesirable consequences of any element of a society. *(p. 18)*

ecofeminism Sociological theory that the ideology that authorizes oppressions such as those based on race, class, gender, sexuality, physical abilities, and species is the same ideology that sanctions the oppression of nature. *(p. 554)*

ecological modernization theory Environmental sociological theory that societies are adapting to the threat of environmental harms by reforming societal practices and policies. *(p. 553)*

economy The social institution that ensures the maintenance of society through the production, distribution, and consumption of goods and services. *(p. 516)*

education The social institution responsible for the systematic transmission of knowledge, skills, and cultural values within a formally organized structure. *(p. 434)*

egalitarian family A family structure in which both partners share power and authority equally. *(p. 142)*

ego According to Sigmund Freud, the rational, reality-oriented component of personality that imposes restrictions on the innate pleasure-seeking drives of the id. *(p. 100)*

elder abuse A term used to describe physical abuse, psychological abuse, financial exploitation, and medical abuse or neglect of people age 65 or older. *(p. 390)*

elite model A view of society in which power in political systems is concentrated in the hands of a small group of elites and the masses are relatively powerless. *(p. 22–11)*

emigration The movement of people out of a geographic area to take up residency elsewhere. *(p. 23-6)*

employment equity A strategy to eliminate the effects of discrimination and to make employment opportunities available to groups who have been excluded. *(p. 337)*

environmental justice The idea that all people and all communities are entitled to equal protection under environmental health laws and regulations. *(p. 561)*

environmental racism The targeting of racial minority communities for toxic waste facilities, the official sanctioning of poisons and pollutants in minority communities, and the systematic exclusion of racial minorities from decision making regarding the production of environmental conditions that affect their lives and livelihoods. *(p. 562)*

environmental sociology The study of the interaction between human society and the physical environment. *(p. 548)*

epidemics Sudden, significant increases in the numbers of people contracting a disease. *(p. 416)*

ethnic group A collection of people distinguished, by others or by themselves, primarily on the basis of cultural or nationality characteristics. *(p. 292)*

ethnic pluralism The coexistence of a variety of distinct racial and ethnic groups within one society. *(p. 304)*

ethnocentrism The tendency to regard one's own culture and group as the standard—and thus superior—whereas all other groups are seen as inferior. *(p. 76)*

ethnomethodology The study of the commonsense knowledge that people use to understand the situations in which they find themselves. *(p. 126)*

experiment A test conducted under controlled conditions in which an investigator tests a hypothesis by manipulating an independent variable and examining its impact on a dependent variable *(p. 35)*.

experimental group Subjects in an experiment who are exposed to the independent variable. *(p. 36)*

explanatory study Research that attempts to explain relationships and to provide information on why certain events do or do not occur. *(p. 31)*

extended family A family unit composed of relatives in addition to parents and children who live in the same household. *(p. 140)*

fad A temporary but widely copied activity enthusiastically followed by large numbers of people. *(p. 581)*

faith Unquestioning belief that does not require proof or scientific evidence. *(p. 456)*

families we choose Social arrangements that include intimate relationships between couples and close familial relationships with other couples, as well as with other adults and children. *(p. 138)*

family A relationship in which people live together with commitment, form an economic unit and care for any young, and consider their identity to be significantly attached to the group. *(p. 139)*

family of orientation The family into which a person is born and in which early socialization usually takes place. *(p. 139)*

family of procreation The family that a person forms by having or adopting children. *(p. 139)*

fashion A currently valued style of behaviour, thinking, or appearance. *(p. 581)*

feminism The belief that women and men are equal and that they should be valued equally and have equal rights. *(p. 341)*

feminist perspectives The sociological approach that focuses on the significance of gender in understanding and explaining inequalities that exist between men and women in the household, in the paid labour force, and in the realms of politics, law, and culture. *(p. 21)*

feminization of poverty The trend in which women are disproportionately represented among individuals living in poverty. *(p. 249)*

fertility The actual level of childbearing for an individual or a population. *(p. 23-4)*

field research The study of social life in its natural setting: observing and interviewing people where they live, work, and play. *(p. 42)*

folkways Informal norms or everyday customs that may be violated without serious consequences within a particular culture. *(p. 70)*

formal education Learning that takes place within an academic setting, such as a school, that has a planned instructional process and teachers who convey specific knowledge, skills, and thinking processes to students. *(p. 434)*

formal organization A highly structured group formed for the purpose of completing certain tasks or achieving specific goals. *(p. 120)*

functional age A term used to describe observable individual attributes, such as physical appearance, mobility, strength, coordination, and mental capacity, that are used to assign people to age categories. *(p. 378)*

functionalist perspectives The sociological approach that views society as a stable, orderly system. *(p. 18)*

Gemeinschaft (guh-MINE-shoft) A traditional society in which social relationships are based on personal bonds of friendship and kinship and on intergenerational stability. *(p. 122)*

gender The culturally and socially constructed differences between females and males found in the meanings, beliefs, and practices associated with "femininity" and "masculinity." *(p. 324)*

gender bias Behaviour that shows favouritism toward one gender over the other. *(p. 334)*

gender identity A person's perception of the self as female or male. *(p. 324)*

gender role Attitudes, behaviour, and activities that are socially defined as appropriate for each sex and are learned through the socialization process. *(p. 324)*

gender socialization The aspect of socialization that contains specific messages and practices concerning the nature of being female or male in a specific group or society. *(p. 96)*

generalized other George Herbert Mead's term for the child's awareness of the demands and expectations of the society as a whole or of the child's subculture. *(p. 99)*

gentrification The process by which members of the middle and upper-middle classes move into the central city area and renovate existing properties. *(p. 23–22)*

gerontology The study of aging and older people. *(p. 381)*

Gesellschaft (guh-ZELL-shoft) A large, urban society in which social bonds are based on impersonal and specialized relationships, with little long-term commitment to the group or consensus on values. *(p. 122)*

global interdependence A relationship in which the lives of all people are closely intertwined and any one nation's problems are part of a larger global problem. *(p. 4)*

goal displacement A process that occurs in organizations when the rules become an end in themselves rather than a means to an end. *(p. 181)*

global village A world in which distances have been shrunk by modern communications technology so that everyone is socially and economically interdependent. *(p. 507)*

gossip Rumours about the personal lives of individuals. *(p. 580)*

government The formal organization that has the legal and political authority to regulate the relationships among members of a society and between the society and those outside its borders. *(p. 22-4)*

green criminology A branch of criminology that encompasses a broad range of behaviours, ranging from acts that are clearly harmful, such as dumping toxic waste into the ocean, to acts that many people consider to be acceptable, such as eating meat or wearing leather shoes. *(p. 567)*

gross national income (GNI) All the goods and services produced in a country in a given year, plus the income earned outside the country by individuals or corporations. *(p. 263)*

groupthink The process by which members of a cohesive group arrive at a decision that many individual members privately believe is unwise. *(p. 176)*

hate racism (or **overt racism**) Racism that may take the form of deliberate and highly personal attacks, including derogatory slurs and name-calling toward members of a racial or ethnic group who are perceived to be "inferior." *(p. 299)*

health The state of complete physical, mental, and social well-being. *(p. 406)*

healthcare Any activity intended to improve health. *(p. 407)*

hidden curriculum The transmission of cultural values and attitudes, such as conformity and obedience to authority, through implied demands found in the rules, routines, and regulations of schools. *(p. 440)*

high-income countries Nations with highly industrialized economies; technologically advanced industrial, administrative, and service occupations; and relatively high levels of national and personal income. *(p. 7)*

high-income economies Countries with an annual per capita gross national income of $US12,477 or more. *(p. 264)*

homogamy The pattern of individuals marrying those who have similar characteristics, such as race/ethnicity, religious background, age, education, or social class. *(p. 147)*

hyperreality A situation in which the distinction between reality and simulation has become blurred. *(p. 497)*

hypotheses Tentative statements of the relationship between two or more concepts or variables. *(p. 30)*

id Sigmund Freud's term for the component of personality that includes all of the individual's basic biological drives and needs that demand immediate gratification. *(p. 100)*

ideal culture The values and standards of behaviour that people in a society profess to hold. *(p. 69)*

illegitimate opportunity structures Circumstances that provide an opportunity for people to acquire through illegitimate activities what they cannot achieve through legitimate channels. *(p. 196)*

immigration The movement of people into a geographic area to take up residency. *(p. 23-6)*

impression management (or presentation of self) A term for people's efforts to present themselves to others in ways that are most favourable to their own interests or image. *(p. 126)*

income The economic gain derived from wages, salaries, income transfers (governmental aid), and ownership of property. *(p. 233)*

independent variable A variable that is presumed to cause or determine a dependent variable. *(p. 33)*

inductive approach Research in which the investigator collects information or data (facts or evidence) and then generates theories from the analysis of that data. *(p. 32)*

industrialization The process by which societies are transformed from dependence on agriculture and handmade products to an emphasis on manufacturing and related industries. *(p. 10)*

infant mortality rate The number of deaths of infants under one year of age per 1000 live births in a given year. *(p. 23–5)*

infertility An inability to conceive after one year of unprotected sexual relations. *(p. 149)*

informal education Learning that occurs in a spontaneous, unplanned way. *(p. 434)*

informal structure Those aspects of participants' day-to-day activities and interactions that ignore, bypass, or do not correspond with the official rules and procedures of the bureaucracy. *(p. 179)*

ingroup A group to which a person belongs and identifies with. *(p. 167)*

institutionalized racism A situation where the established rules, policies, and practices within an institution or organization produce differential treatment of various groups based on race. *(p. 301)*

intergenerational mobility The social movement (upward or downward) experienced by family members from one generation to the next. *(p. 230)*

internal colonialism According to conflict theorists, a situation in which members of a racial or ethnic group are conquered or colonized and forcibly placed under the economic and political control of the dominant group. *(p. 306)*

Internet The network infrastructure that links together the world's millions of computers. *(p. 485)*

interpretive sociology An approach to sociology that examines the meaning that people give to aspects of their social lives *(p. 45)*.

intersex person An individual who is born with a reproductive or sexual anatomy that does not correspond to typical definitions of male or female; in other words, the person's sexual differentiation is ambiguous. *(p. 351)*

interview A research method using a data collection encounter in which an interviewer asks the respondent questions and records the answers. *(p. 40)*

intimate partner violence The physical and nonphysical violence experienced by women and men at the hands of current or former partners. *(p. 367)*

intragenerational mobility The social movement (upward or downward) experienced by individuals within their own lifetime. *(p. 230)*

invasion The process by which a new category of people or type of land use arrives in an area previously occupied by another group or land use. *(p. 23–21)*

invention The process of reshaping existing cultural items into a new form. *(p. 71)*

job deskilling A reduction in the proficiency needed to perform a specific job, which leads to a corresponding reduction in the wages paid for that job. *(p. 252)*

kinship A social network of people based on common ancestry, marriage, or adoption. *(p. 139)*

labelling theory The proposition that deviants are those people who have been successfully labelled as such by others. *(p. 199)*

labour union A group of employees who join together to bargain with an employer over wages, benefits, and working conditions. *(p. 537)*

language A system of symbols that expresses ideas and enables people to think and communicate with one another. *(p. 64)*

latent functions Unintended functions that are hidden and remain unacknowledged by participants. *(p. 18)*

laws Formal, standardized norms that have been enacted by legislatures and are enforced by formal sanctions. *(p. 70)*

life chances Max Weber's term for the extent to which individuals have access to important societal resources, such as food, clothing, shelter, education, and healthcare. *(p. 230)*

life expectancy The average length of time a group of individuals of the same age will live. *(p. 378)*

looking-glass self Charles Horton Cooley's term for the way in which a person's sense of self is derived from the perceptions of others. *(p. 97)*

low-income countries Countries that are primarily agrarian, with little industrialization and low levels of national and personal income. *(p. 8)*

low-income cutoff The income level at which a family may be in "straitened circumstances" because it spends considerably more on the basic necessities of life (food, shelter, and clothing) than does the average family. *(p. 246)*

low-income economies Countries with an annual per capita gross national income of $US1045 or less. *(p. 264)*

lower-middle-income economies Countries with an annual per capita gross national income between $US1046 and $US4125. *(p. 264)*

macrolevel analysis Sociological theory and research that focuses on whole societies, large-scale social structures, and social systems. *(p. 21)*

majority (dominant) group An advantaged group that has the greatest power and resources in a society. *(p. 296)*

manifest functions Open, stated, and intended goals or consequences of activities within an organization or institution. *(p. 18)*

marginal job A position that differs from the employment norms of the society in which it is located. *(p. 533)*

marriage A legally recognized and/or socially approved arrangement between two or more individuals that carries certain rights and obligations and usually involves sexual activity. *(pp. 140, 354)*

mass A large number of people who share an interest in a specific idea or issue but are not in another's immediate physical vicinity. *(p. 575)*

mass behaviour Collective behaviour that takes place when people (who are often geographically separated from one another) respond to the same event in much the same way. *(p. 580)*

mass education Free, public schooling for wide segments of a nation's population. *(p. 434)*

mass hysteria A form of dispersed collective behaviour that occurs when a large number of people react with strong emotions and self-destructive behaviour to a real or perceived threat. *(p. 580)*

mass media Any technologically based means of communicating between large numbers of people distributed widely over space or time. *(p. 484)*

master status A term used to describe the most important status a person occupies. *(p. 116)*

material culture A component of culture that consists of the physical or tangible creations—such as clothing, shelter, and art—that members of a society make, use, and share. *(p. 61)*

matriarchal family A family structure in which authority is held by the eldest female (usually the mother). *(p. 141)*

matriarchy A hierarchical system of social organization in which cultural, political, and economic structures are controlled by women. *(p. 328)*

matrilineal descent A system of tracing descent through the mother's side of the family. *(p. 141)*

means of production Karl Marx's term for tools, land, factories, and money for investment that form the economic basis of a society. *(p. 14)*

mechanical solidarity Émile Durkheim's term for the social cohesion that exists in preindustrial societies, in which there is a minimal division of labour and people feel united by shared values and common social bonds. *(p. 122)*

media literacy The ability to access, analyze, evaluate, and create media in a varity of forms. *(p. 485)*

medicalization The process whereby an object or a condition becomes defined by society as a physical or psychological illness. *(p. 408)*

medicine An institutionalized system for the scientific diagnosis, treatment, and prevention of illness. *(p. 407)*

meritocracy A hierarchy in which all positions are rewarded based on people's ability and credentials. *(p. 252)*

metropolis One or more central cities and their surrounding suburbs that dominate the economic and cultural life of a region. *(p. 23–20)*

microlevel analysis Sociological theory and research that focus on small groups rather than on large-scale social structures. *(p. 21)*

middle-income countries Nations with industrializing economies, particularly in urban areas, and moderate levels of national and personal income. *(p. 8)*

migration The movement of people from one geographic area to another for the purpose of changing residency. *(p. 23–6)*

minority (subordinate) group A group whose members, because of physical or cultural characteristics, are disadvantaged and subjected to unequal treatment and discrimination by the dominant group. *(p. 296)*

mixed economy An economic system that combines elements of a market economy (capitalism) with elements of a command economy (socialism). *(p. 525)*

mob A highly emotional crowd whose members engage in, or are ready to engage in, violence against a specific target, which may be a person, a category of people, or physical property. *(p. 576)*

modernization theory A perspective that links global inequality to different levels of economic development and that suggests that low-income economies can move to middle- and high-income economies by achieving self-sustained economic growth. *(p. 275)*

monarchy A political system in which power resides in one person or family and is passed from generation to generation through lines of inheritance. *(p. 22-8)*

monogamy An intimate relationship with one person at a time. *(pp. 140, 354)*

monotheism A belief in a single, supreme being or god who is responsible for significant events, such as the creation of the world. *(p. 458)*

moral crusades Public campaigns that help generate public and political support for moral entrepreneurs' causes. *(p. 200)*

moral entrepreneurs People or groups who take an active role in trying to have particular behaviours defined as deviant. *(p. 200)*

mores Strongly held norms with moral and ethical connotations that may not be violated without serious consequences in a particular culture. *(p. 70)*

mortality The incidence of death in a population. *(p. 23–5)*

multinational corporations Large companies that are headquartered in one country and have subsidiaries or branches in other countries. *(p. 522)*

network A web of social relationships that link a number of people. *(p. 168)*

network enterprise Separate businesses, which may be companies or parts of companies, join together for specific projects that become the focus of the network. *(p. 188)*

New Ecological Paradigm An alternative sociological approach, which recognizes that the social world does not exist on its own but is embedded in the natural world. *(p. 548)*

new international division of labour theory The perspective that commodity production is being split into fragments that can be assigned to whichever part of the world can provide the most profitable combination of capital and labour. *(p. 279)*

nonmaterial culture A component of culture that consists of the abstract or intangible human creations of society—such as attitudes, beliefs, and values—that influence people's behaviour. *(p. 61)*

nontheistic religion A religion based on a belief in divine spiritual forces, such as sacred principles of thought and conduct, rather than on a god or gods. *(p. 458)*

nonverbal communication The transfer of information between persons without the use of speech. *(p. 129)*

norms Established rules of behaviour or standards of conduct. *(p. 69)*

nuclear family A family made up of one or two parents and their dependent children, all of whom live apart from other relatives. *(p. 140)*

nursing home Any institution that offers medical care for chronically ill older people but is not a hospital. *(p. 398)*

objective Free from distorted subjective (personal or emotional) bias. *(p. 31)*

occupational, or white-collar, crime A term used to describe illegal activities committed by people in the course of their employment or in dealing with their financial affairs. *(p. 207)*

occupations Categories of jobs that involve similar activities at different work sites. *(p. 530)*

oligopoly The situation that exists when several companies overwhelmingly control an entire industry. *(p. 523)*

operational definition An explanation of an abstract concept in terms of observable features that are specific enough to measure the variable. *(p. 34)*

organic solidarity Émile Durkheim's term for the social cohesion that exists in industrial (and perhaps post-industrial) societies, in which people perform specialized tasks and feel united by their mutual dependence. *(p. 122)*

organized crime A business operation that supplies illegal goods and/or services for profit. *(p. 208)*

outgroup A group to which a person does not belong and toward which the person may feel a sense of competitiveness or hostility. *(p. 167)*

overt racism (or **hate racism)** Racism that may take the form of deliberate and highly personal attacks, including derogatory slurs and name-calling toward members of a racial or ethnic group who are perceived to be "inferior." *(p. 299)*

panic A form of crowd behaviour that occurs when a large number of people react to a real or perceived threat with strong emotions and self-destructive behaviour. *(p. 577)*

participant observation A research method in which researchers collect systematic observations while being part of the activities of the group they are studying. *(p. 43)*

patriarchal family A family structure in which authority is held by the eldest male (usually the father). *(p. 141)*

patriarchy A hierarchical system of social organization in which cultural, political, and economic structures are controlled by men. *(p. 328)*

patrilineal descent A system of tracing descent through the father's side of the family. *(p. 141)*

pay equity (comparable worth) The belief that wages ought to reflect the worth of a job, not the gender or race of the worker. *(p. 337)*

peer group A group of people who are linked by common interests, equal social position, and (usually) similar age. *(p. 94)*

peripheral nations According to world-systems analysis, nations that are dependent on core nations for capital, have little or no industrialization, and have uneven patterns of urbanization. *(p. 279)*

personal space The immediate area surrounding a person that the person claims as private. *(p. 131)*

perspective An overall approach to or viewpoint on some subject. *(p. 18)*

pink-collar occupation Relatively low-paying, nonmanual, semiskilled positions primarily held by women. *(p. 240)*

pluralist model An analysis of political systems that views power as widely dispersed throughout many competing interest groups. *(p. 22–10)*

polite racism A term used to describe an attempt to disguise a dislike of others through behaviour that is outwardly nonprejudicial. *(p. 299)*

political crime Illegal or unethical acts involving the usurpation of power by government officials, or illegal or unethical acts perpetrated against a government by outsiders seeking to make a political statement or to undermine or overthrow the government. *(p. 209)*

political socialization The process by which people learn political attitudes, values, and behaviour. *(p. 22–18)*

politics The social institution through which power is acquired and exercised by some people and groups. *(p. 22–4)*

polyamory Intimate relationships that involve mutually acknowledged emotional, sexual, or romantic relationships with multiple partners. *(p. 355)*

polyandry The concurrent marriage of one woman with two or more men. *(pp. 141, 357)*

polygamy The concurrent marriage of a person of one sex with two or more members of the opposite sex. *(pp. 141, 356)*

polygyny The concurrent marriage of one man with two or more women. *(pp. 141, 356)*

polytheism A belief in more than one god. *(p. 458)*

population In a research study, those persons about whom we want to be able to draw conclusions. *(p. 40)*

population composition In demography, the biological and social characteristics of a population. *(p. 23–11)*

population pyramid A graphic representation of the distribution of a population by sex and age. *(p. 23–11)*

positivism A belief that the world can best be understood through scientific inquiry. *(p. 11)*

post-industrial economy An economy that is based on the provision of services rather than goods. *(p. 519)*

postmodern perspectives The sociological approach that attempts to explain social life in contemporary societies that are characterized by post-industrialization, consumerism, and global communications. *(p. 22)*

power According to Max Weber, the ability of people or groups to achieve their goals despite opposition from others. *(p. 238)*

power elite A term devised by C. Wright Mills for a structure composed of leaders at the top of business, the executive branch of the federal government, and the military. *(p. 22–11)*

prejudice A negative attitude based on preconceived notions about members of selected groups. *(p. 297)*

prestige The respect or regard with which a person or status position is regarded by others. *(p. 238)*

preventive medicine Medicine that emphasizes a healthy lifestyle in order to prevent poor health before it occurs. *(p. 407)*

primary deviance A term used to describe the initial act of rule breaking. *(p. 199)*

primary group A small, less specialized group in which members engage in face-to-face, emotion-based interactions over an extended time. *(p. 120)*

primary sector production The sector of the economy that extracts raw materials and natural resources from the environment. *(p. 516)*

primary sex characteristics The genitalia used in the reproductive process. *(p. 350)*

profane A term used to describe the everyday, secular, or "worldly" aspects of life. *(p. 456)*

professions High-status, knowledge-based occupations. *(p. 531)*

proletariat Karl Marx's term for those who must sell their labour because they have no other means to earn a livelihood. *(p. 14)*

propaganda Information provided by individuals or groups that have a vested interest in furthering their own cause or damaging an opposing one. *(p. 582)*

public opinion The political attitudes and beliefs communicated by ordinary citizens to decision makers. *(p. 582)*

questionnaire A research instrument containing a series of items to which subjects respond. *(p. 40)*

race A term used by many people to specify groups of people distinguished by physical characteristics, such as skin colour; also, a category of people who have been singled out as inferior or superior, often on the basis of real or alleged physical characteristics, such as skin colour, hair texture, eye shape, or other subjectively selected attributes. *(p. 292)*

racial prejudice Beliefs that certain racial groups are innately inferior to others or have a disproportionate number of negative traits. *(p. 297)*

racism A set of ideas that implies the superiority of one social group over another on the basis of biological or cultural characteristics, together with the power to put these beliefs into practice in a way that controls, excludes, or exploits minority women and men. *(p. 299)*

rationality The process by which traditional methods of social organization, characterized by informality and spontaneity, are gradually replaced by formal rules and procedures (bureaucracy). *(p. 185)*

rational-legal authority Power legitimized by law or written rules and regulations. *(p. 22–6)*

reactivity The tendency of experiment participants to change their behaviour in response to the presence of the researcher or to the fact that they know they are being studied. *(p. 39)*

real culture The values and standards of behaviour that people actually follow (as contrasted with *ideal culture*). *(p. 69)*

reciprocal socialization The process by which the feelings, thoughts, appearance, and behaviour of individuals who are undergoing socialization also have a direct influence on those agents of socialization who are attempting to influence them. *(p. 93)*

reference group A group that strongly influences a person's behaviour and social attitudes, regardless of whether that individual is a member. *(p. 168)*

relative poverty A level of economic deprivation in which people may be able to afford basic necessities but still are unable to maintain an average standard of living. *(p. 247)*

reliability In sociological research, the extent to which a study or research instrument yields consistent results. *(p. 34)*

religion A system of beliefs, symbols, and rituals, based on some sacred or supernatural realm, that guides human behaviour, gives meaning to life, and unites believers into a community. *(p. 456)*

replication In sociological research, the repetition of the investigation in substantially the same way that it originally was conducted. *(p. 34)*

representative sample A selection where the sample has the essential characteristics of the total population. *(p. 40)*

research methods Specific strategies or techniques for conducting research. *(p. 35)*

resocialization The process of learning a new and different set of attitudes, values, and behaviours from those in one's previous background and experience. *(p. 105)*

riot Violent crowd behaviour that is fuelled by deep-seated emotions but not directed at one specific target. *(p. 576)*

rituals Regularly repeated and carefully prescribed forms of behaviour that symbolize a cherished value or belief. *(p. 456)*

role A set of behavioural expectations associated with a given status. *(p. 117)*

role conflict A situation in which incompatible role demands are placed on a person by two or more statuses held at the same time. *(p. 118)*

role exit A situation in which people disengage from social roles that have been central to their self-identity. *(p. 119)*

role expectation A group's or society's definition of the way a specific role ought to be played. *(p. 117)*

role performance How a person plays a role. *(p. 117)*

role strain The strain experienced by a person when incompatible demands are built into a single status that the person occupies. *(p. 118)*

role-taking The process by which a person mentally assumes the role of another person in order to understand the world from that person's point of view. *(p. 97)*

routinization of charisma A term for the process by which charismatic authority is succeeded by a bureaucracy controlled by a rationally established authority or by a combination of traditional and bureaucratic authority. *(p. 22–6)*

rumour An unsubstantiated report on an issue or subject. *(p. 580)*

sacred A term used to describe those aspects of life that are extraordinary or supernatural. *(p. 456)*

sample The people who are selected from the population to be studied. *(p. 40)*

sanctions Rewards for appropriate behaviour or penalties for inappropriate behaviour. *(p. 69)*

Sapir–Whorf hypothesis The proposition that language shapes its speakers' view of reality. *(p. 65)*

scapegoat A person or group that is incapable of offering resistance to the hostility or aggression of others. *(p. 298)*

secondary analysis A research method in which researchers use existing material and analyze data that originally was collected by others. *(p. 46)*

secondary deviance A term used to describe the process whereby a person who has been labelled deviant accepts that new identity and continues the deviant behaviour. *(p. 199)*

secondary group A larger, more specialized group in which the members engage in more impersonal, goal-oriented relationships for a limited time. *(p. 120)*

secondary sector production The sector of the economy that processes raw materials (from the primary sector) into finished goods. *(p. 518)*

secondary sex characteristics The physical traits (other than reproductive organs) that identify an individual's sex. *(p. 350)*

second shift Arlie Hochschild's term for the domestic work that employed women perform at home after they complete their workday on the job. *(p. 152)*

sect A relatively small religious group that has broken away from another religious organization to renew what it views as the original version of the faith. *(p. 468)*

segregation A term used to describe the spatial and social separation of categories of people by race/ethnicity, class, gender, and/or religion. *(p. 304)*

self-concept The totality of our beliefs and feelings about ourselves. *(p. 97)*

self-fulfilling prophecy A situation in which a false belief or prediction produces behaviour that makes the originally false belief come true. *(p. 125)*

semiperipheral nations According to world-systems analysis, nations that are more developed than peripheral nations but less developed than core nations. *(p. 279)*

sex A term used to describe the biological and anatomical differences between females and males. *(p. 350)*

sexism The subordination of one sex, usually female, based on the assumed superiority of the other sex. *(p. 327)*

sexual double standard The belief that men and women are expected to conform to different standards of sexual conduct. *(p. 370)*

sexuality (sexual) The range of human activities designed to produce erotic response and pleasure. *(p. 351)*

sexual orientation A person's preference for emotional–sexual relationships with members of the opposite sex (heterosexuality), the same sex (homosexuality), both (bisexuality), or neither (asexuality). *(p. 352)*

sexual revolution A term used for the dramatic changes that occurred regarding sexual attitudes, behaviours, and values during the 1960s. *(p. 369)*

sexual scripts Culturally created guidelines that define how, where, with whom, and under what conditions a person is to behave as a sexual being. *(p. 349)*

sick role Patterns of behaviour defined as appropriate for people who are sick. *(p. 407)*

significant others Those persons whose care, affection, and approval are especially desired and who are most important in the development of the self. *(p. 98)*

simple random sample A selection in which everyone in the target population has an equal chance of being chosen. *(p. 40)*

simple supernaturalism The belief that supernatural forces affect people's lives either positively or negatively. *(p. 458)*

slavery An extreme form of stratification in which some people are owned by others. *(p. 230)*

small group A collectivity small enough for all members to be acquainted with one another and to interact simultaneously. *(p. 170)*

social bond theory The proposition that the likelihood of deviant behaviour increases when a person's ties to society are weakened or broken. *(p. 197)*

social change The alteration, modification, or transformation of public policy, culture, or social institutions over time. *(p. 573)*

social construction of reality The process by which our perception of reality is shaped largely by the subjective meaning that we give to an experience. *(p. 125)*

social control Systematic practices developed by social groups to encourage conformity and discourage deviance. *(p. 194)*

social Darwinism The belief that those species of animals (including human beings) best adapted to their environment survive and prosper, whereas those poorly adapted die out. *(p. 13)*

social determinants of health The conditions in which people are born, grow, live, work and age, including the health system. *(p. 414)*

social devaluation A situation in which a person or group is considered to have less social value than other individuals or groups. *(p. 105)*

social facts Émile Durkheim's term for patterned ways of acting, thinking, and feeling that exist outside any one individual. *(p. 13)*

social group A group that consists of two or more people who interact frequently and share a common identity and a feeling of interdependence. *(p. 120)*

social institution A set of organized beliefs and rules that establish how a society will attempt to meet its basic social needs. *(p. 121)*

social interaction The process by which people act toward or respond to other people. *(p. 113)*

socialism An economic system characterized by public ownership of the means of production, the pursuit of collective goals, and centralized decision making. *(p. 524)*

socialization The lifelong process of social interaction through which individuals acquire a self-identity and the physical, mental, and social skills needed for survival in society. *(p. 88)*

social marginality The state of being part insider and part outsider in the social structure. *(p. 115)*

social mobility The movement of individuals or groups from one level in a stratification system to another. *(p. 230)*

social movement An organized group that acts consciously to promote or resist change through collective action. *(p. 582)*

social network A series of social relationships that link an individual to others. *(p. 120)*

social stratification The hierarchical arrangement of large social groups based on their control over basic resources. *(p. 230)*

social structure The complex framework of societal institutions (such as the economy, politics, and religion) and the social practices (such as rules and social roles) that make up a society and that organize and establish limits on people's behavior. *(p. 113)*

societal consensus A situation whereby the majority of members share a common set of values, beliefs, and behavioural expectations. *(p. 18)*

society A large social grouping that shares the same geographical territory and is subject to the same political authority and dominant cultural expectations. *(p. 4)*

sociobiology The systematic study of how biology affects social behaviour. *(p. 89)*

socioeconomic status (SES) A combined measure that attempts to classify individuals, families, or households in terms of indicators, such as income, occupation, and education, to determine class location. *(p. 238)*

sociological imagination C. Wright Mills's term for the ability to see the relationship between individual experiences and the larger society. *(p. 6)*

sociology The systematic study of human society and social interaction. *(p. 4)*

sociology of family The subdiscipline of sociology that attempts to describe and explain patterns of family life and variations in family structure. *(p. 142)*

special interest groups Political coalitions made up of individuals or groups that share a specific interest they wish to protect or advance with the help of the political system. *(p. 22–10)*

split labour market A term used to describe the division of the economy into two areas of employment: a primary sector, or upper tier, composed of higher-paid (usually dominant group) workers in more secure jobs; and a secondary sector, or lower tier, composed of lower-paid (often subordinate group) workers in jobs with little security and hazardous working conditions. *(p. 307)*

state The political entity that possesses a legitimate monopoly over the use of force within its territory to achieve its goals. *(p. 22–4)*

status A socially defined position in a group or society characterized by certain expectations, rights, and duties. *(p. 115)*

status conferral The process of giving prominence to particular individuals by focusing media attention on them. *(p. 488)*

status set A term used to describe all the statuses that a person occupies at a given time. *(p. 116)*

status symbol A material sign that informs others of a person's specific status. *(p. 117)*

stereotype An overgeneralization about the appearance, behaviour, or other characteristics of members of particular groups. *(p. 297)*

stigma According to Erving Goffman, any physical or social attribute or sign that so devalues a person's social identity that it disqualifies that person from full social acceptance. *(p. 115)*

strain theory The proposition that people feel strain when they are exposed to cultural goals that they are unable to obtain because they do not have access to culturally approved means of achieving these goals. *(p. 196)*

street crime All violent crime, certain property crimes, and certain morals crimes. *(p. 207)*

subculture A group of people who share a distinctive set of cultural beliefs and behaviours that differ in some significant way from those of the larger society. *(p. 72)*

subliminal racism A term used to describe an unconscious racism that occurs when there is a conflict of values. *(p. 300)*

succession The process by which a new category of people or type of land use gradually predominates in an area formerly dominated by another group or activity. *(p. 23–22)*

superego Sigmund Freud's term for the human conscience, consisting of the moral and ethical aspects of personality. *(p. 100)*

survey A research method in which a number of respondents are asked identical questions through a systematic questionnaire or interview. *(p. 40)*

symbol Anything that meaningfully represents something else. *(p. 21)*

symbolic interactionist perspectives The sociological approach that views society as the sum of the interactions of individuals and groups. *(p. 21)*

taboos Mores so strong that their violation is considered extremely offensive and even unmentionable. *(p. 70)*

technology The knowledge, techniques, and tools that make it possible for people to transform resources into usable forms, as well as the knowledge and skills required to use them after they are developed. *(p. 61)*

terrorism Acts of serious violence, planned and executed clandestinely and committed to achieve political ends. *(p. 583)*

theism A belief in a god or gods. *(p. 458)*

theory A set of logically interrelated statements that attempts to describe, explain, and (occasionally) predict social events. *(p. 18)*

total institution Erving Goffman's term for a place where people are isolated from the rest of society for a set period of time and come under the control of the officials who run the institution. *(p. 107)*

totalitarian political system A political system in which the state seeks to regulate all aspects of people's public and private lives. *(p. 22–8)*

tracking The assignment of students to specific courses and educational programs based on their test scores, previous grades, or both. *(p. 438)*

traditional authority Power that is legitimized by respect for long-standing custom. *(p. 22–6)*

transgender person An individual whose gender identity (self-identification as woman, man, neither, or both) does not match the person's assigned six (identification by others as male, female, or intersex based on physical/genetic sex). *(p. 352)*

treadmill of accumulation The never-ending process of accumulating more and more profit. *(p. 552)*

treadmill of production A cycle in which resources are taken out of the environment to produce goods, and waste and pollution are deposited back into the environment. *(p. 552)*

triad A group composed of three members. *(p. 170)*

triangulation Using several different research methods, data sources, investigators, and/or theoretical perspectives in the same study. *(p. 49)*

unemployment rate The percentage of unemployed persons in the labour force actively seeking jobs. *(p. 536)*

universal healthcare system System in which all citizens receive medical services paid for through tax revenues. *(p. 423)*

unstructured interview A research method involving an extended, open-ended interaction between an interviewer and an interviewee. *(p. 44)*

upper-middle-income economies Countries with an annual per capita gross national income between $US4126 and $US12,736. *(p. 264)*

urbanization The process by which an increasing proportion of a population lives in cities rather than in rural areas. *(p. 10)*

urban sociology A subfield of sociology that examines social relationships and political and economic structures in the city. *(p. 23–19)*

validity In sociological research, the extent to which a study or research instrument accurately measures what it is supposed to measure. *(p. 34)*

value contradiction Values that conflict with one another or are mutually exclusive. *(p. 69)*

values Collective ideas about what is right or wrong, good or bad, and desirable or undesirable in a particular culture. *(p. 69)*

variable In sociological research, any concept with measurable traits or characteristics that can change or vary from one person, time, situation, or society to another. *(p. 33)*

visible minority Government term used to describe those who are nonwhite, non-Indigenous, or non-Caucasian. This term is used interchangeably with "people of colour" and "racialized minorities." *(p. 296)*

wage gap A term used to describe the disparity between women's and men's earnings. *(p. 336)*

wealth The value of all of a person's or family's economic assets, including income and property, such as buildings, land, farms, houses, factories, and cars, as well as other assets, such as money in bank accounts, corporate stocks, bonds, and insurance policies. *(p. 234)*

working class (proletariat) Karl Marx's term for those who must sell their labour in order to earn enough money to survive. *(p. 236)*

world-systems analysis The perspective that the capitalist world economy is a global system divided into a hierarchy of three major types of nations—core, semiperipheral, and peripheral—in which upward or downward mobility is conditioned by the resources and obstacles that characterize the international system. *(p. 279)*

World Wide Web The computer language that allows us to access information on the Internet. *(p. 485)*

Abella, Irving. 1974. *On Strike: Six Key Labour Struggles in Canada 1919–1949.* Toronto: James Lewis and Samuel.

———, and Harold Troper. 1982. *None Is Too Many.* Toronto: Lester and Orpen Dennys.

Aberle, David F. 1966. *The Peyote Religion Among the Navaho.* Chicago: Aldine.

Aboriginal Affairs and Northern Development Canada. 2013. "Fact Sheet - 2011 National Household Survey Aboriginal Demographics, Educational Attainment and Labour Market Outcomes." Aboriginal Affairs and Northern Development Canada. Retrieved from https://www.aadnc-aandc.gc.ca/eng/1376329205785/1376329233875.

Aboujaoude, Elias. 2010. "Problematic Internet Use: An Overview." *World Psychiatry* 9 (2): 85–90.

Abraham, Carolyn. 2010. "Failing boys and the powder keg of sexual politics. *The Globe and Mail*, December 6.

———. 2012. "Unnatural Selection: Is Evolving Reproductive Technology Ushering in a New Age of Eugenics?" *The Globe and Mail* (January 10). Available: http://www.theglobeandmail.com/life/parenting/pregnancy/pregnancy-trends/unnatural-selection-is-evolving-reproductive-technology-ushering-in-a-new-age-of-eugenics/article2294636/?service=mobile.

Achilles, Rona. 1996. "Assisted Reproduction: The Social Issues." In E.D. Nelson and B.W. Robinson (eds.), *Gender in the 1990s.* Toronto: Thomson Nelson, 346–364.

Adams, Tom. 1991. *Grass Roots: How Ordinary People Are Changing America.* New York: Citadel Press.

Adler, Patricia A., and Peter Adler. 1994. *Constructions of Deviance: Social Power, Context, and Interaction.* Belmont, CA: Wadsworth.

Adorno, Theodor W., Else Frenkel-Brunswik, Daniel Levinson, and Nevitt Sanford. 1950. *The Authoritarian Personality.* New York: Harper & Row.

Africentric Alternative School Support Committee. 2009. *Africentric Education: Commonly Asked Questions.* Retrieved May 10, 2012. Available: http://www.tdsb.on.ca/boardroom/bd_agenda/uploads/generalinfo/080516%20africentric%20q&as.pdf.

AFP. 2013. "Wombs for Rent: Commercial Surrogacy Big Business in India." *The Express Tribune with the International Herald Tribune* (Feb. 25). Retrieved July 7, 2013. Online: http://tribune.com.pk/story/512264/wombs-for-rent-commercial-surrogacy-big-business-in-india.

Agger, Ben. 1993. *Gender, Culture, and Power: Toward a Feminist Postmodern Critical Theory.* Westport, CT: Praeger.

Agnew, Robert. 2012. "Dire Forecast: A Theoretical Model of the Impact of Climate Change on Crime." *Theoretical Criminology* 16 (1): 21–42.

Agyeman J. 2005. *Sustainable Communities and the Challenge of Environmental Justice.* New York: NYU Press.

Ahn, Sun Joo, and Jeremy Bailenson. 2011. "Self-Endorsing Versus Other-Endorsing in Virtual Environments: The Effect on Brand Attitude and Purchase Intention." *Journal of Advertising* 40 (Summer): 93–106.

Aiello, John R., and S.E. Jones. 1971. "Field Study of Proxemic Behavior of Young School Children in Three Subcultural Groups." *Journal of Personality and Social Psychology* 19: 351–356.

Albanese, Catherine L. 1992. *America, Religions and Religion.* Belmont, CA: Wadsworth.

Albas, Cheryl, and Daniel Albas. 1988. "Emotion Work and Emotion Rules: The Case of Exams." *Qualitative Sociology* 11 (4): 259–275.

Alberici, Emma. 2007. "What Causes People to Become Multiple Killers?" Australian Broadcasting Corporation. Accessed 8 June 2014. http://www.abc.net.au/worldtoday/content/2007/s1900575.htm.

Albrecht, Gary L. 1992. *The Disability Business: Rehabilitation in America.* Newbury Park, CA: Sage.

Allahar, Anton. 1989. *Sociology and the Periphery: Theories and Issues.* Toronto: Garamond.

Allen, Andrew. 2010. "Beyond Kentes and Kwanzaa: Reconceptualizing the Africentric school and curriculum using the principles of anti-racism education." *Our Schools/Ourselves*, Vol. 19, no. 23, pp. 327–342, Canadian Centre for Policy Alternatives. Available: https://www.policyalternatives.ca/sites/default/files/uploads/publications/ourselves/docs/OSOS_99_spring.pdf.

Allen, Mary. 2014. "Police Reported Hate Crime in Canada." Statistics Canada Catalogue no. 85-002-X http://www.statcan.gc.ca/pub/85-002-x/2014001/article/14028-eng.pdf.

Altemeyer, Bob. 1981. *Right-Wing Authoritarianism.* Winnipeg: University of Manitoba Press.

———. 1988. *Enemies of Freedom: Understanding Right-Wing Authoritarianism.* San Francisco: Jossey-Bass.

Alzheimer Society of Canada. 2010. *Rising Tide: The Impact of Dementia on Canadian Society.* Ottawa: Alzheimer Society of Canada. Available: http://www.alzheimer.ca/english/media/putyourmind09-RisingTide.htm.

Amato, Paul R. 2005. "The Impact of Family Formation Change on the Cognitive, Social, and Emotional Well-Being of the next Generation." *Marriage and Child Wellbeing* (Fall). Retrieved Feb. 12, 2012. Online: www.princeton.edu/futureofchildren/publications/docs/15_02_05.pdf.

Ambert, Anne-Marie. 2001. *Families in the New Millennium.* Toronto: Allyn and Bacon.

———. 2006. *Changing Families: Relationships in Context* (Canadian ed.). Toronto: Pearson

———. 2009. *Divorce: Facts, Causes and Consequences* (3rd ed.). Ottawa: Vanier Institute of the Family. Retrieved January 17, 2012. Available: http://www.vifamily.ca/sites/default/files/divorce_facts_causes_consequences.pdf.

Amnesty International. 2013. *Matching international commitments with national action: A human rights agenda for Canada.* Amnesty International, 9. Available: http://www.amnesty.ca/sites/default/files/canadaaihra19december12.pdf.

Andersen, Margaret L. 2006. *Thinking about Women: Sociological Perspectives on Sex and Gender*, (7th ed.). Boston: Pearson.

———. 2010. *Thinking About Women: Sociological Perspectives on Sex and Gender.* (9th ed.). Boston: Pearson.

———. 2014. *Thinking About Women: Sociological Perspectives on Sex and Gender.* (10th ed.). Boston: Pearson.

———, and Patricia Hill Collins, eds. 1998. *Race, Class, and Gender: An Anthology* (3rd ed.). Belmont, CA: Wadsworth.

Anderson, Chris. 2008. "The End of Theory: The Data Deluge Makes the Scientific Method Obsolete." *Wired* 16 June 23. Retrieved October 24, 2014. Available: http://archive.wired.com/science/discoveries/magazine/16-07/pb_theory.

Anderson, Elijah. 1990. *Streetwise: Race, Class, and Change in an Urban Community.* Chicago: University of Chicago Press.

Anderson, Eric. 2010. "At Least with Cheating There Is an Attempt at Monogamy: Cheating and Monogamism Among Undergraduate Heterosexual Men." *Journal of Social and Personal Relationships* 27 (7): 851–872. doi:10.1177/0265407510373908.

———. 2012. *The Monogamy Gap: Men, Love and the Reality of Cheating.* New York: Oxford University Press.

Anderson, Karen. 1996. *Sociology: A Critical Introduction.* Toronto: Thomson Nelson.

Anderson, Mark, and Carmen Robertson. 2011. *Seeing Red: A History of Natives in Canadian Newspapers.* Winnipeg: University of Manitoba Press.

Andrews, Lori. 2012. *I Know Who You Are and I Saw What You Did: Social Networks and the Death of Privacy.* New York: The Free Press.

Angier, Natalie. 1993. "'Stopit!' She Said. 'Nomore!'" *New York Times Book Review* (April 25): 12.

Appleton, Lynn M. 1995. "The Gender Regimes in American Cities." In Judith A. Garber and Robyne S. Turner (eds.), *Gender in Urban Research.* Thousand Oaks, CA: Sage, 44–59.

Arendt, Hannah. 1973. *On Revolution.* London: Penguin.

Armstrong, Elizabeth, Laura Hamilton, and Paula England. 2010. "Is Hooking Up Bad for Young Women?" *Contexts* (August 2010, Vol. 9, No. 3), 22–27.

Armstrong, Pat. 1993. "Work and Family Life: Changing Patterns." In G.N. Ramu (ed.), *Marriage and the Family in Canada Today* (2nd ed.). Scarborough, ON: Prentice-Hall, 127–145.

———. 1994. *The Double Ghetto: Canadian Women and Their Segregated Work.* Toronto: McClelland and Stewart.

———, and Hugh Armstrong. 1983. *A Working Majority: What Women Must Do for Pay.* Ottawa: Canadian Government Publishing Centre.

———, and Hugh Armstrong. 1994. *The Double Ghetto: Canadian Women and Their Segregated Work.* Toronto: McClelland and Stewart.

Arnold, Regina A. 1990. "Processes of Victimization and Criminalization of Black Women." *Social Justice* 17 (3): 153–166.

Arquilla, John, and David Ronfeldt. 2001. *Networks and Netwars.* Santa Monica: RAND Corporation.

Asch, Adrienne. 1986. "Will Populism Empower Disabled People?" In Harry G. Boyle and Frank Reissman (eds.), *The New Populism: The Power of Empowerment.* Philadelphia: Temple University Press, 213–228.

———. 2004. "Critical Race Theory, Feminism, and Disability." In Bonnie G. Smith and Beth Hutchison (eds.), *Gendering Disability.* New Brunswick, NJ: Rutgers University Press, 9–44.

Asch, Solomon E. 1955. "Opinions and Social Pressure." *Scientific American* 193 (5): 31–35.

———. 1956. "Studies of Independence and Conformity: A Minority of One Against a Unanimous Majority." *Psychological Monographs* 70 (9) (Whole no. 416).

Ashkenaz, Marissa. 2008. "You Mean You Chose to Be Fat? Body Image in a Virtual World." Retrieved January 2, 2012. Available: http://marissaracecourse.com/2008/06/03/you-mean-you-chose-to-be-fat-body-image-in-a-virtual-world.

Assembly of First Nations. 2008. "Sexual Exploitation/Abuse of First Nations Children." Retrieved May 22, 2009. Available: http://www.afn.ca/cmslib/general/Sex-Ex.pdf.

Association for Canadian Studies. 2012. *In God We Canadians Trust?* Retrieved May 8, 2012. Available: https://docs.google.com/viewer?url=http%3A%2F%2Fwww.acs-aec.ca%2Fpdf%2Fpolls%2FIn%2520God%2520Canadians%2520Trust%2520II.pdf.

Association of Faculties of Medicine of Canada. 2014. *Canadian Medical Education Statistics 2014.* Ottawa: Association of Faculties of Medicine of Canada.

Atchley, Robert C., and Amanda Barusch, eds. 2004. *Social Forces and Aging: An Introduction to Social Gerontology* (10th ed.). Belmont, CA: Wadsworth.

Australian Human Rights Commission. 2013. *Fact or Fiction: Stereotypes of Older Australians*. Sydney: Australian Human Rights Commission.
Avert. 2005. "HIV and AIDS in Uganda." Retrieved July 22, 2005. Available: http://www.avert.org/aidsuganda.htm.
Aw, Dr. James. 2011. "Paging Dr. Smartphone: How Medical Apps are Changing Diagnoses and Treatments." *National Post* (November 8).
Aylward, Carol A. 1999. *Canadian Critical Race Theory: Racism and the Law*. Halifax: Fernwood Publishing.
Babad, Michael. 2012. "Joe Oliver Taints All with Talk of Environmentalists, Radicals." *The Globe and Mail*. January 9. Available: http://www.theglobeandmail.com/report-on-business/top-business-stories/joe-oliver-taints-all-with-talk-of-environmentalists-radicals/article4085710/.
Baber, K. M., and Murray, C. I. 2001. "A Postmodern Feminist Approach to Teaching Human Sexuality." *Family Relations* 50 (1): 23–33.
Babbie, Earl. 2004. *The Practice of Social Research* (10th ed.). Belmont, CA: Wadsworth.
Backstrom, Lars, Paolo Boldi, Marco Rosa, Johan Ugander, and Sebastiano Vigna. 2011. "Four Degrees of Separation." CoRR, abs/1111.4570.
Baikie, Caitlyn. 2012. "Inuit Perspectives on Recent Climate Change." Skeptical Science Blog, 27 September. Available: http://www.skepticalscience.com/Inuit-Climate-Change.html.
Baker, Maureen. 2009. *Families: Changing Trends in Canada* (6th ed). Toronto: McGraw-Hill.
Baker, Robert. 1993. "'Pricks' and 'Chicks': A Plea for 'Persons.'" In Anne Minas (ed.), *Gender Basics: Feminist Perspectives on Women and Men*. Belmont, CA: Wadsworth, 66–68.
Bakht, Natasha. 2012. "Veiled Objections: Facing Public Opposition to the Niqab." In Lori G. Beaman (ed.), *Reasonable Accommodation: Managing Religious Diversity*. Vancouver: UBC Press, 70–108.
Ballantine, Jeanne H., and Floyd M. Hammack. 2012. *The Sociology of Education: A Systematic Analysis* (7th ed.). Upper Saddle River, NJ: Prentice-Hall/Pearson.
Bane, Mary Jo. 1986. "Household Composition and Poverty: Which Comes First?" In Sheldon H. Danzinger and Daniel H. Weinberg (eds.), *Fighting Poverty: What Works and What Doesn't*. Cambridge, MA: Harvard University Press.
Banerjee, Abhijit, Esther Duflo, Nathanael Goldberg, Dean Karlan, Robert Osei, William Pariente, Jeremy Shapiro, Bram Thuysbaert, and Christopher Udry. 2015. "A Multifaceted Program Causes Lasting Progress for the Very Poor: Evidence From Six Countries." *Science*, 15 May: 1260799.
Banner, Lois W. 1993. *In Full Flower: Aging Women, Power, and Sexuality*. New York: Vintage.
Barlow, Maude. 2013. *Blue Future*. New York: New York Press.
———, and Heather-Jane Robertson. 1994. *Class Warfare: The Assault on Canada's Schools*. Toronto: Key Porter Books.
Barnard, Chester. 1938. *The Functions of the Executive*. Cambridge, MA: Harvard University Press.
Barrett, David V. 2001. *The New Believers*. London: Cassell and Company.
Bascaramurty, Dakshana. 2012a. "Boomers, we'll pay for your retirement, but we won't enjoy it." *Globe and Mail* (March 30). Accessed June 14, 2014, online: http://www.theglobeandmail.com/news/national/boomers-well-pay-for-your-retirements-but-we-wont-be-happy-about-it/article4097235/.
———. 2012b. "Ethnic-focused Nursing Homes Put a Canadian Face on Filial Piety." *Globe and Mail* (January 27). Retrieved May 8, 2012. Available: http://www.theglobeandmail.com/news/national/ethnic-focused-nursing-homes-put-a-canadian-face-on-filial-piety/article1359997/?page=all.
Basow, Susan A. 1992. *Gender Stereotypes and Roles* (3rd ed.). Pacific Grove, CA: Brooks/Cole.
Battams, Nathan, Nora Spinks, and Roger Sauvé. 2014. "The Current State of Canadian Family Finances, 2013–2014 Report." Ottawa: Vanier Institute for the Family. Available: http://www.vanierinstitute.ca/family_finances.
Baudrillard, Jean. 1983. *Simulations*. New York: Semiotext.
———. 1995. *Simulacra and Simulation*. Ann Arbor: University of Michigan Press.
———. 1998. *The Consumer Society: Myths and Structures*. London: Sage (orig. pub. 1970).
Bauman, Z. (2001). *The Individualized Society*. Cambridge, UK: Polity Press.
Baym, Nancy. 2010. *Personal Connections in the Digital Age*. Cambridge, UK: Polity.
Baxter, J. 1970. "Interpersonal Spacing in Natural Settings." *Sociology* 36 (3): 444–456.
BBC News. 2012. "FBI Probes Anonymous Intercept of US-UK Hacking Call" (February 3). Retrieved February 4, 2012. Available: http://www.bbc.co.uk/news/world-us-canada-16881582.
Beaman, Lori G. 2012. *Reasonable Accommodation: Managing Religious Diversity*. Vancouver: UBC Press.
Beattie, Sara and Adam Cotter. 2010. "Homicide in Canada." *Juristat*. Ottawa: Statistics Canada.
Beaujot, Roderic. 1991. *Population Change in Canada: The Challenges of Policy Adaptation*. Toronto: McClelland and Stewart.
Becker, Howard S. 1963. *Outsiders: Studies in the Sociology of Deviance*. New York: Free Press.
Beeghley, Leonard. 2000. *The Structure of Social Stratification in the United States* (3rd ed.). Boston: Allyn & Bacon.
Beirne, Piers, and Nigel South. 2007. *Issues in Green Criminology: Confronting Harms Against Environments, Humanity and Other Animals*. Portland: Willan Publishing.
Bell, Michael. 2009. *An Invitation to Environmental Sociology* 3rd ed. Thousand Oaks, CA: Pine Forge Press.
———. 2012. *An Invitation to Environmental Sociology* 4th ed. Thousand Oaks, CA: Pine Forge Press.
Bellan, Ruben. 1978. *Winnipeg First Century: An Economic History*. Winnipeg: Queenston House Publishing.
Belley, Philippe, Marc Frenette, and Lance Lochner. 2011. "Post-Secondary Attendance by Parental Income in the U.S. and Canada: What Role for Financial Aid Policy?" University of Western Ontario, Economic Policy Research Institute Working Papers 20113, University of Western Ontario, Economic Policy Research Institute.
Bennett, Colin, Kevin Haggerty, David Lyon, and Valerie Steeves. 2014. *Transparent Lives: Surveillance in Canada*. Edmonton: Athabasca University Press.
Bergen, Raquel Kennedy. 1993. "Interviewing Survivors of Marital Rape." In Claire M. Renzetti and Raymond M. Lee (eds.), *Researching Sensitive Topics*. Newbury Park: Sage, 97–211.
Berger, Peter. 1963. *Invitation to Sociology: A Humanistic Perspective*. New York: Anchor.
———. 1967. *The Sacred Canopy: Elements of a Sociological Theory of Religion*. New York: Doubleday.
———, and Hansfried Kellner. 1964. "Marriage and the Construction of Reality." *Diogenes* 46: 1–32.
———, and Thomas Luckmann. 1967. *The Social Construction of Reality: A Treatise in the Sociology of Knowledge*. Garden City, NY: Anchor Books.
Bernard, Jessie. 1982. *The Future of Marriage*. New Haven, CT: Yale University Press (orig. pub. 1973).
Best, Joel. 2012. *Damned Lies and Statistics: Untangling Numbers from the Media, Politicians, and Activists*. Berkeley, CA: University of California Press. ProQuest ebrary. Web. 3 January 2015.
Beyerstein, Barry. 1997. "Alternative Medicine: Where's the Evidence?" *Canadian Journal of Public Health* 88 (May/June): 149–150.
Bibby, Reginald W. 1987. *Fragmented Gods: The Poverty and Potential of Religion in Canada*. Toronto: Irwin.
———. 1993. *Unknown Gods: The Ongoing Story of Religion in Canada*. Toronto: Stoddart.
———. 2001. *Canada's Teens: Today, Yesterday, and Tomorrow*. Toronto: Stoddart.
———. 2002. *Restless Gods: The Renaissance of Religion in Canada*. Toronto: Stoddart.
Bielski, Zosia. 2010. "Canada's Teen Birth and Abortion Rate Drops by 36.9 per cent." *Globe and Mail* (May 26). Retrieved April 21, 2011. Available: http://www.theglobeandmail.com/life/parenting/canadas-teen-birth-and-abortion-rate-drops-by-369-per-cent/article571685.
Biles, John, Meyer Burstein, and James Frideres. 2008. *Immigration and Integration in Canada in the Twenty-First Century*. Kingston: McGill-Queen's University Press.
Binks, Georgie. 2003. "Are Older Women Invisible?" CBC News Viewpoint. www.cbc.ca.
Bissoondath, Neil. 1994. *Selling Illusions: The Cult of Multiculturalism in Canada*. Toronto: Penguin.
———. 1998. "No Place Like Home." *New Internationalist*. Issue 305 (September). Retrieved June 3, 2009, 1. Available: http://www.geocities.com/frankie_meehan/NoPlaceLikeHome.html.
Bittle, Steven. 2014. In Deborah Brock, Amanda Glasbeek, and Carmela Murdocca (eds.), *Criminalization, Representation, Regulation: Thinking Differently About Crime*. North York, ON: University of Toronto Press, 357–384.
Bittner, Egon. 1980. *Popular Interests in Psychiatric Remedies: A Study in Social Control*. New York: Ayer.
Blackford, Karen A. 1996. "Families and Parental Disability." In Marion Lynn (ed.), *Voices: Essays on Canadian Families*. Toronto: Thomson Nelson, 161–163.
Blascovich, Jim, and Jeremy Bailenson. 2011. *Reality: Avatars, Eternal Life, New Worlds, and the Dawn of the Virtual Revolution*. New York: HarperCollins.
Blau, Peter M., and Otis Dudley Duncan. 1967. *The American Occupational Structure*. New York: Wiley.
———, and Marshall W. Meyer. 1987. *Bureaucracy in Modern Society* (3rd ed.). New York: Random House.
Blauner, Robert. 1972. *Racial Oppression in America*. New York: Harper & Row.
Bluestone, Barry, and Bennett Harrison. 1982. *The Deindustrialization of America*. New York: Basic Books.
Blumberg, Leonard. 1977. "The Ideology of a Therapeutic Social Movement: Alcoholics Anonymous." *Journal of Studies on Alcohol* 38: 2122–2143.
Blumer, Herbert G. 1946. "Collective Behavior." In Alfred McClung Lee (ed.), *A New Outline of the Principles of Sociology*. New York: Barnes & Noble, 167–219.
———. 1969. *Symbolic Interactionism: Perspective and Method*. Englewood Cliffs, NJ: Prentice-Hall.
———. 1974. "Social Movements." In R. Serge Denisoff (ed.), *The Sociology of Dissent*. New York: Harcourt Brace Jovanovich, 74–90.
BMO Student Survey. 2013. "Students Stressing More Over Finances than Academics." BMO 2012 Student

Survey. Available: http://newsroom.bmo.com/press-releases/students-stressing-more-over-finances-than-academi-tsx-bmo-20120817081292000l.

BMO Wealth Institute. 2014. "Wealth Generation: The Financial Challenges for Generations X & Y," January 2014. Available: http://www.bmo.com/pdf/mf/prospectus/en/BMO_WealthInstitute_QR_Q1_2014_EN_LQ.pdf.

Boesveld, Sarah. 2014. "The end of gender? North American society may be ready for more shades in between male and female." *National Post.* May 30, 2014. Available: news.nationalpost.com/news/Canada/the-end-of-gender-north-american-society-may-be-ready-for-more-shades-in-between-male-and-female.

Bogle, Kathleen. 2008. *Hooking Up: Sex, Dating and Relationships on Campus.* New York: New York University Press.

Bolaria, B. Singh, and P. Li. 1988. *Racial Oppression in Canada* (2nd ed.). Toronto: Garamond.

Boldt, Menno. 1993. *Surviving as Indians: The Challenge of Self-Government.* Toronto: University of Toronto Press.

Bolton, M. Anne. 1995. "Who Can Let You Die?" In Mark Novak (ed.), *Aging in Society: A Canadian Reader.* Toronto: Thomson Nelson, 385–392.

Bolton, M., A. McKay, and M. Schneider. (2010). "Relational influences on condom use discontinuation: A qualitative study of young adult women in dating relationships." *The Canadian Journal of Human Sexuality, 19*(3), 91–104.

Bonaccio, S. 2014. "Barriers to Employment for Individuals with Disabilities." Ottawa: University of Ottawa. Retrieved from: http://saea.uottawa.ca/cyber/images/dtregister/uploads/Barriers%20to%20Employment%20for%20Workers%20with%20Disabilities%20Final.pptx.

Bonacich, Edna. 1972. "A Theory of Ethnic Antagonism: The Split Labor Market." *American Sociological Review* 37: 547–549.

———. 1976. "Advanced Capitalism and Black–White Relations in the United States: A Split Labor Market Interpretation." *American Sociological Review* 41: 34–51.

Bond, Robert, Christopher Fariss, Jason Jones, Adam Kramer, Cameron Marlow, Jaime Settle, and James Fowler. 2012. "A 61-Million-Person Experiment in Social Influence and Political Mobilization." *Nature, 489*: 295–298.

Bonvillain, Nancy. 2001. *Women & Men: Cultural Constructs of Gender* (3rd ed.). Upper Saddle River, NJ: Prentice-Hall.

Bookman, Sonia. 2011. "Media, Consumption, and Everyday Life." In Will Straw, Sandra Gabrielle, and Ira Wagman (eds.). *Intersection of Media and Communications.* Toronto: Emond Montgomery, 267–287.

Bougie, Evelyn. 2010. "Family, Community, and Aboriginal Language among Young First Nations Children Living Off Reserve in Canada." Statistics Canada, *Canadian Social Trends,* Cat. no. 11-008-X. Retrieved April 22, 2102. Available: http://www.statcan.gc.ca/pub/11-008-x/2010002/article/11336-eng.pdf.

Bourdieu, Pierre. 1984. *Distinction: A Social Critique of the Judgement of Taste.* Trans. Richard Nice. Cambridge, MA: Harvard University Press.

———, and Jean-Claude Passeron. 1990. *Reproduction in Education, Society and Culture.* Newbury Park, CA: Sage.

Boyce, Jillian. 2015. "Police-Reported Crime Statistics in Canada, 2014. *Juristat.* Catalogue no. 85-002-X. Ottawa: Statistics Canada.

Boyd, Danah. 2014. *It's Complicated.* New Haven, CT: Yale University Press, 20.

———, and Kate Crawford. 2011. "Six Provocations for Big Data." Paper presented at the Oxford Internet Institute's Symposium on the Dynamics of the Internet and Society. Oxford, UK.

Boyd, Susan and Connie Carter. 2014. *Killer Weed: Marihuana Grow Ops, Media, and Justice.* Toronto: University of Toronto Press.

Brady, David, Regina Baker, and Ryan Finnigan. 2013. "When Unionization Disappears: State-Level Unionization and Working Poverty in the United States. *American Sociological Review, 78: 872–896.*

Braithwaite, Andrea. 2014. "'Seriously, Get Out': Feminists on the Forums and the War(craft) on Women." *New Media & Society* 16: 703–718.

Brannnigan, Augustine. 2004. *The Rise and Fall of Social Psychology: The Use and Misuse of the Experimental Method.* Hawthorne, NY: Aldine.

Braverman, Harry. 1974. *Labor and Monopoly Capital: The Degradation of Work in the Twentieth Century.* New York: Monthly Review Press.

Brennan, Shannon. 2011. "Violent victimization of Aboriginal women in the Canadian provinces, 2009." *Juristat.* Statistics Canada catalogue no. 85-002-X. Available: http://www.statcan.gc.ca/pub/85-002-x/2011001/article/11439-eng.pdf.

Brenner, Susan. 2010. *Cybercrime: Threats from Cyberspace.* Santa Barbara, CA: ABC-CLIO.

Breton, Raymond. 1988. "French–English Relations." In James Curtis and Lorne Tepperman (eds.), *Understanding Canadian Society.* Toronto: McGraw-Hill, 557–585.

Britt, Lory. 1993. "From Shame to Pride: Social Movements and Individual Affect." Paper presented at the 88th Annual Meeting of the American Sociological Association, Miami (August).

Broad, William, John Markoff, and David Sanger. 2011. "Israeli Test on Worm Called Crucial in Iran Nuclear Delay." *New York Times* (January 15). Retrieved: February 4, 2012. Available: http://www.nytimes.com/2011/01/16/world/middleeast/16stuxnet.html?pagewanted=all.

Brodeur, Jean-Paul. 2010. *The Policing Web.* New York: Oxford University Press.

Brooks Gardner, Carol. 1989. "Analyzing Gender in Public Places: Rethinking Goffman's Vision of Everyday Life." *American Sociologist* 20 (Spring): 42–56.

Brotman, S., B. Ryan, Y. Jalbert, and B. Rowe. 2002. "Reclaiming Space—Regaining Health: The Health Care Experiences of Two-Spirit People in Canada." *Journal of Gay and Lesbian Social Services,* 14: 67–87.

Brown, Andrew. 2012. *The Sphere of Seduction: How One-Dimensional Seduction Emerges from the Two-Dimensional Screen.* Unpublished bachelor's thesis. Vancouver: University of British Columbia.

Brown, Robert W. 1954. "Mass Phenomena." In Gardner Lindzey (ed.), *Handbook of Social Psychology,* vol. 2. Reading, MA: Addison-Wesley, 833–873.

Brownlee, Jamie. 2005. *Ruling Canada: Corporate Cohesion and Democracy.* Halifax: Fernwood Publishing.

Brownridge, D.A. 2006a. "Intergenerational transmission and dating violence victimization: Evidence from a sample of female university students in Manitoba." *Canadian Journal of Community Mental Health,* 25(1), 75–93.

———. 2006b. "Partner violence against women with disabilities: Prevalence, risk and explanations." *Violence Against Women,* 12(9), 805–822.

———. 2008. "Understanding the elevated risk of partner violence against Aboriginal women: A comparison of two nationally representative surveys of Canada." *Journal of Family Violence,* 23(5), 353–367.

Bruce, Steve. 2006. "Secularization and the Impotence of Individualized Religion." *The Hedgehog Review.* Spring/Summer: 35–45.

Brynjolfsson, Erik and Andrew McAfee. 2014. *The Second Machine Age.* New York: W.W. Norton.

Buechler, Steven M. 2000. *Social Movements in Advanced Capitalism: The Political Economy and Cultural Construction of Social Activism.* New York: Oxford University Press.

Bullard, Robert B., and Beverly H. Wright. 1992. "The Quest for Environmental Equity: Mobilizing the African-American Community for Social Change." In Riley E. Dunlap and Angela G. Mertig (eds.), *American Environmentalism: The U.S. Environmental Movement, 1970–1990.* New York: Taylor & Francis, 39–49.

Bullard, Robert D. 1990. *Dumping in Dixie: Race, Class, and Environmental Quality.* Boulder, CO: Westview Press.

———. 1993. *Confronting Environmental Racism.* Boston: South End Press.

———. 1996. *Symposium: The legacy of American apartheid and environmental racism.* St. John's J. Legal Comment. 9:445–74

———. (Ed.). 2007. *Growing Smarter: Achieving Livable Communities, Environmental Justice, and Regional Equity.* Cambridge, MA: MIT Press.

Bureau of Labor Statistics. 2014. Union Members–2013. Washington: Bureau of Labor Statistics.

Burger, Jerry M. 2009. "Replicating Milgram: Would People Still Obey Today?" *American Psychologist* 64: 1–11.

Burgess, Ernest W. 1925. "The Growth of the City." In Robert E. Park and Ernest W. Burgess (eds.), *The City.* Chicago: University of Chicago Press, 47–62.

Butler, J. 1999. *Gender Trouble: Feminism and the Subversion of Identity.* New York: Routledge.

Butler, Robert N. 1975. *Why Survive? Being Old in America.* New York: Harper & Row.

Buttel, Frederick H. 2010. "Social institutions and environmental change." *The International Handbook of Environmental Sociology,* 2nd ed., 33–47.

Cable News Network. 1997. "Study: Despair Increases Health Risks in Middle-Aged Men." CNN Website: August 26, 1997. Available: http://www.cnn.com.

Cable, Sherry, and Charles Cable. 1995. *Environmental Problems, Grassroots Solutions: The Politics of Grassroots Environmental Conflict.* New York: St. Martin's Press.

Cain, P.A. 1993. "Feminism and the Limits of Equality." In D.K. Weisberg (ed.), *Feminist Legal Theory: Foundations.* Philadelphia: Temple University Press, 237–247.

California State Auditor. 2010. "California Department of Corrections and Rehabilitation: Inmates Sentence Under the Three Strikes Law and a Small Number of Inmates Receiving Specialty Health Care Represent Significant Costs." *Report 2009-107.2* (May). Retrieved June 7, 2011. Available: www.bsa.ca.gov/reports/summary/2009-107.2.

Calliste, Agnes. 1987. "Sleeping Car Porters in Canada: An Ethically Submerged Split Labour Market." *Canadian Ethnic Studies* 19: 1–20.

Callwood, June. 2005. "On Turning 80." *Chatelaine* (May): 56.

Campaign 2000. 2011. "Revisiting Family Security in Insecure Times." *Report Card on Child and Family Poverty in Canada.* Available: http://www.campaign2000.ca/reportCards/national/2011EnglishReportCard.pdf.

———. 2014. "2014 Report Card on Child and Family Poverty in Canada." Available: http://www.campaign2000.ca/anniversaryreport/CanadaRC2014EN.pdf.

Campbell, Marie, and Frances Gregor. 2004. "Theory 'in' Everyday Life." In William K. Carroll (ed.), *Critical Strategies for Social Research.* Toronto: Canadian Scholars' Press, 170–180.

Canada (Attorney General) v. *Bedford,* 2012 ONCA 186.

Canada (Attorney General) v. *Bedford,* 2013 SCC 72, [2013] 3 S.C.R. 1101.

Canada Centre for Global Security Studies. 2011. *Casting a Wider Net.* Toronto: Munk School of Global Affairs, University of Toronto.

Canadian Bar Association. 2013. *The Future of Legal Services in Canada: Trends and Issues.* Ottawa: Canadian Bar Association.
———. 2015. "Canada's Richest People 2015: The Top 100 Richest Canadians." Retrieved from: http://www.canadianbusiness.com/lists-and-rankings/richest-people/top-100-richest-canadians-2015/.
Canadian Cancer Society. 2015. Breast Cancer Statistics. Available: http://www.cancer.ca/en/cancer-information/cancer-type/breast/statistics/?region=bc.
Canadian Council on Learning. 2007. "Canada Slow to Overcome Limits for Disabled Learners." *Lessons in Learning* (February 27). Retrieved May 6, 2009. Available: http://www.ccl-cca.ca/pdfs/LessonsIn-Learning/Feb-26-07-Canada-slow-to-ov.pdf.
Canadian Council on Social Development. 2009. "A Profile of Economic Security in Canada." Stats and Facts Retrieved June 17, 2009. Available: http://www.ccsd.ca/factsheets/economic_security/poverty/index.htm.
Canadian Diabetes Association. 2015. "Diabetes Organizations React to Auditor General Report on Aboriginal Health Services." Available: https://www.diabetes.ca/newsroom/search-news/diabetes-orgs-react-to-auditor-general-report.
Canadian Federation of Students. 2013. 2013 Fact Sheets. Available: http://cfs-fcee.ca/wp-content/uploads/sites/2/2013/11/Fact-Sheet-fees-2013-Nov-En.pdf.
———. 2014. "Tuition Fees and Access to Education." Available: http://cfs-fcee.ca/the-issues/tuition-fees-and-access-to-education/.
Canadian Institute for Health Information. 2004. *Improving the Health of Canadians.* Retrieved July 22, 2005. Available: http://www.cihi.ca.
———. 2013. *Physicians in Canada, 2013: Summary Report.* Ottawa: Canadian Institute for Health Information.
———. 2014. *National Health Expenditure Trends, 1975 to 2014: Report.* Ottawa: Canadian Institute for Health Information. Available: http://www.cihi.ca/web/resource/en/nhex_2014_report_en.pdf.
Canadian Internet Registration Authority. 2013. Factbook 2013. Available: http://www.cira.ca/factbook/2013/ca-in-canada.html.
Canadian Medical Association. 2010. "Health Care System: Canadians Fear Effect of the 'Silver Tsunami.'" Retrieved May 13, 2012. Available: http://www.cma.ca/advocacy/silver-tsunami.
Canadian Mental Health Association. 2015. "Income." Canadian Mental Health Association Ontario. Retrieved from: https://ontario.cmha.ca/mental-health/services-and-support/income/
Canadian Race Relations Foundation. 2014. "Religion, Racism, and Intergroup Relations in Canada 2014." Available: http://www.crr.ca/en/news-a-events/item/24974-information-handout-on-religion-racism-intergroup-relations-and-integration-results-from-january-2014-survey.
Canadian Race Relations Foundation. 2014. *Report on Canadian Values.* Available: http://www.crr.ca/images/Our_Canada/Report_on_Canadian_Values_Billingual-wCOVER.pdf.
Canadian Safe Schools Network. 2012. "Programs and Resources." Retrieved May 10, 2012. Available: http://www.canadiansafeschools.com/programs/overview.htm.
Canadian Sportfishing Industry Association. 2007. "Federal Animal Cruelty Legislation." Available: http://www.csia.ca/media/FEDERAL_ANIMAL_CRUELTY_LEGISLATION.pdf. Accessed August 18, 2007.
Cancian, Francesca M. 1990. "The Feminization of Love." In C. Carlson (ed.), *Perspectives on the Family: History, Class, and Feminism.* Belmont, CA: Wadsworth.
Cannon, M. 1998. "The Regulation of First Nations Sexuality." *Canadian Journal of Native Studies* 18 (1): 1–18.
Cantor, Muriel G. 1980. *Prime-Time Television: Content and Control.* Newbury Park, CA: Sage.
———. 1987. "Popular Culture and the Portrayal of Women: Content and Control." In Beth B. Hess and Myra Marx Ferree, *Analyzing Gender: A Handbook of Social Science Research.* Newbury Park, CA: Sage, 190–214.
Cantril, Hadley. 1941. *The Psychology of Social Movements.* New York: Wiley.
Cappon, Paul. 2011. *"Exploring the 'Boy Crisis' in Education."* Canadian Council on Learning, Bosch Foundation, 1. Available: http://files.eric.ed.gov/fulltext/ED518173.pdf.
Carmichael, Amy. 2005. "And the Booby Prize Goes to ..." Retrieved January 13, 2006. Available: http://www.theglobe.ca/servlet/story/RTGAM.20050724.wimplants0724/BNStory/National/.
Carrington, Peter, and Robin Fitzgerald. 2012. "Do Police Discriminate Against Minority Youth in Canada?" In Julian Roberts and Michelle Grossman (eds.), *Criminal Justice in Canada* (4th ed.). Toronto: Nelson, 187–198.
Carroll, William K. 2004. *Critical Strategies for Social Research.* Toronto: Canadian Scholars' Press.
Carroll, William, and Jerome Klassen. 2010. "Hollowing Out Corporate Canada? Changes in the Corporate Network Since the 1990s." *Canadian Journal of Sociology* 35: 1–30.
Carter, Jimmy. 2009. "The Words of God do not Justify Cruelty to Women." The Elders. Available: http://theedlders.org/article/words-god-do-not-justify-cruelty-women.
Cassidy, B., R. Lord, and N. Mandell. 2001. "Silenced and Forgotten Women: Race, Poverty, and Disability." In Nancy Mandell (ed.), *Feminist Issues: Race, Class, and Society* (3rd ed.). Toronto: Prentice-Hall, 75–107.
Castellano, Marlene. 2002. "Aboriginal Family Trends: Extended Families, Nuclear Families, Families of the Heart." Ottawa: Vanier Institute of the Family. Retrieved May 12, 2009. Available: http://www.vifamily.ca/library/cft/aboriginal.html.
Castells, Manuel. 1977. *The Urban Question.* London: Edward Arnold (orig. pub. 1972 as *La Question Urbaine,* Paris).
———. 2000a. *The Rise of the Network Society* (2nd ed.). Oxford: Blackwell Publishers.
———. 2000b. "Materials for an Exploratory Theory of the Network Society." *British Journal of Sociology* (January/March): 5–24.
———. 2004. "Informationalism and the Network Society." In Manuel Castells (ed.), *The Network Society.* Cheltenham: Edward Elgar, 3–45.
Castles, Stephen. 1995. "Trois Siècles de Dépopulation Amerindienne." In L. Normandeau and V. Piche (eds.), *Les Populations Amerindienne et Inuit du Canada.* Montreal: Presse de l'Université de Montréal.
Catalyst Canada. 2012. "Women CEOs and Heads of the Financial Post 500." Retrieved February 6, 2012. Available: http://www.catalyst.org/publication/322/women-ceos-and-heads-of-the-financial-post-500.
Catalyst Canada. 2014. "Statistical Overview of Women in the Workplace." March 3, 2014. Available: www.catalyst.org/knowledge/statistical-overview-women-workplace.
Catholic News Agency. 2009. "Benedict XVI Says Church Needs to Proclaim Gospel on the 'Digital Continent.'" Retrieved May 8, 2012. Available: http://www.catholicnewsagency.com/news/benedict_xvi_says_church_needs_to_proclaim_gospel_on_the_digital_continent.
Catton, William, and Riley Dunlap. 1978. "Environmental Sociology: A New Paradigm." *American Sociologist,* 13:41–49.
Cavender, Gray. 1995. "Alternative Theory: Labeling and Critical Perspectives." In Joseph F. Sheley (ed.), *Criminology: A Contemporary Handbook* (2nd ed.). Belmont, CA: Wadsworth, 349–371.
CBC. 2011. "Lost Boy: Mijok Lang." *The Current.* 27 June 2011.
CBC News. 2006. "In Depth: The Hutterites." CBC News In Depth. Retrieved January 20, 2009. Available: http://www.cbc.ca/news/background/hutterites/.
———. 2008. "PM Cites 'Sad Chapter' in Apology for Residential Schools." Retrieved April 9, 2012. Available: http://www.cbc.ca/news/canada/story/2008/06/11/aboriginal-apology.html.
CBS News. 2007. "Outsourced 'Wombs-for-Rent' in India." Retrieved Feb. 9, 2008. Online: www.cbsnews.com/stories/2007/12/31/health/main3658750.shtml.
Ceballos, Gerardo, Paul R. Ehrlich, Anthony D. Barnosky, Andrés García, Robert M. Pringle, and Todd M. Palmer. 2015 "Accelerated modern human–induced species losses: Entering the sixth mass extinction." *Science Advances* 1, no. 5: 1–5.
Center for Media Literacy. 2011. "What Is Media Literacy? A Definition ... and More." Retrieved January 30, 2012. Available: http://www.medialit.org/reading-room/what-media-literacy-definitionand-more.
Center for Strategic and International Studies. 2011. *Global Aging and the Future of Emerging Markets.* Washington: Center for Strategic and International Studies.
Chafetz, Janet Saltzman. 1984. *Sex and Advantage: A Comparative, Macro-Structural Theory of Sex Stratification.* Totowa, NJ: Rowman & Allanheld.
Chalfant, H. Paul, Robert E. Beckley, and C. Eddie Palmer. 1994. *Religion in Contemporary Society* (3rd. ed.). Ithaca, IL: Peacock.
Challenge of Change: A Study of Canada's Criminal Prostitution Laws. 2006. Report of the Standing Committee on Justice and Human Rights. Report of the Subcommittee on Solicitation Laws.
Chandler, Tertius, and Gerald Fox. 1974. *3000 Years of Urban History.* New York: Academic Press.
Chapkis, W. 2000. "Power and Control in the Commercial Sex Trade." In Ronald Weitzer (ed.), *Sex for Sale: Prostitution, Pornography and the Sex Industry.* New York: Routledge, 181–201.
Chawla, R., and S. Uppal. 2012. "Household debt in Canada." Statistics Canada. Component of Statistics Canada Catalogue no. 75-001-X. *Perspectives on Labour and Income.* Available: http://www.statcan.gc.ca/pub/75-001-x/2012002/article/11636-eng.pdf.
Chiasson, Miriam. 2013. "Incommodations, 2008–2013: The Aftermath of the Reasonable Accommodation Crisis." Third of Five Reports prepared by Miriam Chiasson for David Howes and the Centaur Jurisprudence Project, Centre for Human Rights and Legal Pluralism, McGill University, August 2013. Available: http://canadianicon.org/wp-content/uploads/2014/03/TMODPart3-Sources-1.pdf.
Cho, Kyong. 2011. "New Media and Religion: Observations on Research." *Communication Research Trends* (March). Retrieved May 11, 2012. Available: http://findarticles.com/p/articles/mi_7081/is_1_30/ai_n57221190/?tag=content;col1.
Chossudovsky, Michel. 1997. *The Globalization of Poverty.* Penang: Third World Network.
Christakis, Dimitri, Michelle Garrison, Todd Herrenkohl, Kevin Haggerty, Frederick Rivara, Chuan Zhou, and Kimberley Liekweg. 2013. "Modifying Media Content for Preschool Children: A Randomized Control Trial." *Pediatrics* 131: 431–438.
Chu, Kathy. 2012. "Apple Plans Environmental Audits of China Suppliers." *USA Today* (Feb. 20). Available: www.usatoday.com/tech/

news/story/2012-02-20/apple-china-environmental-audits/53167970/1.
Chudnovsky, Anna. 2014. "In My Class, Child Poverty Is No Numbers Game." *The Tyee* (November 25). Available: http://thetyee.ca/Opinion/2014/11/25/Child-Poverty-No-Numbers-Game/.
Chunn, Dorothy E. 2000. "Politicizing the Personal: Feminism, Law, and Public Policy." In Nancy Mandell and Ann Duffy (eds.), *Canadian Families: Diversity, Conflict, and Change* (2nd ed.). Toronto: Harcourt, 225–259.
Church, Elizabeth. 2003. "Kinship and Stepfamilies." In Marion M. Lynn (ed.), *Voices: Essays on Canadian Families* (2nd ed.). Toronto: Thomson Nelson, 55–75.
CIA (Central Intelligence Agency). 2001. *The World Factbook 2001*. Washington, D.C.: Office of Public Affairs.
———. 2004. *The World Factbook 2004*. Washington D.C.: Office of Public Affairs.
———. 2008. *The World Factbook 2004*. Washington D.C.: Office of Public Affairs.
———. 2014. *The World Factbook*. Retrieved November 6, 2014. Available: https://www.cia.gov/library/publications/the-world-factbook/geos/rs.html.
———. 2015. *The World Factbook*. Washington: Central Intelligence Agency.
Cisco. 2014. "Gen Y: New Dawn for Work, Play, Identity." Cisco Connected World Technology Report. Available: http://www.cisco.com/c/dam/en/us/solutions/enterprise/connected-world-technology-report/2012-CCWTR-Chapter1-Global-Results.pdf.
Citizenship and Immigration Canada. 2008. *Facts and Figures 2007—Immigration Overview*. Ottawa: Citizenship and Immigration Canada. Retrieved April 10, 2009. Available: http://www.cic.gc.ca/english/resources/statistics/facts2007/index.asp.
———. 2011. *Facts and Figures 2011 – Immigration Overview*. Ottawa: Citizenship and Immigration Canada.
———. 2014. *2014 Annual Report to Parliament on Immigration*. Ottawa: Citizenship and Immigration Canada.
———. 2015. Facts and Figures 2013 – Immigration overview: Permanent residents. Available: http://www.cic.gc.ca/english/resources/statistics/facts2013/permanent/index.asp#figure1
Clark, Lorenne, and Debra Lewis. 1977. *Rape: the Price of Coercive Sexuality*. Toronto: Women's Press.
Clark, Warren. 1998. "Religious Observance: Marriage and Family." *Canadian Social Trends* (Autumn): 2–7. Ottawa: Statistics Canada.
———. 2003. "Pockets of Belief: Religious Attendance in Canada." *Canadian Social Trends* (Spring): 2–5. Ottawa: Statistics Canada.
———, and Susan Crompton. 2006. "Till Death Do Us Part? The Risk of First and Second Marriage Dissolution." *Canadian Social Trends*. Statistics Canada Cat. no. 11-008. Retrieved May 6, 2009. Available: http://www.statcan.gc.ca/pub/11-008-x/2006001/pdf/9198-eng.pdf.
———, and Grant Schellenberg. 2006. "Who's Religious?" *Canadian Social Trends* (Summer): 2–9. Ottawa: Statistics Canada.
Clayman, Steven E. 1993. "Booing: The Anatomy of a Disaffiliative Response." *American Sociological Review* 58 (1): 110–131.
Clement, Wallace. 1975. *The Canadian Corporate Elite*. Toronto: McClelland and Stewart.
———, and John Myles. 1994. *Relations of Ruling: Class and Gender in Postindustrial Societies*. Montreal: McGill-Queen's University Press.
Cloward, Richard A., and Lloyd E. Ohlin. 1960. *Delinquency and Opportunity: A Theory of Delinquent Gangs*. New York: Free Press.
CMEC (Council of Ministers of Education). 2014. "Education Indicators in Canada: An International Perspective 2013." Catalogue no. 81-604-X, ISSN: 1920-5910. Tourism and the Centre for Education Statistics Division. Available: http://www.cmec.ca/Publicatons/Lists?Publications/Attachments/322/Education-Indicators-Canada-Internationsl-Perspective-2013.pdf.
CNN. 2005. "Leadership Vacuum Stymied Aid Offers." CNN.com. Retrieved September 16, 2005. Available: http://edition.cnn.com/2005/US/09/15/katrina.response/index.html.
Coakley, Jay. 2009. *Sports in Society: Issues and Controversies* (10th ed.). New York: McGraw-Hill.
Coates, Ken. 2015. *#idlenomore and the Remaking of Canada*. Regina: University of Regina Press.
Cohen, Marjorie Griffin. 1993. "Capitalist Development, Industrialization, and Women's Work." In Graham S. Lowe and Harvey J. Krahn (eds.), *Work in Canada*. Toronto: Thomson Nelson, 142–144.
Cohen, Theodore, ed. 2001. *Men and Masculinity: A Text Reader*. Belmont, CA: Wadsworth, 2–3.
Collier, Peter, and David Horowitz. 1987. *The Fords: An American Epic*. New York: Summit Books.
Collins, Randall. 1971. "A Conflict Theory of Sexual Stratification." *Social Problems* 19 (1): 3–21.
———. 1982. *Sociological Insight: An Introduction to Non-Obvious Sociology*. New York: Oxford University Press.
———. 1987. "Interaction Ritual Chains, Power, and Property: The Micro–Macro Connection as an Empirically Based Theoretical Problem." In Jeffrey C. Alexander et al. (eds.), *The Micro–Macro Link*. Berkeley, CA: University of California Press, 193–206.
———. 1997. "An Asian Route to Capitalism: Religious Economy and the Origins of Self-Transforming Growth in Japan." *American Sociological Review* 62: 843–865.
Collins, Rebecca L. 2011. "Content Analysis of Gender Roles in Media: Where Are We Now and Where Should We Go?" *Sex Roles*, 64 (3/4): 290–329.
Collison, Robert. 2006. "Men and the Pursuit of Youthfulness." *Financial Post* (September 30).
Coltrane, Scott. 1992. "The Micropolitics of Gender in Nonindustrial Societies." *Gender and Society* 6: 86–107.
Comack, Elizabeth. 1996. *Women in Trouble*. Halifax: Fernwood Publishing
———. 2009. "Feminism and Criminology." In Rick Linden (ed.), *Criminology: A Canadian Perspective* (6th ed.) Toronto: Nelson, 164–195.
———. 2012. "Feminism and Criminology." In Rick Linden (ed.), *Criminology: A Canadian Perspective* (7th ed.) Toronto: Nelson, 173, 179–216.
———. 2016. "Feminism and Criminology." In Rick Linden (ed.), *Criminology: A Canadian Perspective* (8th ed.). Toronto: Nelson.
———, and Evan Bowness. 2010. "Dealing the Race Card: Public Discourse on the Policing of Winnipeg's Inner-city Communities." *Canadian Journal of Urban Research* 19: 34–50.
Comfort, Alex. 1976. "Age Prejudice in America." *Social Policy* 7 (3): 3–8.
Commission of Inquiry into the Investigation of the Bombing of Air India Flight 182. 2010. *Air India Flight 182: A Canadian Tragedy. Volume One: The Overview*. Ottawa: Public Works and Government Services Canada.
Commission on Systemic Racism in the Ontario Criminal Justice System. 1995. *Report of the Commission on Systemic Racism in the Ontario Criminal Justice System*. Toronto: Queen's Printer for Ontario.
Commonwealth Fund. 2002. "Canadian Adults' Health Care System Views and Experiences." *Commonwealth Fund 2001 International Health Policy Survey*. New York: Commonwealth Fund. Retrieved December 17, 2002. Available: http://www.cmwf.org/programs/international/can_sb_552.pdf.
———. 2007. *2007 International Health Policy Survey in Seven Countries*. New York: Commonwealth Fund. Retrieved April 21, 2009. Available: http://www.commonwealthfund.org/Content/Surveys/2007/2007-International-Health-Policy-Survey-in-Seven-Countries.aspx.
———. 2015. *2014 International Profiles of Health Care Systems*. New York: The Commonwealth Fund.
Condry, Sandra McConnell, John C. Condry Jr., and Lee Wolfram Pogatshnik. 1983. "Sex Differences: A Study of the Ear of the Beholder." *Sex Roles* 9: 697–704.
Conference Board of Canada. 2011. *World Income Inequality: Is the world becoming more unequal?* Retrieved from: http://www.conferenceboard.ca/hcp/hot-topics/worldinequality.aspx.
Conrad, Peter. 1975. "The Discovery of Hyperkinesis." *Social Problems* 23 (October): 12–21.
Conrad, Peter, and Joseph W. Schneider. 1980. "The Medical Control of Deviance: Conquests and Consequences." In Julius A. Roth (ed.), *Research in the Sociology of Health Care: A Research Annual*, vol. 1. Greenwich, CT: Jai Press, 1–53.
———. 1992. *Deviance and Medicalization: From Badness to Sickness*. Philadelphia: Temple University Press.
Cook, Kevin. 2014. *Kitty Genovese: The Murder, the Bystanders, the Crime that Changed America*. New York: W.W. Norton and Company.
Cook, Ramsay. 1995. *Canada, Quebec and the Uses of Nationalism* (2nd ed.). Toronto: McClelland and Stewart.
Cook, Shirley J. 1969. "Canadian Narcotics Legislation, 1908–1923: A Conflict Model Interpretation." *Canadian Review of Sociology and Anthropology* 6 (1): 36–46.
Cooke, Martin, Daniel Beavon, and Mindy McHardy. 2004. *Measuring the Well-Being of Aboriginal People: An Application of the United Nations' Human Development Index to Registered Indians in Canada, 1981–2001*. Ottawa: Indian and Northern Affairs Canada.
Cool, Julie, 2010. *Wage Gap Between Women and Men*. Background Paper: Library of Parliament Publication No 2010-30-E (July 29). Retrieved June 1, 2012. Available: http://www.parl.gc.ca/Content/LOP/ResearchPublications/2010-30-e.pdf.
Cooley, Charles Horton. 1922. *Human Nature and Social Order*. New York: Scribner (orig. pub. 1902).
———. 1962. *Social Organization*. New York: Schocken Books (orig. pub. 1909).
Coontz, Stephanie. 1992. *The Way We Never Were: American Families and the Nostalgia Trap*. New York: Basic Books.
———. 2011. The Sociology of Childhood (3rd ed.). Thousand Oaks, CA: Pine Forge.
Cossman, Brenda. 2004. "Sexuality, Queer Theory, and 'Feminism After': Reading and Rereading the Sexual Subject." *McGill Law Journal* 49:847–851.
Council of Ministers of Education (CMEC). 2014. *Education Indicators in Canada: An International Perspective 2013*. Catalogue no. 81-604-X ISSN: 1920-5910. Tourism and the Centre for Education Statistics Division. Available: http://www.cmec.ca/Publications/Lists/Publications/Attachments/322/Education-Indicators-Canada-International-Perspective-2013.pdf.
Cowgill, Donald O. 1986. *Aging Around the World*. Belmont, CA: Wadsworth.
Craig, E. 2008. "Re-interpreting the Criminal Regulation of Sex Work in Light of R. v. Labaye." *Canadian Criminal Law Review* 12:3, 327.
———. 2009. "Ten Years After Ewanchuk the Art of Seduction Is Alive and Well: An Examination of the Mistaken Belief in Consent Defence." *Canadian Criminal Law Review* 13:3, 247.
Craig, Steve. 1992. "Considering Men and the Media." In Steve Craig (ed.), *Men, Masculinity, and the Media*. Newbury Park, CA: Sage, 1–7.
Craig, Wendy, Debra Pepler, and Dilys Haner. 2012. "Healthy Development

Depends on Healthy Relationships." Paper prepared for the Division of Childhood and Adolescence, Centre for Health Promotion, Public Health Agency of Canada, November 15, 2012.
Crane, B., and Crane-Seeber, J. 2003. "The Four Boxes of Gendered Sexuality: Good Girl/Bad Girl and Tough Guy/Sweet Guy." In Robert Heasley and Betsy Crane, *Sexual Lives: A Reader on the Theories and Realities of Human Sexualities*. New York: McGraw-Hill, 196–216.
Cranswick, Kelly. 2003. *General Social Survey Cycle 16: Caring for an Aging Society.* Ottawa: Statistics Canada.
Creswell, John W. 1998. *Qualitative Inquiry and Research Design: Choosing Among Five Traditions.* Thousand Oaks, CA: Sage.
Crichton, Anne, Ann Robertson, Christine Gordon, and Wendy Farrant. 1997. *Health Care: A Community Concern?* Calgary: University of Calgary Press.
Crompton, Susan. 2005. *Always a Bridesmaid: People Who Don't Intend to Marry.* Canadian Social Trends, Statistics Canada. Cat. no. 11-008. Retrieved January 26, 2012. Available: http://thesurvey.womenshealthdata.ca/pdf_files/7961.pdf.
———. 2011. "Women with Activity Limitations." In *Women in Canada: A Gender-Based Statistical Report*. Ottawa: Statistics Canada. Statistics Canada. Cat. no. 89-502-X. Retrieved September 13, 2011. Available: http://www.statcan.gc.ca/pub/89-503-x/2010001/article/11545-eng.htm.
CRTC. 2011. "Frequently Asked Questions." Ottawa: CRTC. Retrieved September 23, 2011. Available: http://www.crtc.gc.ca/eng/faqs.htm.
CTV. 2005. "Blacks Stopped More Often by Police, Study Finds." Retrieved July 12, 2005. Available: http://www.ctv.ca/servlet/ArticleNews/story/CTVNews/1117145635847_112554835?s_name=&no_ads=.
Cui, D. 2011. Two Multicultural Debates and the Lived Experiences of Chinese-Canadian Youth. *Canadian Ethnic Studies*, 43/44(3–1), 123–143.
Cumming, Elaine C., and William E. Henry. 1961. *Growing Old: The Process of Disengagement.* New York: Basic Books.
Currie, C.L., T.C. Wild, D.P. Schopflocher, L. Laing, and P. Veugelers. 2012. "Racial Discrimination Experienced by Aboriginal University Students in Canada." *Canadian Journal of Psychiatry*, 57(10), 617–625.
Curtis, James, and Ronald D. Lambert. 1994. "Culture." In R. Hagedorn (ed.), *Sociology* (5th ed.). Toronto: Holt Rinehart and Winston, 57–86.
———, Edward Grabb, and Neil Guppy, eds. 1999. *Social Inequality in Canada: Patterns, Problems, and Policies* (3rd ed.). Scarborough, ON: Prentice-Hall.
Curtiss, Susan. 1977. *Genie: A Psycholinguistic Study of a Modern Day "Wild Child."* New York: Academic Press.
Czitrom, D. 1982. *Media and the American Mind: From Morse to McLuhan.* Chapel Hill, NC: University of North Carolina Press.
Dahl, Robert A. 1961. *Who Governs?* New Haven, CT: Yale University Press.
Dahrendorf, Ralf. 1959. *Class and Class Conflict in an Industrial Society.* Stanford, Cal.: Stanford University Press.
Daschuk, James. 2013. *Clearing the Plains: Disease Politics of Starvation and the Loss of Aboriginal Life.* Regina: University of Regina Press.
Dauvergne, Mia. 2003. "Family Violence Against Seniors." *Canadian Social Trends* (Spring): 10–14. Ottawa: Statistics Canada.
David Suzuki Foundation. 2015. "Canada's Emissions." Available: http://www.davidsuzuki.org/issues/climate-change/science/canada-climate-change/canadas-emissions/.
Davies, Scott, and Neil Guppy. 2013. *The Schooled Society: An Introduction to the Sociology of Education* (3rd ed.). Don Mills, ON: Oxford University Press.
———, and Neil Guppy. 2014. *The Schooled Society: An Introduction to the Sociology of Education* (3rd ed.). Don Mills, ON: Oxford University Press.
Davis, K. 1937. "The Sociology of Prostitution." *American Sociological Review* 2 (5): 744–755.
Davis, Fred. 1992. *Fashion, Culture, and Identity.* Chicago: University of Chicago Press.
Davis, Kingsley. 1940. "Extreme Social Isolation of a Child." *American Journal of Sociology* 45 (4): 554–565.
———, and Judith Blake. 1956. "Social Structure and Fertility: An Analytical Framework." *Economic Development and Cultural Change* 4 (April): 211–235.
———, and Wilbert Moore. 1945. "Some Principles of Stratification." *American Sociological Review* 7 (April): 242–249.
Dean, L.M., F.N. Willis, and J.N. la Rocco. 1976. "Invasion of Personal Space as a Function of Age, Sex and Race." *Psychological Reports* 38 (3) (pt. 1): 959–965.
Dear, Michael, and Steven Flusty. 1998. "Postmodern Urbanism." *Annals of the Association of American Geographers* 88 (1): 50–72.
Deetz, Nanette Bradley. 2014. "Idle No More's Sylvia McAdam Reacts to Enbridge Northern Gateway Pipeline." Online. Net. Retrieved 30 November 2014. Available: http://nativenewsonline.net/currents/idle-mores-sylvia-mcadam-reacts-enbridge-northern-gateway-pipeline/.
DeKeseredy, W.S., and M. Dragiewicz. (2014). "Woman abuse in Canada: Sociological reflections on the past, suggestions for the future." *Violence Against Women*, 20(2) 228–244.
Delgado, Richard. 1995. "Introduction." In Richard Delgado (ed.), *Critical Race Theory: The Cutting Edge.* Philadelphia: Temple University Press, xiii–xvi.
Denny, Keith, and Marnie Brownell. 2010. "Taking a Social Determinants Perspective on Children's Health and Development." *Canadian Journal of Public Health*, 101: S4-S7.
Denzin, Norman K. 1989. *The Research Act* (3rd ed.). Englewood Cliffs, NJ: Prentice-Hall.
DePape, B. (2012, Spring). "Power of youth: Youth and community-led activism in Canada." *Our Schools, Our Selves, 21*, 15–21. Retrieved from http://search.proquest.com/docview/1038154846?accountid=15067.
Department of Justice. 2009. *Assisted Human Reproduction Act.* Retrieved May 25, 2009. Available: http://laws.justice.gc.ca/en/A-13.4/.
———. 2011. "About Family Violence in Canada." Retrieved January 23, 2012. Available: http://www.justice.gc.ca/eng/pi/fv-vf/about-aprop.
———. 2014. "Technical Paper: Bill C-36, Protection of Communities and Exploited Persons Act." Department of Justice Canada.
Derber, Charles. 1983. *The Pursuit of Attention: Power and Individualism in Everyday Life.* New York: Oxford University Press.
Deutschmann, Linda B. 2002. *Deviance and Social Control* (3rd ed). Toronto: Thomson Nelson.
Dewing, Michael. 2013. "Canadian Multiculturalism." Library of Parliament. Background Paper. Available: http://www.parl.gc.ca/Content/LOP/ResearchPublications/2009-20-e.pdf.
Dobson-Mitchell, Scott. 2012. "The Many Regrets of a Fourth-Year Student." Retrieved from: http://oncampus.macleans.ca/education/2012/01/27/the-many-regrets-of-a-fourth-year-student.
Dodds, Peter Sheridan, Roby Muhamad, and Duncan J. Watts. 2003. "An Experimental Study of Search in Global Networks." *Science* 301: 827–829.
Dollard, John, Leonard W. Doob, Neil E. Miller, O.H. Mowrer, and Robert R. Sears. 1939. *Frustration and Aggression.* New Haven, CT: Yale University Press.
Domhoff, G. William. 1978. *The Powers That Be: Processes of Ruling Class Domination in America.* New York: Random House.
———. 1983. *Who Rules America Now? A View for the '80s.* Englewood Cliffs, NJ: Prentice-Hall.
———. 2002. *Who Rules America? Power and Politics* (4th ed.). New York: McGraw-Hill.
———. 2005. "The Class-Domination Theory of Power." Retrieved Mar. 23, 2013. Online: www2.ucsc.edu/whorulesamerica/power//class_domination.html.
Donkin, Karissa. 2013. "Social Media Helps Drive Idle No More Movement." *Toronto Star* 11 January. Retrieved: 1 December 2014. Available: http://www.thestar.com/news/canada/2013/01/11/social_media_helps_drive_idle_no_more_movement.html.
Doyle, Aaron. 2006. "How Not to Think About Crime in the Media." *Canadian Journal of Criminology and Criminal Justice.* October: 867–885.
Drolet, Marie. 2011. *Why Has the Gender Wage Gap Narrowed?* Cat no. 75-001-X. Ottawa: Statistics Canada. Retrieved February 6, 2012. Available: http://www.statcan.gc.ca/pub/75-001-x/2011001/pdf/11394-eng.pdf.
Dror, Itiel, David Charlton, and Ailsa Peron. 2006. "Contextual Information Renders Experts Vulnerable to Making Erroneous Identifications." *Forensic Science International*, 156: 74–78.
Drucker, Peter. 1994. "The Age of Social Transformation." *The Atlantic Monthly* (November): 53–80.
Dubois, R. (2012). Student Activism in Canada: Continuing the fight for public education. *Our Schools / Our Selves, 21*(3), 149–157.
Dufur, M., B.A. McKune, J.P. Hoffmann, and S.J. Bahr. 2007. "Adolescent Outcomes in Single Parent, Heterosexual Couple, and Homosexual Couple Families: Findings from a National Survey." Paper presented at the annual meeting of the American Sociological Association, New York. Available: http://bit.ly/XVzJhF.
Dunbar, Polly. 2007. "Wombs to Rent: Childless British Couples Pay Indian Women to Carry Their Babies." *The Daily Mail* (Dec. 8). Retrieved Feb. 9, 2008. Online: www.dailymail.co.uk/pages/live/articles/news/worldnews.html?in_article_id=500601.
Dunlap, Riley. 2010. "The Maturation and Diversification of Environmental Sociology: From Constructivism and Realism to Agnosticism and Pragmatism". In Michael Redclift and Graham Woodgate (eds.), *The International Handbook of Environmental Sociology*, 2nd ed. Cheltenham, UK: Edward Elgar Publishing, 15–32.
———, and William Catton, Jr. 2002. "Which Function(s) of the Environment Do We Study? A Comparison of Environmental and Natural Resource Sociology." *Society and Natural Resources*, 15 (3), 239–249.
Dunlap, Riley E., Frederick H. Buttel, Peter Dickens, and August Gijswijt (eds.) 2002. *Sociological Theory and the Environment: Classical Foundations, Contemporary Insights.* Rowman & Littlefield.
Durkheim, Émile. 1933. *Division of Labor in Society.* Trans. George Simpson. New York: Free Press (orig. pub. 1893).
———. 1956. *Education and Sociology.* Trans. Sherwood D. Fox. Glencoe, IL: Free Press, 28.
———. 1964a. *The Rules of Sociological Method.* Trans. Sarah A. Solovay and John H. Mueller. New York: Free Press (orig. pub. 1895).
———. 1995. *The Elementary Forms of Religious Life.* Trans. Karen E. Fields. New York: Free Press (orig. pub. 1912).
Durlauf, Steven, and Daniel Nagin. 2011. *The Deterrent Effect of Imprisonment.* Retrieved June 9, 2011, 38. Available: www.nber.org/chapters/c12078.pdf.

Durning, Alan. 1993. "Life on the Brink." In William Dan Perdue (ed.), *Systemic Crisis: Problems in Society, Politics, and World Order.* Fort Worth: Harcourt Brace, 274–282.

Dyck, P.R. 1996. *Canadian Politics: Critical Approaches* (2nd ed.) Scarborough, ON: Nelson Canada.

Dyck, Rand. 2000. *Canadian Politics: Critical Approaches* (3rd ed.) Toronto: Nelson Canada.

———. 2004. *Canadian Politics: Critical Approaches* (4th ed.). Toronto: Thomson Nelson.

———. 2008. *Canadian Politics: Critical Approaches* (5th ed.). Toronto: Nelson.

——— , and C. Cochrane. 2013. *Canadian Politics: Critical Approaches* (7th ed.). Toronto: Nelson.

Dye, Thomas R., and Harmon Zeigler. 2006. *The Irony of Democracy: An Uncommon Introduction to American Politics* (13th ed.). Belmont, CA: Wadsworth.

Dyer, Evan. 2014. "Conservatives' Gun Bill, C-42, Fails to Impress Gun Owners." CBC News. 23 November. Available: http://www.cbc.ca/m/touch/news/story/1.2837842.

Ebaugh, Helen Rose Fuchs. 1988. *Becoming an EX: The Process of Role Exit.* Chicago: University of Chicago Press.

The Economist. 1997. "The Anti-Management Guru." May 4.

Economist.com. 2011. "Environmental Activism in China: Poison Protests—A Huge Demonstration Over a Chemical Factory Unnerves Officials" (Aug. 20). Online: www.economist.com/node/21526417.

Edmonston, B., S. M. Lee, and Z. Wu. 2008. *Childless Canadian Couples.* Victoria: Department of Sociology and Population Research Group, University of Victoria. Retrieved January 20, 2012. Available: http://www.horizons.gc.ca/doclib/PA-pwfc2008-Edmonston-eng.pdf.

Edwards, Richard. 1993. "An Education in Interviewing." In C.M. Renzetti and R.M. Lee (eds.), *Researching Sensitive Topics.* Newbury Park, CA: Sage, 181–196.

Edwards, Steven. 2007. "UN labels anti-racism language as racist: 'Visible minorities,' other Canadian terms run afoul of watchdog." *The Ottawa Citizen*, March 8, 2007.

Ehrenreich, Barbara. 2001. *Nickel and Dimed: On (Not) Getting By in America.* New York: Metropolitan.

———. 2011. *Nickel and Dimed: On (Not) Getting By in America* (10th anniversary edition. New York: Picador.

Ehrlich, Paul R., Anne H. Ehrlich, and Gretchen C. Daily. 1995. *The Stork and the Plow: An Equity Answer to the Human Dilemma.* New Haven, CT: Yale University Press.

Eichler, Margrit. 1981. "The Inadequacy of the Monolithic Model of the Family." *Canadian Journal of Sociology* 6: 367–388.

———. 1988. *Nonsexist Research Methods: A Practical Guide.* Boston: Allen & Unwin.

Eisenberg, Mark, Kristian Filion, Arik Azoulay, Anya Brox, Seema Haider, and Louise Pilote. 2005. *Archives of Internal Medicine* 165: 1506–1513.

Eisenstein, Zillah R. 1994. *The Color of Gender: Reimaging Democracy.* Berkeley, CA: University of California Press.

Elections Canada. 2011. "Estimation of Voter Turnout by Age Group and Gender at the 2011 General Election." Ottawa: Elections Canada.

———. 2012. *National Youth Survey Report.* Retrieved April 30, 2012. Available: http://www.elections.ca/content.aspx?section=res&dir=rec/part/nysr&document=bkr&lang=e.

Elkind, David. 1995. "School and Family in the Postmodern World." *Phi Delta Kappan* (September): 8–21.

Ellison, Duncan. 2013. "Bottling Water – One of the Older Professions." *Policy Horizons Canada.* Ottawa: Environment Canada.

Emling, Shelley. 1997. "Haiti Held in Grip of Another Drought." *Austin American-Statesman* (September 19): A17, A18.

Employment and Social Development Canada. 2013a. "Snapshot of Racialized Poverty in Canada." Retrieved from: http://www.esdc.gc.ca/eng/communities/reports/poverty_profile/snapshot.shtml.

———. 2013b. "Indicators of Well-Being in Canada: School Drop-Outs. Available: http://www4.hrsdc.gc.ca/.3ndic.1t.4r@-eng.jsp?iid=32.

———. 2014. "Indicators of Well-Being in Canada, Financial Security- Low Income Incidence." Available at: http://www4.hrsdc.gc.ca/.3ndic.1t.4r@-eng.jsp?iid=23. Data source: Statistics Canada. Table 202-0802 - Persons in low income families, annual, CANSIM (database).

Engels, Friedrich. 1970. *The Origins of the Family, Private Property, and the States.* Ed. Eleanor Burke Leacock. New York: International (orig. pub. 1884).

Environics. 2012. *Focus Canada, 2012.* The Environics Institute. Available: http://www.environicsinstitute.org/uploads/institute-projects/environics%20institute%20-%20focus%20canada%202012%20final%20report.pdf.

———. 2013. *Focus Canada, 2012.* The Environics Institute. Available: http://www.environicsinstitute.org/uploads/institute-projects/environics%20institute%20-%20focus%20canada%202012%20final%20report.pdf.

———. 2014. *Focus Canada 2014: Canadian Public Opinion About Climate Change.* Available: http://www.environicsinstitute.org/uploads/news/focus%20canada%202014%20-%20public%20opinion%20on%20climate%20change%20-%20final%20report%20-%20english%20-%20november%2025-2014.pdf.

———. 2014. *From Boomers to Millennials: How an era can shape financial planning habits.* Available: http://www.newswire.ca/en/story/1098987/from-boomers-to-millennials-how-an-era-can-shape-financial-planning-habits.

Epstein, Cynthia Fuchs. 1988. *Deceptive Distinctions: Sex, Gender, and the Social Order.* New Haven, CT: Yale University Press.

Erickson, Stephane. 2014. "Trinity Western law school has no right to judge its gay students." *The Globe and Mail,* February 21.

Erikson, Eric H. 1963. *Childhood and Society.* New York: Norton.

Erikson, Kai T. 1962. "Notes on the Sociology of Deviance." *Social Problems* 9: 307–314.

———. 1976. *Everything in Its Path: Destruction of Community in the Buffalo Creek Flood.* New York: Simon & Schuster.

———. 1994. *A New Species of Trouble: Explorations in Disaster, Trauma, and Community.* New York: Norton.

Esbensen, Finn-Aage, and David Huizinga. 1993. "Gangs, Drugs, and Delinquency in a Survey of Urban Youth." *Criminology* 31 (4): 565–589.

Esping Anderson, Gosta. 2000. "Two Societies, One Sociology, and No Theory." *British Journal of Sociology* (January/March): 59–77.

Essed, Philomena. 1991. *Understanding Everyday Racism.* Newbury Park, CA: Sage.

Euromonitor International. 2015. *Bottled Water in Canada.* Available: http://www.euromonitor.com/bottled-water-in-canada/report.

Evans, Peter B., and John D. Stephens. 1988. "Development and the World Economy." In Neil J. Smelser (ed.), *Handbook of Sociology.* Newbury Park, CA: Sage, 739–773.

Fahmi, Wael Salah. 2001. " 'Honey, I Shrunk the Space': Planning in the Information Age." Paper presented at the 37th International Planning Congress. Utrecht: The Netherlands, 5.

Fanelli, Carlo. 2014. Climate Change: 'The Greatest Challenge of Our Time.' *Alternate Routes* 25(14):15–31.

Farley, John E. 1992. *Sociology* (2nd ed.). Englewood Cliffs, NJ: Prentice-Hall.

Fausto-Sterling, Anne. 1993. "The Five Sexes: Why male and female are not enough." *The Sciences* (March/April 1993): 20–24.

———. (2000). "The five sexes, revisited." *Sciences (New York)* **40** (4): 18–23.

———. 2002. "The Five Sexes: Why Male and Female Are Not Enough." In C. Williams and A. Stein (eds.), *Sexuality and Gender.* London: Blackwell, 489–473.

FDA. 2011. *FDA 101: Health Fraud Awareness.* Available: http://www.fda.gov/ForConsumers/ConsumerUpdates/ucm235995.htm.

Feagin, Joe R., and Clairece Booher Feagin. 1994. *Social Problems: A Critical Power–Conflict Perspective* (4th ed). Englewood Cliffs, NJ: Prentice Hall.

———, and Clairece Booher Feagin. 2011. *Racial and Ethnic Relations* (9th ed.). Upper Saddle River, NJ: Prentice Hall.

———, and Robert Parker. 1990. *Building American Cities: The Urban Real Estate Game* (2nd ed.). Englewood Cliffs, NJ: Prentice-Hall.

———, David B. Baker, and Clairece B. Feagin. 2006. *Social Problems: A Critical Power–Conflict Perspective* (6th.). Englewood Cliffs, NJ: Prentice-Hall.

Ferguson, Sarah Jane, and John Zhao. 2013. "Education in Canada: Attainment, Field of Study and Location of Study." Statistics Canada Catalogue no. 99-012-X2011001. 2011 National Household Survey Analytical Document. Ottawa.

Fern, Carolyn, and Martha Friendly. 2014. "The state of early childhood education and care in Canada 2012." Moving Childcare Forward Project (a joint initiative of the Childcare Resource and Research Unit, Centre for Work, Families and Well-Being at the University of Guelph, and the Department of Sociology at the University of Manitoba). Available: http://childcarecanada.org/sites/default/files/StateofECEC2012.pdf.

Ferrao, Vincent. 2010. "Paid Work." In Statistics Canada, 2011, *Women in Canada: A Gender-Based Statistical Report* (6th ed.). Cat. no. 89-503-X. Retrieved February 6, 2012. Available: http://www.statcan.gc.ca/pub/89-503-x/2010001/article/11387-eng.pdf.

Findlay, Deborah A., and Leslie J. Miller. 2002. "Through Medical Eyes: The Medicalization of Women's Bodies and Women's Lives." In B. Singh Bolaria and Harley D. Dickinson (eds.), *Health, Illness, and Health Care in Canada* (3rd ed.). Toronto: Nelson, 185–210.

Firestone, Shulamith. 1970. *The Dialectic of Sex.* New York: Morrow.

First Nations Health Authority. 2015. *Drinking Water Advisories.* Available: http://www.fnha.ca/what-we-do/environmental-health.

Fiske, J. (1994) *Media Matters: Everyday Culture and Political Change.* Minneapolis: University of Minnesota Press.

Fjellman, Stephen M. 1992. *Vinyl Leaves: Walt Disney World and America.* Boulder, CO: Westview.

Flanagan, William G. 1999. *Urban Sociology: Images and Structure* (3rd ed.). Needham Heights, MA: Allyn & Bacon.

Flanders, Cait. 2014. "No 'lost' time in taking a gap year." *Globe and Mail.* June 27, 2014. http://www.theglobeandmail.com/news/national/education/no-lost-time-in-taking-a-gap-year/article19308616/.

Fleras, Augie. 2012. *Unequal Relations: An Introduction to Race, Ethnic, and Aboriginal Dynamics in Canada* (7th ed.). Toronto: Pearson Canada.

———, and Jean Leonard Elliott. 1996. *Unequal Relations: An Introduction to Race, Ethnic and Aboriginal Dynamics in Canada* (2nd ed.). Scarborough, ON: Prentice-Hall Canada.

———. 2003. *Unequal Relations: An Introduction to Race, Ethnic and*

Aboriginal Dynamics in Canada (4th ed.). Scarborough, ON: Prentice-Hall Canada.
Flintoff, Corey. 2011. "Selective Abortions Blamed for Girl Shortage in India." NPR.org (April 14). Retrieved February 5, 2012. Online: www.npr.org/2011/04/14/135417647/in-india-number-of-female-children-drops.
Forbes. 2015. "The World's Billionaires." Available: http://www.forbes.com/billionaires/gallery.
Foster, John Bellamy. 2005. "The Treadmill of Accumulation: Schnaiberg's Environment and Marxian Political Economy." *Organization & Environment* (18.1): 7–18.
Foucault, Michel. 1978. *The History of Sexuality, Volume 1: An Introduction*. New York: Pantheon.
———. 1979. *Discipline and Punish: The Birth of the Prison*. New York: Vintage.
———. 1991. "Governmentality." In Buchell et al. (eds.). *The Foucault Effect*. Hemel Hempstead: Harvester Wheatsheaf.
Fox, John, and Michael Ornstein. 1986. "The Canadian State and Corporate Elites in the Post-War Period." *Canadian Review of Sociology and Anthropology* 23: 481–506.
Fox, Mary Frank. 1989. "Women and Higher Education: Gender Differences in the Status of Students and Scholars." In Jo Freeman (ed.), *Women: A Feminist Perspective*. Mountain View, CA: Mayfield, 217–235.
Fox, Susannah. 2011. "The Social Life of Health Information, 2011." Pew Internet and American Life Project. Retrieved February 18, 2012. Available: http://pewinternet.org/Reports/2011/Social-Life-of-Health-Info/Part-1/Section-1.aspx.
Frank, Andre Gunder. 1969. *Latin America: Underdevelopment or Revolution?* New York: Monthly Review Press.
———. 1981. *Reflections on the World Economic Crisis*. New York: Monthly Review Press.
Frankenberg, Ruth. 1993. *White Women, Race Matters: The Social Construction of Whiteness*. Minneapolis, MN: University of Minnesota Press.
Fraser Report. 1985. *Canada, Pornography and Prostitution in Canada, Volume II*. Special Committee on Pornography and Prostitution (Fraser Committee). Ottawa: Department of Supply and Services.
Frederick, Judith A., and Janet E. Fast. 2001. "Enjoying Work: An Effective Strategy in the Struggle to Juggle." *Canadian Social Trends* (Summer): 8–11.
Freedman, E., and J. d'Emilio. 1998. *Intimate Matters: A History of Sexuality in America*. Chicago: University of Chicago Press.
Freedom House. 2014. *2014 Freedom in the World*. Retrieved: 19 November 2014. Available: https://freedomhouse.org/report-types/freedom-world#.VGzYP5NnCaG.
Freeland, Chrystia. 2012. "Social Media Statecraft: A Multiplatform Strategy." *Globe and Mail* (April 5). Retrieved April 27, 2012. Available: http://www.theglobeandmail.com/report-on-business/commentary/chrystia-freeland/social-media-statecraft-a-multiplatform-strategy/article2393250.
Freeman, John G., Matthew King, and William Pickett with Wendy Craig, Frank Elgar, Ian Janssen, and Don Klinger. (2011). *The Health of Canada's Young People: A Mental Health Focus*. Ottawa: Public Health Agency of Canada.
Freidson, Eliot. 1965. "Disability as Social Deviance." In Marvin B. Sussman (ed.), *Sociology and Rehabilitation*. Washington, D.C.: American Sociology Association, 71–99.
Freud, Sigmund. 1924. *A General Introduction to Psychoanalysis* (2nd ed.). New York: Boni & Liveright.
Frideres, James. 1994. "The Future of Our Past: Native Elderly in Canadian Society." In National Advisory Council on Aging, *Aboriginal Seniors' Issues*, 17–37. Cat. no. H71-2/1-15-1994E. Ottawa: Minister of Supply and Services.
———, and René R. Gadacz. 2011. *Aboriginal Peoples in Canada: Contemporary Conflicts* (9th ed.). Paperback. Scarborough, ON: Pearson.
Friedman, Thomas L. 2005. *The World Is Flat*. New York: Farrar, Straus and Giroux.
Fries, Christopher. 2008. "Governing the Health of the Hybrid Self: Integrative Medicine, Neoliberalism, and the Shifting Biopolitics of Subjectivity." *Health Sociology Review* 17 (December): 353–367.
Fukuyama, Francis. 2005. "The Calvinist Manifesto." *The New York Times Review of Books*, 2. Retrieved March 13, 2005. Available: www.nytimes.com/200/03/13/books/review/013/FUKUYA.
Gaard, Greta. 1993. *Ecofeminism: Women, Animals and Nature*. Philadelphia: Temple University Press.
Gabriels, Katleen, K. Poels, and J. Braeckman. 2013. "Morality and involvement in social virtual worlds: The Intensity of moral emotions in response to virtual versus real life cheating." *New Media and Society* 16: 451-469.
Gadd, Jane. 1998. "Young Men Across Canada Earning Less, Report Says." *Globe and Mail* (July 29): A5.
Gaetz, Stephen. (2014). *A Safe and Decent Place to Live: Towards a Housing First Framework for Youth*. Toronto: The Homeless Hub Press.
Gagnon, John H., and *William* Simon. 1973. *Sexual Conduct: The Social Sources of Human Sexuality (*paperback ed.). Chicago: Aldine.
Galarneau, Diane, and Marian Radulescu. 2009. "Employment Among the Disabled." *Perspectives on Labour and Income* 19 (5). Retrieved June 18, 2009. Available: http://www.statcan.gc.ca/pub/75-001-x/2009105/pdf/10865-eng.pdf.
———, and Thao Sohn. 2014. "Long Term Trends in Unionization." Ottawa: Statistics Canada.
Gann, R. 2000. "Postmodern Perspectives on Race and Racism: Help or Hindrance?" Retrieved September 2, 2005. Paper for the Political Studies Association-UK, 10–13 April 2000. Ebsco host database.
Gans, Herbert. 1974. *Popular Culture and High Culture: An Analysis and Evaluation of Tastes*. New York: Basic Books.
———. 1982. *The Urban Villagers: Group and Class in the Life of Italian Americans* (updated and expanded edition; orig. pub. 1962). New York: Free Press.
Gardner, Sue. 2014. "Why Women Are Leaving the Tech Industry in Droves." *LA Times*, 5 December.
Garfinkel, Harold. 1967. *Studies in Ethnomethodology*. Englewood Cliffs, NJ: Prentice-Hall.
Garson, Barbara. 1989. *The Electronic Sweatshop: How Computers Are Transforming the Office of the Future into the Factory of the Past*. New York: Penguin.
Gaskell, Jane. 2009. "Feminist Approaches to the Sociology of Education in Canada." In Cynthia Levine-Rasky (ed.), *Canadian Perspectives on the Sociology of Education*. Toronto: Oxford University Press, 17–29.
Gee, Ellen. 2000. "Voodoo Demography, Population Aging, and Social Policy." In E.M. Gee and G.M. Guttman (eds.), *The Overselling of Population Aging: Apocalyptic Demography, Intergenerational Challenges, and Social Policy*. Don Mills, ON: Oxford University Press.
"General Facts on Sweden." 2005. Retrieved July 23, 2005. Available: http://www.finansforbundet.se/Resource.phx/plaza/content/material/internationellteu.htx.pdf.material.3.pdf.
George, Anne, Andrew Jin, Mariana Brussoni, and Christopher Lalonde. 2015. "Is the Injury Gap Closing Between the Aboriginal and General Populations of British Columbia?" *Health Reports* 26, Number 1.
Gerber, Linda. 1990. "Multiple Jeopardy: A Socioeconomic Comparison of Women Among the Indian, Metis, and Inuit Peoples of Canada." *Canadian Ethnic Studies* 22 (3): 22–34.
Gerds, Jenna. 2011. "Famous for Being Famous: Celebrity Socialites and the Framework of Fame." In Robin DeRosa (ed.), *Simulation in Media and Culture: Believing the Hype*. Lanham, MD: Lexington Books, 8–24.
Gereffi, Gary. 1994. "The International Economy and Economic Development." In Neil J. Smelser and Richard Swedberg (eds.), *The Handbook of Economic Sociology*. Princeton, NJ: Princeton University Press, 206–233.
Gerschenkron, Alexander. 1962. *Economic Backwardness in Historical Perspective*. Cambridge, MA: Harvard University Press.
Gibbs, Jennifer, Nicole Ellison, and Chih-Hui Lai. 2011. "First Comes Love, Then Comes Google: An Investigation of Uncertainty Reduction Strategies and Self-Disclosure in Online Dating." *Communication Research* 38 (1): 70–100.
Giddens, Anthony. 1996. *Introduction to Sociology* (2nd ed.). New York: W.W. Norton & Co.
Gidengil, Elisabeth, Neil Nevitte, Andre Blais, Patrick Fournier, and Joanna Everitt. 2004. "Why Johnny Won't Vote." *Globe and Mail* (August 4): A11.
Gilbert, Dennis L. 2011. *The American Class Structure in an Age of Growing Inequality* (8th ed.). Thousand Oaks, CA: Sage.
Gilbert, Nicolas, Nathalie Auger, Russell Wilkins, and Michael Kramer. 2013. "Neighbourhood Income and Neonatal, Postneonatal and Sudden Infant Death Syndrome (SIDS) Mortality in Canada, 1991–2005." *Canadian Journal of Public Health*, 104: 187–192.
Gill, Indermit, and Homi Kharas. 2007. *An East Asian Renaissance: Ideas for Economic Growth*. Washington: The World Bank.
Gilligan, Carol. 1982. *In a Different Voice: Psychological Theory and Women's Development*. Cambridge, Mass.: Harvard University Press.
Gillis, A.R. 1995. "Urbanization." In Robert J. Brym (ed.), *New Society: Sociology for the 21st Century*. Toronto: Harcourt Brace and Company, 13.1–13.40.
Ginsberg, Jeremy, Matthew Mohebbi, Rajan Patel, Lynnette Brammer, Mark Smolinski, and Larry Brilliant. 2009. "Detecting Flu Epidemics Using Search Engine Query Data." *Nature* 457 (February 19): 1012–1014.
Girard, April, Suzanne Day, and Laureen Snider. 2010. "Tracking Environmental Crime Through CEPA: Canada's Environment Cops of Industry's Best Friend?" *Canadian Journal of Sociology* 35 (2): 219–241.
Glaser, Barney, and Anselm Strauss. 1967. *The Discovery of Grounded Theory*. Chicago: Aldine.
Gleick, Peter. 2010. *Bottled and Sold: The Story Behind Our Obsession with Bottled Water*. Washington: Island Press.
Glenny, Misha. 2011. *DarkMarket: Cyberthieves, Cybercops and You*. New York: Albert A. Knopf.
Global Health Council. 2002. *Health: A Key to Prosperity. Success Stories in Developing Countries*. Retrieved July 22, 2005. Available: http://www.globalhealth.org/sources/view.php3?id=390.
Global Media Monitoring Project. 2010. *Who Makes the News*. London: World Association for Christian Communication. Retrieved January 12, 2012. Available: http://www.whomakesthenews.org.
Globe and Mail, Report on Business Magazine. 1990. October: B80.
Godin-Beers, Monique, and Cinderina Williams. 1994. "Report of the Spallumcheen Child Welfare Program." Research study prepared for the Royal Commission on Aboriginal Peoples (unpublished), cited in Castellano, Marlene. 2002. "Aboriginal Family Trends: Extended Families, Nuclear Families, Families of

the Heart." Ottawa: Vanier Institute of the Family. Retrieved May 12, 2009. Available: http://www.vifamily.ca/library/cft/aboriginal.html.

Godley, J., and L. McLaren. 2010. "Socioeconomic Status and Body Mass Index in Canada: Exploring Measures and Mechanisms SES and BMI in Canada." *Canadian Review of Sociology, 47*(4), 381–403. doi:10.1111/j.1755-618X.2010.01244.x.

Goffman, Erving. 1956. "The Nature of Deference and Demeanor." *American Anthropologist* 58: 473–502.

———. 1959. *The Presentation of Self in Everyday Life.* New York: Doubleday.

———. 1961a. *Asylums: Essays on the Social Situation of Mental Patients and Other Inmates.* Chicago: Aldine.

———. 1961b. *Encounters: Two Studies in the Sociology of Interaction.* Indianapolis, IN: Bobbs-Merrill.

———. 1963a. *Behavior in Public Places: Notes on the Social Structure of Gatherings.* New York: Free Press.

———. 1963b. *Stigma: Notes on the Management of Spoiled Identity.* Englewood Cliffs, NJ: Prentice-Hall.

———. 1967. *Interaction Ritual: Essays on Face to Face Behavior.* Garden City, NY: Anchor Books.

———. 1971. *Relations in Public.* New York: Basic Books.

———. 1974. *Frame Analysis: An Essay on the Organization of Experience.* Boston: Northeastern University Press.

———. 1979. *Gender Advertisements.* Cambridge, MA: Harvard University Press.

Goldberg, Robert A. 1991. *Grassroots Resistance: Social Movements in Twentieth Century America.* Belmont, CA: Wadsworth.

Goodman, Peter S. 1996. "The High Cost of Sneakers." *Austin American-Statesman* (July 7): F1, F6.

Gordon, David. 1973. "Capitalism, Class, and Crime in America." *Crime and Delinquency* 19: 163–186.

Gordon, Milton. 1964. *Assimilation in American Life: The Role of Race, Religion, and National Origins.* New York: Oxford University Press.

Gorey, Kevin, Eric J. Holowaty, Gordon Fehringer, Ethan Laukkanen, Agnes Moskowitz, David J. Webster, and Nancy L. Richter. 1997. "An International Comparison of Cancer Survival: Toronto, Ontario, and Detroit, Michigan, Metropolitan Areas." *American Journal of Public Health* 87: 1156–1163.

Gorkoff, Kelly, and Jane Runner (eds.). 2003. *Being Heard: The Experiences of Young Women in Prostitution.* Halifax and Winnipeg: Fernwood Publishing.

Gottdiener, Mark. 1985. *The Social Production of Urban Space.* Austin, TX: University of Texas Press.

Gough, K. 1975. "The Origin of the Family." In Rayna R. Reiter (ed.), *Toward an Anthropology of Women.* New York: Monthly Review Press, 69–70.

Gould, Kenneth Alan, and Tammy L. Lewis. 2009. *Twenty Lessons in Environmental Sociology.* Oxford University Press.

———. 2015. *Twenty Lessons in Environmental Sociology* (2nd ed.). Kenneth Alan Gould and Tammy L. Lewis (eds.). Oxford University Press.

Graham, John, and Francois Leveque. 2010. *First Nations Communities in Distress: Dealing With Causes Not Symptoms.* Ottawa: Institute on Governance. Available: http://iog.ca/wp-content/uploads/2012/12/2010_April_First-Nation-Communities-in-Distress.pdf.

Grahame, Peter. 2004. "Ethnography, Institutions, and the Problematic of the Everyday World." In William K. Carroll (ed.), *Critical Strategies for Social Research.* Toronto: Canadian Scholars' Press, 181–190.

Granovetter, Mark. 1995. *Getting a Job: A Study in Contacts and Careers* (2nd ed.). Chicago: University of Chicago Press.

Grant, Karen. 1993. "Health and Health Care." In Peter S. Li and B. Singh Bolaria (eds.), *Contemporary Sociology: Critical Perspectives.* Toronto: Copp-Clark Pitman, 394–409.

Grant, Tavia. 2014. "The 15-Hour Workweek: Canada's Part-Time Problem. *The Globe and Mail,* 4 October.

Gratton, Bruce. 1986. "The New History of the Aged." In David Van Tassel and Paul N. Stearns (eds.), *Old Age in a Bureaucratic Society.* Westport, CT: Greenwood Press, 3–29.

Gray, Jeff. 2014. "University of Toronto's Next Lawyer: A Computer Program Named Ross." *The Globe and Mail,* 11 December.

Green, Emma. 2014. "Kicked Out of Heaven for Wanting Women Priests." *The Atlantic,* June 24.

Greenspan, Edward. 1982. "The Role of the Defence Lawyer in Sentencing." In Craig L. Boydell and Ingrid Connidis (eds.), *The Canadian Criminal Justice System.* Toronto: Holt, Rinehart and Winston, 200–210.

Gregg, Allan. 2006. "Identity Crisis." *The Walrus* (March).

Guha-sapir, Debarati, Phillipe Hoyois, and Regina Below. 2014. *Annual Disaster Statistical Review 2013: The Numbers and Trends.* Brussels: CRED.

Gusfield, Joseph. 1963. *Symbolic Crusade: Status Politics and the American Temperance Movement.* Urbana, IL: University of Illinois Press.

Guyatt, Gordon, P.J. Devereaux, Joel Lexchin, Samuel Stone, Armine Yalnizyan, and David Himmelstein. 2007. "A Systematic Review of Studies Comparing Health Outcomes in Canada and the United States." *Open Medicine* 1. Available: http://www.cnbc.com/id/100840148.

Haas, Jack. 1977. "Learning Real Feelings: A Study of High Steel Ironworkers' Reactions to Fear and Danger." *Sociology of Work and Occupations* 4 (May): 147–170.

Hadden, Jeffrey. 1987. "Toward Desacralizing Secularization Theory." *Social Forces* 65: 587–611.

Hadden, Richard W. 1997. *Sociological Theory: An Introduction to the Classical Tradition.* Peterborough, ON: Broadview.

Hahn, Harlan. 1987. "Civil Rights for Disabled Americans: The Foundation of a Political Agenda." In Alan Gartner and Tom Joe (eds.), *Images of the Disabled, Disabling Images.* New York: Praeger, 181–203.

Haines, Valerie A. 1997. "Spencer and His Critics." In Charles Camic (ed.), *Reclaiming the Sociological Classics: The State of the Scholarship.* Malden, MA: Blackwell, 81–111.

Halberstadt, Amy G., and Martha B. Saitta. 1987. "Gender, Nonverbal Behavior, and Perceived Dominance: A Test of the Theory." *Journal of Personality and Social Psychology* 53: 257–272.

Hall, Edward. 1966. *The Hidden Dimension.* New York: Anchor/Doubleday.

Hall, Evan. 2014. "Conservatives' Gun Bill, C-42, Fails to Impress Gun Owners." CBC News. Retrieved: 21 November 2014. Available: http://www.cbc.ca/news/politics/conservatives-gun-bill-c-42-fails-to-impress-gun-owners-1.2837842.

Hall, Michael, David Lasby, Glenn Gumulka, and Catherine Tryon. 2006. *Caring Canadians, Involved Canadians: Highlights from the 2004 Canada Survey of Giving, Volunteering and Participating.* Ottawa: Statistics Canada.

Hall, Peter M. 1972. "A Symbolic Interactionist Analysis of Politics." *Sociological Inquiry* 42: 35–75.

Hall, Stuart. 1982. "The Rediscovery of 'Ideology': Return of the Repressed in Media Studies." In Michael Gurevitch, Trevor Bennett, James Curran, and J. Woollacott (eds.), *Culture, Society, and the Media.* London: Methuen, 56–90.

Hallinan, Maureen. 2005. "Should Your School Eliminate Tracking? The History of Tracking and Detracking in America's Schools." *Education Matters,* 1–2.

Hamilton, Allen C., and C. Murray Sinclair. 1991. *Report of the Aboriginal Justice Inquiry of Manitoba,* Vol. 1. Winnipeg: Queen's Printer, 91.

Hango, D. 2013. "Gender differences in science, technology, engineering, mathematics and computer science (STEM) programs at university." Statistics Canada. December 2013. Catalogue no. 75-006-X. Available: http://www.statcan.gc.ca/pub/75-006-x/2013001/article/11874-eng.pdf.

Hannigan, John. 1998. *Fantasy City: Pleasure and Profit in the Postmodern Metropolis.* London: Routledge.

Hannigan, John. 2006. *Environmental Sociology, Second Edition.* New York: Routledge, 29.

Harlow, Harry F., and Margaret Kuenne Harlow. 1962. "Social Deprivation in Monkeys." *Scientific American* 207 (5): 137–146.

———. 1977. "Effects of Various Mother–Infant Relationships on Rhesus Monkey Behaviors." In Brian M. Foss (ed.), *Determinants of Infant Behavior,* vol. 4. London: Methuen, 15–36.

Harman, Lesley. 1989. *When a Hostel Becomes a Home: Experiences of Women.* Toronto: Garamond Press, 42, 91.

Harper, Miheala, and Andrew Ploeg. 2011. "Drafting the Hyperreal: Ownership, Agency, Responsibility in Fantasy Sports." In Robin DeRosa (ed.), *Simulation in Media and Culture: Believing the Hype.* Lanham, Md.: Lexington Books, 151–161.

Harris, Chauncey D., and Edward L. Ullman. 1945. "The Nature of Cities." *Annals of the Academy of Political and Social Sciences* (November): 7–17.

Harris, Marvin. 1974. *Cows, Pigs, Wars, and Witches.* New York: Random House.

———. 1985. *Good to Eat: Riddles of Food and Culture.* New York: Simon & Schuster.

Hartmann, Heidi. 1976. "Capitalism, Patriarchy, and Job Segregation by Sex." *Signs: Journal of Women in Culture and Society* 1 (Spring): 137–169.

———. 1981. "The Unhappy Marriage of Marxism and Feminism." In Lydia Sargent (ed.), *Women and Revolution.* Boston: South End Press.

Hartnagel, Timothy F. 2004. "Correlates of Criminal Behaviour." In Rick Linden (ed.), *Criminology: A Canadian Perspective* (5th ed.). Toronto: Nelson.

———. 2012. "Correlates of Criminal Behaviour." In Rick Linden (ed.), *Criminology: A Canadian Perspective* (7th ed.) Toronto: Nelson, 133–178.

Havighurst, Robert J., Bernice L. Neugarten, and Sheldon S. Tobin. 1968. "Disengagement and Patterns of Aging." In Bernice L. Neugarten (ed.), *Middle Age and Aging.* Chicago: University of Chicago Press, 161–172.

Haviland, William A. 1993. *Cultural Anthropology* (7th ed.). Orlando, FL: Harcourt Brace Jovanovich.

Haynes, Jeff. 1997. "Religion, Secularisation and Politics: A Postmodern Conspectus." *Third World Quarterly* 18: 709–728.

Health Canada. 2005. *A Statistical Profile on the Health of First Nations in Canada: Highlights.* Ottawa: Health Canada.

———. 2009. *A Statistical Profile on the Health of First Nations in Canada: Self-rated Health and Selected Conditions, 2002 to 2005.* Ottawa: Health Canada.

———. 2015. *Drinking Water Advisories in First Nations Communities.* Available: http://www.hc-sc.gc.ca/fniah-spnia/promotion/public-publique/water-dwa-eau-aqep-eng.php#more.

Health Nexus. 2015a. "How does poverty affect your health?" Retrieved from: http://en.healthnexus.ca/topics-tools/health-equity-topics/poverty.

Health Nexus. 2015b. "I'm Still Hungry: Child and Family in Ontario," 35. Retrieved from: http://www.beststart.org/resources/anti_poverty/pdf/child_poverty_guide_rev.pdf.

Health and Welfare Canada. 1998. *Active Health Report: The Active Health Report*

on Seniors. Ottawa: Minister of Supply and Services.
Hébert, Benoît-Paul, and May Luong. 2008. "Bridge Employment." *Perspectives on Labour and Income* (November): 5–12.
Heldman, C., and L. Wade. 2010. "Hook-up culture: Setting a new research agenda." *Sexuality Research and Social Policy*. 7:323–333.
HelpAge International. 2013. *Global AgeWatch Index 2013*. Retrieved 6 November 2014. Available: http://www.helpage.org/global-agewatch/.
Henley, Nancy. 1977. *Body Politics: Power, Sex, and Nonverbal Communication*. Englewood Cliffs, NJ: Prentice-Hall.
Henry, Frances, and Carol Tator. 2000. *The Colour of Democracy: Racism in Canadian Society* (2nd ed). Toronto: Harcourt Canada.
———, and Carol Tator. 2006. *The Colour of Democracy: Racism in Canadian Society* (3rd ed.). Toronto: Thomson Nelson.
———, and Carol Tator. 2009. *The Colour of Democracy: Racism in Canadian Society* (4th ed.). Toronto: Nelson.
Herbenick, Debby, and Aleta Baldwin. 2014. "It's Complicated: What Each of Facebook's 51 New Gender Options Means." *The Daily Beast*, Feb. 15, 2014. http://www.thedailybeast.com/articles/2014/02/15/the-complete-glossary-of-facebook-s-51-gender-options.html.
Heritage, John. 1984. *Garfinkel and Ethnomethodology*. Cambridge, MA: Polity.
Herman, Nancy. 1996. "'Mixed Nutters,' 'Looney Tuners,' and 'Daffy Ducks.'" In Earl Rubington and Martin S. Weinberg (eds.), *Deviance: The Interactionist Perspective* (6th ed.). Boston: Allyn & Bacon, 254–266, 310, 323.
Herman, Norman, and Noam Chomsky. 1998. *Manufacturing Consent*. London: Vintage.
Heshka, Stanley, and Yona Nelson. 1972. "Interpersonal Speaking Distances as a Function of Age, Sex, and Relationship." *Sociometry* 35 (4): 491–498.
Hettne, Bjorn. 1995. *Development Theory and the Three Worlds* (2nd ed.). Essex, UK: Longman.
Hier, S.E., and K. Walby. 2006. "Competing Analytical Paradigms in the Sociological Study of Racism in Canada." *Canadian Ethnic Studies*, *38*(1), 83–104.
Higgitt, Nancy, Susan Wingert, and Janice Ristock. (2003). *Voices from the Margins: Experiences of Street-Involved Youth In Winnipeg*. Winnipeg Inner-city Research Alliance. Available: https://www.uwinnipeg.ca/index/cms-filesystem-action?file=pdfs/media-releases/voices-report.pdf.
Hill, Darryl. 2005. "Coming to Terms: Using Technology to Know Identity." *Social Problems* 9 (3): 24–52.
The Hindu. 1998. "The Idea of Human Development." (October 25): 25.
Hirschi, Travis. 1969. *Causes of Delinquency*. Berkeley, CA: University of California Press.
———, and Michael Gottfredson. 1983. "Age and the Explanation of Crime." *American Journal of Sociology* 89 (3): 552–584.
Hochschild, Arlie Russell. 1983. *The Managed Heart: Commercialization of Human Feeling*. Berkeley, CA: University of California Press.
———. 1989. *The Second Shift: Working Parents and the Revolution at Home*. New York: Viking/Penguin.
———. 2003. *The Commercialization of Intimate Life: Notes from Home and Work*. Berkeley, CA: University of California Press.
———. 2012. *The Second Shift: Working Families and the Revolution at Home*. New York: Penguin.
Hodgetts, Darrin, Kerry Chamberlain, and Graeme Bassett. 2003. "Between Television and the Audience: Negotiating Representations of Ageing." *Health: An Interdisciplinary Journal for the Social Study of Health, Illness and Medicine* 7 (4): 417–438.
Hodkinson, Paul. 2011. *Media, Culture and Society: An Introduction*. London: Sage.
Hodson, Randy, and Teresa A. Sullivan. 2002. *The Social Organization of Work* (3rd ed.). Belmont, CA: Wadsworth.
Hoecker-Drysdale, Susan. 1992. *Harriet Martineau: First Woman Sociologist*. Oxford, UK: Berg.
Hoffman, Bruce. 1995. "'Holy Terror': The Implications of Terrorism Motivated by a Religious Imperative." *Studies in Conflict and Terrorism* 18: 271–284.
Holmes, Mark. 1998. *The Reformation of Canada's Schools: Breaking the Barriers to Parental Choice*. Montreal: McGill-Queen's University Press.
Holmes, Morgan. 2002. "Rethinking the Meaning and Management of Intersexuality." *Sexualities* 5 (2): 159–180.
Holmes, M. Morgan, Linda Mooney, David Knox, and Caroline Schacht. 2016. *Understanding Social Problems* (5th Canadian ed.). Toronto: Nelson.
Homer-Dixon, Thomas. 1993. *Environmental Scarcity and Global Security*. Foreign Policy Association, Headline Series, Number 300. Ephrata, PA: Science Press.
Horkheimer, Max, and Theodor W. Adorno. 1972. *Dialectic of Enlightenment*. New York: Herder and Herder.
Houle, René, and LahouraiaYssaad. 2010. "Recognition of Newcomers' Foreign Credentials and Work Experience." *Perspectives* (September): 18–33. Ottawa: Statistics Canada.
Hounshell, David. 1984. *From the American System to Mass Production, 1800–1932: The Development of Manufacturing Technology in the United States*. Baltimore: Johns Hopkins University Press.
Howard, Alison, and Jessica Edge. 2013. "Enough for All: Household Food Security in Canada by Conference Board of Canada." Available: http://www.conferenceboard.ca/temp/43b73fe8-0424-4af3-95b8-689a26e00ddf/14-058_enoughforall_cfic.pdf.
Howard, Ross. 1998. "No Way Out for Despairing Port Hardy." *Globe and Mail* (June 15): A4.
Hoyt, Homer. 1939. *The Structure and Growth of Residential Neighborhoods in American Cities*. Washington, D.C.: Federal Housing Administration.
Hubbard, Ben. 2014. "The Franchising of Al Qaeda." *The New York Times*. 25 January.
Hughes, Everett C. 1945. "Dilemmas and Contradictions of Status." *American Journal of Sociology* 50: 353–359.
Hugill, David. 2010. *Missing Women, Missing News: Covering Crisis in Vancouver's Downtown Eastside*. Halifax/Winnipeg: Fernwood.
Hull, Gloria T., Patricia Bell-Scott, and Barbara Smith. 1982. *All the Women Are White, All the Blacks Are Men, But Some of Us Are Brave*. Old Westbury, NY: Feminist.
Humphreys, Laud. 1970. *Tearoom Trade: Impersonal Sex in Public Places*. Chicago: Aldine.
Hunt, Charles W. 1989. "Migrant Labor and Sexually Transmitted Diseases: AIDS in Africa." *Journal of Health and Social Behaviour* 30: 353–73.
Hurd Clarke, Laura. 2001. "Older women's bodies and the self: The construction of identity in later life." *Canadian Review of Sociology and Anthropology*, 38 (4), 441–464.
Ilyniak, Natalia. 2014. "Mercury Poisoning in Grassy Narrows: Environmental Injustice, Colonialism, and Capitalist Expansion in Canada." *McGill Sociological Review*, Volume 4 (February 2014): 43–66. Available: http://www.mcgill.ca/msr/msr-volume-4/mercury-poisoning-grassy-narrows.
Index Mundi. 2015. "Russia-Mortality Rate." Available: http://www.indexmundi.com/facts/russia/mortality-rate.
Innis, Harold. 1984. *The Fur Trade in Canada*. Toronto: University of Toronto Press (orig. pub. 1930).
Institut National d'Études Demographiques. 1995. From Julie DaVanzo and David Adamson. 1997. "Russia's Demographic 'Crisis': How Real Is It?" *Rand Issue Paper*, July 1997. Santa Monica, CA: Rand Center for Russian and Eurasian Studies.
Internet World Stats. 2015. "Internet Users in the World: Distribution by World Regions – 2014 Q2." Available: http://www.internetworldstats.com/stats.htm.
Inter-Parliamentary Union. 2014. "Women in National Parliaments, Situation as of 1 Oct 2014." Retrieved November 21, 2014. Available: http://www.ipu.org/wmn-e/classif.htm.
Intersex Society of North America. 2011. "What is intersex?" Available: www.isna.org/faq/what_is_intersex.
IPCC (Intergovernmental Panel on Climate Change). 2015. *Climate Change 2014: Synthesis Report*. Geneva: World Meteorological Organization. Available: https://docs.google.com/viewer?url=http%3A%2F%2Fwww.ipcc.ch%2Fpdf%2Fassessment-report%2Far5%2Fsyr%2FSYR_AR5_FINAL_full.pdf.
Ito, Aki. 2014. "Machines That Can Learn Could Replace Half of American Jobs In the Next Decade or Two." *National Post*. 14 March.
Ivanova, I. (2013). "Time to Rethink the Way We Fund Higher Education." *Our Schools / Our Selves*, *22*(2), 143–147.
IVF.ca. 2012. "Frequently Asked Questions." Retrieved January 24, 2012. Availble: http://www.ivf.ca/faq.htm.
Jackson, John D., Greg Nielsen, and Yon Hsu. 2011. *Mediated Society: A Critical Sociology of Media*. Toronto: Oxford University Press.
Jacques, Peter J., Riley E. Dunlap, and Mark Freeman. 2008. "The Organisation of Denial: Conservative Think Tanks and Environmental Scepticism." *Environmental Politics* 17.3: 349–385.
James, Carl E. 1998. "'Up to No Good': Black on the Streets and Encountering the Police." In Vic Satzewich (ed.), *Racism and Social Inequality in Canada*. Toronto: Thompson Educational Publishing, 157–176.
———. 2005. *Possibilities and Limitations: Multicultural Policies and Programs in Canada*. Halifax: Fernwood.
———. 2010. *Seeing Ourselves: Exploring Race, Ethnicity and Culture* (4th ed.). Toronto: Thompson Books.
Jameson, Fredric. 1984. "Postmodernism, or, the Cultural Logic of Late Capitalism." *New Left Review* 146: 59–92.
Jamieson, L. 2004. "Intimacy, Negotiated Non-Monogamy and the Limits of the Couple." In J. Duncombe, K. Harrison, G. Allan, and D. Marsden (eds.), *The State of Affairs*. Mahwah, NJ: Lawrence Erlbaum, 35–57.
Janis, Irving. 1972. *Victims of Groupthink*. Boston: Houghton Mifflin.
———. 1989. *Crucial Decisions: Leadership in Policymaking and Crisis Management*. New York: Free Press.
Jankowski, Martin Sanchez. 1991. *Islands in the Street: Gangs and American Urban Society*. Berkeley, CA: University of California Press.
Jha, Prabhat, Richard Peto, Witold Zatroski, Jillian Boreham, Martin Jarvis, and Alan Lopez. 2006. "Social inequalities in male mortality, and in male mortality from smoking: Indirect estimation from national death rates in England and Wales, Poland, and North America." *Lancet* 368: 367–370.
Jiwani, J., and M. Young. 2006. "Missing and Murdered Women: Reproducing Marginality in News Discourse." *Canadian Journal of Communication* 31 (4): 895–917.
Jochelson, Richard. 2009. "After Labaye: The Harm Test of Obscenity, the New Judicial Vacuum and the Relevance of Familiar Voices." *Alberta Law Review* 46 (3): 741–768.
Johns Hopkins. 1998. "Can Religion Be Good Medicine?" *The Johns Hopkins Medical Letter* (November 3).

Johnson, Holly. 1996. *Dangerous Domains: Violence Against Women in Canada.* Toronto: Nelson Canada.

———, and Myrna Dawson. 2011. *Violence Against Women in Canada: Research and Policy Perspectives.* Toronto: Oxford University Press.

Johnson, Matthew. 2013. "Connected, Mobile and Social: The Online Lives of Canadian Youth." Vanier Institute for the Family. *Transitions Volume 43 (3).* Available: http://www.vanierinstitute.ca/research-topic_communications-and-media#.VKsA31yqa9A.

Johnstone, Ronald L. 1997. *Religion in Society: A Sociology of Religion* (5th ed.). Saddle River, NJ: Prentice-Hall.

Jones, Melanie, and Tannis Smith. 2011. "Violence against Aboriginal Women and Child Welfare Connections Paper and Annotated Bibliography." Ontario Native Women's Association. Available: http://www.onwa.ca/upload/documents/violence-against-women-and-child-welfare-paper.pdf.

Jones, Meredith. 2004. "Architecture of the Body: Cosmetic Surgery and Postmodern Space." *Space and Culture* 7 (1): 90–101.

Juergensmeyer, Mark. 2003. *Terror in the Mind of God.* Berkeley, CA: University of California Press, 431.

Jussim, Lee. 2013. "Teacher Expectations." Education.com. Retrieved Apr. 13, 2013. Online: www.education.com/reference/article/teacher-expectations.

Kail, Ben, Jill Quadagno, and Jennifer Keene. 2009. "The political economy perspective of aging." In Vern Bengston, Daphna Gans, Norella Putney, and Merril Silverstein (eds.), *Handbook of Theories of Aging,* 2nd ed. New York: Springer, 555–571.

Kaklenberg, Susan, and Michelle Hein. 2010. "Progression on Nickelodeon? Gender-Role Stereotypes in Toy Commercials." *Sex Roles,* 62 (11): 830–847.

Kamkwamba, William, and Bryan Mealer. 2009. *The Boy Who Harnessed the Wind.* New York: HarperCollins.

Kanter, Rosabeth Moss. 1977. *Men and Women of the Corporation.* New York: Basic Books.

———. 1983. *The Change Masters: Innovation and Entrepreneurship in the American Corporation.* New York: Simon & Schuster.

Kapadia, Kamal. 2008. *Developments After a Disaster: The Tsunami, Poverty, Conflict and Reconstruction in Sri Lanka.* Unpublished PhD Dissertation, University of California at Berkeley, 197; 268; 338–339.

Karabanow, Jeff. 2008. "Getting off the Street: Exploring the Processes of Young People's Street Exits" *American Behavioral Scientist* 51: 772, 783.

Karabanow, J., A. Carson, and P. Clement, (2010). *Leaving the Streets: Stories of Canadian Youth.* Fernwood Publishing Ltd: Halifax, NS.

Karabanow, J., J. Hughes, J. Ticknor, S. Kidd, and D. Patterson. (2010). *The Economics of Being Young and Poor: How Homeless Youth Survive in Neo-liberal Times. Journal of Sociology and Social Welfare, 37*(4), 39-63.

Karp, David A., and William C. Yoels. 1976. "The College Classroom: Some Observations on the Meanings of Student Participation." *Sociology and Social Research* 60: 421–439.

Kata, Anna. 2011. "Anti-Vaccine Activists, Web 2.0, and the Postmodern Paradigm—An Overview of Tactics and Tropes Used Online by the Anti-Vaccination Movement." *Vaccine Special Issue.* Online (December 13).

Katz, Stephen. 1999. *Old Age as Lifestyle in an Active Society.* Doreen B. Townsend Center Occasional Papers. Berkeley, CA: University of California, 3. Retrieved May 30, 2005. Available: http://townsendcenter.berkeley.edu/pubs/OP19_Katz.pdf.

Katzer, Jeffrey, Kenneth H. Cook, and Wayne W. Crouch. 1991. *Evaluating Information: A Guide for Users of Social Science Research.* New York: McGraw-Hill.

Kaufman, Tracy L. 1996. *Out of Reach: Can America Pay the Rent?* Washington, DC: National Low Income Housing Coalition.

Kay, Barbara. 2011. "Multiculturalism Is Not a Quebec Value." *National Post.* January 15, 2011. Available: http://news.nationalpost.com/full-comment/barbara-kay-multiculturalism-is-not-a-quebec-value.

Kedrowski, Karen. 2010. "Women's Health Activism in Canada: The Cases of Breast Cancer and Breastfeeding." Paper presented to the Canadian Political Science Association Meeting, Montreal.

Keen, Judy. 2010. "Wis. Law Lets Residents Challenge Race-Based Mascots." *USA Today* (Oct. 7). Available: www.usatoday.com/news/nation/2010-10-06-Kewaunee-school-mascot_N.htm.

Kellner, Douglas. 2011. "Cultural Studies, Multiculturalism, and Media Culture." In Gail Hines and Jean Humez (eds.), *Gender, Race, and Class in Media.* Los Angeles: Sage, 7.

Kellett, Anthony. 2004. "Terrorism in Canada, 1960–1992." In Jeffrey Ian Ross (ed.), *Violence in Canada: Sociopolitical Perspectives* (2nd ed.). New Brunswick, N.J.: Transaction Press, 284–312.

Kelman, Steven. 1991. "Sweden Sour? Downsizing the 'Third Way.'" *New Republic* (July 29): 19–23.

Kemp, Alice Abel. 1994. *Women's Work: Degraded and Devalued.* Englewood Cliffs, NJ: Prentice-Hall.

Kendall, Diana. 2002. *The Power of Good Deeds: Privileged Women and the Social Reproduction of the Upper Class.* Lanham, MD: Rowman & Littlefield.

———. 2011. *Framing Class: Media Representations of Wealth and Poverty in the United States* (2nd ed.). Lanham, MD: Rowman & Littlefield.

Kennedy, K. 2012. "Acting Out: Discussions with young feminists on political engagement." *Our Schools / Our Selves, 21*(3), 23–29.

Kennedy, Leslie W. 1983. *The Urban Kaleidoscope: Canadian Perspectives.* Toronto: McGraw-Hill Ryerson.

Kennedy, Paul. 1993. *Preparing for the Twenty-First Century.* New York: Random House.

Kenny, Charles. 2003. "Development's False Divide." *Foreign Policy* (January/February): 76–77.

Kettle, John. 1998. "Death Still Looks Like a Healthy Business." *Globe and Mail* (May 7): B15.

Khan, S. (1993). Canadian Muslim Women and Shari'a Law: A Feminist Response to "Oh! Canada!" *Canadian Journal of Women and the Law,* 6(1), 52–65.

Khayatt, Didi. 1994. "The Boundaries of Identity at the Intersection of Race, Class and Gender." *Canadian Woman Studies* 14 (Spring).

Kidd, S.A., L. Davidson "'You have to adapt because you have no other choice': The stories of strength and resilience of 208 homeless youth in New York City and Toronto." *Journal of Community Psychology,* 35 (2007), pp. 219–23.

Kidd, S. A., J. Karabanow, J. Hughes, and T. Frederick. (2013). "Brief report: Youth pathways out of homelessness – Preliminary findings." *Journal Of Adolescence, 36*(6), 1035–1037. doi:10.1016/j.adolescence.2013.08.009.

Killian, Lewis. 1984. "Organization, Rationality, and Spontaneity in the Civil Rights Movement." *American Sociological Review* 49: 770–783.

King, Gary, Robert O. Keohane, and Sidney Verba. 1994. *Designing Social Inquiry: Scientific Inference in Qualitative Research.* Princeton, NJ: Princeton University Press.

King, Samantha. 2006. Pink *Ribbons, Inc.: Breast Cancer and the Culture of Philanthropy.* Minneapolis, MN: University of Minnesota Press.

Kinsey, A., C. Pomeroy, and C. Martin. 1948. *Sexual Behavior in the Human Male.* Philadelphia: W.B. Sanders.

Kinsey, A., C. Pomeroy, W. Gebhard, and C. Marin. 1953. *Sexual Behaviour in the Human Female.* Philadelphia: W.B. Sanders.

Kinsey Institute. 2012. "Kinsey's Homosexual-Heterosexual Rating Scale." Retrieved August 1, 2012. Available: http://www.iub.edu/~kinsey/research/ak-hhscale.html.

Kirby, Sandra, and Kate McKenna. 1989. *Experience Research Social Change: Methods from the Margins.* Toronto: Garamond.

Kirk, Ruth. 1986. *Wisdom of the Elders: Native Traditions on the Northwest Coast.* Vancouver: Douglas and Mclntyre.

Klein, Naomi. 2000. *No Logo.* Toronto: Vintage Canada.

Kleinfeld, Judith S. 2002. "The Small World Problem." *Society* (January/February): 61–66.

Klesse, C. 2006. "Polyamory and Its 'Others': Contesting the Terms of Non-Monogamy." *Sexualities* 9 (5): 565–583. Available: http://sites.middlebury.edu/sexandsociety/files/2015/01/polyamory4-klesse-polyamory.pdf.

Klockars, Carl B. 1979. "The Contemporary Crises of Marxist Criminology." *Criminology* 16: 477–515.

Knoll, James. 2010. "The 'Pseudocommando' Mass Murderer: Part II, The Language of Revenge." *Journal of the American Academy of Psychiatry and the Law."* Online 38: 263–272.

Knox, Paul L., and Peter J. Taylor, eds. 1995. *World Cities in a World-System.* Cambridge, England: Cambridge University Press.

Kohl, Beth. 2007. "On Indian Surrogates." The Huffington Post (Oct. 30). Retrieved Feb. 9, 2008. Online: www.huffingtonpost.com/beth-kohl/on-indian-surrogates_b_70425.html.

Kohlberg, Lawrence. 1969. "Stage and Sequence: The Cognitive-Developmental Approach to Socialization." In David A. Goslin, *Handbook of Socialization Theory and Research.* Chicago: Rand McNally, 347–480.

———. 1981. *The Philosophy of Moral Development: Moral Stages and the Idea of Justice,* Vol. 1: *Essays on Moral Development.* San Francisco: Harper & Row.

Kohn, Melvin L., Atsushi Naoi, Carrie Schoenbach, Carmi Schooler, and Kazimierz M. Slomczynski. 1990. "Position in the Class Structure and Psychological Functioning in the United States, Japan, and Poland." *American Journal of Sociology* 95: 964–1008.

Kolata, Gina. 2006. "Old But Not Frail: A Matter of Heart and Head." *The New York Times* (October 5).

Kome, Penney. 2002. "Canada Court Tells Parliament to OK Gay Marriages." Women's eNews. Retrieved August 23, 2003, 1. Available: http://www.womensnews.org/article/cfm/dyn/aid/987/context/archive.

Komljenovic, D. (2013). Supreme Court Win Ensures a Future for Special Needs Education. *Our Schools / Our Selves, 22*(2), 21–23.

Koopmans, Ruud. 1999. "Political. Opportunity. Structure. Some Splitting to Balance the Lumping." *Sociological Forum* (Mar.): 93–105.

Korhonen, Pekka. 1994. "The Theory of the Flying Geese Pattern of Development and Its Interpretations." *Journal of Peace Research* 31: 93–108.

Korkeila, Jyrki. 2010. "Problematic Use in Context." *World Psychiatry* 9 (2): 94–95.

Korte, Charles, and Stanley Milgram. 1970. "Acquaintance Networks Between Racial Groups: Application of the Small World Method." *Journal of Personality and Social Psychology* 15: 101–108.

Krahn, Harvey J., Graham Lowe, and Karen Hughes. 2007. *Work, Industry and Canadian Society* (5th ed.). Toronto: Nelson.

———, Graham Lowe, and Karen Hughes. 2011. *Work, Industry and Canadian Society* (6th ed.). Toronto: Nelson.

———, Karen Hughes, and Graham Lowe. 2015. *Work, Industry, and*

Canadian Society (7th ed.). Toronto: Nelson.
———, and Alison Taylor. 2007. "Streaming in the 10th Grade in Four Canadian Provinces in 2000." *Education Matters*. Statistics Canada Cat. no. 81-004-XIE. Retrieved April 14, 2009. Available: http://www.statcan.gc.ca/pub/81-004-x/2007002/9994-eng.htm.
Kramer, Adam, Jamie Guillory, and Jeffrey Hancock. 2014. "Experimental Evidence of Massive-Scale Emotional Contagion Through Social Networks." *PNAS* 111:8788-8790.
Kraus, Krystalline. 2013. "Grassy Narrows wants justice for destructive logging and mercury poisoning." *Rabble.ca*. Retrieved http://rabble.ca/news/2013/11/grassy-narrows-wants-justice-destructive-logging-and-mercury-poisoning.
Kreager, D.A., and J. Staff. 2009. "The sexual double standard and adolescent peer acceptance." *Social Psychology Quarterly*, *72*(2), 143–164.
Krebs, Valdis E. 2002. "Mapping Networks of Terrorist Cells." *Connections* 24 (3): 43–52.
Kristof, Nicholas D. 2006. "Looking for Islam's Luthers." *New York Times* (Oct. 15): A22.
———. 2012a. "After Recess: Change the World." *New York Times* (February 4).
———. 2012b. "Africa on the Rise." *New York Times* (July 1): SR11.
Kshetri, Nir. 2010. *The Global Cybercrime Industry*. Berlin: Springer-Verlag.
Kubes, Danielle. 2015. "The spending diaries: What three millennials spend their money on and why." *Financial Post* (May 25).
Kuhn, Bob. 2014. "A Message from President Kuhn About the TWO School of Law." Retrieved: 12 November 2014. Available: http://www.twu.ca/academics/school-of-law/news/2014/003-msg-from-president.html.
Kunz, Jean L., Anne Milan, and Sylvain Schetagne. 2000. *Unequal Access: A Canadian Profile of Racial Differences in Education, Employment and Income*. Ottawa: Canadian Race Relations Foundation.
Kurtz, Lester. 1995. *Gods in the Global Village: The World's Religions in Sociological Perspective*. Thousand Oaks, CA: Sage.
Kvale, Steinar. 1996. *Interviews: An Introduction to Qualitative Research Interviewing*. Thousand Oaks, CA: Sage.
Laberge, Danielle. 1991. "Women's Criminality, Criminal Women, Criminalized Women? Questions in and for a Feminist Perspective." *Journal of Human Justice* 2 (2): 37–56.
Ladd, E.C., Jr. 1966. *Negro Political Leadership in the South*. Ithaca, NY: Cornell University Press.
Lamanna, Mary Ann, and Agnes Riedmann. 2003. *Marriages and Families: Making Choices and Facing Change* (8th ed.). Belmont, CA: Wadsworth.
———. 2012. *Marriages, Families, and Relationships: Making Choices in a Diverse Society* (11th ed.). Belmont, CA: Wadsworth/Cengage.
Lancet. 2012. "Living with Grief." Volume 379 (February 18): 589.
Lancet and University College London Institute for Global Health Commission. 2009. "Managing the Effects of Climate Change." *Lancet* Vol. 373, 1693–1733. Available: http://www.thelancet.com/pdfs/journals/lancet/PIIS0140-6736(09)60935-1.pdf.
Langdon, Steven. 1999. *Global Poverty, Democracy and North-South Change*. Toronto: Garamond Press, 41.
Langton, Jerry. 2013. *The Notorious Bacon Brothers: Inside Gang Warfare on Vancouver Streets*. Mississauga, ON: John Wiley.
Lareau, Annette. 2011. *Unequal Childhoods: Class, Race, and Family Life* (2nd ed.). Berkeley, CA: University of California Press.
Larsen, Janet. 2011. "Cancer Now Leading Cause of Death in China." Earth Policy Institute (May 25). Online: www.earth-policy.org/plan_b_updates/2011/update96.
Latané, Bibb, and John M. Darley. 1970. *The Unresponsive Bystander: Why Doesn't He Help?* New York: Appleton Century Crofts.
Laird, Kristin. 2009. "TD's Old Men Are Grumpy About Mortgages." *Marketer News* (April 1). Available: http://www.marketingmag.ca/english/news/marketer/article.jsp?content=20090331_173938_7252.
Laxer, Gordon. 1989. *Open for Business: The Roots of Foreign Ownership in Canada*. Don Mills, ON: Oxford University Press.
Lazar, Shira. 2012. "How the White House Became a Social-Media Powerhouse." *The Daily Dose* (February 8).
Lazarsfeld, Paul, and Robert Merton. 1948. "Mass Communication, Popular Taste and Organized Social Action." In L. Bryson (ed.), *The Communication of Ideas*. New York: Harper and Brothers, 95–118.
LeBlanc, Cathie. 2011. "Fear and Loathing in *Second Life*: Body Surveillance in the Online Community." In Robin DeRosa (ed.), *Simulation in Media and Culture: Believing the Hype*. Lanham, MD: Lexington Books, 113–119.
Le Bon, Gustave. 1960. *The Crowd: A Study of the Popular Mind*. New York: Viking (orig. pub. 1895).
Leiss, William, Stephen Kline, Sut Jhally, and Jacqueline Botterill. 2006. *Social Communication in Advertising: Consumption in the Mediated Marketplace* (3rd ed.) New York: Routledge.
Lemaire, Brad. 2012. "Wal-Mart Gets China Approval for Stake Increase in Yihaodian." Proactiveinvestors.com. Retrieved Sept. 15, 2012. Online: www.proactiveinvestors.com/companies/news/33255/walmart-gets-china-approval-for-stake-increase-in-yihaodian-33255.html.
Lemert, Charles. 1997. *Postmodernism Is Not What You Think*. Malden, MA: Blackwell.
Lemert, Edwin M. 1951. *Social Pathology*. New York: McGraw-Hill.
Lengermann, Patricia Madoo, and Jill Niebrugge-Brantley. 1998. *The Women Founders: Sociology and Social Theory, 1830–1930*. New York: McGraw-Hill.
Lenski, Gerhard, Jean Lenski, and Patrick Nolan. 1991. *Human Societies: An Introduction to Macrosociology* (6th ed.). New York: McGraw-Hill.
Leo, Geoff. 2015. "HIV Rates on Sask. Reserves Higher Than Some African Nations." CBC News. 3 June.
LeShan, Eda. 1994. *I Want More of Everything*. New York: New Market Press.
Levin, William C. 1988. "Age Stereotyping: College Student Evaluations." *Research on Aging* 10 (1): 134–148.
Levine, Nancy E., and Joan B. Silk. 1997. "Why Polyandry Fails: Sources of Instability in Polyandrous Marriages." *Current Anthropology* (June): 375–399.
Lewis, Kevin, Marco Gonzalez, and Jason Kaufman. 2012. "Social Selection and Peer Influence in an Online Social Network." *Proceedings of the National Academy of Sciences of the United States of America* 109: 68–72.
Lewis, Paul. 1998. "Marx's Stock Resurges on a 150-Year Tip." *New York Times* (June 27): A17, A19.
Lewis-Thornton, Rae. 1994. "Facing AIDS." *Essence* (December): 63–130.
Levitt, Kari. 1970. *Silent Surrender: The Multinational Corporation in Canada*. Toronto: Macmillan of Canada.
———. 1997. *Dying Hard: The Ravages of Industrial Carnage*. Toronto: Oxford University Press.
Li, Peter S. 2003. *Cultural Diversity in Canada: The Social Construction of Racial Differences*. Available: http://www.justice.gc.ca/eng/rp-pr/csj-sjc/jsp-sjp/rp02_8-dr02_8/rp02_8.pdf.
Liebow, Elliot. 1993. *Tell Them Who I Am: The Lives of Homeless Women*. New York: Free Press.
Lim, Chaeyoon and Robert Putnam. 2010. "Religion, Social Networks, and Life Satisfaction." *American Sociological Review*. 75: 914–933.
Lin, Ken-Hou and Jennifer Lundquist. 2013. "Mate Selection in Cyberspace: The Intersection of Race, Gender, and Education." *American Journal of Sociology* 119: 183–215.
Linden, Rick. 1994. "Deviance and Crime." In Lorne Tepperman, James E. Curtis, and R.J. Richardson (eds.), *The Social World* (3rd ed.). Whitby, ON: McGraw-Hill Ryerson, 188–226.
———. 2009. *Criminology: A Canadian Perspective* (6th ed.). Toronto: Nelson.
———. 2016. *Criminology: A Canadian Perspective* (8th ed.). Toronto: Nelson.
———, and Raymond C. Currie. 1977. "Religiosity and Drug Use: A Test of Social Control Theory." *Canadian Journal of Criminology and Corrections* 19: 346–355.
———, and Cathy Fillmore. 1981. "A Comparative Study of Delinquency Involvement." *Canadian Review of Sociology and Anthropology* 18: 343–361.
———, and Dan Koenig. 2016. "Deterrence, Routine Activity, and Rational Choice Theories." In Rick Linden (ed.), *Criminology: A Canadian Perspective* (8th ed.). Toronto: Nelson.
Lindsay, Colin. 2002. *Poverty Profile, 1999*. Ottawa: Statistics Canada, 108–110.
———, and Marcia Almey. 2006. *Women in Canada* (5th ed.). Ottawa: Statistics Canada.
Linton, Ralph. 1936. *The Study of Man*. New York: Appleton-Century-Crofts.
Lipton, Eric, Christopher Drew, Scott Shane, and David Rohde. 2005. "Breakdowns Marked Path from Hurricane to Anarchy." NYTimes.com (September 11). Retrieved September 12, 2005. Available: http://www.nytimes.com/2005/09/11/national/nationalspecial/11response.html?pagewanted=1&ei=5070&en=b1231d972456e252&ex=1126670400.
Livingston, D.W. 2014. 1. Class, Race and Gender Differences in Schooling. *Our Schools / Our Selves, 23*(2), 9–39.
Lofland, John. 1993. "Collective Behavior: The Elementary Forms." In Russell L. Curtis, Jr., and Benigno E. Aguirre (eds.), *Collective Behavior and Social Movements*. Boston: Allyn & Bacon, 70–75.
———, and Rodney Stark. 1965. "Becoming a World-Saver: A Theory of Conversion to a Deviant Perspective." *American Sociological Review* 30 (6): 862–875.
Lofland, Lyn. 1973. *A World of Strangers: Order and Action in Urban Public Space*. New York: Basic Books.
Logical Outcomes. 2014. *"This Issue Has Been With Us for Ages:" A Community-Based Assessment of Police Contact Carding in 31 Division*. Toronto: Logical Outcomes.
Lorber, Judith. 1994. *Paradoxes of Gender*. New Haven, CT: Yale University Press.
Lothian, Jane. 1990. *Status Offenders in Manitoba: Hidden Systems of Social Control*. Unpublished master's thesis, University of Manitoba.
Low, Setha. 2003. *Behind the Gates: Life, Security, and the Pursuit of Happiness in Fortress America*. New York: Routledge.
Lui, Emma. 2015. "On Notice For a Drinking Water Crisis in Canada." The Council of Canadians. Available: http://canadians.org/sites/default/files/publications/report-drinking-water-0315.pdf.
Lummis, C. Douglas. 1992. "Equality." In Wolfgang Sachs (ed.), *The Development Dictionary*. Atlantic Highlands, NJ: Zed Books, 38–52.
Lupul, M.R. 1988. "Ukrainians: The Fifth Cultural Wheel in Canada." In Ian H. Angus (ed.), *Ethnicity in a Technological Age*. Edmonton: Canadian Institute of Ukrainian Studies, University of Alberta, 177–192.
Luxton, Meg. 1995. "Two Hands for the Clock: Changing Patterns of Gendered Division of Labour in the Home." In E.D. Nelson and B.W. Robinson

(eds.), *Gender in the 1990s.* Toronto: Thomson Nelson, 288–301.
———. 2012. "Changing Families: New Understandings." *Contemporary Family Trends.* Vanier Institute of the Family, 2.
———, and J. Corman. 2001. *Getting By in Hard Times: Gendered Labour at Home and on the Job.* Toronto: University of Toronto Press.
Lynch, Michael J., and Paul Stretesky. 2007. "Green Criminology in the United States." In Piers Beirne and Nigel South (eds.), *Issues in Green Criminology: Confronting Harms Against Environments, Humanity and Other Animals.* Portland: Willan Publishing, 248–269.
Lyons, H., P. Giordano, W. Manning, and M. Longmore. 2011. "Identity, Peer Relationships, and Adolescent Girls' Sexual Behavior: An Exploration of the Contemporary Double Standard." *Journal of Sex Research, 48*(5), 437–449.
Lyons, John. 1998. "The Way We Live: Central Plains." *Winnipeg Free Press* (June 7): B3.
Lyotard, Jean-Francois. 1984. *The Postmodern Condition.* Manchester, UK: Manchester University Press.
Maass, Peter, and Megha Rajagopalan. 2012. "That's No Phone. That's My Tracker." *New York Times* (July 13). Retrieved July 23, 2012. Available: http://www.nytimes.com/2012/07/15/sunday-review/thats-not-my-phone-its-my-tracker.html.
MacDonald, D. (2014). *Outrageous Fortune: Documenting Canada's Wealth Gap.* Canadian Centre for Policy Alternatives. ISBN 978-1-77125-116-7. Retrieved from: https://www.policyalternatives.ca/sites/default/files/uploads/publications/National%20Office/2014/04/Outrageous_Fortune.pdf.
Macdonald, David, and Erika Shaker. 2012. *Eduflation and the High Cost of Learning.* September 2012, CCPA.
MacDonald, Marci. 1996. "The New Spirituality." *Maclean's* (October 10): 44–48.
Mackie, Marlene. 1995. "Gender in the Family: Changing Patterns." In Nancy Mandell and Ann Duffy (eds.), *Canadian Families: Diversity, Conflict, and Change* (2nd ed.). Toronto: Harcourt Brace, 17–43.
MacKinnon, Catherine. 1982. "Feminism, Marxism, Method and the State: An Agenda for Theory." In N.O. Keohane et al. (eds), *Feminist Theory: A Critique of Ideology.* Chicago: University of Chicago Press, 1–30.
Maclean's Campus Online. 2012. Scott Dobson-Mitchell. "The Many Regrets of a Fourth-Year Student." Available: http://oncampus.macleans.ca/education/2012/01/27/the-many-regrets-of-a-fourth-year-student.
MacLeod, Linda. 1980. *Wife Battering in Canada: The Vicious Circle.* Ottawa: Canadian Council on the Status of Women.
Macmillan, Ross, Annette Nierobisz, and Sandy Welsh. 2000. "Experiencing the Streets: Harassment and Perceptions of Safety Among Women." *Journal of Research in Crime and Delinquency* 37 (3).
Malenfant, Eric, André Lebel, and Laurent Martel. 2010. *Projections of the Diversity of the Canadian Population, 2006–2031.* Ottawa: Statistics Canada.
Malinowski, Bronislaw. 1922. *Argonauts of the Western Pacific.* New York: Dutton.
Mandell, Nancy, ed. 2001. *Feminist Issues: Race, Class, and Sexuality* (3rd ed.). Toronto: Prentice-Hall.
———, and Ann Duffy, eds. 2011. Canadian Families: Diversity Conflict and Change (4th ed.). Toronto: Nelson.
Mangan, Dan. 2013. "Medical Bills Are the Biggest Cause of US Bankruptcies: Study." CNBC, 25 June. Available: http://www.cnbc.com/id/100840148.
Mann, Patricia S. 1994. *Micro-Politics: Agency in Postfeminist Era.* Minneapolis, MN: University of Minnesota Press.
Mantell, David Mark. 1971. "The Potential for Violence in Germany." *Journal of Social Issues* 27 (4): 101–112.
Marchak, Patricia. 1975. *Ideological Perspectives on Canadian Society.* Toronto: McGraw-Hill.
Marger, Martin N. 1987. *Elites and Masses: An Introduction to Political Sociology* (2nd ed.). Belmont, CA: Wadsworth.
———. 1994. *Race and Ethnic Relations: American and Global Perspectives.* Belmont, CA: Wadsworth.
———. 2009. *Race and Ethnic Relations: American and Global Perspectives* (8th ed.). Belmont, CA: Wadsworth, Cengage Learning.
———. 2012. *Race and Ethnic Relations: American and Global Perspectives* (9th ed.). Belmont, CA: Wadsworth, Cengage.
Marion, Russ, and Mary Uhl-Bien. 2003. "Complexity Theory and Al-Qaeda: Examining Complex Leadership." *Emergence* 5 (1): 54–76.
Mark, Kristen P., Erick Janssen, and Robin R. Milhausen. 2011. "Infidelity in Heterosexual Couples: Demographic, Interpersonal, and Personality-Related Predictors of Extradyadic Sex." *Archives of Sexual Behavior* 40 (5): 971–982.
MarketingCharts. 2013. "18-24-Year-Old Smartphone Owners Send and Receive Almost 4K Texts per Month." Available: http://www.marketingcharts.com/online/18-24-year-old-smartphone-owners-send-and-receive-almost-4k-texts-per-month-27993/.
Marquardt, Elizabeth. 2006. "The Revolution in Parenthood: The Emerging Global Clash Between Adult Rights and Children's Needs." Institute for American Values. Retrieved May 22, 2009. Available: http://www.americanvalues.org/pdfs/parenthood.pdf.
Marshall, Gordon, ed. 1998. *The Concise Oxford Dictionary of Sociology* (2nd ed.). New York: Oxford University Press.
Marshall, Katherine. 1995. "Dual Earners: Who's Responsible for Housework?" In E.D. Nelson and B.W. Robinson, *Gender in the 1990s.* Toronto: Thomson Nelson, 302–308.
———. 2011. "Generational Change in Paid and Unpaid Work." *Canadian Social Trends 82.* Statistics Canada Cat. no. 11-008-X. Retrieved January 20, 2011, 13. Available: http://www.statcan.gc.ca/pub/11-008-x/2011002/article/11520-eng.pdf.
Marshall, S.L.A. 1947. *Men Against Fire.* New York: Morrow.
Martin, Nick. 1996. "Aboriginal Speech Dying." *Winnipeg Free Press* (March 29): A8.
Martin, Vivian B. 2008. "Media Bias: Going Beyond Fair and Balanced." *Scientific American.* Retrieved October 17, 2011. Available: http://www.scientificamerican.com/article.cfm?id=media-bias-presidential-election.
Martineau, Harriet. 1962. *Society in America* (edited, abridged). Garden City, NY: Doubleday (orig. pub. 1837).
Martinussen, John. 1997. *Society, State and Market: A Guide to Competing Theories of Development.* Halifax: Fernwood Books.
Marx, Karl. 1967. *Capital: A Critique of Political Economy.* Friedrich Engels (ed.). New York: International Publishers (orig. pub. 1867).
———. 1970. *The German Ideology,* Part 1. C.J. Arthur (ed). New York: International (orig. pub. 1845–1846).
———, and Friedrich Engels. 1967. *The Communist Manifesto.* New York: Pantheon (orig. pub. 1848).
Mascarenhas, Michael. 2012. *Where the Waters Divide: Neoliberalism, White Privilege, and Environmental Racism in Canada.* Toronto: Lexington Books.
———. 2015. "Environmental Inequality and Environmental Justice." In Kenneth Alan Gould and Tammy L. Lewis (eds.), *Twenty Lessons in Environmental Sociology* (2nd ed.). Oxford University Press, 161–178.
Mason Lee, Robert. 1992. *Death and Deliverance.* Toronto: Macfarlane Walter & Ross.
———. 1998. "I'll Be Home for Christmas." *Globe and Mail* (December 24).
Mastro, Dana, and Susanah Stern. 2006. "Race and Gender in Advertising: A Look at Sexualized Images in Prime-Time Commercials." In Tom Reichert and Jacqueline Lambiase (eds.), *Sex in Consumer Culture: The Erotic Content of Media and Marketing.* Mahwah, NJ: Lawrence Erlbaum, 281–299.
Maticka-Tyndale, E. 2008. "Sexuality and Sexual Health of Canadian Adolescents: Yesterday, Today and Tomorrow." *Canadian Journal of Human Sexuality,* 17 (3): 85–95.
Matthews, Christopher. 2013. "How Does One Fake Tweet Cause a Stock Market Crash?" *Time* (April 24). Retrieved May 23, 2013. Online: http://business.time.com/2013/04/24/how-does-one-fake-tweet-cause-a-stock-market-crash/#ixzz2U8tGJinY.
Maynard, Rona. 1987. "How Do You Like Your Job?" *Globe and Mail Report on Business Magazine* (November): 120–25.
McAdam, Doug. 1996. "Conceptual Origins, Current Problems, Future Directions." In Doug McAdam, John McCarthy, and Meyer N. Zald (eds.), *Comparative Perspectives on Social Movements.* New York: Cambridge University Press, 23–40.
McClelland, Mac. 2012. "I Was a Warehouse Wage Slave." *Mother Jones* (March/April), 6.
McCormick, Chris. 1995. *Constructing Danger: The Mis/Representation of Crime in the News.* Halifax: Fernwood Publishing.
McCormick, Susan. 2015. "The Sociology of Environmental Health." In Kenneth Alan Gould and Tammy L. Lewis (eds.), *Twenty Lessons in Environmental Sociology* (2nd ed.). Oxford University Press, 179–190.
McDonald's. 2014. Frequently Asked Questions. Retrieved: 3 December 2014. Available: http://www.mcdonalds.ca/ca/en/contact_us/faq.html.
McDuling, John. 2015. "Why Investors are Pouring Millions into Fantasy Sports." *Quartz* April 6. Available: http://qz.com/377305/why-investors-are-pouring-millions-into-fantasy-sports/.
McEachern, William A. 1994. *Economics: A Contemporary Introduction.* Cincinnati: South-Western.
McGee, Reece. 1975. *Points of Departure.* Hinsdale, IL: Dryden Press.
McGovern, Celeste. 1995. "Dr. Death Speaks." *Alberta Report/Western Report* 22 (January 9): 33.
McGuire, Meredith B. 1992. *Religion: The Social Context* (2nd ed.). Belmont, CA: Wadsworth.
———. 1997. *Religion: The Social Context* (4th ed.). Belmont, CA: Wadsworth.
McIntosh, Mary. 1968. "The Homosexual Role." *Social Problems* 16 (2): 182–192.
———. 1978. "The State and the Oppression of Women." In Annette Kuhn and Ann Marie Wolpe (eds.), *Feminism and Materialism.* London: Routledge and Kegan Paul.
McInturff, Kate. 2013. *Closing Canada's Gender Gap: Year 2240 Here We Come.* Canadian Centre for Policy Alternatives. Available: http://www.policyalternatives.ca/sites/default/files/uploads/publications/National%20Office/2013/04/Closing_Canadas_Gender_Gap_0.pdf.
McIntyre, M. 2011. "Rape Victim Inviting: So No Jail." *Winnipeg Free Press* (February 24). Retrieved July 9, 2012. Available: http://www.winnipegfreepress.com/breakingnews/rape-victim-inviting-so-no-jail--rape-victim-inviting-so-no-jail-116801578.html.
McKay, A. 2006. "Trends in Teen Pregnancy in Canada with Comparisons to U.S.A. and England/Wales." *Canadian Journal of Human Sexuality* 15 (3–4): 157–161.

———. (2009). "Sexual Health Education in the Schools: Questions and Answers," 3rd ed. Sex Information and Education Council of Canada (SIECCAN). *The Canadian Journal of Human Sexuality*, 18 (1–2), 47–60.
———. 2010. *Sexual health education in the schools: Questions & answers*. Sex Information and Education Council of Canada.
———. 2012. "Trends in Canadian National And Provincial/Territorial Teen Pregnancy Rates: 2001–2010," *The Canadian Journal of Human Sexuality*, 21 (3–4), 161–175.
———, and M. Barrett. (2008). "Rising reported rates of chlamydia among young women in Canada: What do they tell us about trends in the actual prevalence of the infection?" *The Canadian Journal of Human Sexuality*, 17, 61–69.
———, and M. Barrett. (2010). "Trends in teen pregnancy from 1996–2006: A comparison of Canada, Sweden, U.S.A. and England/Wales. *The Canadian Journal of Human Sexuality*, 19, 43–52.
McKenzie, Roderick D. 1925. "The Ecological Approach to the Study of the Human Community." In Robert Park, Ernest Burgess, and Roderick D. McKenzie (eds.), *The City*. Chicago: University of Chicago Press.
McKnight, Peter. 2012. "The Funhouse Mirror: Media Representations of Crime and Justice." In Julian Roberts and Michelle Grossman (eds.), *Criminal Justice in Canada*. Toronto: Nelson, 44–54.
McLuhan, Marshall. 1964. *Understanding Media: The Extensions of Man*. New York: McGraw-Hill.
———, Quentin Fiore, and Shepard Fairey. 1967. *The Medium Is the Massage*. New York: Bantam Books.
McMahon, Tamsin. 2014. "Seniors and the Generation Spending Gap." *Maclean's*. Retrieved November 6, 2014. Available: http://www.macleans.ca/society/life/seniors-and-the-generation-spending-gap/.
McNish, Jacquie, and Sinclair Stewart. 2004. *Wrong Way: The Fall of Conrad Black*. Toronto: Viking Canada.
McPherson, Barry D. 1998. *Aging as a Social Process: An Introduction to Individual and Population Aging*. Toronto: Harcourt Brace.
McSpotlight. 1999. "McDonald's and Employment." Retrieved September 7, 1999. Available: http://www.mcspotlight.org/issues/rants/employment.html.
McSpotlight. 2014. "The Issues: Employment." Retrieved: 3 December 2014. Available: http://www.mcspotlight.org/issues/employment/index.html.
Mead, George Herbert. 1962. *Mind, Self, and Society*. Chicago: University of Chicago Press (orig. pub. 1934).
Meisel, Dr. Zachary. 2011. "Googling Symptoms Helps Patients and Doctors." *Time Health* (January 19). Retrieved May 14, 2012. Available: http://www.time.com/time/health/article/0,8599,2043125,00.html.
Meisner, Natalie. 2014. *Double Pregnant: Two Lesbians Make a Family*. Fernwood Press.
Melchers, Ronald. 2003. "Do Toronto Police Engage in Racial Profiling?" *Canadian Journal of Criminology and Criminal Justice* 45 (July): 347–366.
Merton, Robert King. 1938. "Social Structure and Anomie." *American Sociological Review* 3 (6): 672–682.
———. 1949. "Discrimination and the American Creed." In Robert M. MacIver (ed.), *Discrimination and National Welfare*. New York: Harper & Row, 99–126.
———. 1968. *Social Theory and Social Structure* (enlarged ed.). New York: Free Press.
Messner, Michael A. 2000. "Barbie Girls Versus Sea Monsters: Children Constructing Gender." In Margaret L. Andersen (ed.), *Thinking About Women: Sociological Perspectives on Sex and Gender* (7th ed.). Boston: Pearson, 765–784.
Mian, C. 2013. "Bullying." *Our Schools / Our Selves, 22*(4), 145–162.
Michael, Robert T., John H. Gagnon, Edward O. Laumann, and Gina Kolata. 1994. *Sex in America*. Boston: Little, Brown.
Michelson, William H. 1994. "Cities and Urbanization." In Lorne Tepperman, James Curtis, and R.J. Richardson (eds.), *The Social World* (3rd ed.). Toronto: McGraw-Hill, 672–709.
Mickleburgh, Rod. 2011. "Anti–Wall Street Protests Take Off Thanks to a Canadian Idea." *Globe and Mail* (October 4). Retrieved January 11, 2012. Available: http://www.theglobeandmail.com/news/world/americas/article2191364.ece.
Miki, Roy, and Cassandra Kobayashi. 1991. *Justice in Our Own Time: The Japanese Canadian Redress Settlement*. Vancouver: Talonbooks.
Milan, Anne. 2005. "Willing to Participate: Political Engagement of Young Adults." *Canadian Social Trends* (Winter): 2–7.
———. 2013. "Marital Status: Overview, 2011." Statistics Canada. Available: http://www.statcan.gc.ca/pub/91-209-x/2013001/article/11788/fig/fig7-eng.htm.
———, Leslie-Anne Keown, and Covadonga Robles Urquijo. 2011. "Families, Living Arrangements and Unpaid Work." In *Women in Canada: A Gender-Based Statistical Report*, Statistics Canada Cat. no. 89-503X.
———, and Mireille Vezina. 2011. "Senior Women." Component of Statistics Canada Catalogue no. 89-503-X, *Women in Canada: A Gender-Based Statistical Report*. Ottawa: Statistics Canada.
———, Irene Wong, and Mireille Vezina. 2014. "Emerging Trends in Living Arrangements and Conjugal Unions for Current and Future Seniors." Ottawa: Statistics Canada. Retrieved 7 November 2014. Available: http://www.statcan.gc.ca/pub/75-006-x/2014001/article/11904-eng.htm.
Milgram, Stanley. 1963. "Behavioral Study of Obedience." *Journal of Abnormal and Social Psychology* 67: 371–378.
———. 1967. "The Small World Problem." *Psychology Today* 2: 60–67.
———. 1974. *Obedience to Authority*. New York: Harper & Row.
Milhausen, R.R., and E.S. Herold. 2001. "Does the sexual double standard still exist? Perceptions of university women." *Journal of Sex Research, 36*, 361–368.
Milhausen, Robin R., and Kristen P. Mark. 2009. "Infidelity in Committed Relationships." In Harry T. Reis and Susan Sprecher (eds.), *Encyclopedia of Human Relationships*. Thousand Oaks, CA: Sage.
Miliband, Ralph. 1969. *The State in Capitalist Society*. New York: Basic Books.
Miller, Grant. 2008. "Women's Suffrage, Political Responsiveness, and Child Survival in American History." *The Quarterly Journal of Economics* 123 (3): 1287–1327.
Mills, C. Wright. 1956. *White Collar*. New York: Oxford University Press.
———. 1959a. *The Sociological Imagination*. London: Oxford University Press.
———. 1959b. *The Power Elite*. Fair Lawn, NJ: Oxford University Press.
Misztal, Barbara A. 1993. "Understanding Political Change in Eastern Europe: A Sociological Perspective." *Sociology* 27 (3): 451–471.
Mitchell, Kaitlyn. 2014. "Standing up for Canadians' Access to Clean Drinking Water." *Ecojustice*. Summer, 2014, 3. Available: http://www.ecojustice.ca/wp-content/uploads/2013/10/Ecojustice_news_Summer_2014_V6_ONLINE_.pdf.
———. 2015. "Environmental Racism Remains a Reality in Canada." *Huffington Post*, May 7, 2015. Available: http://www.huffingtonpost.ca/ecojustice/environmental-racism-canadal_b_7224904.html.
Mol, Arthur P.J., and Gert Spaargaren. 2000. "Ecological modernisation theory in debate: a review." *Environmental politics* 9.1, 17–49.
Money, John, and Anke A. Ehrhardt. 1972. *Man and Woman, Boy and Girl*. Baltimore: Johns Hopkins University Press.
Moody, Harry. 1998. *Aging: Concepts and Controversy* (2nd ed.). Thousand Oaks, CA: Pine Forge Press.
Moore, Patricia, with C.P. Conn. 1985. *Disguised*. Waco, TX: Word Books.
Morgan, Steven, and Colleen Cunningham. 2011. "Population Aging and the Determinants of Hospital, Medical and Pharmaceutical Care in British Columbia, 1996 to 2006." *Healthcare Policy* 7 (1): 68–79.
Morris, David B. 1998. *Illness and Culture in the Postmodern Age*. Berkeley, CA: University of California Press.
Morissette, René, Garnett Picot, and Yuqian Lu. 2013. *The Evolution of Canadian Wages over the Last Three Decades*. Analytical Studies Branch Research Paper Series. Statistics Canada. Catalogue no. 11F0019M – No. 347. Available: http://www.statcan.gc.ca/pub/11f0019m/11f0019m2013347-eng.pdf.
———, Grant Schellenberg, and Anick Johnson. 2005. "Diverging Trends in Unionization." *Perspectives on Labour and Income* 6 (April): 5–12. Statistics Canada Cat. no. 75-001-XIE. Ottawa: Statistics Canada.
———, and Xuelin Zhang. 2006. "Revisiting Wealth Inequality." *Perspectives on Labour and Income* (December). Statistics Canada Cat. no. 75-001-XIE. Retrieved June 17, 2009. Available: http://www.statcan.gc.ca/pub/75-001-x/11206/9543-eng.pdf.
Multani v. Commission scolaire Marguerite-Bourgeoys, [2006] 1 S.C.R. 256, 2006 SCC6.
Murdock, George P. 1945. "The Common Denominator of Cultures." In Ralph Linton (ed.), *The Science of Man in the World Crisis*. New York: Columbia University Press, 123–142.
Murphy, Christopher, and Curtis Clarke. 2005. "Policing Communities and Communities of Policing: A Comparative Study of Policing and Security in Two Canadian Communities." In Dennis Cooley (ed.), *Re-Imagining Policing in Canada*. Toronto: University of Toronto Press, 209–259.
Murphy, Emily F. 1922. *The Black Candle*. Toronto: Thomas Allen.
Murphy, Raymond. 2004. "Disaster of Sustainability: The Dance of Agents with Nature's Actants." *Canadian Review of Sociology and Anthropology* 41: 249–266.
Murphy, Robert E., Jessica Scheer, Yolanda Murphy, and Richard Mack. 1988. "Physical Disability and Social Liminality: A Study in Rituals of Adversity." *Social Science and Medicine* 26: 235–242.
Myles, John. 1999. "Demography or Democracy? The 'Crisis' of Old-Age Security." In Curtis, James E., Edward Grabb, and Neil Guppy (eds.), *Social Inequality in Canada: Patterns, Problems, Policies* (3rd ed.). Scarborough, ON: Prentice Hall.
Nader, George A. 1976. *Cities of Canada*, Vol. 2. Profiles of Fifteen Metropolitan Centres. Toronto: Macmillan of Canada.
Naiman, Joanne. 2008. *How Societies Work: Class, Power, and Change in a Canadian Context* (4th ed.). Halifax: Fernwood.
Nairne, Doug. 1998. "Good Samaritan Feels That He Was Victimized Twice." *Winnipeg Free Press* (June 4): A4.
Nanda, Serena. 2000. *Gender Diversity: Cross Cultural Variations*. Prospect Heights, IL: Waveland, 20.
Natanson, Maurice. 1963. "A Study in Philosophy and the Social Sciences." In Maurice Natanson (ed.), *Philosophy of the Social Science: A Reader*. New York: Random House, 271–285.
National Centre for School Statistics 2012 Indicators of School Crime and

Safety: 2011 (Feb.). Retrieved Apr. 13, 2013. Online: www.bjs.gov/content/pub/pdf/iscs11.pdf.

National Post. 2013. "'When do I get to be a boy?' Transgender youth can't remember a time when he was happy to be a girl." September 3rd, 2013. Available at: http://news.nationalpost.com/news/canada/when-do-i-get-to-be-a-boy-transgender-youth-cant-remember-a-time-when-he-was-happy-to-be-a-girl.

Navarrette, Ruben, Jr. 1997. "A Darker Shade of Crimson." In Diana Kendall (ed.), *Race, Class, and Gender in a Diverse Society.* Boston: Allyn & Bacon, 274–279. Reprinted from Ruben Navarrette Jr., *A Darker Shade of Crimson.* New York: Bantam, 1993.

Negroponte, Nicholas. 1995. *Being Digital.* New York: Alfred A. Knopf.

Nelson, Adie. 2006. *Gender in Canada* (3rd ed.). Toronto: Pearson.

Nemeth, Mary, Nora Underwood, and John Howse. 1993. "God Is Alive." *Maclean's* (April 12): 32–36.

Nevitte, Neil. 2000. "Value Change and Reorientations in Citizen-State Relations." *Canadian Public Policy* 26 (Supplement): 73–94.

New York Times. 2002. "Text: Senate Judiciary Committee Hearing, June 6, 2002." Retrieved June 9, 2002. Available: http://www.nytimes.com/2002/06/06../06TEXT-INQ2.html.

———. 2005. "The Missing Condoms." Retrieved September 4, 2005. Available: http://www.nytimes.com/2005/09/04/opinion/04sun2.html.

NHS in Brief National Household Survey (NHS). 2011. Available: http://www12.statcan.gc.ca/nhs-enm/2011/as-sa/99-012-x/99-012-x2011003_3-eng.pdf.

Nielsen, Joyce McCarl. 1990. *Sex and Gender in Society: Perspectives on Stratification* (2nd ed.). Prospect Heights, IL: Waveland Press.

Nielsen, T.M. 2005. "Streets, Strangers, and Solidarity." In Bruce Ravelli (ed.), *Exploring Canadian Sociology: A Reader.* Toronto: Pearson, 89–98.

Noel, Donald L. 1972. *The Origins of American Slavery and Racism.* Columbus, OH: Merrill.

Nolen, Stephanie. 2007. "Swaziland: The Economics of an Epidemic." *Globe and Mail* (December 22): A15.

Norris, Pippa, and Ronald Inglehart. 2004. *Sacred and Secular: Religion and Politics Worldwide.* Cambridge: Cambridge University Press.

Northcott, Herbert, C. 1982. "The Best Years of Your Life." *Canadian Journal on Aging* 1: 72–78.

Novak, Mark. 1993. *Aging and Society: A Canadian Perspective.* Toronto: Thomson Nelson.

———. 1995. "Successful Aging." In *Aging and Society: A Canadian Reader.* Toronto: Thomson Nelson, 125.

———. 1997. *Aging and Society* (3rd ed.). Toronto: Thomson Nelson.

———, Lori Campbell, and Herbert Northcott. 2013. *Aging and Society: A Canadian Perspective.* Toronto: Nelson.

O'Connor, The Honourable Dennis R. 2002. *Report of the Walkerton Inquiry: Part One: Summary.* Toronto: Queen's Printer for Ontario.

O'Connor, James. 1973. *The Fiscal Crisis of the State.* New York: St. Martin's Press.

O'Donnell, V., and S. Wallace. 2011. "First Nations, Métis and Inuit Women in Canada: A Gender-based Statistical Report." Statistics Canada. Catalogue no. 89-503-X. Available: http://www.statcan.gc.ca/pub/89-503-x/2010001/article/11442-eng.pdf.

OECD (Organisation for Economic Co-operation and Development). 2012. Education At a Glance: OECD Indicators. OECD Publishing, 2012. Available: http://www.keepeek.com/Digital-Asset-Management/oecd/education/education-at-a-glance-2012_eag-2012-en#page1.

OECD. 2013. *Education At a Glance: OECD Indicators.* OECD Publishing, 2013. Available: http://www.keepeek.com/Digital-Asset-Management/oecd/education/education-at-a-glance-2013_eag-2013-en#page3

———. 2013. *Health at a Glance 2013: OECD Indicators.* Paris: OECD. Available: http://www.oecd.org/els/health-systems/Health-at-a-Glance-2013.pdf.

———, and Statistics Canada. 2000. *Literacy in the Information Age: Final Report of the International Adult Literacy Survey.* Ottawa: Statistics Canada.

Office of the Auditor General of Canada. 2015. *2013 Spring Report of the Auditor General of Canada.* Available: http://www.oag-bvg.gc.ca/internet/English/parl_oag_201304_05_e_38190.html.

Office of the Correctional Investigator. 2014. *Annual Report of the Office of the Correctional Investigator 2013–2014.* Ottawa: Office of the Correctional Investigator.

Ogburn, William F. 1966. *Social Change with Respect to Culture and Original Nature.* New York: Dell (orig. pub. 1922).

Ogden, Russel D. 1994. *Euthanasia and Assisted Suicide in Persons with Acquired Immunodeficiency Syndrome (AIDS) or Human Immunodeficiency Virus (HIV).* Pitt Meadows, BC: Perreault Goedman.

Ogrodnik, Lucie. 2007. *Seniors as Victims of Crime: 2004 and 2005.* Ottawa: Canadian Centre for Justice Statistics.

Oliver, Michael. 1990. *The Politics of Disablement: A Sociological Approach.* New York: St. Martin's Press.

Olmstead, S., K. Pasley, and F. Fincham. (2013). "Hooking up and penetrative hookups: Correlates that differentiate college men." *Archives of Sexual Behavior*, 42(4), 573–583.

Ontario Federation of Teaching Parents. 2014. "Homeschooling FAQs." Available: http://ontariohomeschool.org/homeschooling-faq/.

O'Rand, A.M., and J.C. Henretta. 1999. *Age and inequality: Diverse pathways through later life.* Boulder, CO: Westview Press.

Orbach, Susie. 1978. *Fat Is a Feminist Issue.* New York: Paddington.

O'Reilly, Terry. 2011. *The Age of Persuasion,* Season 5, Episode 10. Podcast. 2011.

Oreopoulos, Philip, 2009. *Why Do Skilled Immigrants Struggle in the Labor Market? A Field Experiment with 6000 Résumés.* Metropolis British Columbia. Centre of Excellence for Research on Immigration and Diversity Working Paper Series No. 09-03 (May). Retrieved June 19, 2012. Available: http://mbc.metropolis.net/assets/uploads/files/wp/2009/WP09-03.pdf.

Oriola, Temitope. 2016. "Correlates of Criminal Behaviour." In Rick Linden (ed.), *Criminology: A Canadian Perspective* (8th ed.). Toronto: Nelson, 119–154.

Osborne, Ken. 1999. *Education: A Guide to the Canadian School Debate—Or, Who Wants What and Why.* Toronto: Penguin Books.

Owen, J., G. Rhoades, S. Stanley, and F. Fincham. (2010). "Hooking up among college students: Demographic and psychosocial correlates." *Archives of Sexual Behavior*, 39(3), 653–663.

Oxfam. 2001. *Rigged Trade and Not Much Aid: How Rich Countries Help to Keep the Least Developed Countries Poor.* London: Oxfam.

———. 2014. *Working for the Few: Political Capture and Economic Inequality.* Oxford: Oxfam. Available: https://www.oxfam.org/en/research/working-few.

Page, Charles H. 1946. "Bureaucracy's Other Face." *Social Forces* 25 (October): 89–94.

Palangkaraya, Alfons, and Jongsay Yong. 2009. "Population Ageing and Its Implications on Aggregate Health Care Demand: Empirical Evidence From 22 OECD Countries." *International Journal of Health Finance and Economics* 9: 391–402.

Palmore, Erdman. 1981. *Social Patterns in Normal Aging: Findings from the Duke Longitudinal Study.* Durham, NC: Duke University Press.

Palys, Ted. 1997. *Research Decisions: Quantitative and Qualitative Perspectives.* Toronto: Harcourt Brace.

Pampel, Fred. 2011. "Cohort Change, Diffusion, and Support for Gender Egalitarianism in Cross-National Perspective." *Demographic Research,* 25: 667–694. Retrieved Feb. 11, 2012. Online: www.demographic-research.org/volumes/vol25/21/25-21.pdf.

Parenti, Michael. 1994. *Land of Idols: Political Mythology in America.* New York: St. Martin's Press.

———. 1996. *Democracy for the Few* (5th ed.). New York: St. Martin's Press.

———. 2001. "Monopoly Media Manipulation." *Michael Parenti Political Archive.* Retrieved October 17, 2011. Available: http://www.michaelparenti.org/MonopolyMedia.html.

Park, Jungwee. 2011. "Retirement, Health and Employment Among Those 55 Plus." *Perspectives on Labour and Income* 31 (January): 3–12.

Park, Robert E. 1915. "The City: Suggestions for the Investigation of Human Behavior in the City." *American Journal of Sociology* 20: 577–612.

———. 1928. "Human Migration and the Marginal Man." *American Journal of Sociology* 33.

———. 1936. "Human Ecology." *American Journal of Sociology* 42: 1–15.

———, and Ernest W. Burgess. 1921. *Human Ecology.* Chicago: University of Chicago Press.

Parliament of Canada. 2006. *Final Report on the Canadian News Media.* Ottawa: Standing Senate Committee on Transportation and Communications.

Parrish, Dee Anna. 1990. *Abused: A Guide to Recovery for Adult Survivors of Emotional/Physical Child Abuse.* Barrytown, NY: Station Hill Press.

Parsons, Talcott. 1951. *The Social System.* Glencoe, IL: Free Press.

———. 1955. "The American Family: Its Relations to Personality and to the Social Structure." In Talcott Parsons and Robert F. Bales (eds.), *Family, Socialization and Interaction Process.* Glencoe, IL: Free Press, 3–33.

———. 1960. "Toward a Healthy Maturity." *Journal of Health and Social Behavior* 1: 163–173.

———, and Edward A. Shils, eds. 1951. *Toward a General Theory of Action.* Cambridge, MA: Harvard University Press.

Patten, Melanie. 2014. "Nova Scotia Ban on Trinity Grads Would Be 'Prejudice' University Head Says". *The Globe and Mail.* March 4.

Patterson, Christopher, and Elizabeth Podnieks. 1995. "A Guide to the Diagnosis and Treatment of Elder Abuse." In Mark Novak (ed.), *Aging and Society: A Canadian Reader.* Toronto: Thomson Nelson.

Patterson, Richard. 2015. "Can Behavioral Tools Improve Online Student Outcomes? Experimental Evidence from a Massive Open Online Course." Ithaca, NY: Cornell University. Available: https://www.ilr.cornell.edu/sites/ilr.cornell.edu/files/cheri_wp165_0.pdf.

Pavlik, John V., and Shawn McIntosh. 2011. *Converging Media: A New Introduction to Mass Communication* (2nd ed.). New York: Oxford University Press.

Pearce, Diana. 1978. "The Feminization of Poverty: Women, Work, and Welfare." *Urban and Social Change Review* 11 (1/2): 28–36.

Pearson, Judy C. 1985. *Gender and Communication.* Dubuque, IA: Brown.

Peikoff, Tannis. 2000. *Anglican Missionaries and Governing the Self: An Encounter with Aboriginal Peoples in Western Canada.* Unpublished Ph.D. dissertation, University of Manitoba.

Pellow, David N., and Holly Nyseth Brehm. 2013. "An Environmental Sociology for the Twenty-First Century." *American Review of Sociology* 39 220–250.

Pendakur, K., and Pendakur, R. (2013). Aboriginal Income Disparity in Canada. Aboriginal Affairs and Northern Development. Available: https://

www.aadnc-aandc.gc.ca/DAM/DAM-INTER-HQ-AI/STAGING/texte-text/rs_re_brief_incomedisparity-PDF_1378400531873_eng.pdf.

Pentland, Alex. 2014. *Social Physics*. New York: The Penguin Press.

People's Daily Online. 2012. "China's River Pollution a Threat to People's Lives" (Feb. 17). Online: http://english.people.com.cn/90882/7732438.html.

Perreault, Samuel. 2009. "Violent victimization of Aboriginal people in the Canadian provinces 2009." *Juristat*. Statistics Canada catalogue no. 85-002-X. Available: http://www.statcan.gc.ca/pub/85-002-x/2011001/article/11415-eng.pdf.

———, and Shannon Brennan. 2010. "Criminal Victimization in Canada 2009." *Juristat* 30. Ottawa: Statistics Canada.

Perry, David C., and Alfred J. Watkins, eds. 1977. *The Rise of the Sunbelt Cities*. Beverly Hills, CA: Sage.

Petersen, John L. 1994. *The Road to 2015: Profiles of the Future*. Corte Madera, CA: Waite Group Press.

Pew Research Center. 2013. "Canada's Changing Religious Landscape." Retrieved: 15 November 2014. Available: http://www.pewforum.org/2013/06/27/canadas-changing-religious-landscape/.

———. 2013. "The Global Divide on Homosexuality: Greater Acceptance in More Secular and Affluent Countries." *Pew Research Center*. June 4: http://www.pewglobal.org/files/2013/06/Pew-Global-Attitudes-Homosexuality-Report-FINAL-JUNE-4-2013.pdf.

Phoenix, A. and A.Woollett. 1991. "Motherhood, Social Construction, Politics, and Psychology." In A. Pheonix, A Woollett, and E. Lloyd (eds.), *Motherhood: Meanings, Practices and Ideologies*. London: Sage, 7.

Piaget, Jean. 1954. *The Construction of Reality in the Child.* Trans. Margaret Cook. New York: Basic Books.

Picot, Garnett, and John Myles. 2004. "Income Inequality and Low Income in Canada." *Horizons* 7 (December): 9–18.

Pierce, Jennifer. 1995. *Gender Trials: Emotional Lives in Contemporary Law Firms*. Berkeley, CA: University of California Press.

Pines, Maya. 1981. "The Civilizing of Genie." *Psychology Today* 15 (September): 28–29, 31–32, 34.

Plumwood, Val. 1993. *Feminism and the Mastery of Nature*. New York: Routledge.

Polanyi, Karl. 1944. *The Great Transformation: The Political and Economic Origins of Our Time*. New York: Beacon.

Polivka, Larry. 2000. "Postmodern Aging and the Loss of Meaning." *Journal of Aging and Identity* 5: 225–235.

Popoff, Wilfied. 1996. "One Day You're Family; the Next Day You're Fired." *Globe and Mail* (March 14): A22.

Population Reference Bureau. 2014. 2014 World Population Data Sheet. Washington: Population Reference Bureau.

Porter, Jody. 2015. "Ontario Premier Kathleen Wynne Won't Commit to Grassy Narrows Mercury Cleanup." CBC News, 16 June. Available: http://www.cbc.ca/news/canada/thunder-bay/ontario-premier-kathleen-wynne-won-t-commit-to-grassy-narrows-mercury-cleanup-1.3114673.

Porter, John. 1965. *The Vertical Mosaic: An Analysis of Social Class and Power in Canada*. Toronto: University of Toronto Press.

Postone, Moishe. 1997. "Rethinking Marx (in a Post-Marxist World)." In Charles Camic (ed.), *Reclaiming the Sociological Classics: The State of the Scholarship*. Malden, Mass.: Blackwell, 45–80.

Potvin, Maryse. 2013. *Social and Media Discourse in the Reasonable Accommodations Debate*. Available: http://canada.metropolis.net/pdfs/odc_vol7_maryse_potvin_e.pdf.

Preece, Melady. 2004. "When Lone Parents Marry: The Challenge of Stepfamily Relationships." *Transition* (Winter 2003–2004). Ottawa: Vanier Institute of the Family. Available: http://www.vifamily.ca/library/transition/334/334.htm.

Pryor, John, and Kathleen McKinney (eds.). 1991. "Sexual Harassment." *Basic and Applied Social Psychology* 17 (4).

———. 1995. "Research Advances in Sexual Harassment: Introduction and Overview." *Basic and Applied Social Psychology* 17: 421–424.

Public Health Agency of Canada. 2007. *HIV/AIDS Epi Update*. Ottawa: Public Health Agency of Canada. Retrieved April 17, 2009. Available: www.phac-aspc.gc.ca/aids-sida/publication/epi/pdf/epi2007_e.pdf.

———. 2010. *HIV/AIDS Epi Updates – July 2010*. Retrieved February 19, 2012. Available: http://www.phac-aspc.gc.ca/aids-sida/publication/epi/2010/8-eng.php.

———. 2014. *HIV/AIDS Epi Updates*. Ottawa: Centre for Communicable Diseases and Infection Control.

———. 2015. *Sexually Transmitted Infections (STI): Sexual Health Facts and Information for the Public*. Available: http://www.phac-aspc.gc.ca/std-mts/faq-eng.php.

Public Safety Canada. 2013. *Corrections and Conditional Release: Statistical Overview*. Ottawa: Public Safety Canada.

Purcell, Kristen. 2012. "Teens 2012: Truth, Trends, and Myths About Teen Online Behavior." Pew Internet & American Life Project. Retrieved Mar. 7, 2013. Online: http://pewinternet.org/-/media//Files/Presentations/2012/July/KPurcell%20ACT%20Conf_PDF.pdf.

Quadagno, Jill S. 1984. "Welfare Capitalism and the Social Security Act of 1935." *American Sociological Review* 49: 632–647.

Quan, Douglas. 2014. "Justin Bourque: Terrorism Charges Were Once Pondered Against Man Who Shot Dead Three Mounties." *Postmedia*, 31 October. Retrieved: 25 November 2014. Available: http://www.canada.com/news/Justin+Bourque+Terrorism+charges+were+once+pondered+against+shot+dead+three+Mounties/10343213/story.html.

Queen, Stuart A., and David B. Carpenter. 1953. *The American City*. New York: McGraw-Hill.

Quinney, Richard. 1979. *Class, State, and Crime*. New York: McKay.

———. 1980. *Class, State, and Crime* (2nd ed.). New York: Longman.

Quinton, Rhonda. 1989. "Liability of Search and Rescuers." Unpublished paper. Faculty of Law, University of Victoria.

Radcliffe-Brown, A.R. 1952. *Structure and Function in Primitive Society*. New York: Free Press.

Rankin, Jim, and Betsy Powell. 2008. "Why the Difference for 'Non-Whites'?" *Toronto Star* (July 21): A6.

Ranson, Dave. 2008. "ADISQ vs. Heri: Can the CRTC Control Internet Radio Content?" *Alberta Law Review Online Supplement* (October 28). Retrieved September 23, 2011. Available: http://ualbertalaw.typepad.com/alr_supplement/2008/10/index.html.

Razack, Sherene H. 1998. *Looking White People in the Eye*. Toronto: University of Toronto Press.

Redmond, Ashley. 2014."Gap Years Aren't Just for Brits Anymore." *Huffington Post*. Available: http://www.huffingtonpost.ca/ashley-redmond/gap-year-canada_b_5948708.html.

Reed, Christopher. 1998. "No Fingerprints Puts Man Under Society's Thumb." *Globe and Mail* (April 23): A11.

Reiman, Jeffrey H. 1979. *The Rich Get Richer and the Poor Get Prison*. New York: Wiley.

———. 1984. *The Rich Get Richer and the Poor Get Prison* (2nd ed.). New York: Wiley.

Reinharz, Shulamit. 1992. *Feminist Methods in Social Research*. New York: Oxford University Press.

Reitz, J.G. 2010. "Selecting immigrants for the short term: Is it smart in the long run?" *Policy Options, 31*(7), 12–16.

———. 2012. "The distinctiveness of Canadian immigration experience." *Patterns of Prejudice, 46*(5), 518–538. doi:10.1080/0031322X.2012.718168.

Renzetti, Claire M., and Daniel J. Curran. 1995. *Women, Men, and Society* (3rd ed.). Boston: Allyn & Bacon.

Renzetti, Elizabeth. 2015. "Remember the U.S. Ebola Crisis? The Only Epidemic Was Fear-mongering." *The Globe and Mail*, 27 February.

Reskin, Barbara F., and Irene Padavic. 1994. *Women and Men at Work*. Thousand Oaks, CA: Pine Forge Press.

Rhoten, Diana, and Wayne Lutters. 2009. "Virtual Worlds for Virtual Organizing." In W.S. Bainbridge (ed.), *Online Worlds: Convergence of the Real and the Virtual*. London: Springer-Verlag, 175–186.

Riahi, A., and T. McSorley. (2013). "Another Layer of Colonialism: Resource extraction, toxic pollution and First Nations." *Canadian Dimension, 47*(6), 34–35.

Rice, G. 2013. "Feasibly Free." *Our Schools / Our Selves, 22*(3), 57–61.

Rich, A. 1980. "Compulsory Heterosexuality and Lesbian Existence." *Signs* 5 (4): 631–660.

Richardson, Laurel. 1993. "Inequalities of Power, Property, and Prestige." In Virginia Cyrus (ed.), *Experiencing Race, Class, and Gender in the United States*. Mountain View, CA: Mayfield, 229–236.

Rigler, David. 1993. "Letters: A Psychologist Portrayed in a Book About an Abused Child Speaks Out for the First Time in 22 Years." *New York Times Book Review* (June 13): 35.

Riley, Matilda, Anne Foner, and John Riley. 1999. "The Aging and Society Paradigm." In Merril Silverstein, Vern Bengston, and Warner Schaie (eds.), *Handbook of Theories in Aging*. New York: Springer, 327–343.

Rinehart, James W. 1996. *The Tyranny of Work: Alienation and the Labour Process* (3rd ed.). Toronto: Harcourt Brace.

Risman, B., and P. Schwartz. 2002. "After the sexual revolution: Gender politics in teen dating." *Contexts, 1*(1), 16–24.

Ritzer, George. 1993. *The McDonaldization of Society: An Investigation into the Changing Character of Contemporary Social Life*. Thousand Oaks, CA: Pine Forge Press.

———. 1997. *Postmodern Society Theory*. New York: McGraw-Hill.

———. 1998. *The McDonaldization Thesis*. London: Sage.

———. 1999. *Enchanting a Disenchanted World: Revolutionizing the Means of Consumption*. Thousand Oaks, CA: Pine Forge.

———. 2000. *The McDonaldization of Society*. Thousand Oaks, CA: Pine Forge.

———. 2004. An Introduction to McDonaldization. *The McDonaldization of Society, Revised New Century Edition*. Thousand Oaks, CA: Sage.

———. 2011. *Sociological Theory* (8th ed.). New York: McGraw-Hill.

Rivers, Madeline. 2003. "Shock and confusion with love." In Mary Boenke (ed.), *Transforming Families: Real Stories about Transgendered Loved Ones* (2nd ed.). Oak Knoll Press, 59–60.

Roberts, J. Timmons. 2009. "Climate Change: Why the Old Approaches Aren't Working." In *Twenty Lessons in Environmental Sociology*, 1st ed. Kenneth Alan Gould and Tammy L. Lewis (eds.). Oxford University Press, 191–208.

Roberts, Keith A. 1995b. *Religion in Sociological Perspective*. Belmont, CA: Wadsworth, 360.

Robnett, Rachael D., and Joshue E. Susskind. 2010. "Who Cares About Being Gentle? The Impact of Social Identity and the Gender of One's Friends on Children's Display of Same-Gender Favoritism." Sex Roles, 64 (1/2): 90–102.

Rodriguez, Havidan, and Russell Dynes. 2006. *Finding and Framing Katrina: The Social Construction of Disaster.* Understanding Katrina. Social Science Research Council. Retrieved February 1, 2009. Available: http://understandingkatrina.ssrc.org/Dynes_Rodriguez/.

———, Joseph Trainor, and Enrico Quarantelli. 2006. "Rising to the Challenges of a Catastrophe: The Emergent and Prosocial Behavior Following Hurricane Katrina." *Annals of the American Academy of Political and Social Sciences* 604 (March): 82–101.

Roethlisberger, Fritz J., and William J. Dickson. 1939. *Management and the Worker.* Cambridge, MA: Harvard University Press.

Rollins, Judith. 1985. *Between Women: Domestics and Their Employers.* Philadelphia: Temple University Press.

Romaniuc, Anatole. 1994. Statistics Canada. "Fertility: Fewer children, older moms." Canadian Megatrends. *The Daily*, 121–22. Available: http://www.statcan.gc.ca/pub/11-630-x/11-630-x2014002-eng.htm.

———. 1994. "Fertility in Canada: Retrospective and Prospective." In Frank Trovato and Carl F. Grindstaff (eds.), *Perspectives on Canada's Population.* Toronto: Oxford University Press, 214–229.

Roof, Wade Clark. 1993. *A Generation of Seekers: The Spiritual Journeys of the Baby Boom Generation.* San Francisco: HarperSanFrancisco.

Roos, Noralou, Evelyn Forget, and Gerard Beirne. 2004. "Health Care User Fees: Clinic Charges are Wrong Way to Go." *Winnipeg Free Press* (January 20): A11.

Rosenbloom, Stephanie. 2011. "Love, Lies and What They Learned." *The New York Times* (November 13): ST1.

Rosenfeld, Michael J., and Reuben J. Thomas. 2012. "Searching for a Mate: The Rise of the Internet as a Social Intermediary." *American Sociological Review,* 77 (4): 523–547.

Rosenthal, Naomi, Meryl Fingrutd, Michele Ethier, Roberta Karant, and David McDonald. 1985. "Social Movements and Network Analysis: A Case Study of Nineteenth-Century Women's Reform in New York State." *American Journal of Sociology* 90: 1022–1054.

Rosenthal, Robert, and Lenore Jacobson. 1968. *Pygmalion in the Classroom: Teacher Expectation and Student's Intellectual Development.* New York: Holt, Rinehart, and Winston.

Roshanafshar, Shirin, and Emma Hawkins. 2015. "Food Insecurity in Canada." Statistics Canada, Health at a Glance. Available: http://www.statcan.gc.ca/pub/82-624-x/2015001/article/14138-eng.pdf.

Ross, Becki. 2010. "Sex and (Evacuation from) the City: The Moral and Legal Regulation of Sex Workers in Vancouver's West End, 1975–1985." *Sexualities* 13 (2): 197–218.

Ross, David P., Katherine Scott, and Peter Smith. 2000. *The Canadian Fact Book on Poverty, 2000.* Ottawa: Canadian Council on Social Development.

Ross, Rupert. 1996. *Returning to the Teachings: Exploring Aboriginal Justice.* Toronto: Penguin Books.

Rostow, Walt W. 1971. *The Stages of Economic Growth: A Non-Communist Manifesto* (2nd ed.). Cambridge: Cambridge University Press (orig. pub. 1960).

———. 1978. *The World Economy: History and Prospect.* Austin, TX: University of Texas Press.

Rotermann, Michelle. 2008. *Trends in Teen Sexual Behaviour and Condom Use.* Statistics Canada Cat. no. 82-003-X. Retrieved April 21, 2012. Available: http://www.statcan.gc.ca/pub/82-003-x/2008003/article/10664-eng.pdf.

———. 2013. "Sexual behaviour and condom use of 15- to 24-year-olds in 2003 and 2009/2010." Statistics Canada Health Reports. Vol 23 (1) catalogue 82-003-x. http://www.statcan.gc.ca/pub/82-003-x/2012001/article/11632-eng.htm.

Roth, Guenther. 1988. "Marianne Weber and Her Circle." In Marianne Weber, *Max Weber: A Biography.* New Brunswick, NJ: Transaction.

Royal Commission on Aboriginal Peoples. 1995. *Choosing Life: Special Report on Suicide Among Aboriginal Peoples.* Ottawa: Canada. Communications Group Publishing.

Rozanova, Julia. 2010. "Discourse of Successful Aging in The Globe & Mail: Insights from Critical Gerontology." *Journal of Aging Studies,* 24, 213–222.

Rudder, Christian. 2014. "We Experiment on Human Beings!" *oktrends* 28 July. Retrieved: 20 October, 2014. Available: http://blog.okcupid.com/index.php/we-experiment-on-human-beings/.

Rukavina, S. 2015. "Quebec judge wouldn't hear case of woman wearing hijab." CBC online: http://www.cbc.ca/news/canada/montreal/quebec-judge-wouldn-t-hear-case-of-woman-wearing-hijab-1.2974282.

Rushkoff, Douglas. 2010. *Program or Be Programmed: Ten Commands for a Digital Age.* New York: OR Books.

Russell, Diana E.H. 1975. *The Politics of Rape: The Victim's Perspective.* 1st ed. New York: Stein & Day.

———. 2003. *The Politics of Rape: The Victim's Perspective.* (3rd ed.). Lincoln, NB: iUniverse.

Ruthven, Malise. 2004. *Fundamentalism: The Search for Meaning.* Oxford: Oxford University Press.

Rymer, Russ. 1993. *Genie: An Abused Child's Flight from Silence.* New York: HarperCollins.

Sachs, Jeffrey D. 2012. "Occupy Global Capitalism." In Janet Byrne (ed.), *The Occupy Handbook.* New York: Back Bay Books/Little, Brown, 462–474.

Sadker, David, and Karen Zittleman. 2009. *Still Failing at Fairness: How Gender Bias Cheats Boys and Girls in Schools.* New York: Simon and Schuster.

Samara Democracy Report. 2012. "Who's the Boss? Canadians' Views of Their Democracy." Retrieved: 27 November 2014. Available: http://www.samaracanada.com/research/current-research/who's-the-boss-.

Samovar, Larry A., and Richard E. Porter. 1991. *Communication Between Cultures.* Belmont, CA: Wadsworth.

Sandals, Leah. 2007. "'Public Space Protection'—But for Which 'Public'?" Retrieved February 24, 2007. Available: http://spacing.ca/wire/?p=1466.

Sanger, David, and Nicole Perlroth. 2015. "Bank Hackers Steal Millions via Malware." *The New York Times,* Feb. 14.

Sapir, Edward. 1961. *Culture, Language and Personality.* Berkeley, CA: University of California Press.

Sargent, Margaret. 1987. *Sociology for Australians* (2nd ed.). Melbourne, Australia: Longman Cheshire.

Satzewich, Vic, ed. 1998. *Racism and Social Inequality in Canada: Concepts, Controversies and Strategies for Resistance.* Toronto: Thompson Educational Publishing.

Satzewich, V., and N. Liodakis. 2007. *Race and Ethnicity in Canada.* Toronto: Oxford University Press.

———. 2010. "Aboriginal and Non-Aboriginal Relations." *Race and Ethnicity in Canada: A Critical Introduction,* 2nd ed. Don Mills, ON: Oxford University Press.

Saulnier, Beth. 1998. "Small World." *Cornell Magazine On//line.* Available: http://cornell-magazine.cornell.edu/Archive/JulyAugust98/JulyWorld.html.

Saunders, Eileen. 1999. "Theoretical Approaches to the Study of Women." In James Curtis, Edward Grabb, and Neil Guppy (eds.), *Social Inequality in Canada: Patterns and Policies.* Scarborough, ON: Prentice-Hall, 168–185.

Sauvé, Roger. 2008. "The Current State of Canadian Family Finances, 2007 Report." Ottawa: Vanier Institute for the Family. Retrieved June 17, 2009. Available: http://www.vifamily.ca/library/cft/famfin07.pdf.

Schafer, A. 1998. *Down and Out in Winnipeg and Toronto: The Ethics of Legislating Against Panhandling.* Ottawa: Institute of Social Policy.

Schama, Simon. 1989. *Citizens: A Chronicle of the French Revolution.* New York: Knopf.

Schellenberg, Grant, and Helene Maheux. 2007. "Immigrants' Perspectives on Their First Four Years in Canada: Highlights from Three Waves of the Longitudinal Survey of Immigrants to Canada." *Canadian Social Trends* (April): 1–35.

———, and Cynthia Silver. 2004. "You Can't Always Get What You Want: Retirement Preferences and Experiences." *Canadian Social Trends* (Winter): 2–7.

Schiller, Herbert. 1992. "A Quarter Century Retrospective." In Herbert Schiller (ed.), *Mass Communications and American Empire* (2nd ed.). Boulder, CO: Westview Press.

Schnaiberg, Allan. 1980. *The Environment: From Surplus to Scarcity.* New York: Oxford University Press.

Schneider, Beth E., and Valerie Jenness. 2005. "Social Control, Civil Liberties, and Women's Sexuality." In Tracey L. Steele (ed.), *Sex, Self, and Society: The Social Content of Sexuality.* Belmont, CA: Wadsworth Press, 2005. 388–401.

Scholastic Parent & Child. 2007. "How and When to Praise." Retrieved January 3, 2008. Available: http://content.scholastic.com/browse/article.jsp?id=2064.

Schor, Juliet B. 1999. *The Overspent American: Upscaling, Downshifting, and the New Consumer.* New York: HarperPerennial.

Schrock, Douglas, Daphne Holden, and Lori Reid. 2004. "Creating Emotional Resonance: Interpersonal Emotion Work and Motivational Framing in a Transgender Community." *Social Problems* 51 (1): 61–81.

Schur, Edwin M. 1965. *Crimes Without Victims: Deviant Behavior and Public Policy.* Englewood Cliffs, NJ: Prentice-Hall.

———. 1983. *Labeling Women Deviant: Gender, Stigma, and Social Control.* Philadelphia: Temple University Press.

Schwartz, M.D., W.S. DeKeseredy, D. Tait, and S. Alvi. (2001). "Male peer support and a feminist routing activities theory: Understanding sexual assault on the college campus." *Justice Quarterly,* 18(3), 623–649.

Scott, Robert A. 1969. *The Making of Blind Men: A Study of Adult Socialization.* New York: Russell Sage Foundation.

Scully, Diana. 1990. *Understanding Sexual Violence: A Study of Convicted Rapists.* Boston: Unwin Hyman.

Segall, Alexander, and Christopher Fries. 2011. *Pursuing Health and Wellness.* Toronto: Oxford University Press.

Sen, Amartya. 1990. "More Than 100 Million Women Are Missing". *New York Review of Books,* 20 December.

Senate Committee on Legal and Constitutional Affairs. 2006. "Evidence." (December 4).

Serazio, Michael. 2008. "Virtual Sports Consumption, Authentic Brotherhood: The Reality of Fantasy Football." In Lawrence W. Hugengerg, Paul Haridakis, and Adam Earnheardt (eds.), *Sports Mania: Essays on Fandom and the Media in the 21st Century.* Jefferson, NC: McFarland, 229–242.

Sev'er, Aysan. 2011. "Marriage-Go-Around: Divorce and Remarriage in Canada." In Nancy Mandell and Ann Duffy (eds.), *Canadian Families: Diversity, Conflict and Change* (4th ed.). Toronto: Nelson Education Ltd., 243–273.

Shadd, Adrienne. 1991. "Institutionalized Racism and Canadian History: Notes of a Black Canadian." In Ormond McKague (ed.), *Racism in Canada.* Saskatoon: Fifth House, 1–5, 11.

Shaker, E. 2013. "Jeopardy, Jackpot, or Wheel of (Mis)Fortune?" *Our Schools / Our Selves, 23*(1), 131–133.

Shaker, Erika, and David Macdonald with Nigel Wodrich. 2013. "Degrees of Uncertainty Navigating the Changing Terrain of University Finance." Canadian Centre for Policy

Alternatives, September 2013. Available: http://www.policyalternatives.ca/sites/default/files/uploads/publications/National%20Office/2013/09/Degrees_of_Uncertainty.pdf.

Shapiro, Ari, and Melissa Block. 2009. "U.S. Busts Largest-Ever ID-Theft Scheme." National Public Radio (August 17). Retrieved February 4, 2012. Available: http://www.npr.org/templates/story/story.php?storyId=111964002.

Shapiro, Eve. 2010. *Gender Circuits: Bodies and Identities in a Technological Age*. New York: Routledge.

Shapiro, Joseph P. 1993. *No Pity: People with Disabilities Forging a New Civil Rights Movement*. Toronto: Time Books/Random House.

Sharma, Monica, and James Tulloch. 1997. "Commentary: Unfinished Business." *Progress of Nations 1996*. New York: The United Nations. Retrieved December 7, 2002, 1. Available: http://www.unicef.org/pon96/heunfini.htm.

Shilts, Randy. 1988. *And the Band Played On: Politics, People, and the AIDS Epidemic*. New York: Penguin.

Shirpak, K., E. Maticka-Tyndale, and M. Chinichian. 2007. "Iranian Immigrants' Perceptions of Sexuality in Canada: A symbolic Interactionist Approach." *Canadian Journal of Human Sexuality* 16 (3–4): 113–128.

Shkilnyk, Anastasia M. 1985. *A Poison Stronger Than Love: The Destruction of an Ojibwa Community*. New Haven, CT: Yale University Press.

Silverman, Rachel. 2012. "Big Firms Try Crowdsourcing." *Wall Street Journal* (January 17).

Silverman, Robert, and Leslie Kennedy. 1993. *Deadly Deeds: Murder in Canada*. Toronto: Thomson Nelson.

Simmel, Georg. 1904. "Fashion." *American Journal of Sociology* 62 (May 1957): 541–558.

Simmel, Georg. 1950. *The Sociology of Georg Simmel*. Trans. Kurt Wolff. Glencoe, IL: Free Press (orig. written 1902–1917).

———. 1990. *The Philosophy of Money*. Ed. David Frisby. New York: Routledge (orig. pub. 1907).

Simon, David R., and D. Stanley Eitzen. 1993. *Elite Deviance* (4th ed.). Boston: Allyn & Bacon.

Simon, W., and J. Gagnon. 1986. "Sexual Scripts: Permanence and Change." *Archives of Sexual Behavior* 15 (2): 97–120.

Sinclair, Murray. 2013. Opening Remarks by Justice Murray Sinclair, Quebec National Event. Montreal. Truth and Reconciliation Commission.

Sinha, M. 2014. *Child Care in Canada*. Statistics Canada Catalogue no. 89-652-X http://www.statcan.gc.ca/pub/89-652-x/89-652-x2014005-eng.pdf.

Sinha, Maire. 2013. "Measuring violence against women: Statistical trends." Statistics Canada catalogue no. 85-002-X. *Juristat*. Available: http://www.statcan.gc.ca/pub/85-002-x/2013001/article/11766-eng.pdf.

Sjoberg, Gideon. 1960. *The Preindustrial City: Past and Present*. New York: The Free Press.

Skype, 2013. "Thanks for Making Skype a Part of Your Daily Lives – 2 Billion Minutes a Day." Available: http://blogs.skype.com/2013/04/03/thanks-for-making-skype-a-part-of-your-daily-lives-2-billion-minutes-a-day/.

Sloan, R.P., E. Bagiella, and T. Powell. 1999. "Religion, Spirituality, and Medicine." *The Lancet* 353: 664–667.

Smandych, Russell. 1985. "Marxism and the Creation of Law: Re-Examining the Origins of Canadian Anti-Combines Legislation." In Thomas Fleming (ed.), *The New Criminologies in Canada: State, Crime and Control*. Toronto: Oxford University Press, 87–99.

———, and Rodney Kueneman. (2010). "The Canadian-Alberta Tar Sands: A Case Study of State-Corporate Environmental Crime." In Rob White (ed.), *Global Environmental Harm: Criminological Perspectives*. Cullompton, Devon: Willan Publishing, 87–109.

Smelser, Neil J. 1963. *Theory of Collective Behaviour*. New York: Free Press.

———. 1988. "Social Structure." In Neil J. Smelser (ed.), *Handbook of Sociology*. Newbury Park, CA: Sage, 103–129.

Smith, Adam. 1976. *An Inquiry into the Nature and Causes of the Wealth of Nations*. Ed. Roy H. Campbell and Andrew S. Skinner. Oxford, England: Clarendon Press (orig. pub. 1776).

Smith, Allen C., III, and Sheryl Kleinman. 1989. "Managing Emotions in Medical School: Students' Contacts with the Living and the Dead." *Social Science Quarterly* 52 (1): 56–69.

Smith, Dorothy. 1974. "Women's Perspective as a Radical Critique of Sociology." *Sociological Inquiry* 44: 7–13.

———. 1985. "Women, Class and Family." In Varda Burstyn and Dorothy Smith (eds.), *Women, Class and the State*. Toronto: Garamond.

———. 1987. *The Everyday World as Problematic: A Feminist Sociology*. Toronto: University of Toronto Press.

Smith, K. 2008. "Symposium Introduction: Mitigating, Adapting, and Suffering: How Much of Each?" *Annual Review of Public Health* 29:11–15.

Smith, M. 2011. Against Ecological Sovereignty: Ethics, Biopolitics, and Saving the Natural World. Minneapolis: University of Minnesota Press.

Smyke, Patricia. 1991. *Women and Health*. Atlantic Highlands, NJ: Zed Books.

Smylie, Janet, and Paul Adomako. 2009. *Indigenous Children's Health Report: Health Assessment in Action*. Toronto: St. Michael's Hospital.

Snow, David A. 2003. "Observations and Comments on Gusfield's Journey." *Symbolic Interaction* 26: 141–149.

———, and Leon Anderson. 1991. "Researching the Homeless: The Characteristic Features and Virtues of the Case Study." In Joe R. Feagin, Anthony M. Orum, and Gideon Sjoberg (eds.), *A Case for the Case Study*. Chapel Hill, NC: University of North Carolina Press, 148–173.

———, Louis A. Zurcher, and Robert Peters. 1981. "Victory Celebrations as Theater: A Dramaturgical Approach to Crowd Behavior." *Symbolic Interaction* 4 (1): 21–41.

Snowdon, Anne. 2012. "Strengthening Communities for Canadian Children with Disabilities." Paper Presented at the Annual Conference of the Sandbox Project.

Solyom, Catherine. 2012. "Debate over Interview Material Could Be Key to Luka Rocco Magnotta Murder Trial." *Montreal Gazette* (August 1).

Spain, Daphne. 2002. "What Happened to Gender Relations on the Way from Chicago to Los Angeles?" *City & Community* 1 (June): 155–169.

Spain, Daphne. 2014. "Gender and Urban Space." *Annual Review of Sociology*. 40: 581–598.

Speilmann, Roger. 2002. *You're So Fat!: Exploring Ojibwe Discourse*. Toronto: University of Toronto Press, 51. Available: https://books.google.ca/books?id=5-cB3BL6HdQC&pg=PA51&lpg=PA51&dq=Our+native+language+embodies+a+value+system+about+how+we+ought+to+live+and+relate+to+each+other&source=bl&ots=zIbCnslsvb&sig=NwTADbIot56QbWwVF9XdnwiSGh0&hl=en&sa=X&ei=i2_aVM-UCMe-ggTvs4GIDA&ved=0CB8Q6AEwAA#v=onepage&q=Our%20native%20language%20embodies%20a%20value%20system%20about%20how%20we%20ought%20to%20live%20and%20relate%20to%20each%20other&f=false.

Stark, Rodney. 1992. *Sociology* (4th ed.). Belmont, CA: Wadsworth.

Stark, Rodney. 1998. *Sociology* (7th ed.). Belmont, CA: Wadsworth.

Stark, Rodney. 1999. "Secularization: R.I.P." *Sociology of Religion* 60: 249–273.

———. 2004. *Exploring the Religious Life*. Baltimore: The Johns Hopkins Press.

———. 2007. *Discovering God: The Origin of the Great Religions and the Evolution of Belief*. New York: HarperOne.

———, Daniel P. Doyle, and Lori Kent. 1982. "Religion and Delinquency: The Ecology of a 'Lost' Relationship." *Journal of Research in Crime and Delinquency* 19: 4–24.

Statistics Canada. 1994. *Women in the Labour Force*. Ottawa: Ministry of Industry, Science, and Technology.

———. 1997. "1996 Census: Immigration and Citizenship." *The Daily* (November 4). Cat. no. 11-001E.

———. 2002a. "Geographic Units: Census Metropolitan Area and Census Agglomeration." Retrieved July 15, 2005. Available: http://www12.statcan.ca/english/census01/Products/Reference/dict/geo009.htm.

———. 2002b. "Impact of Income and Mortality in Urban Canada." *The Daily* (September 26).

———. 2003. "Religions in Canada." *2001 Census: Analysis Series*. Cat. no. 96F0030XIE2001015. Ottawa: Ministry of Industry.

———. 2004. "Corporations Returns Act (CRA), major financial variables." CANSIM Table 179-0004. *The Daily*, November 2, 2004.

———. 2005. "Health Reports: The Use of Alternative Health Care." *The Daily* (March 15).

———. 2007a. "Immigration in Canada: A Portrait of the Foreign-Born Population, 2006 Census: Driver of Population Growth." Ottawa: Statistics Canada. Retrieved April 9, 2009. Available: http://www12.statcan.ca/english/census06/analysis/immcit/canada_foreign.cfm.

———. 2007b. "Participation and Activity Limitation Survey 2001." *The Daily* (December 3).

———. 2008. Against the Flow: Which Households Drink Bottled Water? *EnviroStats*. Available: http://www.statcan.gc.ca/pub/16-002-x/2008002/article/10620-eng.htm.

———. 2010a. Canada Survey of Giving, Volunteering and Participating. Table 1.1 "Donor Rate and Distribution of Donations, by Personal and Economic Characteristics, Population Aged 15 and Older, Canada."

———. 2010b. Healthy People, Healthy Places: Long-Term Unemployment. Ottawa: Statistics Canada.

———. 2010c. Gap in Life Expectancy Projected to Decrease Between Aboriginal People and the Total Canadian Population. Ottawa: Statistics Canada 89-645-X.

———. 2011a. Immigrant Languages in Canada. 98-314-X2011003. Available: http://www12.statcan.gc.ca/census-recensement/2011/as-sa/98-314-x/98-314-x2011003_2-eng.pdf.

———. 2011b. Proportion of Aboriginal identity population, First Nations people, Métis and Inuit for selected Aboriginal language indicators, Canada, 2011, National Household Survey, 2011. Available: http://www12.statcan.gc.ca/nhs-enm/2011/as-sa/99-011-x/2011003/tbl/tbl01-eng.cfm, 2011 Proportion of Aboriginal identity population, First Nations people, Métis and Inuit for selected Aboriginal language indicators, Canada, 2011, National Household Survey, 2011 http://www12.statcan.gc.ca/nhs-enm/2011/as-sa/99-011-x/2011003/tbl/tbl01-eng.cfm.

———. 2011c. HRSDC calculations based on Statistics Canada. Estimates of population, by age group and sex for July 1, Canada, provinces and territories, annual (CANSIM Table 051-0001); and Statistics Canada. Projected population, by projection, scenario, sex and age group as of July 1, Canada, provinces and territories, annual (CANSIM Table 052-0005). Ottawa: Statistics Canada, 2011.

———. 2011d. National Household Survey, Statistics Canada Catalogue no. 99-014-X2011041. Available: http://www12.statcan.gc.ca/nhs-enm/2011/dp-pd/dt-td/Rp-eng.cfm?LANG=E&APATH=3&DETAIL=0&DIM=0&FL=A&FREE=0&GC=0&GID=0&GK=0&GRP=0&PID=106746&PRID=0&P

TYPE=105277&S=0&SHOWALL=0&SUB=0&Temporal=2013&THEME=98&VID=0&VNAMEE&VNAMEF).

———. 2011e. "Deaths." *The Daily* (September 27).

———. 2011f. "Women in Canada: A Gender-Based Statistical Report," 6th ed. Catalogue no. 89-503-XIE2010001.

———. 2012a. Linguistic Characteristics of Canadians Language, 2011 Census of Population October 2012 Catalogue no. 98-314-X2011001 http://www12.statcan.gc.ca/census-recensement/2011/as-sa/98-314-x/98-314-x2011001-eng.pdf

———. 2012b. Immigrant languages in Canada. Language, 2011 Census of Population Catalogue no. 98-314-X2011003 http://www12.statcan.gc.ca/census-recensement/2011/as-sa/98-314-x/98-314-x2011003_2-eng.pdf.

———. 2012c. Population by language spoken most often and regularly at home, age groups (total), for Canada, provinces and territories, Highlight Tables, http://www12.statcan.gc.ca/census-recensement/2011/dp-pd/hlt-fst/lang/Pages/Highlight.cfm?TabID=1&Lang=E&PRCode=01&Age=1&tableID=403&queryID=1.

———. 2012d. Canadian households in 2011: Type and growth. Families, households and marital status. 2011 Census of Population Catalogue no. 98-312-X2011003 http://www12.statcan.gc.ca/census-recensement/2011/as-sa/98-312-x/98-312-x2011003_2-eng.pdf.

———. 2012e. Portrait of Families and Living Arrangements in Canada. Families, households and marital status. 2011 Census of Population Catalogue no. 98-312-X2011001 http://www12.statcan.gc.ca/census-recensement/2011/as-sa/98-312-x/98-312-x2011001-eng.pdf.

———. 2012f. Living arrangements of young adults aged 20 to 29 in Canada. Families, households and marital status. 2011 Census of Population Catalogue no. 98-312-X2011003 http://www12.statcan.gc.ca/census-recensement/2011/as-sa/98-312-x/98-312-x2011003_3-eng.pdf.

———. 2012g. "Canadian Income Survey." *The Daily.* CANSIM table 206-0001. Retrieved from: http://www.statcan.gc.ca/daily-quotidien/141210/t141210a001-eng.htm.

———. 2012h. "University Tuition Fees, 2012–13." *The Daily* (September 12).

———. 2012i. "Deaths, 2009." *The Daily* (May 31).

———. 2013a. Immigration and Ethnocultural Diversity in Canada National Household Survey, 2011. Catalogue no. 99-010-X2011001 ISBN: 978-1-100-22197-7. Available: http://www12.statcan.gc.ca/nhs-enm/2011/as-sa/99-010-x/99-010-x2011001-eng.pdf.

———. 2013b. National Household Survey: Immigration and Ethnocultural Diversity – Religion (108), Immigrant Status and Period of Immigration (11), Age Groups (10) and Sex (3) for the Population in Private Households of Canada, Provinces, Territories, Census Metropolitan Areas and Census Agglomerations, 2011 National Household Survey, National Household Survey year 2011. Available: http://www12.statcan.gc.ca/nhs-enm/2011/dp-pd/dt-td/Rp-eng.cfm?LANG=E&APATH=3&DETAIL=0&DIM=0&FL=A&FREE=0&GC=0&GID=0&GK=0&GRP=1&PID=105399&PRID=0&PTYPE=105277&S=0&SHOWALL=0&SUB=0&Temporal=2013&THEME=95&VID=0&VNAMEE=&VNAMEF=http://www12.statcan.gc.ca/nhs-enm/2011/dp-pd/dt-td/Rp-eng.cfm?LANG=E&APATH=3&DETAIL=0&DIM=0&FL=A&FREE=0&GC=0&GID=0&GK=0&GRP=0&PID=105396&PRID=0&PTYPE=105277&S=0&SHOWALL=0&SUB=0&Temporal=2013&THEME=95&VID=0&VNAMEE=&VNAMEF=.

———. 2013c. "Low Income in Canada: A Multi-line and Multi-index Perspective." Series no. 75F0002M. Retrieved from: http://www.statcan.gc.ca/pub/75f0002m/2012001/chap3-eng.htm.

———. 2013d. Aboriginal Peoples in Canada: First Nations People, Métis and Inuit in Canada National Household Survey, 2011. Catalogue no. 99-011-X2011001. Available: http://www12.statcan.gc.ca/nhs-enm/2011/as-sa/99-011-x/99-011-x2011001-eng.pdf.

———. 2013e. "Live births by age of mother, Canada, provinces and territories," Table 102-4503, http://www5.statcan.gc.ca/cansim/a26?lang=eng&id=1024503&p2=46

———. 2013f. *Life Tables, Canada, Provinces and Territories 2009 to 2011.* Retrieved 31 October, 2014. Available: http://www.statcan.gc.ca/pub/84-537-x/84-537-x2013005-eng.htm.

———. 2013g. *Health at a Glance: Ninety Years of Change in Life Expectancy.* Ottawa: Statistics Canada. Available: http://www.statcan.gc.ca/pub/82-624-x/2014001/article/14009-eng.htm#n3.

———. 2013h. Education in Canada: Attainment, Field of Study and Location of Study. Analytical Document, National Household Survey, 2011. Catalogue no. 99-012-X2011001.

———. 2013i. Summary Elementary and Secondary School Indicators for Canada, the Provinces and Territories, 2006/2007 to 2010/2011 Tourism and Centre for Education Statistics Division Catalogue no. 81595M — No. 099 ISSN 1711-831X ISBN 978-1-100-21490-0 http://www.statcan.gc.ca/pub/81-595-m/81-595-m2013099-eng.pdf

———. 2013j. The educational attainment of Aboriginal peoples in Canada. Catalogue no. 99-012-X2011003.

———. 2013k. "University Tuition Fees, 2013–2014." *The Daily*, September 12, 2013. http://www.statcan.gc.ca/daily-quotidien/130912/dq130912b-eng.pdf.

———. 2013l. "Changing Labour Market Conditions for Young Canadians." *The Daily*, 4 July. Ottawa: Statistics Canada.

———. 2013m. "The Education and Employment Experiences of First Nations People Living Off Reserve, Inuit and Métis: Selected Findings From the 2012 Aboriginal Peoples Survey." *The Daily.* 25 November 2013.

———. 2013n. "Crude Birth Rate, Age-Specific and Total Fertility Rates (Live Births), Canada, Provinces and Territories." CANSIM Table 102-4505.

———. 2013o. "Deaths and Mortality Rates, by Age Group and Sex, Canada, Provinces and Territories." CANSIM Table 102-0504.

———. 2013p, "Live births by age of mother, Canada, provinces and territories," Table 102-4503, http://www5.statcan.gc.ca/cansim/a26?lang=eng&id=1024503&p2=46.

———. 2014a. "Study: Persons with Disabilities and Employment." *The Daily*, 3 December. Ottawa: Statistics Canada.

———. 2014b. 2011 National Household Survey: Data Tables.

———. 2014c. Full-Time and Part-Time Employment by Sex and Age Group. CANSIM Table 282-0002. Ottawa: Statistics Canada.

———. 2014d. "The Underground Economy in Canada, 2011." *The Daily.* 30 January.

———. 2014e. Immigration and Ethnocultural Diversity in Canada. Ottawa: Statistics Canada 99-010-X.

———. 2014f. 2011 National Household Survey: Data Tables. Ottawa: Statistics Canada.

———. 2014g. "Canada's Population Estimates: Subprovincial Areas, July 1, 2013. *The Daily*, 26 February.

———. 2015a. "Average Household Expenditure." CANSIM, table 203-0021 and Catalogue no. 62F0026M.

———. 2015b. "Family violence in Canada: A statistical profile, 2013." Catalogue no. 85-002-X http://www.statcan.gc.ca/pub/85-002-x/2014001/article/14114-eng.pdf.

———. 2015c. "Avoidable Mortality Among First Nations People: A Cohort Analysis." *The Daily*, August 19.

———. 2015d. *Canadian Survey on Disability, 2012.* Ottawa: Statistics Canada.

———. 2015e. "Labour Force Characteristics by Age and Sex – Seasonally Adjusted." *The Daily* (13 March).

Steinem, Gloria. 2011. "Sex, Lies, and Advertising." In Gail Dines and Jean Humez (eds.). *Gender, Race, and Class in Media: A Critical Reader* (3rd ed.). Los Angeles: Sage, 235–241 (orig. pub. in *Ms.* Magazine, July/August 1990).

Steptoe, Andrew, Aparna Shankar, Panayotes Demakakos, and Jane Wardle. 2013. "Social Isolation, Loneliness, and All-Cause Mortality in Older Men and Women." *PNAS*, March 25.

Stevens, Angi Becker. 2012. "Polyamory: Rebooting Our Definitions of Love and Family." *Role/Reboot* (February 7), 3. Retrieved August 1, 2012. Available: http://www.rolereboot.org/family/details/2012-02-polyamory-rebooting-our-definitions-of-love-and-fami.

Stevenson, Seth. 2014. "The Smartphone Game Money Pit." *Slate.* 10 November. Available: http://www.slate.com/articles/technology/technology/2014/11/clash_of_clans_why_i_spent_real_nonfantasy_dollars_to_unlock_a_flock_of.html.

Stiegelbauer, S.M. 1996. "What is an Elder? What do Elders do? First Nation Elders as Teachers in Culture-Based Urban Organizations." *The Canadian Journal of Native Studies* 1:37–66.

Stier, Deborah S., and Judith A. Hall. 1984. "Gender Differences in Touch: An Empirical and Theoretical Review." *Journal of Personality and Social Psychology* 47 (2): 440–459.

Stobert, Susan, and Kelly Cranswick. 2004. "Looking After Seniors: Who Does What for Whom?" *Canadian Social Trends* (Autumn): 2–6.

———, and Anna Kemeny. 2003. "Child Free by Choice." *Canadian Social Trends.* Statistics Canada Cat. no. 11-008. Retrieved May 19, 2009. Available: http://www.statcan.gc.ca/pub/11-008-x/2003001/article/6528-eng.pdf.

Stone, Leroy O. 1967. *Urban Development in Canada: 1961 Census Monograph.* Ottawa: Queen's Printer.

Strapagiel, Lauren. 2015. "Stephen Harper's 'anti-woman' niqab comment mocked on Twitter." (March 11). *National Post.* Available: http://news.nationalpost.com/2015/03/11/stephen-harpers-anti-woman-niqab-comment-mocked-on-twitter-with-dresscodepm-hashtag/.

Straw, Will. 2011. "Dimensions of Media: Time and Space, Storage and Transmission." In Will Straw, Sandra Gabrielle, and Ira Wagman (eds.), *Intersection of Media and Communications.* Toronto: Emond Montgomery, 37–52.

Strogatz, Steven H., and Duncan J. Watts. 1998. "Collective Dynamics of 'Small-World' Networks." *Nature* 393: 440–442.

Suek, Beverly. 2009. "Radical Retirement Communities Rising." *Utne Reader*, November-December. Retrieved: 2 November 2014. Available: http://www.utne.com/print.aspx?id={F4F3B291-550B-4BCF-9040-D9E12A67553B}#axzz3HvsdDGcF.

Sumner, William G. 1959. *Folkways.* New York: Dover (orig. pub. 1906).

Surrogacy in Canada Online. 2015. Surrogacy in Canada. Available: http://surrogacy.ca/surrogacy-in-canada/surrogacy-in-canada.html.

Sutherland, Edwin H. 1939. *Principles of Criminology.* Philadelphia: Lippincott.

Swidler, Ann. 1986. "Culture in Action: Symbols and Strategies." *American Sociological Review* 51 (April): 273–286.

Sykes, Katie. 2015. "Rethinking the Application of Canadian Criminal Law to Factory Farming." In Peter Sankoff, Vaughan Black, and Katie Sykes (eds.),

Canadian Perspectives on Animals and the Law. Toronto: Irwin Law, 33–56.
Synge, Jane. 1980. "Work and Family Support Patterns of the Aged in the Early Twentieth Century." In Victor Marshall (ed.), *Aging in Canada: Social Perspectives.* Toronto: Fitzhenry and Whiteside.
Tannen, Deborah. 1993. "Commencement Address, State University of New York at Binghamton." Reprinted in *Chronicle of Higher Education* (June 9): B5.
Taras, David. 2015. *Digital Mosaic.* Toronto: University of Toronto Press.
Taylor, Angus. 2015. "Philosophy and the Case for Animals." In Peter Sankoff, Vaughan Black, and Katie Sykes (eds.), *Canadian Perspectives on Animals and the Law.* Toronto: Irwin Law, 11–29.
Taylor, Catherine, and Janice Ristock. 2011. "LGBTQ Families in Canada: Private Lives and Public Discourse." In Nancy Mandell and Ann Duffy (eds.), *Canadian Families: Diversity, Conflict and Change,* 4th ed. Toronto: Nelson, 125–163.
Taylor, C., and T. Peter. 2011. "' We Are Not Aliens, We're People, and We Have Rights.' Canadian Human Rights Discourse and High School Climate for LGBTQ Students." *Canadian Review of Sociology* 48 (3): 275–312. doi:10.1111/j.1755-618X.2011.01266.x.
Tcherni, M., A. Davies, G. Lopes, and A. Lizotte. 2015. "The Dark Figure of Online Property Crime: Is Cyberspace Hiding a Crime Wave?" *Justice Quarterly,* DOI: 10.1080/07418825.2014.994658.
Tedds, Lindsay. 2005. "The Underground Economy in Canada." In Chris Bajada and Friedrich Schneider (eds.), *Size, Causes and Consequences of the Underground Economy.* Farnham, U.K.: Ashgate, 157–178.
Terkel, Studs. 1996. *Coming of Age: The Story of Our Century by Those Who've Lived It.* New York: St. Martin's Griffin.
theadventuresofiman.com. 2007. "The Adventures of Iman." Retrieved March 18, 2007. Available: http://www.theadventuresofiman.com/AboutIman.asp.
Thomas, D. 1992. *Criminality Among the Foreign Born: Analysis of Federal Prison Population.* Ottawa: Immigration and Employment Canada.
Thomas, Mark, and Marcelo Giugale. 2014. "African Debt and Debt Relief." In Celestin Monga and Justin Lin (eds.), *The Oxford Handbook of Africa and Economics: Policies and Practices.* Oxford, UK: Oxford University Press.
Thomas, William I., and Dorothy Swaine Thomas. 1928. *The Child in America.* New York: Knopf.
Thorne, Barrie, Cheris Kramarae, and Nancy Henley. 1983. *Language, Gender, and Society.* Rowley, MA: Newbury House.
Tiefer, Leonore. 2005. "Unnatural acts." In T.L. Steele (ed.), *Sex, Self, and Society: The Social Construction of Sexuality.* Belmont, CA: Thomson/Wadsworth, 23–27.
Tieleman, Bill. 2015. "Nestle Pays $2.25 to Bottle and Sell a Million Litres of BC Water." TheTyee.ca. Available: http://thetyee.ca/Opinion/2015/02/24/Nestle-Pays-Nothing-to-Bottle-Water/.
Tiemeyer, Phil. 2013. *Plane Queer: Labor, Sexuality, and AIDS in the History of Male Flight Attendants.* Oakland, CA: University of California Press.
Tilly, Charles, ed. 1975. *The Formation of National States in Western Europe.* Princeton, NJ: Princeton University Press.
Tirrito, Terry. 2003. *Aging in the New Millennium.* Columbia, SC: University of South Carolina Press.
Tiryakian, Edward A. 1978. "Émile Durkheim." In Tom Bottomore and Robert Nisbet (eds.), *A History of Sociological Analysis.* New York: Basic Books, 187–236.
Titchkosky, Tanya. 2003. *Disability, Self and Society.* Toronto: University of Toronto Press.
Tjepkema, Michael, Russell Wilkins, and Andrea Long. 2013. "Cause-Specific Mortality by Income Adequacy in Canada: A 16-year Follow-up Study." *Health Reports* 24 (7): 14–22.
Toffler, Alvin. 1980. *The Third Wave.* New York: Bantam.
Tokunaga, Robert, and Stephen Rains. 2010. "An Evaluation of Two Characterizations of the Relationships between Problematic Internet Use, Time Spent Using the Internet, and Psychosocial Problems." *Human Communication Research* 36 (October): 512–545.
Toma, Catalina, and Jeffrey Hancock. 2010. "Looks and Lies: The Role of Physical Attractiveness in Online Dating Self-Presentation and Deception." *Communication Research* 37 (93): 335–351.
Tomlinson, Kathy. 2014a. "McDonald's Foreign Worker Practices Face Growing Investigation." Retrieved 15 December 2014. Available: http://www.cbc.ca/news/canada/british-columbia/mcdonald-s-foreign-worker-practices-face-growing-investigation-1.2607365.
———. 2014b. "McDonald's Foreign Workers Call it 'Slavery.'" Retrieved 15 December 2014. Available: http://www.cbc.ca/news/canada/edmonton/mcdonald-s-foreign-workers-call-it-slavery-1.2612659.
Tong, Rosemarie. 1989. *Feminist Thought: A Comprehensive Introduction.* Boulder, CO: Westview Press.
Tönnies, Ferdinand. 1940. *Fundamental Concepts of Sociology (Gemeinschaft und Gesellschaft).* Trans. Charles P. Loomis. New York: American Book Company (orig. pub. 1887).
———. 1963. *Community and Society* (Gemeinschaft *and* Gesellschaft). New York: Harper & Row (orig. pub. 1887).
Tonry, Michael. 2009. "The Mostly Unintended Effects of Mandatory Minimum Penalties: Two Centuries of Consistent Findings." In Michael Tonry (ed.), *Crime and Justice: A Review of Research, Volume 38.* Chicago: University of Chicago Press, 65–114.
Trevithick, Alan. 1997. "On a Panhuman Preference for Moandry: Is Polyandry an Exception?" *Journal of Comparative Family Studies* (September): 154–184.
Trocmé, Nico, Bruce MacLaurin, Barbara Fallon, Joanne Daciuk, Diane Billingsley, Marc Tourigny, Micheline Mayer, John Wright, Ken Barter, Gale Burford, Joe Hornick, Richard Sullivan, and Brad McKenzie. 2010. *The Canadian Incidence Study of Reported Child Abuse and Neglect, 2008: Major Findings.* Canadian Public Health Agency. Available: http://www.phac-aspc.gc.ca/cm-vee/csca-ecve/2008/cis-eci-04-eng.php.
Troeltsch, Ernst. 1960. *The Social Teachings of the Christian Churches,* Vols. 1 and 2. Trans. O. Wyon. New York: Harper & Row (orig. pub. 1931).
Trottier, Helen, Laurent Martel, Christian Houle, Jean-Marie Berthelot, and Jacques Legare. 2000. "Living at Home or in an Institution: What Makes the Difference for Seniors?" *Health Reports* 11 (Spring): 49–61.
Truth and Reconciliation Commission of Canada. 2012. Interim Report. Winnipeg: Truth and Reconciliation Commission.
Tuggle, Justin L. and Malcolm D. Holmes. 2000. "Blowing Smoke: Status Politics and the Smoking Ban." In Patricia A. Adler and Peter Adler (eds.), *Constructions of Deviance.* Belmont, CA: Wadsworth, 159–168.
Turcotte, Martin. 2011. *Women and Health.* Ottawa: Statistics Canada.
———. 2012. "Charitable Giving by Canadians." *Canadian Social Trends.* April: 18–36. Ottawa: Statistics Canada.
———. 2014. "Persons with Disabilities and Employment." Ottawa: Statistics Canada.
Turkle, Sherry. 2011. *Alone Together: Why We Expect More from Technology and Less from Each Other.* New York: Basic Books.
Turner, Jonathan, Leonard Beeghley, and Charles H. Powers. 1995. *The Emergence of Sociological Theory* (3rd ed.). Belmont, CA: Wadsworth.
———. 1998. *The Emergence of Sociological Theory* (4th ed.). Belmont, CA: Wadsworth.
Turner, Jonathan H., Royce Singleton, Jr., and David Musick. 1984. *Oppression: A Socio-History of Black–White Relations in America.* Chicago: Nelson-Hall (reprinted 1987).
Turner, Ralph H., and Lewis M. Killian. 1993. "The Field of Collective Behavior." In Russell L. Curtis, Jr., and Benigno E. Aguirre (eds.), *Collective Behavior and Social Movements.* Boston: Allyn & Bacon, 5–20.
Turner, Robin Lanette, and Diana Pei Wu. 2002. *Environmental Justice and Environmental Racism: An Annotated Bibliography and General Overview, Focusing on U.S. Literature, 1996–2002.* Berkeley, CA: Institute of International Studies, University of California.
UN Food and Agriculture Organization. 2014. *The State of Food Insecurity in the World.* Rome: United Nations.
UNAIDS. 2014. *World Aids Day 2014 Report – Fact Sheet.*
UNICEF. 2010. "Child Survival and Development'. Retrieved February 20, 2012. Available: http://www.unicef.org/media/media_45485.html.
———. 2014. "United Nations Children's Fund. Hidden in Plain Sight: A Statistical Analysis of Violence Against Children." New York: UNICEF. Available: http://files.unicef.org/publications/files/Hidden_in_plain_sight_statistical_analysis_EN_3_Sept_2014.pdf.
———. 2015. *The State of the World's Children Report 2015 Statistical Tables.* New York: UNICEF.
UNICEF Canada. (2009). *Aboriginal Children's Health: Leaving No Child Behind.* Canadian Supplement to The State of the World's Children 2009. Retrieved March 13, 2010, from www.unicef.ca/portal/SmartDefault.aspx?at=2063.
UNICEF Innocenti Research Centre. 2000. "A League Table of Child Poverty in Rich Nations." (Innocenti Report Card No. 1, June). Florence, Italy: UNICEF Innocenti Research Centre.
United Nations. 1998. *Human Development Report, 1998.* New York: Oxford University Press.
———. 2007. Meeting of the United Nations Committee on the Elimination of Racial Discrimination. 28 February 2007. Para. 50.
———. 2009. *World Population Prospects: The 2008 Revision.* New York: United Nations Population Division.
———. 2014. Concise Report on the World Population Situation in 2014. New York: United Nations.
———. 2015. "World Population Estimates and Projections." Retrieved on July 15, 2015. http://www.un.org/en/development/desa/population/.
United Nations Development Programme. 2003. *Human Development Report, 2003.* New York: Oxford University Press. Retrieved July 20, 2005. Available: http://hdr.undp.org/reports/global/2003/.
———. 2005. *Human Development Report, 2005.* New York: Oxford University Press. Retrieved December 30, 2005, 31. Available: http://hdr.undp.org/reports/global/2005/.
———. 2010. *Human Development Report, 2010.* New York: Palgrave Macmillan.
———. 2013. Life Expectancy at Birth (Years). Available: http://hdr.undp.org/en/69206
United Nations Human Development Programme. 2014. *Human Development Report 2014.* New York: United Nations.
Uppal, Sharanjit. 2011. "Unionization 2011." *Perspectives on Labour and Income* (Winter): 3–12.
———, Dafna Kohen, and Saeeda Khan. 2007. "Educational Services and the Disabled Child." *Education Matters: Insights on Education, Learning and Training in Canada* 3 (5).

Health Analysis and Measurement Group, Statistics Canada. Retrieved April 24, 2009. Available: http://www.statcan.gc.ca/pub/81-004-x/2006005/9588-eng.htm.

———, and Sébastien LaRochelle-Côté. 2012. "Factors Associated with Voting." *Perspectives on Labour and Income* (Spring): 4–15.

Ursel, Jane. 1996. *Submission to the Commission of Inquiry into the Deaths of Rhonds LaVoie and Roy Lavoie.* Winnipeg.

U.S. Census Bureau. 2000. *World Population Profile 2000.* Washington, U.S. Census Bureau.

———. 2014. *Health Insurance Coverage in the United States: 2013.* Washington: U.S. Government Printing Office.

van der Meulen, Emily, and Elya Maria Durisin. 2008. "Why Decriminalize? How Canada's Municipal and Federal Regulations Increase Sex Workers' Vulnerability." *Canadian Journal of Women and the Law* 20(2): 289–311.

Vander Ploeg, Casey. 2008. *Big Cities and the Census.* Calgary: Canada West Foundation.

Vanier Institute of the Family. 2009. "Fathers Matter." *Fascinating Families* 8. Retrieved May 10, 2009. Available: http://www.vifamily.ca/families/issue8.pdf.

———. 2011. "Four in Ten Marriages End In Divorce." *Fascinating Families* 41. Retrieved January 12, 2012. Available: http://www.vifamily.ca/media/node/945/attachments/2011-09-15_FASFAM_4-in-10-marriages-end-in-divorce_ENG.pdf.

———. 2012. "Family Roles and Responsibilities." *Transition.* Summer 2012 Vol. 42, No. 2. Available: http://www.vanierinstitute.ca/include/get.php?nodeid=2231.

———. 2013. "Same-Sex Families Raising Children." *Fascinating Families.* March 13 issue 51. Available: http://www.vanierinstitute.ca/include/get.php?nodeid=2817.

Veblen, Thorstein. 1967. *The Theory of the Leisure Class.* New York: Viking (orig. pub. 1899).

Vecsey, Christopher. 1987. "Grassy Narrows Reserve: Mercury Pollution, Social Disruption, and Natural Resources: A Question of Autonomy." *American Indian Quarterly* 11(4): 287–314.

Vezina, Mireille, and Susan Crompton. 2012. "Volunteering in Canada." *Canadian Social Trends.* April: 37–55. Ottawa: Statistics Canada.

Vigen, Tyler. n.d. *Spurious Correlations.* Available: http://www.tylervigen.com/view_correlation?id=805.

Volti, Rudi. 2008. *An Introduction to the Sociology of Work and Occupations.* Los Angeles: Pine Forge Press.

Voyageur, Cora. 2011. "Out in the Open: Elected Female Leadership in Canada's First Nations Community." *Canadian Review of Sociology* 48 (February): 67–85.

Wade, Lisa. 2011. "The Promise and Peril of Hookup Culture." Franklin and Marshall College. Available: http://thesocietypages.org/socimages/files/2011/06/FM-TALK-Wade-Hooking-Up-and-Hook-Up-Culture.pdf, 10.

Wadud, Amina. 2002. "A'ishah's Legacy: Amina Wadud Looks at the Struggle for Women's Rights Within Islam." *New Internationalist* (May). Retrieved March 18, 2007. Available: http://newint.org/-features/2002/05/01/aishahs-legacy.

Waine, Marie. 2014. "Private School Costs Across Canada." *Canadian Living.* Available: http://www.canadianliving.com/moms/kids/private_school_costs_across_canada.php.

Waldram, James B., D. Ann Herring, and T. Kue Young. 1995. *Aboriginal Health in Canada: Historical, Cultural, and Epidemiological Perspectives.* Toronto: University of Toronto Press.

Waldron, Ingrid. 1994. "What Do We Know About the Causes of Sex Differences in Mortality? A Review of the Literature." In Peter Conrad and Rochelle Kern (eds.), *The Sociology of Health and Illness: Critical Perspectives.* New York: St. Martin's Press, 42–54.

Walker, Danna L. 2007. "The Longest Day: Could a Class of College Students Survive Without iPods, Cellphones, Computers and TV from One Sunrise to the Next?" *Washington Post Magazine* (August 5): W20. Retrieved December 29, 2011. Available: http://www.washingtonpost.com/wp-dyn/content/article/2007/08/01/AR2007080101720.html.

Walker, James W. St. G. 1997. *"Race": Rights and the Law in the Supreme Court of Canada.* Waterloo: Wilfrid Laurier Press.

Wallace, Walter L. 1971. *The Logic of Science in Sociology.* New York: Aldine de Gruyter.

Wallerstein, Immanuel. 1979. *The Capitalist World-Economy.* Cambridge, UK: Cambridge University Press.

———. 1984. *The Politics of the World Economy.* Cambridge, UK: Cambridge University Press.

———. 1991. *Unthinking Social Science: The Limits of Nineteenth-Century Paradigms.* Cambridge, UK: Polity Press.

———. 2004. *World Systems Analysis: An Introduction.* Durham, NC: Duke University Press.

Walmart Corporation. 2012. "Walmart Facts." Retrieved Sept. 15, 2012. Online: http://news.walmart.com/walmart-facts.

Walmart.com. 2012. Walmart Annual Report. Retrieved Sept. 15, 2012. Online: www.walmartstores.com/sites/annual-report/2012/WalMart_AR.pdf.

Wang, Feng. 2010. "China's Population Destiny: The Looming Crisis." Washington: The Brookings Institute. Online.

Ward, Margaret, and Marc Belanger. 2011. *The Family Dynamic: A Canadian Perspective* (5th ed.). Toronto: Nelson.

Warren, F.J., and D.S. Lemmen. 2014. *Canada in a Changing Climate: Sector Perspectives on Impacts and Adaptation.* Government of Canada, Ottawa, ON, 286.

Waters, Malcolm. 1995. *Globalization.* London and New York: Routledge.

Wathen, N. 2012. "Health Impacts of Violent Victimization on Women and their Children." Department of Justice Canada. Catalogue no. J2-377/2013E-PDF. Available: http://www.justice.gc.ca/eng/rp-pr/cj-jp/fv-vf/rr12_12/index.html.

Weaver, A.D., K.L. MacKeigan, and H.A. MacDonald. 2011. "Experiences and Perceptions of Young Adults in Friends with Benefits Relationships: A Qualitative Study." *Canadian Journal of Human Sexuality* 20 (1/2): 41–53.

Weber, Max. 1947. *The Theory of Social and Economic Organization.* Trans. A.M. Henderson and Talcott Parsons; ed. Talcott Parsons. New York: Oxford University Press.

———. 1963. *The Sociology of Religion.* Trans. E. Fischoff. Boston: Beacon Press (orig. pub. 1922).

———. 1968. *Economy and Society: An Outline of Interpretive Sociology.* Trans. G. Roth and G. Wittich. New York: Bedminster Press (orig. pub. 1922).

———. 1976. *The Protestant Ethic and the Spirit of Capitalism.* Trans. Talcott Parsons. Introduction by Anthony Giddens. New York: Scribner (orig. pub. 1904–1905).

Weber, Peter. 2014. "Confused By All the Facebook Genders? Here's What They Mean." *The Week,* Feb. 21, 2014. Available: http://www.slate.com/blogs/lexicon_valley/2014/02/21/gender_facebook_now_has_56_categories_to_choose_from_including_cisgender.html.

Weeks, John R. 2002. *Population: An Introduction to Concepts and Issues* (8th ed). Belmont, CA: Wadsworth/Thomson Learning.

Wei, Chen, and Liu Jinju. 2009. "Future Population Trends in China: 2005–2050." Centre of Policy Studies/IMPACT Centre Working Papers, Victoria University.

Weigel, Russell, and P.W. Howes. 1985. "Conceptions of Racial Prejudice: Symbolic Racism Revisited." *Journal of Social Issues* 41: 124–132.

Weiskel, Timothy. 1994. "Vicious Circles." *Harvard International Review* 16: 12–20.

Weitz, Rose. 1995. *A Sociology of Health, Illness, and Health Care.* Belmont, CA: Wadsworth.

———. 1996. *The Sociology of Health, Illness, and Health Care: A Critical Approach.* Belmont, CA: Wadsworth.

Weitzer, R. 2005. "New Directions in Research on Prostitution." *Crime, Law and Social Change* 43: 211–235.

Welch, Mary Agnes. 2015. "Diagnosing Poverty: New Stats Show People in North End Die 16 Years Earlier". *Winnipeg Free Press,* June 6.

Wendell, Susan. 1995. "Toward a Feminist Theory of Disability." In E.D. Nelson and B.W. Robinson (eds.), *Gender in the 1990s.* Toronto: Thomson Nelson, 455–465.

Wente, Margaret. 2006. "In the Best Interests of the Child?" *Globe and Mail* (September 30): A21.

———. "Sixty-five's Not 60, and Other Boomer Revelations." *The Globe and Mail,* 9 February.

West, Candice, and Don H. Zimmerman. 1991. "Doing Gender." In J. Lorber and Susan A. Farrell (eds.), *The Social Construction of Gender.* London: Sage Publications, 13–37.

Wharton, Amy. 2000. "Feminism at Work." *The Annals of the American Academy of Political and Social Science* 571 (September): 167–182.

Whitaker, Barbara. 1997. "Earning It; If You Can't Beat Dilbert, Hire Him." *New York Times* (June 29): C12.

White, Rob. 2007. "Green Criminology and the Pursuit of Social and Ecological Justice." In Piers Beirne and Nigel South (eds.), *Issues in Green Criminology: Confronting Harms Against Environments, Humanity and Other Animals.* Portland, OR: Willan Publishing, 32–54.

Whorf, Benjamin Lee. 1956. *Language, Thought and Reality.* John B. Carroll (ed.). Cambridge, MA: MIT Press.

Whyte, William Foote. 1989. "Advancing Scientific Knowledge Through Participatory Action Research." *Sociological Forum* 4: 367–386.

Whyte, William H., Jr. 1957. *The Organization Man.* Garden City, NY: Anchor.

Wilkins-Laflamme, Sarah. 2015. "How Unreligious Are the Religious 'Nones'? Religious Dynamics of the Unaffiliated in Canada. *Canadian Journal of Sociology.*

Williams, Cara. 2011. "Economic Well-being." In Statistics Canada, 2011, *Women in Canada: A Gender-Based Statistical Report* (6th ed.). Cat. no, 89-503-XIE2010001. Available: http: //www.statcan.gc.ca/pub/75-001-x/2011001/pdf/11394-eng.pdf.

Williams, Christine L. 1989. *Gender Differences at Work.* Berkeley, CA: University of California Press.

Williams, Colin. 1994 *Called Unto Liberty: On Language and Nationalism.* Multilingual Matters: Clevedon, Avon.

Williams, Robin M., Jr. 1970. *American Society: A Sociological Interpretation* (3rd ed.). New York: Knopf.

Williamson, Robert C., Alice Duffy Rinehart, and Thomas O. Blank. 1992. *Early Retirement: Promises and Pitfalls.* New York: Plenum Press.

Willows, N., P. Veugelers, K. Raine, and S. Kuhle. 2012 *Associations Between Household Food Insecurity and Health Outcomes in the Aboriginal Population (Excluding Reserves).* Ottawa: Statistics Canada. Cat. no. 82-003-X. www.statcan.gc.ca/pub/82- 003-x/2011002/article/11435-eng.htm (accessed May 12, 2012).

Wilson, Daniel, and David MacDonald. 2010. "The Income Gap Between Aboriginal Peoples and the Rest of Canada." Canadian Centre for Policy Alternatives. Available: http://www.policyalternatives.ca/sites/default/files/uploads/publications/reports/docs/Aboriginal%20Income%20Gap.pdf.

Wilson, David, ed. 1997. "Globalization and the Changing U.S. City." *Annals of*

the American Academy of Political and Social Sciences, special issue, 551 (May).
Wilson, Edward O. 1975. *Sociobiology: A New Synthesis.* Cambridge, MA: Harvard University Press.
Wilson, Elizabeth. 1991. *The Sphinx in the City: Urban Life, the Control of Disorder, and Women.* Berkeley, CA: University of California Press.
Wilson, William Julius. 1996. *When Work Disappears: The World of the New Urban Poor.* New York: Knopf.
Wilton, Katherine. 2014. "Luka Magnotta Case: Montreal Police Cannot View Research Video, Judge Rules." *The Gazette* (January 22).
Wingert, S., N. Higgitt, and J. Ristock, (2005). "Voices from the Margins: Understanding Street Youth in Winnipeg." *Canadian Journal of Urban Research, 14*(1), 54–80.
Wirth, Louis. 1938. "Urbanism as a Way of Life." *American Journal of Sociology* 40: 1–24.
Wise, Tim. 2008a. *Speaking Treason Fluently: Angry Racist Reflections From an Angry White Male.* Soft Skull Press.
———. 2008b. "Explaining White Privilege, (Or, Your Defense Mechanism Is Showing)." Available: http://www.timwise.org/2008/09/explaining-white-privilege-or-your-defense-mechanism-is-showing/.
Wolf, Daniel. 1996. "A Bloody Biker War." *Maclean's* (January 15): 10–11.
Wolfe, Jeanne M. 1992. "Canada's Livable Cities." *Social Policy* 23: 56–63.
Wolfe, Robert, and Lisa Sharp. 2005. "Vaccination or Immunization? The Impact of Search Terms on the Internet." *Journal of Health Communication* 10: 539–553.
Wollstonecraft, Mary 1974. *A Vindication of the Rights of Woman.* New York: Garland (orig pub. 1797).
Wong, Josephine P. 2006. "Age of consent to sexual activity in Canada: Background to proposed new legislation on 'age of protection.'" *The Canadian Journal of Human Sexuality,* Vol. 15 (3–4), 163.
Wong, L., and S. Guo. (2011). "Introduction: Multiculturalism Turns 40: Reflections on the Canadian Policy." *Canadian Ethnic Studies, 43*(1/2), 1–3.
Wood, Daniel B. 2002. "As Homelessness Grows, Even Havens Toughen Up." *Christian Science Monitor* (Nov. 21). Retrieved July 1, 2003. (Online http://www.csmontiro.com/2002/1121/p01s04-ussc.htm.)
Wood, Darryl S., and Curt T. Griffiths. 1996. "Patterns of Aboriginal Crime." In Robert A. Silverman, James J. Teevan, and Vincent F. Sacco (eds.), *Crime in Canadian Society* (5th ed.). Toronto: Harcourt Brace and Company, 222–223.
Wood, Julia T. 1999. *Gendered Lives: Communication, Gender, and Culture* (3rd ed.). Belmont, CA: Wadsworth.
Woodward, Kathleen. 1991. *Aging and Its Discontents: Freud and Other Fictions.* Bloomington: Indiana University Press.
World Bank. 2003. *World Development Indicators 2003.* New York: Author. Retrieved August 8, 2005. Available: http://www.worldbank.org/data/wdi2003.
———. 2004. "Millennium Development Goals: Global Data Monitoring System. Promote Gender Equality and Empower Women." New York: Author. Retrieved August 12, 2005. Available: http://ddp-ext.worldbank.org/ext/MDG/gdmis.do.
———. 2006. *World Development Report 2006.* New York: World Bank.
———. 2008c. *World Development Indicators 2008.* Retrieved May 9, 2009. Available: http://siteresources.worldbank.org/DATASTATISTICS/Resources/front.pdf.
———. 2010. *World Development Report 2010.* Washington: World Bank.
———. 2011. *World Development Report 2011.* Washington: World Bank, 2, 61.
———. 2012. *World Development Report 2012.* "Gender Equality and Development." Washington: World Bank, 14, 122, 129.
———. 2015. PovcalNet: An Online Analysis Tool for Global Poverty Monitoring. Available: http://iresearch.worldbank.org/PovcalNet/.
———. 2015a. "Poverty Overview." Available: http://www.worldbank.org/en/topic/poverty/overview.
———. 2015b. "Country and Lending Groups." Available: http://data.worldbank.org/about/country-and-lending-groups.
World Economic Forum. 2012. *Global Population Ageing: Peril or Promise?* Geneva: World Economic Forum.
World Health Organization. 2012a. *World Health Statistics 2012.* Geneva: World Health Organization.
———. 2012b. "Social Determinants of Health". Retrieved February 22, 2012. Available: http://www.who.int/social_determinants/en.
———. 2014. "Maternal Mortality." Fact Sheet No. 348. Available: http://www.who.int/mediacentre/factsheets/fs348/en/.
———. 2015a. "Infant Mortality." Global Health Observatory Data. Available: http://www.who.int/gho/child_health/mortality/neonatal_infant_text/en/.
———. 2015b. "Life Expectancy." Global Health Observatory Data. Available: http://www.who.int/gho/mortality_burden_disease/life_tables/en/.
"World Wealth Report." 2015. Capgemini Financial Services Analysis. Retrieved from: https://www.worldwealthreport.com/reports/population/north_america/canada.
Worldwatch Institute. 2014. "Agricultural Subsidies Remain a Staple in the Industrial World. Available: http://www.worldwatch.org/agricultural-subsidies-remain-staple-industrial-world-0.
Wortley Scot. 2009. "Introduction. The immigration–crime connection: Competing theoretical perspectives." *Journal of International Migration & Integration* 10(4): 349–358.
———, and Julian Tanner. 2003. "Data, Denials, and Confusion: The Racial Profiling Debate in Toronto." *Canadian Journal of Criminology and Criminal Justice* 45 (July): 367–390.
Wotherspoon, Terry. 2009. *The Sociology of Education in Canada.* Toronto: Oxford University Press.
———. 2014. *The Sociology of Education in Canada: Critical Perspectives,* 4th ed. Oxford University Press.
Wouters, Cas. 1989. "The Sociology of Emotions and Flight Attendants: Hochschild's Managed Heart." *Theory, Culture & Society* 6: 95–123.
Wright, Charles. 1959. *Mass Communication: A Sociological Perspective.* New York: Random House.
Wright, Erik Olin. 1978. "Race, Class, and Income Inequality." *American Journal of Sociology* 83 (6): 1397.
———. 1979. *Class Structure and Income Determination.* New York: Academic Press.
———. 1985. *Class.* London: Verso.
———. 1997. *Class Counts: Comparative Studies in Class Analysis.* Cambridge, UK: Cambridge University Press.
———. 2010. *Envisioning Real Utopias.* London: Verso.
Wuthnow, Robert. 1996. *Poor Richard's Principle: Recovering the American Dream Through the Moral Dimension of Work, Business, and Money.* Princeton, NJ: Princeton University Press.
Yee, Jessica. 2010. "Sustainable Justice Through Knowledge Transfer: Sex Education and Youth." *Canadian Woman Studies.* Fall 2009/Winter 2010: 22–26.
Yee, Nick, and Jeremy Bailenson. 2007. "The Proteus Effect: the Effect of Transformed Self-Representation on Behavior." *Human Communication Research* 33 (July): 271–290.
———, Jeremy Bailenson, and Nicolas Duchenault. 2009. "The Proteus Effect: Implications of Transformed Digital Self-Representation on Online and Offline Behavior." *Communication Research* 36 (March): 285–312.
Yinger, J. Milton. 1960. "Contraculture and Subculture." *American Sociological Review* 25 (October): 625–635.
———. 1982. *Countercultures: The Promise and Peril of a World Turned Upside Down.* New York: Free Press.
Young, A. 2008. "The State Is Still in the Bedrooms of the Nation: The Control and Regulation of Sexuality in Canadian Criminal Law." *Canadian Journal of Human Sexuality* 17 (4): 203–220.
Young, Nathan. 2015. *Environmental Sociology for the Twenty-First Century.* Don Mills, ON: Oxford University Press.
Zakaria, Fareed. 2011. "Fareed's Take: The Role of Social Media in Revolutions." CNN. Retrieved October 12, 2011. Available: http://globalpublicsquare.blogs.cnn.com/2011/03/27/the-role-of-social-media-in-revolutions.
Zehr, Mary Ann. 2006. "Public Schools Fare Well Against Private Schools in Study." *Education Week.* Retrieved April 22, 2009. Available: http://www.edweek.org/login.html?source=http://www.edweek.org/ew/articles/2006/07/26/43private.h25.html&destination=http://www.edweek.org/ew/articles/2006/07/26/43private.h25.html&levelId=2100.
Zeidenberg, Jerry. 1990. "The Just-in-Time Workforce." *Small Business* (May): 31–34.
Zeitlin, Irving M. 1997. *Ideology and Development of Sociological Theory* (6th ed.). Upper Saddle River, NJ: Prentice-Hall.
Zernike, Kate. 2010. "A Young and Unlikely Activist Who Got to the Tea Party Early." *New York Times* (Feb. 28): A1, A19.
———, and Megan Thee-Brenan. 2010. "Discontent's Demography: Who Backs the Tea Party?" *New York Times* (Apr. 15): A1, A17.
Zgodzinski, Rose. 1996. "Where Immigrants Come From." *Globe and Mail* (June 20).
Zhang, Xuelin. 2010. *Low Income Measurement in Canada: What Do Different Lines and Indexes Tell Us?* Statistics Canada, Income Research Series, Cat. no. 75F0002M—No. 3. Retrieved April 12, 2012. Available: http://www.statcan.gc.ca/pub/75f0002m/75f0002m2010003-eng.pdf.
Zipp, John F. 1985. "Perceived Representativeness and Voting: An Assessment of the Impact of 'Choices' vs. 'Echoes.'" *American Political Science Review* 60 (3): 738–759.
Zuckerman, Ethan. 2011. "Cute Cats and the Arab Spring: The 2011 Vancouver Human Rights Lecture." *Ideas.* CBC Radio (December 9, 2011). Available: http://www.cbc.ca/ideas/episodes/2012/02/24/the-vancouver-human-rights-lecture---cute-cats-and-the-arab-spring.

NAME INDEX

Note: Page numbers that begin with "21" and "22" (e.g. 21-11, 22-15–22-16) refer to online chapters 22 and 23, found at www.nelson.com/student

SUBJECT INDEX

Note: Page numbers that begin with "21" and "22" (e.g. 21-11, 22-15–22-16) refer to online chapters 22 and 23, found at www.nelson.com/student